# MACHINE TOOL OPERATIONS

## ABOUT THE AUTHORS

**STEVE F. KRAR** majored in Machine Shop Practice and spent fifteen years in the trade, first as a machinist and finally as a tool and diemaker. After this period he entered Teachers' College and graduated from the University of Toronto with a Specialist's Certificate in Machine Shop Practice. During his nineteen years of teaching, Mr. Krar was active in Vocational and Technical education and served on the executive of many educational organizations. For ten years he was on the summer staff of the University of Toronto, involved in teacher training programs. Active in machine tool associations, Steve Krar has been a senior member of the Society of Manufacturing Engineers for over 25 years. He is a co-author of the McGraw-Hill text *Technology of Machine Tools,* Second Edition. He is also co-author of the following McGraw-Hill Ryerson Ltd. publications: *Machine Shop Training, Machine Shop Operations,* and the overhead transparency kits: *Machine Tools, Measurement and Layout, Threads and Testing Equipment,* and *Cutting Tools.*

**JOSEPH E. ST. AMAND** served his apprenticeship in general machine shop work, jigs and fixtures, and tool and die work. After twelve years in industry he entered Teachers' College and graduated with a Specialist's Certificate in Machine Shop Practice from the University of Toronto. During his thirty-three years of teaching, Joe St. Amand served on several educational committees which were involved with technical research, courses, and standardized examinations. For more than twenty-five years he has been active in chapter work for the Society of Manufacturing Engineers. He is a co-author of the McGraw-Hill text *Technology of Machine Tools,* Second Edition. He is also co-author of the texts: *Machine Shop Training, Machine Shop Operations,* and Machine Shop Transparency Kits: *Machine Tools, Measurement and Layout, Threads and Testing Equipment,* and *Cutting Tools.*

**J. WILLIAM OSWALD** served an apprenticeship in machine shop, and after sixteen years in the trade attended Teachers' College at the University of Toronto. After graduation, he received a Specialist's Certificate in Machine Shop Practice and taught machine shop work for twenty-five years. During this time he attended several up-grading courses in the operation of the latest machine shop and testing equipment. For several years Mr. Oswald served on the teacher-training staffs at the University of Toronto and the University of Western Ontario. He had also worked with various technical educational committees and organizations. He is a co-author of the McGraw-Hill text *Technology of Machine Tools,* Second Edition. Mr. Oswald is also a co-author of *Machine Shop Operations,* and the overhead transparency kits: *Machine Tools, Measurement and Layout, Threads and Testing Equipment,* and *Cutting Tools.*

# MACHINE TOOL OPERATIONS

S. F. Krar   J. W. Oswald   J. E. St. Amand



GLENCOE

Macmillan/McGraw-Hill

Lake Forest, Illinois   Columbus, Ohio
Mission Hills, California   Peoria, Illinois

Sponsoring Editor: Myrna Breskin
Editing Supervisors: Evelyn Belov and Iris Cohen
Design and Art Supervisors: Sheila Granda and Patricia Lowy
Production Supervisor: Priscilla Taguer

Cover Photographer: Gary Gladstone/The Image Bank
Cover Designer: Patricia Lowy

**Library of Congress Cataloging in Publication Data**

Krar, Stephen F.
Machine tool operations.

Includes index.
1. Machine-tools. 2. Machine-shop practice.
I. Oswald, James William. II. St. Amand, J. E.
III. Title.
TJ1185.K6677 670.42'3 81-18587
ISBN 0-07-035430-8 AACR2

**Machine Tool Operations**

Imprint 1991


4 5 6 7 8 9 10 11 12 13 14 15 A-HAL 00 99 98 97 96 95 94 93 92 91

ISBN 0-07-035430-8

# CONTENTS

# PREFACE

After many years of machine shop and teaching experience, the authors decided to produce a basic text which would be easily understood and followed by students, make the teacher's job more effective, and prepare the student to enter the world of work more effectively. Students, teachers, and industry were surveyed in order to meet the needs of all three and this text is the result of their input and suggestions.

*Machine Tool Operations* has been written in unit form to make it suitable for general classroom use as well as for individualized instruction. Each unit is introduced with a set of objectives, along with the related theory and the sequence for performing each operation in easy-to-follow point form. The main body of the text, which includes the theory and operations sequence, is located in the two right-hand columns of each page. A left-hand column contains some tables, artwork, captions, and important notes, which should help to clarify the theory or operations presented.

This text covers all the basic machine tools, measurement and inspection systems, layout and bench work, metallurgy and heat treating, and new machining processes. A new section on job planning should be especially useful to beginning students and machine shop workers. It outlines the sequence of operations which should be followed to produce various parts and should provide valuable information to the beginner. Safety, although discussed in a separate unit, is also discussed throughout the text when it applies to the use of hand tools or the operation of machines.

After completing a unit, the student should be able to complete the operations to the required standard. If the student has any difficulty in understanding the material or performing the operation, the unit should be reviewed and the teacher consulted for assistance. At the end of each unit is a series of review questions designed to test the student's understanding of the unit. These questions may be used by the teacher as class or homework assignments or as a test. A teacher's guide for these units is available. It contains answers in brief point form, which makes marking assignments easier.

Throughout the text, dual dimensioning (U.S. Customary: inch; metric: SI) is used in order to make the transition from one system of measurement to another as easy as possible. The need for a working knowledge of the metric system of measurement cannot be overstressed, since most of the countries in the world are already on or in the process of converting to the metric system.

*The changeover period.* Although both the United States and Canada are now committed to conversion to SI as rapidly as possible, it is likely to be some years before all machine tools and measuring devices are redesigned or converted. The change to the metric system in the machine shop trade will be gradual because of the long-life expectancy of the costly machine tools and measuring equipment involved. Moreover, many manufacturers will continue to make inch-pound-sized products as long as their customers request them. It is probable, therefore, that people involved in the machine shop trade will have to be familiar with both the metric and the inch-pound systems during the long changeover period.

The situations that faced the authors of this book at the time of publication were:

**1.** Students using the text must be prepared to work in a dual-measurement system world for some years yet.
**2.** Many measuring tools were not available in metric sizes, and information was not obtainable about when and in what sizes some items would become available.

To accommodate these two principles, the following policy has been adopted throughout this book. This policy should enable you to work effectively in both systems now, while permitting an easy transition to full metric as the new materials and tools become available.

**1.** Where general measurements or references to quantity are not related specifically to inch-pound standards, tools, or products, only SI units are given.
**2.** Where the student may be exposed to equipment designed to both metric and inch-pound standards, separate information is given on both types of equipment in exact dimensions.
**3.** Where only inch-pound standards, tools, or products exist at the present time, inch measurements are given with a soft conversion to metric provided in parentheses.

In order to be successful in the machine shop trade, a person should be neat, develop sound work habits, and have a good knowledge of mathematics and blueprint reading as they relate to machine shop work. Owing to the ever-changing technology, it is wise to keep abreast of new developments by reading specialized texts, trade literature, and articles related to this exciting line of work.

*Steve F. Krar*
*Joseph E. St. Amand*
*J. William Oswald*

# ACKNOWLEDGMENTS

The authors wish to express their sincere appreciation to Alice H. Krar for typing and checking the manuscript for this text. Her assistance has been invaluable in the preparation of this text.

We wish to acknowledge the contribution of Harold Bacsu, Henry Bickle, and the many teachers, students, and industrial personnel who gave freely of their time to offer helpful suggestions to make this text as up to date, clear, and concise as possible. The authors sincerely hope that the quality of this text justifies the countless hours spent by many people throughout the country.

A sincere vote of thanks is due to the firms who provided the authors with the opportunity of researching up-to-date industrial practices and new machining techniques. We are also grateful to the following firms who were kind enough to supply technical information and illustrations for this text.

Allen, Chas. G. & Co.
American Chain and Cable Co. Inc., Wilson Instrument Division
''American Machinist''
American Superior Electric Co. Ltd.
Ametek Testing Equipment
Armstrong Bros. Tool Co.
Ash Precision Equipment Inc.
Atlas Press Co., Clausing Division
Avco-Bay State Abrasive Company
Bendix Corporation, Automation Group
Bethlehem Steel Corporation
Boston Gear Works
Brown & Sharpe Manufacturing Co.
Buffalo Forge Co.
Butterfield Division, Union Twist Drill Co.
Carborundum Company
Cincinnati Milacron Inc.
Cincinnati Shaper Co.
Cleveland Tapping Machine Co.
Cleveland Twist Drill (Canada) Ltd.
Colchester Lathe & Tool Co.
Coleman Engineering Co. Inc.
Covel Manufacturing Co.
Delmar Publishers Inc.
Delta File Works
DeVlieg Machine Co.
Dillon, W. C. and Co. Inc.
DoALL Company
Elliott Machine Tools
Enco Manufacturing Co.
Ex-Cell-O Corp.
Federal Products Corp.
General Electric Co. Ltd.
General Motors of Canada
Greenfield Tap and Die Co.
Hones, Charles H. Inc.
Inland Steel Co.
Jacobs Manufacturing Co.
Jones and Lamson Division of Waterbury Farrel
Kaiser Steel Corp.
Kostel Enterprises Ltd.
LeBlond, R. K. Machine Tool Co.
Lionite Abrasives Ltd.
Mahr Gage Co. Inc.
Mobil Oil Corporation
Morse Twist Drill and Machine Co.
National Broach & Machine Division, Lear Siegler Inc.
Neill, James & Co. (Sheffield) Ltd.
Nicholson File Co. of Canada Ltd.
Norton Company of Canada Ltd.
Powder Metallurgy Parts Manufacturers' Association
Pratt & Whitney Co. Inc., Machine Tool Division
Rockford Machine Tool Co.
Shore Instrument & Mfg. Co. Inc.
Slocomb, J. T. Co.
South Bend Lathe Inc.
Standard-Modern Tool Co. Ltd.
Stanley Tools Division, Stanley Works
Starrett, L. S. Co.
Sun Oil Co.
Taft-Peirce Manufacturing Co.
Taper Micrometer Corp.
Union Carbide Corp., Linde Division
United States Steel Corporation
Weldon Tool Co.
Whitman & Barnes
Wickman, A. C. Ltd.
Wilkie Brothers Foundation
Williams, A. R., Machinery Co. Ltd.
Williams, J. H. & Co.

# SECTION 

# MEASUREMENT SYSTEMS AND PRECISION MEASUREMENT

With the introduction of space-age technology, machines have been developed which are capable of producing workpieces to extremely fine tolerances. As a result, measuring tools and equipment had to be upgraded to measure the closer tolerances accurately. Therefore, measurement has assumed a very important role in machine shop and manufacturing.

The need for accurate measurement was necessary because of interchangeable manufacture, where parts produced in one plant may be assembled with parts from another plant or even from another country. This type of manufacturing and assembly requires that all parts are made to accurate size with very close tolerances. Interchangeable manufacture, world trade, and the need for high precision have, in turn, led to the need for a standardized system of international measurement. As a result, in 1960 the International System of Units (SI) was developed.

# UNIT 1

# MEASUREMENT SYSTEMS

Currently, two major systems of measurement are used in the world: the metric (decimal) system, and the inch-pound [US Customary System (USCS)]. Over 90 percent of the world's population uses some form of the metric system; the inch-pound system is the one that has been traditionally used in North America.

## OBJECTIVES

The aim of this unit is to enable you to:

**1.** Identify the various measurement systems
**2.** Understand the relationship between the inch and metric measurement systems
**3.** Know when precision and semi-precision tools should be used

## METRIC (DECIMAL) SYSTEMS

The United States is now involved in adopting the metric system known as SI (short for the French "Système International"). SI is the most advanced and easiest to use system of measurement—one which all countries are likely to adopt in time. (See N-1.) In SI, the base unit of length is the meter (m). All other linear units are directly related to the meter by a factor of 10. In order to convert from a smaller to a larger unit, or vice versa, divide or multiply by 10, 100, etc. For example, refer to the table below.

N-1 The basic unit of length measurement in the SI system is the meter.

N-2 The basic unit of length in the inch-pound system is the inch.

N-3 The inch-pound system is more complex than the SI system.

## INCH-POUND SYSTEMS

The inch-pound system (USCS) has for many years been the standard of North American industry. (See N-2.) In this system the base unit of length is the inch. Other linear units are related to the base unit by odd and unusual factors. The inch can be divided by halves, quarters, eighths, sixteenths, thirty-seconds, sixty-fourths, tenths, hundredths, thousandths, ten-thousandths, etc. Some linear units larger than the inch are listed along with the metric equivalent in meters:

| | | | | |
|---|---|---|---|---|
| 1 foot | = | 12 inches | = | 0.3048 m |
| 1 yard | = | 36 inches | = | 0.9144 m |
| 1 rod | = | 198 inches | = | 5.0292 m |
| 1 mile | = | 63,360 inches | = | 1609.344 m |

It can readily be seen that the inch-pound system is far more complex than SI. In the inch-pound system, mass is measured in pounds, temperature in degrees Fahrenheit, volume in cubic inches or fluid ounces, force in foot-pounds, pressure in pounds per square inch, and area in square feet and inches. Other quantities have similarly complex units. (See N-3.)

## STYLE DIFFERENCES BETWEEN SI AND THE INCH-POUND SYSTEM

Because SI is an international language, all countries adopting it should follow the approved style. The approved SI style is followed throughout this text wherever standards, tools, products, or processes are described in metric terms. The approved SI style is also followed as much as possible where the inch-pound system is used. The student should be aware of one particular difference between SI and inch-pound styles. In SI, a zero is placed in front

| Prefix Meaning | Multiplier | Symbol |
|---|---|---|
| micro (one-millionth) | 0.000 001 | μ |
| milli (one-thousandth) | 0.001 | m |
| centi (one-hundredth) | 0.01 | c |
| deci (one-tenth) | 0.1 | d |
| deca (ten) | 10 | da |
| hecto (one hundred) | 100 | h |
| kilo (one thousand) | 1 000 | k |
| mega (one million) | 1 000 000 | M |

of the decimal marker when no other figure is there (e.g., 0.25). The zero is not always required in the inch-pound system and is not used in this text where inch-pound terms are shown.

## CLASSES OF MEASURING TOOLS

There are two types of measuring tools—*basic, or semiprecision,* and *precision* measuring tools. The accuracy of the workpiece required will determine the type of measuring tools which must be used, so the student should pay particular attention to the accuracy required for the job. For instance, if a part is to be 6¼ inches (in.) [158.7 millimeters (mm)] long, there is no need to make it 6.250 in. [158.75 mm] long, since extra time will be required to machine it to this measurement. The use of a steel rule in this case would be accurate enough. On the other hand, if a part is to be 1.000 in. [25.40 mm] in diameter, a micrometer should be used for measurement and not an outside caliper and a rule.

A general procedure may be applied for machining tolerances when using the inch system. If the measurement is in full inches or fractions of an inch, the measurement may be made with a steel rule. If the measurement is stated as a decimal, however, a precision measuring tool must be used. For instance, if a measurement is given as 3 in. or 3³⁄₁₆ in., a rule and/or outside caliper may be used. If the measurement is 3.000 or 3.187, a precision measuring tool, such as a micrometer, must be used. (See N-4 and N-5.)

N-4 The tolerance for fractional dimensions is generally ± 1⁄64 in.

N-5 The tolerance for decimal dimensions is generally ± .002 in. (unless otherwise stated)

For *metric measurements* given in whole or half millimeters, a rule may be used for measuring. For sizes given in decimals, it is advisable to use a precision measuring tool.

## CARE OF TOOLS

The proper use of the correct measuring tool or instrument plays an important part in the quality and accuracy of the finished product. Precision measuring tools are expensive, so they should be handled with the greatest of care to preserve their accuracy. Always keep in mind that an inaccurate measuring tool is worse than no tool at all. Good-quality hardened tools will withstand a lifetime of wear with a reasonable amount of care. Good machinists or toolmakers take pride in keeping their tools accurate and in good condition.

Never place measuring tools in chips or where other tools may be dropped on them. After tools have been used, wipe them clean and give them a light dressing of oil to prevent rusting. Measuring tools should be stored in separate boxes or cases to avoid accidental scratches or nicks which can obscure graduation lines and cause the tool to be inaccurate.

## REVIEW QUESTIONS

**1.** What two measurement systems are used in the world?

**2.** Name the basic unit of length in the SI system.

**3.** What advantages has the SI system over the inch system of measurement?

**4.** If dimensions on a drawing are given in fractions, what tolerance is generally allowed?

**5.** If dimensions are given in decimals, what tolerance is generally allowed?

**6.** List three things that will prolong the life of any measuring tool.

# UNIT 2

# MEASURING WITH RULES

The most common measuring tool used in machine shop work is the steel rule. It is used where fast measurements, which do not require a high degree of accuracy, must be made. Inch steel rules are graduated in fractions or decimals; metric rules are usually graduated in millimeters or half millimeters. The accuracy of the measurement taken depends on the condition and the proper use of the rule.

# OBJECTIVES

This unit will enable you to:

1. Recognize the common types of steel rules and their uses
2. Read inch, decimal, and metric rules
3. Measure round and flat workpieces to within 1/64 in. (0.39 mm) of accuracy
4. Measure from the 1-in. or 10-mm line when accuracy is required

# STEEL RULES

Steel rules are made in a wide variety of types and sizes to suit the shape or size of a section or the entire length of a part. Rules graduated in inch, decimal, or metric sizes are available to suit the requirements of the part being produced and measured.

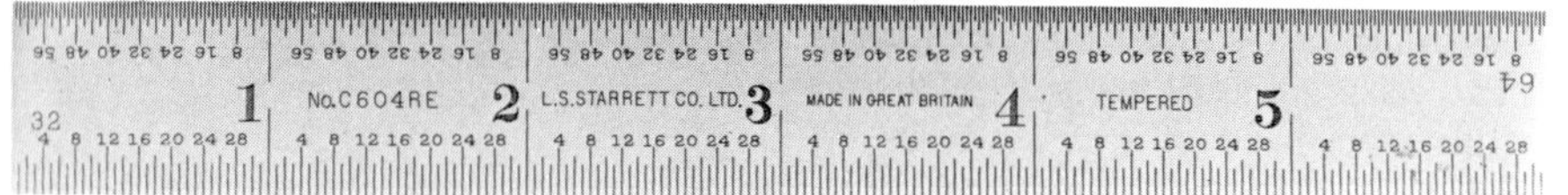


**Fig. 2-1** A steel rule with no. 4 graduations showing 8ths and 16ths. (*Courtesy The L. S. Starrett Company.*)

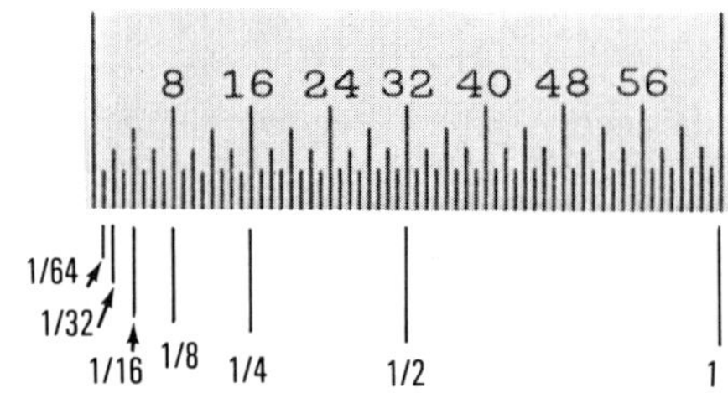


**Fig. 2-2** A steel rule with no. 4 graduations showing 32ds and 64ths. (*Courtesy Kostel Enterprises Ltd.*)

**Fig. 2-3** A flexible steel rule. (*Courtesy The L. S. Starrett Company.*)

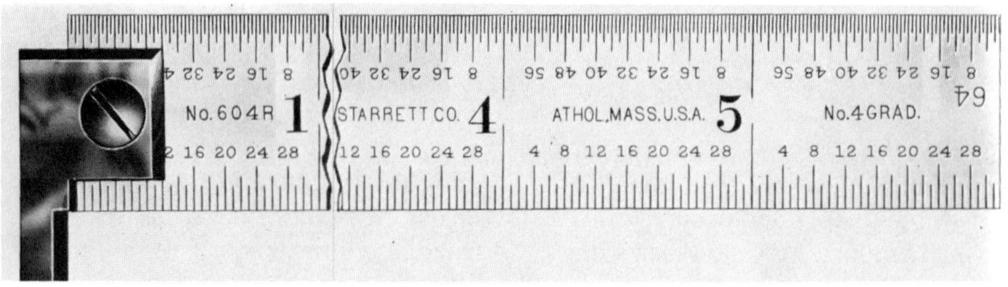


**Fig. 2-4** A hook rule is used to make accurate measurements from a shoulder, step, or edge. (*Courtesy The L. S. Starrett Company.*)

## Inch Steel Rules

The *spring-tempered rule* (Fig. 2-1) with the #4 graduation is the most common rule found in a machine shop. It is graduated in 8ths and 16ths on one side, and in 32ds and 64ths on the other side. Every eighth of an inch is numbered to make it easier to read the 32ds and 64ths graduations (Fig. 2-2).

The *flexible rule,* Fig. 2-3, is similar to the spring-tempered rule but narrower. It is designed to be used where the stiffness of a spring-tempered rule would not permit proper measurement.

The *hook rule,* Fig. 2-4, is used to make accurate measurements from a shoulder, step, or edge of a workpiece. It may also be used for measuring circular pieces and for setting inside calipers to a size.

*Short-length rules,* Fig. 2-5, are used to measure small openings and hard-to-reach locations where standard-length rules could not be used. The five small rules, ranging from 1/4 to 1 in. in length, can be interchanged in the holder. This allows the short rules to be firmly held while making a measurement.

*Decimal rules,* Fig. 2-6A, are generally used when measurements smaller than 1/64 in. are required. Since linear dimensions are sometimes given on a drawing in decimals, these rules are particularly useful to the machinist. The most common graduations found on decimal rules are: .100 (1/10 in.), .050 (1/20 in.), .020 (1/50 in.), and .010 (1/100 in.). Figure 2-6B shows an enlarged sketch of a rule graduated in .010 (1/100 in.).

*Metric rules,* Fig. 2-7, are generally graduated in full millimeters (mm) and half millimeters (0.5 mm). A comparison of the inch and metric dimensions for the most common fractions is shown below.

| Inch | Metric, mm |
|---|---|
| 1/32 | 0.79 |
| 1/16 | 1.58 |
| 1/8 | 3.17 |
| 1/4 | 6.35 |
| 1/2 | 12.70 |
| 1 | 25.40 |

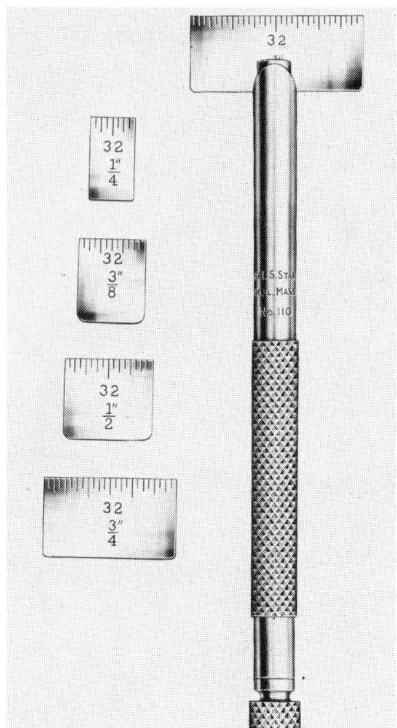

**Fig. 2-5** Short-length rules are used to measure narrow spaces. (*Courtesy the L. S. Starrett Company.*)

N-1 For accurate measurements use the 1-in. or 10-mm graduation on the rule since the end may be worn.

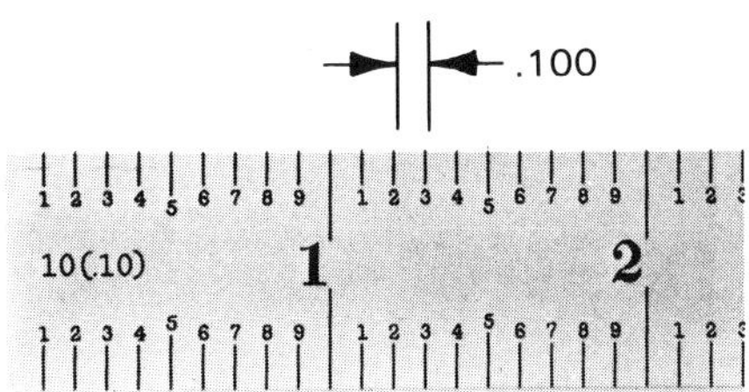


**Fig. 2-6A** A decimal rule showing .100 graduations.

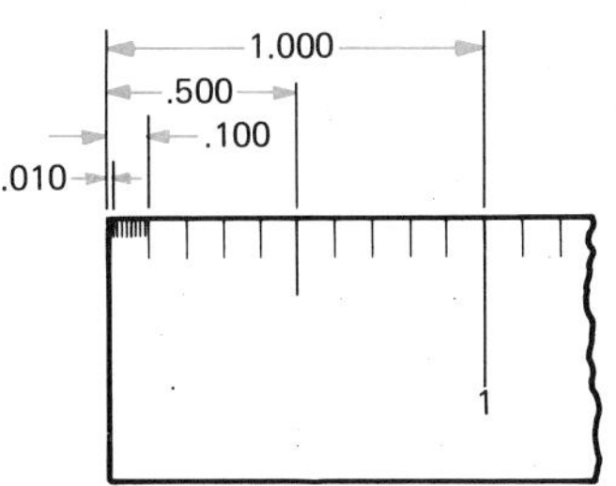


**Fig. 2-6B** Decimal graduations on a rule.

N-2 A magnifying glass will help make reading a decimal rule easier.

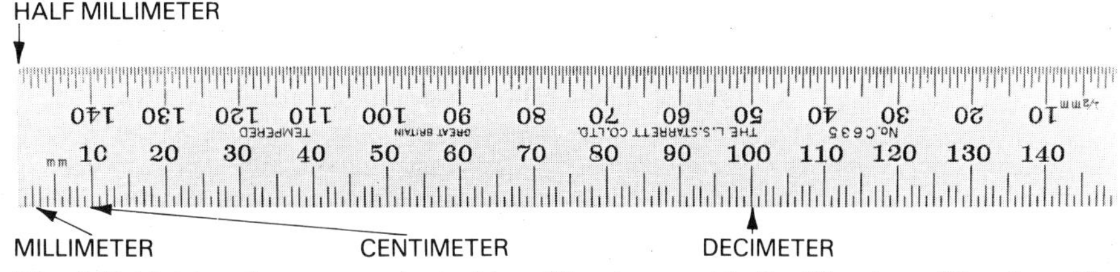


**Fig. 2-7** Metric rules are graduated in millimeters and half millimeters. (*Courtesy The L. S. Starrett Company.*)

# READING FRACTIONAL, DECIMAL, AND METRIC RULES

Although rules are not classed as precision measuring tools, reasonable accuracy is possible if care is taken when measurements are being made.

## Reading a Fractional Rule

When measuring with a steel rule, proceed as follows (see N-1):

1. Note the number of full inches.
2. Add to this the fraction past the last full inch mark.

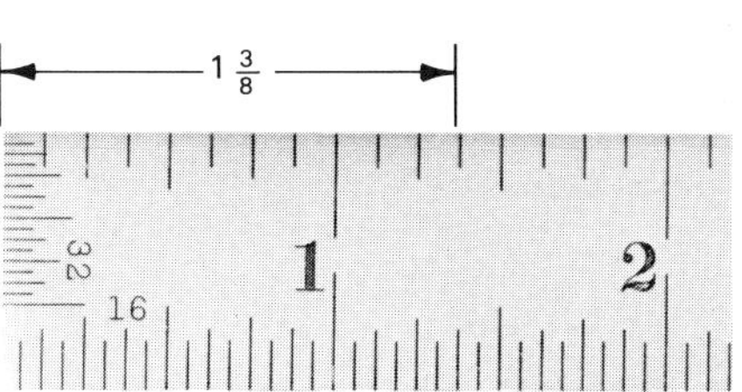


**Fig. 2-8** A reading of $1\frac{3}{8}$ in. on the $\frac{1}{8}$th scale.

Reading is:

1 in. + (3 × $\frac{1}{8}$ in.) = 1 + $\frac{3}{8}$, or $1\frac{3}{8}$ in.

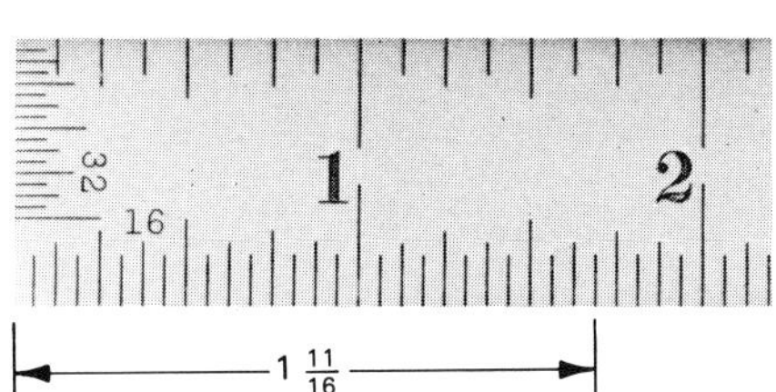


**Fig. 2-9** A reading of $1\frac{11}{16}$ in. on the $\frac{1}{16}$th scale.

Reading is:

1 + (11 × $\frac{1}{16}$) = 1 + $\frac{11}{16}$, or $1\frac{11}{16}$ in.

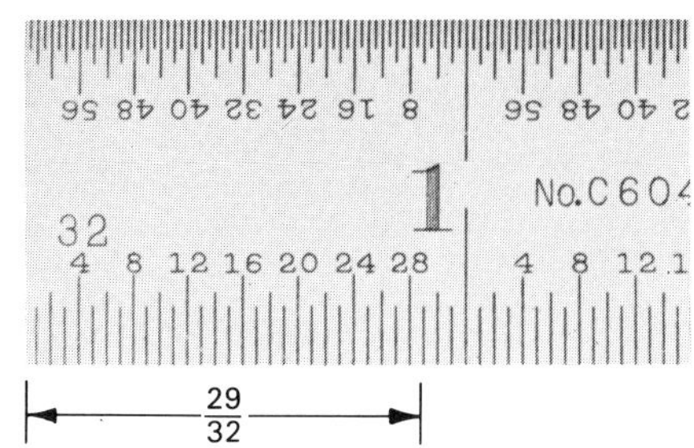


**Fig. 2-10** A reading of $\frac{29}{32}$ in. on the $\frac{1}{32}$nd scale.

Reading is:

one mark past the 28 graduation on the thirty-second scale

Reading is:

28 + $\frac{1}{32}$ = $\frac{29}{32}$ in.

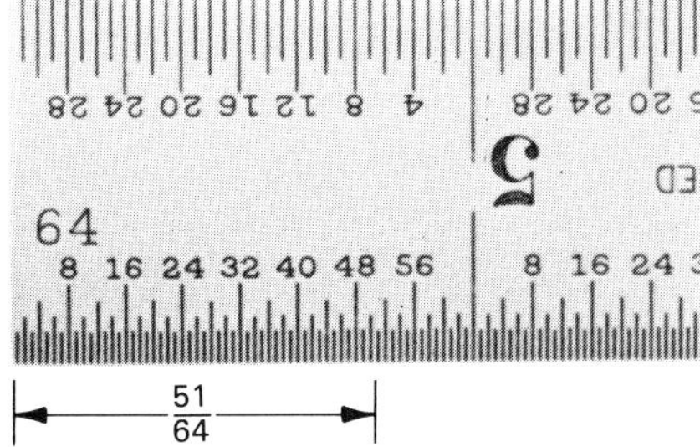


**Fig. 2-11** A reading of $\frac{51}{64}$ in. on the $\frac{1}{64}$th scale.

Reading is:

three marks past the forty-eight graduation on the sixty-fourth scale

Reading is:

48 + $\frac{3}{64}$ = $\frac{51}{64}$ in.

## Reading a Decimal Rule

Decimal rules are used when the dimensions on a drawing are shown in decimals. Accuracy of .010 or better is possible if care is used when reading and using the rule. (See N-2.) In Fig. 2-12 the reading is:

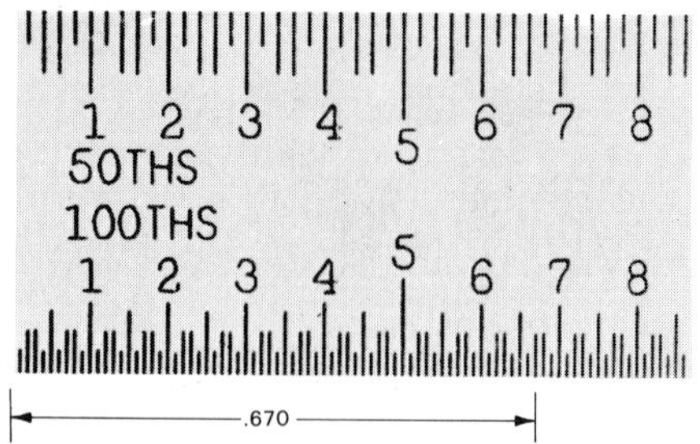


Fig. 2-12 A reading of .670 on the 100ths (.010) scale.

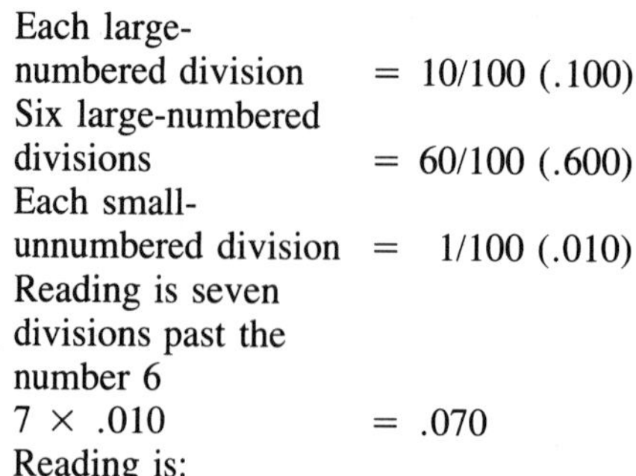

| | | |
|---|---|---|
| Each large-numbered division | = | 10/100 (.100) |
| Six large-numbered divisions | = | 60/100 (.600) |
| Each small-unnumbered division | = | 1/100 (.010) |
| Reading is seven divisions past the number 6 | | |
| 7 × .010 | = | .070 |
| Reading is: | | |
| .600 + .070 | = | .670 |

## Reading a Metric Rule

Metric rules are usually graduated in millimeters and half millimeters and are used for making metric readings which do not require great accuracy. These rules are available in lengths from 150 mm to 1 m.

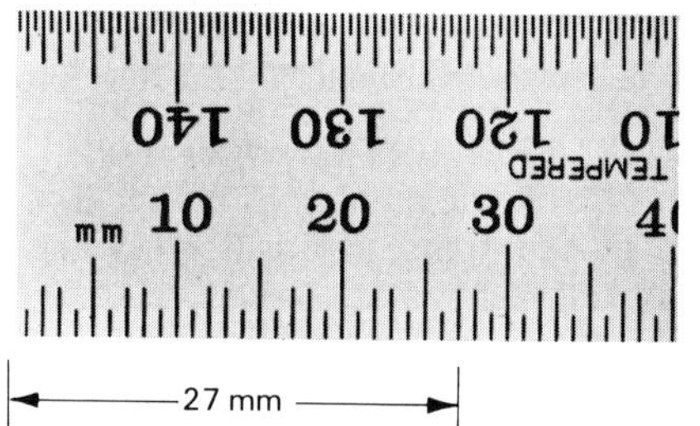


Fig. 2-13 A reading of 27 mm on the millimeter scale. (*Courtesy Kostel Enterprises Ltd.*)

N-3 Before measuring, remove the burrs from the workpiece with a file.

N-4 The rule must always be held parallel to the edge for accurate measurements.

### How to Read a Metric Rule

When measuring with a metric rule, proceed as follows (see Fig. 2-13):

**1.** Note the number of main divisions showing. Each main division has a value of 10 mm.

Reading is: 2 × 10 = 20 mm

**2.** Add the number of lines showing past the 20-mm line. Each small line has a value of 1 mm.

Reading is: 7 × 1 = 7 mm

Total reading 27 mm

## MEASURING WITH RULES

The size of a workpiece may be checked with reasonable accuracy by several methods using a standard steel rule. The following methods may be used for measuring with a rule.

### If the End of the Rule Is in Good Condition and Not Worn

**1.** Place the end of the work and the end of the rule against a flat shoulder or block, Fig. 2-14.

**2.** Read the size of the work by noting the line on the rule which is directly over the end of the work. In this case the reading is 2$\frac{3}{16}$ in.

### If the End of the Rule Is Worn or Damaged

**1.** Remove all burrs from the workpiece. (See N-3.)

**2.** Place the rule *A* on the work so that the center of the 1-in. (or 10-mm) line is over the end *B* of the workpiece, Fig. 2-15A, and parallel to the edge *C*. If the rule is not placed parallel to the edge of the workpiece, a longer measurement will be indicated, Fig. 2-15B. (See N-4.)

**3.** Note the reading on the rule directly over the edge *D* of the workpiece.

**4.** The reading is 4$\frac{13}{16}$ in.; since the reading started at 1 in., the actual length is 4$\frac{13}{16}$ − 1 or 3$\frac{13}{16}$ in.

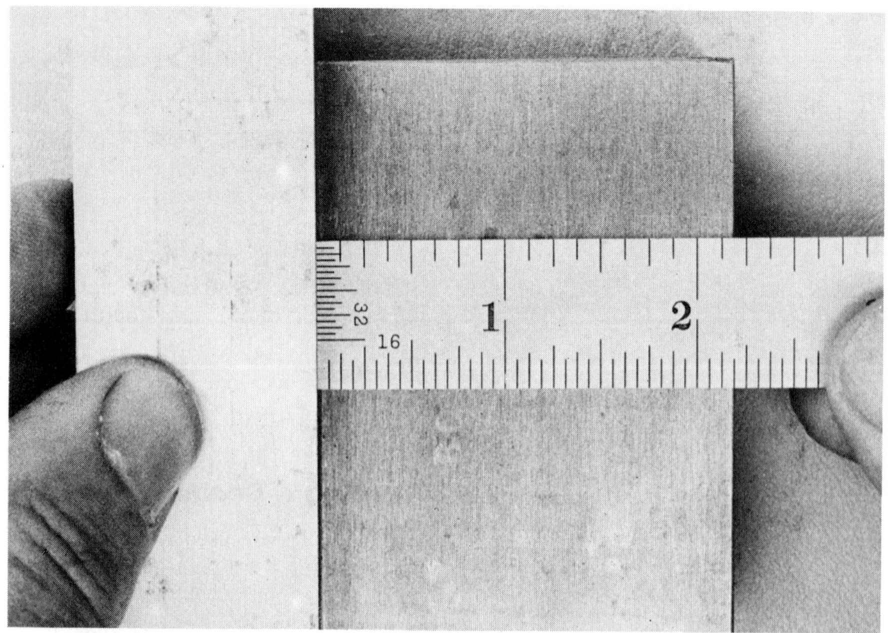

Fig. 2-14 When the end of the rule is in good condition, butt the rule and the work against a shoulder. (*Courtesy Kostel Enterprises Ltd.*)

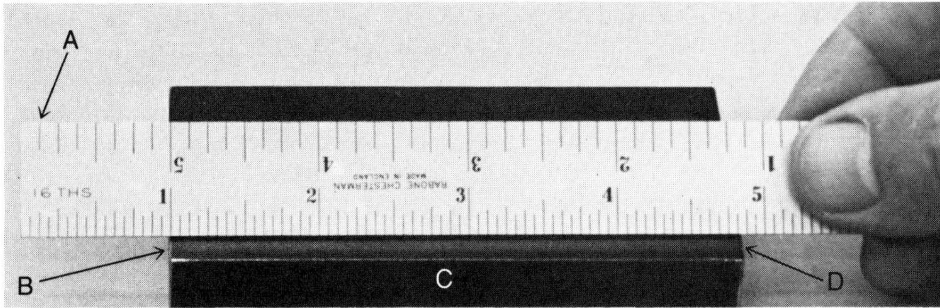


Fig. 2-15A If the end of the rule is worn, place the 1 in. (or 10-mm) line even with the edge of the work.

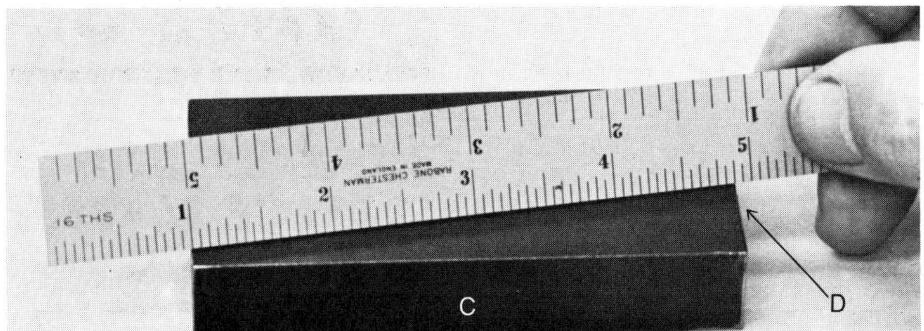


Fig. 2-15B If rule is not parallel to edge C, inaccurate measurement will result.

## When Measuring Round Stock

**1.** Remove all burrs from the end of the workpiece with a file.

**2.** Hold the workpiece in one hand.

**3.** Hold the rule in the other hand with the 1-in. graduation line even with the edge of the workpiece. Be sure that the edge of the rule is on the centerline of the work, Fig. 2-16. (See N-5.)

**4.** Place the thumb against the workpiece to steady the rule.

**5.** Note the reading on rule ($2\frac{1}{4}$ in.).

**6.** Deduct 1 in. from the reading; the diameter of this workpiece is $1\frac{1}{4}$ in. (See N-6.)

N-5 If the rule is not on the centerline, an inaccurate measurement will be made.

N-6 Remember to deduct 1 in. from the measurement when starting at the 1-in. line.

N-7 Do not deduct 1 in. when measuring from the end of a rule.

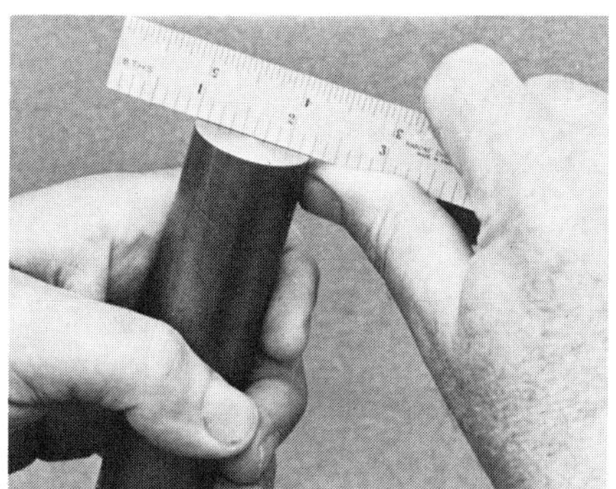

Fig. 2-16 Round work is measured quickly and easily with a rule. (*Courtesy Kostel Enterprises Ltd.*)

## A Hook Rule Is Very Convenient for Measuring the Size of a Workpiece

**1.** Remove the burrs from the workpiece.

**2.** Place the hook over the end of the work, Fig. 2-17.

**3.** Hold the edge of the rule parallel to the edge of the work. When measuring round work, hold the rule on the centerline of the workpiece.

**4.** Note the rule graduation which is directly on the edge of the workpiece. In this case the reading is $2\frac{3}{4}$ in.

*Note:* This is actual size, since the measurement was taken from the end of the rule. Do not deduct 1 in. (See N-7.)

Fig. 2-17 A hook rule provides an accurate means of measuring with a rule. (*Courtesy Kostel Enterprises Ltd.*)

## REVIEW QUESTIONS

**1.** Name four types of steel rules used in machine shop work.
**2.** Describe the most common type of steel rule.
**3.** For what purpose are the following rules used?
(a) Hook rules
(b) Short-length rules
(c) Decimal rules
**4.** How are the following rules usually graduated?
(a) Decimal rules
(b) Metric rules
**5.** What are the readings on the following rules?

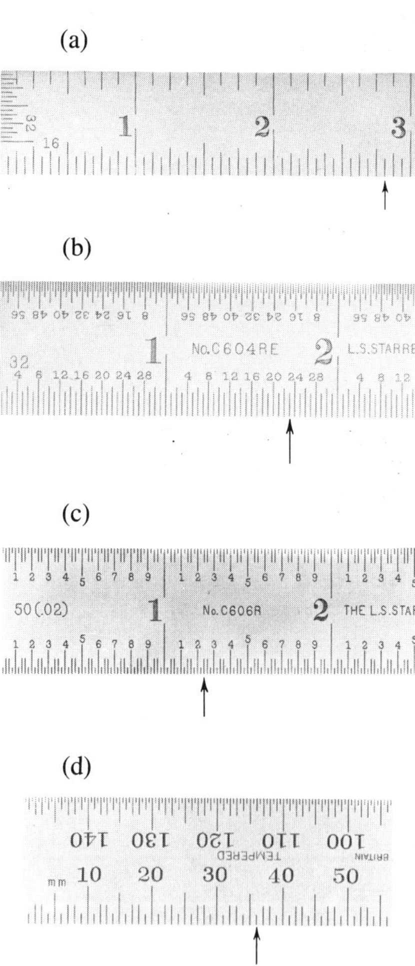

**6.** Name three methods of measuring a rectangular workpiece 4½ in. long.
**7.** Why must a rule be held parallel to the edge of a rectangular workpiece?
**8.** List two precautions that should be observed when measuring a round workpiece.
**9.** Why does a hook rule provide a more accurate means of measuring a piece of work?

# UNIT 3

# CALIPERS, SQUARES, AND SURFACE PLATES

Because of the wide variety of work encountered in machine shops, many tools are required for measuring, testing, and setting up the work. Since these three operations generally determine the accuracy of the part produced, it is important that care be used for these operations. The accuracy obtained with tools such as outside and inside calipers, squares, and surface plates depends on the care given these tools and the skill developed by the operator.

## OBJECTIVES

After completing this unit, you should be able to:

**1.** Measure to within .010-in. [0.25-mm] accuracy when setting a caliper to a rule
**2.** Measure to within .005-in. [0.12-mm] accuracy when setting a caliper to a gage
**3.** Identify and use four types of squares
**4.** Identify and preserve the accuracy of two types of surface plates

## MEASURING WITH CALIPERS

The most common calipers used in a machine shop are the *outside* and *inside* calipers. Both of these are made in the *spring-joint* and *firm-joint* types. The spring-joint calipers are more commonly used since they can easily be adjusted to size.

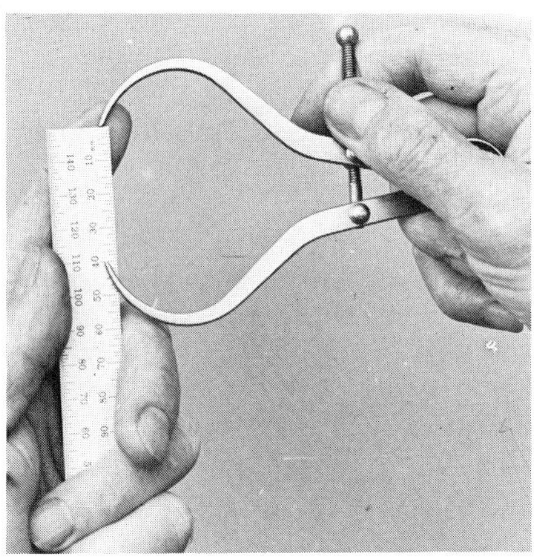

Fig. 3-1 Setting an outside caliper to a rule. (*Courtesy Kostel Enterprises Ltd.*)

N-1 Always keep the caliper legs parallel to the edge of a rule for accurate settings.

N-2 Be sure that the caliper leg splits the desired graduation line on a rule.

N-3 Wait until the machine is fully stopped before measuring work.

N-4 Never force the caliper over the work.

## Outside Calipers

Outside spring calipers are used to measure either round or flat work. They consist of two curved legs, a spring, and an adjusting nut, Fig. 3-1. Since the caliper cannot be read directly, it must be set to a steel rule or to a standard-sized gage. When the caliper is set to a rule, the accuracy obtained will depend on the condition of the rule and the skill of the operator. An accuracy of .010 to .012 in. [0.25 to 0.31 mm] is possible if care is taken in setting and using calipers. If calipers are set to a gage, an experienced machinist can attain an accuracy of .002 to .003 in. [0.05 to 0.06 mm].

### Setting the Caliper to a Rule

**1.** Hold the rule in one hand so that the forefinger extends slightly beyond the edge of the rule, Fig. 3-1.

**2.** Place one leg of the caliper over the end of the rule and support it with the end of the forefinger. (See N-1.)

**3.** Turn the adjusting nut with the thumb and forefinger of the other hand until the end of the "free" caliper leg splits the desired graduation line. (See N-2.)

### Checking the Work Size

**1.** Hold the caliper spring lightly between the thumb and the forefinger, Fig. 3-2. (See N-3 and N-4.)

**2.** Place the caliper on the work so that the line between the legs of the calipers is at right angles to the centerline of the work, Fig. 3-2.

**3.** Note that the diameter is correct when the caliper just slides over the work with its own weight.

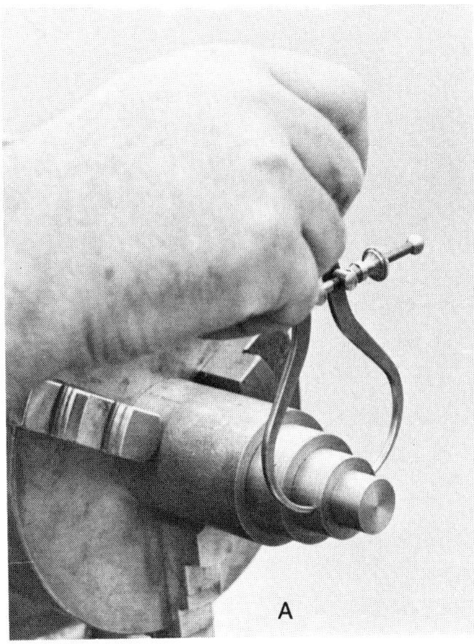


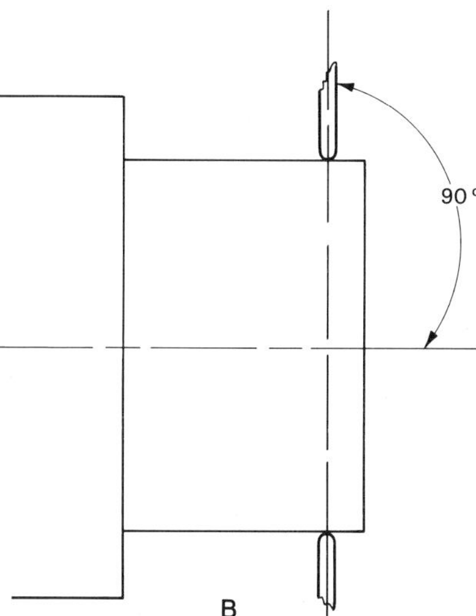


Fig. 3-2 Checking a diameter with an outside caliper. (*Courtesy Kostel Enterprises Ltd.*)

## Inside Calipers

Inside calipers are used to measure the diameter of holes or the width of keyways and slots.

### Setting Inside Calipers to a Rule

Inside calipers set to a rule may be used to measure a piece of work to a rough size only. When checking a workpiece to a finish size, the calipers should be set to a gage or a micrometer.

**1.** Butt the end of a rule and one leg of the caliper against a shoulder or flat surface, Fig. 3-3.

**2.** Keep the ends of both legs parallel to the edge of the rule.

**3.** With the thumb and forefinger, turn the adjusting nut until the free leg splits the desired graduation line on a rule.

### Measuring Inside Diameters

**1.** Place one leg of the caliper at the bottom of the hole. Keep the leg from moving by holding it in place with the forefinger of one hand, Fig. 3-4.

**2.** Adjust the caliper with the other hand until it cannot be moved sideways and a slight drag is felt on the free leg as it is moved in and out of the hole.

N-5 The amount of "drag" on the caliper leg must be the same as the drag felt in the hole.

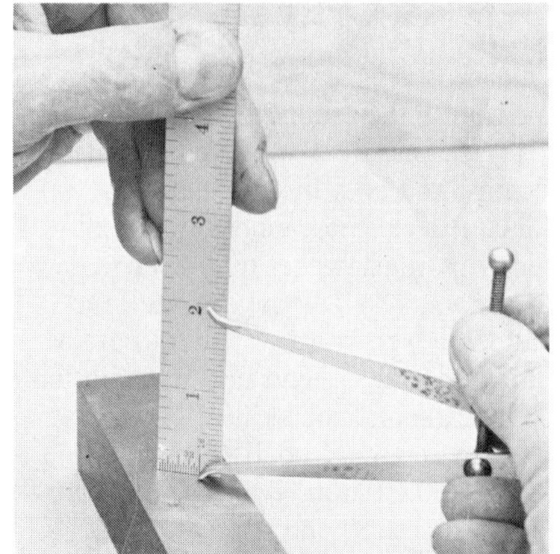
**Fig. 3-3** Setting an inside caliper to a rule. (*Courtesy Kostel Enterprises Ltd.*)

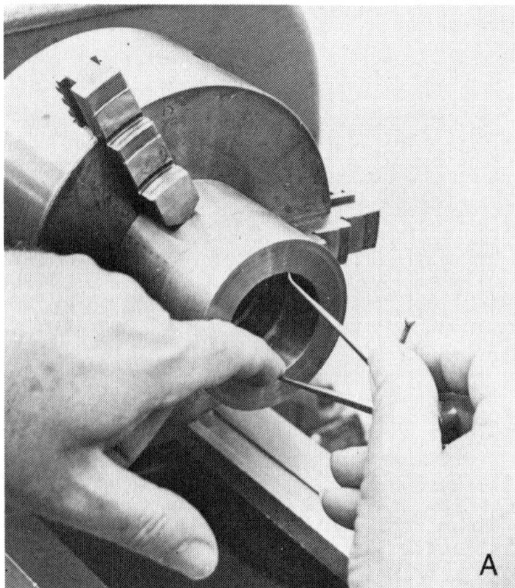


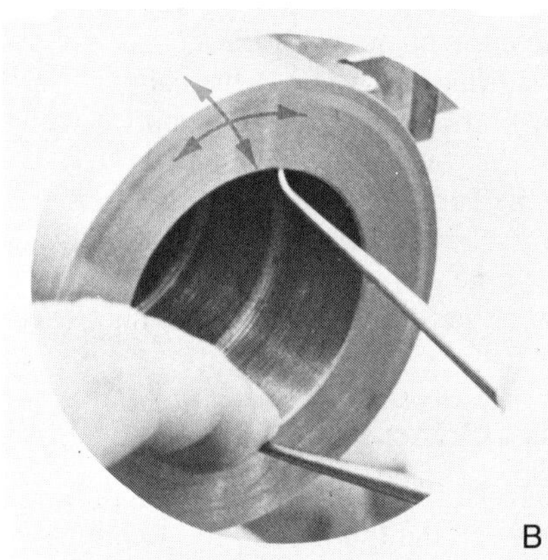


**Fig. 3-4** Adjusting an inside caliper to the size of a hole. (*Courtesy Kostel Enterprises Ltd.*)

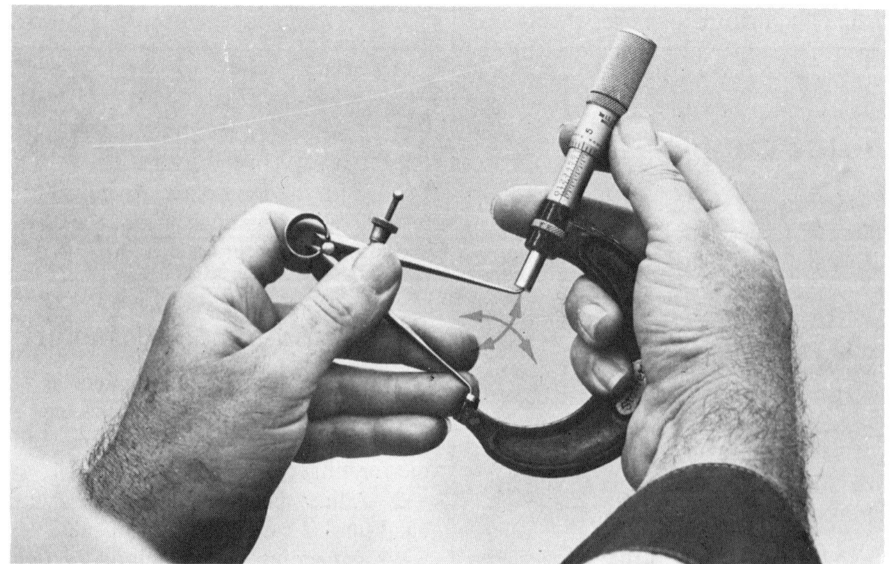
**Fig. 3-5** Inside caliper setting being transferred to a micrometer. (*Courtesy Kostel Enterprises Ltd.*)

**3.** Hold the micrometer in one hand so that it can be adjusted with the thumb and forefinger, Fig. 3-5.

**4.** Place one leg of the caliper on the micrometer anvil and hold it in position with one finger.

**5.** Move the other leg across the micrometer spindle and adjust micrometer until a slight drag is felt. (See N-5.)

**6.** Remove the thumb and forefinger from the micrometer before a reading is made.

## PRECISION SQUARES AND SURFACE PLATES

### Precision Squares

Precision squares are used chiefly for inspection and setup purposes. They are hardened and accurately ground and must be handled carefully to preserve their accuracy. A great variety of squares is manufactured for specific purposes. The squares are all variations of the solid square or the adjustable square, however.

The solid square, Fig. 3-6, sometimes called the "master precision square," is used where extreme accuracy is required. It is made up of two parts, the beam and the blade, both of which are usually hardened and ground. Three uses of a solid square are:

**1.** To check a surface for flatness

**2.** To determine if two surfaces are at right angles (square) to each other

**3.** To check the accuracy of other squares

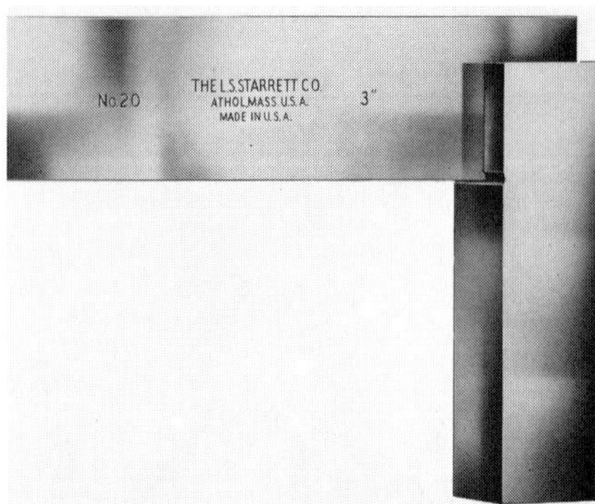


**Fig. 3-6** The better-model solid squares are used as master squares. (*Courtesy The L. S. Starrett Company.*)

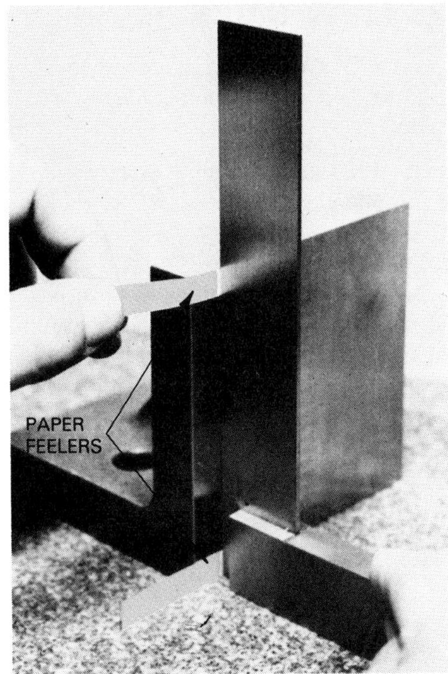


**Fig. 3-7** Squareness may be checked by using paper strips between the work and the blade of the square. (*Courtesy Kostel Enterprises Ltd.*)

N-6 Use approximately .002-in.-thick and ½-in.-wide paper feelers when testing for squareness.

N-7 Handle all precision tools with extreme care.

## Beveled-Edge Squares

The better-quality standard squares used in inspection have a beveled-edge blade which is hardened and ground. The beveled edge allows the blade to make a line contact with the work, thereby permitting a more accurate check. Two methods of using a beveled-edge square for checking purposes are illustrated in Figs. 3-7 and 3-8. In Fig. 3-7 if the work is square (90°), both pieces of paper will be tight between the square and the work. In Fig. 3-8, the light is shut out only where the blade makes line contact with the surface of the work. Light shows through where the blade of the square does not make line contact with the surface being checked. (See N-6.)

**Fig. 3-8** Light shows through where the blade does not make line contact with the work.

**Fig. 3-9** Checking work with a toolmaker's surface plate square.

## Toolmaker's Surface-Plate Square

The toolmaker's surface-plate square (Fig. 3-9) provides a convenient method of checking work for squareness on a surface plate. Since it is of one-piece construction, there is little chance of any inaccuracy developing, as is the case with a blade and beam square.

## Care of Squares

In order to preserve the accuracy of squares, it is important that they be treated with the same care shown toward other precision instruments. The following points should be observed.

**1.** Never drop a square; the accuracy can be destroyed.
**2.** Keep it clean at all times.
**3.** Cover the square with a light film of oil when not in use to prevent rust.
**4.** Store squares in a special individual box or compartment to prevent nicks and burrs. (See N-7.)

## Surface Plates

A surface plate is a rigid block of granite or cast iron, the flat surface of which is used as a reference plane for layout, setup, and inspection work. Surface plates generally have a three-point suspension to prevent rocking when mounted on an uneven surface.

*Cast-iron plates,* Fig. 3-10, are well ribbed and supported to resist deflection under heavy loads. They are made

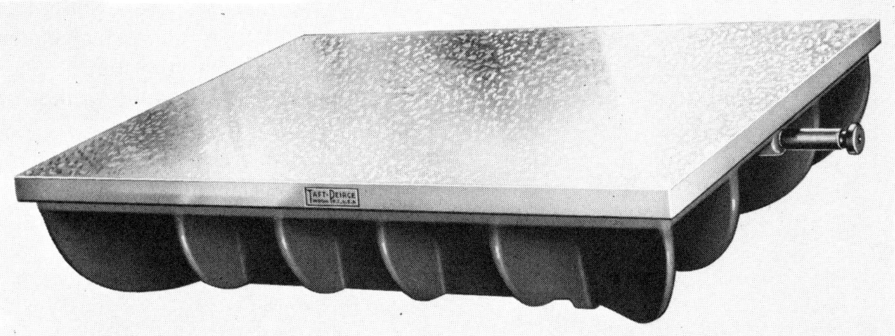

**Fig. 3-10** Cast-iron surface plates are costly and require much care. (*Courtesy Taft-Peirce Mfg. Co.*)

**Fig. 3-11** The surface of better grades of granite surface plates are flat and accurate to within .0001 in. (0.0025 mm).

of close-grained cast iron, which has high strength and good wear-resistance qualities. After a cast-iron surface plate has been machined, its surface must be scraped by hand to a flat plane. This operation is long and tedious; therefore, the cost of these plates is high.

*Granite surface plates,* Fig. 3-11, have many advantages over the cast-iron types and are replacing them in many shops. They may be manufactured from grey, pink, or black granite and can be obtained in several degrees of accuracy. Extremely flat finishes on granite plates are produced by lapping.

The advantages of granite plates are:

**1.** They are not appreciably affected by temperature change. Granite will not burr as does cast iron; therefore the accuracy is not impaired.
**2.** They are nonmagnetic.
**3.** They are rustproof.

### Care of Surface Plates

**1.** Keep surface plates clean at all times and wipe them with a *clean, dry* cloth before using.
**2.** Clean them occasionally with solvent or surface-plate cleaner to remove any film.
**3.** Protect them with a wooden cover when not in use.
**4.** Use parallels whenever possible to prevent damage to the surface of the plates by rough parts or castings.
**5.** Remove burrs from the workpiece before placing it on the plate.
**6.** Slide heavy parts onto the plate rather than place them directly on the plate, since a part might fall and damage the plate.
**7.** Remove all burrs from cast-iron plates by honing.
**8.** When they are not in regular use, cover cast-iron plates with a thin film of oil to prevent rusting. (See N-8.)

N-8 Apply a light film of oil or rust preventative to cast-iron surface plates when not in use.

## REVIEW QUESTIONS

**1.** State two factors affecting the accuracy of outside calipers set to a rule.

**2.** What accuracy may be expected when outside calipers are:
(a) Set to a rule?
(b) Set to a gage?

**3.** List two precautions to observe when checking the workpiece size with outside calipers.

**4.** Describe how inside calipers should be adjusted when measuring an inside diameter.

**5.** Why is a beveled-edge square better for checking workpieces than the standard square?

**6.** List three points to observe in the proper care of squares.

**7.** Why are granite plates considered superior to cast-iron surface plate?

**8.** What precaution should be taken when setting up a rough casting on a surface plate?

**9.** What should be done when a cast-iron surface plate is not being used?

# UNIT 4

# MEASURING WITH MICROMETERS

The *micrometer caliper,* usually called the *micrometer,* is the most commonly used measuring instrument when accuracy is required. The standard inch, vernier, and metric micrometers are available in a wide variety of sizes and shapes to suit workpieces. Since the majority of work in a machine shop is measured with a micrometer, it is very important to be able to read and use micrometers properly in order to measure and machine work accurately.

## OBJECTIVES

Upon completion of this unit you should be able to:

**1.** Identify the most common types of outside micrometers and their use
**2.** Measure the size of a variety of objects to within an accuracy of .001 in.
**3.** Read vernier micrometers to within .0001 in.
**4.** Measure the size of a variety of objects to within an accuracy of 0.01 mm

The *standard inch micrometer* measures accurately to one-thousandth of an inch (.001 in.). Most micrometers have a range of 1 in. and each is classified by the largest size it can measure. That is, a 1-in. micrometer will measure only from 0 to 1 in. A 3-in. micrometer can measure any workpiece between 2 and 3 in. in size. Since many phases of modern manufacturing require greater accuracy, the *vernier micrometer,* capable of measurements of one ten-thousandths of an inch (.0001 in.), is being used to an increasingly greater extent. The *standard metric micrometer,* Fig. 4-1, measures in hundredths of a millimeter, while the *vernier metric micrometer* measures up to two-thousandths of a millimeter.

## Types of Outside Micrometers

Outside micrometers are available in a large variety of types and sizes to suit various purposes. The sizes of micrometers range from 1 in. [25 mm] up to 60 in. [1500 mm]. The most common size measures from 0 to 1.000 in. or from 0 to 25 mm. There are generally two types of outside micrometers:

**1.** Those where the size of the frame or the method of reading them has been changed to suit various applications
**2.** Those where the shape of the anvil, spindle, or both have been changed to suit various workpieces.

## Micrometer Spindles

Micrometers used where severe abrasive conditions exist generally have *tungsten carbide measuring faces* Fig. 4-2. The hardness of this surface pre-

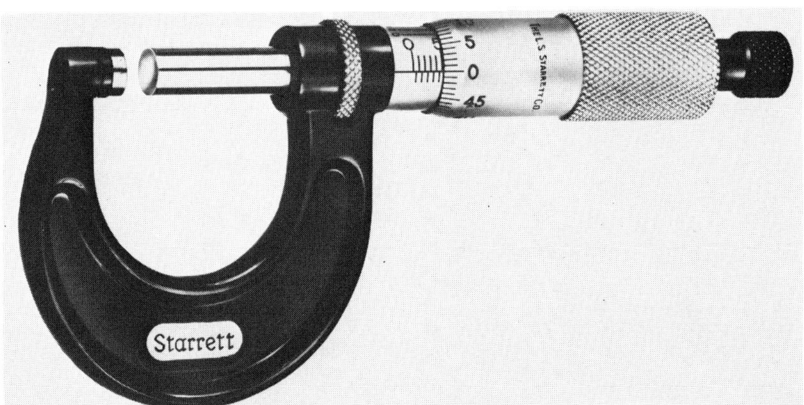


**Fig. 4-1** A standard metric micrometer. (*Courtesy The L. S. Starrett Company.*)

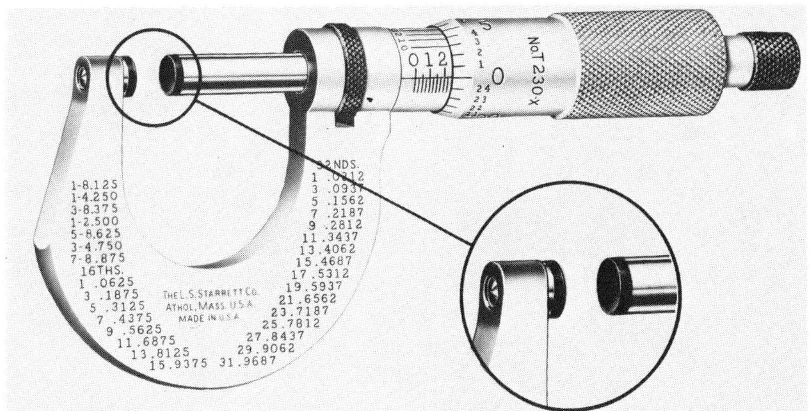

**Fig. 4-2** Carbide measuring faces prevent wear. (*Courtesy The L. S. Starrett Company.*)

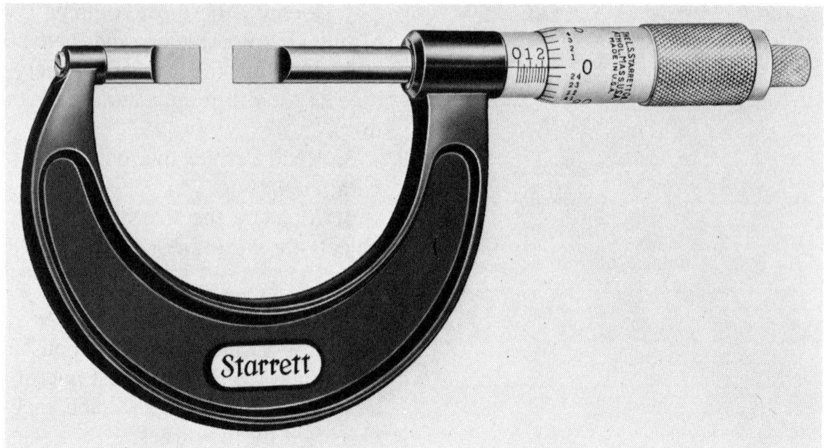


**Fig. 4-3** Blade-type anvils allow measuring in narrow places. (*Courtesy The L. S. Starrett Company.*)

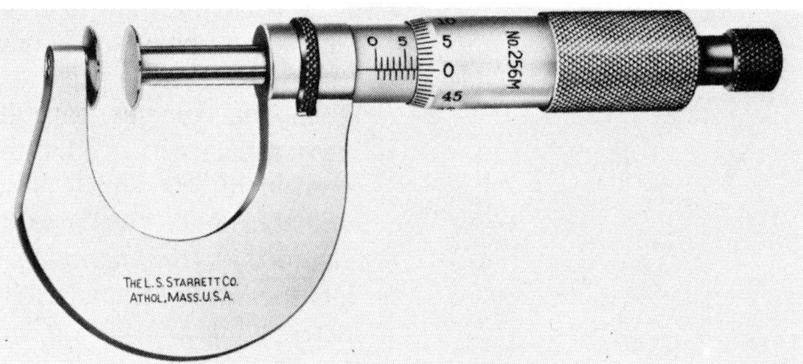


**Fig. 4-4** Disc-type anvils allow the measuring of closely spaced sections on a machine part. (*Courtesy The L. S. Starrett Company.*)

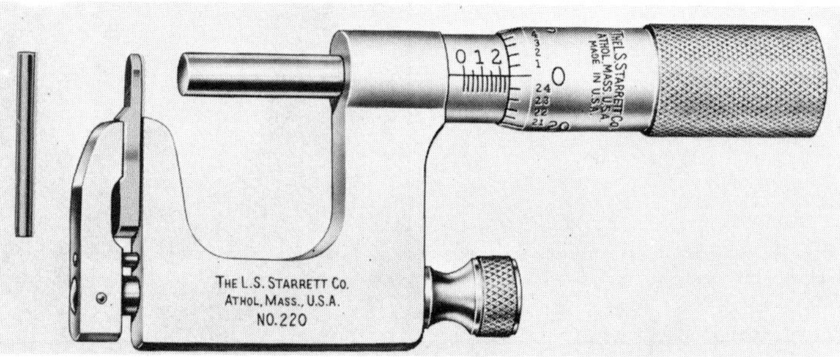


**Fig. 4-5** The Mul-T-Anvil micrometer is used for measuring tubing and distances from a slot to an edge. (*Courtesy The L. S. Starrett Company.*)

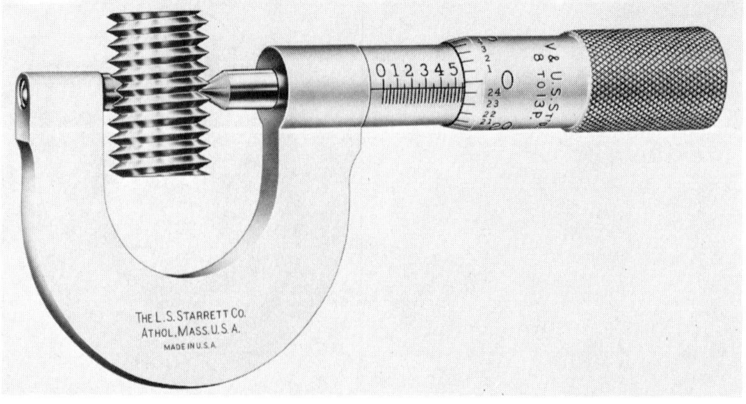


**Fig. 4-6** The screw thread micrometer measures the pitch diameter of a thread. (*Courtesy The L. S. Starrett Company.*)

**Fig. 4-7** Tubular frame micrometers are used for measuring large work. (*Courtesy The L. S. Starrett Company.*)

vents the measuring faces from wearing and thereby ensures the accuracy of the micrometer.

Micrometers with *blade-type anvils,* Fig. 4-3, are used for the accurate measurement of circular-formed tools as well as the diameter and depth of narrow grooves, slots, recesses, keyways, and shallow depths.

*Disc-type micrometers,* Fig. 4-4, are used to measure narrow grooves, slots, and recesses down to .015-in. width. The large-diameter flats on the anvil and spindle measuring faces make this micrometer useful for measuring paper, plastics, etc.

The *Mul-T-Anvil micrometer,* Fig. 4-5, comes equipped with a round and a flat anvil which are interchangeable. The round (rod) anvil is used for measuring the wall thickness of tubing, cylinders, etc., and for measuring from a hole to an edge. The flat anvil is used to measure the distance from inside of slots and grooves to an edge.

The *screw-thread micrometer,* Fig. 4-6, is used to accurately measure the pitch diameter of a thread. The micrometer has a pointed spindle and a double-V anvil to contact the screw thread being measured.

The *tubular frame micrometer,* Fig. 4-7, is made for easier, faster precision measuring of large outside diameters up to 60 in. The hollow tubular frame is made of special steel to give it extreme rigidity and the lightest possible weight. Interchangeable anvils give each micrometer a range of 6 in.

The *direct-reading micrometer,* Fig. 4-8, has the graduations on the thimble and barrel as in a standard micrometer

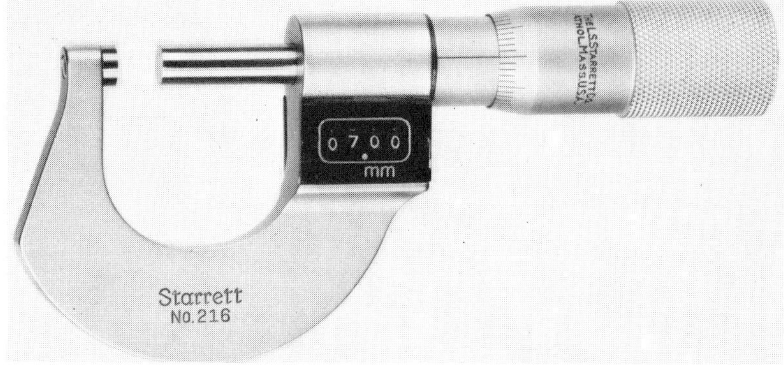


**Fig. 4-8** The direct-reading micrometer has a digital readout built into the frame. (*Courtesy The L. S. Starrett Company.*)

Fig. 4-9 The indicating micrometer can be used as a comparator. (*Courtesy Federal Products Corporation.*)

in addition to a digital readout built into the frame. The exact micrometer reading at any point within its range is shown in the readout. Some micrometers combine both the standard inch reading on the thimble and barrel with the readout given in millimeters.

The *indicating micrometer,* Fig. 4-9, uses an indicating dial and a movable anvil to permit accurate measurements to ten-thousandths of an inch [0.002 mm]. This micrometer may be used as a comparator by setting it to a particular size with gage blocks or a standard and locking the spindle. The tolerance arms are then set to the required limits and each piece of work can be compared with the micrometer setting.

## PARTS OF A MICROMETER

Regardless of the type or size of an outside micrometer, they all contain certain basic parts, Fig. 4-10.

- The U-shaped *frame* holds all the micrometer parts together.
- The *anvil* is the fixed measuring face attached to one end of the frame.
- The *spindle* is the movable measuring face. Turning the thimble moves the spindle, increasing or decreasing the distance between the anvil and the spindle face.
- The *sleeve* (barrel) holds the spindle and is graduated into equal divisions, each line having a value of .025 in. for inch micrometers.
- The *thimble* has equally spaced divisions around its circumference, each having a value of .001 in. for inch micrometers.
- The *friction thimble* on the end of the thimble is used to ensure an accurate measurement, and to prevent too much pressure from being applied to the micrometer. On some micrometers, a *ratchet stop* serves the same purpose.

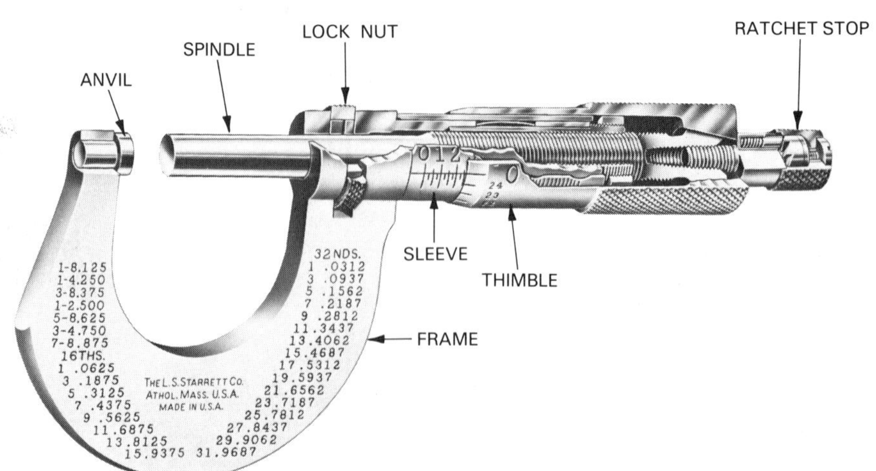


Fig. 4-10 A cutaway view of a standard micrometer. (*Courtesy The L. S. Starrett Company.*)

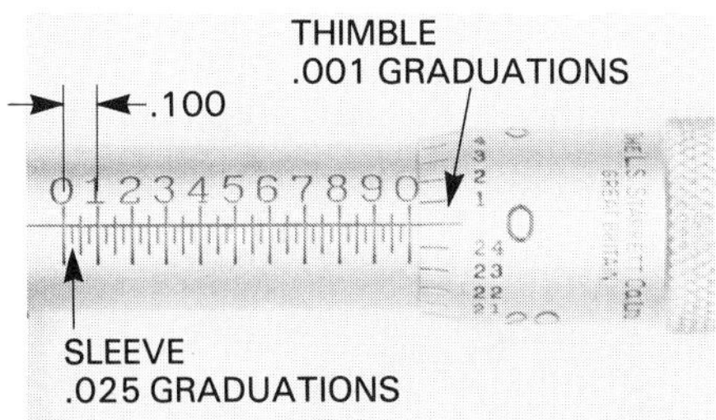


**Fig. 4-11** The graduations on a standard inch micrometer. (*Courtesy Kostel Enterprises Ltd.*)

N-1 Micrometers are precision instruments which must be handled with care.

N-2 Always test the accuracy of a micrometer before using it to measure work.

## CARE OF THE MICROMETER

Everyone should be impressed with the fact that any instrument capable of measuring to within .001 in. or to 0.01 mm must be treated with a great deal of care and not abused, otherwise its accuracy will be damaged. The following points should be kept in mind when using micrometers:

1. Never drop a micrometer; always place it down gently in a clean place.
2. Never place tools or other materials on a micrometer.
3. Never lay a micrometer in steel chips or grinding dust; never handle it with oily hands.
4. Never attempt to use a micrometer on moving work.
5. Keep the micrometer clean and accurately adjusted.

## STANDARD INCH MICROMETER

### Principle

There are 40 threads per inch on the spindle of the inch micrometer. Therefore, one complete revolution of the spindle will either increase or decrease the distance between the measuring faces by 1/40 (.025) in. The 1-in. distance marked on the micrometer sleeve (Fig. 4-11) is divided into 40 equal divisions, each of which equals 1/40 (.025) in. (See N-1.)

If the micrometer is closed until the measuring faces just touch, the zero line on the thimble should line up with the index line on the sleeve (barrel). If the thimble (Fig. 4-11) is revolved counterclockwise one complete revolution, note that one line appears on the sleeve. Each line on the sleeve indicates .025 in. Thus, if three lines are showing on the sleeve (or barrel), the micrometer would open 3 × .025, or .075 in.

Every *fourth* line on the sleeve is longer than the others and is numbered to permit easy reading. Each numbered line indicates a distance of .100 in. For example, number 4 showing on the sleeve indicates a distance between the measuring faces of 4 × .100, or .400 in.

The thimble has 25 equal divisions about its circumference, each of these divisions representing .001 in.

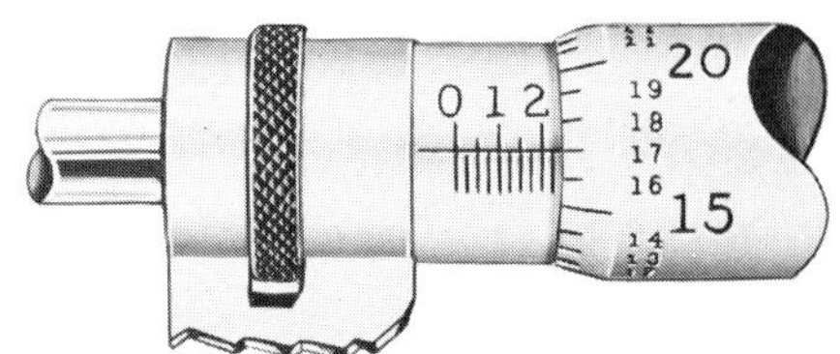


**Fig. 4-12** An inch micrometer showing a reading of .242 in. (*Courtesy The L. S. Starrett Company.*)

### To Read an Inch Micrometer

To read an inch micrometer, note the last number showing on the sleeve and multiply this by .100; then multiply the number of small lines visible past that number by .025 and add the number of divisions on the thimble from zero to the line that coincides with the center, or index line on the sleeve.

In Fig. 4-12, for example:

| | |
|---|---|
| Number 2 is shown on the sleeve | 2 × .100 = .200 in. |
| 1 line is visible past #2 | 1 × .025 = .025 in. |
| 17 divisions past the zero on the thimble | 17 × .001 = .017 in. |
| Total reading | .242 in. |

## TESTING THE ACCURACY OF MICROMETERS

The accuracy of a micrometer should be tested periodically to ensure that the work produced is the size required. Always make sure that both measuring faces are clean before a micrometer is checked for accuracy. (See N-2.)

To test a micrometer, first clean the measuring faces and then turn the thimble using the friction thimble or ratchet stop until the measuring faces contact each other. If the zero line on the thimble coincides with the center (index) line on the sleeve, the micrometer is accurate. Micrometers can also be checked for accuracy by measuring a gage block or known standard (Fig. 4-13). The reading of the micrometer must be the same as the gage block or standard. Any micrometer which is not accurate should be adjusted by a qualified person.

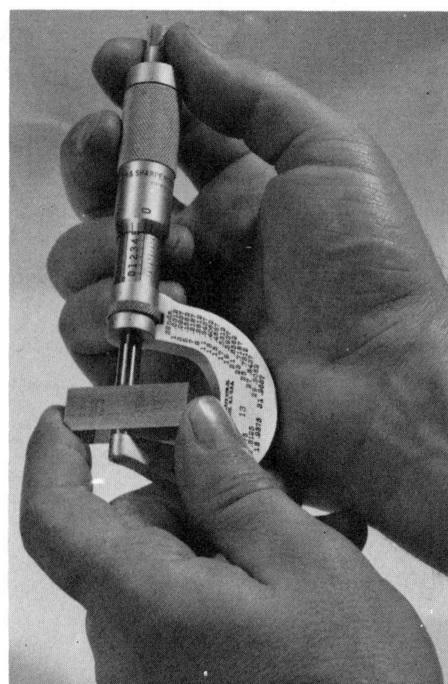

Fig. 4-13 Checking the accuracy of a micrometer with a gage block. (*Courtesy The L. S. Starrett Company.*)

N-3 For the most accurate reading, use the micrometer ratchet stop or friction thimble.

N-4 When measuring with a micrometer, move it slightly to the left and right while turning the thimble. This helps to align the micrometer correctly on the workpiece.

N-5 Vernier inch micrometers measure to within .0001 in. of accuracy.

## MEASURING WITH A MICROMETER

When measuring with a micrometer, make sure that the measuring faces are clean and that the micrometer is held squarely across the work. The thimble should never be adjusted too tightly or the micrometer may be damaged. The correct tension or "feel" may be checked by closing the spindle on a hair or piece of paper until it pulls snugly between the anvil and spindle. It is wise to use the friction thimble or ratchet stop whenever possible in order to get the correct tension and measurement. (See N-3.)

Figure 4-14 shows how a micrometer should be held in order to accurately measure a piece of work held in the hand. Carefully note the position of the fingers. The micrometer is held against the palm of the hand by the little or third finger, and the thumb and forefinger are used to turn the thimble. (See N-4.)

The proper way to hold a micrometer when measuring work in a machine is illustrated in Fig. 4-15. Hold the top of the micrometer frame in the right hand, place the micrometer over the work, and turn the thimble with the thumb and forefinger of the left hand.

## MEASURING WITH A VERNIER MICROMETER

### The Vernier Micrometer

The tolerances required in industry are often specified in sizes much smaller than .001 in. (one one-thousandths of an inch). They are often given in .0001 in. (one ten-thousandths of an inch) or even smaller. The vernier micrometer provides an easy and accurate way to measure to within .0001 in., which is about one-thirtieth the thickness of a sheet in this book.

In addition to the graduations found on a standard micrometer, the *vernier inch micrometer* (Fig. 4-16) has a vernier scale on the sleeve. This vernier scale consists of 10 divisions which run parallel to and above the index line. Note that these 10 divisions on the sleeve occupy the same distance as 9 divisions (.009) on the thimble. One division on the vernier scale, therefore, represents $\frac{1}{10} \times .009$, or .0009 in. Since one graduation on the thimble represents .001 or .0010 in., the difference between one thimble division and one vernier scale division represents .0010 − .0009, or .0001. Therefore, each division on the vernier scale has a value of .0001 in.

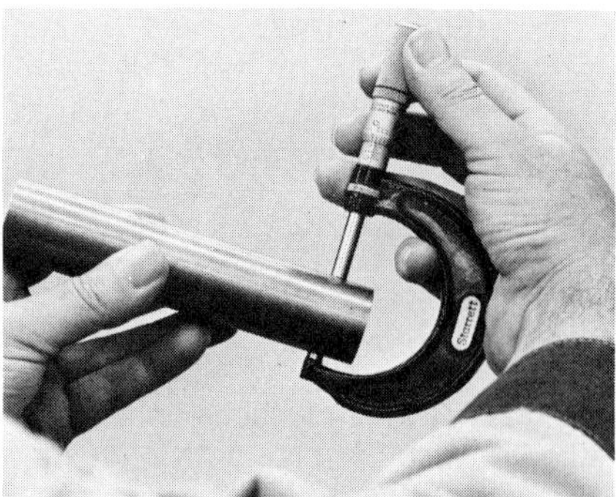

Fig. 4-14 The correct way to hold a micrometer when measuring a workpiece held in the hand. (*Courtesy Kostel Enterprises Ltd.*)

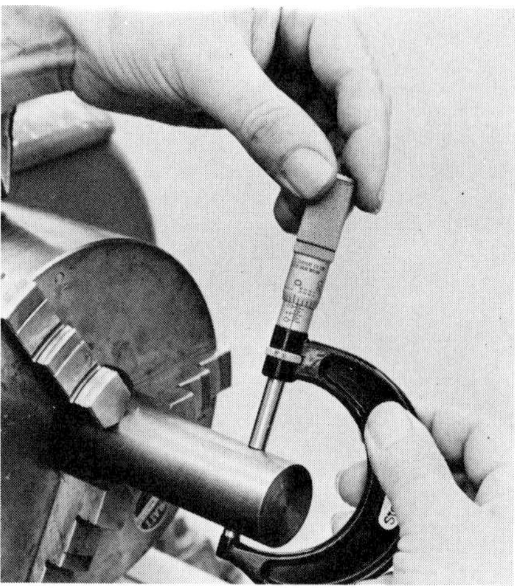

Fig. 4-15 The correct way to hold a micrometer when measuring work in a machine. (*Courtesy Kostel Enterprises Ltd.*)

### To Read a Vernier Micrometer

**1.** Read the micrometer as you would a standard micrometer.

**2.** Note the line on the vernier scale that coincides with a line on the thimble. This line will indicate the number of ten-thousandths that must be added to the above reading. (See N-5.)

The vernier micrometer is read as follows (refer to Fig. 4-17):

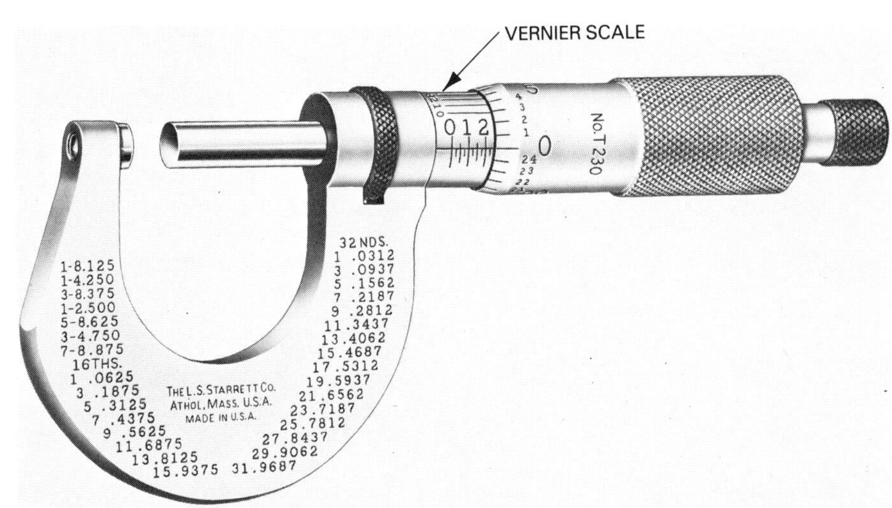


**Fig. 4-16** A vernier micrometer caliper. (*Courtesy The L. S. Starrett Company.*)

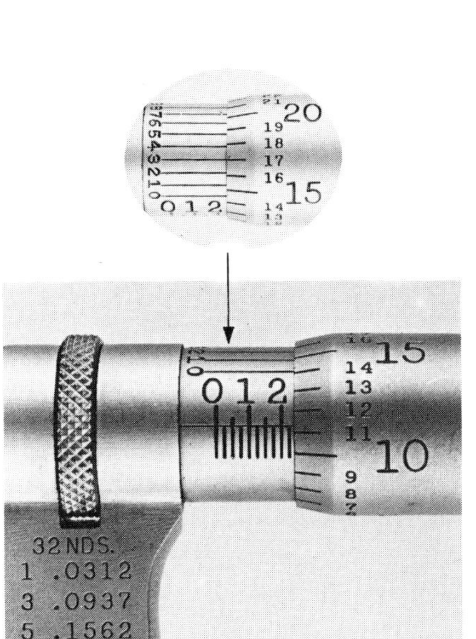


**Fig. 4-17** Vernier micrometer reading of .2363. (*Courtesy Kostel Enterprises Ltd.*)

| | | |
|---|---|---|
| Number 2 is the last visible number | 2 × .100 = | .200 |
| One small line is visible past #2 | 1 × .025 = | .025 |
| Number 11 on the thimble is just past the index line | 11 × .001 = | .011 |
| Number 3 line on the vernier scale coincides with a line on the thimble | 3 × .0001 = | .0003 |
| Total reading | | .2363 |

## MEASURING WITH A METRIC MICROMETER

With the proposed changeover to the metric system in industry, it is important that the student be familiar with metric measurement and understand its relationship to the inch system, where 1 in. = 25.40 mm.

N-6 Metric micrometers measure in millimeters.

### Description of a Metric Micrometer

The pitch of the micrometer screw on metric micrometers is 0.5 (0.50) mm. Therefore, one complete revolution of the spindle will either increase or decrease the distance between the measuring faces 0.50 mm. The graduations on the sleeve (barrel) above the index line are in whole millimeters (from 0 to 25). The graduations below the index line represent 0.50 mm, the pitch of the micrometer thread (Fig. 4-18). Since a complete turn of the micrometer spindle will move the measuring faces 0.50 mm, two complete turns are required to move the spindle 1 mm. (See N-6.)

The circumference of the thimble is graduated into 50 equal divisions, with every fifth line being numbered. Since one revolution of the thimble advances the spindle 0.50 mm, each graduation on the thimble equals $\frac{1}{50} \times 0.50$ mm, or 0.01 mm.

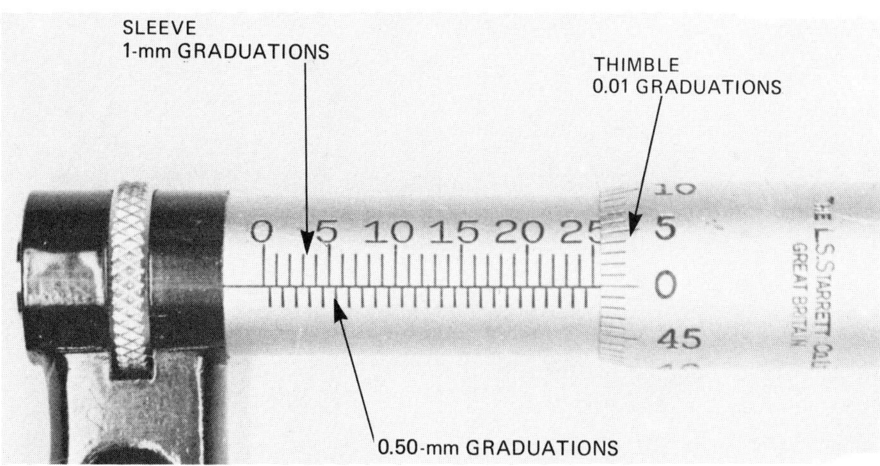


**Fig. 4-18** The graduations on a metric micrometer. (*Courtesy Kostel Enterprises Ltd.*)

### To Read a Metric Micrometer

**1.** Note the last number showing on the sleeve; multiply this by 1 mm.

**2.** Note the number of lines (above and below the index line) which show past the numbered line; multiply these by 0.50 mm.

**3.** Add the number of the line on the thimble that coincides with the index line.

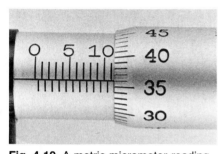

**Fig. 4-19** A metric micrometer reading of 11.36 mm. (*Courtesy Kostel Enterprises Ltd.*)

In Fig. 4-19, for example

| | | |
|---|---|---|
| Number 10 is showing on the sleeve | 10 × 1 | = 10.00 mm |
| There are 2 lines past the number 10 on the sleeve | 2 × 0.50 = | 1.00 mm |
| Number 36 on the thimble is opposite the index line | 36 × 0.01 = | 0.36 mm |
| Total reading | | = 11.36 mm |

## REVIEW QUESTIONS

**1.** What is the difference between a standard micrometer and a vernier inch micrometer?

**2.** What is the difference between a direct reading micrometer and an indicating micrometer?

**3.** What is the purpose of the following?

(a) Frame
(b) Sleeve
(c) Thimble
(d) Friction thimble

### Inch Micrometer

**4.** How many threads per inch are there on a standard inch micrometer?

**5.** What is the value of

(a) Each line on the sleeve
(b) Each numbered line on the sleeve
(c) Each line on the thimble

**6.** Why should a micrometer be tested frequently for accuracy?

**7.** What precaution should be taken

(a) Before measuring with a micrometer
(b) When measuring with a micrometer

**8.** Read the standard micrometer settings.

(a)

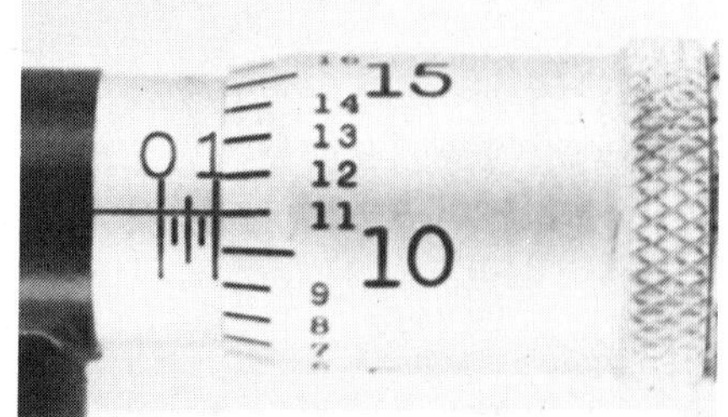


(b)

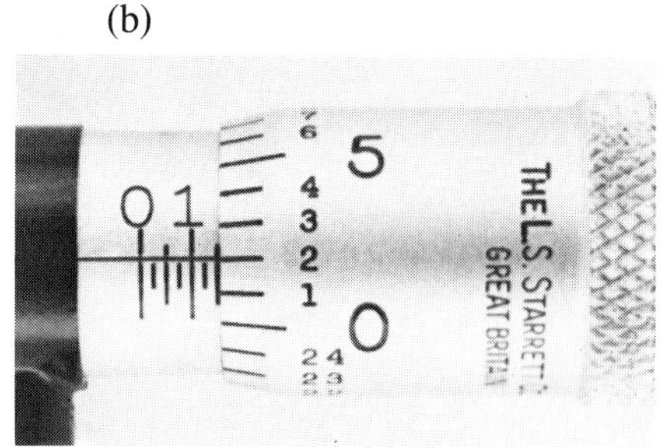


(c)

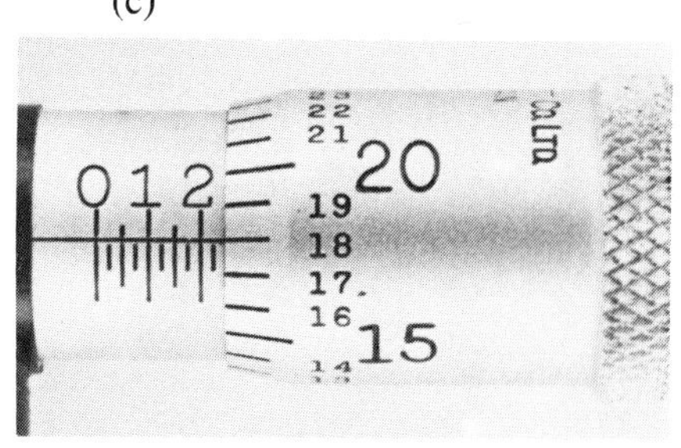


(d)

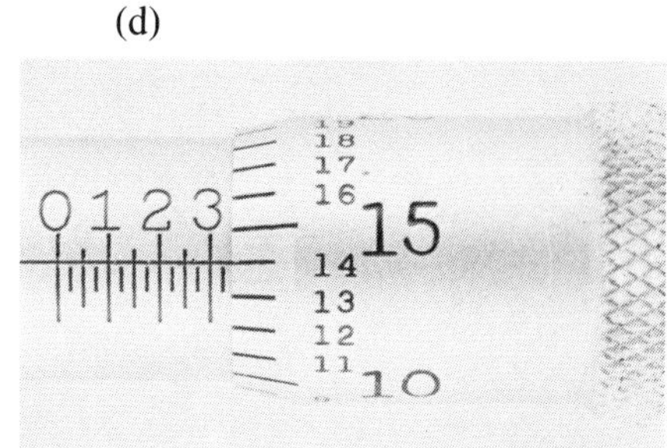


### Vernier Micrometers

**9.** What is the value of each division on the vernier micrometer scale?

**10.** Read the following vernier micrometer settings.

(a)

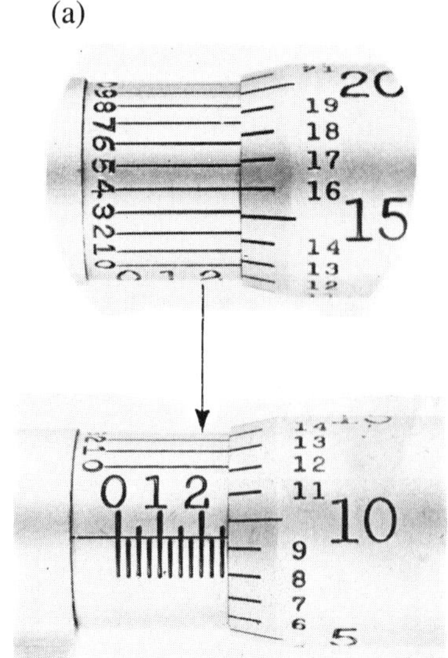

(b)

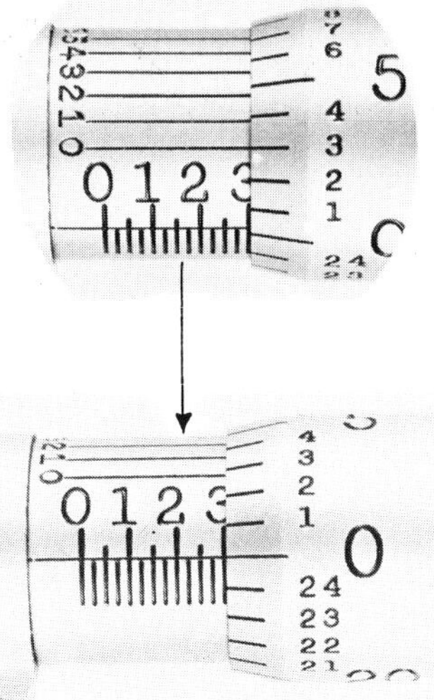

(c)

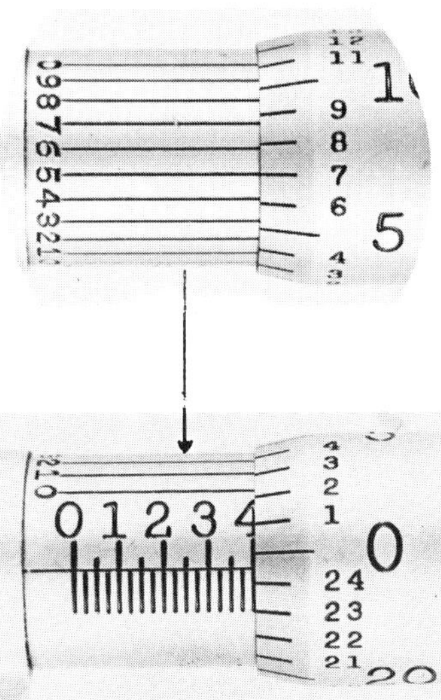

(d)

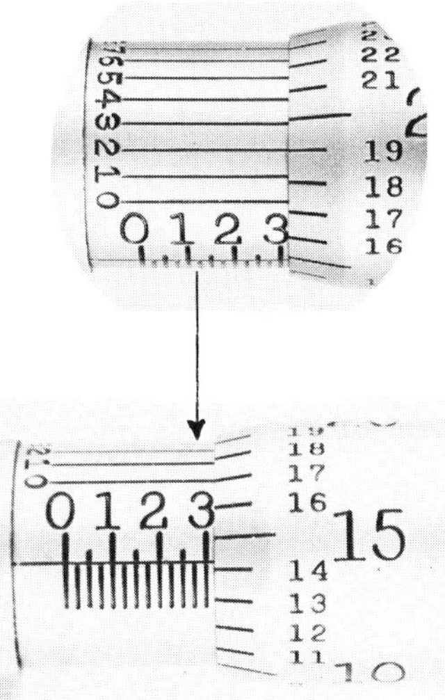

## Metric Micrometers

**11.** What is the pitch of a metric micrometer thread?

**12.** How many complete turns of the thimble are required to move the spindle one large division?

**13.** (a) How many divisions are there on the metric micrometer thimble?
(b) What is the value of each of these divisions?

**14.** What is the reading of each of the following metric micrometer settings?

(a)

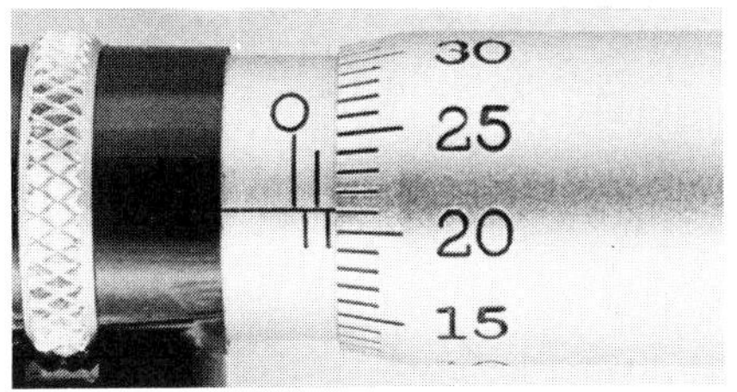


(b)

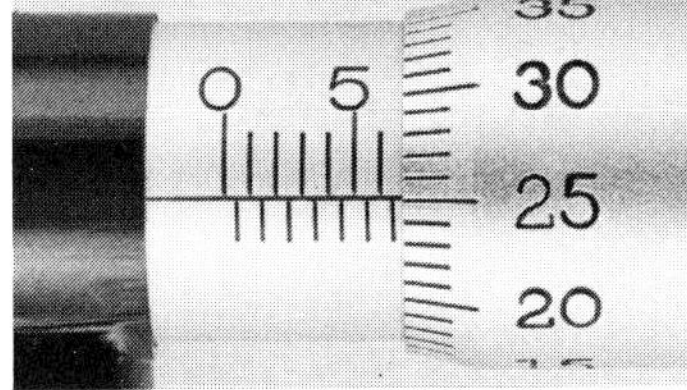


(c)

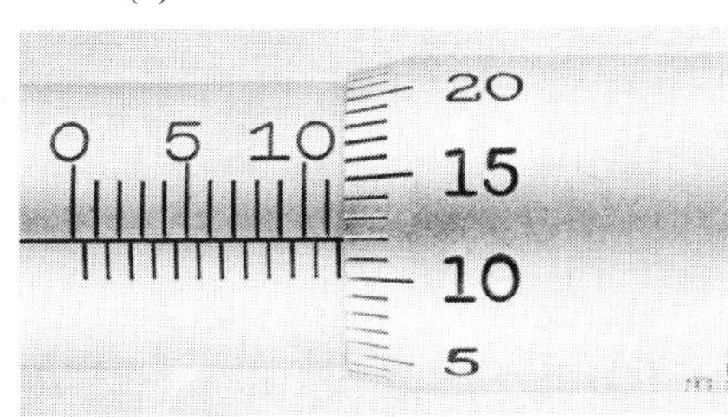


(d)

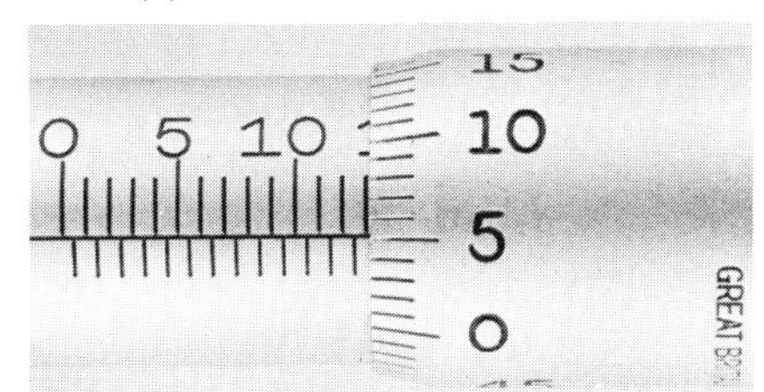

# UNIT 5

# VERNIER CALIPERS

Vernier calipers are precision measuring instruments used to make accurate measurements to within .001 in. for inch tools or to within 0.02 mm for metric tools. The bar and movable jaw are graduated on both sides, one side for taking outside measurements and the other side for taking inside measurements. Vernier calipers are available in inch and in metric graduations, and some types have both scales, Fig. 5-6.

## OBJECTIVES

Upon completion of this unit, you should be able to:

**1.** Measure a variety of workpieces to an accuracy of plus or minus .001 in. with a 25-division inch vernier caliper

**2.** Measure a variety of workpieces to an accuracy of plus or minus .001 in. with a 50-division vernier caliper

**3.** Measure a variety of workpieces to an accuracy of plus or minus 0.02 mm with a metric vernier caliper

## PARTS OF THE VERNIER CALIPER

The parts of the vernier caliper remain the same regardless of the measurement system for which the instrument is designed.

The L-shaped frame consists of a *bar,* containing the *main scale* graduations and the *fixed jaw*. The *movable jaw,* which slides along the bar, contains the *vernier scale*. Adjustments for size are made by means of an *adjusting nut;* the readings may be locked in place by means of *clamp screws*.

Most bars are graduated on both sides—one for inside measurements, the other for outside measurements. The ends of the jaws are designed to permit inside and outside measurement. Inch vernier calipers are manufactured with 25- and 50-division vernier scales. Vernier calipers, graduated in millimeters, are available for taking metric measurements.

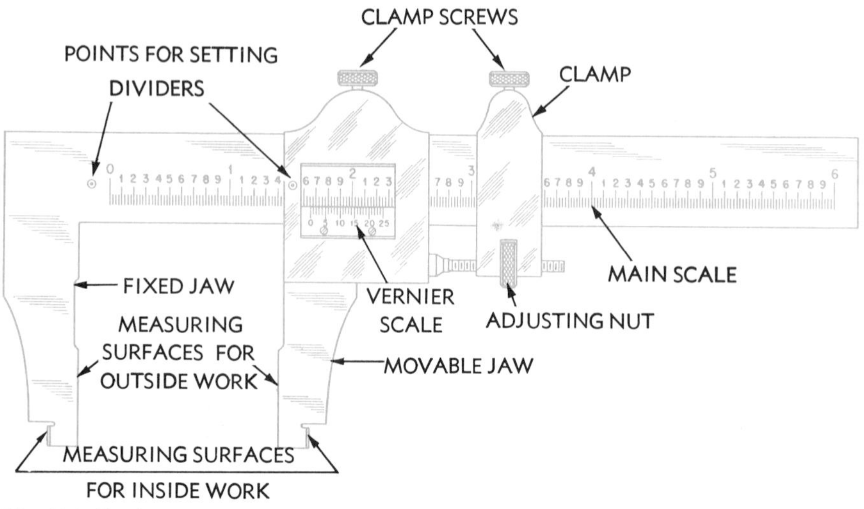

**Fig. 5-1** Setting a vernier to the workpiece.

## THE 25-DIVISION VERNIER CALIPER

### Principle

The bar of this vernier caliper is graduated the same as a micrometer sleeve. Each inch is divided into 40 equal dimensions each having a value of .025 in. Every fourth line—which represents .100, or 1/10, in.—is numbered.

The vernier scale on the movable jaw has 25 equal divisions each representing .001. The 25 divisions on the vernier scale, which are .600 in. in length, are equal to 24 divisions on the main scale. The difference between one division on the main scale and one division on the vernier scale is .025 − .024 = .001. (See N-1.)

N-1 Each division on the vernier scale has a value of .001.

Only one line on the vernier scale will line up exactly with a line on the main scale for any reading. The number of this line is multiplied by .001 to give the reading of the vernier caliper.

### Measuring with a 25-Division Vernier Caliper

**1.** Remove all burrs from the workpiece and clean the surfaces to be measured.

**2.** Open the jaws so that they will pass over the workpiece.

**3.** Close the jaws against the work and lock the right-hand clamp screw, Fig. 5-1.

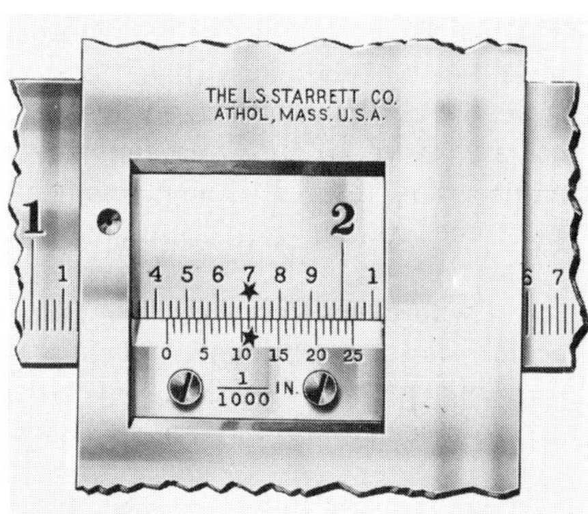


**Fig. 5-2** A 25-division vernier caliper reading of 1.436. (*Courtesy The L. S. Starrett Company.*)

N-2 The jaws of the vernier caliper must be in line with each other for accurate measurements.

N-3 Recheck "feel" of caliper on work before removing the vernier caliper.

N-4 When reading the vernier scale, be sure to hold the bar of the vernier at 90° to your line of sight.

**4.** Turn the adjusting nut until the jaws *just* touch the work surfaces. Be sure that the jaws of the caliper are in line by holding the fixed jaw in place and slightly moving the bar sideways and vertically while setting the adjusting nut. (See N-2.)

**5.** Lock the clamp screw on the movable jaw. (See N-3.)

**6.** Read the measurement as follows, for Fig. 5-2:

| | | |
|---|---|---|
| (a) Note the large number above the main scale on the bar to the left of the vernier scale. This represents the number of whole inches. For this reading it is | 1 inch = | 1.000 |
| (b) Note the small number on the bar to the left of the zero (0) on the vernier scale and multiply this by .100. In this case it is number 4. | 4 × .100 = | .400 |
| (c) Note how many graduations are showing on the bar between the last number and the zero (0) on the vernier scale. Multiply this by .025. In this case it is 1. | 1 × .025 = | .025 |
| (d) Note the line on the vernier scale which matches with a line on the bar. Multiply this number by .001. In this case it is the eleventh line. | 11 × .001 = | .011 |
| Total reading: | | 1.436 in. |

## THE 50-DIVISION VERNIER CALIPER

Construction and use of the 50-division vernier caliper is the same as that for the 25-division caliper. The only difference is in the graduations on the main and vernier scales. The 50-division vernier caliper is becoming more popular than the 25-division model because it is easier to read the vernier scale on the 50-division caliper.

### Principle

Each line on the bar of the vernier represents .050 in. Every second line is numbered and represents .100 in. The vernier scale on the movable jaw has 50 equal divisions, each having a value of .001 in. The 50 divisions on the vernier scale occupy the same space as 49 divisions on the bar. Therefore only one line of the vernier scale will line up exactly with a line on the bar at any one setting.

To read a 50-division vernier, note how many inches—hundred thousandths (.100) and fifty thousandths (.050)—the zero mark on the movable jaw is past the zero mark on the bar. To this total add the number of thousandths (.001) indicated by the line on the vernier scale which exactly coincides with a line on the bar. (See N-4.)

In Fig. 5-3, for example:

| | | |
|---|---|---|
| The large number 1 on the bar | = | 1.000 in. |
| The small number 4 past the number 1 | 4 × .100 = | .400 in. |
| One line visible past number 4 | 1 × .050 = | .050 in. |
| The fourteenth line on the | | |

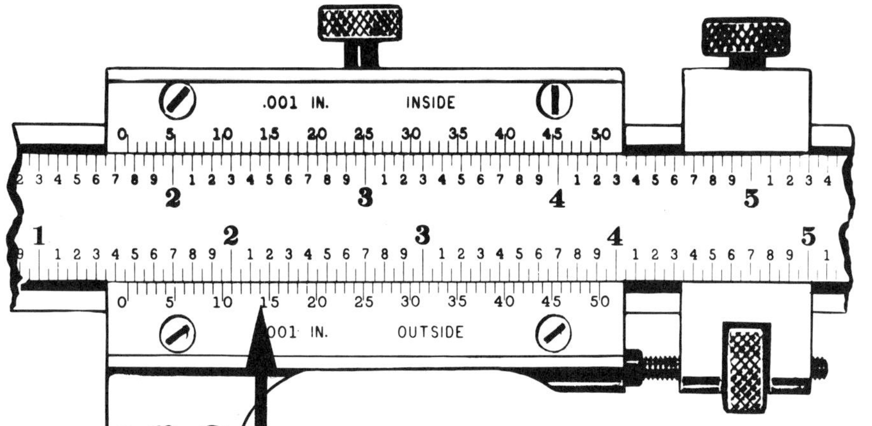


**Fig. 5-3** A 50-division vernier caliper reading of 1.464 in. (*Courtesy The L. S. Starrett Company.*)

| | | |
|---|---|---|
| vernier scale coincides with a line on the bar | 14 × .001 = | .014 in. |
| Total reading | | 1.464 in. |

# METRIC VERNIER CALIPERS

## Measuring with a Metric Vernier Caliper

The main scale on the bar is graduated in millimeters and every main division is numbered. Each numbered division has a value of 10 mm; for example, number 1 represents 10 mm, number 2 represents 20 mm, etc. There are 50 graduations on the sliding, or vernier scale, with every fifth one being numbered. These 50 graduations occupy the same space as 49 graduations on the main scale (49 mm), Fig. 5-4.

Therefore,

$$1 \text{ vernier division} = \frac{49}{50} \text{ mm} = 0.98 \text{ mm}$$

The difference between one main scale division and one vernier scale division

$$= 1 - 0.98 = 0.02 \text{ mm}$$

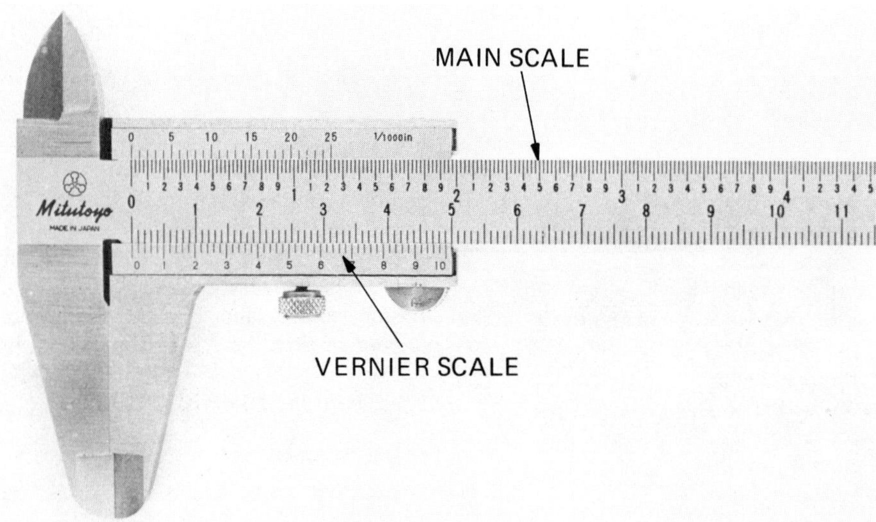


**Fig. 5-4** The graduations on a metric vernier caliper.

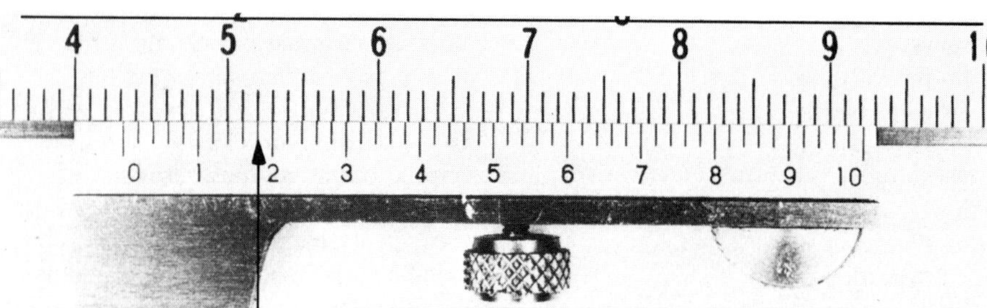

**Fig. 5-5** A metric vernier caliper reading of 43.18 mm. (*Courtesy Kostel Enterprises Ltd.*)

## To Read a Metric Vernier Caliper

**1.** The last numbered division on the bar to the left of the vernier scale represents the number of millimeters multiplied by 10.

**2.** Note how many full graduations are showing between this numbered division and the zero on the vernier scale. Multiply this number by 1 mm.

**3.** Find the line on the vernier scale which coincides with a line on the bar. Multiply this number by 0.02 mm.

In Fig. 5-5, for example

| | | |
|---|---|---|
| The large number 4 graduation on the bar | 4 × 10 mm = | 40 mm |
| Three full lines past the number 4 graduation | 3 × 1 mm = | 3 mm |
| The ninth line on the vernier scale coincides with a line on the bar | 9 × 0.02 = | 0.18 mm |
| Total reading | | 43.18 mm |

The changeover to the metric system will be gradual. For some time many firms will be using dual dimensioning, that is, both inch and metric dimensions. Because of this, some tool manufacturers produce vernier calipers which are graduated in both millimeters and inches, Fig. 5-6.

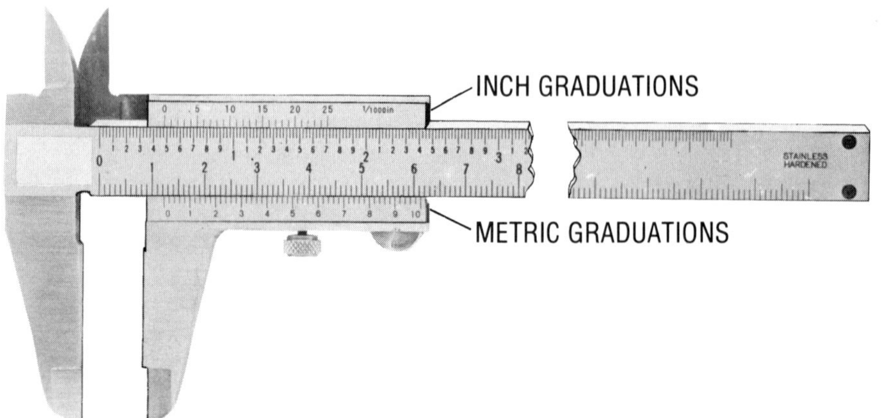


**Fig. 5-6** A vernier caliper with both metric and inch graduations.

## REVIEW QUESTIONS

### Inch Vernier Calipers

**1.** How are vernier calipers graduated in order to make inside and outside measurements?

**2.** How are the following parts of a 50-division inch vernier caliper graduated?

(a) The bar

(b) The sliding jaw

**3.** What is the *value* of one division on any inch vernier caliper?

**4.** What are the following inch vernier caliper settings?

(a)

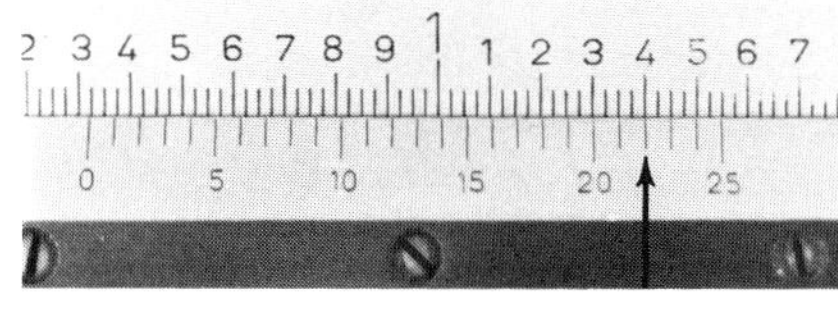

(b)

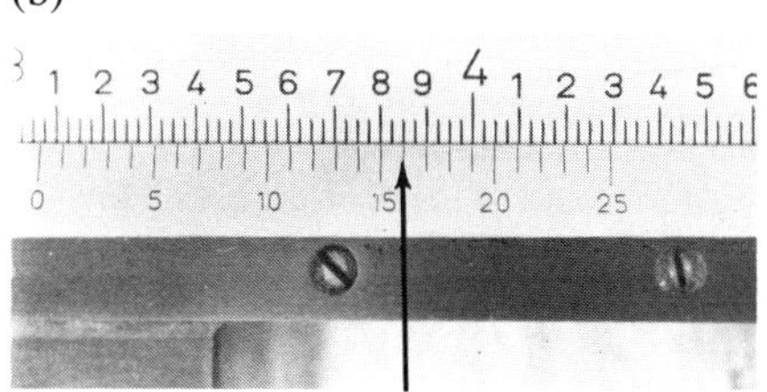

(c)

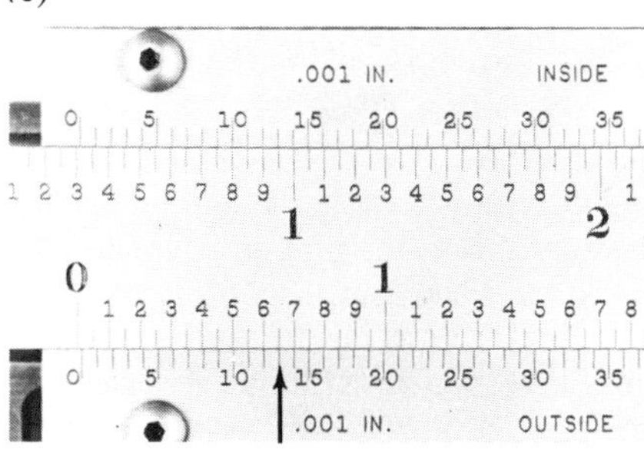


(d)

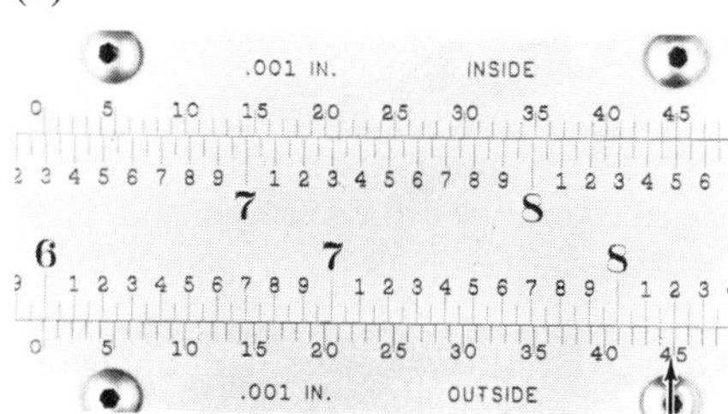


### Metric Vernier Calipers

**5.** What is the relationship between the graduations on the main scale and the vernier scale on a metric vernier caliper?

**6.** What is the value of each?

(a) Numbered division on the main scale

(b) Small division on the main scale

(c) Division on the vernier scale

**7.** What is the reading of the following metric vernier caliper settings?

(a)

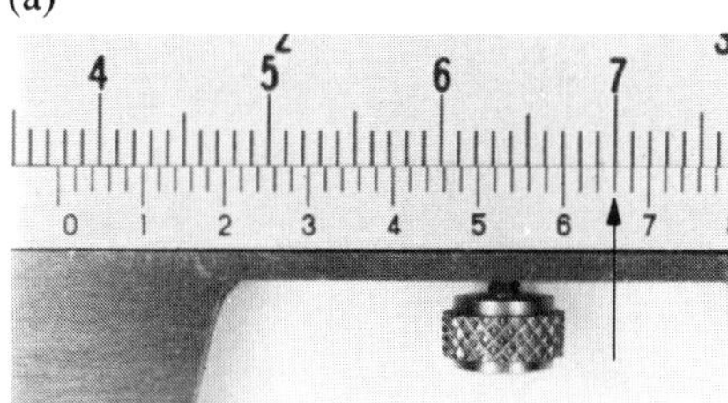

(b)

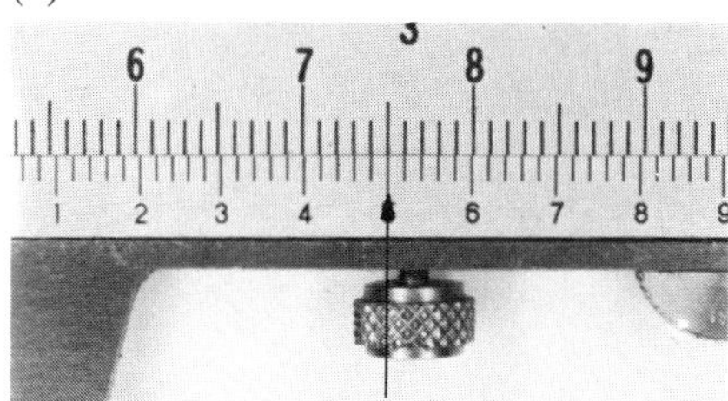

(c)

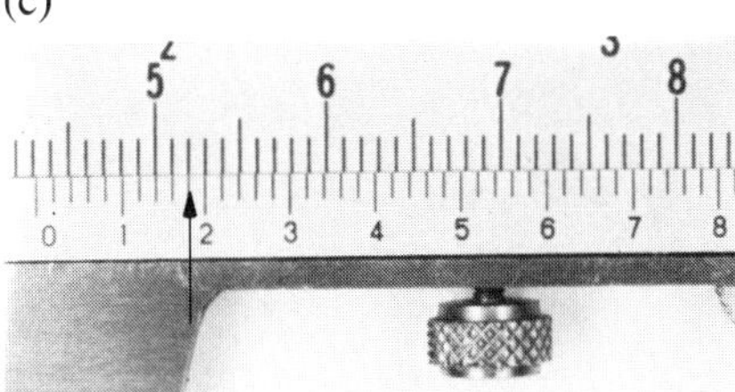

(d)

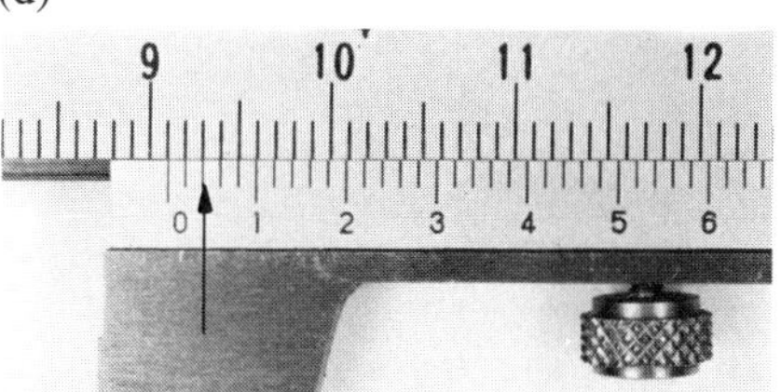

# UNIT 6

## FIXED GAGES

Fixed gages are used in inspection because they provide a quick means of checking a specific dimension. These gages must be easy to use and accurately finished to the required tolerance. They are generally finished to

Fig. 6-1 A cylindrical plug gage is used to check hole sizes.

Fig. 6-2 Checking a hole size with a plug gage.

one-tenth the tolerance they are designed to control. For example, if the tolerance of a piece being checked is to be maintained at .001 in. (0.02 mm), then the gage must be finished to within .0001 in. (0.002 mm) of the required size.

## OBJECTIVES

After completing this unit, you should be able to:

**1.** Identify the common types of ring and plug gages
**2.** Know which gage should be used when checking a particular workpiece for size
**3.** Measure workpieces accurately with ring and plug gages

## CYLINDRICAL PLUG GAGES

Plain cylindrical gages (Fig. 6-1) are used for checking the inside diameter of a straight hole and are generally of the "go" and "no-go" variety. This type consists of a handle and a plug on each end ground and/or lapped to a specific size. The smaller diameter plug, or the "go" gage, checks the lower limit of the hole or the smallest size permissible. The larger diameter plug, or the "no-go" gage, checks the upper limit of the hole or the largest size permissible. (See N-1.)

The dimensions of these gages are usually stamped on the handle at each end adjacent to the plug gage. "Go" gages are made longer than the "no-go" end for easy identification. Sometimes a groove is cut on the handle near the "no-go" gage to distinguish it from the "go" end.

N-1 Plug gages are used to check the diameter of a hole.

N-2 Do not force or turn a plug gage.

N-3 Ring gages are used to check external diameters.

### To Use a Cylindrical Plug Gage

**1.** Select a plug gage of the correct size and tolerance for the hole being checked.
**2.** Clean both ends of the gage and the hole in the workpiece with a clean, dry cloth.
**3.** Check the gage (both ends) and the workpiece for nicks and burrs.
**4.** Wipe both ends of the gage with an oily cloth to deposit a thin film of oil on the surfaces.
**5.** Start the "go" gage squarely into the hole (Fig. 6-2). If the hole is within the limits, the gage will enter easily. The plug should enter the hole for the full length, and there should be no excessive play between the plug and the part. *Note:* If the gage enters only partway, there is a taper in the hole. Excessive play or looseness in one direction indicates that the hole is elliptical (out of round). (See N-2.)
**6.** After the hole has been checked with the "go" gage, it should be checked with the "no-go" end. This gage should not begin to enter the hole. An entry of more than 1⁄16 in. (1.58 mm) indicates an oversized, bell-mouth, or tapered hole.

## PLAIN RING GAGES

Plain ring gages, used to check the outside diameter of pieces, are ground and lapped internally to the desired size. The size is stamped on the side of the gage (Fig. 6-3). The outside diameter is knurled, and the "no-go" gage is identified by an annular groove on the knurled surface. The precautions and procedure regarding the use of a ring gage are similar to those outlined for a plug gage and should be followed carefully. (See N-3.)

## TAPER PLUG GAGES

Taper plug gages (Fig. 6-4), made with standard or special tapers, are used to check the size of the hole and the accuracy of the taper. The gage must slide into the hole for a prescribed depth and fit snugly in the hole. An incorrect taper is evidenced by a wobble between the plug gage and the hole.

### To Check an Internal Taper Using a Taper Plug Gage

**1.** Select the proper taper gage for the hole being checked.
**2.** Wipe the gage and the hole with a clean, dry cloth.
**3.** Check both the gage and hole for nicks and burrs.
**4.** Apply a thin coating of Prussian blue to the surface of the plug gage.

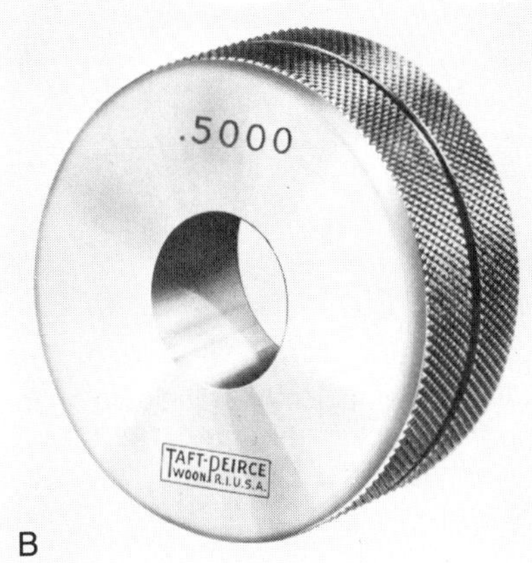


**Fig. 6-3** Plain ring gages are used to check the diameter of round work. (A) "go" gage; (B) "no-go" gage.

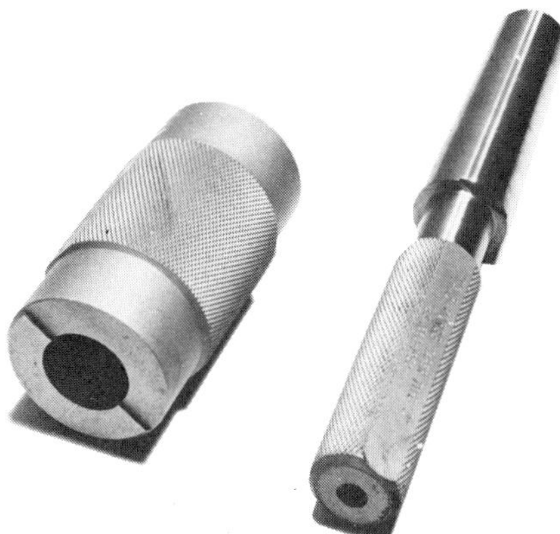

**Fig. 6-4** Taper ring and plug gages. (*Courtesy Kostel Enterprises Ltd.*)

N-4 Some gages have the limit steps ground at the large end; others are ground on the small end.

**5.** Insert the plug gage into the hole as far as it will go (Fig. 6-5).

**6.** Maintaining light-end pressure on the plug gage, rotate it in a counter-clockwise direction for approximately *one-quarter turn*.

**7.** Check the diameter of the hole. A proper size is indicated when the edge of the workpiece lies between the limit steps or lines on the gage. (See N-4.)

**8.** Check the taper of the hole by attempting to move the gage radially in the hole. Movement or play at the large end indicates excessive taper; movement at the small end indicates insufficient taper.

**9.** Remove the gage from the hole to see if the bluing has rubbed off evenly along the length of the gage, a result which would indicate a proper fit. A poor fit is evident if the bluing has been rubbed off more at one end than the other.

## TAPER RING GAGES

Taper ring gages (Fig. 6-4) are used to check both the accuracy and the outside diameter of the taper. Ring gages often have scribed lines or a step ground on the small end to indicate the "go" and "no-go" dimensions.

For a taper ring gage, the precautions and procedures are similar to those outlined for a taper plug gage. However, when work which has not been ground or polished is being checked, three equally spaced chalk lines around the circumference and extending for the full length of the tapered section may be used to indicate the accuracy of the taper (Fig. 6-6). If the work has been ground or polished, it is advisable to use three thin lines of Prussian blue.

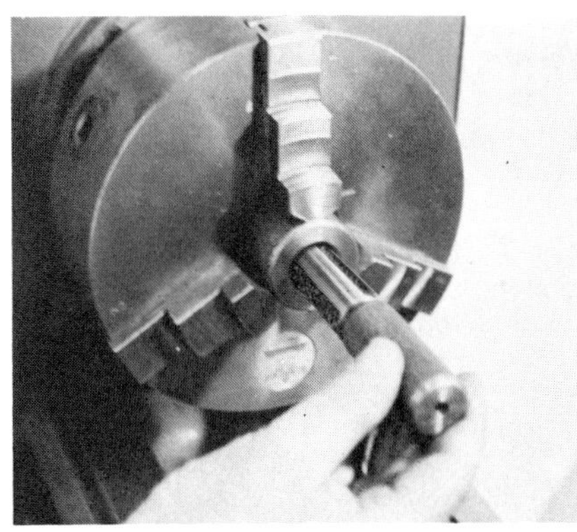

**Fig. 6-5** Checking a tapered hole with a tapered plug gage.

### Care of Plug and Ring Gages

In order to preserve the accuracy and life of gages, observe the following points.

**1.** Store gages in divided wooden trays to protect them from being nicked or burred.

**2.** Check them frequently for size and accuracy.

**3.** Correctly align gages with the workpiece to prevent binding.

**4.** Do not force or twist a plug or ring gage. Forcing or twisting will cause excessive wear.

**5.** Clean the gage and workpiece thoroughly before checking the part.

**6.** Use a light film of oil on the gage to help prevent binding.

**7.** Have gages and work at room temperature to ensure accuracy and prevent damage to the gage.

## THREAD PLUG GAGES

Internal threads are checked with thread plug gages of the "go" and "no-go" variety and employ the same

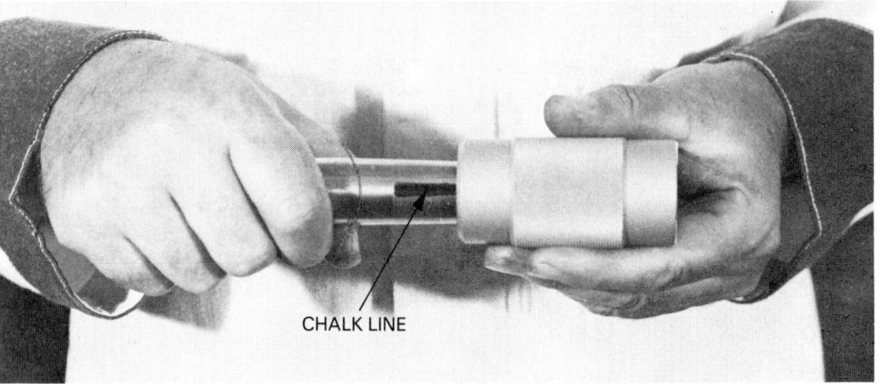


**Fig. 6-6** Checking the accuracy of a taper using chalk lines.

Fig. 6-7 Thread plug gages are used to check the size and accuracy of internal threads. (*Courtesy Taft-Peirce Mfg. Co.*)

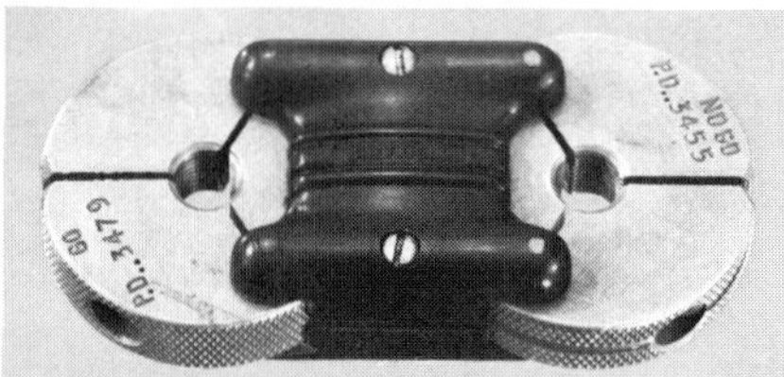

Fig. 6-8 A "go" and "no-go" gage in a holder. (*Courtesy Bendix Corp.*)

principle as cylindrical plug gages (Fig. 6-7). When being used, a little oil should be applied to the thread, and the gage should never be forced into the thread.

## THREAD RING GAGES

These gages, used to check the accuracy of an external thread, have a threaded hole in the center with three radial slots and a set screw to permit small adjustment. The outside diameter is knurled, and the "no-go" gage is identified by an annular groove cut on the knurled surface. Both the "go" and "no-go" gages are generally assembled in one holder for checking the part easily (Fig. 6-8).

## SNAP GAGES

Snap gages are a quick means of checking diameters and threads to within certain limits by comparing the part size to the preset dimension of the snap gage.

Snap gages are generally C-shaped and are usually adjustable to the maximum and minimum limits of the part being checked. Several types of snap gages are shown in Fig. 6-9 A to C. When in use, the work should slide into the "go" gage but not into the "no-go" gage. Snap gages may have anvils or rolls for measuring the work.

Another type of snap gage which permits the work to be checked to .0001 in. (.002 mm) is the dial-indicator adjustable snap gage (Fig. 6-10). This type of gage is more reliable than the solid type, since it overcomes the differences in operator "feel." The size of the part is read directly from the indicator after the gage has been set to the required size.

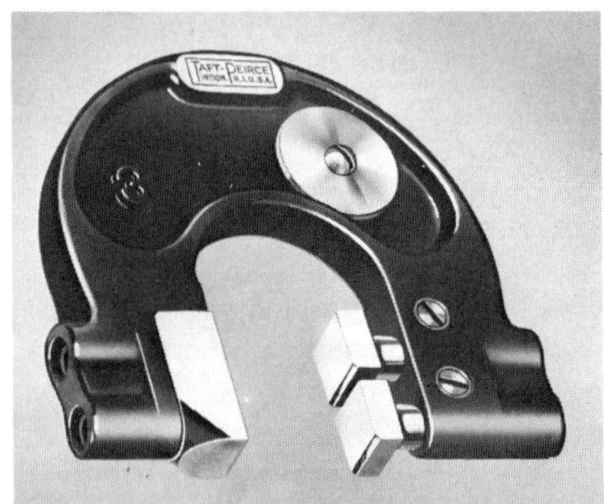

Fig. 6-9A An adjustable snap gage with anvils for checking the work. (*Courtesy Taft-Peirce Mfg. Co.*)

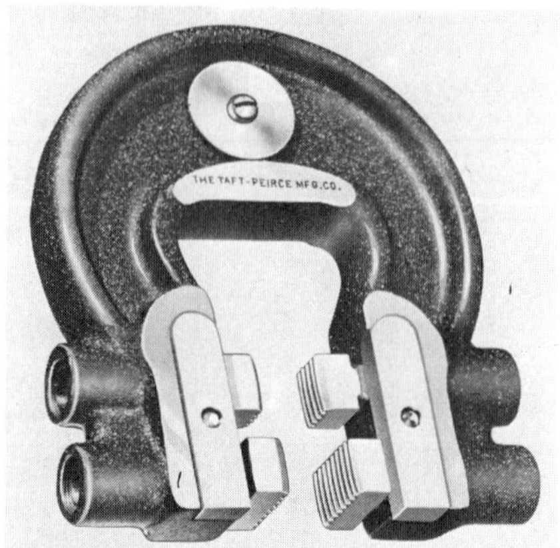

Fig. 6-9B A thread snap gage. (*Courtesy Taft-Peirce Mfg. Co.*)

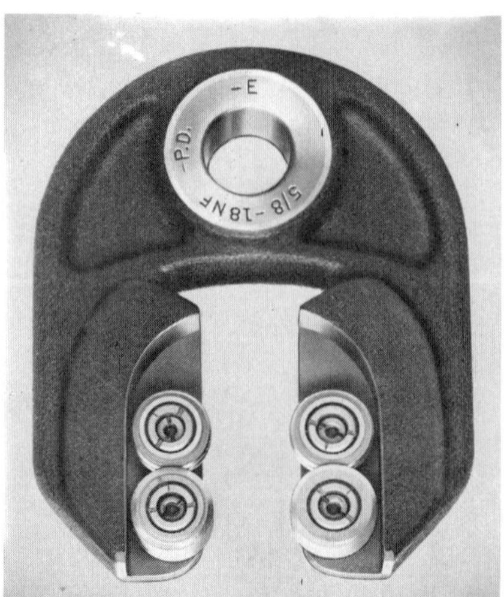

Fig. 6-9C A thread roll gage. (*Courtesy Bendix Corp.*)

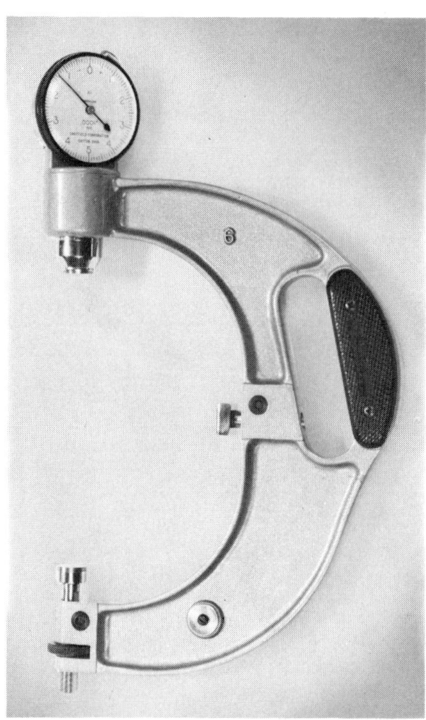

Fig. 6-10 Dial indicator adjustable snap gage. (*Courtesy Taft-Peirce Mfg. Co.*)

## REVIEW QUESTIONS

### Fixed Gages

**1.** What purpose do fixed gages serve in industry?

### Plug and Ring Gages

**2.** When using a plug gage, what limit is checked with the
 (a) "No-go" gage
 (b) "Go" gage

**3.** How may the "go" plug gage be identified?

**4.** How is the "no-go" ring gage identified?

### Taper Plug and Ring Gages

**5.** When checking a taper with a plug gage and bluing, how will the following be indicated?
 (a) Proper fit
 (b) Poor fit

**6.** What procedure should be used when checking a taper, which has not been polished, with a ring gage?

**7.** What two precautions must be observed when checking a taper with a taper plug or ring gage?

**8.** List five points to observe in the proper care of gages.

### Thread Plug and Ring Gages

**9.** What two precautions should be observed when using a thread plug gage?

**10.** How may thread ring gages be adjusted?

### Snap Gages

**11.** What is the purpose of a snap gage?

**12.** What is the advantage of the dial-indicator snap gage over a standard snap gage?

# UNIT 7

# INSIDE, DEPTH, AND HEIGHT MEASUREMENT

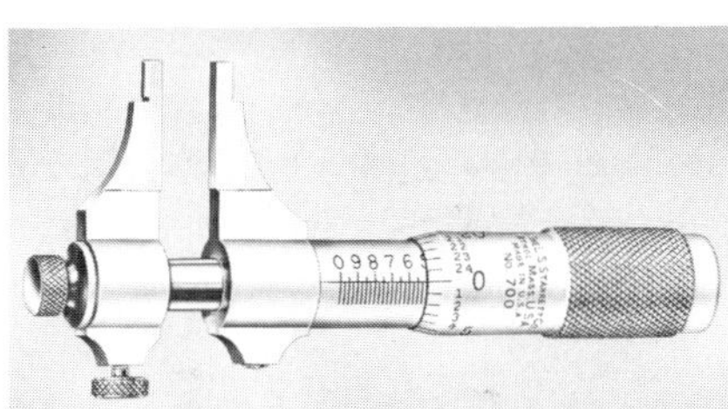

**Fig. 7-1** An inside micrometer caliper with reversed readings. (*Courtesy The L. S. Starrett Company.*)

Because of the wide variety of measurements which are required in machine shop work, it is important that many types of precision measuring tools, each designed for a specific purpose, be available. Measuring tools are available to enable the machinist to measure inside diameters, depths, and heights to a high degree of accuracy. Direct-reading measuring instruments are the most commonly used and generally the most accurate; however, sometimes because of the shape or size of the part, transfer-type instruments must be used.

## OBJECTIVES

Upon completion of this unit, you should be able to:

**1.** Identify and know the purpose of the common direct-reading and transfer-type inside-measuring instruments

**2.** Measure holes to within .001-in. (0.02-mm) accuracy using inside micrometers and inside micrometer calipers

**3.** Measure and record the depths of slots or grooves to an accuracy of plus or minus .001 in. (0.02 mm) with a micrometer depth gage

**4.** Use a vernier height gage to measure heights to an accuracy of plus or minus .001 in. (0.02 mm)

## INSIDE-MEASURING INSTRUMENTS

There is a variety of instruments used to measure the size of internal diameters. All inside-measuring instruments fall into two categories:

**1.** Direct-reading
**2.** Transfer-type

With the *direct-reading instruments* the size of the hole can be read on the instrument being used to measure the hole. The most common direct-reading inside-measuring instruments are the inside micrometer caliper, inside micrometer, and the Intrimik.

The *transfer-type instrument* is set to the diameter of the hole and this setting is then checked for size by an outside micrometer. The most common transfer-type instruments are: inside calipers, small-hole gages, and telescopic gages.

### Direct-Reading Instruments

### Inside Micrometer Calipers

The inside micrometer caliper (Fig. 7-1) is designed for measuring holes, slots, grooves, etc., from .200 to 2.500. The nibs or ends of the jaws are hardened and ground to a small radius to permit the accurate measurement of holes. A locking nut provided with this micrometer can be used to lock the micrometer at any desired size.

The inside micrometer caliper employs the same principle as a standard micrometer; however, the barrel readings on some calipers are reversed (as shown in Fig. 7-1). Extreme care must be taken in reading this type of instrument. Other inside micrometer calipers have the reading on the spindle and are read in the same manner as a

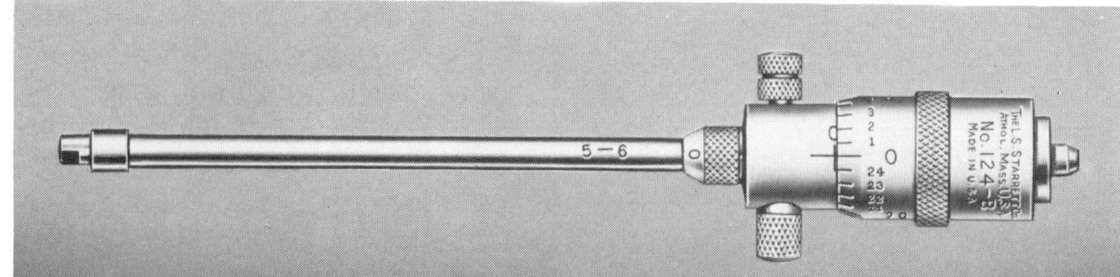


**Fig. 7-2** An inside micrometer set.

**Fig. 7-3** Using a larger tubular inside micrometer. (*"Fundamentals of Dimensional Metrology,"* Delmar Publishers Inc.)

standard outside micrometer. Inside micrometer calipers are special-purpose tools and are not used in mass-production measurement.

## To Use an Inside Micrometer Caliper

**1.** Adjust the jaws to slightly less than the diameter to be measured.

**2.** Hold the fixed jaw against one side of the hole, and adjust the movable jaw until the proper "feel" is obtained. *Note:* Move the movable jaw back and forth to ensure that the measurement taken is across the diameter. (See N-1.)

**3.** Set the lock nut, remove the instrument, and check the reading.

## Inside Micrometers

For internal measurements larger than 1½ in., inside micrometers (Fig. 7-2) are used. The inside micrometer set consists of a micrometer head, having a range of ½ or 1 in., several extension rods of different lengths which may be inserted into the head, and a ½-in. spacing collar. These sets cover a range from 1½ to over 100 in. Sets that are used for the larger sizes generally have hollow tubes, rather than rods, for greater rigidity.

The inside micrometer is read in the same manner as the standard micrometer. Since there is no locking nut on the inside micrometer, the thimble nut is adjusted to a tighter fit on the spindle thread to prevent a change in the setting while it is being removed from the hole.

## To Measure with an Inside Micrometer

**1.** Measure the size of the hole with a rule.

**2.** Insert the correct extension rod after having carefully cleaned the shoulders of the rod and the micrometer head.

**3.** Align the zero marks on the rod and micrometer head, and lock the rod firmly in place with the knurled set screw.

**4.** Adjust the micrometer to slightly less than the diameter to be measured.

**5.** Hold the head in a fixed position and adjust the micrometer to the hole size while moving the rod end in the direction of the arrows (Fig. 7-3). (See N-2.)

**6.** When a micrometer is properly adjusted to size, there should be a slight drag or "feel" when the rod end is moved past the centerline of the hole.

**7.** Carefully remove the micrometer and note the reading.

**8.** To this reading, add the length of the extension rod and collar.

## Intrimik

A difficulty encountered in measuring hole sizes with instruments employing only two measuring faces is that of properly measuring the diameter and not a chord of the circle. An instrument which eliminates this problem is the Intrimik (Fig. 7-4).

N-1 "Feel" is a slight drag between the workpiece and the caliper nibs when making a measurement.

N-2 Never force the micrometer into a hole.

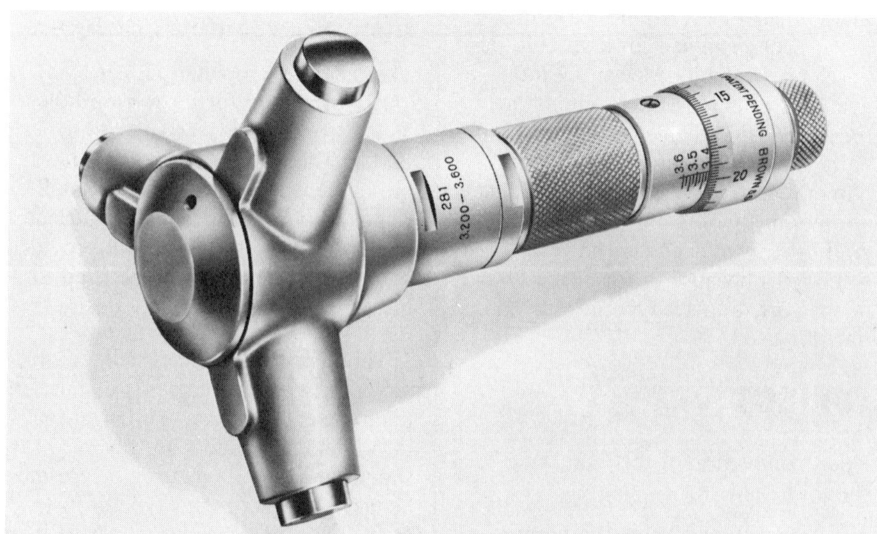


**Fig. 7-4** The three contact points of an Intrimik gives a very accurate internal measurement reading. (*Courtesy Brown & Sharpe Mfg. Co.*)

The Intrimik consists of a head with three contact points spaced 120° apart: this head is attached to a micrometer-type body. The contact points are forced out to contact the inside of the hole by means of a tapered or conical plug attached to the micrometer spindle (Fig. 7-4). The construction of a head with three contact points permits the Intrimik to be self-centering and self-aligning. It is more accurate because it provides a direct reading, eliminating the necessity of transferring measurements to determine hole size as with telescope or small-hole gages.

The range of these instruments is from .275 to 12.000, and the accuracy varies between .0001 and .0005, depending on the head used. The accuracy of the Intrimik should be checked periodically with a setting ring or master ring gage.

## TRANSFER-TYPE INSTRUMENTS

*Transfer measurement* is the method whereby a size is taken with an instrument which is not capable of giving a direct reading. The measurement is then determined by measuring the setting of the instrument with a micrometer or other direct-reading instrument. Only small-hole gages and telescope gages will be covered in this unit.

## SMALL-HOLE GAGES

Small-hole gages are available in sets of four, covering a range from ⅛ to ½ in. They are manufactured in two types (Figs. 7-5 and 7-6).

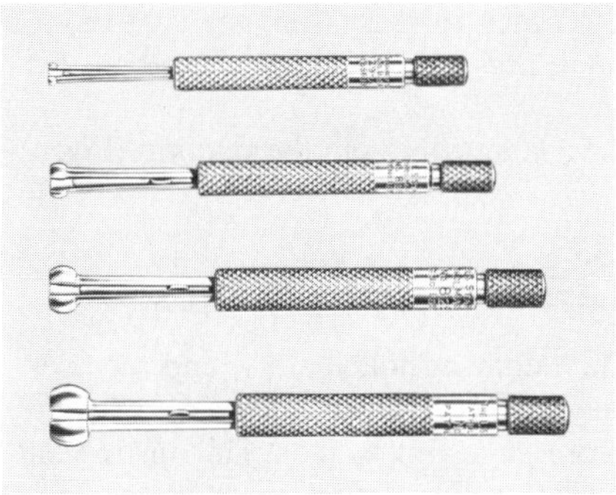

**Fig. 7-5** Small-hole gages with hardened balls.

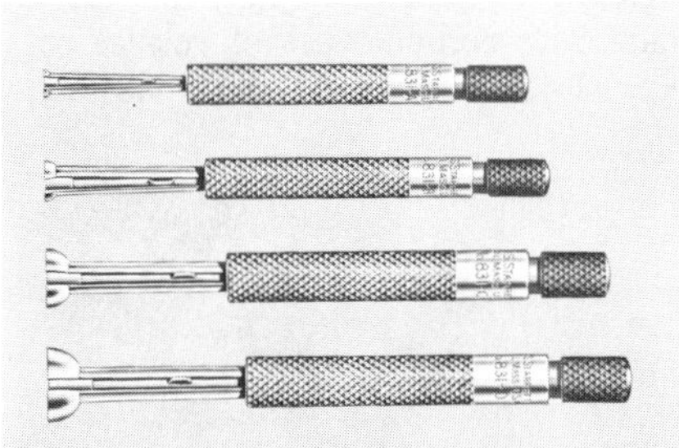

**Fig. 7-6** Small-hole gages with flat bottoms. (*Courtesy The L. S. Starrett Company.*)

The small-hole gages shown in Fig. 7-5 have a small, round end, or ball, and are used for measuring holes, slots, grooves, and recesses which are too small for inside calipers or telescope gages. Those shown in Fig. 7-6 have a flat bottom and are used for similar purposes. The flat bottom permits the measurement of shallow slots, recesses, and holes which is not possible with the rounded type.

Both types are of similar construction and are adjusted to size by turning the knurled knob on the top. This draws up the tapered plunger, causing the two halves of the ball to open up and contact the hole.

### To Use a Small-Hole Gage

Small-hole gages require extreme care in setting since it is easy to get an incorrect setting when checking the diameter of a hole.

**1.** Measure the hole to be checked with a rule.

**2.** Select the proper small-hole gage.

**3.** Clean the hole and the gage.

**4.** Adjust the gage until it is slightly smaller than the hole and insert into the hole.

**5.** Adjust the gage until it can be felt just touching the sides of the hole or slot.

**6.** Swing the handle back and forth, and adjust the knurled end until the proper "feel" is obtained across the widest dimension of the ball.

**7.** Remove the gage and check the size with an outside micrometer. *Note:* It is important that while transferring the measurement, the same "feel" is obtained as when adjusting the gage to the hole. (See N-3.)

N-3 Always use the micrometer ratchet or friction thimble.

### Telescope Gages

Telescope gages (Fig. 7-7) are used for obtaining the size of holes, slots, and recesses from 5⁄16 to 6 in. They are T-shaped instruments, each consisting of a pair of telescoping tubes or plungers connected to a handle. The plungers are spring-loaded to force them apart. The knurled knob on the end of the handle locks the plungers into position when turned in a clockwise direction. (*Note:* In some sets, only one plunger moves.)

### To Measure Using a Telescope Gage

**1.** Measure the hole size and select the proper gage.

**2.** Clean the gage and the hole.

**3.** Depress the plungers until they are slightly smaller than the hole diameter and clamp them in this position.

**4.** Insert it into the hole and with the handle tilted upward slightly, release the plungers.

**5.** Lightly snug up the knurled knob.

**6.** Hold the bottom leg of the telescope gage in position with one hand.

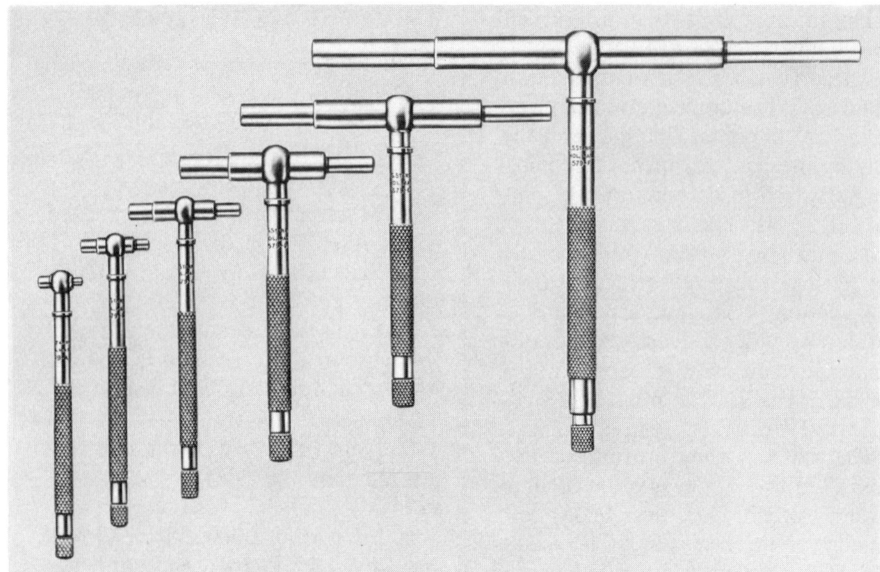

Fig. 7-7 A set of telescope gages. (*Courtesy The L. S. Starrett Company.*)

**7.** Move the handle downward through the center while slightly moving the top leg from side to side.

**8.** Tighten the plungers in position.

**9.** Recheck the "feel" on the gage by testing it in the hole again.

**10.** Check the gage size with outside micrometers, maintaining the same "feel" as in the hole.

## DEPTH MEASUREMENT

N-4 Always clean the work and the measuring instrument before checking a size.

Although a wide variety of measuring tools can be used for measuring depths, the two most commonly used for accurate measurement are the micrometer depth gage and the vernier depth gage.

### Micrometer Depth Gage

Micrometer depth gages are used for measuring depth of blind holes, slots, recesses, and projections. Each gage consists of a flat base attached to a micrometer sleeve. An extension rod of the required length fits through the sleeve and protrudes through the base (Fig. 7-8). This rod is held in position by a threaded cap on the top of the thimble.

Micrometer extension rods are available in various lengths, providing a range up to 9 in. (228.6 mm), while the micrometer screw has a range of ½ or 1 in. Metric depth micrometer screws have a range of 25 mm.

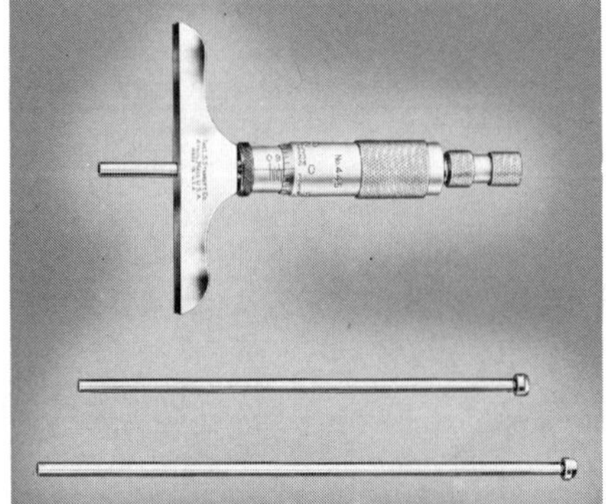

Fig. 7-8 A micrometer depth gage and extension rods. (*Courtesy The L. S. Starrett Company.*)

Depth micrometers are available with both round or flat rods which are not interchangeable with other depth micrometers. The accuracy of these micrometers is controlled by a nut on the end of each extension rod which can be adjusted if necessary.

#### To Measure with a Micrometer Depth Gage

**1.** Remove burrs from the edge of the hole and the face of the workpiece.

**2.** Hold the micrometer base firmly against the surface of the work (Fig. 7-9). (See N-4.)

**3.** Rotate the thimble lightly with the tip of one finger in a clockwise direction until the bottom of the extension rod touches the bottom of the hole or recess.

**4.** Recheck the micrometer setting a few times to make sure that not too much pressure was applied in the setting.

**5.** Carefully note the reading (Fig. 7-10).

### Vernier Depth Gages

The depth of holes, slots, and recesses may also be measured by a vernier depth gage (Fig. 7-11). This instrument is read in the same manner as a standard vernier caliper. Figure 7-11 illustrates how the toolmaker's button may be set up with this instrument.

Depth measurements may also be made with certain types of vernier or dial calipers which are provided with

Fig. 7-9 Measuring the depth of a shoulder. (*Courtesy The L. S. Starrett Company.*)

Fig. 7-10 Graduations on a depth micrometer are reversed to those on an outside micrometer. (*Courtesy The L. S. Starrett Company.*)

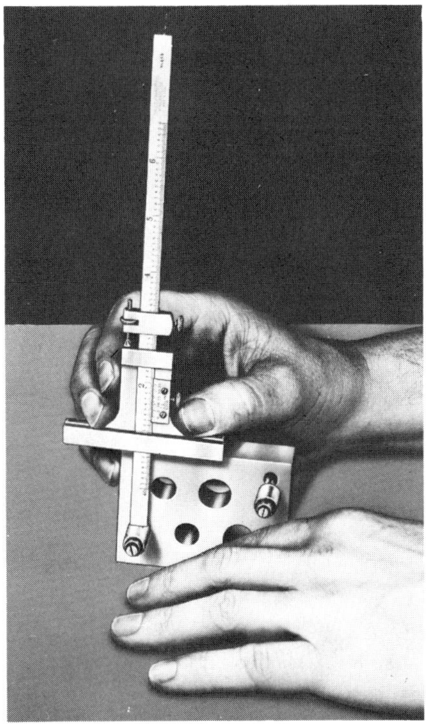

Fig. 7-11 A vernier depth gage can be used to measure the depth of holes and slots and also for checking toolmakers' buttons. (*Courtesy The L. S. Starrett Company.*)

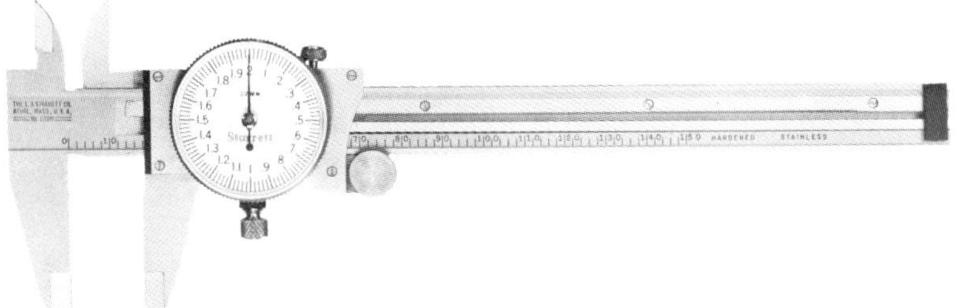

Fig. 7-12 A dial caliper with a thin blade for depth measurement. (*Courtesy The L. S. Starrett Company.*)

a thin sliding blade or depth gage attached to the movable jaw (Fig. 7-12). The blade protrudes from the end of the bar opposite the sliding jaw. The caliper is placed vertically over the depth to be measured, and the end of the bar is held against the shoulder while the blade is inserted into the hole to be measured. Depth readings are identical to standard vernier readings.

## HEIGHT MEASUREMENT

Accurate height measurements are very important in layout and inspection work. With the proper attachments, the vernier height gage is a very useful and versatile instrument for both these purposes.

### Vernier Height Gage

The vernier height gage (Fig. 7-13) is a fine precision instrument used in toolrooms and inspection departments on layout and jig and fixture work to measure and mark off distances accurately. These instruments are available in a variety of sizes from 12 to 72 in., and from 300 to 700 mm. They can be accurately set at any height to within one-thousandth of an inch (0.02 mm). Basically, a vernier height gage is a vernier caliper with a hardened, ground, and lapped base in lieu of a fixed jaw and is always used with a surface plate or an accurate flat surface. The sliding jaw assembly can be raised or lowered to any position along the beam. Fine adjustments are made by means of a fine adjustment screw or knob. It is read in the same manner as a vernier caliper.

The vernier height gage is particularly suited to accurate layout work and may be used for this purpose if a scriber is mounted on the movable jaw (Fig. 7-13). The scriber height may be set either by means of the vernier scale or by setting the scriber to a gage block buildup.

The offset scriber (Fig. 7-14) is a vernier height gage attachment which permits the setting of heights from the face of the surface plate. When using this attachment, it is not necessary to consider the height of the base or the width of the scriber and clamp.

A depth gage attachment may be fastened to the movable jaw, permitting the measurement of height differences which may be difficult to measure by other methods.

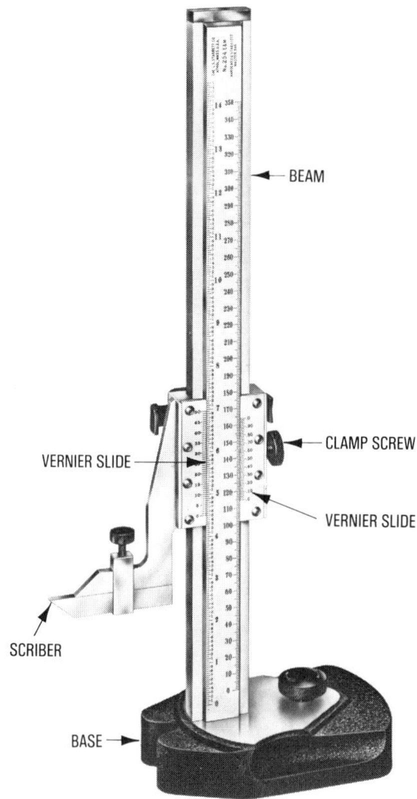


Fig. 7-13 The main parts of a vernier height gage. (*Courtesy The L. S. Starrett Company.*)

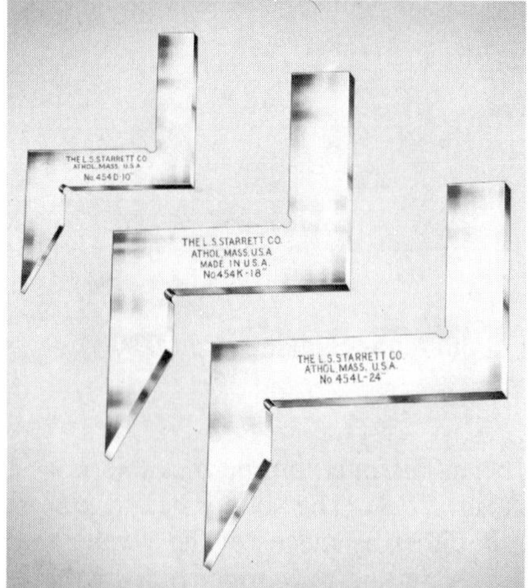


**Fig. 7-14** Offset scribers permit a direct-reading dimension from the face of the surface plate. (*Courtesy The L. S. Starrett Company.*)

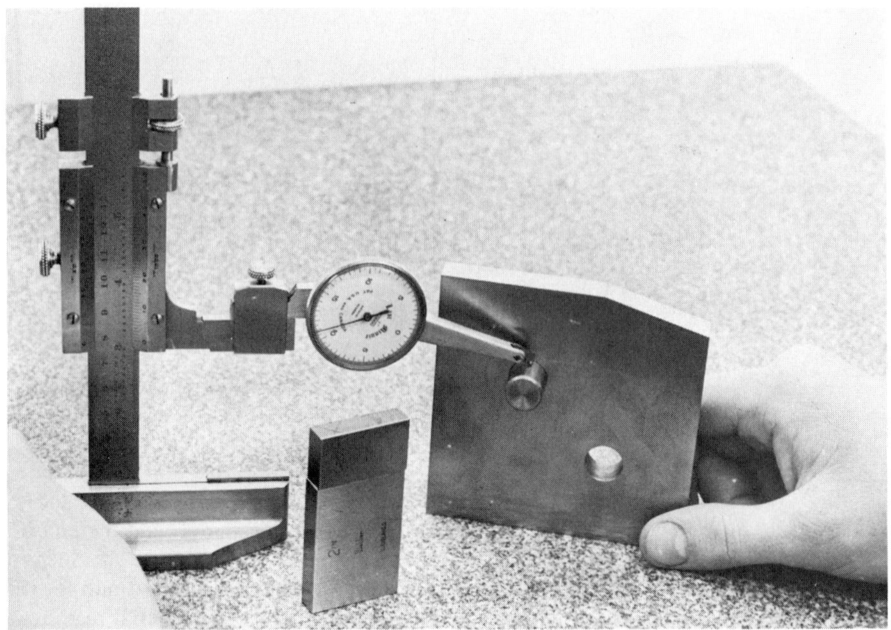

**Fig. 7-15** Using the height gage and dial indicator to check a height. (*Courtesy Kostel Enterprises Ltd.*)

Another important use for the vernier height gage is in inspection work. A dial indicator may be fastened to the movable jaw of the height gage (Fig. 7-15), and distances between holes or surfaces can be checked to within an accuracy of .001 in. (0.02 mm) on the vernier scale. If greater accuracy (less than .001 in., or 0.02 mm) is required, the indicator may be used in conjunction with the gage blocks.

In Fig. 7-15, the height gage is being used to check the location of reamed holes in relation to the edges of the plate and to each other.

### To Measure with a Vernier Height Gage and Dial Indicator

**1.** Thoroughly clean the surface plate, height gage base, and work surface.

**2.** Place the work on the surface plate and clamp it against an angle plate if required.

**3.** Mount the dial indicator on the movable jaw of the height gage.

**4.** Adjust the movable jaw until the indicator almost touches the surface plate.

**5.** Lock the upper slide, and use the fine adjustment knob or screw to move the indicator until the needle registers about one-quarter turn.

**6.** Set the indicator dial to zero.

**7.** Record the reading of the vernier height gage.

**8.** Adjust the vernier height gage until the indicator registers zero on the top of the plug as shown in Fig. 7-15. Record this vernier height gage reading.

**9.** From this reading, subtract the initial reading plus half the diameter of the plug. This will indicate the distance from the surface plate to the center of the hole.

**10.** Check the other hole heights using the same procedure.

## REVIEW QUESTIONS

### Inside-Measuring Instruments

**1.** How is a wide range of measurements possible with an inside micrometer?

**2.** What precaution must be taken when:

(a) Assembling the inside micrometer and an extension rod

(b) Using the inside micrometer

**3.** What is the correct "feel" with an inside micrometer?

**4.** Name two types of small-hole gages and state the purpose of each.

**5.** What precaution should be taken when transferring the measurement to a direct-reading instrument?

6. Describe a telescope gage.
7. Briefly outline the procedure to measure a hole size using a telescope gage.

### Depth Measurement

8. How is the extension rod attached to the micrometer depth gage?
9. How is the workpiece prepared before measuring the depth of a hole or slot with a micrometer depth gage?
10. How can you ensure that too much pressure is not applied when checking the depth?
11. How does the reading of a depth micrometer differ from that of a standard outside micrometer?

### Height Measurement

12. State two important applications of the vernier height gage.
13. What is the advantage of using an offset scriber for setting a height?
14. What accessory is required to *accurately* check the height of a workpiece with a vernier height gage?

# UNIT 8

# GAGE BLOCKS

Gage blocks, the accepted standard of accuracy in the machine industry, have provided industry with a means of maintaining sizes to specific standards of tolerances. This feature has resulted in higher rates of production and has made interchangeable manufacture possible.

N-1 Gage blocks are only accurate to size at 68°F (20°C).

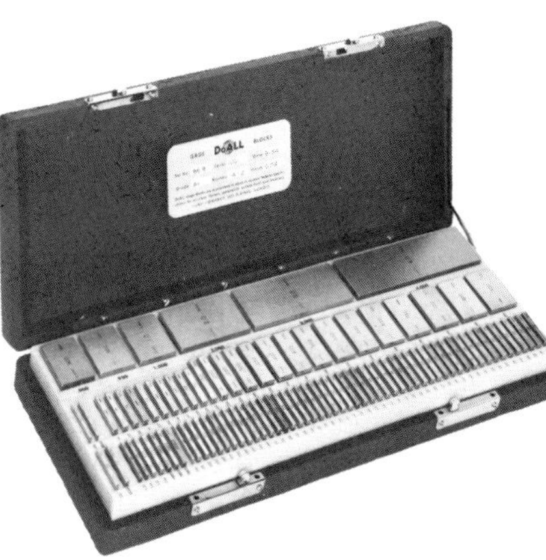

**Fig. 8-1** An 83-piece set of gage blocks. (*Courtesy The DoAll Company.*)

## OBJECTIVES

On completion of this unit you will:

1. Know the care required to maintain the accuracy of inch and metric gage blocks
2. Be aware of the use and application of gage blocks
3. Be able to calculate and make gage-block buildups
4. Be able to ''wring'' the blocks properly to make any buildup

## GAGE-BLOCK MANUFACTURE

Gage blocks are rectangular blocks of hardened and ground alloy steel which have been stabilized by alternate cycles of heating and freezing to remove the strain in the metal. The two opposite measuring surfaces are lapped and polished to an optically flat surface and to a specific size accurate to within a few millionths of an inch. The size of each block is marked on one of its surfaces. Great care is exercised in their manufacture; the final sizing is made under ideal conditions where the temperature is maintained at 68°F (20°C). Therefore, even a standard inch is accurate to size *only* when measured at the standard temperature of 68°F (20°C). (See N-1.)

### Uses

Industry has found gage blocks to be invaluable tools. Because of their extreme accuracy, they are used for the following purposes.

1. To check the accuracy of fixed gages
2. To set adjustable gages, such as micrometers and vernier calipers
3. To set comparators, dial indicators, and height gages to exact dimensions
4. To set sine bars and sine plates for angular setups
5. To function in machine tool setups
6. To measure and inspect the accuracy of finished parts in inspection rooms

## GAGE-BLOCK SETS

Gage blocks are manufactured in a variety of sets that vary from a few blocks to as many as 115 in a set. The most commonly used is the 83-piece set, Fig. 8-1, from which it is possible to make over 120,000 different measurements ranging from one hundred-thousandths of an inch to over 25 in. The blocks which make up an 83-piece set are listed in Table 8-1 on page 36.

Two *wear* blocks, furnished with an 83-piece set, may be either .050 or .100 in. They should be used at each end of a combination to protect the

N-2 Wear blocks preserve the accuracy of the other blocks in a set.

N-3 Gage block temperatures over 68°F (20°C) will cause inaccurate measurements.

other blocks and preserve their accuracy, especially if the blocks will be in contact with hard surfaces or abrasives. During use, it is considered good practice always to expose the same face of the wear block to the work surface so that all the wear will be on one surface and the wringing quality of the other surface will be preserved. (See N-2.)

**TABLE 8-1 SIZES IN AN 83-PIECE SET OF INCH STANDARD GAGE BLOCKS**

| First: .0001-in. series—9 blocks | | | | | | | | | |
|---|---|---|---|---|---|---|---|---|---|
| .1001 | .1002 | .1003 | .1004 | .1005 | .1006 | .1007 | .1008 | .1009 | |
| **Second: .001-in. series—49 blocks** | | | | | | | | | |
| .101 | .102 | .103 | .104 | .105 | .106 | .107 | .108 | .109 | |
| .110 | .111 | .112 | .113 | .114 | .115 | .116 | .117 | .118 | |
| .119 | .120 | .121 | .122 | .123 | .124 | .125 | .126 | .127 | |
| .128 | .129 | .130 | .131 | .132 | .133 | .134 | .135 | .136 | |
| .137 | .138 | .139 | .140 | .141 | .142 | .143 | .144 | .145 | |
| .146 | .147 | .148 | .149 | | | | | | |
| **Third: .050-in. series—19 blocks** | | | | | | | | | |
| .050 | .100 | .150 | .200 | .250 | .300 | .350 | .400 | .450 | .500 |
| .550 | .600 | .650 | .700 | .750 | .800 | .850 | .900 | .950 | |
| **Fourth: 1.000-in. series—4 blocks** | | | | | | | | | |
| 1.000 | 2.000 | 3.000 | 4.000 | | | | | | |
| **Two .050-in. wear blocks** | | | | | | | | | |

**TABLE 8-2 SIZES IN AN 88-PIECE SET OF METRIC GAGE BLOCKS**

| 0.001-mm series—9 blocks | | | | | | | | |
|---|---|---|---|---|---|---|---|---|
| 1.001 | 1.002 | 1.003 | 1.004 | 1.005 | 1.006 | 1.007 | 1.008 | 1.009 |
| **0.01-mm series—49 blocks** | | | | | | | | |
| 1.01 | 1.02 | 1.03 | 1.04 | 1.05 | 1.06 | 1.07 | 1.08 | 1.09 |
| 1.10 | 1.11 | 1.12 | 1.13 | 1.14 | 1.15 | 1.16 | 1.17 | 1.18 |
| 1.19 | 1.20 | 1.21 | 1.22 | 1.23 | 1.24 | 1.25 | 1.26 | 1.27 |
| 1.28 | 1.29 | 1.30 | 1.31 | 1.32 | 1.33 | 1.34 | 1.35 | 1.36 |
| 1.37 | 1.38 | 1.39 | 1.40 | 1.41 | 1.42 | 1.43 | 1.44 | 1.45 |
| 1.46 | 1.47 | 1.48 | 1.49 | | | | | |
| **0.5-mm series—1 block** | | | | | | | | |
| 0.5 | | | | | | | | |
| **0.5-mm series—18 blocks** | | | | | | | | |
| 1 | 1.5 | 2 | 2.5 | 3 | 3.5 | 4 | 4.5 | 5 |
| 5.5 | 6 | 6.5 | 7 | 7.5 | 8 | 8.5 | 9 | 9.5 |
| **10-mm series—9 blocks** | | | | | | | | |
| 10 | 20 | 30 | 40 | 50 | 60 | 70 | 80 | 90 |
| **Two 2-mm wear blocks** | | | | | | | | |

## METRIC GAGE BLOCKS

Metric gage blocks are supplied by most manufacturers in sets of 47-, 88-, and 113-piece blocks. The most common set is the 88-piece set (Table 8-2). Each of these sets contains a pair of 2-mm wear blocks. The blocks are used at each end of the buildup to prolong the accuracy of the other blocks in the set.

## ACCURACY

Gage blocks in the inch and metric standards are manufactured in three common degrees of accuracy, depending on the purpose for which they are used.

**1.** The class AA set, commonly called a laboratory or master set, is accurate to ±.000 002 in. in the inch standard. The metric gage block set is accurate to ±0.000 05 mm. These gage blocks are used in temperature-controlled laboratories as references to compare or check the accuracy of working gages.

**2.** The class A set is used for inspection purposes and is accurate to ±.000 004 in. in the inch standard. The metric gage block set is accurate to +0.000 15 mm and −0.000 05 mm.

**3.** The class B set, commonly called the working set, is accurate to ±.000 008 in. in the inch standard. The metric gage block set is accurate to +0.000 25 mm and −0.000 15 mm. These blocks are used in the shop for machine tool setups, layout work, and measurement.

## THE EFFECT OF TEMPERATURE

Gage blocks have been calibrated at 68°F (20°C); the human body temperature is around 98°F (37°C). A 1°F (0.5°C) rise in temperature will cause a 4-in. (100-mm) stack of gage blocks to expand approximately .000 025 in. (0.0006 mm); therefore it is important that these blocks be handled as little as possible. (See N-3.) The following suggestions are offered to eliminate as much error as possible from temperature change.

1. Handle gage blocks only when required.
2. Hold them for as little time as possible.
3. Hold them between the tips of the fingers so that the area of contact is small.
4. Have the work and gage blocks at the same temperature.

| Step | Procedure | | Check |
|---|---|---|---|
| 1 | Write the dimension required on the paper | 1.6428 | |
| 2 | Deduct the size of two wear blocks: 2 × .050 in. | .100 | .100 |
| | Remainder | 1.5428 | |
| 3 | Use a block which will eliminate the right-hand digit | .1008 | .1008 |
| | Remainder | 1.4420 | |
| 4 | Use a block which will eliminate the right-hand digit and at the same time bring the digit to the left of it to a zero (0) or a five (5) | .142 | .142 |
| | Remainder | 1.300 | |
| 5 | Continue to eliminate the digits from the right to the left until the dimension required is attained | .300 | .300 |
| | Remainder | 1.000 | |
| 6 | Use a 1.000-in. block | 1.000 | 1.000 |
| | Remainder | .000 | 1.6428 |

## GAGE-BLOCK BUILDUPS

When the blocks required to make up a dimension are being calculated, the following procedure should be followed to save time and reduce the chance of error. When calculating a buildup, use as few blocks as possible. (See N-4.)

N-4 Always use as few blocks as possible when making a buildup.

To eliminate the possibility of error in subtraction while making a buildup, it is good practice to use two columns for this calculation. As the following example illustrates, the left-hand column is used to subtract the gage blocks from the original dimension, and the right-hand column is used as a check column. To make any buildup, proceed as in the table above, where the buildup for 1.6428 is used.

## METRIC BUILDUP

For a metric buildup of 27.781 mm, proceed as shown in the table below.

| Step | | Procedure | Check |
|---|---|---|---|
| | | 27.781 | |
| 1 | Two wear blocks: 2 × 2 mm | 4.000 | 4.000 |
| | | 23.781 | |
| 2 | Use 1.001 mm | 1.001 | 1.001 |
| | | 22.780 | |
| 3 | Use 1.080 mm | 1.080 | 1.080 |
| | | 21.700 | |
| 4 | Use 1.700 mm | 1.700 | 1.700 |
| | | 20.000 | |
| 5 | Use 20.000 mm | 20.000 | 20.000 |
| | | 0.000 | 27.781 |

### Care of Gage Blocks

1. Gage blocks should always be protected from dust and dirt by being kept in a closed case when not in use.
2. Gage blocks should not be handled unnecessarily, since they absorb the heat of the hand. Should this occur, the gage blocks must be permitted to settle down to room temperature before use.
3. Fingering of lapped surfaces should be avoided to prevent tarnishing and rusting.
4. Care should be taken that gage blocks are not dropped or their lapped surfaces scratched.
5. Immediately after use, each block should be cleaned, oiled, and replaced in its case.
6. Before gage blocks are wrung together, their faces must be free of oil and dust.
7. Gage blocks should never be left wrung together for any length of time. The slight moisture between the blocks can cause rusting which will permanently damage the block.

### To Wring Blocks Together

It is important when wringing blocks together to take care not to damage them. The sequence of movement to wring blocks together correctly is illustrated in Fig. 8-2.

1. Clean the blocks with a clean, soft cloth.
2. Wipe each of the contacting surfaces on the clean palm of the hand or wrist.
3. Place the end of one block over the end of another block as shown in Fig. 8-2.
4. While applying pressure on the

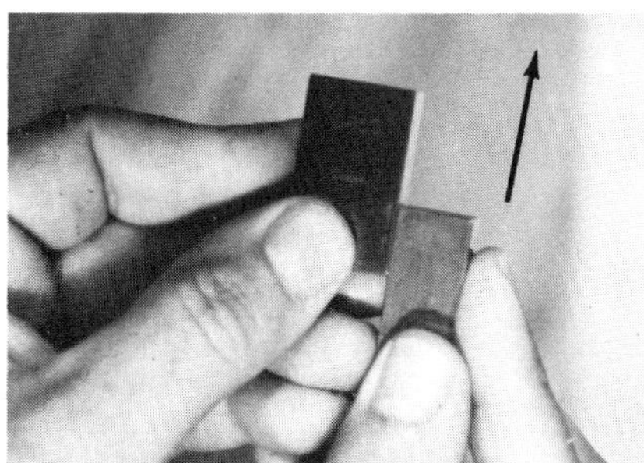

**Fig. 8-2** The procedure for wringing gage blocks.

two blocks, slide one block over the other.

**5.** If the blocks do not adhere to each other, it is generally because the blocks have not been thoroughly cleaned.

## REVIEW QUESTIONS

**1.** How are gage blocks stabilized, and why is this necessary?

**2.** State five general uses for gage blocks.

**3.** For what purpose are wear blocks used?

**4.** Where should wear blocks be placed in a buildup?

**5.** State the difference between a master set and a working set of gage blocks.

**6.** What two precautions are necessary when handling gage blocks in order that the effect of heat on the blocks is minimal?

**7.** List five precautions necessary for the proper care of gage blocks.

**8.** Calculate the gage blocks required for the following buildups. Two wear blocks (.050 in. or 2 mm) are used in each buildup.

| | |
|---|---|
| (a) 2.1743 | (d) 32.079 mm |
| (b) 6.2937 | (e) 74.213 mm |
| (c) 7.8923 | (f) 89.694 mm |

# UNIT 9

# ANGULAR MEASURING INSTRUMENTS

The ability to accurately measure angular surfaces is very important in machine shop work. The type of instrument which should be used to measure an angular surface depends on the accuracy required. The most commonly used tools for accurately laying out or measuring angles are the universal bevel protractor and the sine bar.

## ANGULAR MEASUREMENT UNITS

The angular unit of measurement in the inch system of measurement is the *degree* (°). Each circle consists of 360 degrees and these degrees are subdivided into smaller units of minutes and seconds for more accurate measurement.

| | | |
|---|---|---|
| 1 circle | = 360° | (360 degrees) |
| 1 degree | = 60′ | (60 minutes of an arc) |
| 1 minute | = 60″ | (60 seconds of an arc) |

The angular unit of measurement in the metric system is the *radian* (rad). A radian is the length of an arc on the circumference of a circle which is equal in length to the radius of the circle. Therefore, radians can be converted to degrees as follows:

$$2\pi r = 360°$$
$$1 \text{ radian} = \frac{360}{2\pi}$$
$$= 57.2957795131°$$
*or*
$$57°17'44'' \text{ (approximate)}$$
$$1° = 0.0174532925 \text{ rad}$$

If an angle is expressed in radian measure and a metric measuring tool is not available, it can easily be converted to degrees by using the equivalents shown.

## OBJECTIVES

Upon completion of this unit, you will be able to:

**1.** Identify and know when to use common angular measuring tools

**2.** Measure and record angular measurements to within an accuracy of plus or minus 5′ of a degree with a universal bevel protractor

**3.** Measure and record angular measurements to an accuracy of less than 5′ of a degree with a sine bar and dial indicator

## MEASURING ANGLES WITH A UNIVERSAL BEVEL PROTRACTOR

Precise angular measurements and setups are important phases of machine shop work. If an accuracy of up to 0.5° is required, a bevel protractor head and rule may be used. For an accuracy of one-twelfth of a degree, or 5′, a universal bevel protractor should be used.

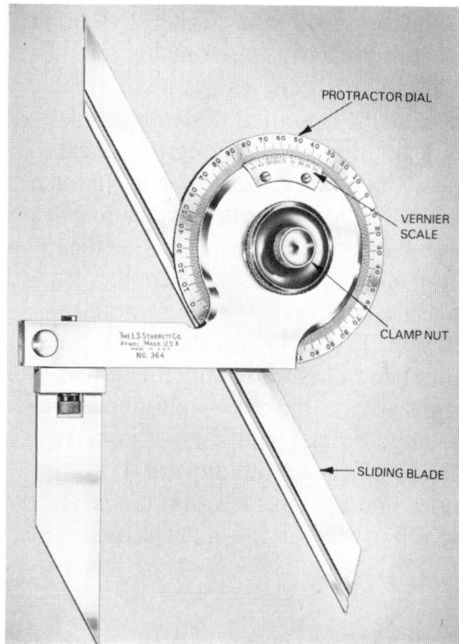


Fig. 9-1 The universal bevel protractor can measure angles to within 5 min of a degree. (*Courtesy The L. S. Starrett Company.*)

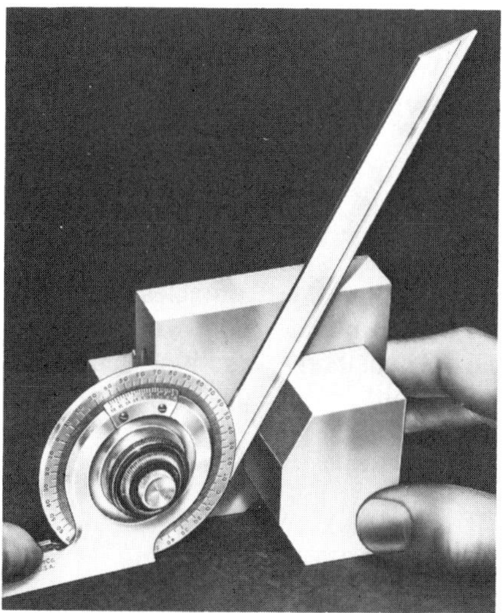

Fig. 9-2 Measuring an obtuse angle using a universal bevel protractor. (*Courtesy The L. S. Starrett Company.*)

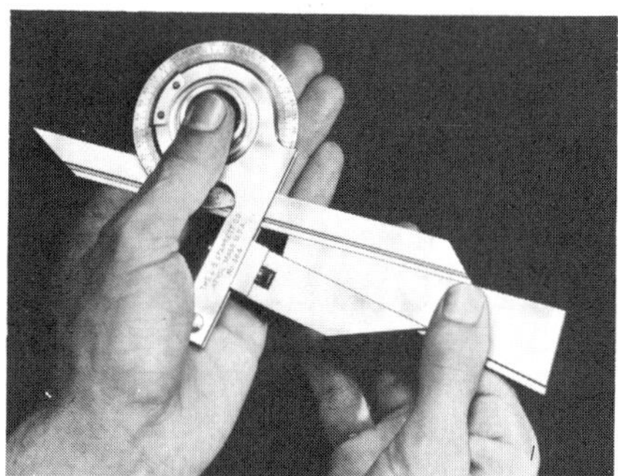

Fig. 9-3 Measuring an acute angle with a universal bevel protractor. (*Courtesy The L. S. Starrett Company.*)

### The Universal Bevel Protractor

The universal bevel protractor (Fig. 9-1) is a precision instrument capable of measuring angles to within 5′, or 1/12° (0.083°). It consists of a base to which a vernier scale is attached. A protractor dial, graduated in degrees with every tenth degree being numbered, is mounted onto the circular section of the base. A sliding blade is fitted into this dial; it may be extended in either direction and set at any angle to the base. The blade and the dial are rotated as a unit. Fine adjustments are obtained with a small knurled-headed pinion which when turned engages with a gear attached to the blade mount. The protractor dial may be locked in any desired position by means of the dial clamp nut.

The vernier protractor (Fig. 9-2) is being used to measure an obtuse angle, or an angle greater than 90° but less than 180°. An acute angle attachment is fastened to the vernier protractor to measure angles less than 90° (Fig. 9-3).

The vernier protractor employs the following principle. The protractor dial, or main scale, is divided into two arcs of 180°. Each arc is divided into two quadrants of 90° and has graduations from zero to 90° to the left and right of the "0" line.

The vernier scale is divided into 12 spaces on each side of the "0" line which occupy the same space as 23° on the protractor dial. By simple calculation, it is easy to prove that one vernier space is 5′ or 1/12° less than two graduations on the main scale. If the zero on the vernier scale coincides with a line on the main scale, the reading will be in degrees only. However, if any other line on the vernier scale coincides with a line on the main scale, the number of vernier graduations beyond the zero should be multiplied by 5 and added to the number of full degrees indicated on the protractor dial.

### To Read a Vernier Protractor

**1.** Note the number of whole degrees between the zero on the main scale and the zero on the vernier scale.

**2.** Proceeding in the same direction beyond the zero on the vernier scale, note which vernier line coincides with a main scale line.

**3.** Multiply this number by 5 and add it to the number of degrees on the protractor dial.

In Fig. 9-4, the angular reading is determined as follows. The number of degrees indicated on the main scale is 50 plus. The fourth line on the vernier scale to the left of the zero coincides with a line on the main scale. Therefore, the reading is:

| | | |
|---|---|---|
| Number of full degrees | = | 50° |
| Value of vernier scale | | |
| 4 × 5′ | = | 20′ |
| Reading | = | 50°20′ |

*Note:* A double check of the reading would locate the vernier scale line on the other side of zero which coincides with a protractor scale line. This line should always equal the complement of 60′. In Fig. 9-4, the 40′ line to the right of the zero coincides with a line on the protractor scale. This reading when added to the 20′ on the left of the scale is equal to 60′, or 1°.

## MEASURING ANGLES WITH A SINE BAR

### The Sine Bar

A *sine bar,* Fig. 9-5, is used when the accuracy of an angle must be checked to less than 5′ (0.083° or 1/12°) or work must be located to a given angle within close limits. The sine bar consists of a steel bar with two cylinders of equal diameter secured near the ends. The centers of these cylinders are on a line exactly parallel with the edge of the bar. The distance between the centers of these lapped cylinders is usually 5 or 10 in. These bars are generally made of stabilized tool steel, hardened, ground, and lapped to extreme accuracy. They are used on surface plates, and any desired angle can be set by raising one end of the bar to a predetermined height with gage blocks.

Sine bars are generally made 5 in. or multiples of 5 in. in length. That is, the lapped cylinders are 5 in. ±.0002 or 10 in. ±.00025 between centers. The face of the sine bar is accurate within .00005 in 5 in. In theory, the sine bar merely becomes the *hypotenuse* of a right-angle triangle. The gage-block buildup forms the *side opposite,* while the face of the surface plate forms the *side adjacent* in the triangle.

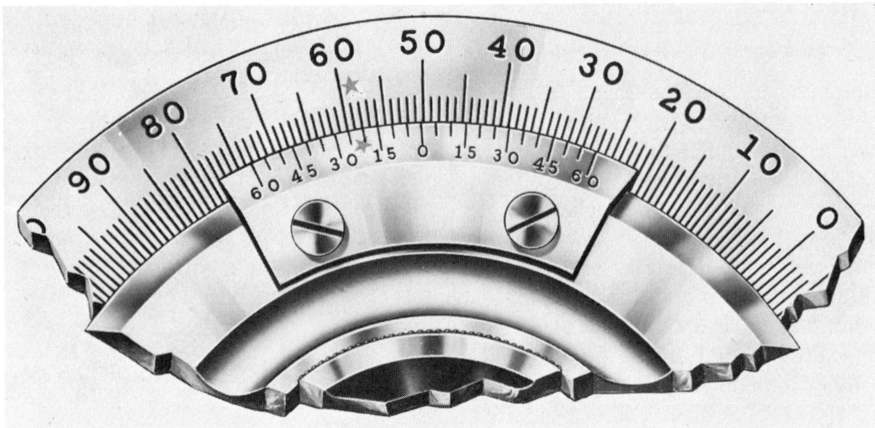

Fig. 9-4 Vernier protractor reading of 50° 20′ (50.33°). (*Courtesy The L. S. Starrett Company.*)

Fig. 9-5 A 5-in. sine bar and gage block buildup is used to set up work to an angle.

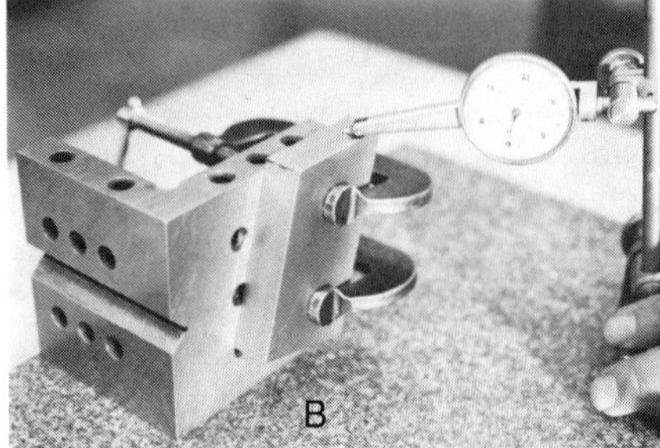

Fig. 9-6 Setting up for an angle greater than 60°. (A) Set the sine bar to the complement of the angle. (B) Turn the angle plate 90° on its side.

Using trigonometry, it is possible to calculate the side opposite, or gage-block buildup, for any angle between 0 and 90° as follows.

$$\text{Sine of given angle} = \frac{\text{side opposite}}{\text{hypotenuse}} = \frac{\text{gage-block buildup}}{\text{length of sine bar}}$$

When using a 5-in. sine bar, this would become

$$\text{Sine of the angle} = \frac{\text{buildup}}{5}$$

Therefore, by transposition, the gage-block buildup for any required angle with a 5-in. sine bar is as follows.

$$\text{Buildup} = 5 \times \text{sine of the required angle}$$

**Example:**
Calculate the gage-block buildup required to set a 5-in. sine bar to an angle of 30°.

$$\begin{aligned}\text{Buildup} &= 5 \text{ sine } 30 \\ &= 5 \times .500 \\ &= 2.500\end{aligned}$$

*Note:* This formula is applied only to angles up to 60°. When an angle greater than 60° is to be checked, it is better to set up the work using the complement of the angle (Fig. 9-6A). The angle plate is then turned 90° to produce the correct angle (Fig. 9-6B). The reason is that when the sine bar is in a near-horizontal position, a small change in the height of the buildup will produce a smaller change in the angle than when the sine bar is in the near-vertical position. Angles can thus be measured more accurately when the sine bar is nearly horizontal.

When small angles are to be checked, it is sometimes impossible to get a buildup small enough to place under one end of the sine bar. In such situations, it will be necessary to place gage blocks under both rolls of the sine bar, having a net difference in measurement equal to the required buildup. For example, the buildup required for 2° is .1745 in. Since it is impossible to make this buildup with the gage blocks available, it is necessary to place the buildup for 1.1745 under one roll and a 1.000 block under the other roll, giving a net difference of .1745.

## REVIEW QUESTIONS

### Universal Bevel Protractor

**1.** Name five parts of a universal bevel protractor.
**2.** What is the accuracy of a universal bevel protractor?
**3.** What is the value of each division on the vernier scale?

### Sine Bar

**4.** Describe the construction and principle of a sine bar.
**5.** What are the accuracies of the 5- and the 10-in. sine bars?
**6.** Calculate the gage-block buildup for the following angles using a 5-in. sine bar.
(a) 7°40′
(b) 25°50′
(c) 40°10′

# UNIT 10

# COMPARISON MEASURING INSTRUMENTS

Manufacturing processes are so precise that component parts can be made in various plants and then shipped to one location for assembly. To make this possible, it is important that each part is made to within certain limits or tolerances to ensure that the part will fit and perform properly. Inspection and quality control ensures that only properly sized parts are used for the final assembly of the finished part.

The inspection of parts can be done quickly and accurately by comparison measurement. This consists of comparing the size of the part to a known standard or master of the dimension required. *Comparators* are gages that incorporate some means of amplification to compare the part size to a set standard. The most common comparison instruments are dial indicators, mechanical and optical comparators, and electronic and air gages.

## OBJECTIVES

After completing this unit, you should be able to:

**1.** Know the purpose of the most common dial indicators
**2.** Check dimensions using a dial indicator and vernier height gage
**3.** Understand the principle and construction of the mechanical, optical, or reed-type comparator
**4.** Be able to set up and use a reed-type comparator

## DIAL INDICATORS

Dial indicators are used to compare sizes and measurements to a known standard and to check the alignment of machine tools, fixtures, and workpieces prior to machining.

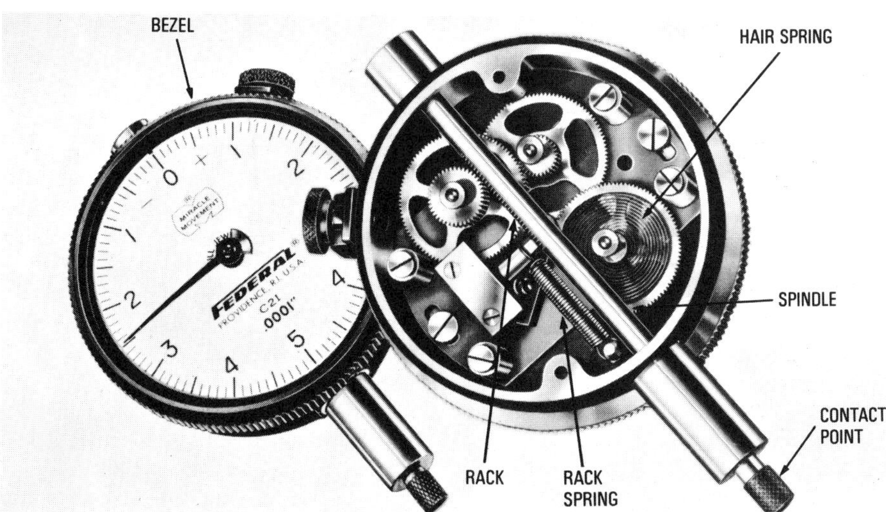


**Fig. 10-1** A balanced-type dial test indicator showing the internal mechanism. (*Courtesy Federal Products Corporation.*)

Many types of dial indicators operate on a gear and rack principle (Fig. 10-1). A rack cut on the plunger or spindle is in mesh with a pinion which in turn is connected with a gear train. Any movement of the spindle is then amplified and transmitted to a hand or pointer over a graduated dial. Inch-designed dials may be graduated in thousandths of an inch or less. The dial, attached to a bezel, may be adjusted to and locked in any position.

During use, the contact point on the end of the spindle bears against the work and is held in constant engagement with the work surface by the rack spring. A hair spring is attached to the gear that meshes with the center pinion. This flat spiral takes up the backlash from the gear train and prevents any lost motion from affecting the accuracy of the gage.

Dial indicators are generally available in two types: the continuous-reading dial indicator and the dial test indicator.

The continuous-reading dial indicator (Fig. 10-2), numbered clockwise for 360°, is available in two types: the regular-range and the long-range indicator. The regular-range dial indicator has only about 2½ revolutions of travel. It is generally used for comparison measurement and setup purposes. The long-range dial indicator (Fig. 10-2) is often used to indicate table travel or cutting-tool movement on machine tools. It has a second smaller hand that indicates the number of revolutions that the large hand has traveled. These indicators usually have a range of 1 in.

Dial test indicators (Fig. 10-1) may have a *balanced-type dial,* that is, one which reads both to the right and left from zero and indicates a plus or minus value. Indicators of this type have a total spindle travel of only 2½ revolutions. These instruments may be equipped with tolerance pointers to indicate the permissible variation of the part being measured.

Perpendicular dial test indicators (Fig. 10-3) or back-plunger indicators have the spindle at right angles (90°) to the dial. They are used extensively in setting up lathe work and for machine table alignment.

The universal dial test indicator (Fig. 10-4) has a contact point that may be set at several positions through a 180° arc. This type of indicator may be conveniently used to check internal and

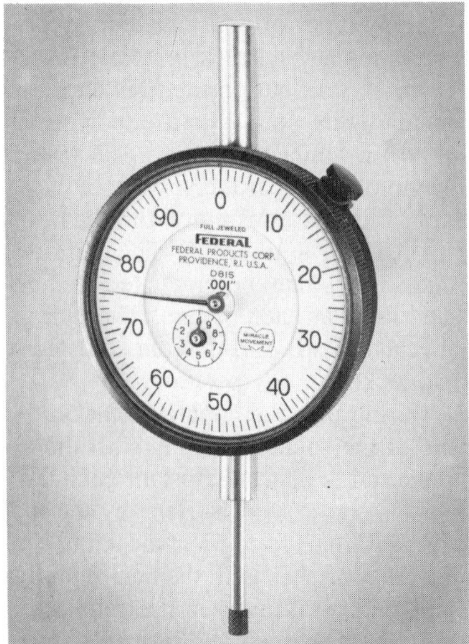


**Fig. 10-2** A long-range continuous-reading dial indicator is often used on machine tools. (*Courtesy Federal Products Corporation.*)

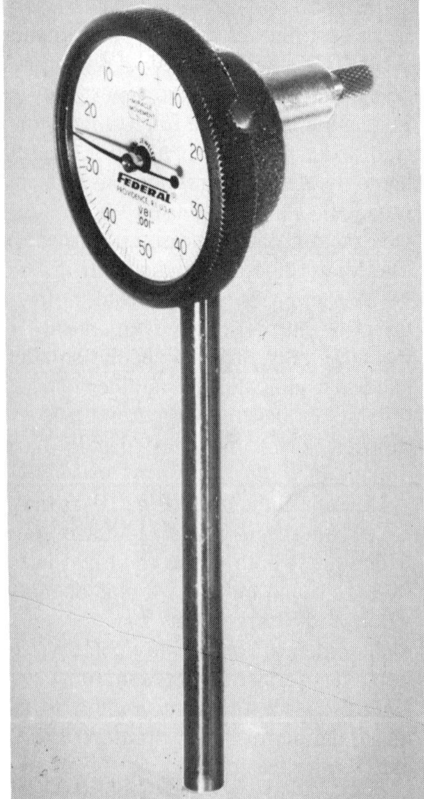


**Fig. 10-3** A perpendicular dial test indicator is used for checking machine-part alignment and work runout. (*Courtesy Federal Products Corporation.*)

N-1 The surface plate, height gage base, and the gage block must be clean.

external surfaces. Figures 10-4 and 10-6 illustrate typical applications of this type of indicator.

Metric dial indicators (Fig. 10-5) are available in both the balanced- and continuous-reading types. The type used for inspection purposes is usually graduated in 0.002 mm and has a range of 0.5 mm. The regular indicators are usually graduated in 0.01 mm and have a range up to 25 mm.

### To Measure with a Dial Test Indicator and Height Gage

**1.** Clean the face of the surface plate and the vernier height gage. (See N-1.)

**2.** Mount the dial test indicator on the movable jaw of the height gage (Fig. 10-6).

**3.** Lower the movable jaw until the indicator point just touches the top of a gage block resting on the surface plate.

**4.** Tighten the upper locking screw on the vernier and loosen the lower locking screw.

**5.** Carefully turn the adjusting nut until the indicator needle registers approximately one-quarter turn.

**6.** Turn the bezel to set the indicator to zero.

**7.** Note the reading on the vernier and record it on a piece of paper.

**8.** Raise the indicator to the height of the first hole to be measured.

**9.** Adjust the vernier until the indicator reads zero.

**10.** Note the vernier reading again and record it.

**11.** Subtract the first reading from the second and add the height of the gage block.

**12.** Proceed in this manner to record the location of all other holes.

## COMPARATORS

A comparator may be classified as any instrument used to compare the size of a workpiece with a known standard. The simplest form of comparator is a dial indicator mounted on a surface gage. All comparators are provided with some means of amplification by which variations from the basic dimensions are noted easily. A small difference in the size of the workpiece will

**Fig. 10-4** A universal dial test indicator being used to center a workpiece with a spindle. (*Courtesy Federal Products Corporation.*)

show up as a much larger space on the gage through some form of magnification.

### Types of Comparators

There are several types of comparators, each of which uses a different principle to measure or compare the size of a workpiece to a known standard or gage. Some of the more common types of comparators used in machine shop are as follows.

*Mechanical comparators* (Fig. 10-7) use several principles, the most common being the gear and rack or a system of levers, both of which operate on the same principle as some dial indicators.

*Optical comparators* (Fig. 10-8) project an enlarged shadow of the part being measured onto a screen where it can be measured or compared with a standard. These comparators are particularly suited for measuring small and odd-shaped workpieces.

*Electronic gages* (Fig. 10-9) can change the amplification ratio to suit the tolerance desired. These comparators are capable of measuring to thousandths (.001), ten-thousandths (.0001), or hundred thousandths (.00001) of an inch and can be set to the desired range by a selector switch. The units may also be used as electronic height gages by mounting a rectangular gaging head on a height-gage stand. This method is particularly suited to the checking of soft, highly polished surfaces because of the light gaging pressure required.

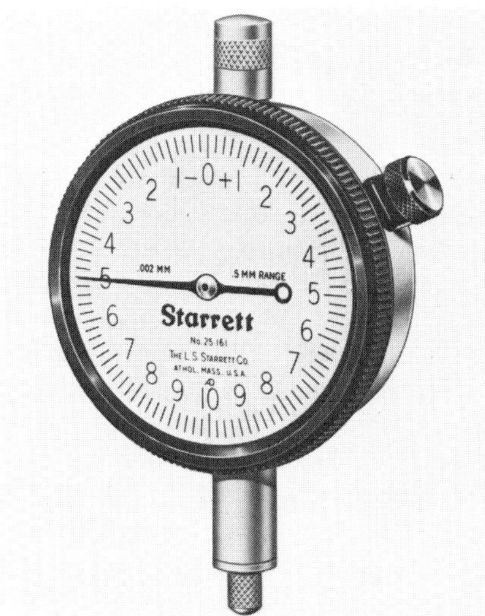


**Fig. 10-5** A metric indicator with a balanced dial. *(Courtesy The L. S. Starrett Company.)*

**Fig. 10-6** Checking measurements with a dial test indicator and height gage. *(Courtesy Federal Products Corporation.)*

**Fig. 10-7** A mechanical comparator being set to a gage block. Note the tolerance pointers that indicate the upper and lower limits.

Height gages with electronic digital readouts may be used as a comparator for inspection work by attaching an indicator to the sliding jaw. The Digit-Hite electronic gage is capable of giving both inch and metric readings to within .0001 in., or 0.003 mm. This instrument has several refinements:

**1.** A memory feature which makes it possible to hold readings taken when the readout is not visible from the operating position

**2.** A comparative measuring feature which shows the amount of deviation of a part from a set standard

## Mechanical-Optical Comparators

The *mechanical-optical comparator* (Fig. 10-10), or the *reed*-type comparator, combines a reed mechanism with a light beam to cast a shadow on a magnified scale to indicate the dimensional variation of the part. It consists of a base and a column as well as a gaging head which contains the reed mechanism and light source.

**Fig. 10-8** Small and intricate parts can be checked easily with an optical comparator.

## The Reed Mechanism

Figure 10-11A illustrates the principle of the reed mechanism. A fixed steel block *A* and a movable block *B* have two pieces of spring steel, or reeds, attached to them. The upper ends of the reeds are joined and connected to a pointer. Since block *A* is fixed, any movement of the spindle attached to block *B* will move this block up or down, causing the pointer to move a much greater distance to the right or left, Fig. 10-11B.

A light beam passing through an aperture illuminates the scale, Fig. 10-12. The pointer, with a *target* attached, is located so that a movement of block *B* will cause the target to interrupt the light beam, casting a highly amplified shadow on the scale. The movement of the target is much greater than the movement of the spindle. Also, the shadow cast on the scale will be larger than the movement of the target. Therefore, the measurement on the scale will be much greater than the movement of the spindle.

The scales for these instruments are graduated in plus and minus with zero being in the center of the scale. The value for each graduation is marked on the scale of every machine.

## To Measure with a Reed Comparator

**1.** Raise the gaging head above the required height and clean the anvil and master thoroughly.

**2.** Place the master gage or gage-block buildup on the anvil.

**3.** Carefully lower the gaging head until the end of the spindle *just touches the master*. *Note:* A shadow will begin to appear on the left side of the scale.

**4.** Clamp the gaging head to the column.

**5.** Turn the adjusting sleeve until the shadow coincides with the zero on the scale.

**6.** Remove the gage blocks and carefully slide the workpiece between the anvil and the spindle.

**7.** Note the reading. If the shadow is to the right of the zero, the part is oversize; if to the left, it is undersize.

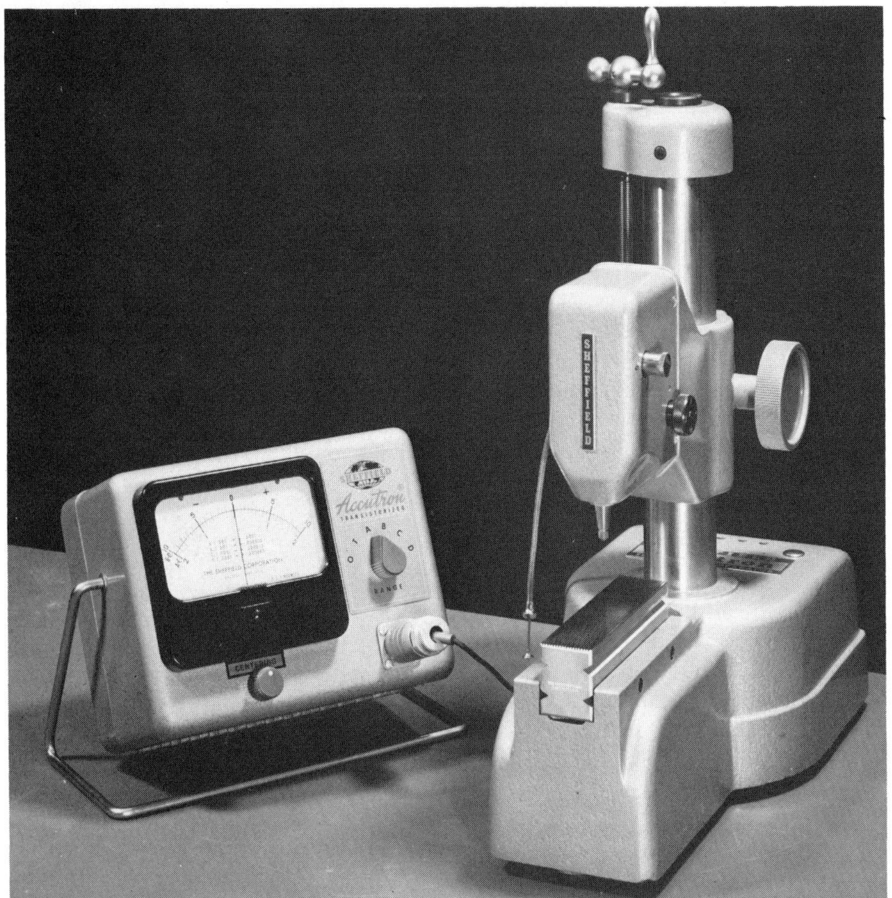

Fig. 10-9 The value of the scale divisions on an electronic comparator can be readily changed to suit the accuracy required. (*Courtesy Bendix Corp.*)

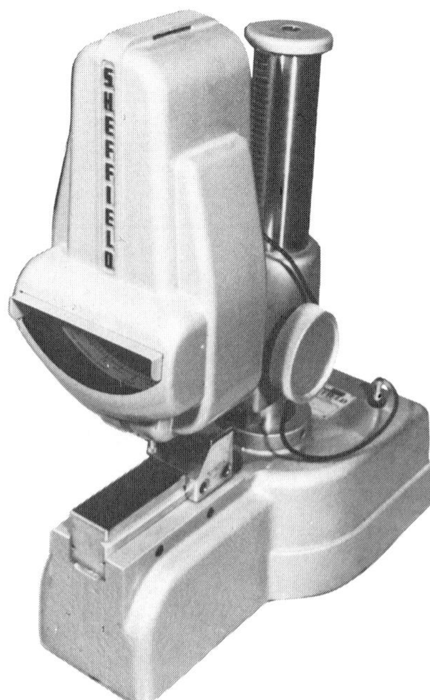

Fig. 10-10 The reed-type comparator uses mechanical and optical amplification to produce a highly magnified and easy-to-read measurement. (*Courtesy Bendix Corp.*)

## AIR GAGES OR PNEUMATIC COMPARATORS

*Air gaging,* a form of comparison measurement, is used to compare workpiece dimensions with those of a master gage by means of air pressure or flow.

Air gages are of two types: the *flow,* or *column, type* (Fig. 10-13), which indicates air velocity; and the *pressure type* (Fig. 10-15), which indicates air pressure in the system.

### Column-Type Air Gage

After air has been passed through a filter and a regulator, it is supplied to the gage at about 10 lb/in.$^2$ (68.9 kPa) (Fig. 10-14). The air flows through a transparent tapered tube in which a float is suspended as a result of this

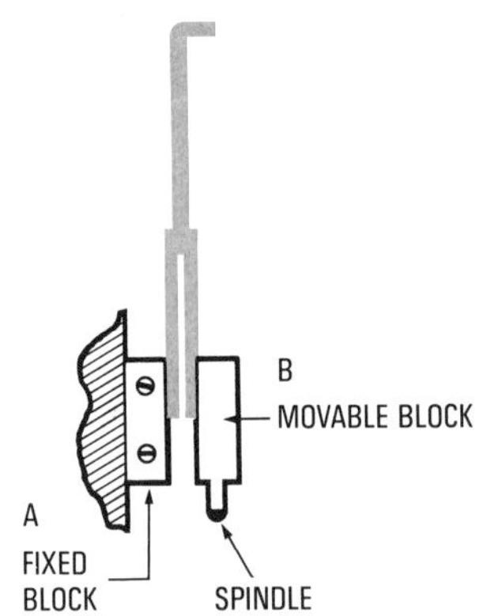


Fig. 10-11A A schematic view of the reed mechanism.

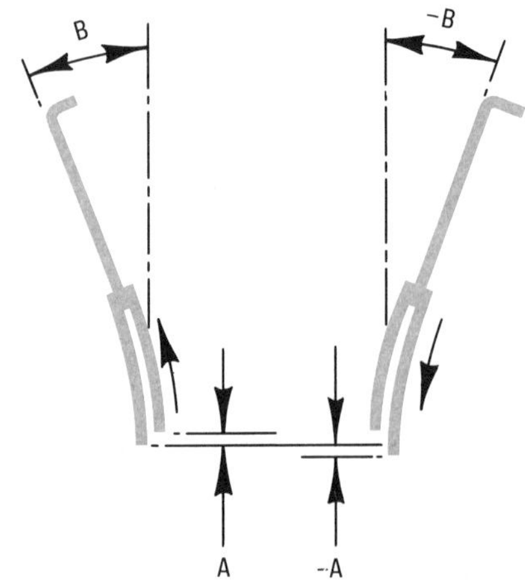


Fig. 10-11B Any upward movement of the spindle will cause the pointer to move to the left, and visa versa.

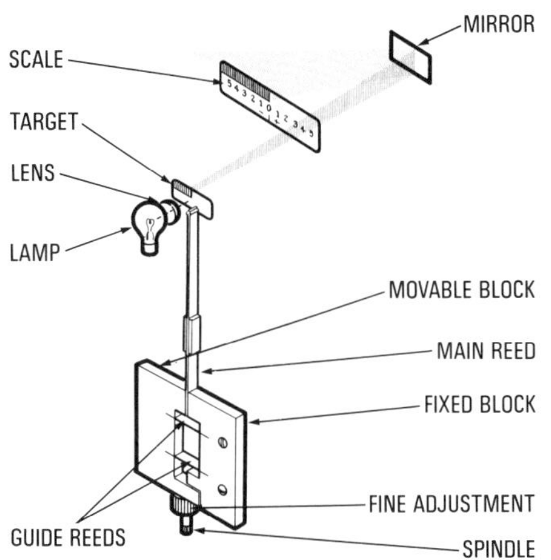


**Fig. 10-12** The light beam amplifies the target movement.

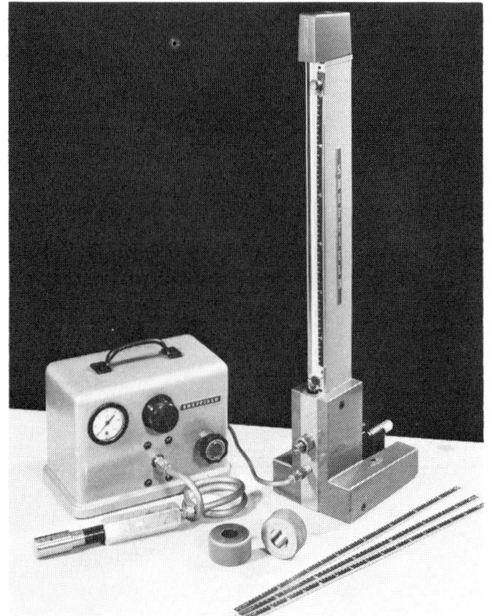

**Fig. 10-13** A column-type air gage. Note the gaging head and the master gage. (*Courtesy Bendix Corp.*)

airflow. The top of this tube is connected to the gaging head by a plastic tube.

The air flowing through the gage exhausts through the passages in the gaging head into the clearance between the head and the workpiece. The rate of flow is proportional to the clearance and is indicated by the position of the float in the column. The gage is set to a master and the float is then positioned by means of an adjusting knob. The upper and lower limits for the workpiece are then set with limit indicators on the scale. If the hole in the workpiece is larger than the hole size of the master, more air will flow through the gaging head, and the float will rise higher in the tube. Conversely, if the hole is smaller than the master, the float will fall in the tube. Amplification from 1000:1 to 40,000:1 may be obtained with this type of gage. Snap-, ring-, and plug-type gaging heads may be fitted to this type of gaging device.

## Pressure-Type Air Gage

In the pressure-type air gage (Fig. 10-15), air passes through a filter and regulator and is then divided into two channels (Fig. 10-16). The air in the *reference channel* escapes to the atmosphere through a zero setting valve. The air in the *measuring channel* escapes to the atmosphere through the gage head jets. The two channels are connected by an extremely accurate differential pressure meter.

The master is placed over the gaging spindle and the zero setting valve is adjusted until the gage needle indicates zero. Any deviation in the workpiece size from the master size changes the reading. If the workpiece is too large, more air will escape through the gaging plug; therefore, pressure in the measuring channel will be less and the dial gage hand will move counterclockwise, indicating how much the piece is oversize. A diameter smaller than the master gage indicates a reading on the right side of the dial. Amplification from 2500:1 to 20,000:1 may be obtained with this type of gage. Pressure-type air gages may also use plug, ring, or snap gaging heads for a wide variety of measuring jobs.

Air gages are widely used since they

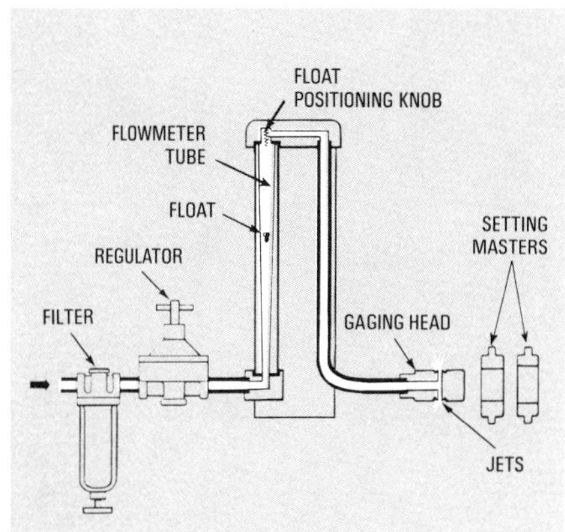


**Fig. 10-14** Principle of the column-type air gage.

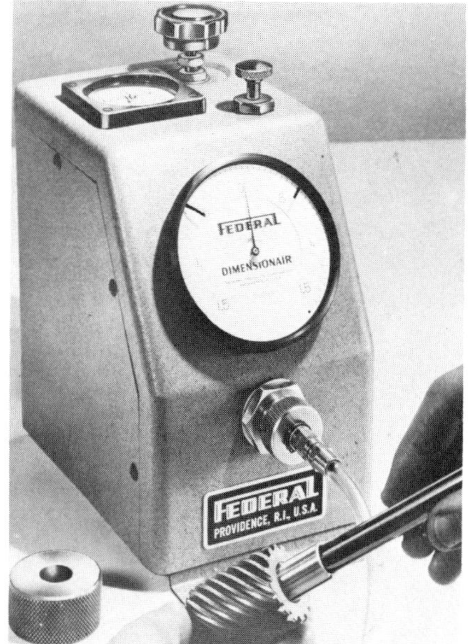


**Fig. 10-15** Gaging a hole using a pressure-type air gage.

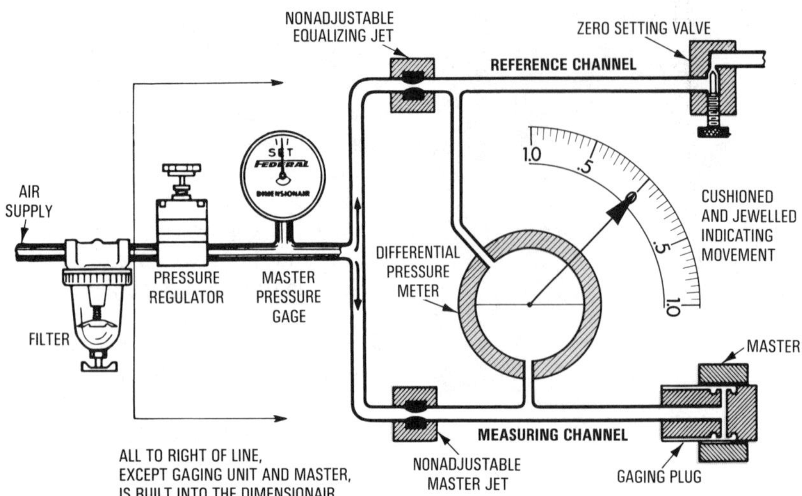


**Fig. 10-16** Operation of the pressure-type air-gaging system. (*Courtesy Federal Products Corporation.*)

have several advantages over other types of gages.

**1.** Holes may be checked for taper, out-of-roundness, concentricity, and irregularity more easily than with mechanical gages.

**2.** The gage does not touch the workpiece; therefore there is little chance to mar the finish.

**3.** Gaging heads last longer than fixed gages since wear is reduced between the head and the workpiece.

**4.** Less operator skill is required to use this type of gaging equipment than other types.

**5.** Gages may be used at a machine or bench.

**6.** More than one diameter may be checked at the same time.

## REVIEW QUESTIONS

**1.** Name two general uses for dial indicators.

**2.** Describe the gear and rack principle found in many types of dial indicators.

**3.** Name two types of dials found on dial indicators and describe each one.

**4.** What is the difference between a regular-range dial indicator and a long-range indicator?

**5.** How are metric dial indicators usually graduated?

**6.** Define a comparator.

**7.** Why is high amplification necessary in any comparison measurement process?

**8.** What is the advantage of an optical comparator?

**9.** Briefly describe the principle of a reed-type comparator.

**10.** How is such great amplification possible with a mechanical-optical comparator?

**11.** List the steps required to set up a reed-type comparator to measure 1.000.

**12.** In the column-type air gage, what size is indicated if

(a) The float is above the zero line

(b) The float is below the zero line

**13.** What are the names of the two channels found in a pressure-type air gage?

**14.** How are the following indicated on a pressure-type air gage?

(a) Oversize

(b) Undersize

# SECTION 2

## LAYOUT WORK

A very important function of any machinist is being able to do accurate layout work. Before any work is machined or holes drilled, it is necessary to first lay out the workpiece to indicate the amount of material that is to be removed and to locate the position of the holes. The dimensions or specifications, which represent the actual size of the finished workpiece, are usually found on a blueprint prepared by a drafter or tool designer. A competent machinist must be able to read and understand blueprints, select and use the proper layout tools, and transfer the dimensions from the blueprint to the metal workpiece.

# UNIT 11

# BASIC LAYOUT TOOLS

Laying out is the process of scribing (marking) center points, circles, arcs, or straight lines on metal to indicate the position of holes to be drilled or the amount of material to be removed on a shaper or milling machine. All layouts should be made from a *base line* or a *machined edge* to ensure the accuracy of the layout and the correct position of dimensions in relation to each other. The accuracy of a finished job is generally determined by the accuracy and amount of care used while laying out. If the part does not have to be precise, time should not be spent in making a precision layout. (See N-1.)

N-1 Keep the layout as simple as the workpiece requirements permit.

## OBJECTIVES

After completing this unit, you should be able to:

**1.** Understand the need for accurate layout work

**2.** Know the purpose of layout tools such as scribers, punches, dividers, squares, and accessories

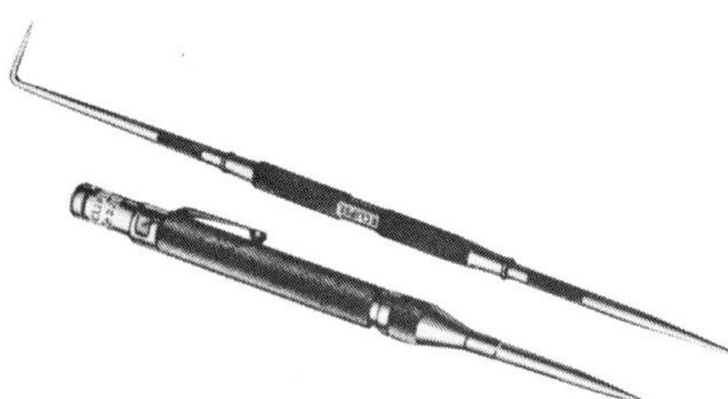

**Fig. 11-1** A pocket and a double end scriber. (*Courtesy "Eclipse" Scribers.*)

## SCRIBER

The *scriber* is a tool used for marking layout lines on metal surfaces. Scribers are made of tool steel with hardened and tempered points. In order for a scriber to mark fine, clear layout lines, it is important that its point be sharp. Never perform any layout work with a dull scriber. Two of the most common types of scribers are illustrated in Fig. 11-1. (See N-2.)

N-2 Always check that the scriber point is sharp before using it for layout work.

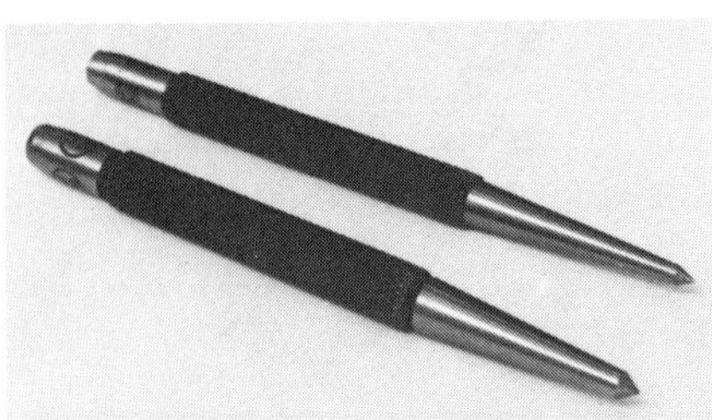

**Fig. 11-2** A prick punch (top) and a center punch (bottom) are used for layout work. (*Courtesy Kostel Enterprises Ltd.*)

## PRICK PUNCH

A *prick punch* (Fig. 11-2) is a layout instrument made of tool steel, with both ends hardened and tempered. Its point is ground to an angle of from 30 to 60°. It is used to make small indentations along layout lines, to mark centers for drilled holes, and also centers for divider points. Punch marks are sometimes called *witness marks*, because the indentations will still remain if the layout lines should be rubbed off the surface of the work.

## CENTER PUNCH

The *center punch* (Fig. 11-2) is similar to a prick punch but its point is ground to an angle of approximately 90°. A center punch is used to enlarge the prick punch marks so that a drill may be started easily and accurately.

The following points should be kept in mind while using either a center or a prick punch:

**1.** Always make sure the point of the punch is *sharp*.

**2.** Hold the punch at a 45° angle and place the point on the layout line.

**3.** Bring the punch to a vertical position and tap it gently with a light hammer.

**4.** Examine the position of the punch mark and correct it if necessary.

**5.** If the center-punch-mark location is correct, the punch mark should be enlarged to permit the point of the drill to start accurately.

## DIVIDER

The *divider* (Fig. 11-3) is a tool with hardened steel points used for transferring measurements, comparing distances, and scribing arcs and circles. Dividers are adjustable and are classified according to size by the maximum opening between the two points. (See N-3.)

N-3 Sharp divider points will produce clear, accurate layout lines.

To set the divider to a size, place one point in the 1-in. or 10-mm line of a steel rule and adjust the other point

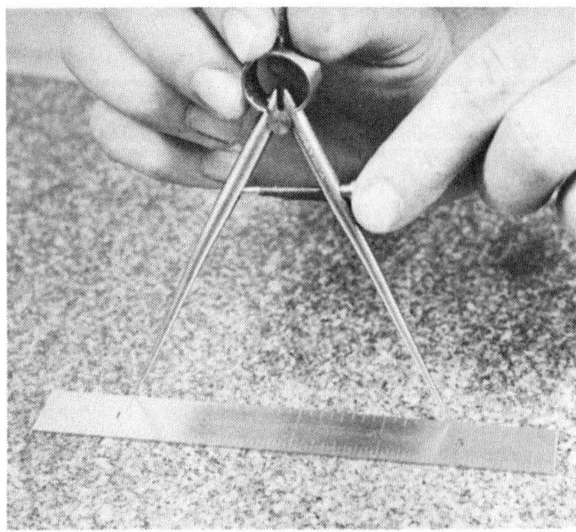

**Fig. 11-3** Dividers are set more accurately by starting at the 1 in. or 1 cm mark.

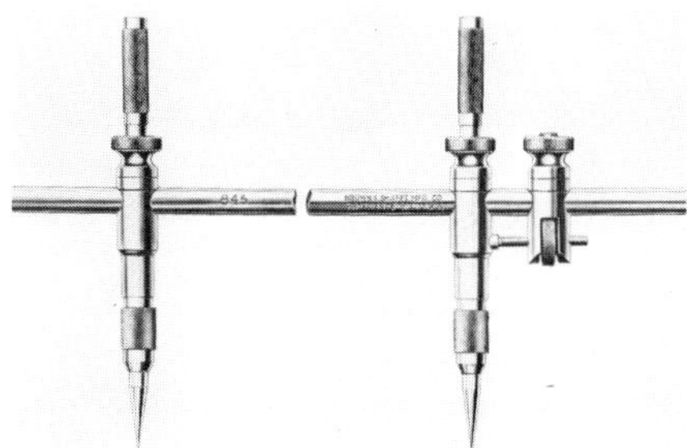

**Fig. 11-4** Trammels are used for work beyond the range of the divider.

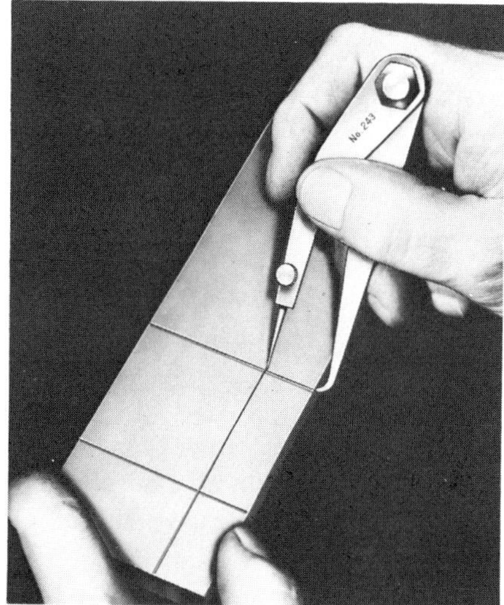

**Fig. 11-5** The hermaphrodite caliper. (*Courtesy The L. S. Starrett Company.*)

N-4 Handle squares with care to preserve their accuracy.

until it splits the graduation line the correct distance away. The graduation lines on a rule are V-shaped, and often a divider may be set more exactly by "feeling" that the point is in the graduation line than by seeing. If an extremely accurate size is needed, check the setting with a magnifying glass.

*Trammels* (Fig. 11-4) are used for transferring measurements and scribing circles and radii beyond the range of the divider. When a circle is laid out from a hole, a ball attachment may be substituted for one of the scriber points. This ball attachment fits into the hole and serves as the center point when arcs are scribed.

## HERMAPHRODITE CALIPER

The *hermaphrodite caliper* is a caliper having one bent and one straight leg. The straight leg has a point so that it can be used to scribe lines. Hermaphrodite calipers can be used to locate centers of round pieces, and also to scribe lines parallel with an edge or shoulder (Fig. 11-5).

To set this tool, place the bent leg on the end of a rule and adjust until the pointed leg is at the desired graduation.

## SQUARES

Squares are used to lay out lines at right angles (90°) or parallel to a machined edge, to test the accuracy of surfaces which must be at 90° to each other, and to set up work for machining. (See N-4.)

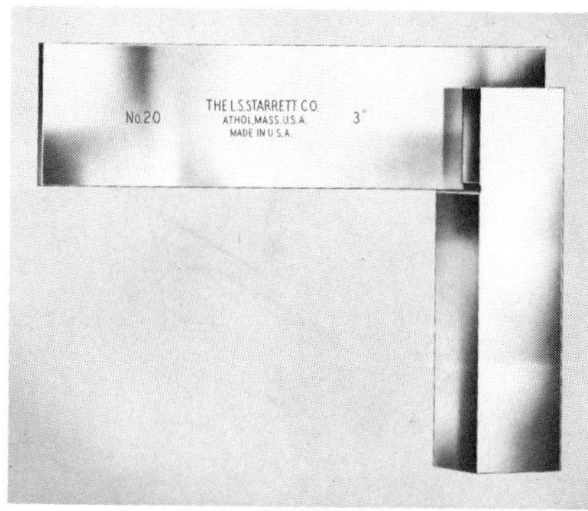


**Fig. 11-6** The solid square. (*Courtesy The L. S. Starrett Company.*)

*Adjustable squares* are used for general-purpose work; the *solid square* (Fig. 11-6) is used where greater accuracy is required. The solid square is made up of two parts, the beam and the blade, which are usually hardened and ground. Extremely accurate solid squares are called master squares, used only for checking the accuracy of other squares.

## COMBINATION SET

The combination set (Fig. 11-7) is one of the most useful and versatile tools in a machine shop. The set consists of four principal parts: steel rule, square head, bevel protractor, and center head.

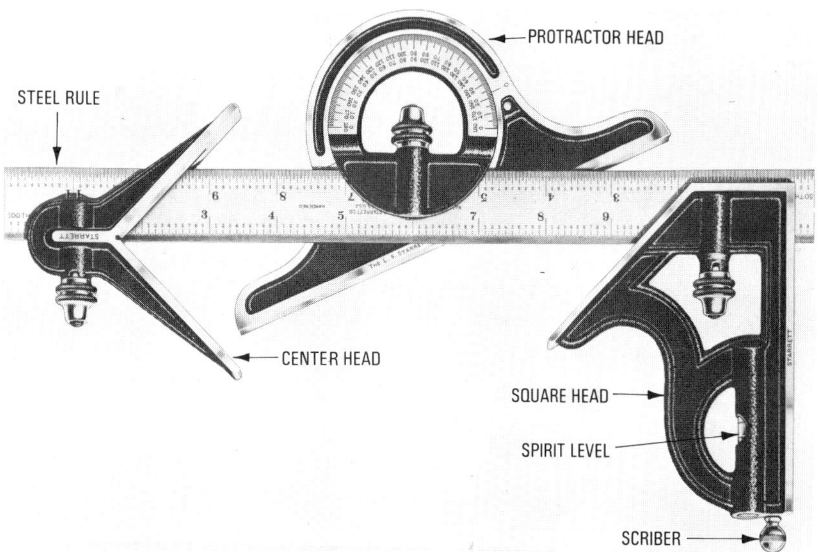


**Fig. 11-7** The main parts of a combination set (*Courtesy The L. S. Starrett Company.*)

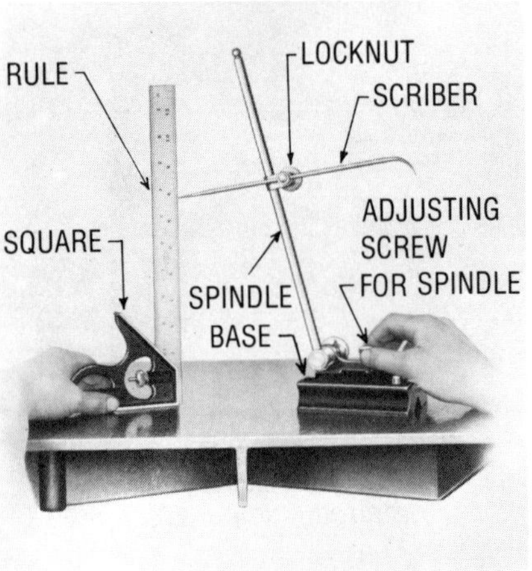

**Fig. 11-8** The parts of a surface gage.

**Fig. 11-9** A surface plate provides a true flat surface for layout work. (*Courtesy Taft-Peirce Mfg. Co.*)

N-5 The center head should be used only on stock which is true in shape.

N-6 To maintain the accuracy of a surface plate and the layout, the top should be kept clean and free of burrs.

### Steel Rule

The steel rule or blade may be fitted to the center head, the bevel protractor, or to the square head. Sometimes it is used separately as a straightedge or for measuring. Inch-combination set rules are usually graduated in 8ths and 16ths on one side and in 32ds and 64ths on the other side. Metric-combination set rules are usually graduated in millimeters and half millimeters.

### Square Head

The square head and rule, or *combination square,* is used to lay out lines parallel and at right angles to an edge. It may also be used as a depth gage or for checking 45 and 90° angles. The square head can be moved along to any position on the rule.

### Bevel Protractor

The bevel protractor is used to lay out and check angles. The protractor can be adjusted from 0 to 180°. On some protractors, the scale is graduated from 0 to 90° from both right and left. The accuracy of this type of protractor is ±0.5°.

### Center Head

The center head forms a center square when clamped to the rule. It can be used for locating centers of round, square, and octagonal stock. (See N-5.)

## SURFACE GAGE

The surface gage (Fig. 11-8) is an instrument used on a surface plate or any flat surface for scribing lines in layout work. It consists of a heavy *base* and an upright *spindle* to which a *scriber* is clamped. The base of the surface gage has a V groove which allows it to be used on cylindrical work as well as on flat surfaces. There are also pins in the base, which may be pushed down so that the surface gage can be used against the edge of a surface plate or a slot. A surface gage can also be used as a height gage and for leveling work in a vise. A surface gage may be set to a size or dimension by using a combination square, which is generally accurate enough for most layout work (Fig. 11-8).

## LAYOUT ACCESSORIES

In addition to the tools used to mark the workpiece, certain accessories are helpful and even necessary in layout work.

### Surface Plates

*Surface plates* (Fig. 11-9) are used where fine accuracy is required in layout work. They are made from a good grade of cast-iron, granite, or ceramic material. Because of their flat surface, surface plates provide a reference or starting point for layout operations. Some granite surface plates are lapped flat to within .0001 in. (0.0025 mm). Besides being used for accurate layout work, they are also used for inspecting gages, jigs, and fixtures. (See N-6.)

### Angle Plates

An *angle plate* (Fig. 11-10) is a precision L-shaped tool made of cast iron or hardened steel, machined to an accurate 90° angle, with all working surfaces and edges ground square and parallel. Angle plates are used to hold work parallel and at right angles to a surface. C-clamps are generally used to fasten work to an angle plate; however, some angle plates are provided with slots and tapped holes for this purpose.

**Fig. 11-10** Work may be clamped to an angle plate for layout purposes. Note the parallel under the workpiece. (*Courtesy Kostel Enterprises Ltd.*)

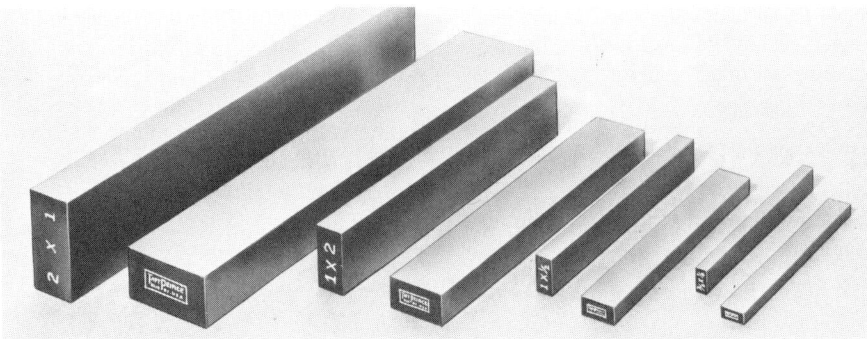


**Fig. 11-11** Parallels are available in a variety of shapes and sizes. (*Courtesy Taft-Peirce Mfg. Co.*)

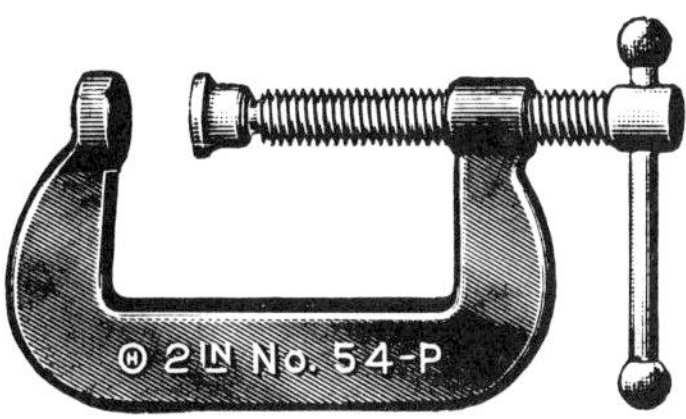


**Fig. 11-12** A C-clamp

### Parallels

*Parallels* (Fig. 11-11) are used in layout work to raise the work to a suitable height and provide a solid seat. When used under work on a surface plate, the work surface is held parallel to the top of the surface plate.

### Clamps

*C-clamps* (Fig. 11-12) are clamps made in the shape of a "C." They are used to fasten work to an angle plate or a drill press table and also for holding two or more pieces of metal together.

*Parallel clamps* or *toolmaker's clamps* (Fig. 11-13) consist of two jaws held together with adjusting screws. They are used for the same purpose as C-clamps.

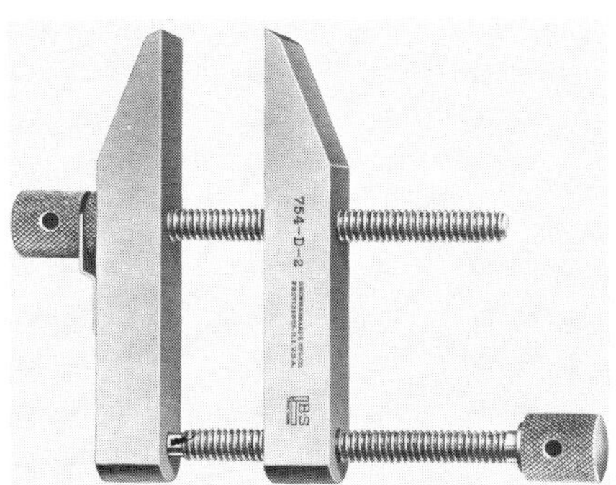


**Fig. 11-13** A parallel or toolmaker's clamp. (*Courtesy Brown & Sharpe Mfg. Co.*)

### V-Blocks

V-blocks (Fig. 11-14) are made of cast iron or heat-treated steel in a wide range of sizes. They are machined with a 90° V-shaped slot on the top and bottom. Some V-blocks are provided with U-shaped clamps to hold the work securely. They are used for laying out round work and for holding it for machining.

**Fig. 11-14** V-blocks are used to hold round work for laying out or machining. (*Courtesy Kostel Enterprises Ltd.*)

## REVIEW QUESTIONS

**1.** What is the purpose of laying out a workpiece?

**2.** Why should all layouts be made from a base line or a machined edge?

**3.** What should be checked before using a scriber?

**4.** What is the main difference between a prick punch and a center punch?

**5.** State three uses of a divider.

**6.** State three uses of squares.

**7.** Name the main parts of the combination set.

**8.** State four uses of the combination square.

**9.** State two uses for a surface gage.

**10.** How is a surface gage generally set to size?

**11.** Name three materials from which surface plates are made.

**12.** What is the purpose of an angle plate?

**13.** For what purpose are parallels used?

**14.** For what purpose are V-blocks used?

# UNIT 12

# BASIC LAYOUT OPERATIONS

Usually the first operations performed in layout work are preparing the work surface and laying out straight lines. Since the accuracy of the layout plays a big part in the accuracy of the finished workpiece, great care should be used in the layout operation.

Unless a high degree of accuracy is required in layout work, the combination set is a very versatile and valuable tool. The *square head* can be used to lay out horizontal and parallel straight lines, the *bevel protractor* can be used for angular lines, and the *center head* can be used to lay out the centers of round, square, or octagonal work. Other common layout tools are the surface gage and the height gage.

## OBJECTIVES

After completing this unit, you should be able to:

1. Know how to prepare the work surface for layout
2. Lay out straight lines using
   (a) The combination square and scriber
   (b) The surface gage and surface plate
   (c) The vernier height gage and surface plate
3. Locate center holes by
   (a) Hermaphrodite calipers
   (b) Center head
   (c) Surface gage

## PREPARING SURFACES FOR LAYOUT

Layout lines must be plain to see. Therefore the surface of the work must first be coated with some type of layout material so that scribed lines will stand out in sharp contrast. Before any type of layout material is applied to work, the surface must be clean and free from surface scale, grease, or oil; otherwise the layout material will not stick to it. There are numerous ways of coating surfaces for layout, a few of which are listed below. (See N-1.)

**1.** A commercial layout dye or bluing is generally used to coat work surfaces. It is inexpensive and fast drying. Layout dye is generally applied with a brush, or sprayed on with an aerosol can.

**2.** Chalk may be rubbed into the rough surface of castings.

**3.** A copper sulphate solution, called blue vitriol, can be used for coating machined surfaces. (See N-2.)

**4.** Certain metals may be heated to a blue color. When the metal is cooled, scribed layout lines may be clearly seen against the blue background.

**Fig. 12-1** Layout out lines parallel to a machined edge. (*Courtesy Kostel Enterprises Ltd.*)

N-1 When copper sulphate is used, the surface of the workpiece must be absolutely clean and free from grease and finger marks.

N-2 Apply blue vitriol only to ferrous metals (those which contain iron in some quantity).

N-3 Since indistinct or double lines are useless, always make sure the scriber point is sharp.

## TO LAY OUT PARALLEL LINES TO AN EDGE

### Combination-Square Method

**1.** If possible, hold the work in a vise to prevent it from moving during the layout operation.

**2.** Remove all burrs from the edge of the work with a file.

**3.** Extend the steel rule the desired distance beyond the body of the square. Always be sure that only *half the graduation line* can be seen.

**4.** Hold the base of the square tightly against a *machined edge* with the rule flat on the work surface.

**5.** Hold the scriber at a *slight angle* (Fig. 12-1) to keep the point against the end of the rule and tilt the top of the scriber in the direction in which the line is to be scribed. (See N-3.)

**6.** Draw a sharp line along the end of the rule.

**7.** Move the square a short distance and again scribe along the end of the rule.

**8.** Continue moving the square and scribing until the line is complete.

**9.** Check the location of the scribed line.

### To Lay Out Lines at Right Angles (90°) to an Edge

Lines at *right angles,* or the location of holes, may be laid out by placing the body of the square against a machined edge of the work that is at right angles (90°) to the first edge, Fig. 12-2.

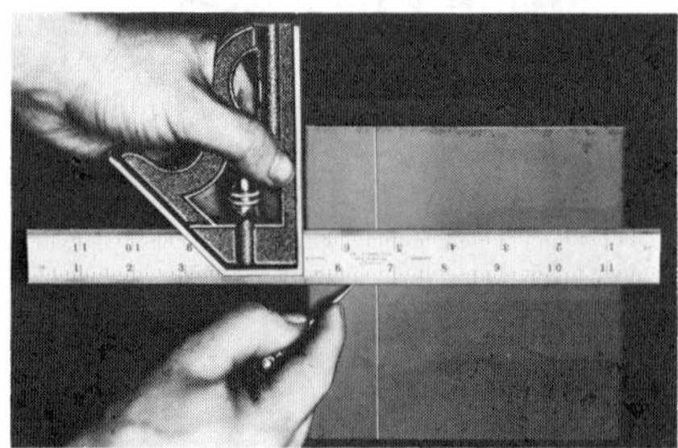

**Fig. 12-2** Laying out intersecting lines (lines at 90°).

**1.** Remove all burrs from the edges of the work with a file.

**2.** Slide the square along the work until the edge of the rule is at the desired distance.

**3.** Hold the base of the square tightly against the machined edge with the blade of the rule flat on the work surface.

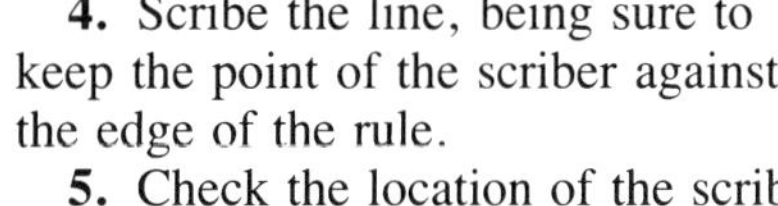

**4.** Scribe the line, being sure to keep the point of the scriber against the edge of the rule.

**5.** Check the location of the scribed line.

## Surface Gage Method

When lines are scribed with a surface gage, it is important that the surface plate, the base of the surface gage, the base of the square, and the work be clean.

**Fig. 12-3** A surface gage being set to a dimension using a combination square.

## To Set a Surface Gage to a Dimension (Fig. 12-3)

**1.** Thoroughly clean the top of the surface plate.

**2.** Set a combination square on the surface plate.

**3.** Loosen the combination square lock nut and be sure that the end of the rule is down against the surface plate.

**4.** Tighten the lock nut.

**5.** Set the surface gage on the surface plate.

**6.** Loosen the scriber clamp nut and adjust the scriber so that it is approximately at the desired dimension.

**7.** Tighten the scriber clamp nut securely to lock the scriber in position.

**8.** Turn the surface gage thumb screw until the scriber point is in the center of the desired graduation on the rule.

## To Lay Out Horizontal Lines Using a Surface Gage (Fig. 12-4)

**1.** Set the surface gage to the dimension required.

**2.** Place the edge of the work from which the line is to be scribed on the surface plate.

**3.** Hold the surface gage *down* on the surface plate.

**4.** Draw the surface gage along the work in the direction of the arrow to scribe the line. *Pushing* will cause the scriber point to dig into the work and cause an inaccurate layout.

**5.** Reset the surface gage to the combination-square rule for each dimension.

**6.** Lay out all lines which are parallel to the edge resting on the surface plate.

## Vernier Height Gage Method

When layout lines must be accurate to within .001 in. (0.02 mm), a vernier height gage should be used, Fig. 12-5A. (See Unit 7 for information on the height gage.) The same precautions regarding cleanliness and scribing as outlined for the surface gage must be observed when scribing lines with a vernier height gage.

The following points should be observed when a surface or a height gage is used for laying out lines.

**Fig. 12-4** Using a surface gage and surface plate to lay out parallel lines.

**Fig. 12-5A** Scribing an accurate layout line with a vernier height gage. Note how the gage is drawn along at about a 45° angle.

**Fig. 12-5B** Turn the angle plate 90° to scribe intersecting lines.

**1.** Thoroughly clean the surface plate and the base of the surface or height gage.

**2.** Hold the gage so that the scriber point is approximately 45° to the surface being laid out (Figs. 12-4 and 12-5A).

**3.** Press down firmly on the surface or height gage base.

**4.** Always pull the gage to maintain an accurate layout line. Pushing tends to make the scriber point dig in and tip the gage, producing inaccurate layout lines.

**5.** If the work is first fastened to an angle plate, all horizontal lines can be laid out as outlined. The angle plate can then be set on its edge (90° to the base) and the vertical lines can be laid out (Fig. 12-5B).

## LAYING OUT CENTER HOLES

When a workpiece is to be machined between centers on a lathe, it is sometimes necessary to lay out and drill the center holes prior to turning the work on a lathe. Generally it is more accurate and convenient to drill lathe center holes in small round stock on the lathe. This eliminates the need for laying out the centers.

### To Locate the Centers of Round Stock

#### Using Hermaphrodite Calipers

**1.** Place the work in a vise and remove the burrs or sharp edges.

**2.** Apply layout dye to both ends.

**3.** Set the hermaphrodite caliper to the approximate radius of the work.

**4.** With the thumb, hold the bent leg just below the edge of the work and scribe an arc.

**5.** Move the bent leg a quarter of a turn and scribe an arc. Repeat until four arcs are obtained on each end of the work, Fig. 12-6.

**6.** Center-punch the center of the arcs and test the accuracy of the centers.

**7.** Repeat steps 3 to 6 on the other end of the workpiece.

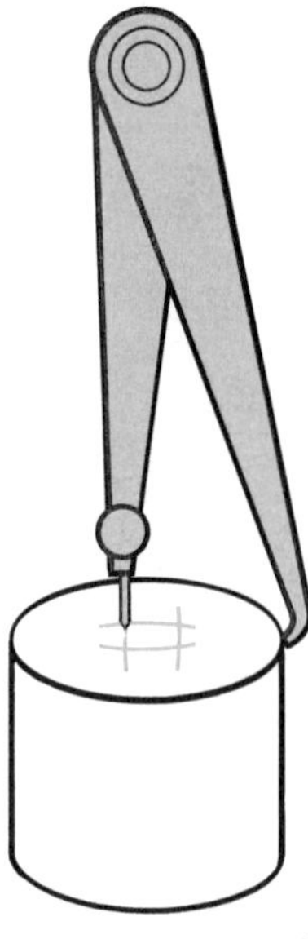

**Fig. 12-6** Using hermaphrodite calipers to center a piece of round stock. The center of the four arcs will be the center of the stock.

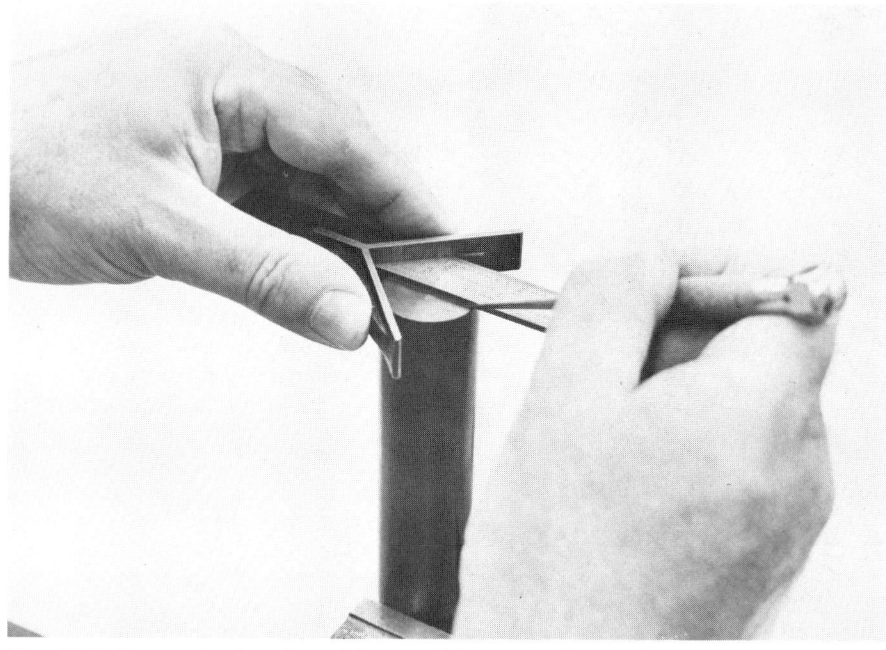

**Fig. 12-7** The center head provides a quick means of centering round, square, or octagonal workpieces. (*Courtesy Kostel Enterprises Ltd.*)

N-4 This tool can be used effectively only when the stock being centered is true in shape (round, square, octagonal) and cut off at a 90° angle.

## Using the Center Head

**1.** Place the work in a vise and remove the burrs or sharp edges with a file.

**2.** Apply layout dye to both ends of the work.

**3.** Hold the center head firmly against the work with the rule flat on the end, Fig. 12-7.

**4.** Hold a *sharp scriber* at an angle so that its point touches the edge of the rule.

**5.** Scribe a line along the edge of the rule.

**6.** Rotate the center head one-quarter of a turn and scribe a second line. (See N-4.)

**7.** *Lightly* center-punch where the two lines cross.

**8.** Repeat steps 3 to 8 on the other end of the work.

**9.** Test the accuracy of the center layout.

**Fig. 12-8** Locating the center of a round workpiece using a V-block and surface gage. (*Courtesy Kostel Enterprises Ltd.*)

## Using a Surface Gage

**1.** Remove the burrs from the ends of the workpiece and apply layout dye.

**2.** Place the workpiece in a V-block which is on a surface plate, Fig. 12-8.

**3.** Set the scriber point of the surface gage to the approximate center of the workpiece.

**4.** Scribe four lines across the end of the workpiece by rotating the work 90° for each line.

**5.** Lightly center-punch the center of the four intersecting lines.

**6.** Test the center for accuracy by the divider or lathe center method described below.

## To Check the Accuracy of the Center Layout

Before any metal is cut or a hole is drilled, it is considered good practice to check the accuracy of the layout. This is especially true when laying out lathe center holes, since an error in layout will cause the center hole to be drilled off center. Two common meth-

Fig. 12-9 Checking the accuracy of the center layout with dividers. (*Courtesy Kostel Enterprises Ltd.*)

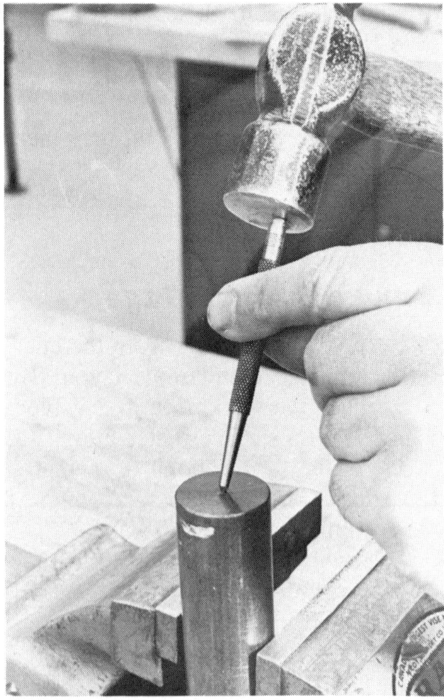

Fig. 12-10 Moving the center punch mark over to the center. (*Courtesy Kostel Enterprises Ltd.*)

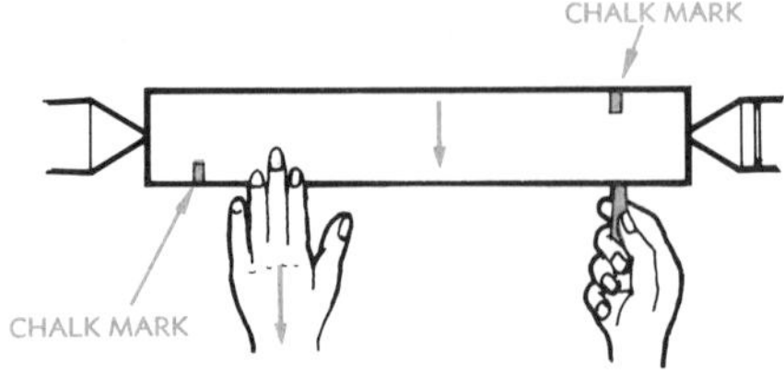


Fig. 12-11 Testing the accuracy of the center layout between lathe centers.

ods are used to check the center hole layout: with a pair of dividers, and by rotating the work between lathe centers.

### Divider Method (Fig. 12-9)

**1.** Place one leg of the divider in the light center punch mark.

**2.** Adjust the divider so that the other leg is on a line and set exactly to the edge of the work (Fig. 12-9).

**3.** Revolve the divider one-half turn (to the other end of the same line) and check if the leg of the divider is in the same relation to the edge of the work as in step 2.

**4.** If the divider leg is not in the same relation to the work edge at both ends of the line, lightly move the center punch mark to correct the error (Fig. 12-10).

**5.** Repeat step 4 on the other scribed line until the punch mark is exactly on center.

**6.** Deepen the center punch mark in preparation for drilling the center hole.

**7.** Repeat steps 1 to 6 on the other end of the workpiece.

### Lathe Center Method (Fig. 12-11)

**1.** Deepen the center punch marks at both ends and mount the work between lathe centers.

**2.** Adjust the lathe center tension so that the work can spin freely.

**3.** Hold a piece of chalk in one hand which is steadied against any part of the lathe (e.g., compound rest).

**4.** Spin the work by hand and bring the chalk close to the work surface until it lightly marks the high spot (Fig. 12-11).

**5.** Repeat step 4 on the other end of the work.

**6.** Fasten the work in a vise, tip the center punch, and with a hammer move the center punch mark toward the chalk mark (Fig. 12-10).

**7.** Retest the workpiece between centers and repeat steps 3 to 6 until the work runs true.

## REVIEW QUESTIONS

**1.** What is the purpose of a layout solution?

**2.** Why must the work surface be clean before applying a layout material?

**3.** How is the edge of the workpiece prepared before any layout work is done?

**4.** How far should the rule extend beyond the body of the square when setting it to a size?

**5.** What precautions must be observed when using a scriber?

**6.** How may intersecting layout lines or lines at right angles be scribed when using a surface or height gage?

**7.** Briefly describe how to locate the centers of a round workpiece using hermaphrodite calipers.

**8.** List the steps required to lay out the center of a round workpiece using a surface gage.

**9.** Briefly explain how the center layout may be tested with dividers.

**10.** What must be done to the center punch mark on the end of a workpiece before testing the accuracy of the center layout in a lathe?

# UNIT 13

# LAYOUT WORK

It is often necessary for a machinist to lay out a workpiece so that the layout lines will provide a guide during the machining process. Most layouts consist of scribing straight lines, circles, arcs, and angles on the surface of a workpiece. There are many ways to lay out work and the method used will depend on the accuracy required and the tools available.

## OBJECTIVES

After completing this unit you will know how to:

**1.** Layout angles by three methods:
   (a) The rule
   (b) The bevel protractor
   (c) The universal bevel protractor

**2.** Choose the proper method of angular layout depending on the accuracy required

**3.** Lay out circles and arcs using dividers and trammels

**4.** Lay out hole locations using test circles as proof circles

## METHODS OF LAYING OUT ANGLES

Angular layout involves the use of straight layout lines, and for most purposes, these lines may be positioned by means of a rule, a bevel protractor, or a universal bevel protractor.

### Angular Layout Using a Rule

This form of angular layout is usually not shown as an angle on the drawing. It merely shows the distance from a corner along two adjacent sides.

Examine Fig. 13-1. The corner *A* is the only one that does not show an angular dimension. Angular layouts of this kind are common and are performed as follows:

**1.** Remove all burrs and clean the face of the workpiece.
**2.** Apply layout dye to the surface to be laid out.
**3.** With a rule, or adjustable square, measure 1 in. from the corner along the two sides and scribe a line on each edge, Fig. 13-2.
**4.** Join these two points with the edge of the rule, Fig. 13-2.
**5.** Scribe a line along the edge of the rule.
**6.** Lightly prick-punch along this line at intervals of ¼ to ⅜ in. (6 to 9 mm).

N-1 Be sure to start at the 1-in. line of the rule when setting a divider to size.

### Angular Layout Using an Adjustable Square

The angle at corner *B*, Fig. 13-1, is indicated in degrees and no tolerances are shown. This automatically implies that ±0.5° is satisfactory. The 45° angle on the body of the adjustable square is accurate enough to be used for laying out this angle.

**1.** Scribe a short line 1 in. from the corner along the end.
**2.** Place the blade in the adjustable square and lock it in position.
**3.** Place the body of the square against the end and slide it along until the edge of the rule is over the mark on the workpiece, Fig. 13-3.
**4.** Hold it firmly in this position and scribe a line along the edge of the rule.
**5.** Lightly prick-punch along the layout line at intervals of about to ⅜ in. (6 to 9 mm).

### Angular Layout Using a Bevel Protractor

The angle at the corner *C* is shown in degrees. This indicates that a bevel protractor can be used for laying out this angle.

**1.** Scribe a short line 2 in. from the end along the side.
**2.** Set the bevel protractor to 30°.
**3.** Repeat steps 3, 4, and 5 as outlined for the adjustable square, Fig. 13-4.

### Laying Out an External Radius

An external convex radius is generally found on the corners or edge of a workpiece. The radius shown in Fig. 13-1 (corner *D*) requires that a quadrant or a quarter of a circle, having a 1-in. radius, must be laid out on the workpiece.

**1.** Set an adjustable square to 1 in. (See N-1.)
**2.** Locate the center of the quadrant 1 in. from the edge and the end, Fig. 13-5, by scribing intersecting lines at this point.
**3.** Prick-punch the intersection of these lines.
**4.** Set the dividers to 1 in.
**5.** Scribe an arc which should just touch the two adjacent edges of the

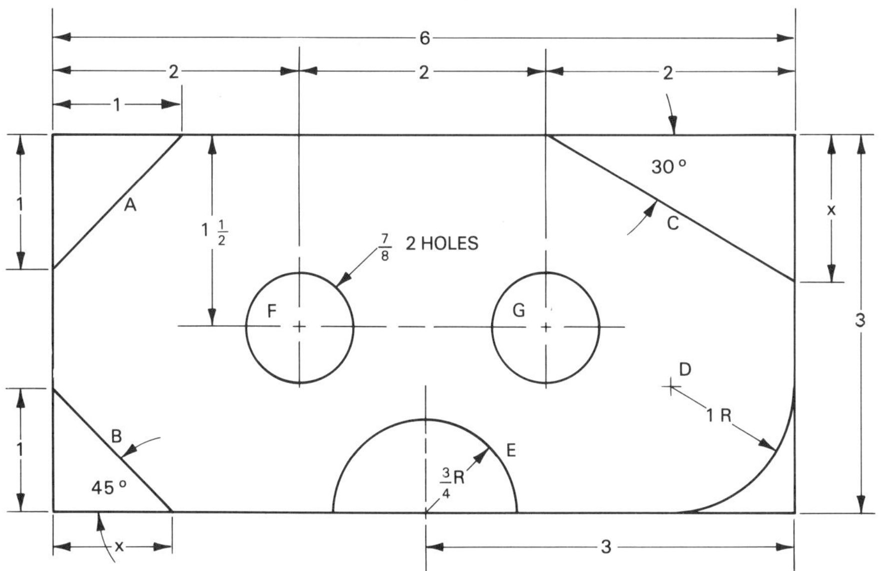

Fig. 13-1 A typical layout.

N-2 For large radii beyond the range of dividers, trammels should be used.

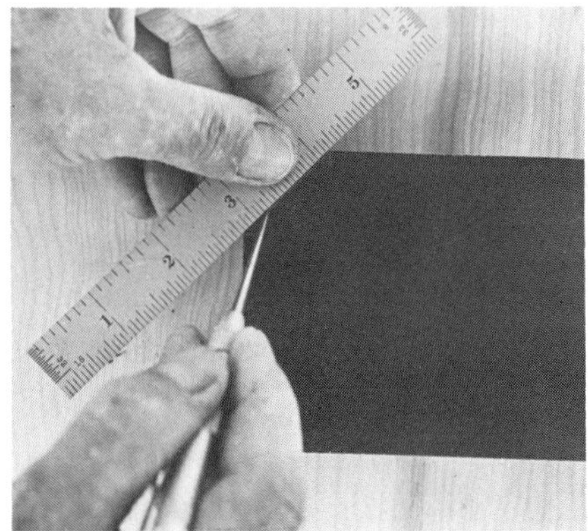

**Fig. 13-2** Laying out an angle using a rule. (*Courtesy Kostel Enterprises Ltd.*)

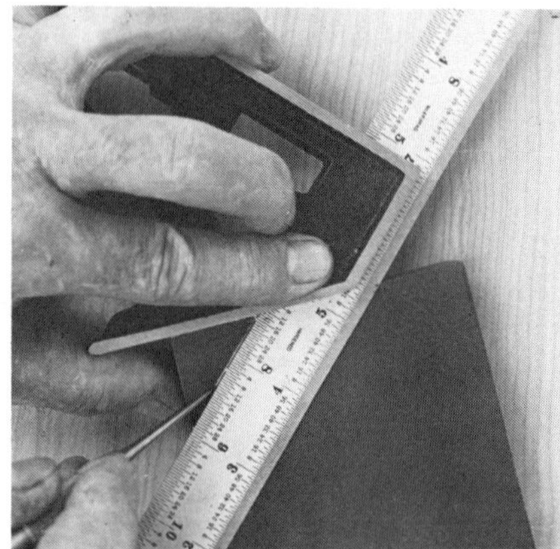

**Fig. 13-3** The adjustable square is commonly used for laying out 45° angles. (*Courtesy Kostel Enterprises Ltd.*)

workpiece at a point 1 in. along each side. (See N-2.)

**6.** Lightly prick-punch along this line.

## Laying Out an Internal Radius

Laying out an internal or concave radius, corner *E* (Fig. 13-1), with dividers is a little more difficult than an external radius, since it is necessary to position the point of the divider on the edge of the workpiece. A block of metal, or wood, held against the edge of the work will make the job much easier. If the workpiece is steel or cast iron, a magnet placed on the edge, Fig. 13-6, will work well.

**1.** Clean the edge of the workpiece and remove all burrs.

**2.** Mark off the position of the center of the internal radius 3 in. from the corner.

**3.** Set the dividers to ¾ in.

**4.** Place a magnet on the edge of the work at the marked position, Fig. 13-6A, or hold a flat metal or wooden block against the edge of the work at this point, Fig. 13-6B.

**5.** Place the point of the divider on the mark and in the small groove between the magnet and the work.

**6.** Scribe a semicircle, being careful not to let the point of the divider slip out of position.

**7.** Lightly prick-punch the layout line.

## Laying Out Hole Locations Using Test and Proof Circles

When hole locations must be accurate, the position of the hole must first be laid out and then test and proof circles should be scribed to indicate the exact position of the hole. Test circles are scribed on the workpiece to show the machinist the relation of the drill point to the location of the hole. This permits the drill-point location to be moved over if necessary, before the drill cuts to the full diameter.

**1.** Place the workpiece on edge against an angle plate.

**2.** With a surface gage, scribe a center 2 in. from one side of the workpiece.

**3.** Rotate the workpiece 90° and

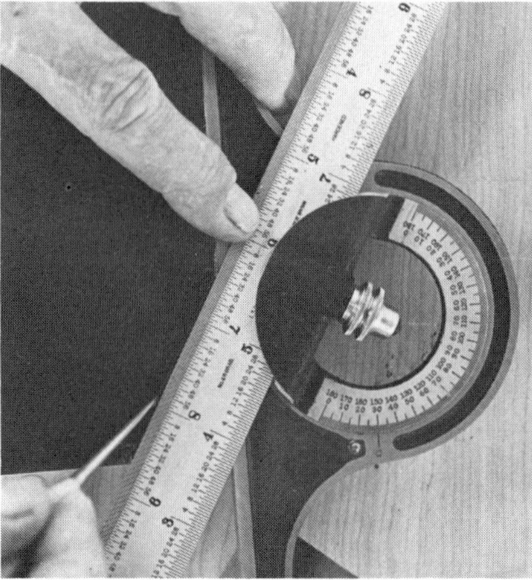

**Fig. 13-4** A bevel protractor is commonly used for laying out angles. (*Courtesy Kostel Enterprises Ltd.*)

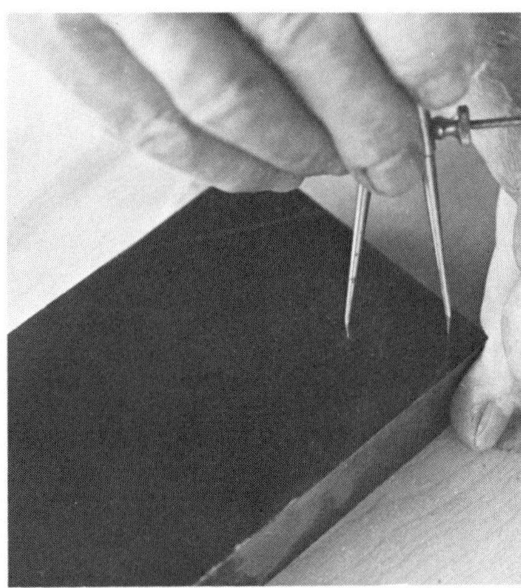

**Fig. 13-5** Scribing an external radius with a pair of dividers. (*Courtesy Kostel Enterprises Ltd.*)

scribe another line intersecting the first.

**4.** Set the surface gage to 4 in. and scribe another line intersecting the first line.

**5.** Prick-punch the intersections of these lines.

**6.** Set the dividers to 7⁄16 in. and scribe two circles ⅞ in. in diameter. These are the proof circles and indicate the location of the holes, Fig. 13-7A and B.

**7.** Set the dividers to ⅜ in. and using the same center points, scribe a circle inside each of the other circles. These are the test circles and will indicate whether the drill point is central with the proof circle before it reaches the full diameter.

**8.** Place four light punch marks on both sets of circles at the intersection of

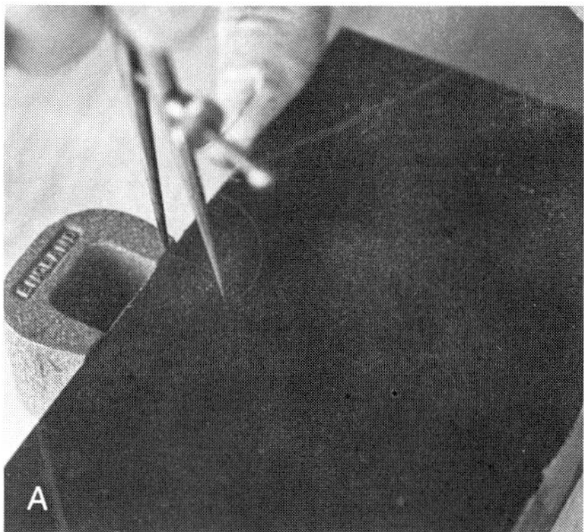


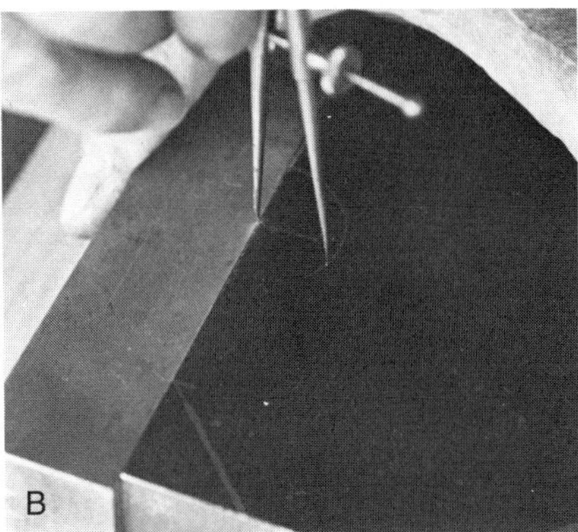


**Fig. 13-6** (**A**) Magnet. (**B**) Metal or wooden block. (*Courtesy Kostel Enterprises Ltd.*) Care must be taken to hold the divider point on the edge of the workpiece when scribing an internal radius.

N-3 The punch or witness marks in the proof circle will indicate whether the hole has been drilled in the proper location.

N-4 It is much easier to connect a line to circles or arcs than it is to connect a circle or an arc to a straight line.

N-5 All measurements for any location must be taken from the base line or finished edge.

the lines if the circles are under ¾-in. (19-mm) diameter. If the circles are over ¾-in. (19-mm) diameter, eight equally spaced punch marks should be made. (See N-3.)

## LAYOUT OPERATIONS

Although the layout required will naturally not be the same for each workpiece, there are certain procedures which should be followed in any layout. The job discussed below is intended to acquaint the reader with layout methods and procedures.

When laying out a workpiece, it is recommended that lines be laid out in the following order.

1. Horizontal and centerlines
2. Vertical and intersecting lines
3. Arcs and circles
4. Straight lines connecting arcs and circles (See N-4.)
5. Angular layout lines

### To Lay Out Hole Locations, Slots, and Radii

1. Study Fig. 13-8, and select the proper stock.
2. Cut off the stock, allowing enough material to square the ends if required.
3. Remove all burrs.
4. Clean the surface thoroughly, and apply layout dye.
5. Place a suitable angle plate on a surface plate. *Note:* Clean both plates.
6. Clamp the work to the angle plate with a finished edge of the part against the surface plate. This is known as the reference surface. Leave one end of the angle plate protruding beyond the workpiece. (See N-5.)
7. With the surface gage set to the proper height, scribe a centerline for the full length of the workpiece (Fig. 13-9A).
8. Using the centerline as a reference, set the surface gage as required, and scribe the centerlines for all hole and radii locations (Fig. 13-9A).
9. Turn the angle plate 90° and scribe the base line at the bottom of the workpiece (Fig. 13-9B).
10. Using the base line as a reference line, locate and scribe the other centerlines for each hole or arc (Fig. 13-9B).
11. Locate the starting point for the angular layout (Fig. 13-9B).
12. Remove the workpiece from the angle plate.

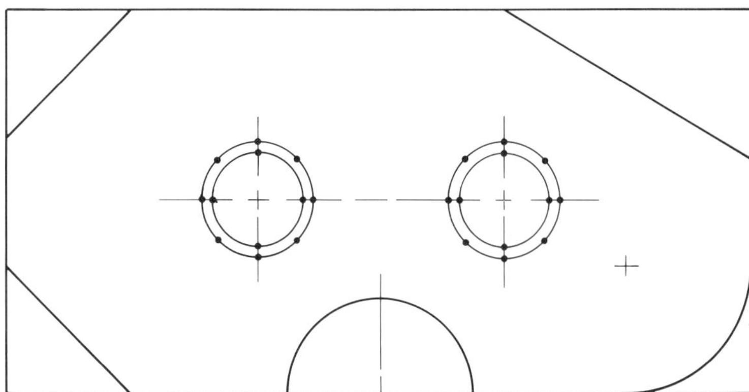

**Fig. 13-7A** Layout showing test and proof circles.

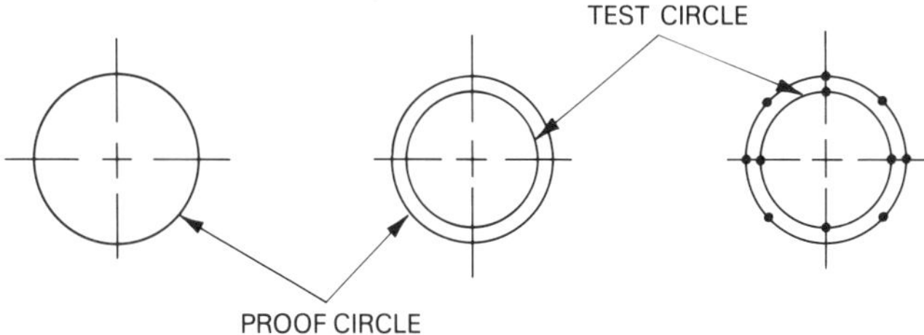


**Fig. 13-7B** Steps in laying out test and proof circles.

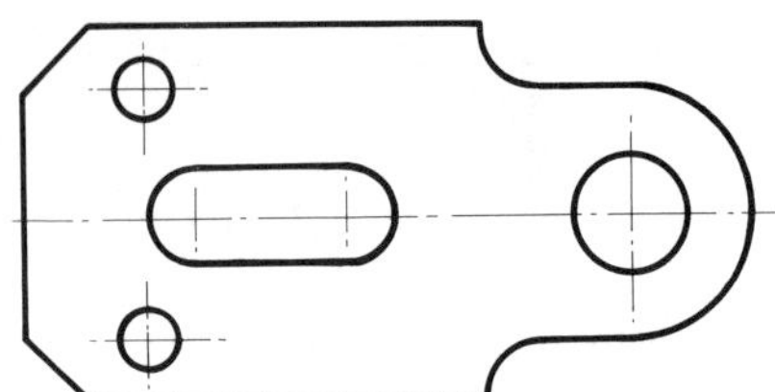

**Fig. 13-8** A typical layout exercise.

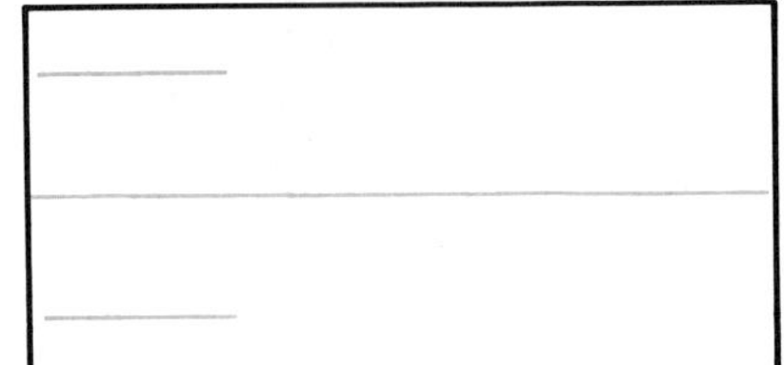

**Fig. 13-9A** Lines scribed parallel to the base.

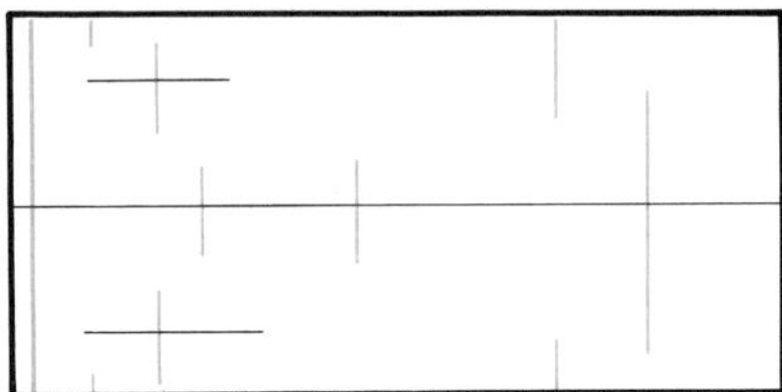

**Fig. 13-9B** Lines scribed at 90°.

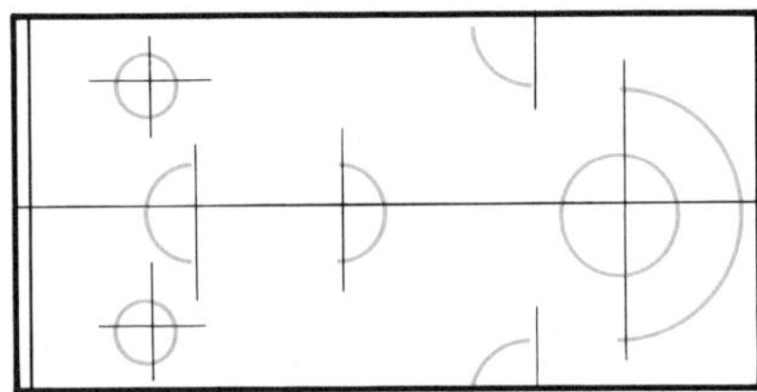

**Fig. 13-9C** Arcs and circles scribed.

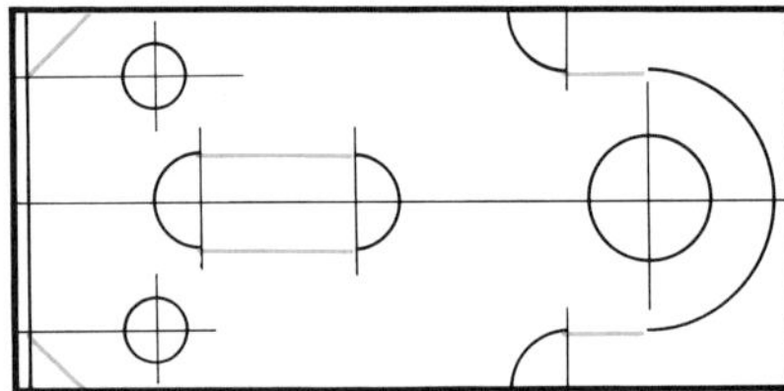

**Fig. 13-9D** Arcs and circles connected by straight lines.

**13.** Carefully prick-punch the center of all hole or radii locations.

**14.** Using a divider set as required, scribe all circles and arcs (Fig. 13-9C).

**15.** Scribe any lines required to connect the arcs or circles (Fig. 13-9D).

**16.** Draw in the angular lines (Fig. 13-9D).

## REVIEW QUESTIONS

**1.** What instrument should be used to lay out an angle if it is shown on a drawing in

(a) Degrees and no tolerance given

(b) Degrees and minutes

**2.** How should any workpiece be prepared before applying layout dye?

**3.** Why are layout lines prick-punched after they are scribed?

**4.** How far apart should the prick-punch marks be located?

**5.** What is a quadrant?

**6.** For what purpose are trammels used?

**7.** How may the point of the divider be held on the edge of the workpiece when laying out an internal radius?

**8.** Name another tool that may be used for laying out smaller radii.

**9.** What is the purpose of

(a) A test circle

(b) A proof circle

**10.** List the order in which layout lines should be scribed on a workpiece.

# SECTION

## SAFETY

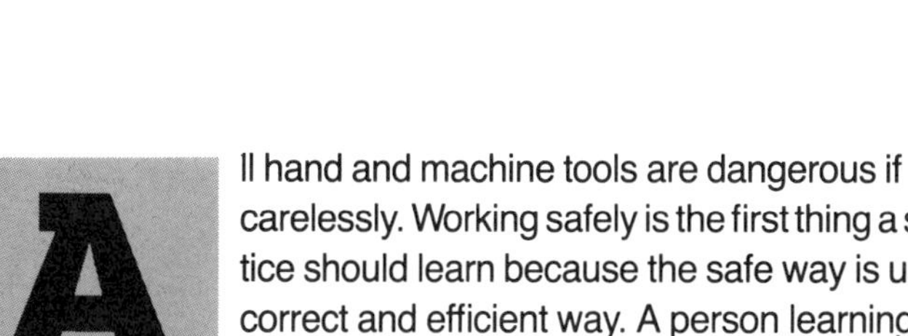

All hand and machine tools are dangerous if used improperly or carelessly. Working safely is the first thing a student or apprentice should learn because the safe way is usually the most correct and efficient way. A person learning to operate machine tools *must first* learn the safety regulations and precautions for each tool or machine. Far too many accidents are caused by carelessness in work habits, or by horseplay. It is easier and much more sensible to develop safe work habits than to suffer the consequences of an accident.

# UNIT 14

# CAUSES OF ACCIDENTS

The safety programs initiated by accident-prevention associations, safety councils, governmental agencies, and industrial firms are constantly attempting to reduce the number of accidents. *Nevertheless,* each year accidents which could have been avoided result not only in millions of dollars' worth of lost time and production, but also in a great deal of pain and many lasting physical handicaps. Modern machine tools are equipped with safety features, but it is still the operator's responsibility to use these machines wisely and safely. (See N-1.)

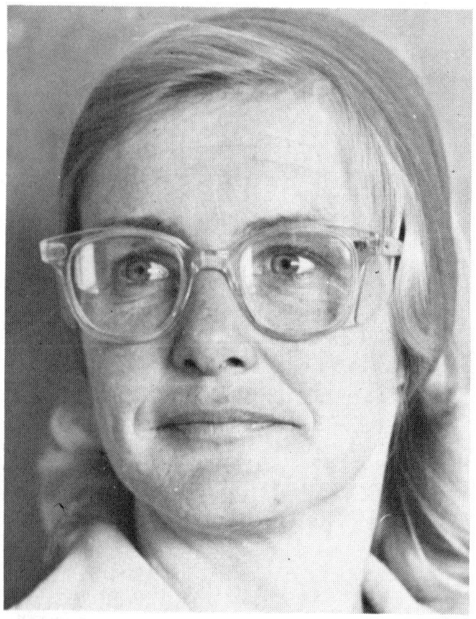

**Fig. 14-1** Safety glasses with side shields offer good eye protection.

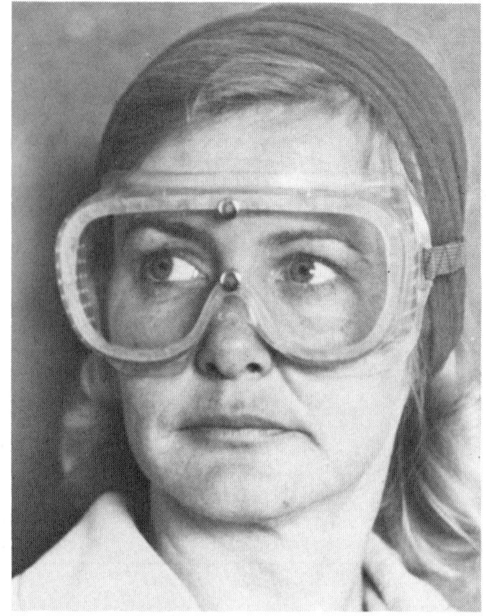

**Fig. 14-2** Safety goggles protect prescription lenses from scratching or damage.

N-1 Safety is everyone's business and responsibility.

N-2 Accidents don't just happen; they are caused.

## OBJECTIVES

Upon completion of this unit you should be able to:

**1.** Identify the four general categories of accidents in a machine shop
**2.** Be aware of the hazards in each category
**3.** Know how to prevent accidents in each category of accidents

## ACCIDENT PREVENTION

It takes many years of experience to become a skilled machinist and one moment of carelessness can quickly put an end to a career. (See N-2.) Since accidents can be avoided, a person learning the machine shop trade must first develop safe work habits. A safe worker should

**1.** Be neat and tidy at all times
**2.** Develop personal responsibility
**3.** Learn to consider the welfare of fellow workers
**4.** Derive satisfaction from performing work accurately and safely

Safety in a machine shop may be divided into four general categories: *personal grooming, housekeeping, handling tools and material,* and *operating machine tools.* Although it is impossible to list every safety rule or unsafe practice that a person may encounter in each of these areas, some general rules are offered.

## PERSONAL GROOMING

Many accidents in a machine shop are caused by careless personal habits. Lack of proper eye care, loose or improper clothing, personal grooming, and horseplay are the major causes of accidents in this category.

### 1. Wear Approved Safety Glasses or Goggles at All Times

It is very important to protect the eyes when in a machine shop. Metal cuttings produced on machine tools often fly a great distance and can cause serious eye injury. Some persons feel that it is necessary to wear eye protection only when operating machine tools. However, since many handcutting tools are made of hardened steel which can break or shatter due to the pressure or force applied, it is wise to wear safety glasses at all times in a machine shop.

Although there are many different types of safety glasses, the ones that offer the best eye protection are the safety glasses with side shields, Fig. 14-1, or the safety goggles, Fig. 14-2, for those who wear prescription glasses.

### 2. Never Wear Loose Clothing around Any Machine

Since loose clothing can become caught in revolving parts of machinery, Fig. 14-3, it is recommended that ties be removed and shirt sleeves rolled up to the elbow. Sweaters with long or fuzzy hair should not be worn in a machine shop.

### 3. Wear Approved Footwear at All Times

Work in a machine shop involves the cutting of metal into razor-sharp chips, therefore it is important to wear at least solid leather shoes. Tennis shoes or sandals do not offer any protection for the feet and should never be worn.

**Fig. 14-3** Loose clothing is dangerous around machinery of any kind. (*Courtesy The Clausing Corporation.*)

**Fig. 14-4** Jewelry such as rings, watches, bracelets, etc., should never be worn in a machine shop. (*Courtesy The Clausing Corporation.*)

**Fig. 14-5** Oil and grease on floors can cause serious falls.

Since there is always the possibility of dropping a piece of steel or machine accessory on the foot, it is recommended that safety shoes with steel toe shields be worn in a machine shop. In most industries, safety shoes must be worn on the job.

### 4. Remove All Rings, Watches, or Bracelets

Jewelry should never be worn in a machine shop since these items can be caught in a machine and draw a finger or the entire hand in, Fig. 14-4. If a ring is worn and a heavy piece of steel should fall on the hand, the ring could be bent and cut the finger. This can be very painful until the bent ring is removed.

### 5. Long Hair Must Be Protected by a Hair Net or an Approved Protective Shop Cap

One of the most common accidents is that of long, unprotected hair becoming caught in the revolving drill or drill press spindle. If long hair becomes caught in the revolving part of any machine, it can cause serious head injury.

### 6. Avoid Horseplay at All Times

Since tools are sharp and machines are made of steel which is hard, horseplay must be avoided in a machine shop. An accidental slip or fall can cause a serious cut or body injury. In many industries, horseplay or playing practical jokes are often grounds for dismissal because so many accidents have resulted from these in the past.

### 7. Never Use Compressed Air to Clean Machines or Clothing

This is a dangerous practice because sharp metal chips can fly quite a distance when blown by compressed air. This can be a hazard to the person cleaning the machine and also to anyone else in the shop.

If compressed air is blown on clothing or the skin, it can force dirt or germs into the skin.

## HOUSEKEEPING

Poor housekeeping has resulted in countless numbers of accidents in a machine shop. Good housekeeping will not only provide safe working conditions, but also improve the efficiency of the job.

### 1. Keep the Floor around a Machine Free of Tools or Stock

Tripping over material which is lying on the floor, especially round bars, can cause dangerous falls. Be sure that the area around a machine is clear so that the operator can move around safely.

### 2. Keep the Floor Free of Oil and Grease

Whenever oil, grease, or cutting fluids are spilled on the floor, be sure to remove them as quickly as possible to prevent dangerous falls, Fig. 14-5. Be sure to store oily rags in an approved container.

### 3. Sweep Up the Metal Chips on the Floor Frequently

Metal chips are sharp and quickly become embedded in the soles of shoes. This causes the soles of the shoes to become very slippery and can cause dangerous falls, especially if a person walks on a concrete or terrazzo floor.

### 4. Always Keep the Machine Clean

A good machinist takes pride in keeping his or her machine clean. This is a reflection on the machinist as a person and provides pleasant, safe working conditions. Always use a brush (Fig. 14-6) to remove metal chips from machines. Oily surfaces which can be dangerous should be cleaned with a cloth.

### 5. Never Place Tools or Materials on a Machine Table

Cluttering up a machine with tools or material creates unsafe working conditions. Use a bench or table near the machine for this purpose.

Fig. 14-6 Always use a brush for removing steel chips from a machine.

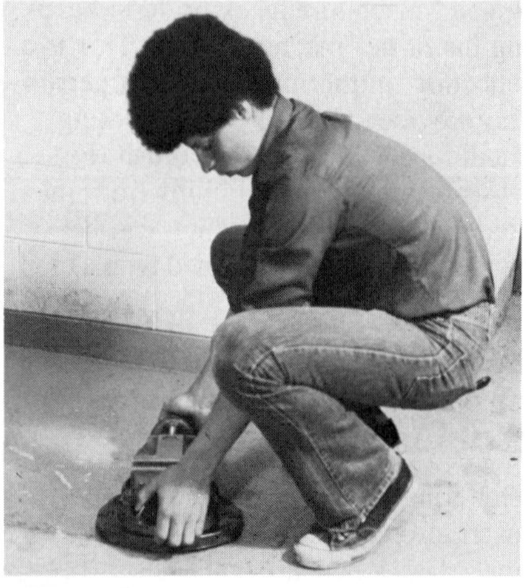

Fig. 14-7 Use the proper lifting procedure to avoid back injury.

### 6. Always Stop a Machine before Attempting to Clean It

Attempting to clean a machine which is operating is both dangerous and very foolish. A rag can easily be caught in the revolving work or part of a machine and draw the entire hand in. It is wise to shut off the main electric switch for a machine being cleaned so that it cannot be started accidentally.

## HANDLING TOOLS AND MATERIALS

Many serious cuts are caused by handling sharp tools or materials improperly. Permanent back injury can result from using improper lifting techniques or attempting to lift something that is too heavy.

### 1. Always Remove the Burrs or Sharp Edges from Workpieces

Whenever a part is to be removed from a machine, be sure to remove the burrs or sharp edges with a file. Get in the habit of always removing burrs from any material; it will save a lot of painful cuts.

### 2. Never Handle Sharp Tools or Cutters by Hand

When handling sharp tools such as milling cutters, end mills, etc., it is wise to use a rag to avoid cuts if they should have a tendency to slip.

### 3. Use Proper Lifting Techniques whenever Lifting Tools or Materials

Many back injuries, which could have been avoided, have been caused when lifting incorrectly or when trying to lift too heavy a load. If large or heavy objects must be lifted, get some help or use some form of lifting equipment. Never try to "show off" how strong you are; it could result in permanent back injury.

The approved method of lifting material is as follows:

**1.** Squat down, bending your knees, but be sure to keep your back straight, Fig. 14-7.

**2.** Grasp the material securely and use your leg muscles to raise the load. *Keep your back straight;* bending the back puts an excessive strain on it.

**3.** Lower the material to the floor by bending the knees.

## OPERATING MACHINE TOOLS

Since different cutting tools and machining procedures are used on various machine tools, the safety precautions for each may vary. The safety precautions which pertain to each machine will be covered in a separate unit in the section which contains a discussion of each machine tool. The following are some general safety rules which should be observed when operating any machine tool.

**1.** Never attempt to operate any machine tool until you fully understand its mechanism and know how to stop it quickly.

**2.** Be sure that all the safety guards are in place before starting any machine tool.

**3.** Never wear loose clothing or jewelry around machine tools.

**4.** Always wear approved safety glasses in a machine shop, especially when operating machine tools.

**5.** Keep your hands away from moving parts of the work or machine.

**6.** Stop the machine before attempting to clean it or measure the size of the workpiece.

## REVIEW QUESTIONS

**1.** What is one of the first things that a person starting the machine shop trade must learn?

**2.** List four qualities of a safe worker.

**3.** Why is it important that safety glasses be worn at all times?

**4.** Name three things which might get caught in revolving work or machine parts.

**5.** What three ways can falling accidents be avoided in a machine shop?

**6.** Explain how sharp cutters and workpieces should be handled.

**7.** What is the correct procedure for lifting material?

**8.** List three safety rules which should be observed when operating machine tools.

# SECTION 4

## BENCHWORK

Hand tools are essential for those operations in machine shop work that cannot be done efficiently or economically by machine tools. Operations such as sawing, filing, polishing, chipping, tapping, and threading must be mastered by an apprentice learning the machine trade. A machinist can only acquire skill in the use of hand tools through patience and practice; there is no easy way.

Hand tools must be used with the same care as is given to the more expensive machine tools. A reasonable amount of care will keep tools in safe and good working condition. One sign of a good machinist is the excellent condition of his tools.

# UNIT 15

# HAND TOOLS (NONCUTTING)

Hand tools used in machine shop work are those tools which use hand power as their driving or turning force. These tools fall into two categories: cutting and noncutting. The cutting tools are generally used to remove metal or cut special forms into metal. The noncutting tools are generally used to hold or turn the workpiece. The most common noncutting hand tools are the vise, hammer, screwdriver, wrench, and pliers. It is important that all hand tools be used properly in order to get the best results and to prevent accidents.

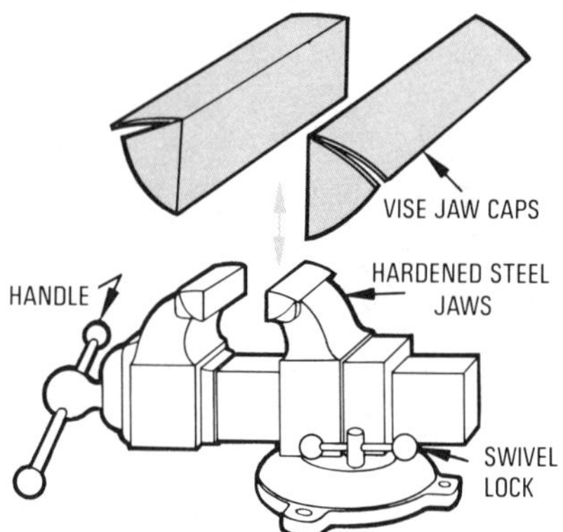


**Fig. 15-1** Vises equipped with a swivel base allow the vise to be turned to any position.

## OBJECTIVES

After completing this unit you should be able to:

**1.** Identify and know the purpose of the machinist's vise

**2.** Identify and know how to use ball-peen and soft-faced hammers

**3.** Follow the correct procedure for using metal stamps

**4.** Identify and use various screwdrivers, wrenches, and pliers correctly

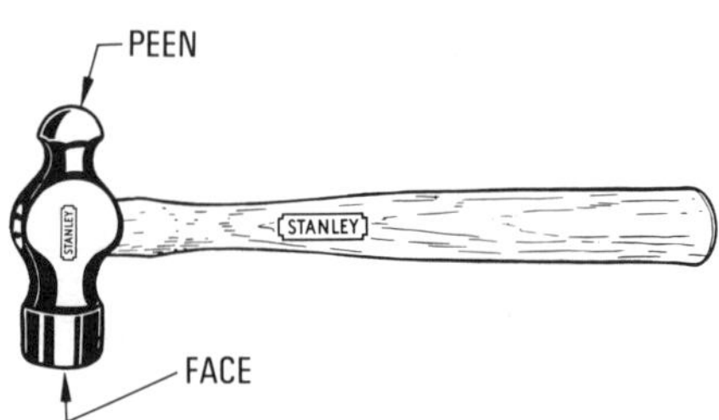


**Fig. 15-2** A ball peen hammer is used in most machine shop work. (*Courtesy Stanley Tools Division of the Stanley Works.*)

## THE MACHINIST'S VISE

The machinist's vise is a work-holding device used to hold work for such operations as sawing, filing, chipping, tapping, threading, etc.

Vises are made in a great variety of sizes, to hold work of many sizes and shapes. Some vises are equipped with a swivel base which allows the vise to be turned to any position (Fig. 15-1) so that it is more convenient to work on the part being held in the vise. To hold finished work, it is wise to put jaws made of aluminum, brass, or copper over the regular jaws to protect the work surface.

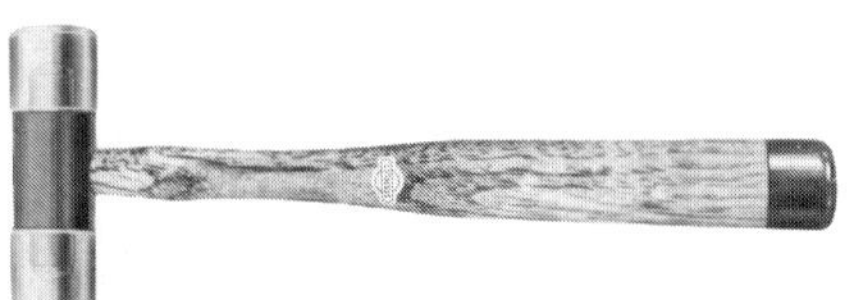

**Fig. 15-3** A soft-faced hammer will not damage the work surface. (*Courtesy Stanley Tools Division of the Stanley Works.*)

N-1 Never strike the face of a hammer head against hardened steel. A chip of metal may fly off and cause a serious injury.

## HAMMERS

The *ball-peen hammer*, or machinist's hammer as it is more commonly called, is the hammer generally used in machine shop work. The rounded top (Fig. 15-2) is called the peen, and the bottom is known as the face. Ball-peen hammers are made in a variety of sizes, with head masses ranging from approximately 2 oz to 3 lb (55 to 1400 g). They are hardened and tempered. The smaller sizes are used for layout work; the larger ones are used for general benchwork.

*Soft-faced hammers* (Fig. 15-3) are used in assembly and setup work because they will not mar the finished surface of work. These hammers have pounding surfaces made of brass, plastic, lead, rawhide, or hard rubber.

When using a hammer, it should be grasped at the end of the handle. This position provides greater striking force and balance for the hammer than if gripped near the head. It also helps to keep the hammer face flat on the work being struck while minimizing the chance of damage to the face of the work. (See N-1.)

Two safety precautions should be observed when using a hammer:

**1.** Be sure that the head is tight on the handle and secured with a proper wedge to keep the handle expanded into the head.

**2.** Never use a hammer with a greasy handle, or when hands are greasy.

## METAL STAMPS

Metal stamps or stencils (Fig. 15-4) are used to mark or identify workpieces. They are made in a variety of sizes with letters or numbers from 1/32 to 1/2 in. (0.8 to 12.7 mm) high. Metal stamps should *never* be used on hardened

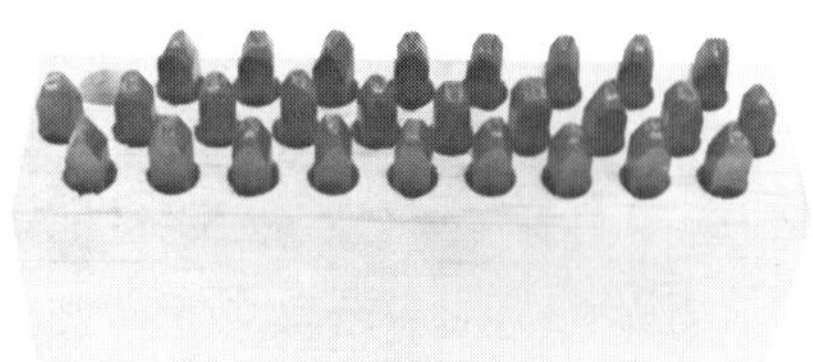

**Fig. 15-4** Metal stamps may be used to identify workpieces. (*Courtesy Kostel Enterprises Ltd.*)

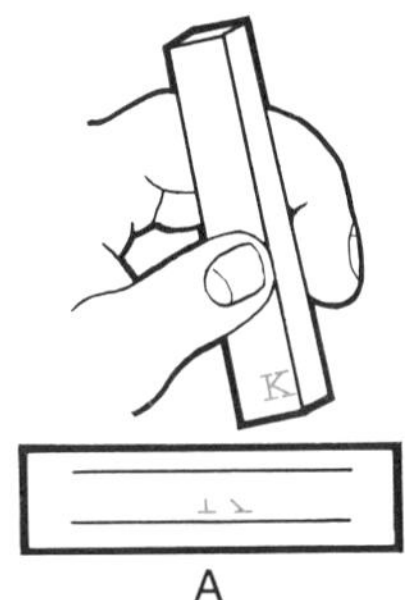

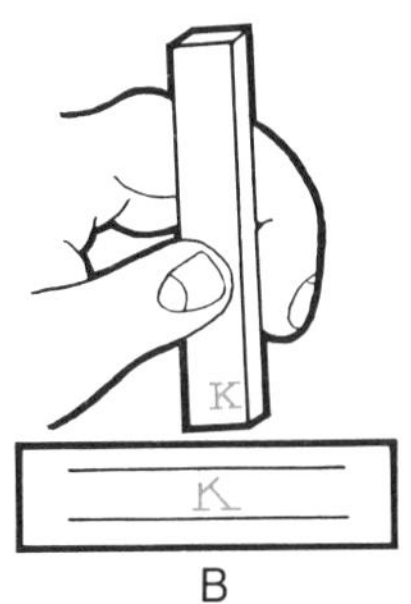

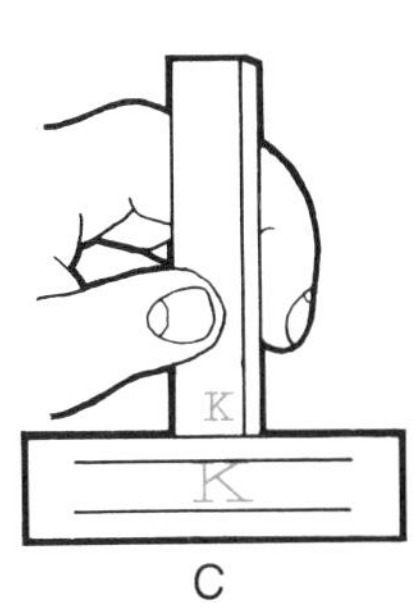


**Fig. 15-5A, B, C** The three-step method is used for large metal stamps.

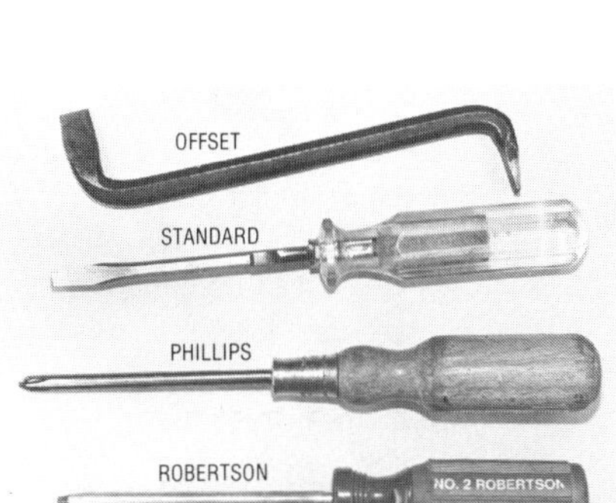


**Fig. 15-6** Various types of commonly used screwdrivers. (*Courtesy Kostel Enterprises Ltd.*)

**Fig. 15-7** A standard screwdriver.

metal, and if used on cast iron or hot rolled steel, the hard outer scale of the metal should first be removed by grinding, chipping, or machining, to prevent damage to the stamps.

Most metal stamps have either the letter (number) of the stamp or a manufacturer's trademark on one side. If this letter (number) or trademark is held so that it faces the operator, the letter (number) will be imprinted right side up on the metal. Those metal stamps without identification of any kind on the side should first be tested, as suggested in N-3, to ensure that the impression in the metal will be right side up. (See N-2.)

### To Use Metal Stamps

**1.** Place the work in a vise or on a bench plate.

**2.** Lay out a base line to indicate where the stamping is to be located.

**3.** Lay out a line for the center of the lettering.

**4.** Stamp the middle letter on the centerline. (See N-3.)

**5.** Stamp all the letters to the right of the centerline and then work from the center to the left, thus balancing the stamping about the centerline.

When using larger-sized metal stamps [¼ in. (6.35 mm) or over] the three-step method is advisable in order to produce clear, sharp impressions.

### To Use the Three-Step Method for Large Metal Stamps

**1.** Place the stamp on the base line inclined toward you and strike it sharply with a medium-sized ball-peen hammer (Fig. 15-5A).

**2.** Replace the stamp in the impression, inclined only slightly, and strike it sharply (Fig. 15-5B).

**3.** Replace the stamp in the impression, hold it *vertically,* and strike it sharply to complete the letter (Fig. 15-5C).

## SCREWDRIVERS

Screwdrivers (Fig. 15-6) are made in a variety of shapes, types, and sizes. The *standard,* or *common,* screwdriver is used on slotted-head screws. It consists of three parts—the *blade,* the *shank,* and the *handle* (Fig. 15-7). Although most shanks are round, those on heavy-duty screwdrivers are generally square. This permits the use of a wrench to turn the screwdriver when extra torque is required.

The *offset* screwdriver is designed for use in confined areas where it is impossible to use a standard screwdriver. The blades on the ends are at right angles to each other. The screw is turned one-quarter of a turn with one end and then one-quarter of a turn with the other end.

Other commonly used screwdrivers are the *Robertson,* which has a square tip or blade, and the *Phillips,* which has a +-shaped point. Both types are made in different sizes to suit the wide range of screw sizes.

### Care of a Screwdriver

**1.** Choose the correct size of screwdriver for the job. If too small a screwdriver is used, both the screw slot and the tip of the driver may become damaged. (See N-4.)

**2.** Do not use the screwdriver as a pry, chisel, or wedge.

**3.** When the point becomes worn or broken, it should be redressed to shape (Fig. 15-8).

N-2 Be sure that the stamp is held so that the letter will be imprinted *right side up.*

N-3 Before stamping a workpiece, test the stamp on a piece of wood to ensure that it will be right side up.

N-4 Always use the correct size screwdriver so that the slot in the head of the screw is not damaged.

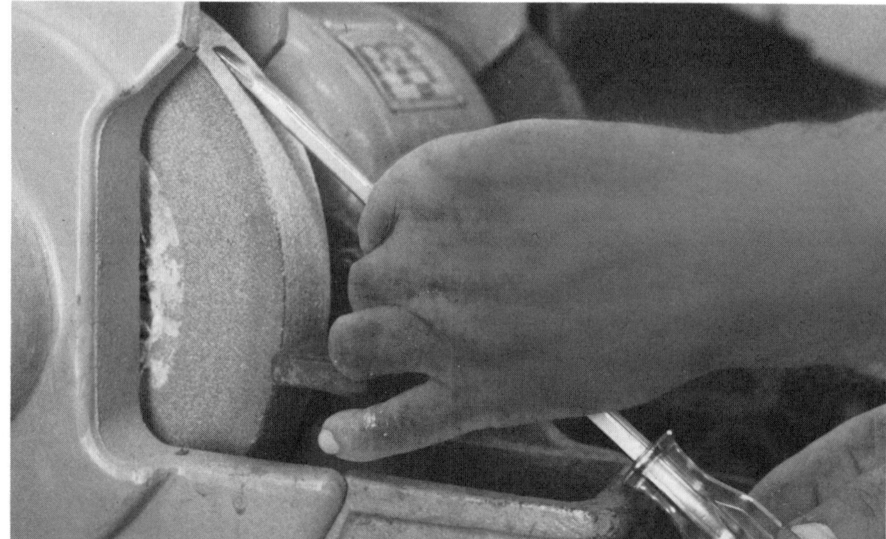

**Fig. 15-8** Regrinding a screwdriver blade.

N-5 When grinding, remove a minimum amount of metal so as not to grind past the hardened zone in the tip. Quench frequently in cold water so as not to draw the temper from the blade.

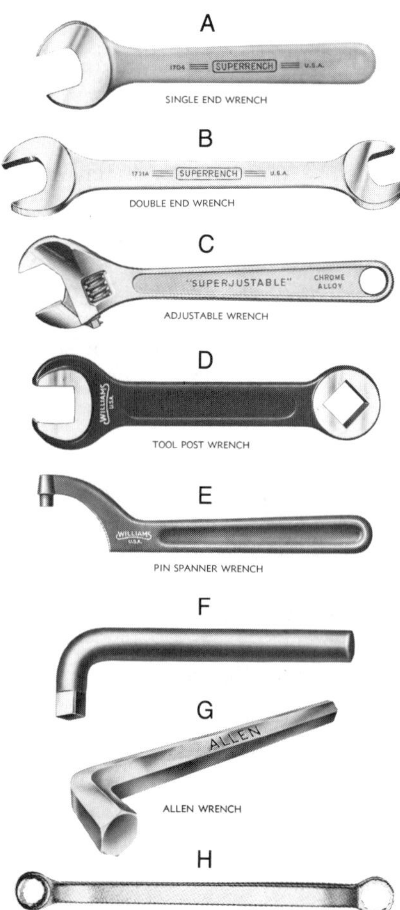


**Fig. 15-9** Various types of common wrenches used in a machine shop. *(A to F: Phototech Studios; H: Courtesy Kostel Enterprises Ltd.)*

## Regrinding a Screwdriver Blade

When regrinding the tip, make the sides of the blade slightly concave by holding the side of the blade tangential to the periphery of the grinding wheel (Fig. 15-8). Grind an equal amount off each side of the blade. This shape will enable the blade to maintain a better grip in the slot. Care should be taken that the original taper, width, and thickness of the tip are retained and that the end is ground square with the centerline of the blade. (See N-5.)

## WRENCHES

Many types of wrenches are used in machine shop work, each being suited for a specific purpose. The name of a wrench is derived from either its shape, its use, or its construction. Various types of wrenches used in a machine shop are illustrated in Fig. 15-9.

A *single-end wrench* is one that fits only one size of bolt, head, or nut. The opening is generally offset at a 15° angle to permit complete rotation of a hexagonal nut in only 30° by turning the wrench over.

A double-end wrench has a different size opening at each end. It is used in the same manner as a single-end wrench.

The *adjustable wrench* is adjustable to fit various size nuts and is particularly useful for odd size nuts. Unfortunately this type of wrench, when not properly adjusted to the flats of the nut, will damage the corners of the nut. Adjustable wrenches should not be used on nuts which are very tight on the bolt. When excess pressure is applied to the handle, the jaws tend to spring, causing damage to the wrench and the corners of the nut. When using an adjustable wrench, the movable jaw should point in the direction of the force being applied, Fig. 15-10.

A *toolpost wrench* is a combination open-end wrench and a box-end wrench. The box end is used on toolpost screws and often on lathe carriage locking screws. In order that the toolpost screw head will not be damaged, it is important that only this type of wrench be used.

The *socket set screw wrench,* commonly called the *Allen wrench,* is hexagonal and fits into the holes in safety set screws or socket-head set screws. They are made of tool steel in various sizes to suit the wide range of screw sizes. They are identified by the distance across the flats. The distance is usually one-half the outside diameter of the set screw thread in which it is used.

The *pin spanner wrench* fits around the circumference of a round nut. The pin on the wrench fits into a hole in the periphery of the nut.

*Box-end* or *twelve-point wrenches* are capable of operating in close quarters. The box-end wrench has twelve notches cut around the inside of the face. This type of wrench completely surrounds the nut and will not slip.

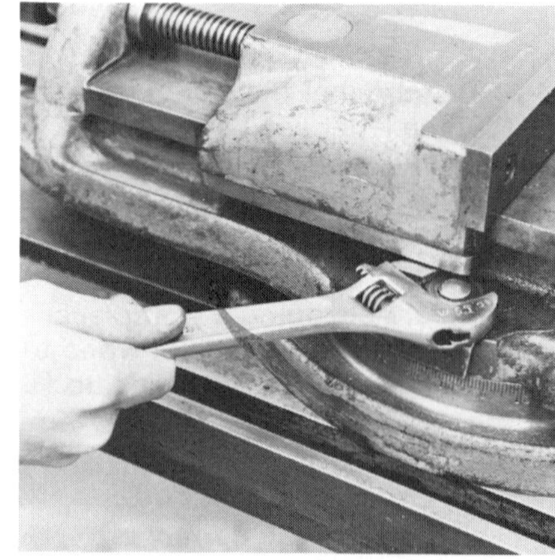

**Fig. 15-10** The correct method of using an adjustable wrench. *(Courtesy Kostel Enterprises Ltd.)*

N-6 Select a wrench that fits the nut or head of the bolt correctly. A loose-fitting wrench will round the corner of a nut and cause the wrench to slip, which may result in an accident.

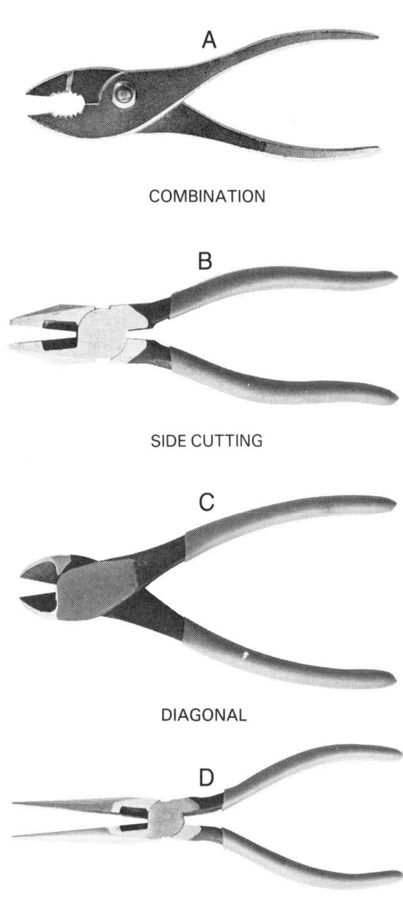


**Fig. 15-11** Various types of pliers. (A) Combination; (B) side cutting; (C) diagonal; (D) needle-nose. (*Courtesy Kostel Enterprises Ltd.*)

### Using Wrenches

**1.** Always select a wrench which fits the nut or bolt properly. A wrench that is too large may slip off the nut and possibly cause an accident. (See N-6.)

**2.** Whenever possible, it is advisable to *pull* rather than push on a wrench in order to avoid injury if the wrench should slip (Fig. 15-10).

**3.** Always be sure that the nut is fully seated in the wrench jaw.

**4.** Use a wrench in the same plane as the nut or bolt head.

**5.** When tightening or loosening a nut, a sharp quick jerk is more effective than a steady pull.

**6.** A drop of oil on the threads, when assembling a bolt and nut, will ensure easier removal later.

## PLIERS

*Pliers,* Fig. 15-11, are hand tools used to grip or hold various parts for assembly or adjustment. They are made in various types and sizes. Pliers should never be used as a wrench because they will slip and damage the corners on a nut or the head of a bolt. Some of the common types are listed below.

*Combination or slip-joint pliers,* Fig. 15-11A, are used for holding round or flat workpieces. The slip joint allows the plier to be quickly adjusted to hold larger workpieces.

*Side cutting pliers,* Fig. 15-11B, are used for gripping flat stock or twisting wire. The cutting side jaw is used for cutting wire.

*Diagonal pliers,* Fig. 15-11C, are special pliers used exclusively for cutting wire.

*Needle-nose pliers,* Fig. 15-11D, are designed with a thin nose or jaws. They are used to reach into small places which would be difficult to reach by hand, and also for forming wire.

## REVIEW QUESTIONS

**1.** How can a finished workpiece surface be protected when holding it in a vise?

**2.** State five materials that may be used for soft-faced hammers.

**3.** List two safety precautions that should be observed when using a hammer.

**4.** What should be done to cast iron or hot-rolled steel before stamping it?

**5.** List the procedure for balancing stamping about a centerline.

**6.** Name three parts of a screwdriver.

**7.** State three important points that should be observed in the care of a screwdriver.

**8.** Describe the procedure for reconditioning the blade of a standard screwdriver.

**9.** List four wrenches used in a machine shop.

**10.** Why should an adjustable wrench not be used on a tight nut?

**11.** List three of the most important hints regarding the use of wrenches.

**12.** Name two types of pliers and state the purpose of each.

# UNIT 16

# USING HACKSAWS AND FILES

Among the many hand cutting tools used in machine shop work, the most common are the hacksaw and the file. The hacksaw can be used to cut work to length or shape and to cut notches or grooves in a workpiece. The file is used to form a metal to shape and size and to improve the surface finish of the workpiece. Since hand cutting tools do remove metal and also produce burrs in the process, it is important to take care in their use to prevent cuts on the hand.

## OBJECTIVES

Upon completion of this unit you should be able to:

**1.** Select and mount the proper blade for various materials

**2.** Saw off work held in a vise

**3.** Identify and know the purpose of the most common files

**4.** File surfaces using the cross-filing and draw-filing methods

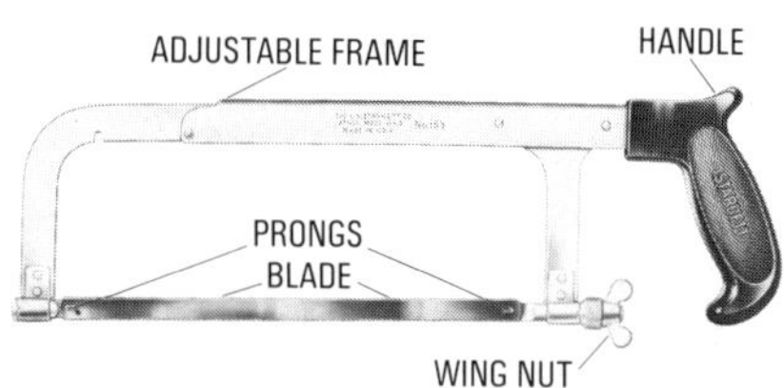


**Fig. 16-1** Parts of a hand hacksaw. (*Courtesy The L. S. Starrett Company.*)

N-1 When mounting a blade in a frame, make sure that the teeth of the blade point away from the handle.

# HAND HACKSAW

The hacksaw is a hand tool used to cut metal. The pistol grip hacksaw (Fig. 16-1) consists of four main parts: handle, frame, blade, and adjusting wing nut. The frame on most hacksaws may be flat or tubular. Some hacksaws have adjustable frames to accommodate various hacksaw blade lengths. Hacksaw blades were available only in inch sizes at the time of publication. Common lengths are 8, 10, and 12 in. (203.2, 254.0, and 304.8 mm).

## Hacksaw Blades

Hacksaw blades are made of high-speed, molybdenum, or tungsten alloy steel that has been hardened and tempered. The saw blades generally used are ½-in. (12.7-mm) wide, and 0.25-in. (0.635-mm) thick. There is a hole at each end of the blade for mounting it on the hacksaw frame.

The distance between each tooth on a blade is called the *pitch*. At the time of publication saw blades were available only in inch pitches. A pitch of 1/18 represents 18 teeth per inch (25.4 mm). The most common blades have 14, 18, 24, or 32 teeth per inch. An 18-tooth blade (18 teeth per inch) is recommended for general use. It is important to use the right pitch for the work being cut. Select a blade as coarse as possible in order to provide plenty of chip clearance and cut through the work quickly. The blade selected should have at least two teeth in contact with the work so that the work cannot jam between the teeth and strip the teeth from the saw blade.

## Mounting a Blade on a Hand Hacksaw

1. Select the proper blade for the job (Fig. 16-2).
2. Adjust the frame to the length of the blade.
3. Place one end of the blade on the back pin (Fig. 16-3). (See N-1.)
4. Place the other end of the blade on the front pin (near the wing nut).
5. Tighten the wing nut until the blade is just snugged up. Do not tighten the blade too much, as the blade may be broken or the hacksaw frame bent.

## To Use a Hand Hacksaw

1. Check that the pitch is proper for the job and be sure that the teeth point away from the handle (Fig. 16-3).
2. Adjust the tension so that the blade is just snugged up.
3. Mark the position of the cut on the workpiece (Fig. 16-4).
4. Mount the stock in the vise so that the cut will be made about ¼ in. (6.35 mm) from the vise jaws.
5. Grip the hacksaw as shown in Fig. 16-5 and assume a comfortable stance.

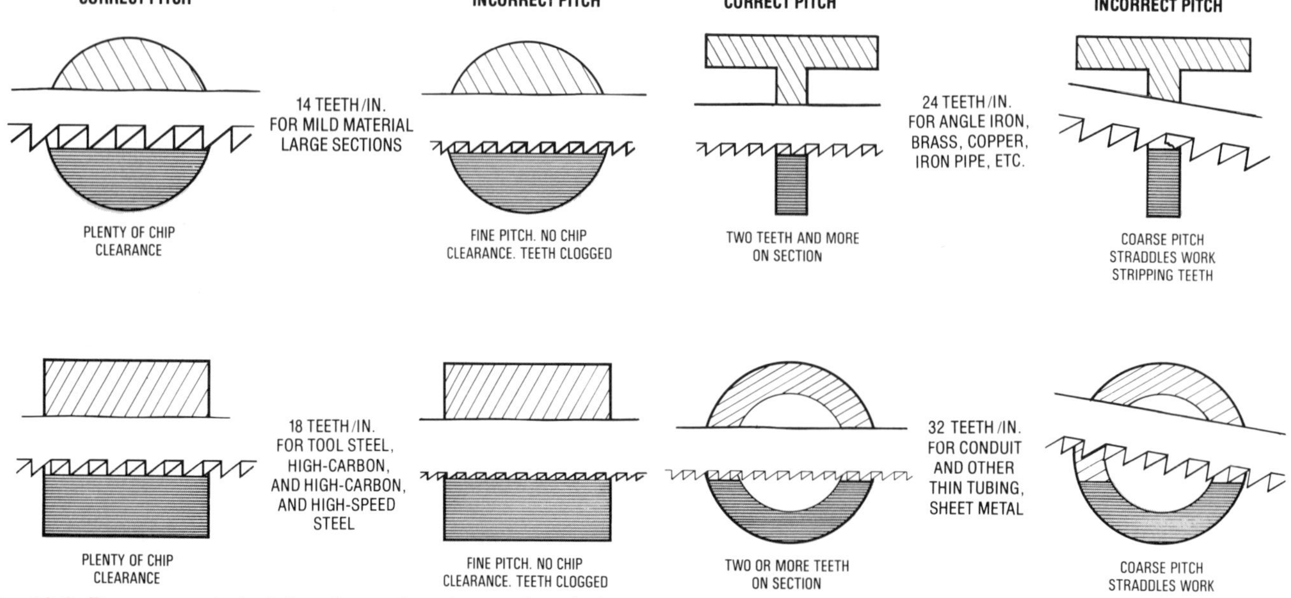


**Fig. 16-2** Recommended pitches for various types of workpieces.

N-2 If the cut does not start in the proper place, file a V-shaped nick at the cutoff mark to guide the saw blade.

N-3 Never start a new blade in an old saw cut because the set on the new blade will cause it to jam and break. Always start a new cut.

**6.** Position the blade on the work just outside the line. (See N-2.)

**7.** Apply pressure on the forward stroke and release it on the return stroke. Use a speed of about 50 strokes per minute.

**8.** When nearing the end of the cut, slow down to control the saw as it breaks through the material.

**9.** If the blade breaks or dulls during a cut, replace it with a new blade. *Do not start a new blade in an old cut. It will jam and break.* (See N-3.)

**10.** To overcome this, revolve the work one-half of a turn in the vise so that the old saw cut is at the bottom, Fig. 16-6.

**11.** Start the new cut opposite the bottom cut and saw through to the old cut.

**12.** When cutting thin material, hold the saw at an angle so that at least two teeth will bear on the work at all times. Sheet metal and other thin material may be clamped between two thin pieces of wood. The cut is then made through all three pieces, as in Fig. 16-7.

## FILES

Files are used in machine shop work when it is impractical to use machine tools. The file is an indispensable tool for removing machine marks, tool and die making, and fitting machine parts.

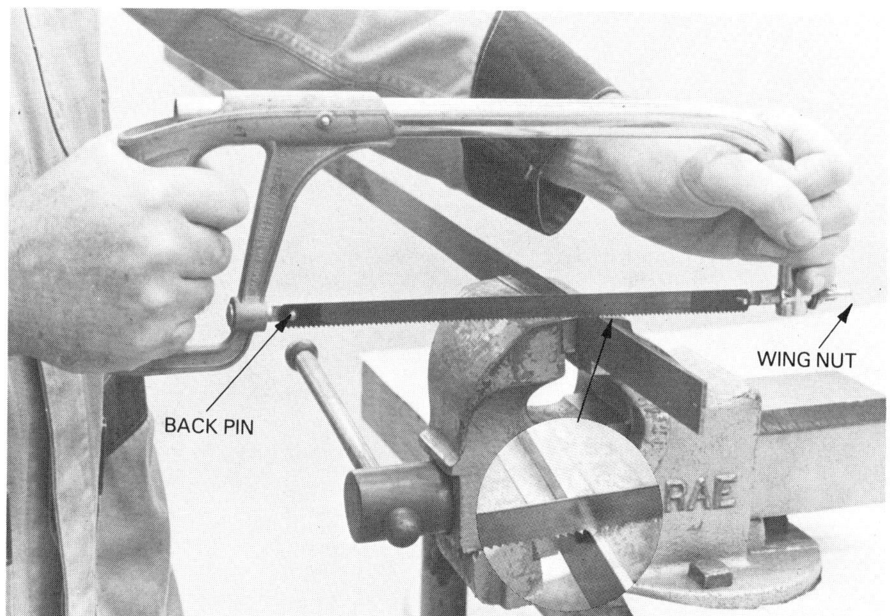


**Fig. 16-3** When mounting a blade on a hand hacksaw, the teeth must point to the front. (*Courtesy Kostel Enterprises Ltd.*)

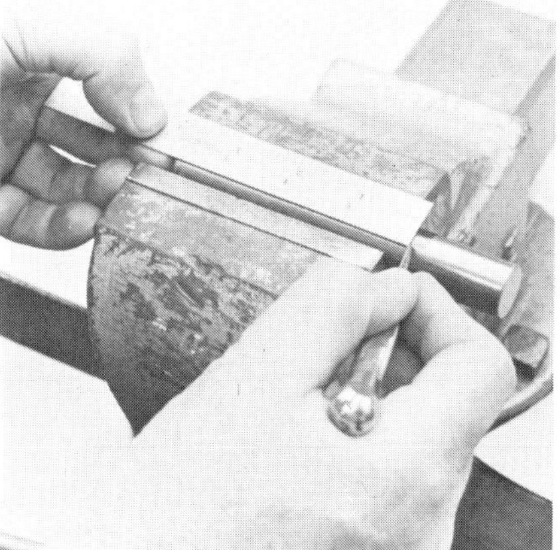

**Fig. 16-4** Marking the length of the workpiece with a scriber. (*Courtesy Kostel Enterprises Ltd.*)

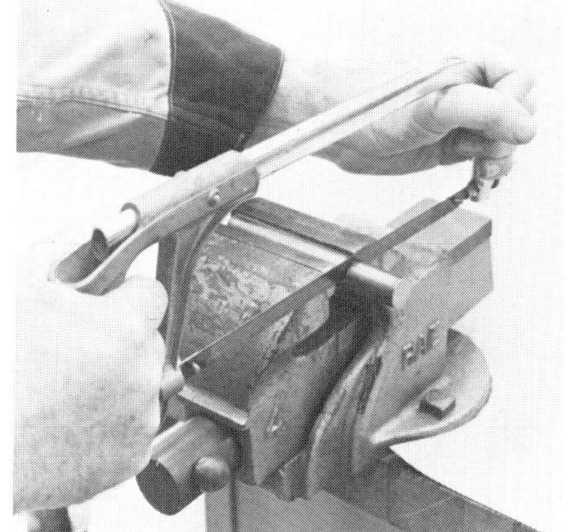

**Fig. 16-5** The correct method of holding a hacksaw. (*Courtesy Kostel Enterprises Ltd.*)

**Fig. 16-6** Start the new cut opposite the old saw cut. (*Courtesy Kostel Enterprises Ltd.*)

**Fig. 16-7** Sawing a thin piece of metal. (*Courtesy Kostel Enterprises Ltd.*)

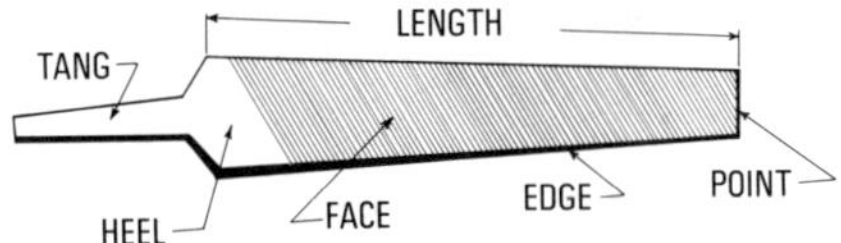


**Fig. 16-8** The parts of a file.

A *file* (Fig. 16-8) is a hand cutting tool with many teeth, used to remove surplus metal and produce finished surfaces. Files are made of high-carbon steel, hardened and tempered. Files are manufactured in a variety of shapes and sizes, each having a specific purpose. Files are divided into two classes, single- and double-cut files.

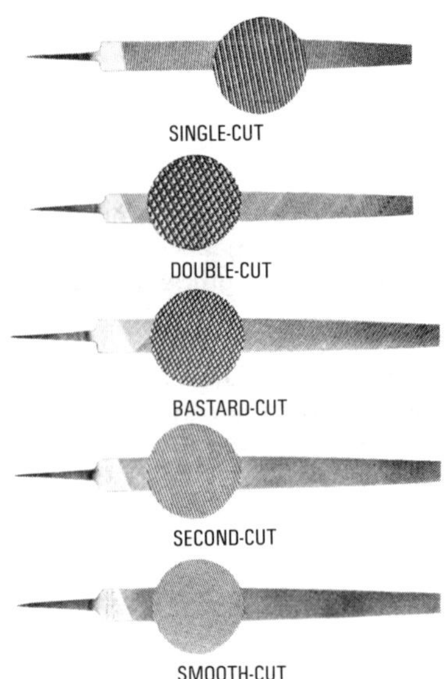


**Fig. 16-9** Single-cut and double-cut files are manufactured in several degrees of coarseness. (*Courtesy Delta File Works.*)

## Single-Cut Files

These files have a single row of parallel teeth across the face at an angle from 65 to 85° (Fig. 16-9). Single-cut files are used when a smooth surface is desired, and when harder metals such as tool steels are to be finished. This group includes the *lathe, mill,* and *saw* files.

## Double-Cut Files

These files have two rows of teeth crossing each other, one row being finer than the other (Fig. 16-9). The two rows crossing each other produce hundreds of sharp cutting teeth which remove metal quickly and make for easy clearing of chips.

## Degrees of Coarseness

Both single-cut and double-cut files are manufactured in various degrees of coarseness. On larger files this is indicated by the terms *rough, coarse, bastard, second-cut, smooth,* and *dead smooth*. The bastard, second-cut, and smooth files are the ones most commonly used in machine shops. On smaller files, the degree of coarseness is indicated by numbers from 00 to 8, number 00 being the coarsest.

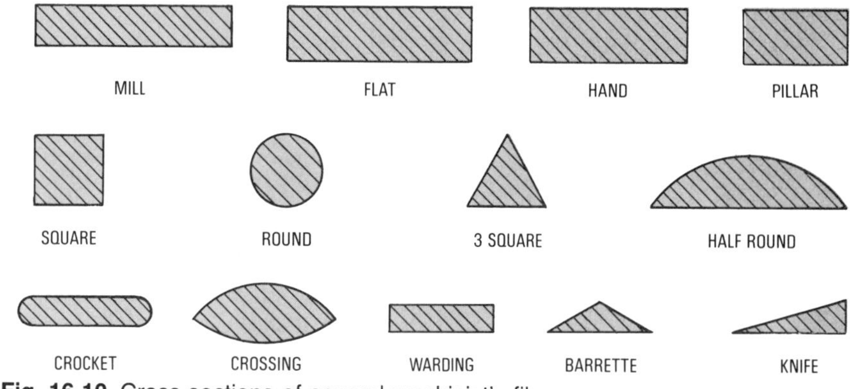


**Fig. 16-10** Cross sections of several machinist's files.

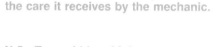

N-4 The life of a file is dependent on the care it receives by the mechanic.

N-5 To avoid hand injury, never use a file without a properly fitted handle.

## Shapes of Files

Files are manufactured in many shapes (Fig. 16-10) and may be identified by their cross section, shape, or special use. The types of files most commonly used in a machine shop are the *hand, flat, round, half-round, square, three-square* (triangular), *pillar, warding* and *knife*.

## Care of Files

Proper care, selection, and use are important if good results are to be obtained when using a file. (See N-4.) In order to preserve the life of a file, the following points should be observed:

**1.** Use a *file card* to keep the file clean and free of chips.

**2.** Do not knock a file on a vise or other metallic surface to clean it.

**3.** Do not apply too much pressure to a new file. This breaks down the cutting edges quickly. Too much pressure also causes ''pinning'' (small particles of metal become wedged between the teeth), which will cause scratches on the surface of the work.

**4.** Never use a file as a pry or hammer. A file is a hardened tool which can snap easily. This may cause small pieces to fly and cause a serious eye injury.

**5.** Always store files where they will not rub together. Hang or store them separately.

## Hints on Using Files

The following general points should be observed when filing.

**1.** *Never use a file without a handle.* This practice can result in a serious injury. If the file should slip, the tang could puncture the hand or arm.

**2.** Always be sure that the handle is tight on the file. (See N-5.)

**3.** To produce a flat surface when cross-filing, the right forearm and the left hand should be held in a horizontal plane (Fig. 16-11). Do not rock the file; push it across the work in a straight line.

**4.** A file cuts only on the forward stroke. When filing, therefore, apply pressure on the forward stroke only and relieve the pressure on the return stroke. Applying pressure on the return stroke tends to dull a file.

**5.** Work to be filed should be held in a vise at about elbow height. Small fine work may be held higher; heavier work, requiring much removal of metal, should be held lower.

**6.** Never rub the hand or fingers across a surface being filed because

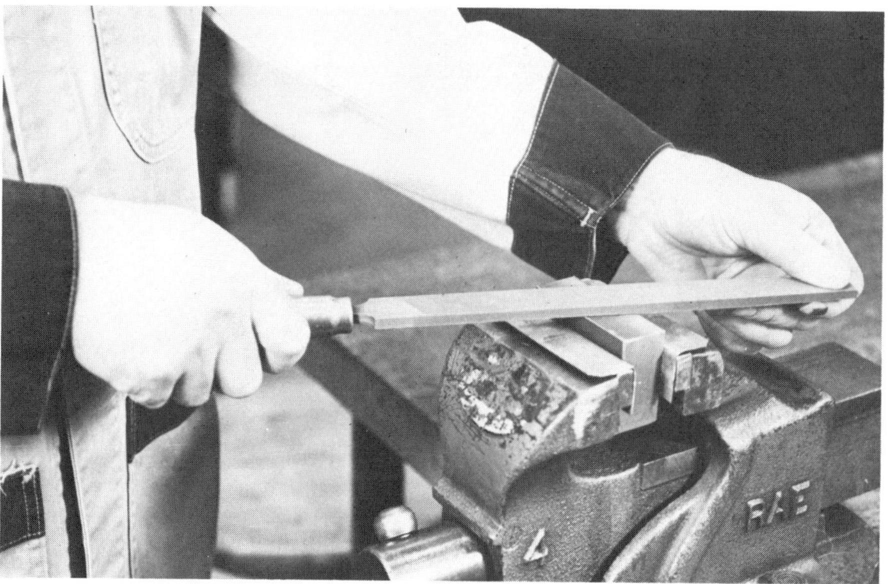

**Fig. 16-11** The correct method of holding a file. (*Courtesy Kostel Enterprises Ltd.*)

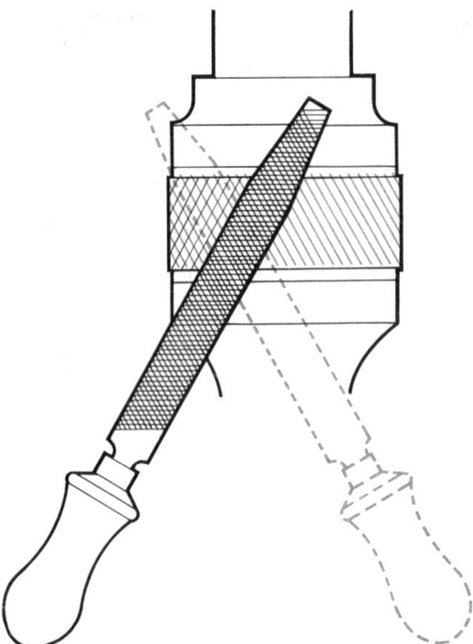

Fig. 16-12 Crossing the file stroke helps to maintain a flat surface.

N-6 A clean file cuts more easily and produces a better finished surface. Use a file card to clean the cuttings from the teeth of a file.

N-7 After draw filing, remove the burrs from the edges of the work to prevent cuts to your fingers or hand.

grease or oil from the hand deposited on the metal causes the file to slide instead of cutting the work. Oil will also cause the filings to clog the file. To prevent clogging, clean the file with a file card and then rub chalk over the surface of the file. (See N-6.)

## Filing Practice

Filing is a skill that can be acquired through patience and practice. The machinist should be able to master both types of filing, i.e., cross filing and draw filing.

*Cross filing* is used when metal is to be removed rapidly or the surface brought flat, prior to finishing by draw filing.

For rough filing, use a double-cut file and cross the stroke at regular intervals to help keep the surface as flat and straight as possible (Fig. 16-12). When finishing, use a single-cut file and take somewhat shorter strokes in order to keep the file flat on the work surface. The pressure on the file should be applied with the fingers of the left hand, which should be kept over the workpiece (Fig. 16-13) during the finishing operation.

The work should be tested occasionally for flatness by laying the edge of a steel rule across its surface (Fig. 16-14A). A steel square should be used to test the squareness of one surface to another (Fig. 16-14B).

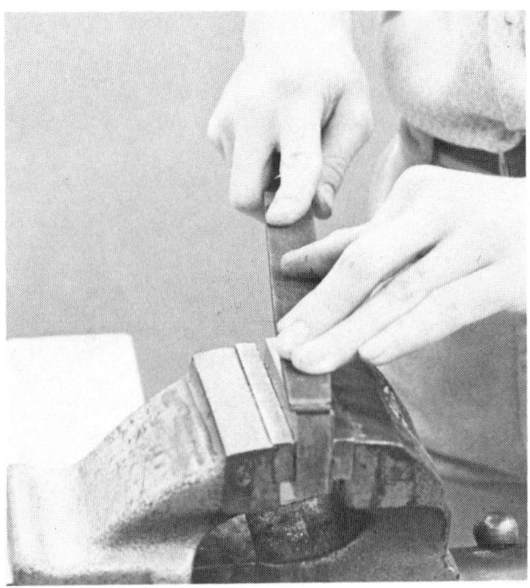

Fig. 16-13 Use only short strokes and maintain pressure on the file directly over the work when finish filing. (*Courtesy Kostel Enterprises Ltd.*)

*Draw filing* is used to produce a straight, square surface with a finer finish than is produced by straight filing. A single-cut file is used and pressure is applied to the file just above the edge of the workpiece (Fig. 16-15). The file is alternately pulled and pushed lengthwise along the surface of the work. The file should be held flat and pressure applied on the forward stroke only. On the return stroke, the file should be slid back without applying pressure but without lifting the file from the work. (See N-7.)

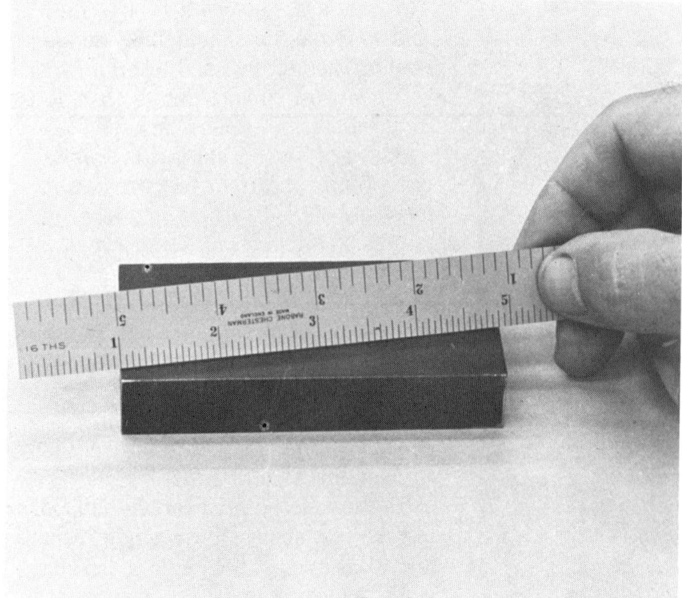

Fig. 16-14A Testing work for flatness. (*Courtesy Kostel Enterprises Ltd.*)

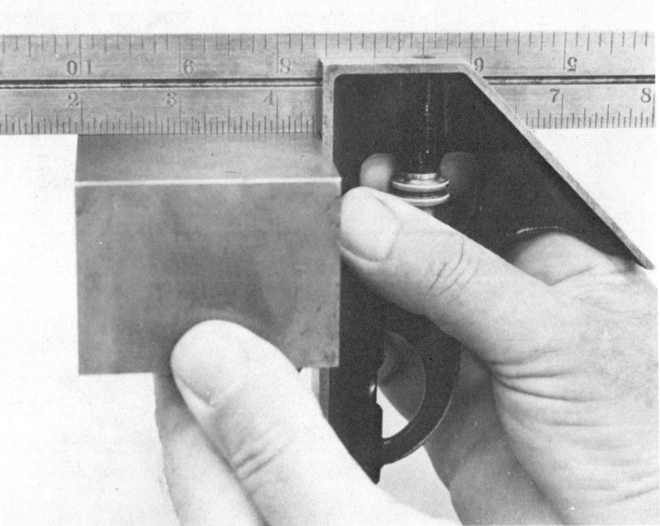

Fig. 16-14B Testing work for squareness. (*Courtesy Kostel Enterprises Ltd.*)

Fig. 16-15 Finger pressure should not extent beyond the edge of the work when draw filing. (*Courtesy Kostel Enterprises Ltd.*)

## REVIEW QUESTIONS

**1.** Of what material are hacksaw blades manufactured?

**2.** What pitch blade is recommended for
 (a) General work
 (b) Tool steel
 (c) Tubing

**3.** When mounting a saw blade on a saw frame, in which direction should the teeth point?

**4.** On what stroke is pressure applied when sawing?

**5.** Why is it advisable to start a new saw cut into the work when a blade breaks or becomes dull?

**6.** Name five parts of a file.

**7.** Of what material are files made?

**8.** State the purpose of
 (a) Single-cut files
 (b) Double-cut files

**9.** Why should a file never be used without a handle?

**10.** On which stroke should pressure be applied to the file? Explain why.

**11.** How can a file be prevented from clogging?

**12.** What tools can be used to test a filed surface for flatness and squareness?

# UNIT 17

# THREAD CUTTING (TAPS AND DIES)

Whenever possible, threads should be cut on machine tools where the accuracy and alignment can be controlled. However, many times it is necessary, due to the size or shape of the workpiece, or because the number of parts required is relatively small, to cut the thread by hand. By using a reasonable amount of care, fairly accurate internal threads can be produced with a tap, while external threads are produced with a die.

## OBJECTIVES

Upon completion of this unit, you should be able to:

**1.** Identify and know the purpose of the three taps in a set

**2.** Calculate the tap drill size for inch and metric threads

**3.** Cut internal threads using a tap

**4.** Identify and know the purpose of various hand dies

**5.** Cut external threads using a die

## HAND TAPS

Taps are cutting tools used to cut internal threads. They are made from high-quality tool steel, hardened and ground. Three or four flutes are cut lengthwise across the threads to form cutting edges, provide clearance for the chips, and admit cutting fluid to lubricate the tap. The end of the shank is square so that a tap wrench can be used to turn the tap into a hole. (See N-1.)

N-1 A tap is a cutting tool used to produce internal threads for fastening parts together.

Hand taps are usually made in sets of three, called *taper, plug,* and *bottoming* (Fig. 17-1).

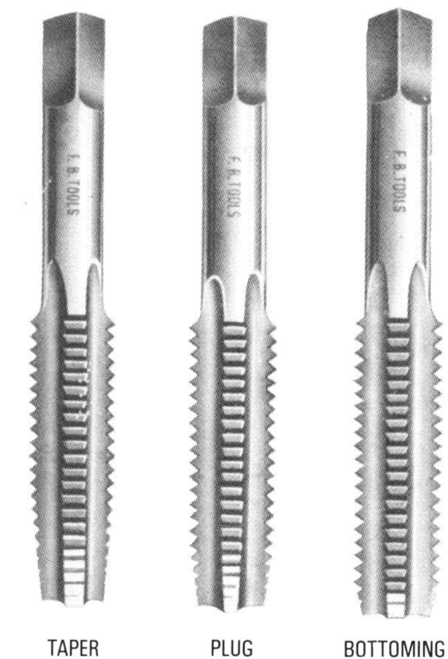


Fig. 17-1 A set of hand taps consists of a taper, plug, and bottoming tap.

N-2 The size, pitch, and thread form are stamped on taps:
1. Inch ½–13 UNC
2. Metric M 2.5 × 0.45

N-3 Always use the proper-size tap wrench for the size of tap being used.

N-4 A tap drill, which is always smaller than the tap diameter, leaves enough material in the hole to produce approximately 75 percent of a full thread.

N-5 Simplified tap drill formula:

Inch T.D.S. $= D - \frac{1}{N}$

Metric T.D.S. $= D - P$

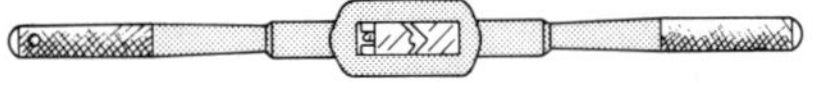

**Fig. 17-2** Double-end adjustable tap wrench.

A *taper tap* is tapered from the end approximately six threads and is used to start a thread easily. It can be used for tapping a hole which goes *through* the work, as well as for starting a *blind* hole (one that does not go through the work).

A *plug tap* is tapered for approximately three threads. Sometimes the plug tap is the only tap used to thread a hole going through a piece of work.

A *bottoming tap* is not tapered but chamfered at the end for one thread. It is used for threading to the bottom of a blind hole. When tapping a blind hole, first use the taper tap, then the plug tap, and complete the hole with a bottoming tap.

## Inch Taps

For inch taps, the major diameter, number of threads per inch, and type of thread are usually found stamped on the shank of a tap. (See N-2.) For example: ½ in.—13 N.C. represents:

| | |
|---|---|
| ½ in. | Major diameter of the tap |
| 13 | Number of threads per inch |
| N.C. | National Coarse (a type of thread) |

## Metric Taps

Metric taps are identified with the letter M followed by the nominal diameter of the thread in millimeters times the pitch in millimeters. A tap with the markings M 2.5 × 0.45 would indicate:

| | |
|---|---|
| M | A metric thread |
| 2.5 | The nominal diameter of the thread in millimeters |
| 0.45 | The pitch of the thread in millimeters |

## Tap Wrenches

Tap wrenches are manufactured in two types and in various sizes to suit the tap being used. (See N-3.)

The *double-end adjustable tap wrench* (Fig. 17-2) is manufactured in several sizes but is generally used for larger taps and in open places where there is room to turn the tap wrench. Because of the greater leverage obtained when using this wrench, it is important that a large wrench not be used on a small tap as small taps are easily broken.

The *adjustable T-handle tap wrench* (Fig. 17-3) is generally used for small taps or in confined areas where it is not possible to use a double-ended tap wrench. When used with very small taps, the body of the wrench should be turned with the thumb and forefinger to advance the tap into the hole. The handle is usually used when threading with larger taps, which will not break as easily as small ones.

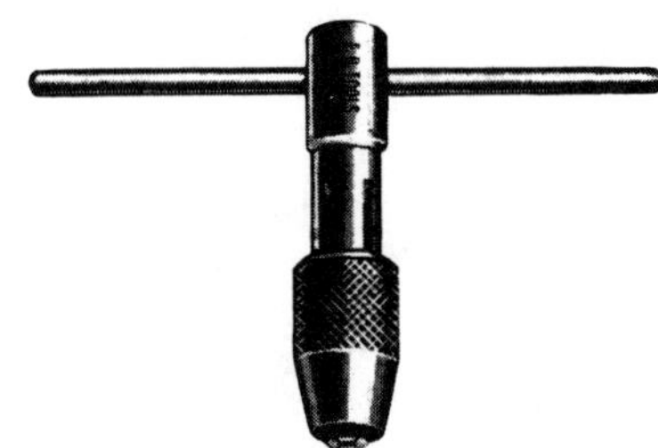

**Fig. 17-3** T-handled tap wrench.

## Tap Drill Size

Before a tap is used, the hole must be drilled to the correct tap drill size, Fig. 17-4. The tap drill size is the size of the drill that should be used to leave the proper amount of material in the hole for a tap to cut a thread. (See N-4.)

## Tap Drill Sizes for Inch Threads

When a chart is not available, the tap drill size for any American National or Unified thread can be easily found by applying this simple formula (see N-5):

$$\text{T.D.S.} = D - \frac{1}{N}$$

= tap drill size
$D$ = major diameter of tap
$N$ = number of threads per inch

**Example:**

Find the tap drill size for a ⅝ in.-11 UNC tap.

$$\begin{aligned}\text{T.D.S.} &= \frac{5}{8} - \frac{1}{11} \\ &= .625 - .091 \\ &= .534 \text{ in.}\end{aligned}$$

The nearest drill size to .534 in. is .531 in. (17/32 in.). Therefore, 17/32 in. is the tap drill size for a ⅝ in.-UNC tap. See Table 5 in the appendix for complete tapping drill chart for American National and Unified threads. For metric threads, refer to Table 6.

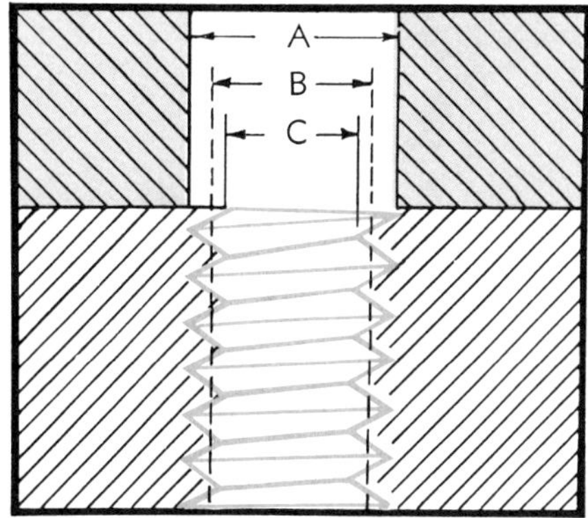


**Fig. 17-4** Cross section of a tapped hole. **(A)** Body size; **(B)** top drill size; **(C)** minor diameter.

**Fig. 17-5** The correct way to start a tap. (*Courtesy Kostel Enterprises Ltd.*)

**Fig. 17-6** Testing a tap for squareness. (*Courtesy Kostel Enterprises Ltd.*)

N-6 Cutting fluid is not required when tapping brass or cast iron.

N-7 Check the tap at two positions 90° to each other for squareness.

N-8 When tapping blind holes use all three taps in order: taper, plug, and then bottoming tap. Before using the bottoming tap, remove all the chips from the hole and be careful not to hit the bottom of the hole with the tap.

## Tap Drill Sizes for Metric Threads

The tap drill sizes for metric threads may be calculated by subtracting the pitch from the nominal diameter.

$$\text{T.D.S.} = D - P$$

For example, calculate the tap drill size for a M 12 × 1.75 thread.

$$\begin{aligned}\text{T.D.S.} &= 12 - 1.75\\ &= 10.25 \text{ mm}\end{aligned}$$

Refer to Table 6 in the appendix for tap drill sizes for metric threads.

## Tapping a Hole

Tapping is the operation of cutting an internal thread using a tap and tap wrench. Because taps are hard and brittle, they are easily broken. *Extreme* care must be used when tapping a hole to prevent breakage. A broken tap in a hole is very difficult to remove and often results in scrapping the work.

### To Tap a Hole by Hand

**1.** Select the correct taps and tap wrench for the job. Be sure that the tap wrench is not too large for the size of tap being used.

**2.** Apply a suitable cutting fluid. (See N-6.)

**3.** Place the tap in the hole as near to vertical as possible.

**4.** Apply equal down pressure on both handles, and turn the tap clockwise (for right-hand thread) for about two turns (Fig. 17-5).

**5.** Remove the tap wrench and check the tap for squareness (Fig. 17-6). (See N-7.)

**6.** If the tap has not entered squarely, remove it from the hole and restart it by applying pressure in the opposite direction from which the tap leans. *Be careful* not to exert too much pressure in the straightening process; otherwise the tap may be broken (Fig. 17-7).

**7.** When a tap has been properly started, feed it into the hole by turning the tap wrench. Down pressure is no longer required since the tap will thread itself into the hole.

**8.** Turn the tap clockwise one-quarter of a turn and then turn it backward about one-half turn to break the chip. This must be done with a steady motion to avoid breaking the tap. (See N-8.)

## Tapping Lubricants

The use of a suitable cutting lubricant when tapping results in longer tool life, better finish, and greater produc-

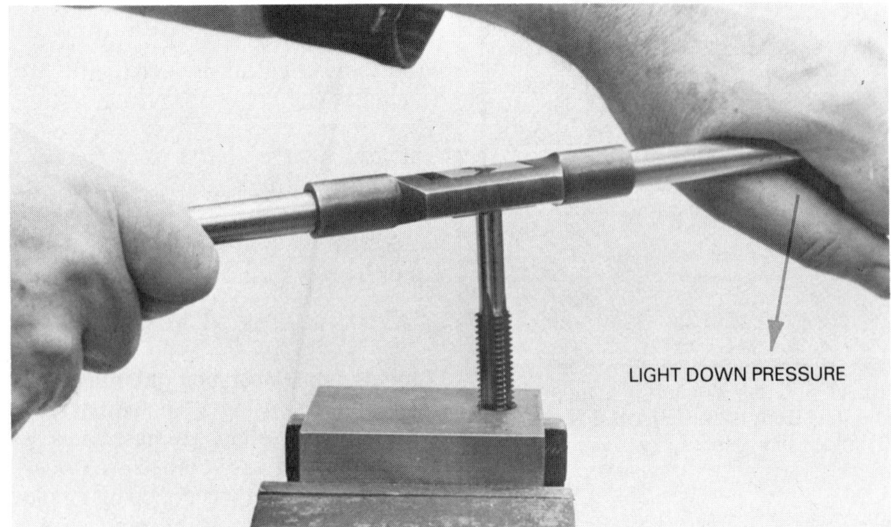


**Fig. 17-7** To straighten a tap in a hole apply pressure, while turning, on the opposite side to which it leans. (*Courtesy Kostel Enterprises Ltd.*)

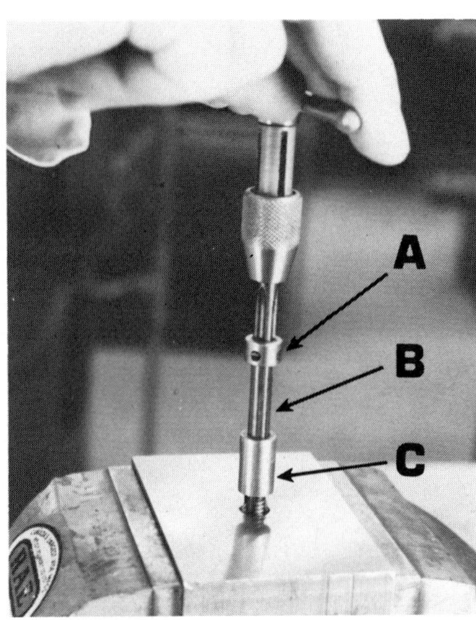


**Fig. 17-8** A tap extractor being used to remove a broken tap.

N-9 Cast-iron cuttings along with a lubricant form a compound which causes the tap to wear. Brass cuttings do not stick to the tap; a lubricant is therefore not required.

N-10 *Remember:* There is no easy way to remove a broken tap. Care should always be taken so that a tap will not be broken.

N-11 Do not force the extractor because this will damage the fingers. It may be necessary to turn the wrench back and forth carefully to free the tap sufficiently to back it out.

tion. The recommended tapping lubricants for the more common metals are listed below. A more complete list will be found in the tables at the back of the book. (See N-9.)

| | |
|---|---|
| Machine steel (hot and cold rolled) | Soluble oil, lard oil |
| Tool steel (carbon and high-speed) | Mineral lard oil Sulphur-base oil |
| Malleable iron | Soluble oil |
| Cast iron | Dry |
| Brass and bronze | Dry |

## Removing Broken Taps

Several methods may be used to remove a broken tap; some may be successful, others will not. (See N-10.)

### Tap Extractor

This tool (Fig. 17-8) has four fingers that slip into the flutes of a broken tap. It is adjustable in order to support the fingers close to the broken tap, even when the broken end is below the surface of the work. A wrench is fitted to the extractor and turned counterclockwise to remove a right-hand tap. Tap extractors are made to fit all sizes of taps.

### Using a Tap Extractor

**1.** Select the proper extractor for the tap to be removed.

**2.** Slide *collar A,* to which the fingers are attached, down *body B* so that the fingers project well below the end of the body.

**3.** Slide the fingers into the flutes of the broken tap, making sure they go down into the hole as far as possible.

**4.** Slide the body down until it rests on top of the broken tap. This will give the maximum support to the fingers. (See N-11.)

**5.** Slide *collar C* down until it rests on top of the work. This also provides support for the fingers.

**6.** Apply a wrench to the square section on the top of the body.

**7.** Turn the wrench *gently* in a counterclockwise direction.

### Drilling

If the broken tap is made of carbon steel, it may be possible to drill it out.

### Procedure

**1.** Heat the broken tap to a bright red color, and allow it to cool *slowly.*

**2.** Center-punch the tap as close to the center as possible.

**3.** Using a drill considerably smaller than the distance between opposite flutes, proceed *carefully* to drill a hole through the broken tap.

**4.** Enlarge this hole to remove as much of the metal between the flutes as possible.

**5.** Collapse the remaining part with a punch and remove the pieces.

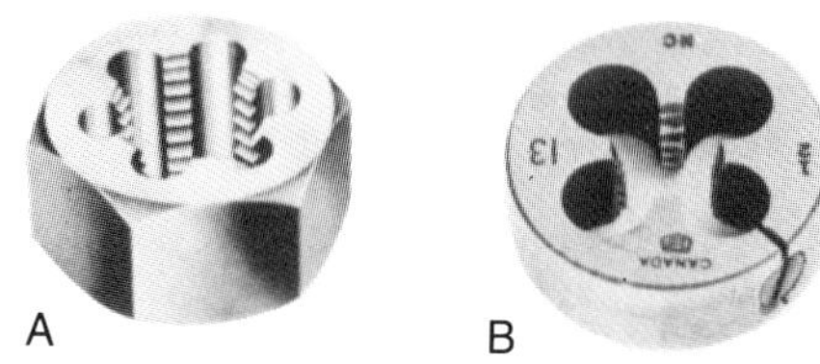


**Fig. 17-9A** Solid die nut.

**Fig. 17-9B** An adjustable round split die.

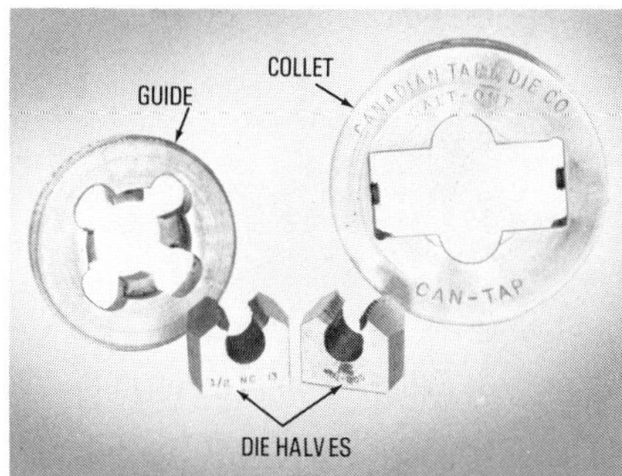


C

**Fig. 17-9C** An adjustable screw plate die.

N-12 Use extreme care when using acid; it can cause serious burns.

N-13 A threading die is a cutting tool used to cut external threads on round material.

### Acid Method

If the broken tap is made of high-speed steel and cannot be removed with a tap extractor, it is sometimes possible to remove it by the acid method.

### Procedure

**1.** Dilute one part nitric acid with five parts water. (See N-12.)

**2.** Inject this mixture into the hole. The acid will act upon the steel and cause the tap to become loose.

**3.** Remove the tap with an extractor or a pair of pliers.

**4.** Wash the remaining acid from the thread with water so that it will not continue to act on the threads.

## THREADING DIES

Threading dies are used to cut external threads on round work. The most common threading dies are the solid, adjustable split, and the adjustable and removable screw plate die.

The *solid die* (Fig. 17-9A) is used for chasing or recutting damaged threads and may be driven by a suitable wrench. It is not adjustable.

The *adjustable split die* (Fig. 17-9B) has an adjusting screw which permits an adjustment over or under the standard depth of thread. This type of die fits into a die stock (Fig. 17-10). (See N-13.)

The *adjustable screw plate die* (Fig. 17-9C) is probably a more efficient die since it provides for greater adjustment than the split die. Two die halves are held securely in a collet by means of a threaded plate which also acts as a guide when threading. The plate, when tightened into the collet, forces the die halves with tapered sides into the tapered slot of the collet. Adjustment is provided by means of two adjusting screws which bear against each die half. The threaded section at the bottom of each die half is tapered to provide for easy starting of the die. Note that the upper side of each die half is stamped with the manufacturer's name, while the lower side of each is stamped with the same serial number. Care should be taken in assembling the die that both serial numbers are facing down. *Never* use two die halves with different serial numbers.

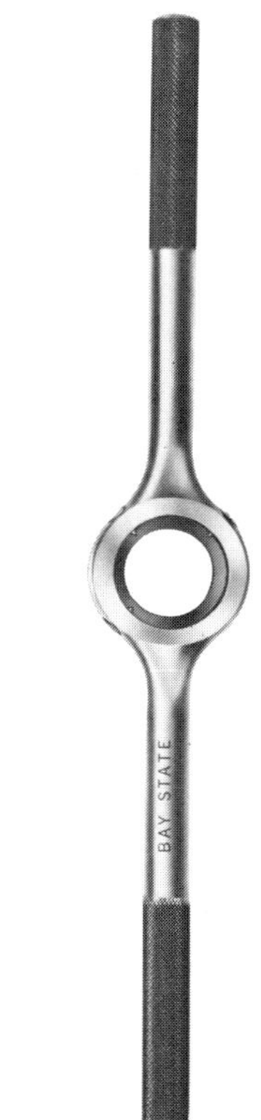


**Fig. 17-10** A die stock is used to turn the die onto the workpiece.

### To Thread with a Hand Die

**1.** Chamfer the end of the workpiece with a file or on the grinder.

**2.** Fasten the work securely in a vise.

**3.** Select the proper die and die stock.

**4.** Lubricate the chamfered end of the workpiece with a suitable cutting lubricant.

**5.** Place the tapered end of the die squarely on the work, Fig. 17-11.

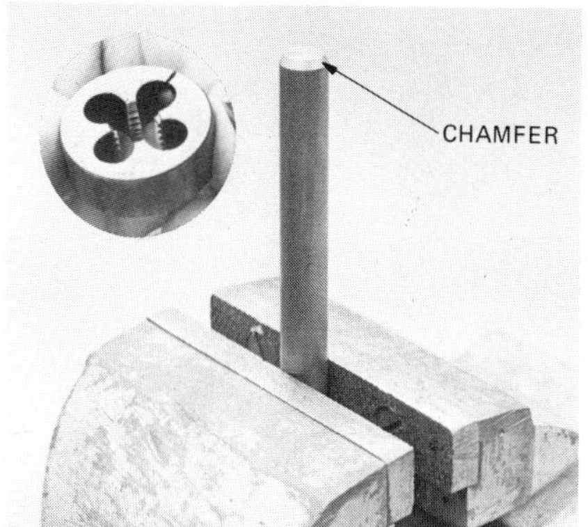


**Fig. 17-11** Hold the die squarely before starting to cut a thread.

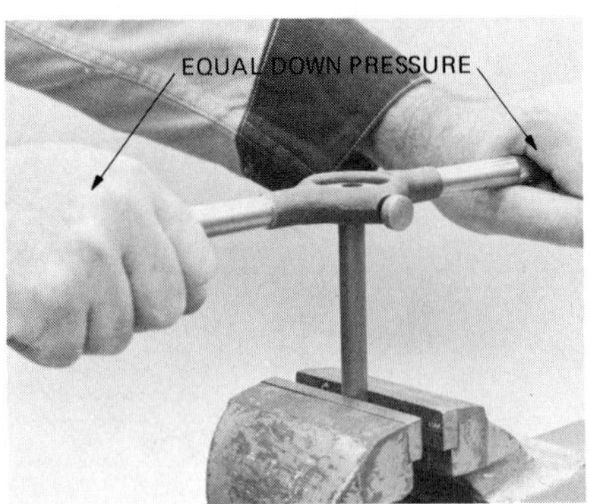


**Fig. 17-12** Apply equal pressure on the handles while starting the die on the workpiece. (*Courtesy Kostel Enterprises Ltd.*)

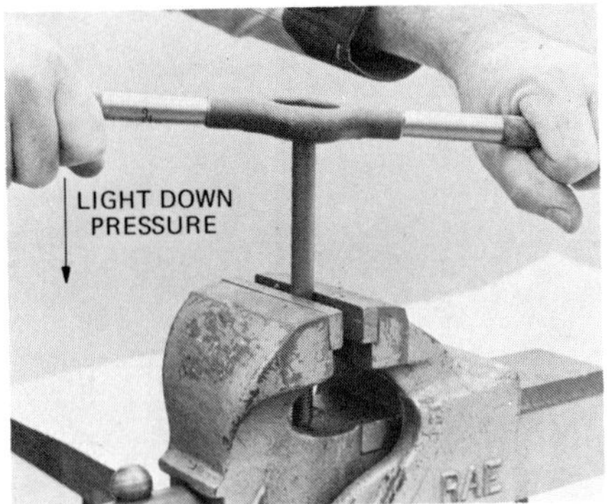


**Fig. 17-13** To square the die on the workpiece apply pressure, while turning, on the opposite side to which it leans. (*Courtesy Kostel Enterprises Ltd.*)

**Fig. 17-14** Place the top of the die (slightly chamfered) in the down position in the die stock. (*Courtesy Kostel Enterprises Ltd.*)

**Fig. 17-15** Thread to within $\frac{1}{16}$ in. (1.59 mm) of the shoulder. (*Courtesy Kostel Enterprises Ltd.*)

**6.** Apply equal pressure on the die-stock handles and turn clockwise several turns, Fig. 17-12.

**7.** Check the die to see that it has started squarely with the work.

**8.** If it is not square, remove the die from the work and restart it squarely by applying pressure on one of the die-stock handles, Fig. 17-13.

**9.** Turn the die forward one turn and then reverse it approximately one-half of a turn to break the chip.

**10.** During the threading process apply cutting fluid frequently. (See N-14.)

**11.** When threading to a shoulder, stop cutting the thread about $\frac{1}{8}$ in. (3.17 mm) from the shoulder, Fig. 17-14.

**12.** Remove the die and reverse it in the die stock with the tapered thread section facing up, Fig. 17-14.

**13.** Start the die on the work and bring it near the bottom of the thread.

**14.** Cut the thread until the die is about $\frac{1}{16}$ in. (1.59 mm) away from the shoulder of the work, Fig. 17-15. (See N-15.)

**15.** Remove the die and test the thread with a nut.

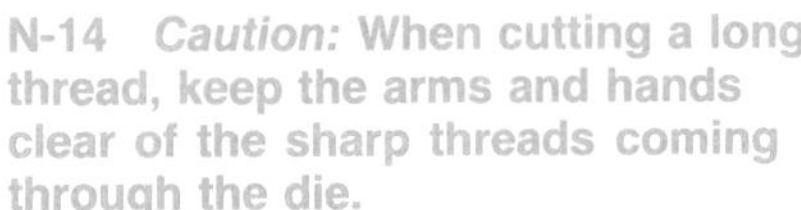
N-14 *Caution:* When cutting a long thread, keep the arms and hands clear of the sharp threads coming through the die.

N-15 *Caution:* If the die hits the shoulder, the work may be bent or the die damaged.

## REVIEW QUESTIONS

1. Define a tap.
2. State three purposes for the flutes on a tap.
3. Name the three taps in a set.
4. State the tap drill formula for
   (a) Inch taps
   (b) Metric taps
5. Calculate the tap drill size for a ¼–20 UNC thread.
6. Explain the procedure for correcting a tap which has not been started squarely.
7. In tapping, why is the tap turned clockwise one-quarter of a turn and then turned counterclockwise?
8. Name three methods of removing broken taps.
9. For what purpose are dies used?
10. Name three types of dies.
11. How should the end of the work be prepared before starting to cut a thread?
12. How is the die started straight on the workpiece?
13. Explain how to correct a die which has not started squarely.
14. How may a thread be cut to a shoulder?
15. What precaution should be observed when threading to a shoulder?
16. What safety precaution should be observed when cutting a long thread on a workpiece?

# UNIT 18

# REAMING AND BROACHING

Hand cutting tools are used to remove metal and bring the workpiecce to the desired shape. Some cutting tools are designed to perform specific operations. *Reamers,* available in a wide variety of types and sizes, are used to bring a hole to size and produce a good surface finish. *Broaches,* generally used in conjunction with an arbor press, are used to produce special shapes in metal. A multitooth cutting tool of the exact size and shape is forced into a hole and it reproduces its shape in the metal.

N-1 A reamer is a cutting tool used to enlarge a drilled or bored hole to accurate dimensions.

## OBJECTIVES

After completing this unit you should be able to:

1. Recognize and know the purpose of the common types of hand reamers
2. Know the purpose of the main parts of a reamer
3. Ream a hole accurately with a hand reamer
4. Understand the principle and purpose of broaching
5. Be able to cut a keyway with a broach using an arbor press

## HAND REAMERS

A hand reamer is a tool used to finish drilled holes accurately and provide a good finish. Reaming is generally performed by machine, but there are times when a hand reamer must be used to finish a hole. Hand reamers, when used properly, will produce holes accurate to size, shape, and finish.

### Standard Reamer Parts

There are several different types of reamers; however, the parts of these reamers are similar. The main parts of a reamer are the *body* and the *shank,* Fig. 18-1A. (See N-1.)

The *body* of the reamer contains several straight or helical grooves called *flutes*. The teeth, or *lands,* are located between these flutes. The *margin* on the top of each land extends from the *body-clearance angle* to the leading edge of each flute. The margin is circle ground with *body clearance* provided behind the back of the margin to reduce the friction while the reamer is cutting.

The *shank,* which is used to turn a hand reamer, has a square end so that a wrench can be used to turn the reamer into the work.

N-2 The expansion reamer is used when a hole must be reamed slightly oversize.

N-3 The adjustable hand reamer is used when a hole must be slightly under- or oversize.

## TYPES OF HAND REAMERS

The *solid hand reamer* (Fig. 18-1B) may be made of carbon steel or high-speed steel. These straight reamers are available in inch sizes of from ⅛ to 1½ in. in diameter and in metric sizes from 1 to 26 mm in diameter. For easy starting, the cutting end of the reamer is ground to a slight taper for a distance equal to the diameter of the reamer. Solid reamers are not adjustable and may have straight or helical flutes. Straight-fluted reamers should not be used on work with a keyway or any other interruption. Since hand reamers are designed to remove only small amounts of metal, no more than .004 to .010 in. (0.10 to 0.25 mm) should be left for reaming, depending on the diameter of the hole. A square on the end of the shank provides the means of driving the reamer with a tap wrench.

The *expansion hand reamer* (Fig. 18-2) is designed to permit an adjustment of approximately .002 in. (0.05 mm) above the nominal diameter. The reamer is made hollow and has slots cut along the length of the cutting section. A tapered threaded plug, fitted into the end of the reamer, provides for limited expansion. If the reamer is expanded too much, it will be easily broken. On *inch expansion hand reamers*, the limit of adjustment is up to .006 in. over the nominal size on reamers up to ½ in. and about .015 in. on reamers over ½ in. The cutting end of the reamer is ground to a slight taper for easy starting. *Metric expansion hand reamers* are available in sizes from 4 to 25 mm. The maximum amount of expansion on these reamers is 1 percent over the nominal size. For example, a 10-mm-diameter reamer can be expanded to 10.1 mm (10 + 1 percent). (See N-2.)

The *adjustable hand reamer* (Fig. 18-3) has tapered slots cut in the body for the entire length. The inner edges of the cutting blades have a corresponding taper so that the blades remain parallel for any setting. The blades are adjusted to size by upper and lower adjusting nuts. The blades on *inch-adjustable hand reamers* have an adjustment range of 1⁄32 in. on the smaller reamers to almost 5⁄16 in. on the larger ones. They are manufactured in sizes ¼ to 3 in. in diameter. (See N-3.)

*Metric-adjustable hand reamers* are available in sizes from #000 (adjustable from 6.4 to 7.2 mm) to #16 (adjustable from 80 to 95 mm).

*Taper reamers* are made to standard tapers and are used to finish tapered holes accurately and smoothly. They may be made with either spiral or straight teeth. Because of its shearing action and its tendency to reduce chat-

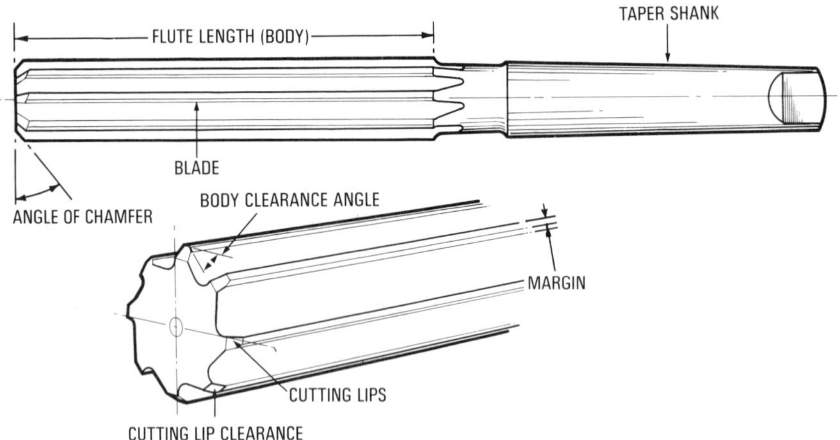


**Fig. 18-1A** The main parts of a reamer.

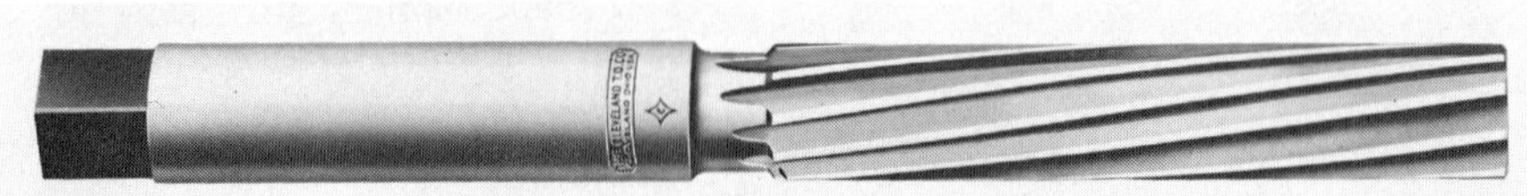

**Fig. 18-1B** A solid hand reamer. (*Courtesy The Cleveland Twist Drill Co.*)

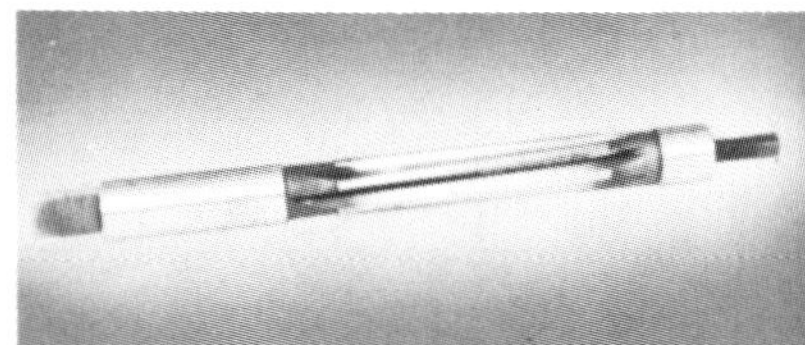

**Fig. 18-2** An expansion hand reamer.

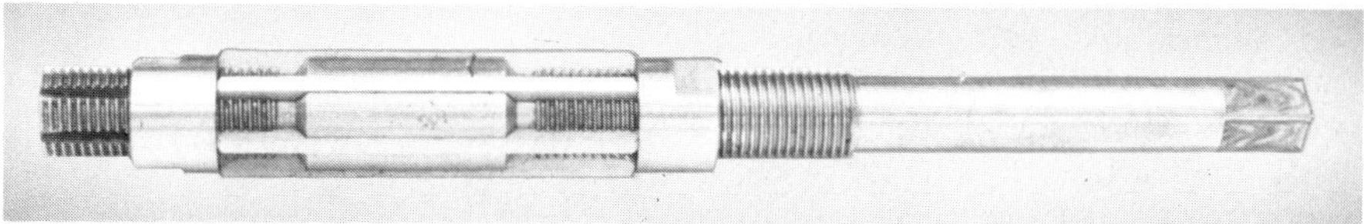

**Fig. 18-3** An adjustable hand reamer.

**Fig. 18-4** A roughing taper reamer.

**Fig. 18-5** A finishing taper reamer.

N-4 Rough out the hole by step drilling before reaming with a tapered reamer.

N-5 The rate of feed should be about one-quarter the diameter of the reamer for each turn.

**Fig. 18-6** For hand reaming, the hole should be .005- to .010-in. (0.12- to 0.25-mm) smaller than the finished hole. (*Courtesy Kostel Enterprises Ltd.*)

ter, the spiral-fluted reamer is superior to the straight one. A *roughing reamer* (Fig. 18-4), with nicks ground at intervals along the teeth, is used for more rapid removal of surplus metal. These nicks or grooves break up the chips into smaller sections; they prevent the tooth from cutting and overloading along its entire length. When a roughing reamer is not available, an old taper reamer is often used prior to finishing the hole with a finishing reamer. (See N-4.)

The *finishing taper reamer* (Fig. 18-5) is used to finish the hole smoothly and to size after the roughing reamer. This reamer, which has either straight or left-hand spiral flutes, is designed to remove only a small amount of metal .010 in. (0.25 mm) from the hole. Since taper reamers do not clear themselves readily, they should be removed frequently from the hole and the chips cleared from the flutes.

## To Ream a Hole with a Straight Reamer

**1.** Check the size of the drilled hole. It should be between .004 to .010 in. (0.10 and 0.25 mm) smaller than the finished hole size (Fig. 18-6).

**2.** Place the end of the reamer into the hole and place the tap wrench on the square end of the reamer.

**3.** Rotate the reamer in a clockwise direction to allow it to align itself with the hole (Fig. 18-7).

**4.** Check the reamer for squareness with the work by testing it with a square at several points around the circumference.

**5.** Brush cutting fluid over the end of the reamer if required.

**6.** Rotate the reamer slowly in a clockwise direction and apply downward pressure. Feed should be fairly rapid and steady to prevent the reamer from chattering. (See N-5.)

## Precautions when Reaming

**1.** Never turn a reamer backward (counterclockwise) because it will dull the cutting teeth.

**2.** Use a cutting lubricant where required.

**3.** Always use a helical-fluted reamer in a hole that has a keyway or oil groove cut in it.

**4.** Never attempt to remove too much material with a hand reamer; about .004 to .010 in. (0.10 to 0.25 mm) should be the maximum.

**5.** Frequently clear a taper reamer (and the hole) of chips.

**6.** Once the reamer is through the workpiece, grasp the tap wrench in the center with one hand.

**7.** Turn the reamer *clockwise* and lift at the same time while removing it from the hole, Fig. 18-8.

# BROACHING

*Broaching* is a process in which a special tapered multitoothed cutter is forced through an opening or along the outside of a piece of work to enlarge or change the shape of the hole or to form the outside to a desired shape.

Broaching was first used for producing internal shapes, such as keyways, splines, and other odd internal shapes (Fig. 18-9). Its application has been extended to exterior surfaces, such as the flat face on automotive engine blocks and cylinder heads. Most broaching is now performed on special machines which either pull or push the broach through or along the material. Hand broaches are used in the machine shop for such operations as keyway cutting. (See N-6.)

N-6 A broach is a multitoothed cutting tool used to produce flat surfaces or various shapes on internal or external surfaces in one pass.

The cutting action of a broach is performed by a series of hardened and ground teeth of the desired shape, each protruding about .003 in. (0.07 mm) farther than the preceding tooth, Fig. 18-10. The last three teeth are generally of the same depth and provide the finish cut.

Broaching has many advantages and an extremely wide range of applications.

**1.** It is possible to machine almost any irregular shape, provided that it is parallel to the broach axis.

**2.** It is rapid: the entire machining process is usually completed in one pass.

**3.** Roughing and finishing cuts are generally combined in the same operation.

**4.** A variety of forms, either internal or external, may be cut simultaneously and the entire width of a surface may be machined in one pass.

## Cutting a Keyway with a Broach

Keyways may be cut by hand in the machine shop quickly and accurately by means of a broach set and an arbor press (Fig. 18-11).

A broach set (Fig. 18-12) covers a wide range of keyways and is a particularly useful piece of equipment when many keyways must be cut. The equipment necessary to cut a keyway is a bushing to suit the hole size in the workpiece, a broach the size of the keyway to be cut, and shims to increase the depth of the cut of the broach.

**1.** Determine the keyway size required for the size of the workpiece.

**2.** Select the proper broach, bushing, and shims.

**3.** Place the workpiece on the arbor press. Use an opening on the base

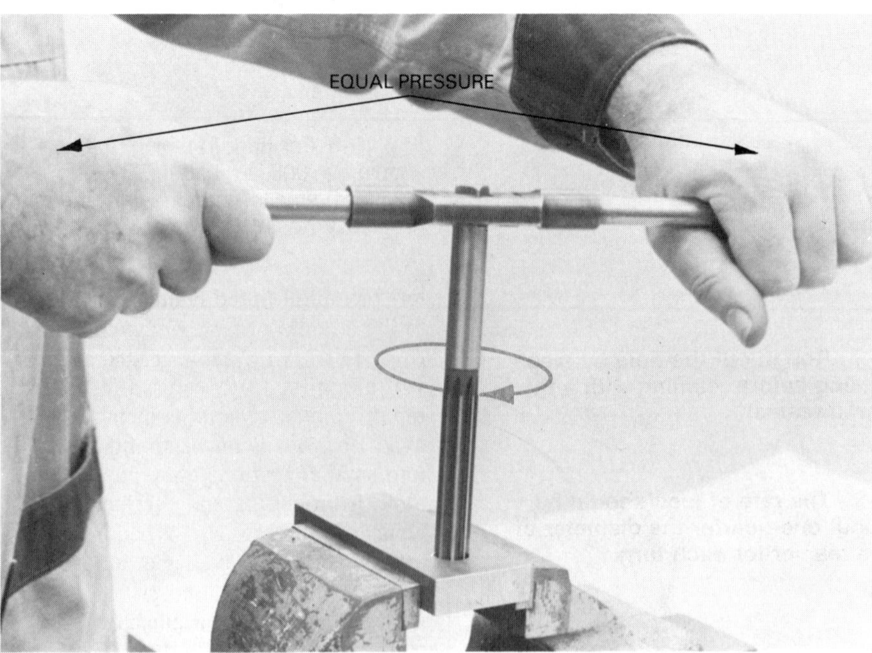


**Fig. 18-7** Start the reamer with equal down pressure on both handles of the tap wrench and rotate in a clockwise direction. (*Courtesy Kostel Enterprises Ltd.*)

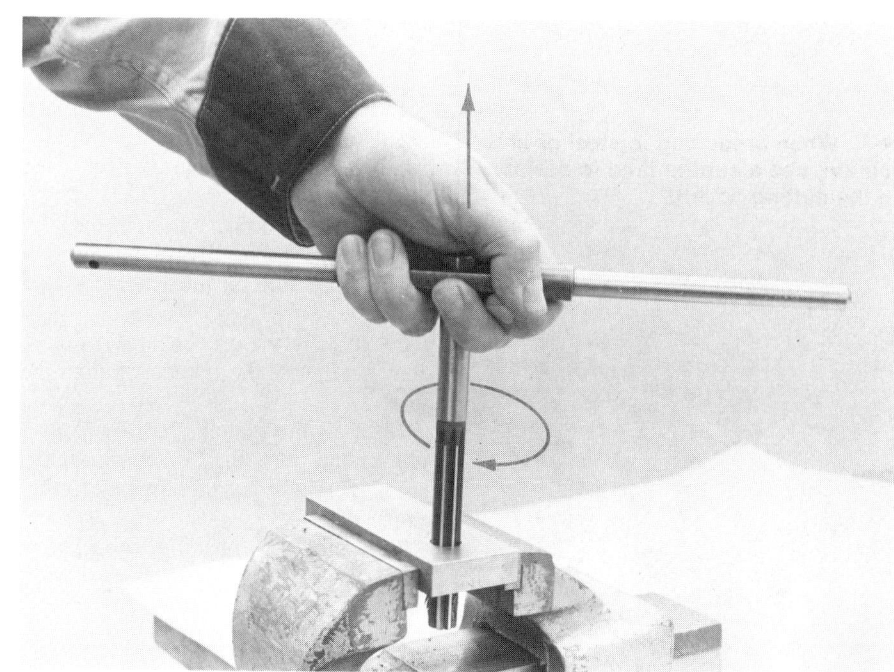

**Fig. 18-8** Remove the reamer by turning clockwise and lifting it upward at the same time. (*Courtesy Kostel Enterprises Ltd.*)

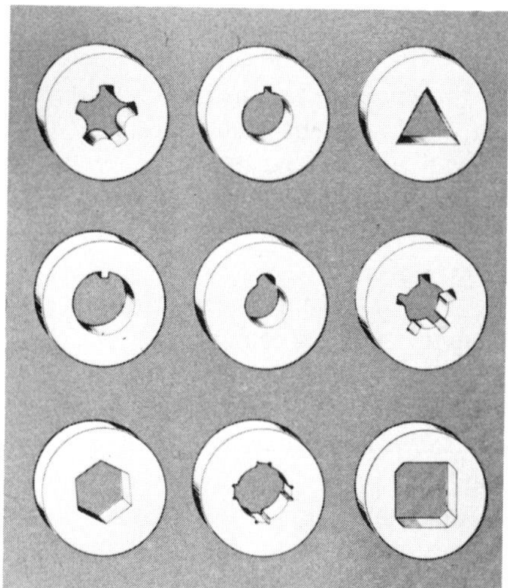

**Fig. 18-9** Examples of internal broaching.

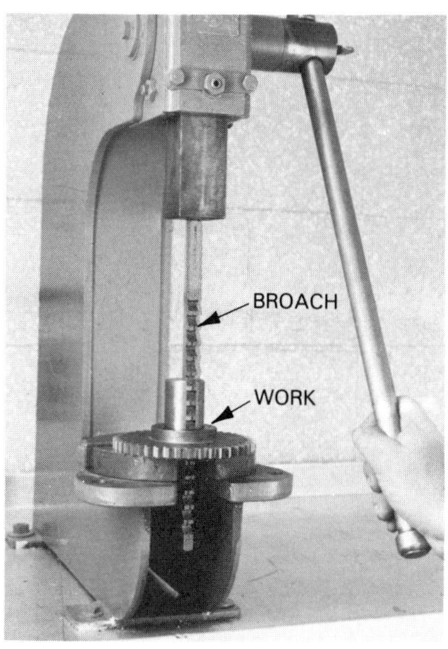


**Fig. 18-11** Using an arbor press to cut a keyway with a broach.

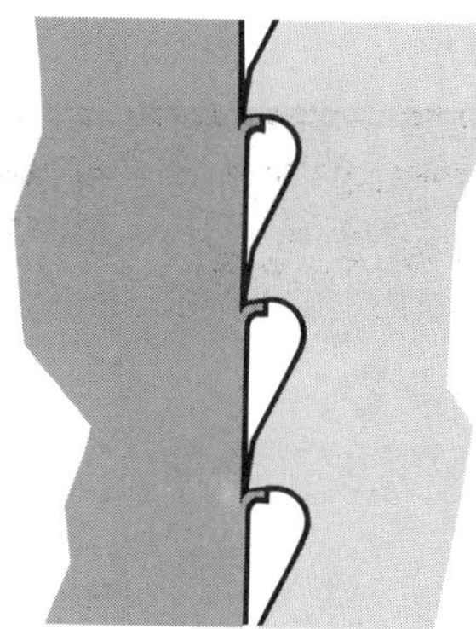


**Fig. 18-10** The cutting action of a broach.

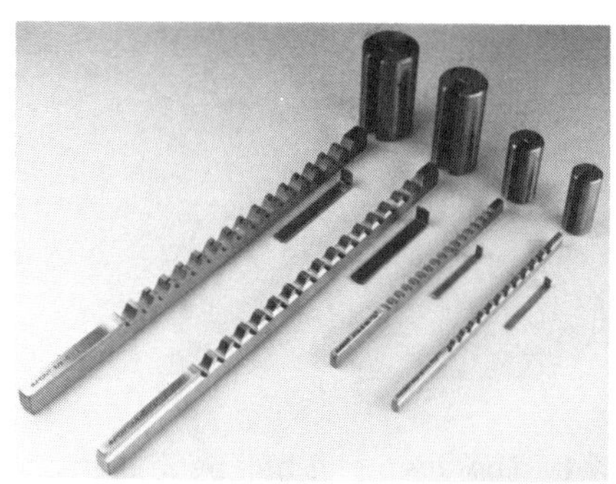

**Fig. 18-12** A broach set for cutting internal keyways.

smaller than the opening in the workpiece so that the bushing will be properly supported.

**4.** Insert the bushing and the broach into the opening. Apply cutting fluid if the workpiece is made of steel. (See N-7.)

N-7 When broaching in steel or aluminum, use a cutting fluid to assist in the cutting action.

**5.** Press the broach through the workpiece (Fig. 18-13), maintaining constant pressure on the arbor-press handle.

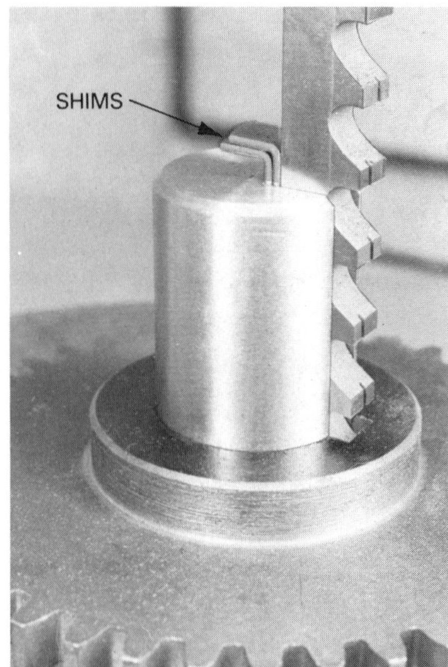

**Fig. 18-13** Two shims are used when making the final pass with the broach.

**6.** Remove the broach, insert one shim, and press the broach through the hole.

**7.** Insert the second shim, if required, and press the broach through again. This will cut the keyway to the proper depth.

**8.** Remove the bushing, broach, and shims.

## REVIEW QUESTIONS

**1.** Define a hand reamer.

**2.** State the purpose of the
 (a) Expansion hand reamer
 (b) Adjustable hand reamer
 (c) Taper reamer

**3.** How much metal should be removed with a hand reamer?

**4.** What direction must the reamer be turned when removing it from a hole?

**5.** What is the main feature of a roughing tapered reamer?

**6.** List three important precautions that should be observed when reaming.

**7.** What is the purpose of broaching?

**8.** Name three operations that may be performed by broaching.

**9.** List four advantages of broaching.

**10.** Briefly describe the procedure for broaching a keyway on an arbor press.

# UNIT 19

# METAL FASTENERS

In machine shop work, many different methods are used to fasten work together. Some of the most common methods such as riveting and fastening with screws and dowel pins, are briefly described.

## OBJECTIVES

After completing this unit you should be able to:

**1.** Identify some of the common metal fasteners

**2.** Fasten pieces of metal together by riveting

**3.** Understand the purpose of dowel and taper pins for alignment purposes

## RIVETING

A rivet is a metal pin made of soft steel, brass, copper, or aluminum, with a head at one end. The two most common types are the round head and the countersunk head (Fig. 19-1). Riveting consists of placing a rivet through holes in two or more pieces of metal and then forming a head on the other end of the rivet with a ball-peen hammer.

### To Rivet

**1.** Drill holes in the metal pieces .015 to .030 in. (0.4 to 0.8 mm) larger than the body of the rivet. If a countersunk rivet is being used, countersink for the head.

**2.** Insert a rivet through the holes, having it extend past the work 1.5 times the diameter of the rivet.

**3.** Place round head rivets in a metal block recessed for the shape of the head (Fig. 19-2). Countersunk rivets may be placed on a flat block.

**4.** Use a ball-peen hammer to form a head on the body of each rivet. (See N-1.)

N-1 The peen of a ball peen hammer is used to form a rivet head.

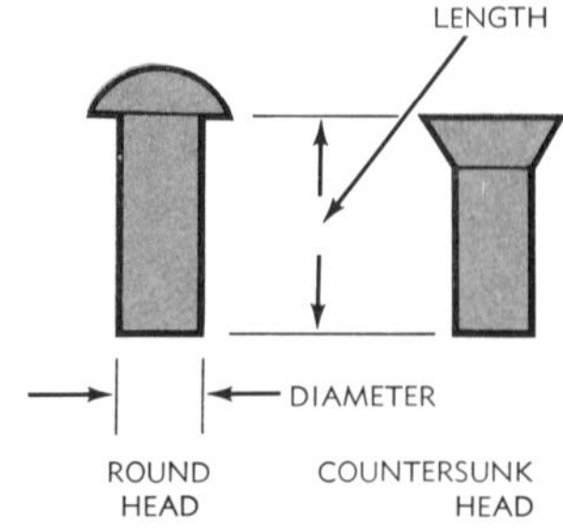

**Fig. 19-1** Rivets are often used to fasten parts together.

## Machine Screws

Machine screws are widely used by the machinist for assembly work. A hole slightly larger than the body size of the screw may be drilled through the workpieces, the screw inserted, and a nut placed on the end of it. Another method of using machine screws is to drill a clearance hole through one piece and drill and tap the other piece to fit the thread of the screw. Some of the more common screws are shown in Fig. 19-3.

Work being fastened by flat head and socket head screws must be recessed so that the head of the screw is flush with the surface of the work. Flat head screws are recessed with an 82° countersink (Fig. 19-4). To recess work for socket head screws (Fig. 19-5) a counterbore or flat bottom drill is used.

## Dowel and Taper Pins

Dowel pins (Fig. 19-6A) are used to locate one piece of work accurately with another. (See N-2.) A hole is drilled and reamed through both workpieces simultaneously and then the dowel pin is forced into the hole to keep both pieces in alignment. A tapered hole must be reamed to accommodate taper pins (Fig. 19-6B). (See N-3.)

N-2 Dowel pin holes should be reamed to a size to provide a push fit for the dowel.

N-3 Tapered pins are sized from a small diameter 7/0 to no. 14 which is about 3/64 to 1¼ in.

## Self-tapping Screws

*Self-tapping, and thread-cutting screws,* Fig. 19-7, are designed to cut a thread and therefore eliminate the need for tapping a hole. These screws are generally used when assembling metal parts, nonferrous materials, and plastics. *Thread-forming screws* produce a thread in a hole by displacing material and not by cutting. Thread-forming screws require more pressure or torque to produce a thread and therefore are more often used to assemble thin workpieces.

Self-tapping screws cut a mating thread in the work which fits snugly to the body of the screw. This close-fitting thread keeps the screw from backing off or coming loose even under vibrating conditions. The following are some advantages of self-tapping screws.

**1.** The assembly of parts costs less because pretapping is not necessary.
**2.** Tapping screws are hardened and have greater strength than machine screws.
**3.** They produce a tight thread that fits close to the screw body, thereby reducing the possibility of loose fits.
**4.** They can be used in steel, aluminum, brass, and fiber glass.

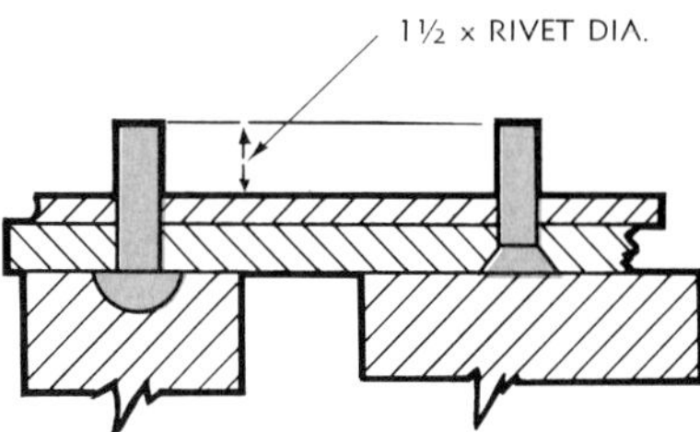


**Fig. 19-2** Work set up for riveting.

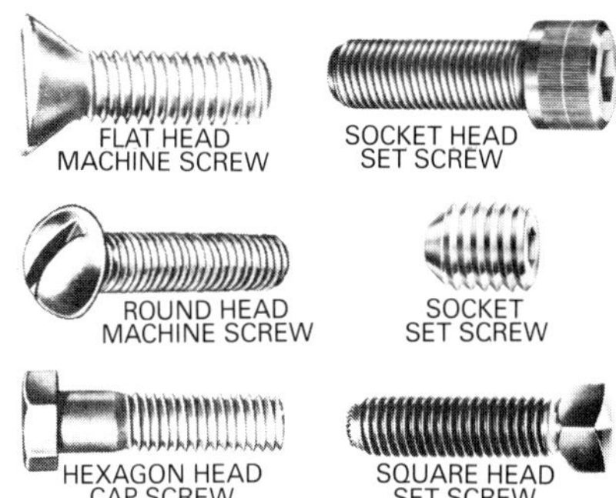


**Fig. 19-3** Thread fasteners are made in a variety of types and sizes.

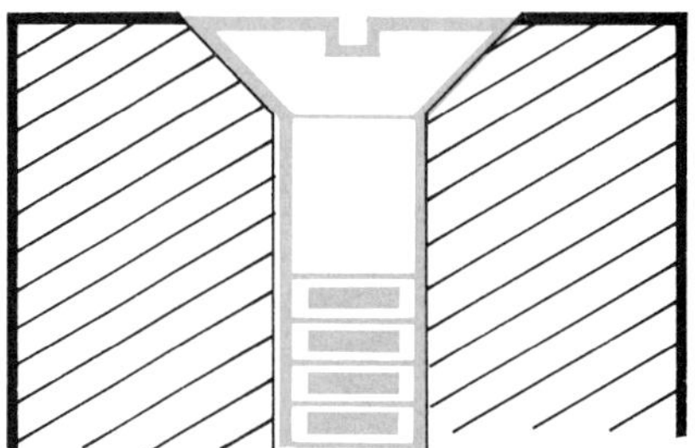

**Fig. 19-4** The work is countersunk properly when the head of the screw is flush with the work surface.

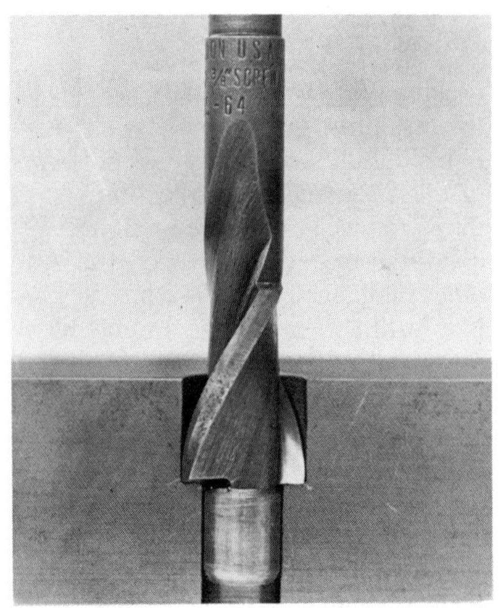

Fig. 19-5 Work must be counterbored when socket head screws are used.

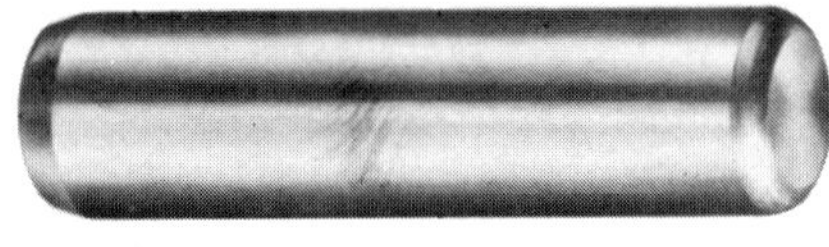

Fig. 19-6 Dowel and tapered pins are used to locate one piece with another.

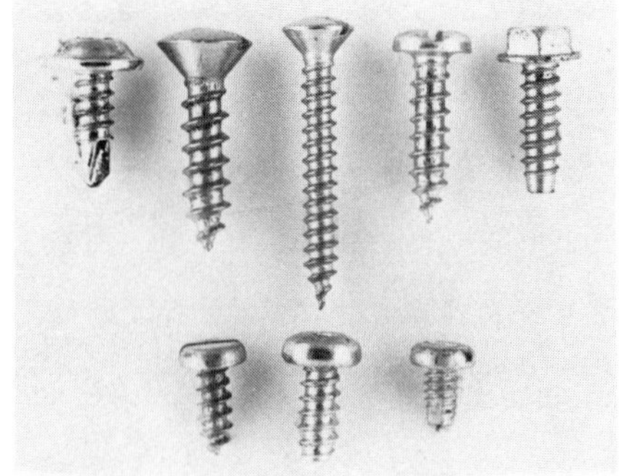

Fig. 19-7 A variety of self-tapping screws. (*Courtesy Kostel Enterprises Ltd.*)

## REVIEW QUESTIONS

**1.** Name two common types of rivets.

**2.** How large should holes for rivets be drilled in a workpiece?

**3.** For what purpose are the following used?

(a) A tapered pin

(b) A self-tapping screw

**4.** Name the tools used to recess the work when using

(a) Flat head screws

(b) Socket head screws

**5.** Briefly describe the procedure for riveting two pieces of metal together.

# SECTION 5

## JOB PLANNING

achine shop work consists of machining a variety of parts (round, flat, contour) and either assembling them into a single unit or using each of them separately to perform some operation. It is important that the sequence of operations be carefully planned in order to produce a part quickly and accurately. Improper planning or following a wrong sequence of operations can often result in spoiled work.

# UNIT 20

# ENGINEERING DRAWINGS

Engineering drawing is the common language by which drafters, tool designers, and engineers indicate to the machinist and toolmaker the physical requirements of a part. Drawings are made up of a variety of lines which represent surfaces, edges, and contours of a workpiece. By adding symbols, dimension lines and sizes, and word notes, the drafter can indicate the exact specifications of each individual part. Students must be familiar with this language of industry so that they can interpret lines and symbols in order to produce the parts quickly and accurately.

A complete product is usually shown on an *assembly drawing* by the drafter. Each part or component of the product is then shown on a *detailed drawing*, which is reproduced in copies called *prints*. The prints are used by the machinist or toolmaker to produce the individual parts which eventually will make up the complete product. Some of the more common drafting lines and symbols will be reviewed briefly. However, it must be assumed that the student already has the basic knowledge of print reading or that a good book on this subject is available for review purposes.

## OBJECTIVES

After completing this unit you will be able to

**1.** Understand the meaning of the various lines used in engineering drawings

**2.** Recognize the various symbols used to convey information

**3.** Read and understand engineering drawings or prints

## TYPES OF DRAWINGS AND LINES

### Orthographic Projection

In order to accurately describe the shape of noncylindrical parts on a drawing or print, the drafter uses the orthographic view or projection method. The orthographic view shows the part from three sides—the front, the top, and the right-hand side, Fig. 20-1. These three views enable the drafter to completely describe a part or

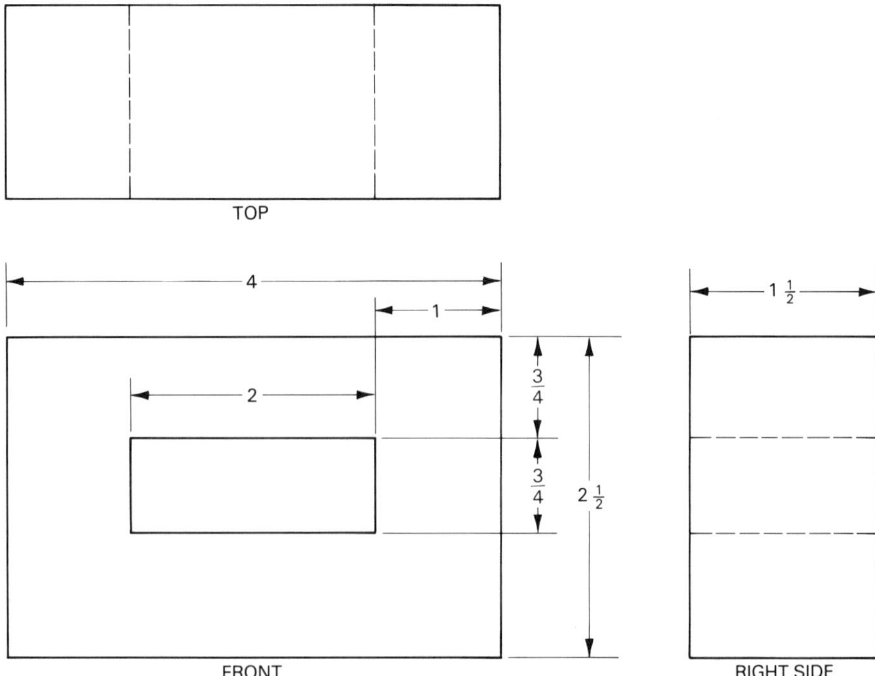


**Fig. 20-1** The three views of orthographic projection make it easier to describe the details of a part.

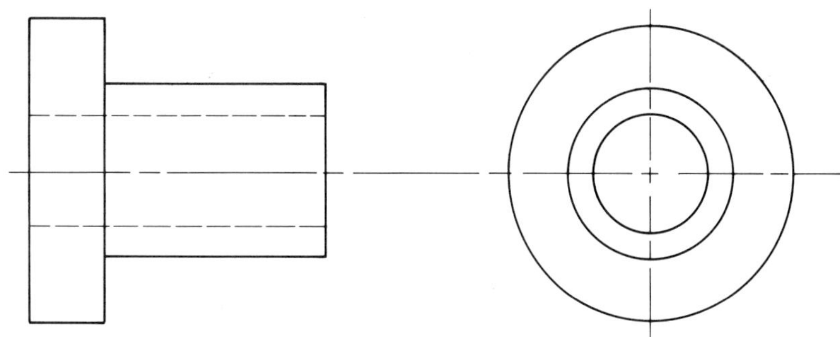

**Fig. 20-2** Cylindrical parts are generally shown in two views.

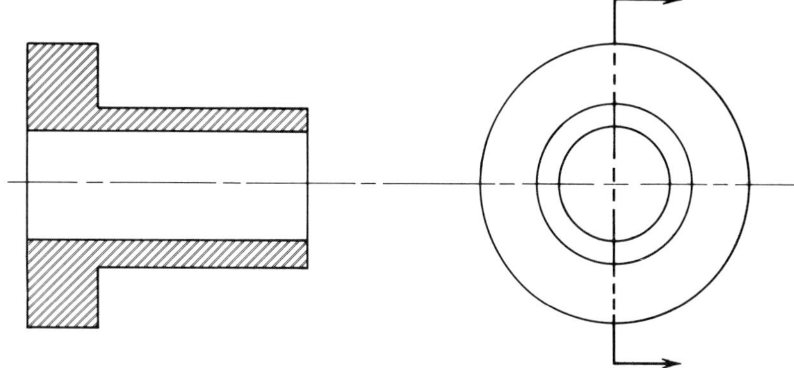

**Fig. 20-3** Section views are used to show complicated interior forms.

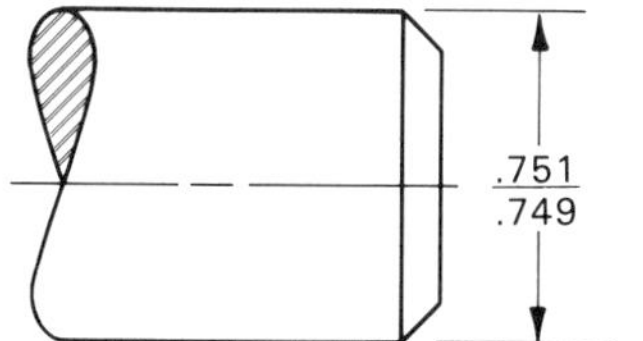


**Fig. 20-4** Limits show the largest and smallest size of a part.

.750 + .001 − .003

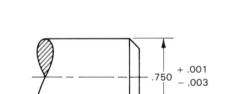

**Fig. 20-5** Tolerance is the permissible variation of a size.

object so that the machinist knows exactly what is required.

Cylindrical parts are generally shown on prints in two views—the front view and the right-side view, Fig. 20-2. However, if a part contains many details, it may be necessary to use the top, bottom, or left-side views to accurately describe the part to the machinist.

### Section Views

In many cases, complicated interior forms are difficult to describe in the usual manner by a drafter. Whenever this occurs a section view, which is obtained by making an imaginary cut through an object, is made. This imaginary cut can be made in a straight line in any direction to best expose the interior contour or form of a part, Fig. 20-3.

### Types of Lines

A wide variety of standard lines are used in engineering drawings in order for the designer to indicate to the machinist exactly what is required. Thick, thin, broken, wavy, and section lines are used on shop or engineering drawings. See Table 20-1 for an example, description, and purpose of some of the more common lines used on shop drawings.

## DRAFTING TERMS AND SYMBOLS

Common drafting terms and symbols are used on shop and engineering drawings in order for the designer to accurately describe each part. If it were not for the universal use of terms, symbols, and abbreviations the designer would have to make extensive notes describing exactly what is required. These not only would be cumbersome but could be misunderstood and therefore result in costly errors. Some of the common drafting terms and symbols are explained.

*Limits* (Fig. 20-4) are the largest and the smallest permissible dimensions of a part (the maximum and minimum dimensions). On a shop drawing both sizes would be given.

**Example:**

.749 smallest dimension
.751 largest dimension

*Tolerance* (Fig. 20-5) is the permissible variation of the size of a part. On a drawing the basic dimension is given plus or minus to variation allowed.

**TABLE 20-1 COMMON LINES USED ON SHOP DRAWINGS**

| Example | Name | Description | Use |
|---|---|---|---|
| a | Object lines | Thick black lines approximately 1/32 in. wide (the width may vary to suit drawing size) | Indicate the visible form or edges of an object |
| b | Hidden lines | Medium weight black lines of 1/8 in. dashes and 1/16 in. spaces | Indicate the hidden contours of an object |
| c | Center lines | Thin lines with alternating long lines and short dashes. Long lines from 1/2 to 3 in. long Short dashes 1/16 to 1/8 in. long, spaces 1/16 in. long | Indicate the centers of holes, cylindrical objects, and other sections |
| $1\frac{1}{2}$ d | Dimension lines | Thin black lines with an arrowhead at each end and a space in the center for a dimension | Indicate the dimension of an object |
| e | Cutting plane lines | Thick black lines make up a series of one long line and two short dashes; arrowheads show the line of sight from where the section is taken | Show where a section is imagined cut |
| f | Cross-section lines | Fine evenly spaced parallel lines at 45°; line spacing is in proportion to the part size | Show the surfaces exposed when a section is cut |

**Example:**

$$.750 \begin{array}{l} +\ .001 \\ -\ .003 \end{array}$$

The tolerance in this case would be .004 (the difference between +.001 oversize and −.003 undersize).

*Allowance* (Fig. 20-6) is the intentional difference in the sizes of mating parts such as the diameter of a shaft and the size of the hole. On a shop drawing both the shaft and the hole are indicated with maximum and minimum sizes in order to produce the best fit.

*Scale size* is necessary because it is impossible to draw parts to the exact size since some drawings would be too large while others may be too small. The scale size of a drawing is generally found in the title block section and it indicates the scale to which the drawing has been made.

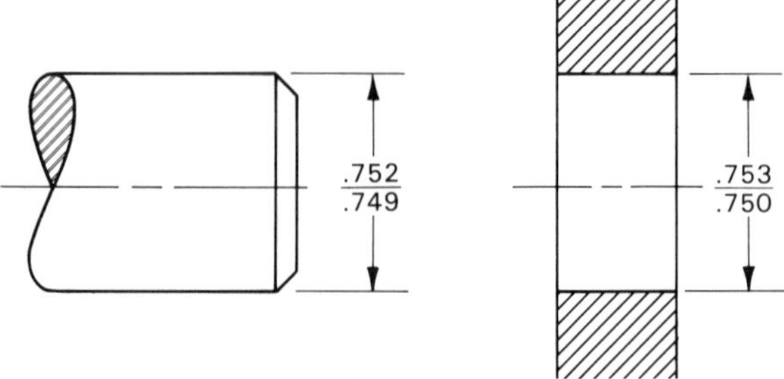


**Fig. 20-6** Allowance is the intentional difference in the sizes of mating parts.

| Common Machine Shop Abbreviations | |
|---|---|
| CBORE | Counterbore |
| CSK | Countersink |
| DIA | Diameter |
| Ø | Diameter |
| HDN | Harden |
| LH | Left hand |
| UNC | Unified national coarse |
| mm | Millimeter |
| NC | National coarse |
| NF | National fine |
| P | Pitch |
| R | Radius |
| RH | Right hand |
| THD | Thread or threads |

| Scale | Definition |
|---|---|
| 1:1 | Drawing is made to the actual size of part or full scale |
| 1:2 | Drawing is made one-half the actual size of the part |
| 2:1 | Drawing is made twice the actual size of the part |

## Symbols

Some of the symbols and abbreviations used on shop drawings indicate the surface finish, type of material, roughness symbols, and common machine shop terms and operations.

### Surface Symbols

The surface-finish mark, used in many cases, indicates which surface of the part must be finished. The number in the $\surd$ indicates the quality of finish required on the surface. In this case the *roughness height* of the surface cannot exceed 40 microinches ($\mu$ in.)

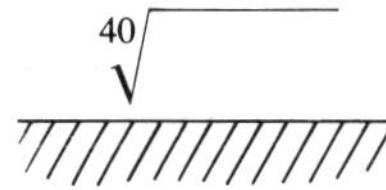


If the surface of a part must be finished to exact specifications, each part of the specification is indicated on the symbol.

0.002
40 $\surd$ 0.001
⊥

| | |
|---|---|
| 40 | Surface finish, in microinches |
| 0.002 | Waviness height, in thousandths |
| 0.001 | Roughness width, in thousandths |
| ⊥ | Machining marks run perpendicular to the boundary of the surface indicated |

The following symbols indicate the direction of lay (marks produced by machining operations on work surfaces).

| | |
|---|---|
| = | Parallel to the boundary line of the surface indicated by the symbol |
| × | Angular in both directions on the surface indicated by the symbol |
| M | Multidirectional |
| C | Approximately circular to the center of the surface indicated by the symbol |
| R | Approximately radial to the center of the surface indicated by the symbol |

### Material Symbols

Represents copper, brass, bronze, etc.

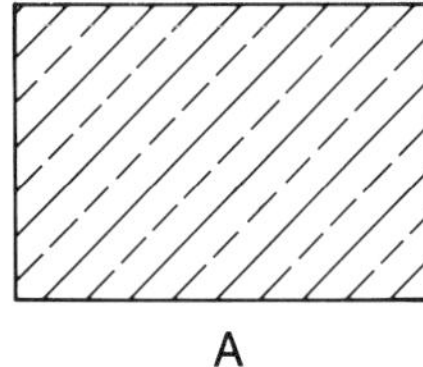

A

Represents aluminum, magnesium, and their alloys

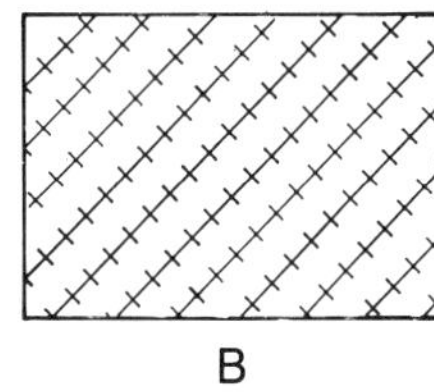

B

Represents steel and wrought iron

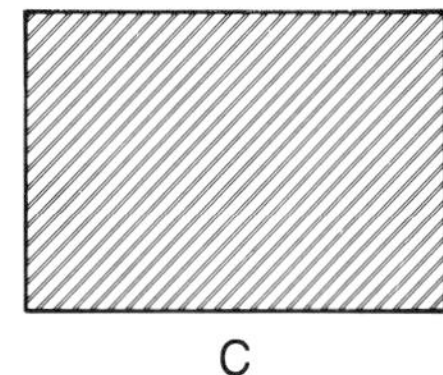

C

Represents cast iron and malleable iron

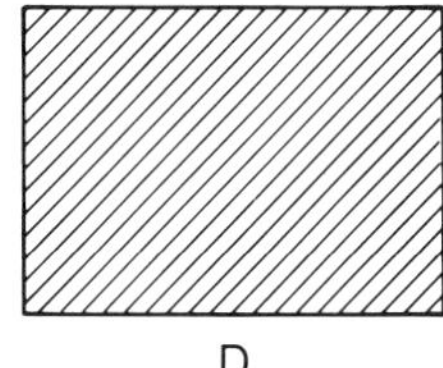

D

**Fig. 20-7** Drafting symbols for various types of materials.

## REVIEW QUESTIONS

**1.** How can a drafter indicate the exact specifications required on a part?

**2.** What is the purpose of
(a) An assembly drawing
(b) A detailed drawing

**3.** What is the purpose of an orthographic view?

**4.** Why are section views shown?

**5.** What lines are used to show
(a) The form of a part
(b) The centers of holes, objects, or sections
(c) The exposed surfaces of where a section is cut

**6.** Define
(a) Limits
(b) Tolerance
(c) Allowance

**7.** How is half scale indicated on an engineering drawing?

**8.** Define each part of the surface-finish symbol.

$$ {}_{60}\sqrt{\frac{0.003}{0.\underline{\underline{00}}2}} $$

---

**9.** What do the following abbreviations mean?
(a) CBORE
(b) HDN
(c) mm
(d) THD

# UNIT 21

# MACHINING PROCEDURES FOR VARIOUS WORKPIECES

Planning the procedures for machining any part is very important so that the part can be machined accurately and as quickly as possible. Many parts have been spoiled because the incorrect sequence has been followed in the machining process. Although it would be impossible to list the exact sequence of operations that would apply to every type and shape of workpiece, some general rules should be followed to machine a part accurately and in the shortest time possible.

## OBJECTIVES

After completing this unit, you should be able to:

**1.** Plan the sequence of operations and machine round work mounted between lathe centers

**2.** Plan the sequence of operations and machine round work mounted in lathe chuck

**3.** Plan the sequence of operations and machine flat workpieces

## MACHINING PROCEDURES FOR ROUND WORK

The largest percentage of work machined in a machine shop is round and is turned to size on a lathe. In industry much of the round work is held in a chuck, while a larger percentage of work in school shops is machined between centers because of the need to reset work more often. In either case it is important to follow the correct machining sequence of operations in order to prevent spoiling work, which so often happens when incorrect procedures are followed.

## GENERAL RULES FOR ROUND WORK

**1.** Rough turn all diameters to within 1/32 in. (0.79 mm) of the size required.

- Machine the largest diameter first and progress to the smallest.
- If the small diameters are rough turned first, it is quite possible that the work would bend when the large diameters are machined.

**2.** Rough turn all steps and shoulders to within 1/32 in. (0.79 mm) of the length required.

- Be sure to measure all lengths from one end of the workpiece.
- If all measurements are not taken from the end of the workpiece, the length of each step would be 1/32 in. (0.79 mm) shorter than required. If four steps were required, the length of the fourth step would be 1/8 in. (3.17 mm) shorter than required [4 × 1/32 in. (4 × 0.79 mm)] and would leave too much material for the finishing operation.

**3.** If any special operations such as knurling or grooving are required, they should be done next.

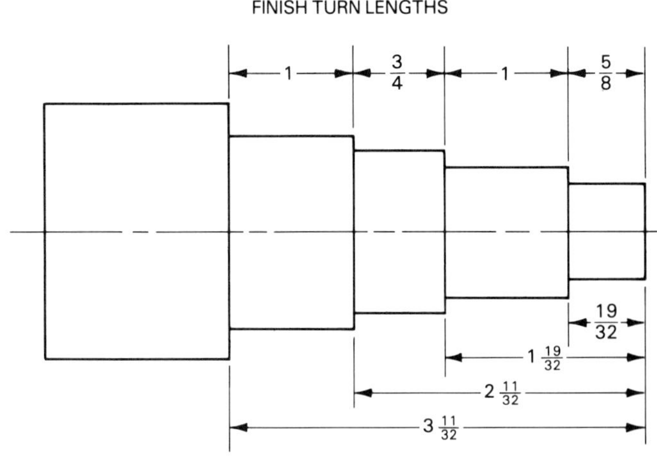


**Fig. 21-1** A sample part showing rough and finish turn lengths.

**4.** Cool the workpiece before starting the finishing operations. (See N-1.)

- Metal expands due to the friction caused by the machining process and any measurements taken while work is hot are incorrect.
- When the workpiece is too cold, the diameters of round work will be smaller than required.

**5.** Finish turn all diameters and lengths.

- Finish the largest diameters first and work down to the smallest diameter.
- Finish the shoulder of one step to the correct length and then cut the diameter to size.

N-1 All finish measurements should be taken when the parts are at room temperature (68°F or 20°C).

N-2 Knurling is a forming operation and takes a certain amount of pressure to form the pattern which might bend the workpiece. Since a knurl is used only for grip, it really does not matter if that diameter is concentric with the other diameters.

## WORKPIECES REQUIRING CENTER HOLES

There are times when it is necessary to machine the entire length of a round workpiece. When this is required, usually on shorter workpieces, center holes are drilled in each end. The workpiece shown in Fig. 21-2 is a typical part which can be machined between the centers on a lathe.

### Machining Sequence

**1.** Cut off a piece of steel ⅛ in. (3.17 mm) longer and ⅛ in. (3.17 mm) larger in diameter than required.

- In this case the diameter of the steel cut off would be 1⅝ in. and its length would be 16⅛ in.

**2.** Hold the work in a three-jaw chuck, face one end square, and then drill the center hole.

**3.** Face the other end to length and then drill the center hole.

**4.** Mount the workpiece between the centers on a lathe.

**5.** Rough turn the largest diameter to within 1/32 in. (0.79 mm) of finish size or 1 17/32 in.

**6.** Finish turn the diameter which is to be knurled. (See N-2.)

**7.** Knurl the 1½-in. diameter.

**8.** Machine the 45° chamfer on the end.

**9.** Reverse the work in the lathe, protecting the knurl from the lathe dog with a piece of soft metal.

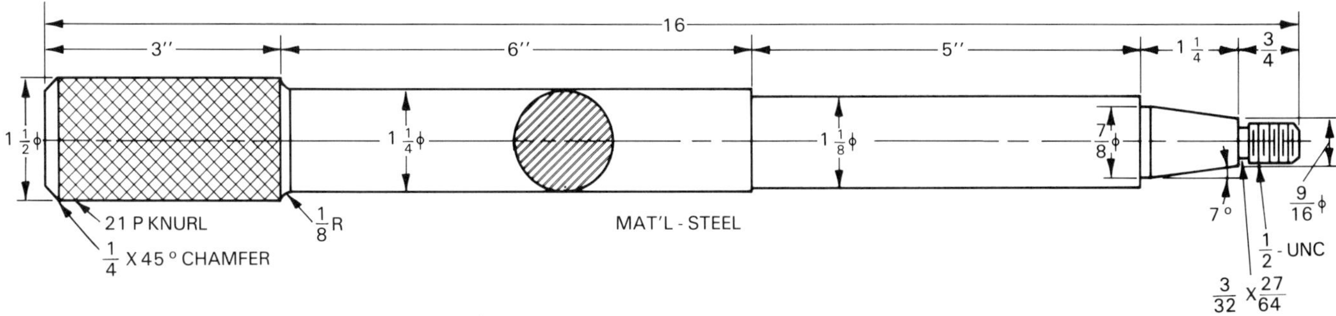


**Fig. 21-2** Sample of a round shaft which can be machined.

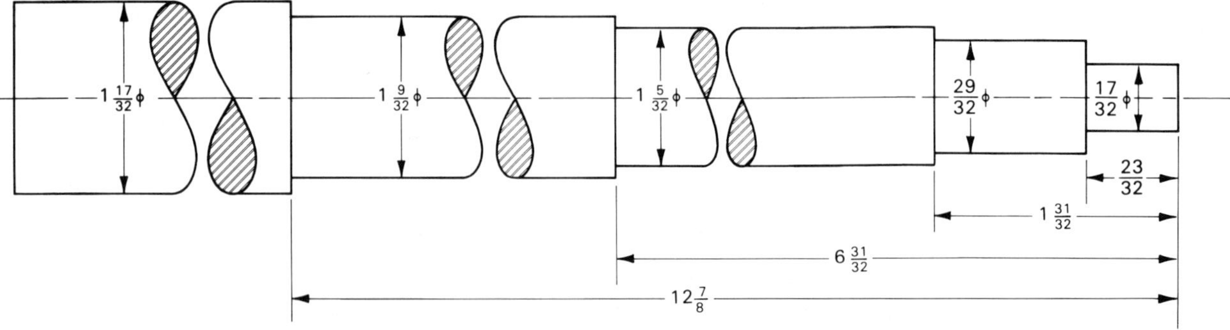


**Fig. 21-3** The rough-turned diameters and lengths on the shaft.

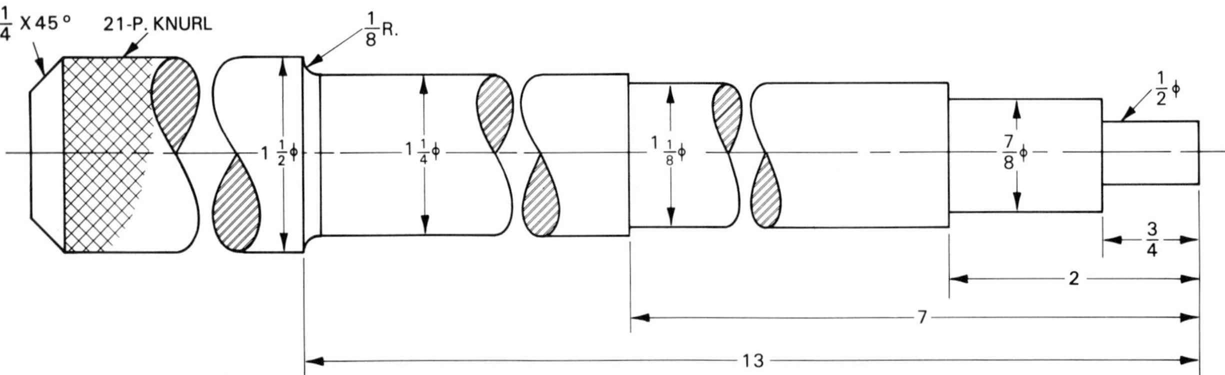


**Fig. 21-4** The shaft finish turned to diameter and length.

N-3 Heat causes material to expand (grow larger), so if diameters are finished to size while the work is hot, the diameters will be undersize when the work cools to room temperature.

N-4 Hone the cutting edge of the finish turning tool for better cutting action and smoother surface finishes.

N-5 Use cutting fluid and a steady infeed to prevent bending the work while grooving.

**10.** Rough turn the 1¼-in. diameters to 1 9/32 in. (Fig. 21-3).

- Be sure to leave the length of this section ⅛ in. (3.17 mm) short (12⅞ in. from the end) to allow for finishing the ⅛-in. (3.17-mm) radius.

**11.** Rough turn the 1⅛-in. diameter to 1 5/32 in.

- Leave the length of this section 1/32 in. (0.79 mm) short (6 31/32 in. from the end) to allow for finishing the shoulder.

**12.** Rough turn the ⅞-in. diameter to 29/32 in.

- Leave the length of this section 1/32 in. (0.79 mm) short (1 31/32 in. from the end) to allow for finishing the shoulder.

**13.** Rough turn the ½-in. (12.7-mm) diameter to 17/32 in.

- Machine the length of this section to 23/32 in.

**14.** Cool the work to room temperature before starting the finishing operations. (See N-3.)

**15.** Finish turn the 1¼-in. diameter to 12⅞ in. from the end.

**16.** Mount a ⅛-in. (3.17-mm) radius tool and finish the corner to the correct length. (See N-4.)

**17.** Finish turn the 1⅛-in. diameter to 7 in. from the end (Fig. 21-4).

**18.** Finish turn the ⅞-in. diameter to 2 in. (50.8 mm) from the end.

**19.** Set the compound rest to 7° and machine the taper to size.

**20.** Finish turn the ½-in. diameter to ¾ in. from the end.

**21.** With a cutoff tool cut the groove at the end of the ½-in. diameter. (See N-5.)

**22.** Chamfer the end of the section to be threaded (Fig. 21-5).

**23.** Set the lathe for threading and cut the thread to size.

## WORKPIECES HELD IN A CHUCK

The procedure for machining the external surfaces of round workpieces held in a chuck (three-jaw, four-jaw, collet, etc.) is basically the same as for machining work held between lathe centers. However, if both external and internal surfaces must be machined on work held in a chuck, the sequence of some operations is changed.

Whenever work is held in a chuck for machining, it is very important that

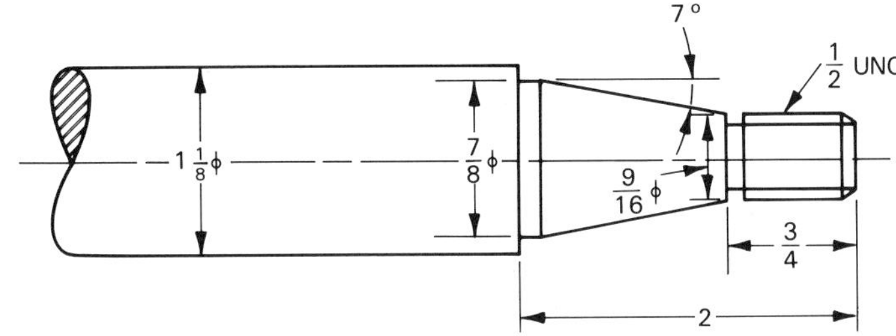


**Fig. 21-5** The special operations completed on the shaft.

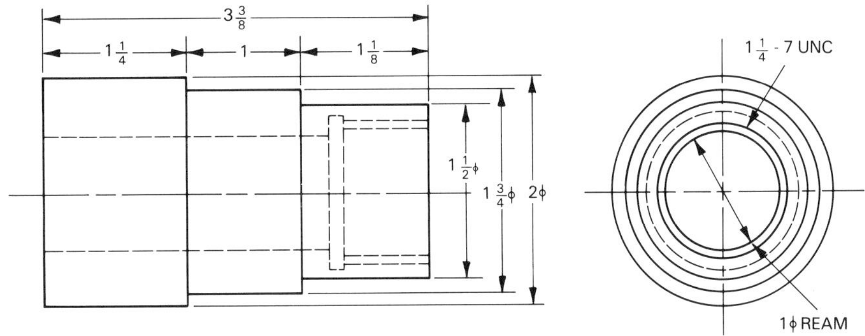


**Fig. 21-6** A round part requiring internal and external machining.

N-6 Work must not extend beyond the chuck jaws more than three times its diameter.

N-7 Use a four-jaw chuck to grip work securely.

N-8 Use aluminum or brass strips between the chuck jaws and finish diameters.

the workpiece be held short for rigidity and prevention of accidents. Never have work extending more than *three times its diameter* beyond the chuck jaws unless it is supported by some means such as a steadyrest. (See N-6.)

## Machining External and Internal Diameters in a Chuck

To machine the part shown in Fig. 21-6, the following sequence of operations is suggested.

**1.** Cut off a piece of steel ⅛ in. (3.17 mm) larger in diameter and ½ in. (12.7 mm) longer than required.

- In this case the rough diameter would be 2⅛ in. (53.98 mm).
- The length would be 3⅞ in. (98.4 mm); the extra length is to allow the piece to be gripped in the chuck.

**2.** Mount and center the workpiece in a four-jaw chuck, gripping only 5⁄16 to ⅜ in. (7.9 to 9.5 mm) of the material in the chuck jaws. (See N-7.)

- A three-jaw chuck would not hold this size workpiece securely enough for the internal and external machining operations.

**3.** Face the end of the work square. Remove only the minimum amount of material required to square the end.

**4.** Rough turn the three external diameters, starting with the largest and progressing to the smallest to within 1⁄32 in. (0.79 mm) of size and length.

**5.** Mount a drill chuck in the tailstock spindle and center drill the work.

**6.** Drill a ½-in. (12.7-mm) diameter hole through the work.

**7.** Mount a 15⁄16-in. (23.81-mm) drill in the tailstock and drill through the work.

**8.** Mount a boring bar in the toolpost and bore the 1-in. (25.4-mm) ream hole to .968-in. (24.58-mm) diameter.

**9.** Bore the 1¼-in.—7 UNC threaded section to the tap drill size, which is 1.107 in. (28.11 mm).

**10.** Cut the groove at the end of the section to be threaded to length and a little deeper than the major diameter of the thread.

**11.** Mount a threading tool in the boring bar and cut the 1¼-in.—7 thread to size.

**12.** Mount a 1-in. (25.4-mm) reamer in the tailstock and cut the hole to size.

**13.** Finish turn the external diameters to diameter and length starting with the largest and working down to the smallest.

**14.** Reverse the workpiece in the chuck and protect the finish diameter with a piece of soft metal between it and the chuck jaws. (See N-8.)

**15.** Face the end surface to the proper length.

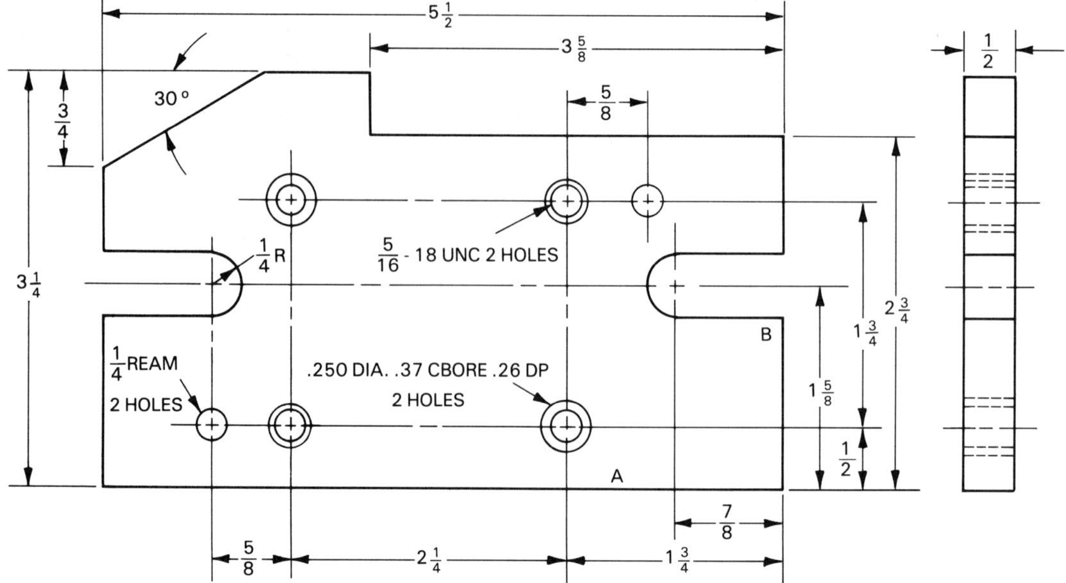


**Fig. 21-7** A typical flat part which must be laid out and machined.

N-9 Allow .010 in. (0.25 mm) oversize on each surface which must be ground.

N-10 Never saw closer than within 1/32 in. (0.79 mm) of the layout line to leave enough material for finish machining.

N-11 Always allow 1/16 in. (1.58 mm) on each surface of the workpiece for machining.

## MACHINING FLAT WORKPIECES

Since there are so many variations in size and shape of flat workpieces, it is difficult to give specific machining rules to follow that would apply to each workpiece. Some general rules are listed, but they may have to be modified to suit particular workpieces.

**1.** Select and cut off the material a little larger than the size required.

**2.** Machine all surfaces to size in a shaper or milling machine. (See N-9.)

**3.** Lay out the physical contours of the part such as angles, steps, radii, etc.

**4.** Lightly prick-punch the layout lines which indicate the surfaces to be cut.

**5.** Remove large sections of the workpiece on a contour bandsaw. (See N-10.)

**6.** Machine all forms such as steps, angles, radii, and grooves.

**7.** Lay out all hole locations.

**8.** Drill all holes and tap those which require threads.

**9.** Ream holes.

**10.** Surface grind any surfaces which are required.

### Operations Sequence for a Sample Flat Part

The part shown in Fig. 21-7 is used only as an example to set forth a sequence of operations which should be followed when machining similar parts. These are not meant to be hard and fast rules, but only guides for the student.

The sequence of operations suggested for the sample part shown in Fig. 21-7 is different from those suggested for machining a block square and parallel as outlined in the milling machine unit because:

**1.** The part is relatively thin and has a large surface area.

**2.** Since at least 1/8 in. (3.17 mm) of work should be above the vise jaws, it would be difficult to use a round bar between the work and movable jaw for machining the large flat surfaces.

**3.** A small inaccuracy (out-of-squareness) on the narrow edge would create a greater error when the large surface was machined.

### Procedure

**1.** Cut off a piece of steel 5/8 in. (15.87 mm) × 3 3/8 in. (85.72 mm) × 5 5/8 in. (142.87 mm) long. (See N-11.)

**2.** In a milling machine, finish one of the largest surfaces (face) first.

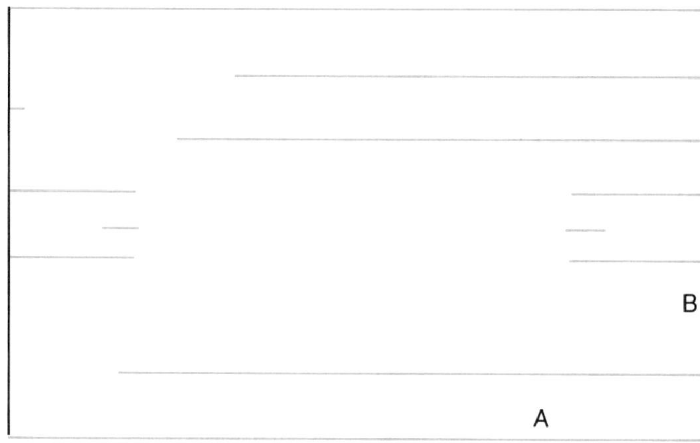

Fig. 21-8 Lay out all horizontal lines using edge A as a reference surface.

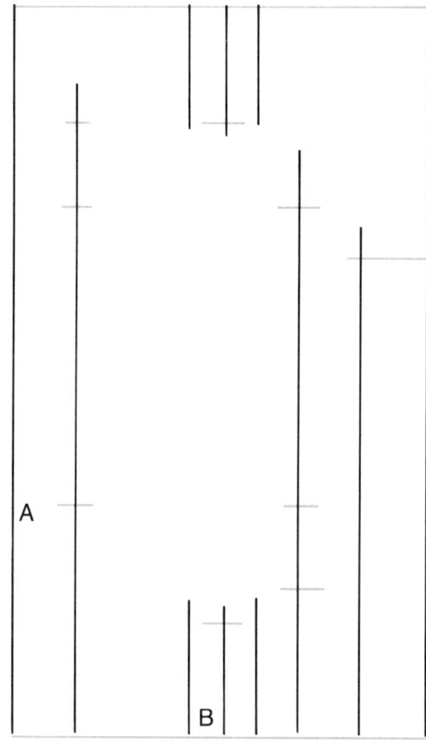

Fig. 21-9 Lay out all vertical lines using edge B as the reference surface.

N-12 Allow .010 in. (0.25 mm) oversize on each surface which must be ground.

N-13 Holes to be tapped should be countersunk slightly larger than the tap diameter to:

1. Protect the top of the thread from damage
2. Remove the burr or sharp edge caused by a tap

**3.** Turn the workpiece over and machine the other face to ½ in. (12.7 mm) thick. (See N-12.)

**4.** Machine one edge square with the face.

**5.** Machine an adjacent edge square (at 90°) with the first edge.

**6.** Place the longest finished edge *A* down in the machine vise and cut the opposite edge to 3¼ in. (82.55 mm) wide.

**7.** Place the narrower finished edge *B* down in the machine vise and cut the opposite edge to 5½ in. (139.7 mm) long.

**8.** With edge *A* as a reference surface, lay out all the horizontal dimensions with an adjustable square, surface gage, or height gage, Fig. 21-8.

**9.** With edge *B* as a reference surface, lay out all the vertical dimensions with an adjustable square, surface gage, or height gage, Fig. 21-9.

**10.** Use a bevel protractor to lay out the 30° angle on the upper-left-hand edge.

**11.** With a divider set to ¼ in. (6.35 mm), draw the arcs for the two center slots.

**12.** With a sharp prick punch, lightly mark all the surfaces to be cut and the centers of all hole locations.

**13.** Center punch and drill ½-in. (12.7-mm) diameter holes for the two center slots.

**14.** On a vertical bandsaw, cut the 30° angle to within 1⁄32 in. (0.79 mm) of the layout line.

**15.** Place the workpiece in a vertical mill and machine the two ½-in. (12.7-mm) slots.

**16.** Machine the step on the top edge of the workpiece.

**17.** Set the work to 30° in the machine vise and finish the 30° angle.

**18.** Center-punch all hole locations.

**19.** Center-drill all hole locations.

**20.** Drill and counterbore the holes for the ¼-in.—20 NC screws.

**21.** Tap drill the 5⁄16-in.—18-thread holes to ¼ in. (6.35 mm).

**22.** Drill the ¼-in. (6.35-mm) ream holes to 15⁄64 in. (5.96 mm).

**23.** Countersink all holes to be tapped slightly larger than their finished size. (See N-13.)

**24.** Ream the ¼-in. (6.35-mm) holes to size.

**25.** Tap the 5⁄16-in.—18 UNC holes.

## REVIEW QUESTIONS

**1.** Why is it not advisable to rough turn small diameters first?

**2.** To what size should work be rough turned?

**3.** Why must all measurements be taken from one end of a workpiece?

**4.** Why is it important that the workpiece be cooled before finish turning?

**5.** Why can a workpiece be bent during the knurling operation?

**6.** Why is the work cut off longer than required when machining in a chuck?

**7.** How far should a 6-in.-long workpiece, 1-in. diameter, extend beyond the chuck jaws?

**8.** How deep should the groove be cut for the internal section to be threaded?

**9.** How is a finish diameter protected from the chuck jaws?

**10.** How much material should be left on a flat surface for grinding?

**11.** When using a bandsaw to remove excess material, how close to the layout line should the cut be layout line should the cut be made?

**12.** When machining flat surfaces what surface should be machined first?

# SECTION

# CUTTING TOOLS AND FLUIDS

Because of the need for longer tool life and better surface finishes brought about by interchangeable manufacture, extensive research was necessary in the area of metal cutting. This resulted in the development of better, more efficient cutting tools and cutting fluids.

# UNIT 22

# METAL CUTTING

For hundreds of years, human beings used tools to cut metals without really understanding the process. It was thought originally that the metal split ahead of the cutting tool similar to the action of an ax splitting wood, Fig. 22-1. This false conception of the cutting process is shown in Fig. 22-2.

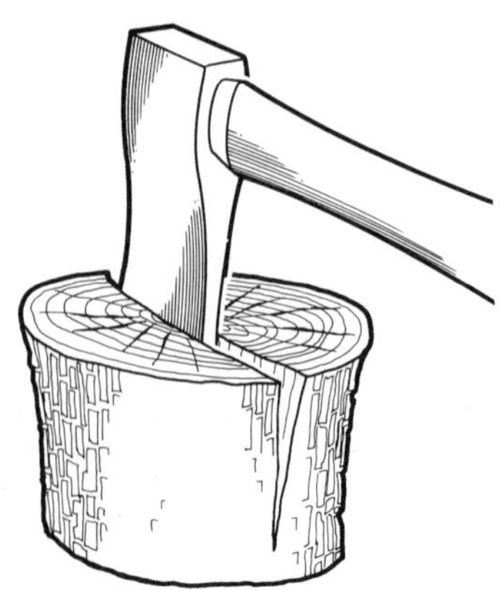

Fig. 22-1 The simile of an ax cutting wood was often used incorrectly to illustrate the action of a cutting tool.

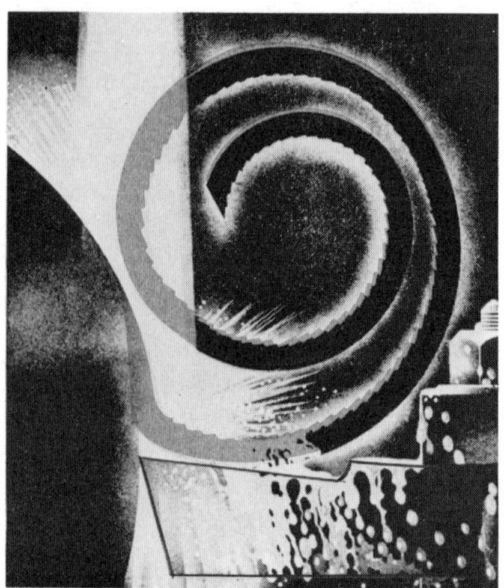

Fig. 22-2 A false conception of the metal-cutting process. (*Courtesy Cincinnati Milacron Inc.*)

## OBJECTIVES

The purpose of this unit is to:

1. Explain the cutting action and terminology used for cutting tools
2. Describe the various types of chips produced when machining metals
3. Discuss the machinability of metals

## METAL-CUTTING TERMINOLOGY

A number of terms resulted from the research conducted on metal cutting, and it may be wise to clearly define these terms.

A *built-up edge* is a layer of compressed metal from the material being cut which adheres to and piles up on the cutting tool edge during a machining operation (Fig. 22-3A).

The *chip-tool interface* is the portion of the face of the cutting tool upon which the chip slides as it is cut from the metal (Fig. 22-3A).

*Plastic deformation* is the deformation of the work material occurring in the shear zone during a cutting action (Fig. 22-3B).

*Plastic flow* is the flow of metal occurring on the shear plane which extends from the cutting tool edge *A* to the corner between the chip and the unmachined work surface *B* in Fig. 22-3B.

A *rupture* is the tear that occurs when brittle materials, such as cast iron, are cut and the chip breaks away from the work surface. This generally occurs when discontinuous or segmented chips are produced.

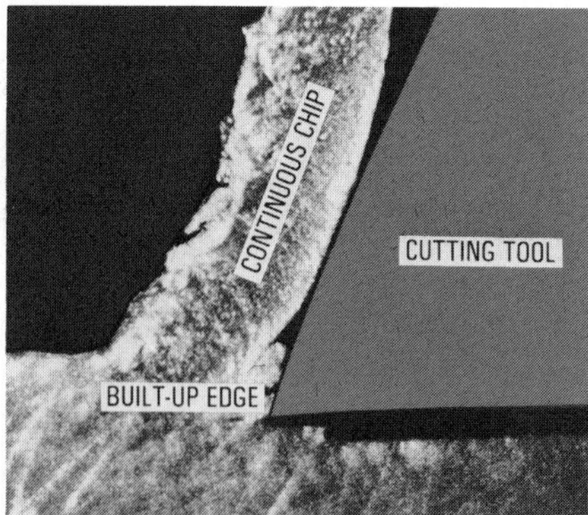


Fig. 22-3A Chip-tool interface. (*Courtesy Cincinnati Milacron Inc.*)

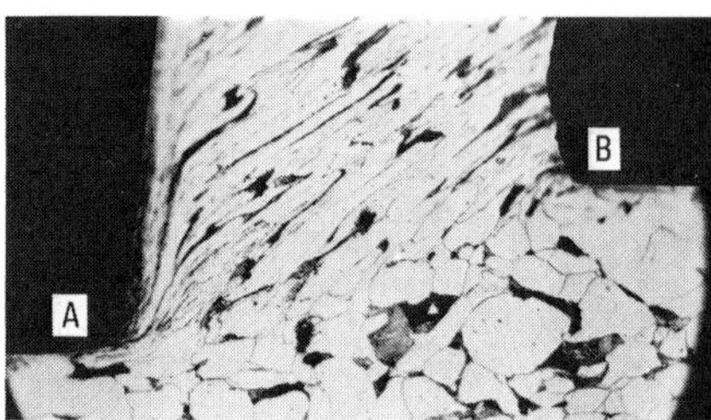


Fig. 22-3B A photomicrograph of a chip showing the crystal elongation and plastic deformation. (*Courtesy Cincinnati Milacron Inc.*)

The *shear angle* or *plane* is the angle of the area of material where plastic deformation is occurring (Fig. 22-3B).

The *shear zone* is the area where plastic deformation of the metal occurs. It is along a plane from the cutting edge of the tool to the original work surface.

## CHIP FORMATION

Machining operations performed on lathes, shapers, milling machines, or similar machine tools, produce chips which fall into three basic types: discontinuous chip, continuous chip, and continuous chip with a built-up edge (Fig. 22-4).

### Type 1: Discontinuous (Segmented) Chip

Discontinuous, or segmented, chips (Fig. 22-5), are produced when brittle metals, such as cast iron and hard bronze, or some ductile metals are cut under poor cutting conditions. As the point of the cutting tool contacts the metal, some compression occurs, as

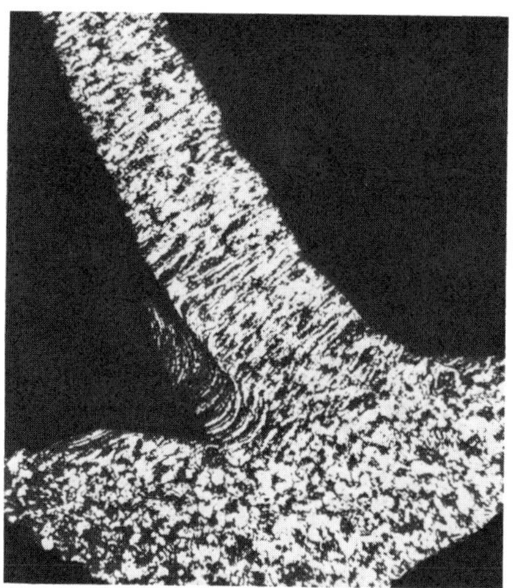

**Fig. 22-4** Continuous chip with a built-up edge. (*Courtesy Cincinnati Milacron Inc.*)

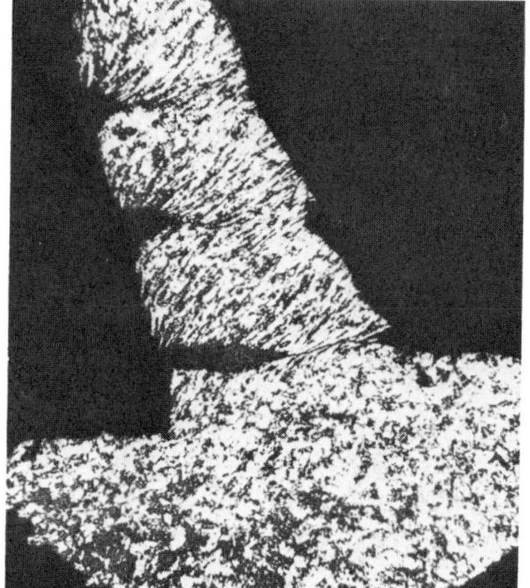

**Fig. 22-5** Discontinuous chip.

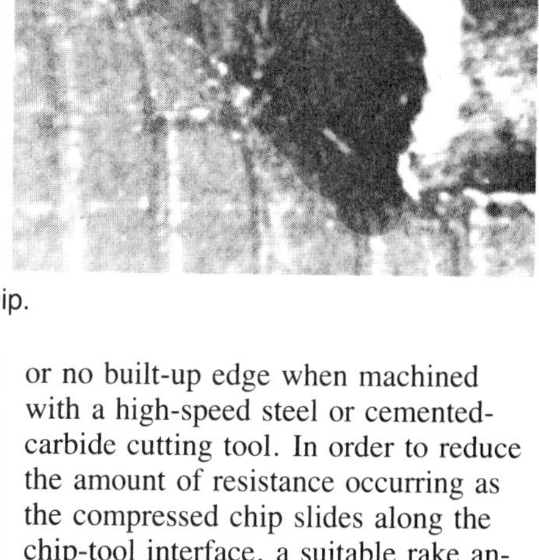

**Fig. 22-6** Formation of a discontinuous chip.

can be noted in Fig. 22-6A and B, and the chip begins flowing along the chip-tool interface. As more stress is applied to brittle metal by the cutting action, the metal compresses until it reaches a point where rupture occurs and the chip separates from the unmachined portion. This cycle is repeated indefinitely during the cutting operation with the rupture of each segment occurring on the shear angle or plane. Generally, as a result of these successive ruptures, a poor surface is produced on the workpiece.

Excessive machine chatter sometimes causes discontinuous chips to be produced when ductile material is cut.

## Type 2: Continuous Chip

The type 2 chip is a continuous ribbon produced when the flow of metal next to the tool face is not greatly retarded by a builtup edge or friction at the chip-tool interface. The continuous ribbon chip is considered ideal for efficient cutting action because it results in better surface finishes.

When ductile materials are cut, plastic flow in the metal takes place by the deformed metal sliding on a great number of crystallographic slip planes. As is the case with the type 1 chip, fractures or ruptures do not occur because of the ductile nature of the metal.

In Fig. 22-3B it can be seen that the crystal structure of the ductile metal is elongated when it is compressed by the action of the cutting tool and as the chip separates from the metal. The process of chip formation occurs in a single plane extending from the cutting tool to the unmachined work surface.

Machine steel generally forms a continuous (unbroken) chip with little or no built-up edge when machined with a high-speed steel or cemented-carbide cutting tool. In order to reduce the amount of resistance occurring as the compressed chip slides along the chip-tool interface, a suitable rake angle is ground on the tool, and cutting fluid is used during the cutting operation. This allows the compressed chip to flow relatively freely along the chip-tool interface. A shiny layer on the back of a continuous-type chip indicates ideal cutting conditions with little resistance to chip flow.

## Type 3: Continuous Chip with a Built-up Edge

Low-carbon machine steel, when cut with a high-speed steel cutting tool without the use of cutting fluids, generally produces a continuous-type chip with a built-up edge.

The metal ahead of the cutting tool is compressed and forms a chip which begins to flow along the chip-tool interface (Fig. 22-7). As a result of the

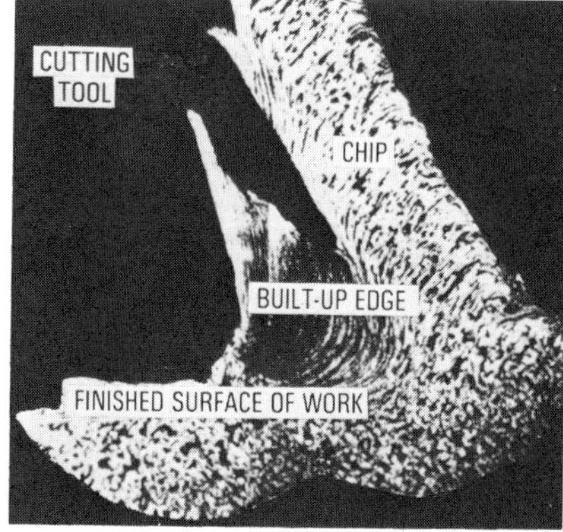


**Fig. 22-7** A type 3 continuous chip with a built-up edge being formed. (*Courtesy Cincinnati Milacron Inc.*)

high temperature, the high pressure, and the high frictional resistance against the flow of the chip along the chip-tool interface, small particles of metal begin adhering to the edge of the cutting tool while the chip shears away. The built-up edge increases in size and becomes more unstable; eventually a point is reached where fragments are torn off. Portions of these fragments which break off stick to both the chip and the workpiece. These fragments adhere to and score the machined surface; the result is a poor surface finish. The continuous chip with the built-up edge, as well as creating a poor surface finish on the workpiece, also shortens the cutting tool life.

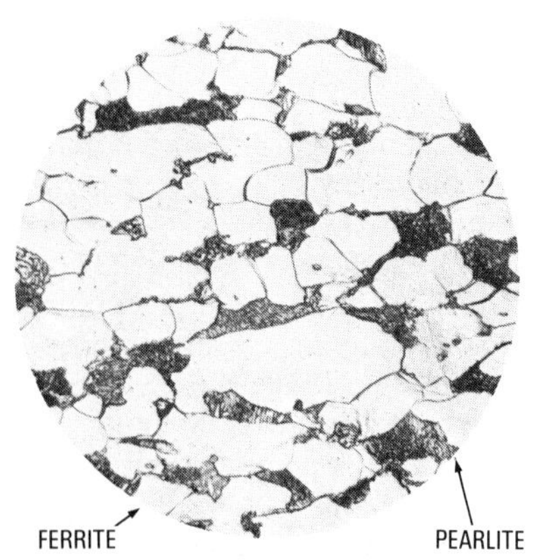


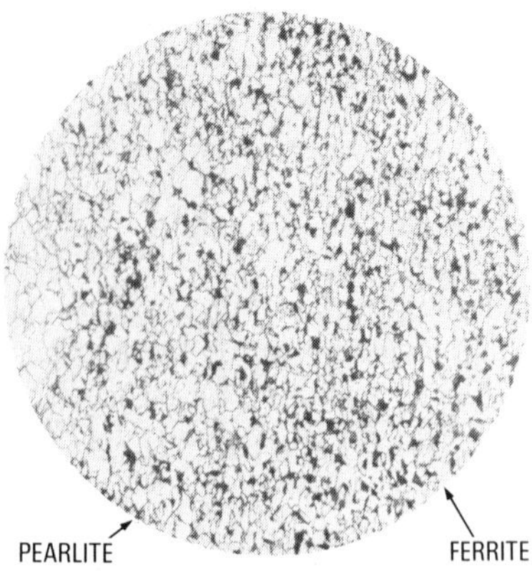


**Fig. 22-8A and B** Photomicrographs indicating undesirable steel microstructures. (*Courtesy Cincinnati Milacron Inc.*)

## MACHINABILITY OF METALS

*Machinability* describes the ease or difficulty with which a metal can be machined. Such factors as cutting-tool life, surface finish produced, and power required must be considered. Machinability has been measured by the length of the cutting-tool life in minutes or by the rate of stock removal in relation to the cutting speed employed, that is, depth of cut. For finish cuts, machinability refers to the life of the cutting tool and the ease with which a good surface finish is produced.

### Grain Structure

The machinability of a metal is affected by its microstructure and will vary if the metal has been annealed. Certain chemical and physical modifications of steel will improve their machinability. Free-machining steels have generally been modified in the following manner by:

1. The addition of sulfur
2. The addition of lead
3. The addition of sodium sulfite
4. Cold working which modifies the ductility

By making these (free-machining) modifications to the steel, three main machining characteristics become evident:

1. Tool life is increased
2. A better surface finish is produced
3. Lower power consumption is required for machining

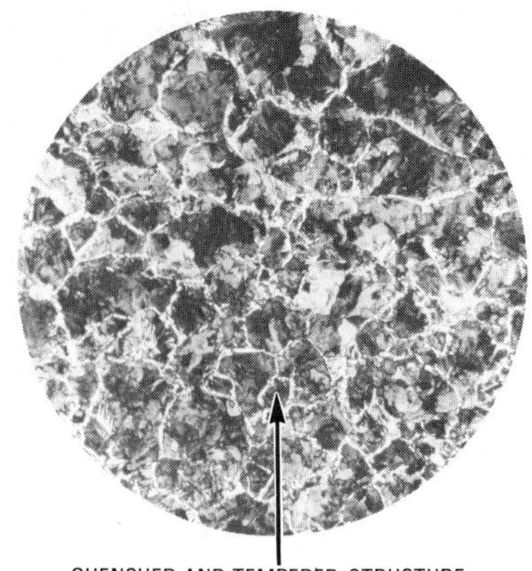


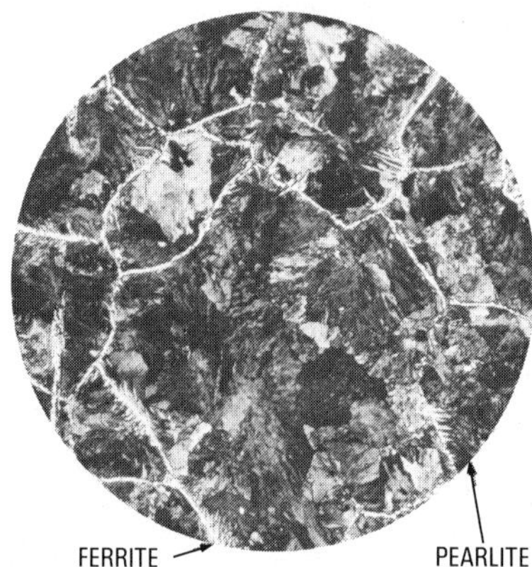


**Fig. 22-9A and B** Photomicrographs showing a desirable microstructure in steel. (*Courtesy Cincinnati Milacron Inc.*)

### Low-Carbon (Machine) Steel

The microstructure of low-carbon steel may have large areas of ferrite (iron) interspersed with small areas of pearlite (Fig. 22-8A and B). Ferrite is soft with high ductility and low strength, while pearlite, a combination of ferrite and iron carbide, has low ductility and high strength. Fig. 22-9A and B illustrates a desirable microstructure in steel because the pearlite is well distributed and is therefore better for machining purposes.

### High-Carbon (Tool) Steel

A greater amount of pearlite is present in high-carbon (tool) steel because of the higher carbon content. The greater the amount of pearlite (low ductility and high strength) present in the steel, the more difficult it becomes to machine the steel efficiently. It is there-

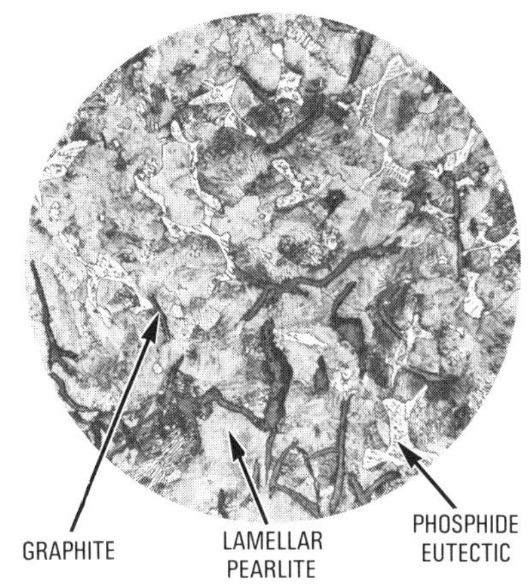


**Fig. 22-10A** The microstructure of white cast iron. (*Courtesy Cincinnati Milacron Inc.*)

**Fig. 22-10B** The microstructure of grey cast iron. (*Courtesy Cincinnati Milacron Inc.*)

N-1 Cutting fluids cool and lubricate the cutting tool and the workpiece.

fore desirable to anneal these steels to alter their microstructure and, as a result, improve their machining qualities.

### Alloy Steel

Alloy steels, which are a combination of two or more metals, are generally a little more difficult to machine than low- or high-carbon steels. In order to improve their machining qualities, combinations of sulfur and lead or sulfur and manganese in proper proportions are sometimes added to alloy steels. The machining of stainless steel, generally difficult because of its work-hardening qualities, can be greatly eased by the addition of selenium.

### Cast Iron

Cast irons, consisting generally of ferrite, iron carbide, and free carbon, form an important group of materials used by industry. The microstructure of cast iron can be controlled by the addition of alloys, the method of casting, the rate of cooling, and by heat treating. *White cast iron* (Fig. 22-10A), cooled rapidly after casting, is usually hard and brittle because of the formation of hard iron carbide. *Grey cast iron* (Fig. 22-10B) is cooled gradually; its structure is composed of compound pearlite, a mixture of fine ferrite and iron carbide, and flakes of graphite. Because of the gradual cooling, it is softer and therefore easier to machine.

## REVIEW QUESTIONS

**1.** What occurs when metal is cut with a cutting tool?

**2.** List three developments which have assisted in the machining of metals.

**3.** Define:

(a) Plastic flow

(b) Chip-tool interface

**4.** Name the type of chip produced when machining

(a) Cast iron

(b) Machine steel

**5.** What additions are added to steel to improve their machinability?

**6.** What operation is recommended in order to improve the machinability of tool steel with a high amount of pearlite?

# UNIT 23

# CUTTING FLUIDS

Cutting fluids are essential in most metal-cutting operations. During a machining process, considerable heat and friction are created by the plastic deformation of metal occurring in the shear zone when the chip slides along the chip-tool interface. This heat and friction causes metal to adhere to the cutting edge of the tool and the tool may break down; the result is a poor finish and inaccurate work.

## OBJECTIVES

This unit will enable you to:

**1.** Know the advantages and characteristics of cutting fluids

**2.** Select the proper cutting fluid to suit various materials and machining operations

## PURPOSE AND ADVANTAGES

During a machining process, considerable heat and friction are created. The correct selection and application of cutting fluids can prevent these results by effectively cooling the work and reducing the friction. Their use can result in the following economic advantages.

**1.** *Reduction of tool costs.* Cutting fluids reduce tool wear; they last longer and require less time in resharpening and resetting. (See N-1.)

**2.** *Increased speed of production.* Because cutting oils help reduce heat and friction, higher cutting speeds can be used for machining operations (Fig. 23-1).

**3.** *Reduction of labor costs.* As cutting tools last longer and require less regrinding when cutting fluids are used, there is less stoppage of production, thereby reducing the labor cost per part.

**4.** *Reduction of power costs.* Since friction is reduced by a cutting fluid, less power is required for machining operations and a corresponding saving in power costs is possible.

**Fig. 23-1** Cutting fluids improve the cutting action and result in higher production.

**Fig. 23-2** Cutting oils are generally mineral oils plus certain additives.

**5.** *Better surface finish.* Since cutting fluids reduce friction and prolong the keen edge of cutting tools, better surface finishes and dimensional accuracy can be obtained.

## CHARACTERISTICS OF A GOOD CUTTING FLUID

For a cutting fluid to function effectively, it should possess the following desirable characteristics.

**1.** *Good cooling capacity,* to reduce the cutting temperature, increase tool life and production, and improve dimensional accuracy

**2.** *Good lubricating qualities,* to prevent metal from adhering to the cutting edge, forming a built-up edge, resulting in a poor surface finish

**3.** *Rust resistance,* which should eliminate stain, rust, or corrosion to the workpiece or machine

**4.** *Stability (long life),* both in storage and in use

**5.** *Resistance to rancidity*

**6.** *Nontoxicity,* so it will not cause skin irritation to the operator

**7.** *Transparency,* so that the operator can clearly see the work during machining

**8.** *Relatively low viscosity,* to permit the chips and dirt to settle quickly

**9.** *Nonflammability,* so it will not burn easily and preferably be noncombustible, as well as not smoke excessively, form gummy deposits which may cause machine slides to become sticky, or clog the circulating system

## TYPES OF CUTTING FLUIDS

The need for a cutting fluid which possesses as many of the desirable characteristics of a good cutting fluid as possible has resulted in the development of many different types. The most commonly used cutting fluids are either aqueous- (water-) based solutions or cutting oils. These fluids fall into three categories: cutting oils, emulsifiable oils, and chemical (synthetic) cutting fluids.

### Cutting Oils

Cutting oils are classified under two types: active or inactive. These terms relate to the oil's chemical activity or ability to react with the metal surface at elevated temperatures to protect it and improve the cutting action.

*Active cutting oils,* which may be dark or transparent, fall into three general categories.

**1.** *Sulfurized mineral oils* contain from 0.5 to 0.8 percent sulfur. They are generally light-colored, transparent, and have good cooling, lubricating, and antiweld properties. They are useful for the cutting of low-carbon steels and tough, ductile metals.

**2.** *Sulfochlorinated mineral oils* contain up to 3 percent sulfur and 1 percent chlorine. These oils prevent excessive built-up edges from forming and prolong the life of the cutting tool. Sulfochlorinated mineral oils are effective in cutting tough low-carbon and chrome-nickel alloy steels.

**3.** *Sulfochlorinated fatty oil blends* contain more sulfur than the other types and are effective cutting fluids for heavy-duty machining.

*Inactive cutting oils,* where the sulfur is so firmly attached to the oil that very little is released to react with the work surface during the cutting action, fall into four general categories.

**1.** *Straight mineral oils,* because of their low viscosity, have faster wetting and penetrating factors. They are used for the machining of nonferrous metals, such as aluminum, brass, and magnesium, where lubricating and cooling properties are not essential. Straight mineral oils are also recommended for use in cutting leaded (free-machining) metals and the tapping and threading of white metal.

**2.** *Fatty oils,* such as lard and sperm oil, once widely used, find limited applications as cutting fluids today. They are generally used for severe cutting operations on tough, nonferrous metals where a sulfurized oil might cause discoloration.

**3.** *Fatty and mineral oil blends* are combinations of fatty and mineral oil, resulting in better wetting and penetrating qualities than straight mineral oils.

**4.** *Sulfurized fatty-mineral oil blends* are made by sulfur being combined with fatty oils and then mixed with certain mineral oils. They are often used on machines when ferrous and nonferrous metals are machined at the same time.

Fig. 23-3 Heavy mixtures of cutting oils tend to smoke.

Fig. 23-4 Water-base cutting fluids should provide rust control.

N-2 Low cutting speeds require a fluid with high friction-reducing qualities. High cutting speeds require a coolant with excellent cooling ability.

N-3 Chemical fluids are synthetic fluids and do not contain emulsifiable oils.

## Emulsifiable (Soluble) Oils

An effective cutting fluid should possess high heat conductivity, and neither mineral nor fatty oils are very effective as coolants. Water is the best cooling medium known; however, used as a cutting fluid, water alone would cause rust and have little lubricating value. By adding a certain percentage of soluble oil to water, it is possible to add rust-resistance and lubrication qualities to the excellent cooling capabilities of water. (See N-2.)

Emulsifiable, or soluble, oils are mineral oils containing a soaplike material (emulsifier) which makes them soluble in water. These emulsifiers break the oil into minute particles and keep them separated in the water for a long period of time. Emulsifiable or soluble oils are supplied in concentrate form. From *one to five parts* of this concentrate are added to *100 parts* of water.

Because of their good cooling and lubricating qualities, soluble oils are used when machining is done at high cutting speeds, at low cutting pressure, and when considerable heat is generated.

## Chemical Cutting Fluids

*Chemical cutting fluids,* sometimes called synthetic fluids, have been widely accepted since they were first introduced for machining purposes. They are stable, preformed emulsions which contain very little oil and mix easily with water. (See N-3.) Chemical cutting fluids depend on chemical agents for lubrication and friction reduction. Some types of chemical cutting fluids contain *extreme-pressure* (EP) lubricants which react with freshly machined metal under the heat and pressure of a cut to form a solid lubricant. Fluids containing extreme-pressure lubricants reduce both the *heat of friction* between the chip and tool face and the *heat caused by plastic deformation* of the metal.

As a result of the chemical agents which are added to the cooling qualities of water, synthetic fluids provide the following advantages.

**1.** Good rust control.
**2.** Resistance for long periods of time to becoming rancid.
**3.** Reduction of the amount of heat generated during a cutting action.
**4.** Excellent cooling qualities.
**5.** Longer durability than cutting or soluble oils.
**6.** They are nonflammable and nonsmoking.
**7.** They are nontoxic.
**8.** Easy separation from the work and chips, which makes them clean to work with.
**9.** Quick settling of grit and fine chips so they are not recirculated in the cooling system.
**10.** No clogging of the machine cooling system due to detergent action of the fluid.

Chemical cutting fluids are manufactured in three types: *true solution fluids, wetting agent types,* and *wetting agent types with extreme-pressure lubricants*.

*Caution:* Although chemical cutting fluids have been widely accepted and used for many types of metal cutting operations, there are certain precautions which should be observed regarding their use. Chemical cutting fluids are generally used on ferrous metals; however, many aluminum alloys can be machined successfully with them. Most chemical cutting fluids are not recommended for use on alloys of magnesium, zinc, cadmium, or lead. Certain types of paint (generally poor quality) may be affected by some chemical cutting fluids which may mar the machine's appearance and allow paint to get into the coolant.

## Special Coolants

Special coolants are specifically made to be used for a special operation in machining such as tapping, honing, thread forming, and metal forming. *Solid-form lubricants* such as graphite, molybdenum disulfide, pastes, soaps, and waxes are usually used in operations where heavy metal forming is performed.

Air is the most common *gaseous coolant,* being present under atmospheric pressure in dry operations, and always present when fluids are used. Sometimes air is compressed to provide better cooling, with a stream directed at the cutting zone to remove heat by forced convection. Important advantages of using inert gases include good cooling ability, a clear view of the operation, elimination of mist, and *noncontamination* of the workpiece, chips, or machine lubricants.

**TABLE 23-1 RECOMMENDED CUTTING FLUIDS FOR VARIOUS MATERIALS**

| Material | Drilling | Reaming | Threading | Turning | Milling |
|---|---|---|---|---|---|
| **Aluminum** | Soluble oil<br>Kerosene<br>Kerosene and lard oil | Soluble oil<br>Kerosene<br>Mineral oil | Soluble oil<br>Kerosene and lard oil | Soluble oil | Soluble oil<br>Lard oil<br>Mineral oil<br>Dry |
| **Brass** | Dry<br>Soluble oil<br>Kerosene and lard oil | Dry<br>Soluble oil | Soluble oil<br>Lard oil | Soluble oil | Dry<br>Soluble oil |
| **Bronze** | Dry<br>Soluble oil<br>Mineral oil<br>Lard oil | Dry<br>Soluble oil<br>Mineral oil<br>Lard oil | Soluble oil<br>Lard oil | Soluble oil | Dry<br>Soluble oil<br>Mineral oil<br>Lard oil |
| **Cast iron** | Dry<br>Air jet<br>Soluble oil | Dry<br>Soluble oil<br>Mineral lard oil | Dry<br>Sulphurized oil<br>Mineral lard oil | Dry<br>Soluble oil | Dry<br>Soluble oil |
| **Copper** | Dry<br>Soluble oil<br>Mineral lard oil<br>Kerosene | Soluble oil<br>Lard oil | Soluble oil<br>Lard oil | Soluble oil | Dry<br>Soluble oil |
| **Malleable iron** | Dry<br>Soda water | Dry<br>Soda water | Lard oil<br>Soda water | Soluble oil | Dry<br>Soda water |
| **Monel metal** | Soluble oil<br>Lard oil | Soluble oil<br>Lard oil | Lard oil | Soluble oil | Soluble oil |
| **Steel alloys** | Soluble oil<br>Sulphurized oil<br>Mineral lard oil | Soluble oil<br>Sulphurized oil<br>Mineral lard oil | Sulphurized oil<br>Lard oil | Soluble oil | Soluble oil<br>Mineral lard oil |
| **Steel, machine** | Soluble oil<br>Sulphurized oil<br>Lard oil<br>Mineral lard oil | Soluble oil<br>Mineral lard oil | Soluble oil<br>Mineral lard oil | Soluble oil | Soluble oil<br>Mineral lard oil |
| **Steel, tool** | Soluble oil<br>Sulphurized oil<br>Mineral lard oil | Soluble oil<br>Sulphurized oil<br>Lard oil | Sulphurized oil<br>Lard oil | Soluble oil | Soluble oil<br>Lard oil |

*Note:* Chemical cutting fluids can be used successfully for most of the cutting operations above. These concentrates are diluted with water in proportions ranging from 1 part cutting fluid to 15 and as high as 100 parts of water, depending on the metal being cut and the type of machining operation. When using chemical cutting fluids, it is wise to follow the manufacturer's recommendations for use and mixture. *(Courtesy Cincinnati Milacron Inc.)*

## APPLICATION OF CUTTING FLUIDS

Cutting tool life and the machining operation are greatly influenced by the way that the cutting fluid is applied. It should be supplied in a copious stream under low pressure so that the work and cutting tool are covered well. The rule of thumb is that the inside diameter of the supply nozzle should be about three-fourths the width of the cutting tool. The fluid should be directed to the area where the chip is being formed to reduce and control the heat created during the cutting action and to prolong tool life.

### Lathe-Type Operations

On horizontal-type turning and boring machines, cutting fluid should be applied to that portion of the cutting tool which is producing the chip. For general turning and facing operations, cutting fluid should be supplied directly over the cutting tool, close to the zone of chip formation (Fig. 23-5). (See N-4.) In heavy-duty turning and facing

N-4 For most operations, cutting fluid should be applied directly to the cutting edge which is producing the chip.

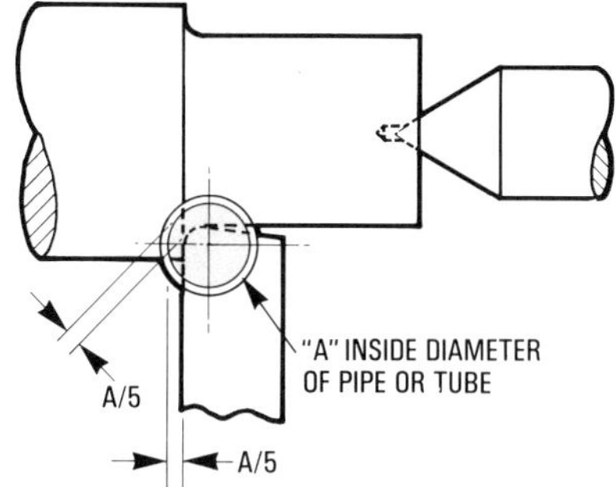


Fig. 23-5 Cutting fluid can be applied with one nozzle for most turning operations. (*Courtesy Cincinnati Milacron Inc.*)

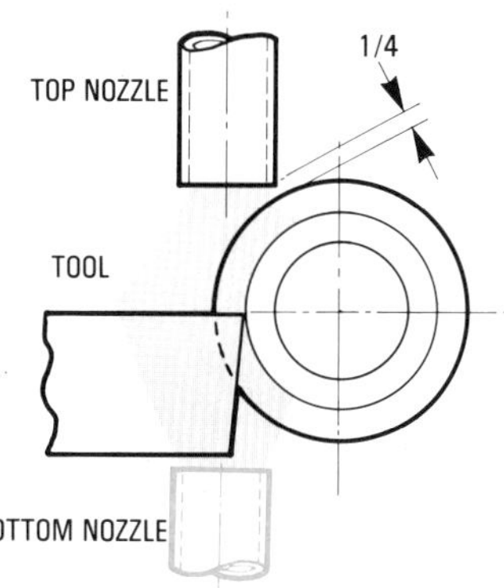


Fig. 23-6 Two nozzles should be used to supply cutting fluid for heavy-duty turning operations. (*Courtesy Cincinnati Milacron Inc.*)

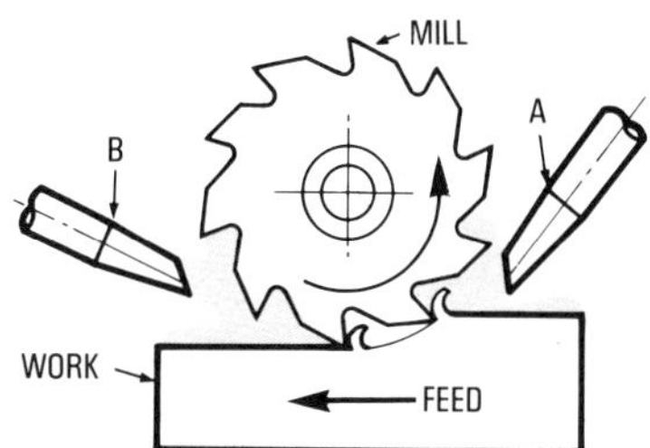


Fig. 23-8 Cutting fluid being supplied to both sides of the cutter in slab milling. (*Courtesy Cincinnati Milacron Inc.*)

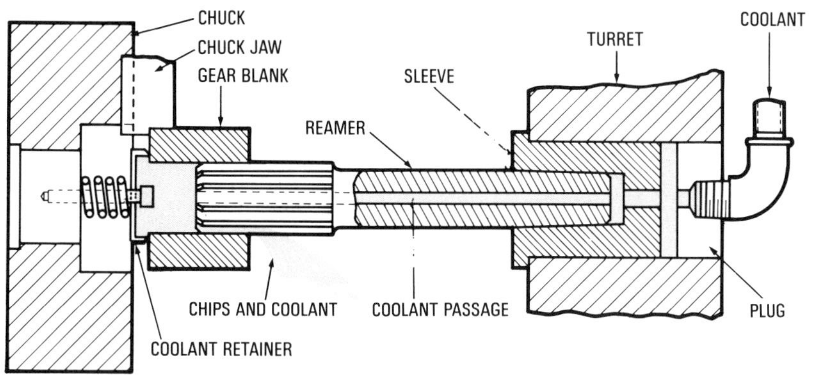


Fig. 23-7 Cutting fluid being supplied through a hole in the center of a reamer. (*Courtesy Cincinnati Milacron Inc.*)

Fig. 23-9 Cutting fluid can be applied with a ring-type distributor for face milling. (*Courtesy Cincinnati Milacron Inc.*)

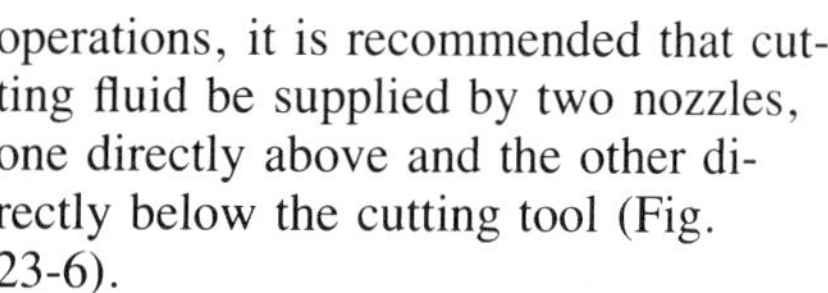

operations, it is recommended that cutting fluid be supplied by two nozzles, one directly above and the other directly below the cutting tool (Fig. 23-6).

## Drilling and Reaming

The most effective method of applying cutting fluids for these operations is to use "oil-feed" drills and hollow-shank reamers. Tools of this type transmit the cutting fluid directly to the cutting edges and at the same time flush the chips out of the hole (Fig. 23-7). When conventional drills and reamers are used, an abundant supply of fluid should be applied to the cutting edges.

## Milling

In *slab milling*, cutting fluid should be directed to both sides of the cutter by fan-shaped nozzles approximately three-fourths the width of the cutter (Fig. 23-8).

For *face milling*, a ring-type distributor (Fig. 23-9) is recommended in order to completely flood the cutter. Keeping each tooth of the cutter immersed in cutting fluid at all times can increase cutter life almost 100 percent.

# REVIEW QUESTIONS

**1.** What is the purpose of a cutting fluid?

**2.** List three advantages of a cutting fluid.

**3.** Name five characteristics of a good coolant.

**4.** Name two types of cutting oils.

**5.** What is the recommended mixture of emulsifiable oils with water?

**6.** List five advantages of chemical cutting fluids.

**7.** Where should the cutting fluid be applied for general turning operations?

# SECTION

# POWER SAWS

There are two common types of power saws used in industry—the *horizontal* and the *vertical* bandsaw. The horizontal bandsaw is generally used for cutting workpieces to rough length from bar stock prior to machining. The vertical bandsaw is usually used for slotting, notching, angular and radius cutting; and it is often used for cutting stock to approximate shape prior to machining in the milling machine or the shaper.

Bandsaws have a continuous cutting action and therefore all teeth in the band perform an equal share of the cutting action. This is more efficient than the reciprocating action of the power hacksaw, where the saw cuts only on the forward stroke and generally on a short section of the blade.

# UNIT 24

# CUTOFF SAWS

Cutoff saws are used to rough cut work to the required length from a longer piece of metal. Two types of cutoff saws are commonly used in machine shop work: the power hacksaw and the horizontal bandsaw. The horizontal bandsaw is becoming the most popular saw because the thin, endless blade has a continuous cutting action which cuts off work quickly. It is important that the correct blade type and pitch be selected to suit the workpiece cross section and material.

## OBJECTIVES

After completing this unit you should be able to:

**1.** Know the function of the main parts of a horizontal bandsaw

**2.** Select the proper saw blade pitch which should be used for cutting various workpiece cross sections

**3.** Install a saw band properly on a horizontal bandsaw

**4.** Use a bandsaw to accurately cut off work to length

The *power hacksaw,* which is a reciprocating type of saw, is usually permanently mounted to the floor. The saw frame and blade travel back and forth, with pressure being applied automatically only on the forward stroke. The power hacksaw finds limited use in machine shop work since this saw cuts only on the forward stroke, which results in considerable wasted motion.

The *horizontal bandsaw,* (Fig. 24-1), has a flexible, "endless" blade which cuts continuously in one direction. The thin blade travels over the rims of two pulley wheels and through roller guide brackets which support the blade and keep it running true. Horizontal bandsaws are available in a wide variety of types and sizes, and are becoming increasingly popular because of their high production and versatility.

### Horizontal Bandsaw Parts

The *saw frame,* hinged at the motor end, has two pulley wheels over which the continuous blade passes.

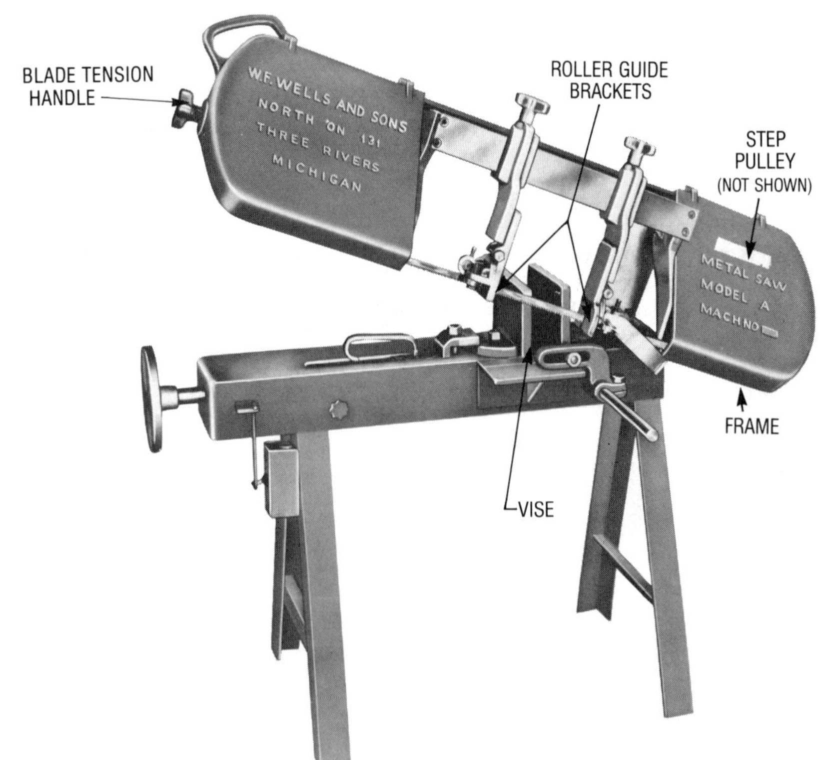


**Fig. 24-1** The main parts of a horizontal bandsaw. (*Courtesy Wells Manufacturing Corporation.*)

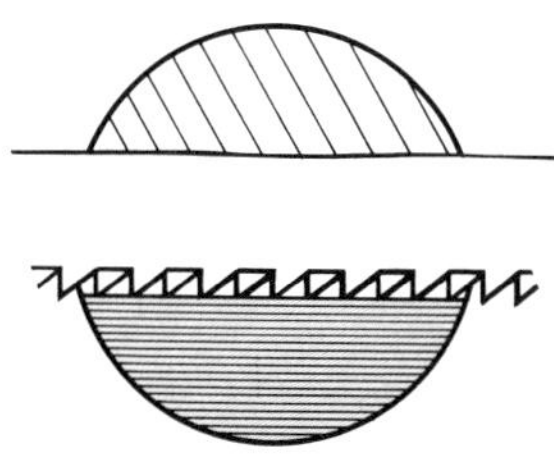


**Fig. 24-2** Use as coarse a pitch blade as possible for the workpiece being cut.

The *step pulleys* are used to vary the speed of the continuous blade to suit the type of material being cut.

The *roller guide brackets* provide rigidity for a section of the blade and prevent it from wandering and twisting. They can be adjusted to accommodate various widths of material. These brackets should be adjusted to just clear the width of the work being cut.

The *blade tension handle* is used to adjust the tension on the saw blade. The blade should be adjusted snugly to prevent it from wandering.

The *vise,* mounted on the table, can be adjusted to hold various sizes of workpieces. It can also be swiveled for making angular cuts on the end of a piece of material.

## SAW BLADES

High-speed tungsten and high-speed molybdenum steel are commonly used in the manufacture of saw blades, and for the power hacksaw they are usually hardened completely. Flexible blades used on bandsaws have only the saw teeth hardened.

Saw blades are manufactured in various degrees of coarseness ranging from 4- to 14-pitch. When cutting large sections, use a coarse or 4-pitch blade which provides the greatest chip clearance and helps to increase tooth penetration. For cutting tool steel and thin material, a 14-pitch blade is recommended. A 10-pitch blade is recommended for general-purpose sawing. *Metric saw blades* are now available in similar sizes but in teeth per 25 millimeters of length rather than teeth per inch. Therefore, the pitch of a blade having 10 teeth per 25 mm would be 25 ÷ 10, or 2.5 mm. Always select a saw blade as coarse as possible, but make sure that *two teeth* of the blade will be in contact with the work at all times. If less than two teeth are in contact with the work, the work can be caught in the tooth space (gullet), which would cause the teeth of the blade to strip or break. (See N-1.)

N-1 Always be sure that at least two saw teeth contact the work at all times.

## SAWING

For the most efficient sawing, it is important that the correct type and pitch of saw blade be selected and that it be run at the proper speed for the material being cut. Use finer-tooth blades when cutting thin cross sections and extra hard materials. Coarser-tooth blades should be used for thick cross sections and material which is soft and stringy. (See N-2.) The blade speed should suit the type and thickness of the material being cut. Too fast a blade speed or excessive feeding pressure will dull the saw teeth quickly and cause an inaccurate cut.

N-2 Use coarse pitch blades for thick sections and fine pitch blades for thin sections.

### Installing a Blade

When replacing a blade, always make sure that the teeth are pointing in the direction of saw travel or toward the motor end of the machine (Fig. 24-3). The blade tension should be tightened to prevent the blade from twisting or wandering during a cut. If it is necessary to replace a blade before a cut is finished, rotate the work one-half of a turn in the vise. This will prevent the new blade from jamming or breaking in the narrower cut made by the worn saw blade.

### To Install a Saw Blade

1. Loosen the blade tension handle.
2. Move the adjustable pulley wheel forward slightly.
3. Mount the new saw band over the two pulleys. (See N-3.)
4. Place the saw blade between the rollers of the guide brackets.
5. Tighten the blade tension handle only enough to hold the blade on the pulleys (Fig. 24-4).
6. Start and quickly stop the machine in order to make the saw blade revolve a turn or two. This will seat the blade on the pulleys.
7. Tighten the blade tension handle as tightly as possible with *one hand.*

N-3 Be sure that the saw teeth are pointing toward the motor end of the machine.

### Sawing Hints

1. Never attempt to mount, measure, or remove work unless the saw is stopped.
2. Guard long material at both ends to prevent anyone from coming in contact with it.
3. Use cutting fluid, whenever possible, to help prolong the life of the saw blade and improve the cutting action.
4. When sawing thin pieces, hold the material flat in the vise to prevent the saw teeth from breaking.

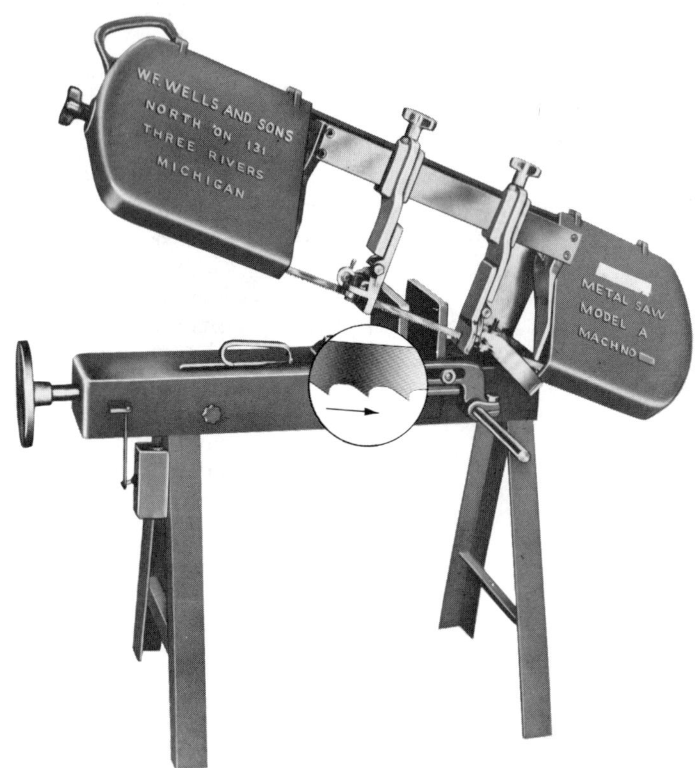


**Fig. 24-3** The saw teeth must point in the direction of the blade travel. (*Courtesy Wells Manufacturing Corporation.*)

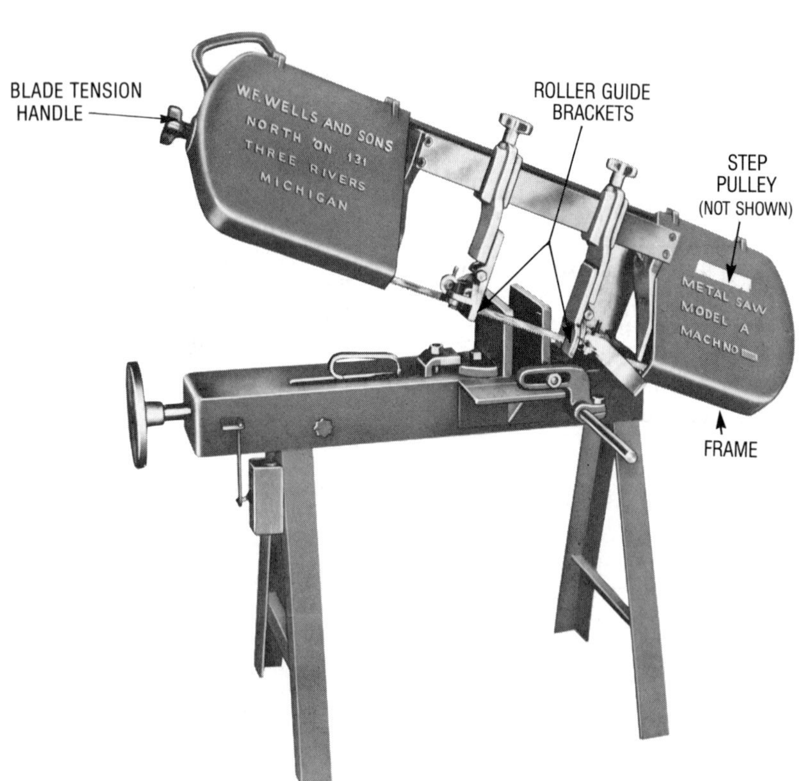


**Fig. 24-4** The saw parts used when installing a new saw band. (*Courtesy Wells Manufacturing Corporation.*)

**Fig. 24-5** The stop gage is used when many pieces of the same length must be cut. (*Courtesy Wells Manufacturing Corporation.*)

**Fig. 24-6** A spacer block should be used to prevent the vise jaw from twisting when holding short pieces. (*Courtesy Kostel Enterprises Ltd.*)

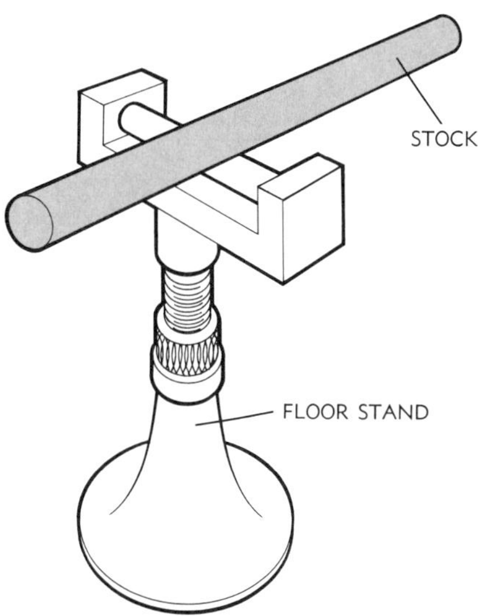


**Fig. 24-7** A floor stand is used to support long workpieces and keep them level.

N-4 Be sure that the machine is stopped when setting the work and saw for cutting.

**5.** Use caution when applying extra pressure to the saw frame, as this generally causes work to be cut out of square.

**6.** When several pieces of the same length are required, set the stop gage which is supplied with most cutoff saws (Fig. 24-5).

**7.** When holding short work in a vise, be sure to place a short piece of the same thickness in the opposite end of the vise. This will prevent the vise from twisting when it is tightened (Fig. 24-6).

## To Saw Work to Length

**1.** Check the solid vise jaw with a square to make sure it is at right angles to the saw blade. (See N-4.)

**2.** Place the material in the vise, supporting long pieces with a floor stand (Fig. 24-7).

**3.** Lower the saw frame until the blade just clears the work. Keep it in this position by engaging the ratchet lever or by closing the hydraulic valve.

**4.** Adjust the roller guide brackets until they just clear both sides of the material to be cut.

**5.** Hold a steel rule against the edge of the saw blade and move the material until the correct length is obtained.

**6.** Always allow 1/16 in. (1.58 mm) for each 1 in. (25.4 mm) of thickness longer than required to compensate for any saw runout (slightly angular cut caused by hard spots in steel or a dull saw blade).

**7.** Tighten the vise and recheck the length from the blade to the end of the material to make sure that the work has not moved.

**8.** Raise the saw frame slightly, release the ratchet lever or open the hydraulic valve and then start the machine.

**9.** Lower the saw frame slowly until the revolving saw blade just touches the work.

**10.** When the cut has been completed, the machine will shut off automatically.

## REVIEW QUESTIONS

**1.** What part on a horizontal bandsaw prevents the blade from wandering and twisting?

**2.** Of what two materials are saw blades generally manufactured?

**3.** What pitch saw blade is recommended for the following?

(a) Large sections
(b) Tool steel
(c) Angle iron
(d) Thin tubing

**4.** Why should at least two teeth of a saw blade be in contact with the work at all times?

**5.** What three things are important for efficient sawing?

**6.** In which direction should the teeth of a saw blade point?

**7.** What part of thc horizontal bandsaw is used to tighten the saw blade on the pulleys?

**8.** Name two functions of cutting fluids when sawing material.

**9.** How should thin material be held in a bandsaw vise?

**10.** What part of the horizontal bandsaw can be used when several pieces of the same length must be cut off?

# UNIT 25

# CONTOUR BANDSAW PARTS AND ACCESSORIES

The contour bandsaw is generally fabricated from steel rather than cast, as are most other machine tools. This machine tool has enabled industry to quickly cut work to size and shape while at the same time removing material in large sections rather than in chips as with other machines.

Contour bandsaws are generally available in two types. Machines with two band carrier pulleys are generally used in school shops and toolrooms. Industrial models which have a larger sawing capacity generally have three band carrier pulleys.

The contour bandsaw has many advantages in the metal-cutting trade. Some of the more common ones are illustrated in Fig. 25-1.

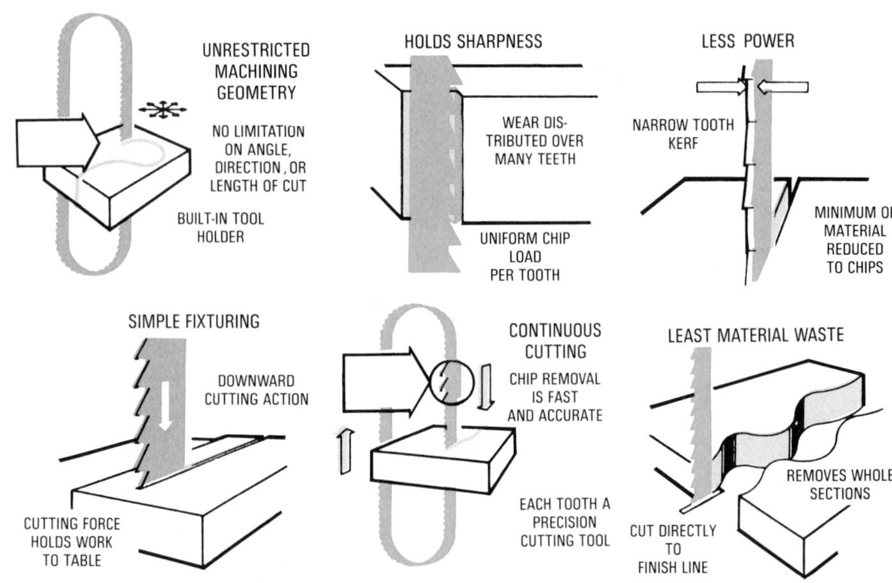


**Fig. 25-1** Some major advantages of the contour-cutting bandsaw. (*Courtesy The DoAll Company.*)

# OBJECTIVES

After completing this unit, you should be able to:

**1.** Name and state the purpose of the main operative parts of a contour bandsaw

**2.** Recognize and state the purpose of three common tooth forms and sets

**3.** Determine the pitch of the saw blade required for each job

**4.** Calculate the length of saw band required for a two-pulley bandsaw

# CONTOUR BANDSAW PARTS AND OPERATIONS

The contour bandsaw is generally fabricated from steel rather than cast, as are most other machine tools. Even though there are a great variety of types and sizes, all contain parts which are basic to all machines.

The basic parts are found on the base, column, or head of the machine, Fig. 25-2.

## Base

The *base* of the contour bandsaw supports the column and houses the assembly that provides the drive for the blade.

- The *lower pulley* is driven by a variable-speed pulley and can be adjusted to various speeds by the *variable-speed handwheel.*
- The *table* is attached to the base by means of a *trunnion.* It can be tilted 45° to the right and 10° to the left for making angular cuts by turning the *table tilt handwheel.* A removable *filler plate slide* and a *center plate* are mounted in the table.
- *The lower saw guide,* attached to the trunnion, supports and guides the blade to keep it from twisting.
- A removable *filler plate slide* and a *center plate* are mounted in the table.

## Column

The column supports the *head,* the left-hand *blade guard, welding unit,* and the *variable-speed handwheel.*

- The variable-speed handwheel is used to control the speed of the bandsaw for various types and sizes of material. (See N-1.)
- The welding unit, attached to the column, is used to weld, anneal, and grind the saw blades.
- The blade tension indicator and the speed indicator are also located in the column.

N-1 The variable-speed handwheel adjusts the speed of the saw band.

## Head

The parts found in the head of a saw

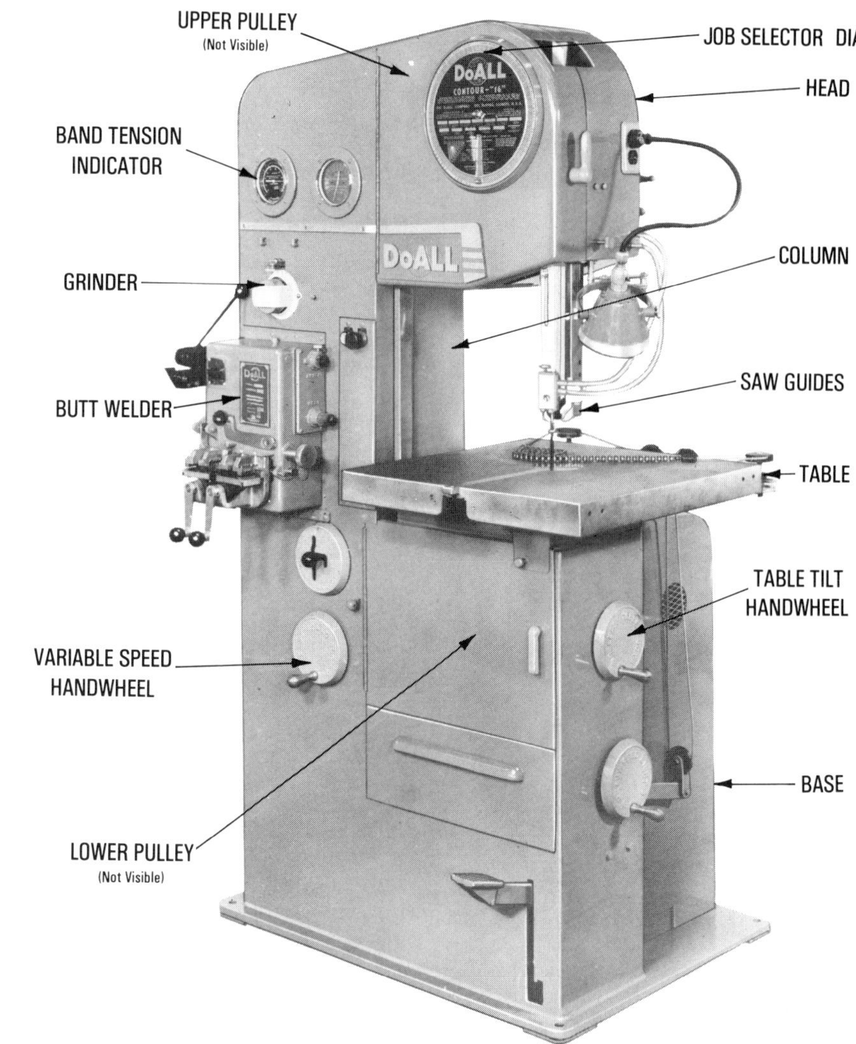


**Fig. 25-2** The main parts of a contour bandsaw. (*Courtesy The DoAll Company.*)

are generally used to guide or support the saw band.

- The *upper saw pulley* supports the saw band, which is adjusted by the tension and tracking controls.
- The *upper saw guide,* attached to the *saw guide post,* supports and guides the saw blade to keep it from twisting. It can be adjusted vertically to accommodate various sizes of work. (See N-2.)
- The *saw guard* and the *air nozzle*, for keeping the area being cut free of chips, are also found in the head.

N-2 The upper and lower saw guides support and guide the saw band.

N-3 Material is removed in one piece instead of in chips.

## BANDSAW OPERATIONS

There are many types of operations which may be performed faster on the contour bandsaw than on any other machine. In addition to saving time, using a contour bandsaw will also result in a saving in material, since large sections of a workpiece can be removed as a solid instead of reduced to steel chips as on conventional machines. (See N-3.) Some more common operations are shown in Fig. 25-3A to D.

### Notching

Sections of metal can be removed in one piece rather than in chips. This can result in the saving of considerable material especially when large sections are cut.

### Slotting

This operation can be done quickly and accurately without the need of expensive fixtures. The cutting action of the

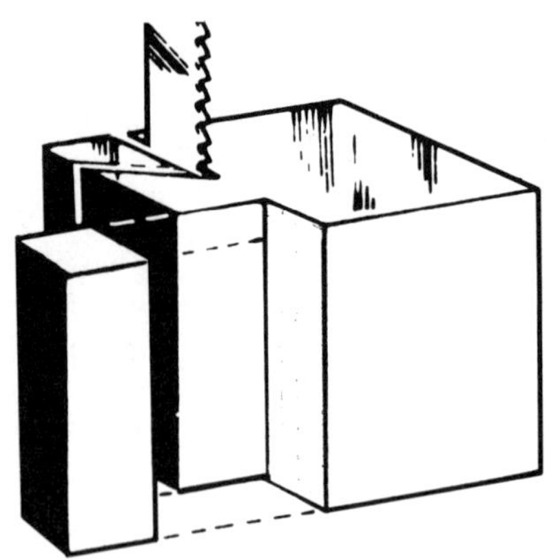

Fig. 25-3A Notching.

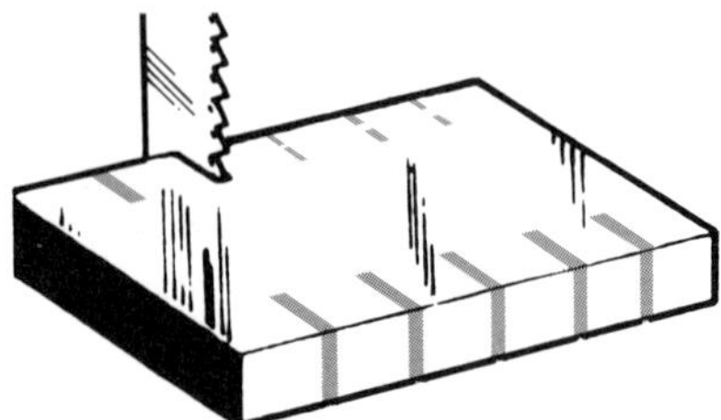

Fig. 25-3B Slotting.

Fig. 25-3C Angular cutting.

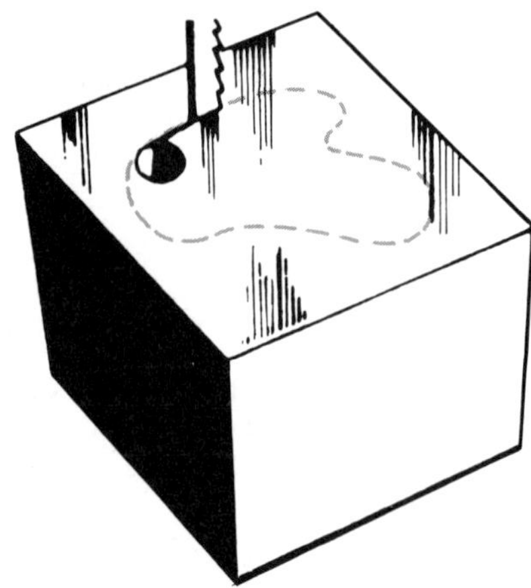

Fig. 25-3D Radius cutting.

N-4 Select a pitch blade where at least two teeth contact the work at all times.

saw blade keeps the work down firmly on the machine table.

### Angular Cutting

The work may be clamped at any angle and fed through the saw. The table may be tilted for compound angles. The need for expensive holding jigs and fixtures is eliminated.

### Radius Cutting

Internal or external contours may be cut easily. Internal sections are generally removed in one piece as shown. Since the saw blade can cut fairly close to layout lines, the time required for finishing operations is reduced.

## SAW-BLADE TYPES AND SELECTION

There are three types of saw blades generally used on bandsaws: carbon-alloy, high-speed steel, and carbide-tipped blades. To obtain the best sawing results, it is important to select the proper blade for a particular job. The size and type of material to be cut generally determines the tooth form, pitch set, and type of saw band which should be used.

### Tooth Forms

Carbon and high-speed steel blades are available in three types of tooth forms (Fig. 25-4).

#### Precision or Regular Tooth

This is the most generally used type of tooth. It has a zero rake angle and about 30° back-clearance angle. It is used when a fine finish and an accurate cut are required.

#### Claw or Hook Tooth

This tooth form has a positive rake on the cutting face and slightly less back clearance than the precision or buttress blade. It has the same general application as the buttress-tooth form. It is faster cutting and longer lasting than the buttress tooth, but will not produce as smooth a finish.

#### Buttress, or Skip Tooth

This tooth form is similar to the precision tooth; however, the teeth are spaced farther apart to provide more chip clearance. The tooth angles are the same as on the precision type. Buttress, or skip-tooth, blades are used to advantage on thick work sections and on deep cuts in soft material.

### Pitch

Each of the tooth forms is available in various pitches, or numbers of teeth per standard reference length. Inch saw-blade pitch is determined by the number of teeth per inch of length, metric blade pitch by the number of teeth per 25 mm, Fig. 25-5.

The thickness of the material to be cut determines the pitch of the blade to be used. When cutting thick materials, use a coarse-pitch blade: thin materials require a fine-pitch blade. It is well to remember when the proper pitch is being selected that there must be at least two teeth in contact with the material being cut. (See N-4.)

### Set

The set of a blade is the amount that the teeth are offset on either side of the center to produce clearance for the back of the band or blade. There are three common set patterns (Fig. 25-6).

*Raker set* has one tooth offset to the right and one to the left; the third tooth is straight. This is the most common pattern and is used for most sawing applications.

Fig. 25-4A Precision saw-tooth form. (*Courtesy The DoAll Company.*)

Fig. 25-4B Buttress saw-tooth form. (*Courtesy The DoAll Company.*)

Fig. 25-4C Claw saw-tooth form. (*Courtesy The DoAll Company.*)

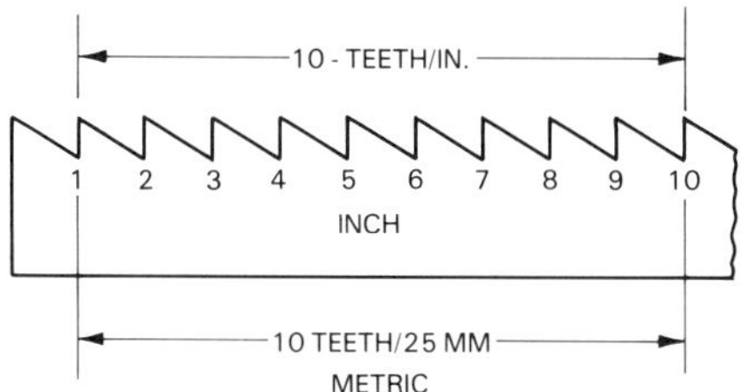


**Fig. 25-5** The pitch of a saw blade.

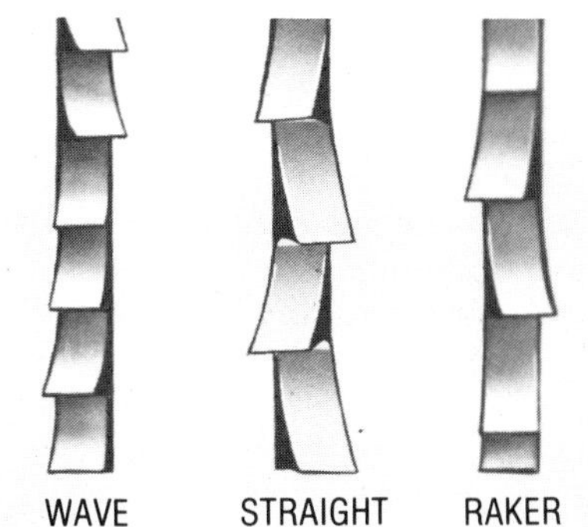


**Fig. 25-6** Common set patterns. (*Courtesy The DoAll Company.*)

N-5 Use the thickest blade possible for strength.

N-6 When measuring the center distance between wheels, make sure the top wheel is lowered approximately 1 in. (25 mm) from its top position. This will allow for stretching when the blade is tensioned.

N-7 Blade length = 2 (C.D.) + P.C.
where C.D. = center distance
P.C. = pulley circumference

*Wave set* has a group of teeth offset to the right and the next group to the left, a pattern which produces a wavelike appearance. Wave set blades are generally used when the cross section of the workpiece changes, such as on structural-steel sections or on pipe.

*Straight set* has one tooth offset to the right and the next to the left. It is used for cutting light nonferrous castings, thin sheet metal, tubing, and Bakelite.

## Width

When making straight, accurate cuts, it is advisable to select a wide blade. Narrow blades are used to cut small radii. Radius charts, advising the proper width blade to use for contour sawing, are generally found on all bandsaws. When selecting a blade for contour cutting, it is advisable to select the widest blade which can cut the smallest radius on the workpiece.

## Gage

The gage is the thickness of the saw blade. This has been standardized according to the width of the blade. Blades up to ½-in. (12.7-mm) wide are .025-in. (0.64-mm) thick; ⅝- and ¾-in. (15.88- and 19.05-mm) blades are .032-in. (0.81-mm) thick; and 1-in. (25.4-mm) blades are .035-in. (0.89-mm) thick. Since thick blades are stronger than thin blades, it is recommended that the thickest blade possible be used for sawing tough material. (See N-5.)

# JOB REQUIREMENTS

The bandsaw operator should be familiar with the various types of blades and be able to select the one which will do the job to the specified requirements of finish and accuracy at the lowest cost. Table 25-1 will serve as a guide to more efficient cutting.

**TABLE 25-1 JOB REQUIREMENT CHART**

| TO INCREASE | TRY ONE OR MORE OF THE FOLLOWING | | | | | | | | |
|---|---|---|---|---|---|---|---|---|---|
| | Faster Tool Velocity (more teeth per min) | Slower Tool Velocity (less teeth per min) | Finer Pitch Band Tool (more teeth & smaller gullets) | Coarser Pitch Band Tool (less teeth & larger gullets) | Slower Feeding Rate (decreases chip load) | Faster Feeding Rate (increases chip load) | Medium Feeding Rate | Claw Tooth (positive rake angle) | Precision & Buttress (0° rake angle) |
| CUTTING RATE | ✓ | | | ✓ | | ✓ | | ✓ | |
| TOOL LIFE | | ✓ | ✓ | | | | ✓ | ✓ | |
| FINISH | ✓ | | ✓ | | ✓ | | | | ✓ |
| ACCURACY | ✓ | | | | ✓ | | | | |

## Blade Length

Metal-cutting saw bands are usually packaged in coils about 100 to 150 ft (30.5 to 152 m) in length. The length required is cut from the coil and its two ends are then welded together.

To calculate the length required for a two-wheel bandsaw, take twice the center distance between each pulley and to it add the circumference of one wheel. This is the total length of the saw band, Fig. 25-7. (See N-6.)

**Example:**

Calculate the length of a saw blade for a bandsaw which has:
(*a*) Two 24-in.-diameter pulleys and a center-to-center distance of 48 in.
(*b*) Two 600-mm-diameter pulleys and a center-to-center distance of 1200 mm.

**Solution (see N-7):**

(*a*) Blade length = 2 (C.D.) + P.C.
= 2 (48) + (24 × 3.1416)
= 96 + 75.4
= 171.4 in.

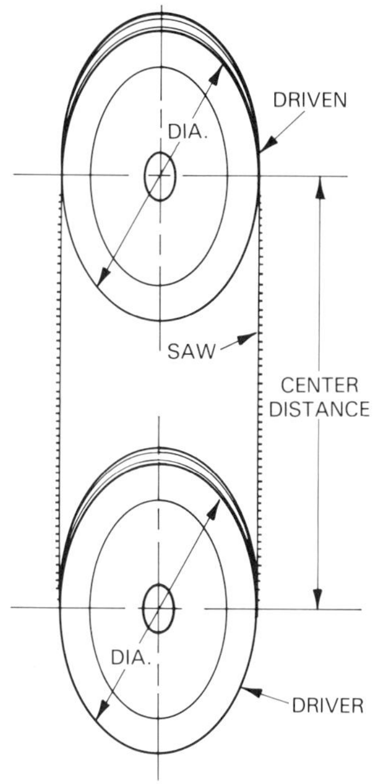


**Fig. 25-7** Calculating the length of a saw band.

(*b*) Blade

length = 2 (C.D.) + P.C.
= 2 (1200) + (600 × 3.1416)
= 2400 + 1885
= 4285 mm

## REVIEW QUESTIONS

**1.** Name two types of contour bandsaws.
**2.** List six advantages of a contour bandsaw.
**3.** How is the speed adjusted on a contour bandsaw?
**4.** What supports and guides the blade to keep it from twisting?
**5.** For what purpose are the following types of saw blades used?
(a) Precision
(b) Claw
(c) Buttress
**6.** What general rule applies when selecting the pitch of a saw blade?
**7.** Describe the following blade sets.
(a) Raker
(b) Wave
(c) Straight
**8.** Calculate the length of saw band required for the following contour bandsaws:
(a) Two 30-in.-diameter pulleys with a center-to-center distance of 50 in.
(b) Two 750-mm-diameter pulleys with a center-to-center distance of 1250 mm.

# UNIT 26

# CONTOUR BANDSAW OPERATIONS

The contour bandsaw provides a machinist with the ability to cut material close to the form required quickly while at the same time removing large sections which can be used for other jobs. The versatility of a bandsaw can be increased by using various attachments and cutting tools. Operations such as sawing, filing, polishing, grinding, and friction and high-speed sawing are all possible on a bandsaw with the proper attachments and cutting tools.

## OBJECTIVES

After completing this unit, you should be able to:

**1.** Set up the machine and saw external sections to within 1/32 in. (0.8 mm) of layout lines
**2.** Saw internal sections to within 1/32 in. (0.8 mm) of layout lines
**3.** Set up a contour bandsaw to file to a layout
**4.** Know the purpose of special cutting tools used on a contour bandsaw

## SAWING EXTERNAL AND INTERNAL SECTIONS

With the proper machine setups and attachments a wide variety of operations can be performed on a contour bandsaw. The most common operations are external and internal cutting. For both operations, it is important that a person be able to select, weld, and mount the correct saw to suit the size and type of workpiece material.

### To Mount a Saw Blade

**1.** Select the correct saw guides for the width of blade being used.
**2.** Mount both upper and lower saw guides on the machine (Fig. 26-1).
**3.** Adjust the upper saw guide until it clears the top of the work by 1/4 in. (6.35 mm).
**4.** Insert the saw plate in the table.
**5.** Place the saw band on both upper and lower pulleys, making sure that the teeth are pointed in the direction of band travel (toward the table). (See N-1.)

N-1 Be sure that the cutting edge of the saw teeth are pointing toward the table.

**6.** Adjust the upper pulley with the tension handwheel (Fig. 26-2) until some tension is registered on the tension gage.
**7.** Set the gearshift lever to neutral and turn the upper pulley by hand to see that the saw blade is tracking properly on the pulley.
**8.** Reengage the gearshift lever and start the machine.

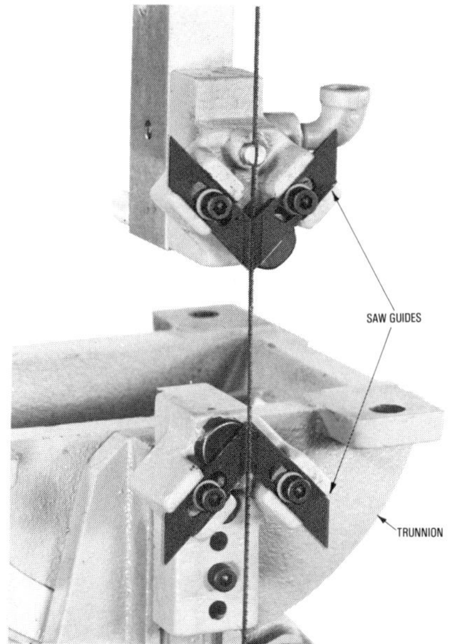


Fig. 26-1 Upper and lower saw guides are used to support and guide the saw band.

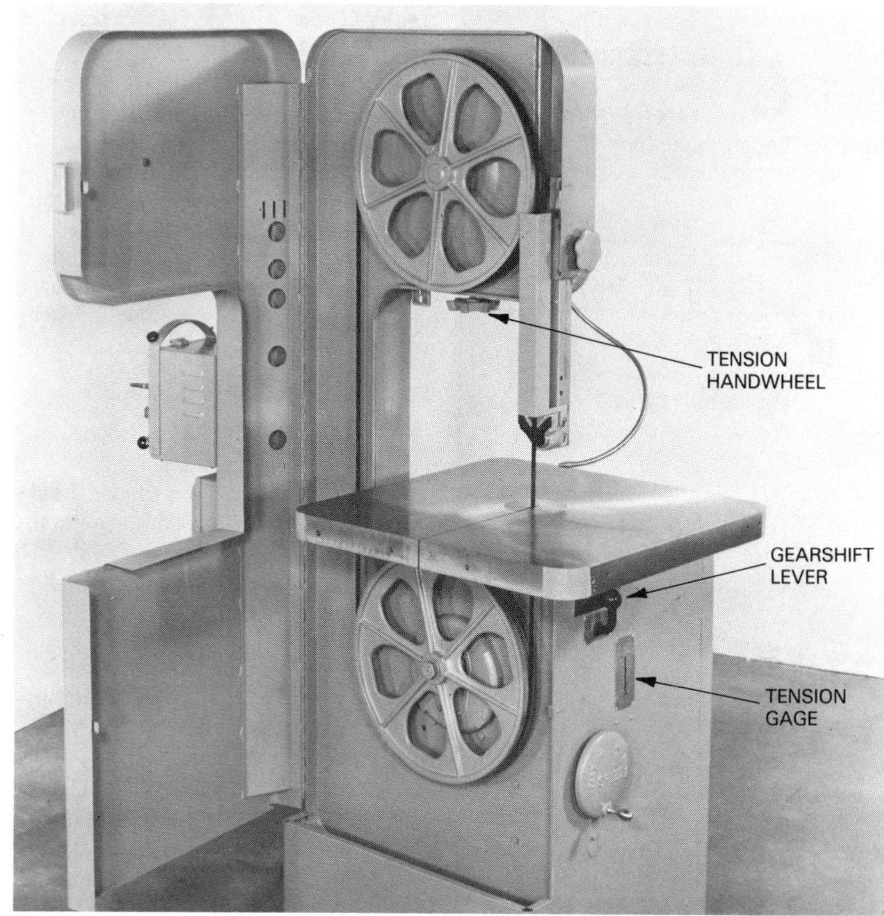


Fig. 26-2 Adjust the tension handwheel until the band tension indicator shows that it is correct for the size of blade being used. (*Courtesy The DoAll Company.*)

Fig. 26-3 The job selector lists cutting speeds for various types and thicknesses of metal. (*Courtesy The DoAll Company.*)

N-2 Keep the fingers well clear of the moving saw blade.

N-3 Use a work-holding jaw or piece of wood when pushing work toward a moving saw blade.

**9.** Adjust the blade to the recommended tension as indicated on the tension gage by using the tension handwheel.

## To Saw to a Layout

**1.** Set the machine to the proper speed for the type of blade and the material being cut. Consult the job selector (Fig. 26-3). (See N-2.)

**2.** Use the work-holding jaw (Fig. 26-4) or a piece of wood to feed the work into the saw. (See N-3.)

**3.** Feed the work into the saw blade steadily. Do not apply too much pressure and crowd the blade.

**4.** Cut to within approximately 1/32 in. (0.8 mm) of the layout lines. This allows material for finishing (Fig. 26-5).

**5.** Never attempt to cut too small a radius with a wide saw. This will damage the blade and the lower drive wheel.

**6.** Drill a hole at every point where sharp turns must be made to allow the workpiece to be turned easily.

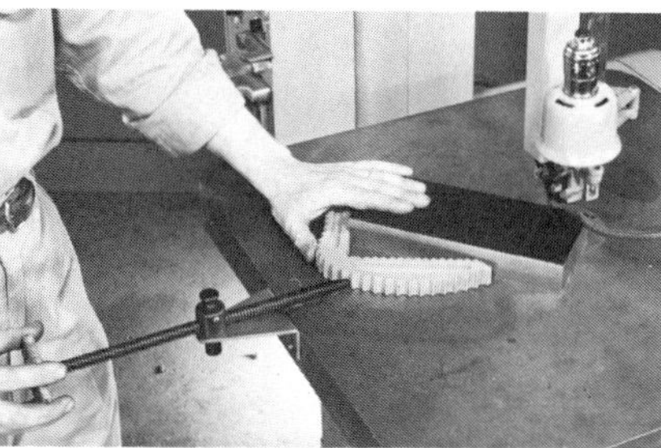

Fig. 26-4 The work-holding jaw is used to feed the work into the saw. (*Courtesy The DoAll Company.*)

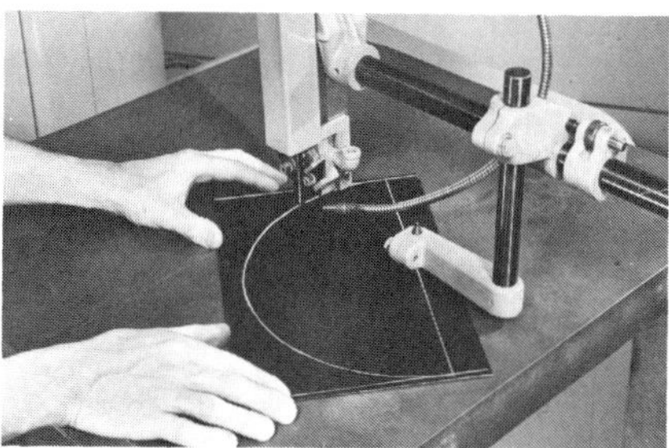

Fig. 26-5 A sample workpiece laid out to show the section which is to be removed. (*"Fundamentals of Band Machining," Delmar Publishers Inc.*)

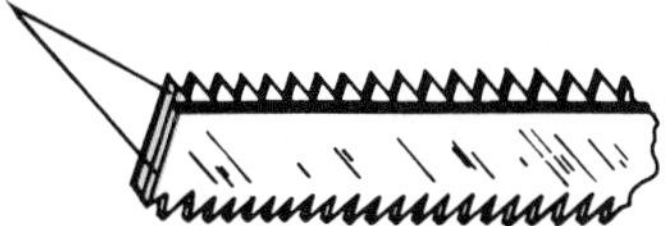

Fig. 26-6 Square both ends of the saw blade at the same time.

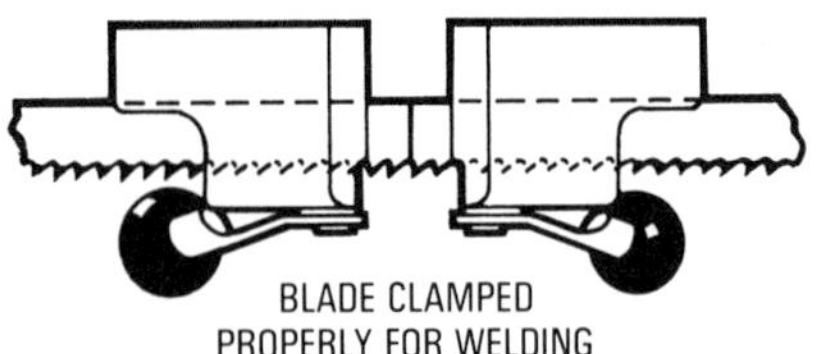


Fig. 26-7 A saw blade properly clamped for welding.

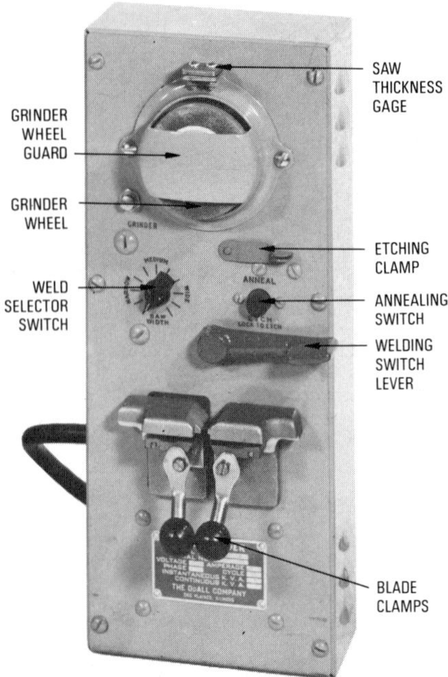


Fig. 26-8 The butt welder is used to weld bandsaw blades. (*Courtesy The DoAll Company.*)

N-4 Do not allow the blade to become too hot when annealing, as it will air-harden on cooling and break easily.

N-5 A hole should be drilled at every point where sharp turns must be made.

N-6 Be sure that the saw teeth are pointed toward the table.

N-7 Always shut off the machine power switch before making any adjustments or installing accessories.

## SAWING INTERNAL SECTIONS

The bandsaw is well suited for removing internal sections of a workpiece. A starting hole must be drilled through the section to be removed to allow the saw blade to be inserted and welded. The saw cut should be made close to the layout line and allow enough material for the finishing operation.

### To Weld a Saw Blade

**1.** Select the proper saw blade for the material to be cut.

**2.** If the ends are not square, hold the saw as in Fig. 26-6 and grind both ends at the same time.

**3.** Insert the saw between the welding jaws, keeping it back against the aligning edge of the jaw.

**4.** Clamp the blade so that both ends are in the center between the two jaws. Make sure that the two ends are square and do not overlap (Fig. 26-7).

**5.** Set the weld selector switch (Fig. 26-8) for the width of the blade being welded.

**6.** Depress the welding lever, holding it down until the weld has cooled.

**7.** Release the stationary jaw clamp and then release the welding lever.

**8.** Clamp the blade in the annealing section of the jaws, having just the saw teeth extending beyond the jaws.

**9.** Press the annealing switch until the saw blade becomes a dull cherry-red color. Repeat this two or three times.

**10.** Allow the saw to cool slowly by intermittently pressing the annealing switch. This prevents the blade from air-hardening. (See N-4.)

**11.** Remove the blade and grind off the weld bead so that it will fit the saw-thickness gage (Fig. 26-8).

**12.** Reanneal the blade to a blue color.

### To Saw Internal Sections

**1.** Drill a hole slightly larger than the width of the saw blade near the edge of the section to be cut. (See N-5.)

**2.** Cut the saw blade and thread it through one of the drilled holes in the workpiece.

Fig. 26-9 Welding a saw blade for sawing internal sections. (*Courtesy The DoAll Company.*)

**3.** Weld the blade and then grind off the weld bead to fit the saw-thickness gage (Fig. 26-9).

**4.** Anneal the weld section to remove the brittleness and prevent the blade from breaking.

**5.** Mount the saw band on the upper and lower pulleys and apply the proper tension for the size of blade. (See N-6.)

**6.** Insert the table filler plate.

**7.** Set the machine to the proper speed for the type and thickness of material being cut.

**8.** Cut out the internal section staying within 1/32 in. (0.8 mm) of the layout line.

**9.** Remove the saw band from the pulleys.

**10.** Cut the blade at the weld point on the cutoff shear.

**11.** Remove the workpiece and the blade.

## FILING ON A CONTOUR BANDSAW

When it is necessary to finish straight or contour forms to the layout lines, it may be done by mounting a band file on a contour bandsaw. The band file consists of a number of short interlocking file segments which are riveted to a steel band (Fig. 26-10). The two ends of the steel band are locked together to form a continuous loop when it is mounted on a bandsaw.

### To Mount a File Band

**1.** Remove the saw blade, upper and lower saw guides, and the table filler plate. (See N-7.)

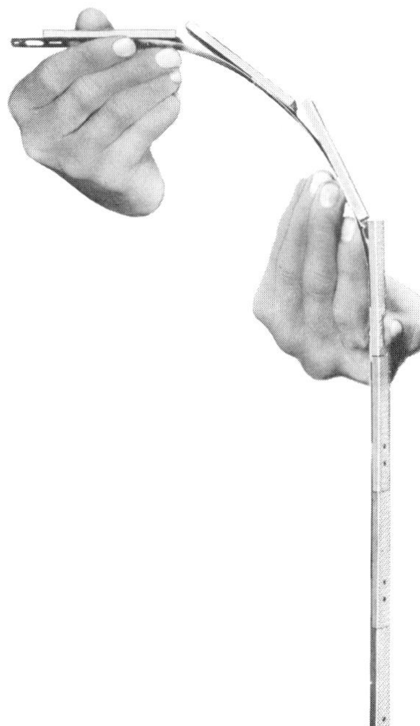

**Fig. 26-10** A band file is made up of a number of individual file segments.

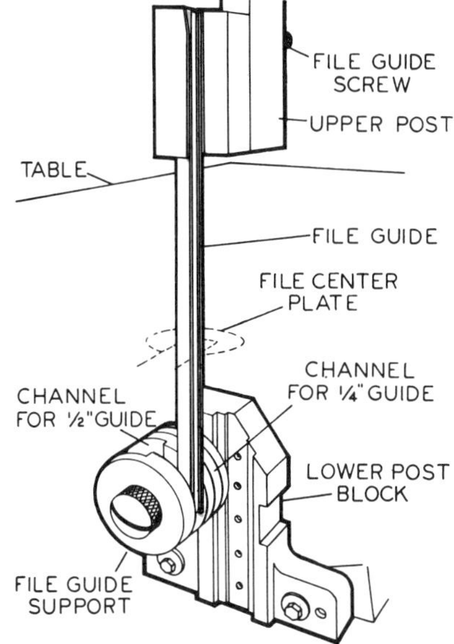


**Fig. 26-11** A file guide and support mounted on the posts of a bandsaw. (*Courtesy The DoAll Company.*)

N-8 Make sure that the *unhinged end* of the file segment is pointing upward.

N-9 Stop the machine before attempting to clean a file.

**2.** Mount the correct file guide support on the lower post block (Fig. 26-11).

**3.** Insert the file plate into the table.

**4.** Mount the file guide to the upper post.

**5.** Lower the upper post until the file guide is below the center of the file guide support.

**6.** Thread the file band upward through the hole in the table center disc. (See N-8.)

**7.** Join the ends of the file band, Fig. 26-12, and mount it on the upper and lower pulleys.

**8.** Adjust the tension control handle to tighten the file band slightly on the machine pulleys.

**9.** Set the speed range transmission in neutral to allow the pulleys to turn freely.

**10.** Revolve the upper pulley by hand and at the same time adjust the tracking controls so that the file band is in the center of the file guide.

**11.** Engage the speed range transmission and start the machine.

**12.** Adjust the file band to the proper tension and set the machine to the proper speed for the material to be filed.

### To File on a Contour Bandsaw

**1.** Consult the job selector, Fig. 26-3, and set the machine to the proper speed. The best filing speeds are between 50 and 100 ft (15 and 30 m) per minute.

**2.** Apply light work pressure to the file band (Fig. 26-12). It not only gives a better finish but prevents the teeth from becoming clogged.

**3.** Keep moving the work sideways against the file to prevent filing grooves in the work.

**4.** Use a file card to keep the file clean. Loaded files cause bumpy filing and scratches in the work. (See N-9.)

## ADDITIONAL BAND TOOLS

Although the saw blade and band file are the most commonly used, several other band tools make this machine particularly versatile.

*Knife-edge blades* (Fig. 26-13) are available with knife, wavy, and scalloped edges, and are used for cutting soft, fibrous materials such as cloth, cardboard, cork, and rubber. Scalloped-edge blades are particularly suited for cutting thin corrugated aluminum. Special guides must be used with knife-edge blades.

**Fig. 26-12** Mounting a file band on a contour bandsaw. (*Courtesy The DoAll Company.*)

**Fig. 26-13** Knife-edge blades are used for cutting soft material.

**Fig. 26-14** Spiral-edge blades have cutting edges around the circumference.

**Fig. 26-15** Intricate contours and patterns can be cut with spiral-edge blades.

*Spiral-edge blades* (Fig. 26-14) are round and have a continuous helical cutting edge around the circumference. This provides a cutting edge of 360° and permits the machining of intricate contours and patterns (Fig. 26-15) without the workpiece having to be turned.

Spiral-edge blades are made in two types: the spring-tempered blade used for plastics and wood, and the all-hard blade used for light metals. These blades are manufactured in diameters of .020, .040, .050, and .074 in. (0.51, 1.02, 1.27, and 1.88 mm). Special guides (Fig. 26-15) are used on the machine with this blade. When spiral blades are welded, sheet copper is used to protect the cutting edges from the welder jaws.

*Polishing bands* are used to remove burrs and provide a good finish to surfaces which have been sawed or filed. They may also be used for sharpening carbide toolbits. A polishing band is a continuous loop of 1-in.- (25.4-mm-) wide abrasive cloth manufactured to a specific length to fit the machine. They are available in several grain sizes in both aluminum oxide and silicon carbide abrasive.

The polishing band is mounted in the same manner as a saw band. The special polishing guide uses the same backup support as the file bands. A special polishing-band center plate is used during band polishing. Most polishing bands are marked with an arrow on the back to indicate the direction in which they should be run.

## REVIEW QUESTIONS

**1.** How close should the saw guides be set to the workpiece?

**2.** In which direction should the teeth of the saw blade point when installing it on the machine?

**3.** How close to the layout lines should the saw cut be made?

**4.** How can the weld of the saw blade be prevented from air-hardening?

**5.** Why is it important that the weld section on a saw blade be annealed?

**6.** What parts must be mounted on the upper and lower posts to support a file band?

**7.** In which direction should the file segment point when it is mounted on a machine?

**8.** What speeds are recommended for filing on a contour bandsaw?

**9.** How can a loaded file be cleaned?

**10.** Name three types of knife-edge blades.

**11.** Name and state the purposes of two types of spiral-edge blades.

**12.** Name two common abrasives used in the manufacture of polishing bands.

# SECTION

# DRILL PRESSES

The drill press, probably the first mechanical device developed in prehistoric times, is one of the most commonly used machines in a machine shop. The main purpose of a drill press is to grip, revolve, and feed a twist drill in order to produce a hole in a piece of metal or other material. Main parts of any drill press include the *spindle,* which holds and revolves the cutting tool, and the *table* upon which the work is held or fastened. The revolving drill or cutting tool is generally fed into the workpiece manually on bench-type drill presses, and either manually or automatically on floor-type drill presses. The variety of cutting tools and attachments which are available allow operations such as drilling, reaming, countersinking, counterboring, tapping, spot-facing, and boring to be performed.

# UNIT 27

## TYPES OF DRILL PRESSES

Drill presses are available in a wide variety of types and sizes to suit industry. (See N-1.) These range from the small hobby-type drill press to the larger, more complex and numerically controlled machines used by industry for production purposes. The most common machines found in a machine shop are the *bench-type sensitive drill press* and the *floor-type drill press*. Other drill presses, such as the upright, post, radial, horizontal, gang, portable, multiple spindle, and numerically controlled types, are variations of the standard machine and are generally designed for specific purposes. (See N-2.)

N-1 The size of a drill press is generally given as the distance from the *edge of the column* to the center of the *drill-press spindle*.

N-2 The vertical capacity of a drill press is measured when the head is in the highest position and the table in its lowest position.

**Fig. 27-2** A floor-type sensitive drill press. (*Courtesy Buffalo Forge Co.*)

## OBJECTIVES

After completing this unit you should be able to

**1.** Recognize and know the purpose of the two categories of sensitive drill presses
**2.** State the purpose of four common drill presses
**3.** Know the function and purpose of the main operative parts of a drill press
**4.** Observe good safety practices required for operating drill presses

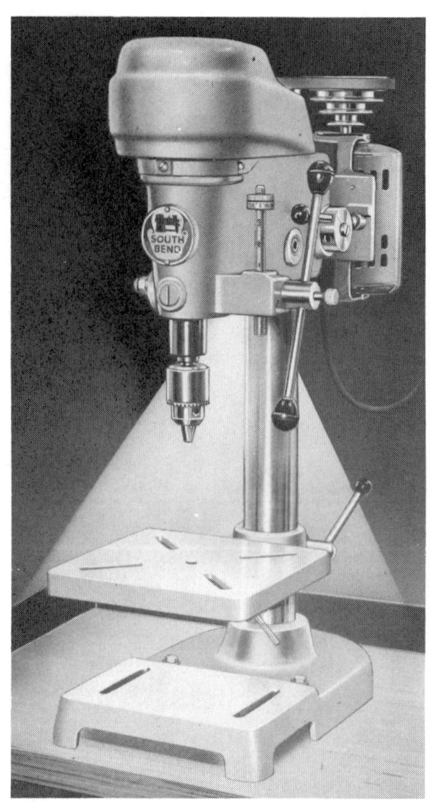


**Fig. 27-1** A bench-type sensitive drill press. (*Courtesy South Bend Lathe, Inc.*)

## SENSITIVE DRILL PRESSES

The simplest and most commonly used drilling machine is the sensitive drill press. The name of this machine is derived from the hand-feed mechanism which enables the operator to "feel" the cutting action and to regulate the down-feed pressure accordingly. Sensitive drill presses fall into two categories: the bench type and the floor type.

The *bench type* (Fig. 27-1) has a short column and a table to support the workpiece. This type of drill press is mounted on a table or bench and is

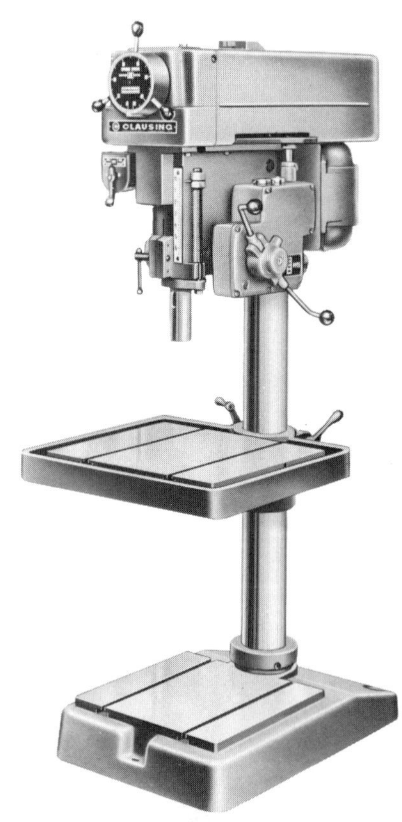


**Fig. 27-3** A standard upright drilling machine with a square or production-type table. (*Courtesy The Clausing Corporation.*)

**Fig. 27-4** A multispindle drilling head. (*Courtesy Cleveland Tapping Machine Co.*)

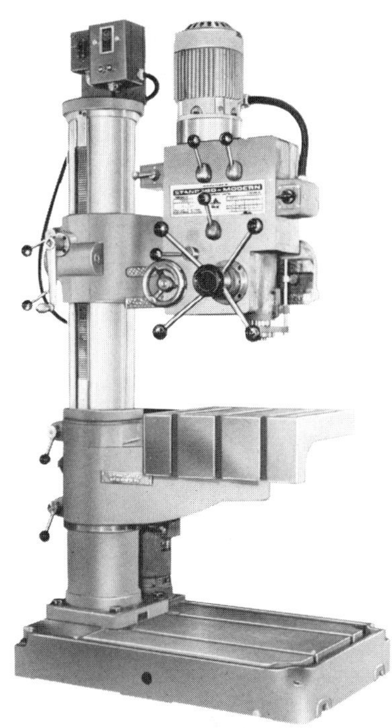

**Fig. 27-5** A radial drill press permits large parts to be drilled. (*Courtesy The A. R. Williams Machinery Co.*)

N-3 Upright drill presses have a gear-drive mechanism.

used for drilling holes in small workpieces. The *floor type* (Fig. 27-2) has a longer column on which the table is adjusted to accommodate longer workpieces. Both bench and floor drill presses are designed to drill holes up to ½-in. (12.7-mm) diameter.

Some specially designed machines of this type are called *super sensitive* or *super speed* because of the high speeds at which they can operate (up to 12,000 r/min). These machines are production drilling machines and are used for drilling small holes less than ¼-in. (6.35-mm) diameter in materials such as copper, brass, aluminum, and other nonferrous metals.

## UPRIGHT DRILLING MACHINES

The *upright drill press* (Fig. 27-3) is similar in design to the sensitive type but is larger and more sturdily built. This type of drill press is more powerful and is used for large-hole drilling and heavier machining operations. It differs in construction from the smaller sensitive machines in that it usually has a gear drive mechanism used to change the spindle speeds and feeds. (See N-3.) The machine may have a mechanical device to raise or lower the work table. Some upright drill presses are equipped with a coolant reservoir built into the base.

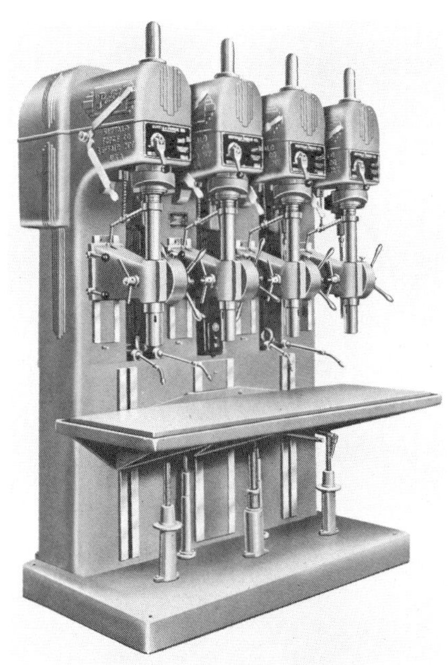

**Fig. 27-6** A gang drilling machine. (*Courtesy Buffalo Forge Co.*)

## MULTIPLE SPINDLE HEAD MACHINES

A *multiple spindle head machine* (Fig. 27-4) is equipped with two or more drilling heads. Each head has several spindles which can be positioned to machine several accurately located holes in a workpiece at one time. All the spindles in one head are set up to perform the same operation—such as drilling. When further operations, such as reaming and tapping, are required, additional heads with the tools in the proper locations may be used. Since most drill heads are used for a specific job, this type of machine is used for production work only.

## THE RADIAL DRILLING MACHINE

When work is too large to be drilled on a sensitive or upright drill press, it can be set up and drilled on a *radial drilling machine* (Fig. 27-5). In this type of drilling machine, the drill can be positioned over the hole location and locked in place without moving the workpiece. Thus, several holes can be drilled in various locations without moving the workpiece.

## GANG DRILLING MACHINE

The *gang drilling machine* (Fig. 27-6), also called a multiple spindle drill, is basically a series of single spindle drills mounted on a long common table. This machine is designed for mass production and is used to perform a number of drill-press operations in sequence. Each separate and individual head may be fitted with a different cutting tool such as a twist drill, a countersink, or a reamer. The work then is moved along the table to each successive work station where another operation is performed. This setup eliminates the need for changing cutting tools for each subsequent operation.

## DRILL-PRESS PARTS

Although drill presses are manufactured in a wide variety of types and sizes, all drilling machines contain certain

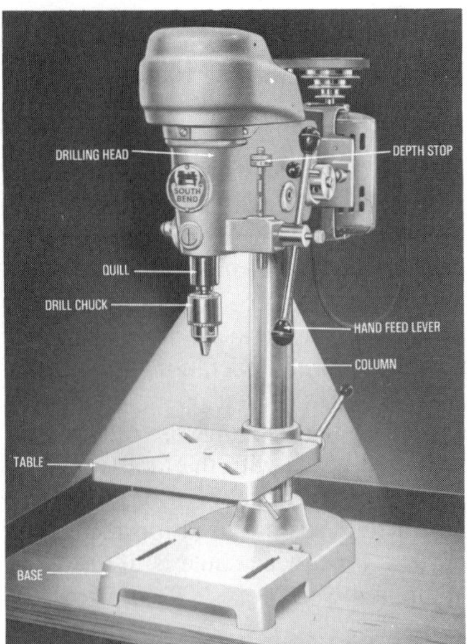


**Fig. 27-7** Parts of a bench-type sensitive drill press. (*Courtesy South Bend Lathe, Inc.*)

basic parts. The main parts on the bench- and floor-type models are *base, column, table,* and *drilling head* (Fig. 27-7). The floor-type model is larger and has a longer column than the bench type.

## Base

The base, usually made of cast iron, provides stability for the machine and also rigid mounting for the column. The base is usually provided with holes so that it may be bolted to a table or bench. The slots or ribs in the base allow the work-holding device or the workpiece to be fastened to the base.

## Column

The column is an accurate cylindrical post which fits into the base. The table, which is fitted to the column, may be raised or lowered to any point between the base and head. The head of the drill press is mounted near the top of the column.

## Table

The table, either round or rectangular in shape, is used to support the workpiece to be machined. The table, whose surface is at 90° to the column, may be raised, lowered, and swiveled around the column. On some models it is possible to tilt the table in either direction for drilling holes on an angle. Slots are provided in most tables to allow jigs, fixtures, or large workpieces to be clamped directly to the table.

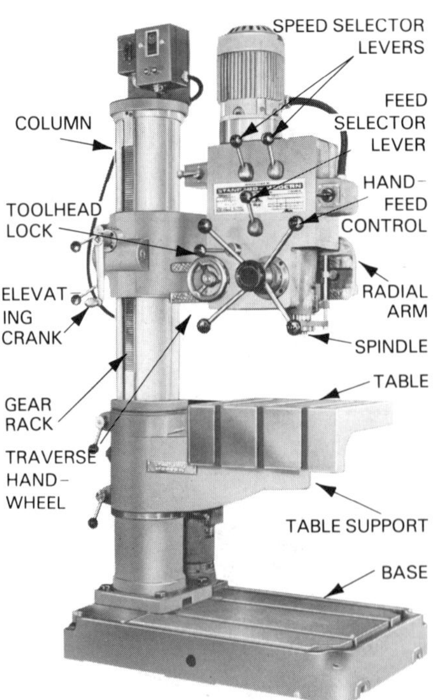


**Fig. 27-8** Parts of a radial drill press. (*Courtesy The A. R. Williams Machinery Co.*)

## Drilling Head

The head, mounted close to the top of the column, contains the mechanism which is used to revolve the cutting tool and advance it into the workpiece. The *spindle,* which is a round shaft that holds and drives the cutting tool, is housed in the *spindle sleeve* or *quill.* The spindle sleeve does not revolve but slides up and down inside the head to provide a downfeed for the cutting tool. The end of the spindle may have a tapered hole to hold taper shank tools, or may be threaded or tapered for attaching a *drill chuck* (Fig. 27-7).

The *hand-feed lever* is used to control the vertical movement of the spindle sleeve and the cutting tool. A *depth stop,* attached to the spindle sleeve, can be set to control the depth that a cutting tool enters the workpiece.

# RADIAL DRILLING MACHINE

The *radial drilling machine* (Fig. 27-8), sometimes called a radial-arm drill, has been developed primarily for the handling of larger workpieces than is possible on upright machines. The advantages of this machine over the upright drill are:

**1.** Larger and heavier work may be machined.

**2.** The drilling head may be easily raised or lowered to accommodate various heights of work.

**3.** The drilling head may be moved rapidly to any desired location, while the workpiece remains clamped in one position, a feature that permits greater production.

**4.** The machine has more power; thus, larger cutting tools can be used.

**5.** On universal models, the head may be swiveled so that holes can be drilled on an angle.

## RADIAL DRILL-PRESS PARTS

The radial drill press (Fig. 27-8) contains all the basic parts found on other types of drilling machines, plus the radial arm. Although the function of each part of the radial drill press is similar to that of other types of drill presses, the size and shape of each part are modified to suit the purpose of the machine.

### Base

The base (Fig. 27-8) is made of heavy-ribbed cast iron or semisteel construction. The base provides rigidity to the machine and supports the vertical column. The base has a number of T-slots for clamping large workpieces or a worktable directly to the base. A coolant trough which surrounds the working surface drains into the coolant reservoir section of the base. Most standard bases have provisions for attaching special base extensions to increase the versatility of the radial drill press.

### Column

The column is a sturdy cylindrical member fitted vertically into the base of the machine. The column supports the radial arm, which can be raised and lowered by engaging the elevating screw. The radial arm can be swung around the column in a complete arc on some models and in a partial arc on other models.

### Radial Arm

The radial arm is attached to the column at 90° to the column surface. The radial arm is of heavy, box-type construction. It can be raised and lowered by means of a power elevating screw. When the *elevating arm lever* (Fig. 27-8) is moved to the ''up'' position, the radial arm rises; moving the lever down lowers the arm. The radial arm is clamped securely to the column when the lever is returned to the center position.

### Drilling Head

The drilling head (Fig. 27-8) is mounted on the radial arm on dovetail guides and can be moved to any position along the length of the arm. The *traverse handwheel* is used to move the drilling head along the radial arm. This handwheel turns a gear that engages into a rack on the arm. The spindle feed and speed are controlled by selector levers that engage the proper gears in the drilling head.

The spindle contains an internal Morse taper to accommodate cutting tools and may be advanced toward the work manually or automatically. Rapid hand feed is accomplished by turning the four-lever feed-control handles; the fine-feed handwheel provides a fine manual feed. The automatic spindle feed is engaged by moving the lever feed-control handles away from the drilling head. On universal models, the drilling head can be swiveled for drilling holes on an angle.

## DRILL-PRESS SAFETY

The drill press is probably the most common machine tool used in industry, school shops, and the home. Because it is so common, good safety practices which can prevent accidents are too often ignored. Before operating a drill press, the operator should become familiar with the safety rules in order to avoid an accident and personal injury.

### Safety Rules

**1.** Never wear any loose clothing or ties around a machine. Roll up the sleeves to above the elbow to prevent getting caught in the machine (Fig. 27-9). (See N-4.)

**2.** Long hair should be protected by a hair net or shop cap to prevent it from becoming caught in the revolving parts on a drill press.

**3.** Never wear rings, watches, or bracelets while working in a machine shop (Fig. 27-10).

**4.** Always wear safety glasses when operating any machine.

**5.** Never attempt to set the speeds, adjust or measure the work until the machine is completely stopped.

**6.** Keep the work area and floor clean and free of oil and grease.

**7.** Never leave a chuck key in a drill chuck at any time.

**8.** Always use a brush to remove chips.

**Fig. 27-9** Loose clothing can be very dangerous around machine tools.

N-4 Always follow good safety practices.

Fig. 27-10 Wearing rings and watches can be dangerous in a machine shop.

**9.** Never attempt to hold work by hand when drilling holes larger than ½ in. (12.7 mm) in diameter. Use a clamp or table stop to prevent the work from spinning.
**10.** Ease up on the drilling pressure as the drill breaks through the workpiece. This will prevent the drill from pulling into the work and breaking.
**11.** Always remove the burrs from a hole which has been drilled.

## REVIEW QUESTIONS

**1.** Name six operations which can be performed on a drill press.
**2.** Name two common types of drill presses.
**3.** How is the size of a drill press generally given?
**4.** For what purpose are the following drill presses used?
(a) Upright
(b) Multispindle head
(c) Radial
(d) Gang
**5.** Name four main parts of a drill press.
**6.** State the purpose of the following drill-press parts:
(a) Spindle
(b) Spindle sleeve
(c) Hand-feed lever
(d) Depth stop
**7.** State three reasons why the radial arm makes the radial drill press so versatile.
**8.** Why is loose clothing and long hair dangerous around a machine?
**9.** Why should drilling pressure be eased as the drill breaks through the work?

# UNIT 28

# TWIST DRILLS

A *twist drill* is an end-cutting tool used to produce a hole in a piece of metal or other material. The most common drill manufactured has two cutting edges (lips) and two straight or helical flutes which provide the cutting edges, admit cutting fluid, and allow the chips to escape during the drilling operation.

## OBJECTIVES

After completing this unit you should be able to:

**1.** Identify and state the purpose of four different types of twist drills
**2.** Identify and state the purpose of four main parts on a standard twist drill
**3.** Know the importance of the point angle and clearance on a twist drill
**4.** Measure drills for size

## TYPES OF DRILLS

A variety of twist-drill styles are manufactured to suit specific drilling operations, types and sizes of material, high production rates, and special applications. The design of drills may vary in the number and width of the flutes, the amount of helix or rake angle of the flutes, or the shape of the land or margin. In addition, the flutes may be straight or helical, and the helix may be a right-hand or left-hand helix. Only commonly used drills are covered in this text. (See N-1.) For special-purpose drills, consult a manufacturer's catalog.

N-1 The jobbers drill, which has two helical flutes, is the most commonly used drill.

Twist drills are manufactured from carbon tool steel, high-speed steel, and cemented carbides. *Carbon-steel drills* are generally used in hobby shops and not recommended for machine shop work since the cutting edges tend to wear down quickly. *High-speed steel drills* are commonly used in machine shop work because they can be operated at twice the speed of carbon-steel

**Fig. 28-1** A general-purpose drill. (*Courtesy The Cleveland Twist Drill Co.*)

**Fig. 28-2** A low-helix drill. (*Courtesy The Cleveland Twist Drill Co.*)

**Fig. 28-3** A high-helix drill. (*Courtesy The Cleveland Twist Drill Co.*)

**Fig. 28-4** A core drill. (*Courtesy The Cleveland Twist Drill Co.*)

**Fig. 28-5** A coolant hole drill. (*Courtesy The Cleveland Twist Drill Co.*)

**Fig. 28-6** A straight flute drill. (*Courtesy The Cleveland Twist Drill Co.*)

N-2 High-speed drills are very common in a machine shop.

drills and the cutting edges can withstand more heat and wear. (See N-2.) *Cemented-carbide drills,* which can be operated much faster than high-speed steel drills, are used to drill hard materials. Cemented-carbide drills have found wide use in production work because they can be operated at high speeds, the cutting edges do not wear rapidly, and they are capable of withstanding higher heat.

The most commonly used drill is the *general-purpose drill,* which has two helical flutes, Fig. 28-1. This drill is designed to give performance on a wide variety of materials, equipment, and job conditions. The general-purpose drill can be made to suit different conditions and materials by varying the point angle and the speeds and feeds used. Straight shank drills are commonly known as *general-purpose jobbers length drills.*

The *low-helix drill,* Fig. 28-2, was developed primarily to drill brass and thin materials. This type of drill is used to drill shallow holes in some aluminum and magnesium alloys. Because of its design, the low helix drill can remove the large volume of chips formed by high rates of penetration when it is used on machines such as turret lathes and screw machines.

The *high-helix drill,* Fig. 28-3, is used to drill deep holes in low-tensile-strength materials, such as aluminum, copper, die-cast materials, and plastics. Chips from these materials tend to jam in the flutes. The wide polished flutes and the high helix help to clear the chips and prevent jamming.

The *core drill,* Fig. 28-4, is used to enlarge previously drilled, cored, or punched holes. The core drill may have three or four flutes. The primary use of this drill is to enlarge cored holes in castings. The advantages of multiflute drills of this type are increased productivity, a better finish, and a greater accuracy with respect to hole size and location.

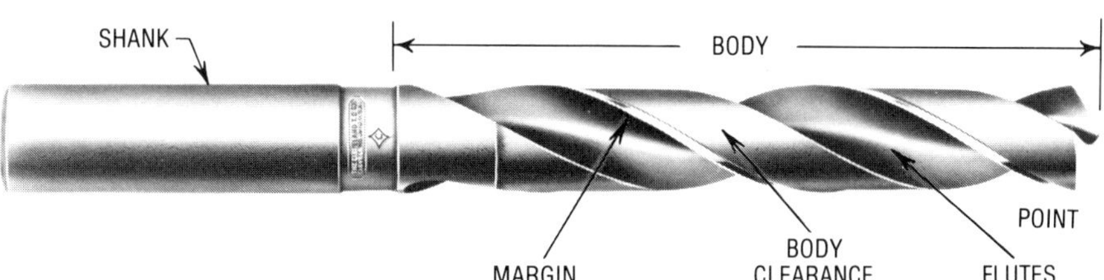



**Fig. 28-7** The main parts of a twist drill. (*Courtesy The Cleveland Twist Drill Co.*)

The *coolant hole drill* (commonly called the oil hole drill), Fig. 28-5, has two holes running through the lands from the shank to the cutting point of the drill. During use, the coolant is forced under high pressure through the holes in the lands to the critical machining area (the chisel edge and point) and serves as a lubricant and coolant. The coolant, which may be oil, water mist, or air, aids in ejecting chips from the hole because of its high pressure and volume. Coolant hole drills are generally used in screw-machine and turret-lathe production work.

The *straight flute drill,* Fig. 28-6, is designed to produce short chips and is recommended for drilling brass, sheet-metal work, and nonferrous materials. The straight flutes of this drill prevent it from digging in (drawing itself into the material) while cutting.

## Twist Drill Parts

A twist drill (Fig. 28-7) may be divided into three main sections: the *shank, body,* and *point.*

### Shank

The shank is the part of the drill which fits into a holding device that revolves the drill. The shanks of twist drills may be either straight or tapered (Fig. 28-7). Straight shanks are generally provided on drills up to ½ in. (12.7 mm) in diameter, while drills over ½-in. (12.7-mm) diameter usually have tapered shanks. Straight-shank drills are held in some type of drill chuck while taper-shank drills fit into the internal taper of the drill press spindle.

The *tang,* at the small end of the tapered shank, is machined flat to fit the slot in the drill press spindle. Its main purpose is to allow the drill to be removed from the spindle with a drift without damaging the shank. The tang may also prevent the shank from turning in the drill press spindle because of a poor fit on the taper or too much drilling pressure.

### Body

The body of a twist drill consists of that portion between the shank and the point. The body contains the *flutes, margin, body clearance,* and *web* of the drill.

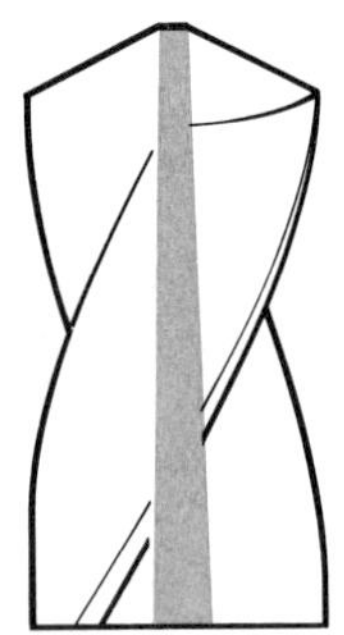

Fig. 28-8 The thickness of the web increases toward the shank to strengthen the drill.

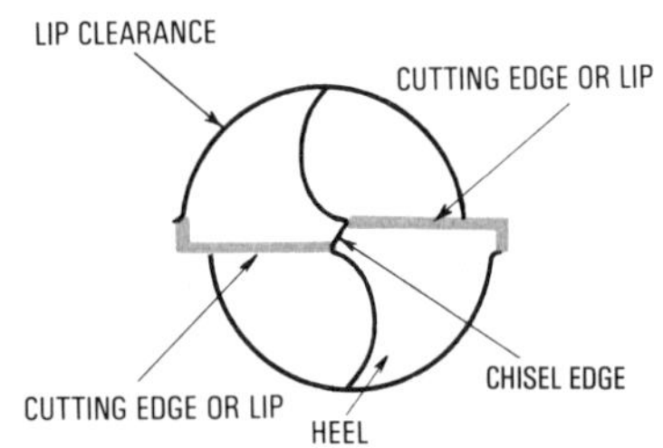


Fig. 28-9 The parts of a twist drill point.

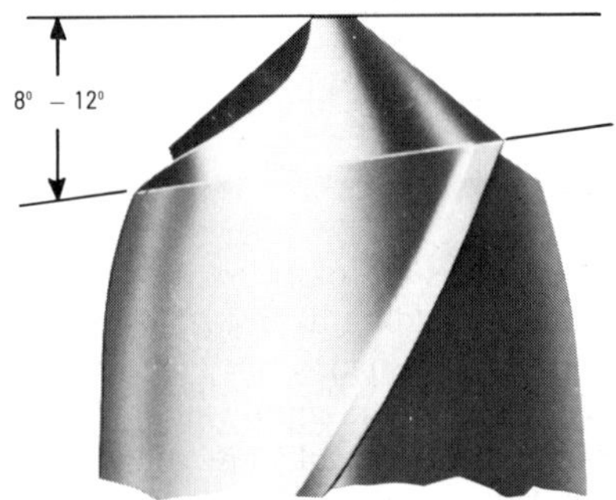


Fig. 28-10 The average lip clearance behind the cutting edge ranges from 8 to 12°.

1. The *flutes* on most drills consist of two or more helical grooves cut along the body of the drill. The flutes form the cutting edges of the drill, admit cutting fluid, and allow chips to escape during the drilling operation.

2. The *margin* is the narrow, raised section on the body immediately next to the flutes. The diameter of the drill is measured across the margin which extends the full length of the flutes. (See N-3.)

3. The *body clearance* is the undercut portion of the body between the margin and the flute.

4. The *web* is the thin metal partition in the center of the drill which extends the full length of the flutes. This part forms the chisel edge at the cutting end of the drill. The web gradually increases in thickness toward the shank to give the drill strength (Fig. 28-8).

## Point

The point of a twist drill consists of the entire cone-shaped cutting end of the drill. The shape and condition of the point are very important to the cutting action of the drill. The drill point consists of the *chisel edge, cutting edges or lips, lip clearance,* and *heel* (Fig. 28-9).

1. The *chisel edge* is that portion which connects the two cutting edges. It is formed by the intersection of the cone-shaped surface of the point. The cutting action of the chisel edge is not very efficient and when drilling holes over ½-in. (12.7-mm) diameter, it is wise to first drill a lead or pilot hole in the workpiece to relieve some of the pressure on the drill point. (See N-4.)

2. The *cutting edges or lips* are formed by the intersection of the flutes and the cone-shaped point. The lips must both be the same length and have the same angle, so that the drill will run true and not cut a hole larger than the size of the drill. (See N-5.)

3. The *lip clearance* is the relief which is ground on the point of the drill extending from the cutting lips back to the heel (Fig. 28-10). Lip clearance allows the lips of the drill to cut into the metal without the heel rubbing. The average lip clearance is from 8 to 12° depending on the type of the material to be drilled.

### Drill-Point Angles and Clearances

For general-purpose drilling, the drill point should be ground to an included angle of 118° (Fig. 28-11A), and the lip clearance should range from 8 to 12°. The drill point for hard materials should be ground to an included angle from 135 to 150° (Fig. 28-11B), and the lip clearance should be from 8 to 10°. For drilling soft materials, the drill point should be ground to an included angle of 90° (Fig. 28-11C) with the lip clearance ranging from 15 to 18°.

## SYSTEMS OF DRILL SIZES

Twist drills are available in both inch and metric sizes. Inch drills are designated by *fractional, number,* and *letter* systems. *Metric* drills are available in various set ranges. The size of straight-shank drills is marked on the shank, while taper-shank drills are generally stamped on the neck between the body and the shank.

1. *Fractional* inch drills are manufactured in sizes from 1/64 to 3½ in. in diameter, varying in steps of 1/64 in. from one size to the next. Drills larger than 3½ in. in diameter must be ordered specially from the manufacturer.

2. *Number* size drills range from the #1 drill (.228 in.) to the #97 drill (.0059 in.). The most common number drill set contains drills from #1 to #60. The large range of sizes enables almost any hole between .0059 in. to .228 in. to be drilled.

N-3 Measure the diameter of a drill across the margin.

N-4 A lead or pilot hole reduces the amount of pressure required to drill large holes.

N-5 The lip angle and length must be the same for a drill to cut properly.

Fig. 28-11 Drill point angles for various materials.

**Fig. 28-12A** Checking a drill for size with a drill gage.

**Fig. 28-12B** Checking a drill for size with a micrometer.

N-6 Measure the diameter of a drill across the margin.

N-7 Use the proper cutting speed for best production and long cutting tool life.

**3.** *Letter* size drills range from A to Z. The letter A drill is the smallest in the set (.234 in.) and the letter Z is the largest (.413 in.).

Metric size drills are available in various sets but are not designated by various systems. The miniature metric drill set ranges from 0.04 to 0.99 mm in steps of 0.01 mm. Straight-shank metric drills are available in sizes from 0.5 to 20 mm ranging in steps of 0.02 to 1 mm, depending on the size. Taper-shank metric drills are available in sizes from 8 to 80 mm.

See the tables in the back of this book for the number, letter, and metric drill size tables.

### Measuring the Size of a Drill

In order to produce a hole to the required size, it is important that the correct size drill is used to drill the hole. It is good practice to always check a drill for size before using it to drill a hole. Drills may be checked for size by two methods: with a drill gage (Fig. 28-12A) and with a micrometer (Fig. 28-12B). When a drill is being checked for size with a micrometer, always be sure that the measurement is taken across the margin of the drill. (See N-6.)

## CUTTING SPEEDS AND FEEDS

The selection of the proper speeds and feeds for the cutting tool to be used and the type of material being drilled are important factors which the operator must consider. These two factors affect the amount of time required to complete an operation (production rate) and how long a cutting tool will perform satisfactorily. Unnecessary time will be wasted if the speed and feed are too low, while the cutting tool will wear quickly if the speed and feed are set too high. (See N-7.)

### Cutting Speed

The speed at which a twist drill should be operated is often referred to as *cutting speed,* surface speed, or peripheral speed. Cutting speed may be defined as the distance in either surface feet or surface meters that a point on the circumference of the drill travels in 1 min. For example, if tool steel has a recommended cutting speed of 85 ft (25.9 m) per minute, the drill-press speed should be set so that a point on the circumference of the drill will travel 85 ft (25.9 m) per minute. The wide range of drill sizes used for drilling holes in the various kinds of metals requires an equally wide range of speeds at which the drills can be efficiently operated. The size of the drill, the material it is made from, and the type of material to be drilled must all be taken into account when determining a safe and efficient speed at which to operate a drill press.

As a result of many years of research, cutting tool and steel manufacturers recommend that various types of metal be machined at certain cutting speeds for the best production rates and the best tool life. The recommended cutting speeds for various materials are listed in Table 28-1.

Whenever reference is made to the *speed* of a drill, the cutting speed in surface feet or in surface meters per minute is implied, and not in revolutions per minute (r/min), unless specifically stated.

### Factors Affecting Drill Speed (r/min)

The calculated drill speed may have to be varied slightly to suit the following factors:

**1.** The type and condition of the machine
**2.** The accuracy and finish of the hole required
**3.** The rigidity of the work setup
**4.** The use of cutting fluid

#### Revolutions per Minute

The number of revolutions necessary to produce the desired cutting speed is called revolutions per minute (r/min). A small drill operating at the same r/min as a larger drill will travel fewer feet or meters per minute and naturally will cut more efficiently at a higher number of r/min.

To find the number of revolutions per minute at which a drill press spindle must be set to obtain a certain cutting speed, the following information must be known:

**TABLE 28-1 CUTTING SPEEDS FOR HIGH-SPEED STEEL DRILLS**

| Size | | Steel casting | | Tool steel | | Cast iron | | Machine steel | | Brass and aluminum | |
|---|---|---|---|---|---|---|---|---|---|---|---|
| | | Cutting speeds | | | | | | | | | |
| | | ft/min 40 | m/min 12 | ft/min 60 | m/min 18 | ft/min 80 | m/min 24 | ft/min 100 | m/min 30 | ft/min 200 | m/min 60 |
| Millimeter | Inch | Revolutions per minute | | | | | | | | | |
| 2 | 1/16 | 2445 | 1910 | 3665 | 2865 | 4890 | 3820 | 6110 | 4775 | 12225 | 9550 |
| 3 | 1/8 | 1220 | 1275 | 1835 | 1910 | 2445 | 2545 | 3055 | 3185 | 6110 | 6365 |
| 4 | 3/16 | 815 | 955 | 1220 | 1430 | 1630 | 1910 | 2035 | 2385 | 4075 | 4775 |
| 5 | 1/4 | 610 | 765 | 915 | 1145 | 1220 | 1530 | 1530 | 1910 | 3055 | 3820 |
| 6 | 5/16 | 490 | 635 | 735 | 955 | 980 | 1275 | 1220 | 1590 | 2445 | 3180 |
| 7 | 3/8 | 405 | 545 | 610 | 820 | 815 | 1090 | 1020 | 1305 | 2035 | 2730 |
| 8 | 7/16 | 350 | 475 | 525 | 715 | 700 | 955 | 875 | 1195 | 1745 | 2390 |
| 9 | 1/2 | 305 | 425 | 460 | 635 | 610 | 850 | 765 | 1060 | 1530 | 2120 |
| 10 | 5/8 | 245 | 350 | 365 | 520 | 490 | 695 | 610 | 870 | 1220 | 1735 |
| 15 | 3/4 | 205 | 255 | 305 | 380 | 405 | 510 | 510 | 635 | 1020 | 1275 |
| 20 | 7/8 | 175 | 190 | 260 | 285 | 350 | 380 | 435 | 475 | 875 | 955 |
| 25 | 1 | 155 | 150 | 230 | 230 | 305 | 305 | 380 | 380 | 765 | 765 |

**1.** The recommended cutting speed of the material to be drilled

**2.** The type of material from which the drill is made

**3.** The diameter of the drill

Apply one of the following formulas to calculate the spindle speed (r/min) at which the drill press should be set.

## INCH DRILLS

N-8
CS = cutting speed of material to be drilled in *feet per minute*
D = diameter of drill in *inches*

$$\text{r/min} = \frac{CS \times 12}{\pi \times D}$$

Since only a few machines are equipped with variable speed drives which allows them to be set to the exact calculated speed, a simplified formula can be used to calculate r/min. (See N-8.) The π (3.1416) on the bottom line of the formula will divide into 12 of the top line approximately four times. This results in a simplified formula which is close enough for most drill presses.

$$\text{r/min} = \frac{CS \times 4}{D}$$

**Example:**

Calculate the r/min at which the drill press should be set to drill a ½-in.-diameter hole in a piece of machine steel. (See Table 28-1 for the cutting speed of machine steel.)

$$\text{r/min} = \frac{CS \times 4}{D}$$

$$= \frac{100 \times 4}{1/2}$$

$$= 800$$

When it is not possible to set the drill press to the exact speed, always set it to the closest speed *under* the calculated speed.

**TABLE 28-2 RECOMMENDED DRILL FEEDS**

| Drill size, inch | Feed per revolution | Drill size, millimeter | Feed per revolution |
|---|---|---|---|
| 1/8 and smaller | .001–.002 | 3 and smaller | 0.02–0.05 |
| 1/8–1/4 | .002–.004 | 3–6 | 0.05–0.10 |
| 1/4–1/2 | .004–.007 | 6–12 | 0.10–0.17 |
| 1/2–1 | .007–.015 | 12–25 | 0.17–0.37 |
| 1–1½ | .015–.025 | 25–38 | 0.37–0.63 |

## METRIC DRILLS

$$\text{r/min} = \frac{\text{cutting speed [meters per minute]}}{\pi \times \text{diameter [millimeters]}}$$

It is necessary to convert the meters in the numerator to millimeters so that both parts of the equation are in the same unit. To accomplish this, multi-

N-9
CS = cutting speed of material to be drilled in *meters per minute*
D = diameter of drill in *millimeters*

ply the cutting speed in meters by 1000 to bring it to millimeters. (See N-9.)

$$\text{r/min} = \frac{CS \times 1000}{\pi \times D}$$

Since only a few machines are equipped with variable-speed drives which allows them to be set to the exact calculated speed, a simplified formula can be used to calculate r/min. The $\pi$ (3.1416) on the bottom line of the formula will divide into 1000 of the top line approximately 320 times. This results in a simplified formula which is close enough for most drill presses.

$$\text{r/min} = \frac{CS \times 320}{D}$$

**Example:**

Calculate the r/min at which a drill press should be set to drill a 19-mm hole in a piece of cast iron. (See Table 28-1.)

$$\text{r/min} = \frac{CS \times 320}{D}$$

$$= \frac{24 \times 320}{19}$$

$$= 404$$

## Feed

Feed is the distance (in hundredths of a millimeter) that a drill advances into the work for each complete revolution. The feed rate is important because it affects the life of the drill and also the rate of production. (See N-10.) Too coarse a feed may cause the cutting edges to break or chip, while too fine a feed causes a drill to chatter which dulls the cutting edges. The recommended feeds per revolution for millimeter and fractional inch size drills are listed in Table 28-2.

N-10 The rate of feed affects production and cutting tool life.

## REVIEW QUESTIONS

**1.** Name three purposes of the flutes on a drill.

**2.** Of what three materials are twist drills manufactured?

**3.** Describe for what purpose each of the following three drills is used?

(a) Low helix
(b) High helix
(c) Straight flute

**4.** Name three main parts of a twist drill.

**5.** State the purpose of each of these parts:

(a) Tang
(b) Flutes
(c) Margin
(d) Web

**6.** What is the recommended drill point angle and lip clearance for a general-purpose drill?

**7.** Name the three systems of inch drill sizes.

**8.** What size are the following drills? #15, 60, letters B, Z.

### Cutting Speed

**9.** Why are speeds and feeds so important?

**10.** List four factors which may affect the calculated drill speed.

**11.** What three factors affect the speed a drill should be run?

**12.** Calculate the revolutions per minute for the following:

(a) ½-in.-diameter drill for machine steel (100 CS)
(b) ¾-in.-diameter drill for cast iron (80 CS)
(c) 6-mm-diameter drill for tool steel (18 CS)
(d) 35-mm-diameter drill for aluminum (60 CS)

# UNIT 29
# DRILL GRINDING

The success of any drilling operation depends largely on the ability of the drill to produce chips which can be ejected readily from the hole. Since there is limited room for the chips in the flutes of a drill, the most desirable chips are those which are broken into relatively small pieces. Coiled chips should be avoided, especially when drilling deep holes, since they tend to pack in the flutes, stop succeeding chips from being ejected, and prevent the cutting fluid from reaching the cutting edges of the drill. (See N-1.) The main factors which affect the type of chips produced are the ductility of the material being drilled, the thickness of

N-1 Avoid coiled chips whenever possible.

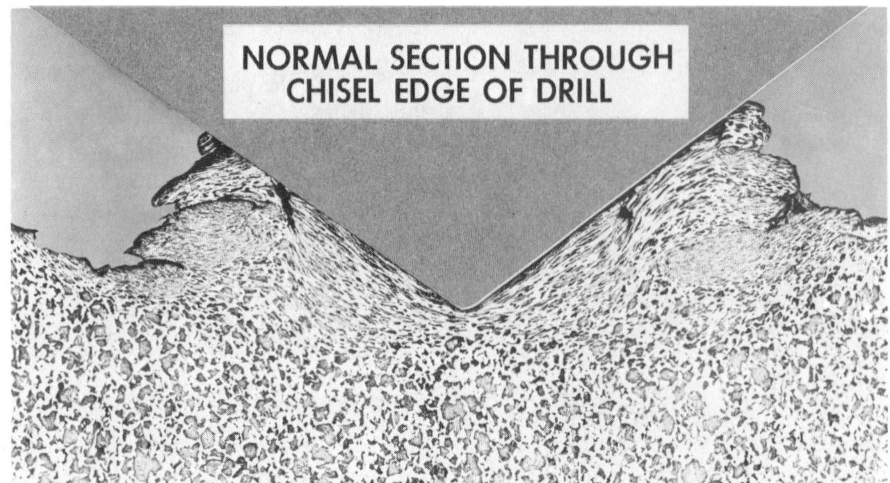


**Fig. 29-1** The photomicrograph shows the metal deformation and flow which occurs under the chisel edge of a twist drill. (*Courtesy National Twist Drill & Tool Co.*)

N-2 The cutting edges must be in good condition for the best drilling results.

the chip or the feed per revolution, and the characteristics of the drill point.

## OBJECTIVES

The purpose of this unit is to enable you to:

1. Understand the cutting action of the twist drill point
2. Know the drill-point angles and clearances required for various materials
3. Recognize the characteristics of a properly ground drill
4. Sharpen a drill properly on a pedestal grinder

## CUTTING ACTION

Two distinct cutting actions occur as a result of the construction of a twist drill: one at the chisel edge and the other at the lips or cutting edges.

The *chisel edge,* although considered a cutting edge, has a negative rake and generally does not produce a very efficient cutting action. As the chisel edge of the drill point is forced against the metal, it acts more like a center punch and tends to indent the metal, (Fig. 29-1). The cutting efficiency of the chisel edge can be improved to suit certain conditions or materials by thinning the web of the drill.

The most effficient cutting action occurs as the *lips* or *cutting edges* of the drill contact the work material. Since the condition of the cutting lips affects the efficiency of the drill, it is important for the machinist to be aware of the characteristics of a drill point. (See N-2.)

**TABLE 29-1 SUGGESTED LIP CLEARANCE ANGLES AT PERIPHERY**

| Drill diameters | | | | |
|---|---|---|---|---|
| Inch | Metric, mm | For general purpose,° | Hard and tough materials,° | Soft and free machining materials,° |
| #97 to 81 | 0.15– 0.33 | 28 | 22 | 30 |
| #80 to 61 | 0.34– 0.99 | 24 | 20 | 26 |
| #60 to 41 | 1 – 2.45 | 21 | 18 | 24 |
| #40 to 31 | 2.5 – 3 | 18 | 16 | 22 |
| #30 to 1/4 in. | 3.3 – 6.35 | 16 | 14 | 20 |
| F to 11/32 in. | 6.5 – 8.37 | 14 | 12 | 18 |
| S to 1/2 in. | 8.8 –12.7 | 12 | 10 | 16 |
| 33/64 to 3/4 in. | 13.1 –19.05 | 10 | 8 | 14 |
| 49/64 in. and up | 19.45 and up | 8 | 7 | 12 |

## DRILL-POINT ANGLES AND CLEARANCES

The angle of the drill point and the clearance behind the cutting lips both affect the size and condition of the hole produced by the drill. If the point angle and lip clearance are varied, the cutting action of a drill can be changed to suit various types of materials and drilling conditions. By changing point angles and lip clearances and by thinning the drill web, the following can be achieved:

- Control of the size and quality of a hole
- Control of the size, shape, and flow of the chips
- Use of various speeds and feeds for more efficient drilling
- Control of the amount of burr produced
- Reduction of the amount of heat generated during drilling
- Increase in the strength of the lips of the drill to reduce wear and prevent them from chipping
- Reduction of the amount of drilling pressure required

### Lip Clearance

*Lip clearance* is the amount that the heel of a drill is ground below the lips or cutting edges of the drill, (Fig. 29-2). Lip clearance is measured at the circumference of the drill and can be changed to suit the material being

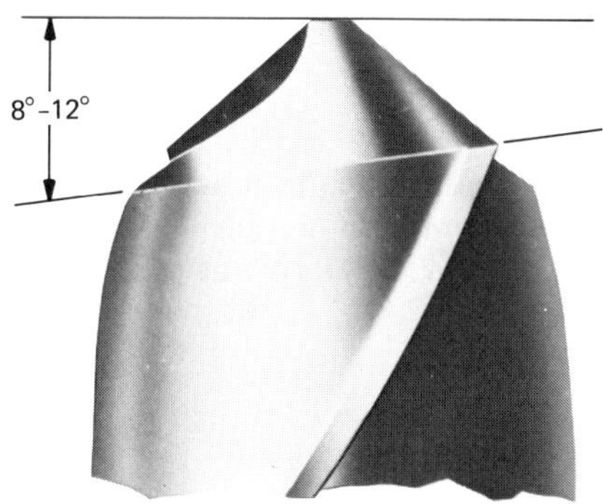


**Fig. 29-2** The lip clearance for general-purpose drills ranges from 8 to 12°. *(Courtesy The Cleveland Twist Drill Co.)*

N-3 Too little lip clearance causes a drill to rub; too much weakens the point.

N-4 Lips with unequal angles will produce oversize holes.

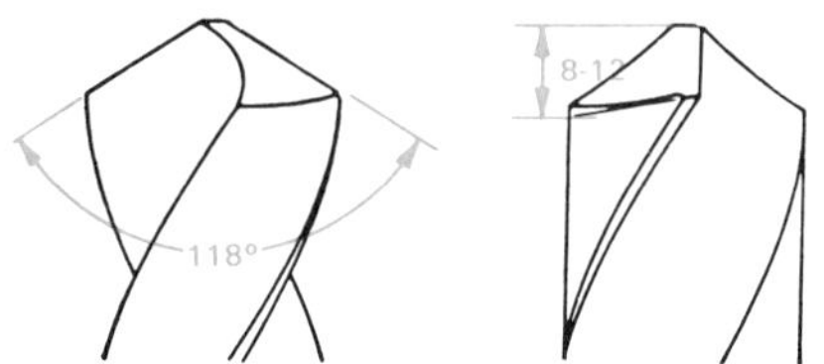


**Fig. 29-3** General-purpose (conventional) drill-point characteristics. *(Courtesy The Cleveland Twist Drill Co.)*

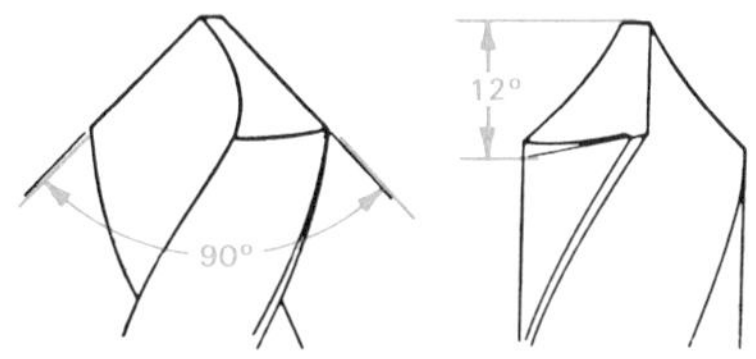


**Fig. 29-4** Long-angle drill-point characteristics. *(Courtesy Greenfield Tap & Die.)*

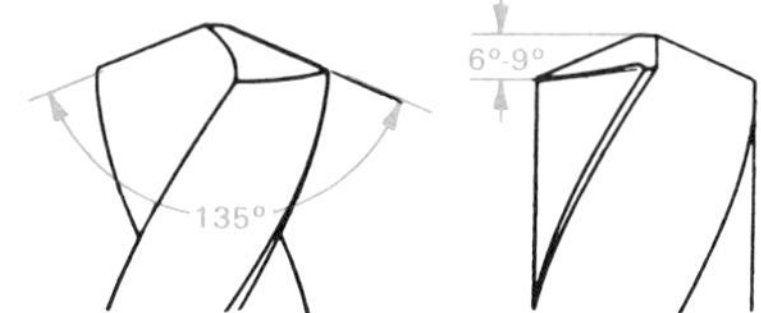


**Fig. 29-5** Flat-angle drill-point characteristics. *(Courtesy Greenfield Tap & Die.)*

drilled and the size of the drill. This clearance allows the cutting edge of the drill to penetrate into the work. Too little or no lip clearance will prevent the drill from cutting, cause it to rub, and create excessive heat which may damage the drill point. Too much lip clearance will weaken the point of the drill and cause the cutting edges to chip and break easily. (See N-3.)

For general-purpose drills, the average lip clearance ranges from 8 to 12°. When hard and tough materials are to be drilled, the lip clearance is reduced (6 to 9°) to provide strength for the cutting edges. A lip clearance of 12 to 15° and higher is provided on low helix drills and those used for drilling soft and nonferrous materials.

Smaller drills, ¼ in. (6.35 mm) or less in diameter, require more lip clearance than the larger drill sizes because their decreased torsional stiffness makes them sensitive to the increased loading resulting from dulling. The higher lip clearances delay dulling and permit easier penetration into the workpiece. For very small drills (no. 40 to no. 97), a lip clearance of 20 to 30° is absolutely necessary to achieve smooth, free-cutting operation. See Table 29-1 for the suggested lip clearance angles for various sizes of drills and work materials.

### Drill-Point Angles

Three common drill points are used to drill most types of materials: the *conventional* or *general-purpose point,* the *long-angle point,* and the *flat-angle point*. Variations of these basic drill points are available to suit the hardness and texture of the work material or the machine conditions. Regardless of the drill point used, it is important that the angle and length of both lips be the same: if they are not, the drill will not cut properly. (See N-4.)

The *conventional,* or *general-purpose,* drill point, Fig. 29-3, has an included angle of 118°. It is the most commonly used drill point and should give satisfactory drilling results for a wide range of materials. The general-purpose drill should be provided with an average lip clearance of 8 to 12° for best results. Consult Table 29-1 for the exact lip-clearance angles to suit the type of material and the size of drill required.

The *long-angle* drill point, (Fig. 29-4), has an included angle ranging from 60 to 90°. This point is commonly used on low helix drills for drilling soft plastics, nonferrous metals, soft cast iron, fibers, and wood. The lip clearance for long-angle drill points is usually 12° but may go as high as 15° for some types of materials.

The *flat-angle* drill point, (Fig. 29-5), is used to drill hard and tough materials. The included angle, which ranges between 135 and 150°, provides a shorter cutting edge which requires less drilling power. Thus, the strain on the drill is reduced. The flat angle also helps to strengthen the point of the drill. The 6 to 9° lip clearance on flat-angle point drills provides as much support as possible for the cutting edges.

## DRILL GRINDING

The cutting efficiency of a drill is determined by the characteristics and condition of the point of the drill. Most types of new drills are provided with a general-purpose point (118° point angle and an 8 to 12° lip clearance). As a drill is used, the cutting edges may wear and become chipped or the drill may break due to the conditions of the drilling process. Drills are generally resharpened by hand. However, small drill-point grinders or drill-sharpening attachments are inexpensive and readily available, and they provide much more consistent quality than hand grinding.

To ensure that a drill will perform satisfactorily, it is recommended that the drill point be examined carefully before the drill is mounted in the drill press. A properly ground drill should have the following characteristics:

- The length of both cutting lips should be the same. Lips of unequal length will force the drill point off center, cause one lip to do more cutting than the other lip, and produce an oversize hole (Fig. 29-6A).
- The angle of both lips should be the same. If the angles are unequal, the drill will cut an oversize hole because one lip will do more cutting than the other (Fig. 29-6B).
- The lips should be free from nicks or wear.
- There should be no sign of wear on the margin.

If the drill does not meet these requirements, it should be resharpened. If

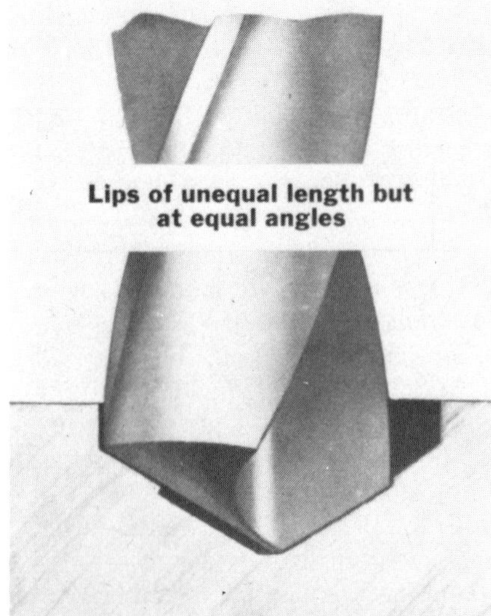


**Fig. 29-6A** An incorrect point with lips of unequal length. (*Courtesy The Cleveland Twist Drill Co.*)

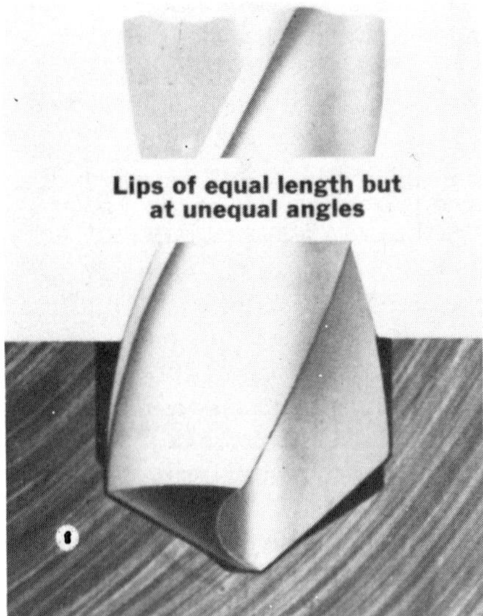


**Fig. 29-6B** Lips with unequal angles produce oversized holes. (*Courtesy The Cleveland Twist Drill Co.*)

N-5 Never use a drill which is worn or improperly ground.

N-6 A worn margin will cause a drill to jam in a hole.

N-7 Always wear safety glasses when operating any machine tools.

the drill is not resharpened, it will give poor service, produce inaccurate holes, and may break due to excessive drilling strain. (See N-5.)

While a drill is being used, there will be definite signs to indicate that the drill is not cutting properly and should be resharpened. If the drill is not resharpened at the first sign of dullness, it will require extra power and thrust to force the slightly dulled drill into the work. This causes more heat to be generated at the cutting lips and results in a faster rate of wear. When any of the following conditions arise while a drill is in use, it should be examined and reconditioned.

- The shape and color of the chips change.
- More drilling pressure is required to force the drill into the work.
- The drill discolors (turns blue) due to the excessive heat generated while drilling.
- The top of the hole is out-of-round and a poor finish is produced in the hole.
- The drill chatters when it contacts the metal.
- The drill squeals and may jam in the hole.
- There is an excessive burr left around the drilled hole.

# DRILL RECONDITIONING

Drills, like other cutting tools, should not be allowed to become so dull that they cannot cut. Overdulling of any metal-cutting tool generally results in poor production rates, inaccurate work, and the shortening of the life of the tool. Premature dulling of a drill may be caused by any one of a number of factors.

- The drill speed may be too high for the type of hardness of the material being cut.
- The feed may be too heavy and overload the cutting lips.
- The feed may be too light and cause the lips to scrape rather than cut.
- There may be hard spots or scale on the work surface.
- The drill or work may not be supported adequately, resulting in springing and chatter.
- The drill point is incorrect for the material being drilled.
- The grinding finish on the lips is poor.

*At the first sign of wear,* a drill should be reground to remove the wear and restore the cutting lips to their original efficiency. Three separate steps are necessary to recondition dull or worn drills:

1. Remove the worn section.
2. Regrind the point.
3. Thin the web.

## Remove the Worn Section

A drill starts to wear as soon as it is put into operation. The first wear occurs at the corners of the drill (Fig. 29-7A) beginning with a slight rounding of the corners. As the roundness of the corner increases, the wear travels back along the margin. This results in the loss of size and a negative back-taper on the tool (Fig. 29-7B). To recondition a drill properly, any sign of wear on the lips or margin must be removed, as well as any burned sections of the drill. (See N-6.)

## Regrind the Drill Point

Drills may be sharpened while being held by hand or in a special drill-sharpening attachment or grinder. Regardless of how a drill is held, three factors are to be considered when the drill point is sharpened:

1. The length and angle of the lips
2. The lip clearance
3. The location of the chisel edge at the center of the drill

Grinding a drill by hand requires practice. An apprentice should make every attempt to become skilled at this operation.

### Sharpening a Drill by Hand

The procedures for grinding any drill held by hand are basically the same; therefore, the following procedure is for the general-purpose drill point only (118° point angle with 8 to 12° lip clearance).

1. Wear approved safety glasses even though the grinder may be equipped with eye shields. (See N-7.)

2. Examine the periphery of the grinding wheel. Dress the wheel if it is

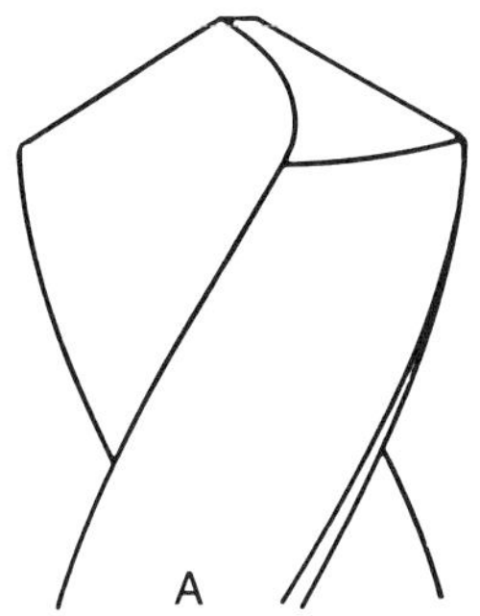


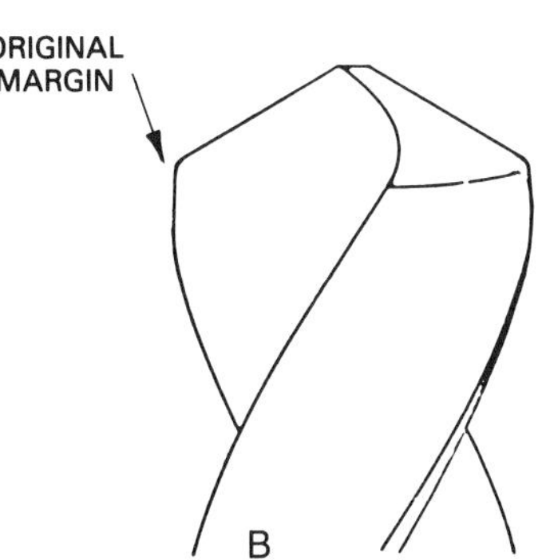


**Fig. 29-7** Most wear occurs first at the corners and then along the margin of the drill. (A) Corners rounded; (B) worn margin.

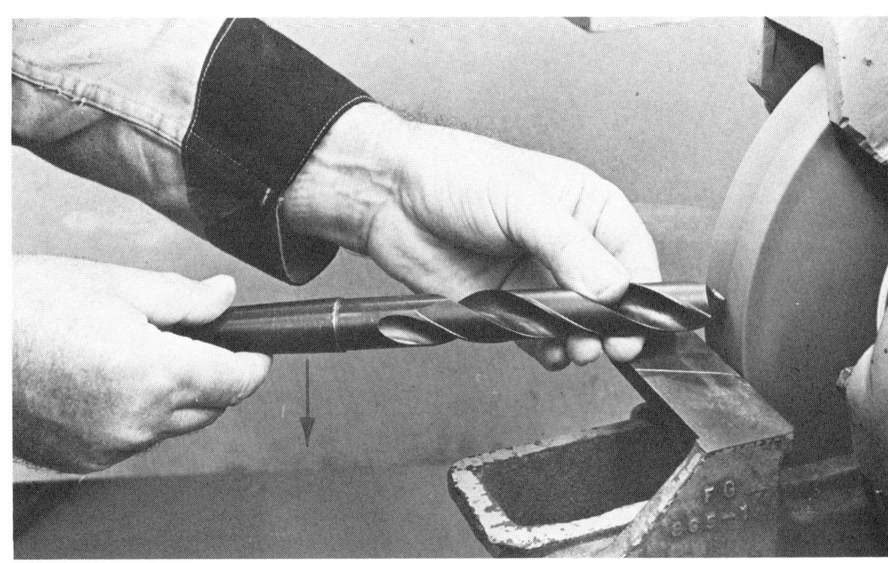

**Fig. 29-8** The drill shank should be held slightly lower than the point. (*Courtesy Kostel Enterprises Ltd.*)

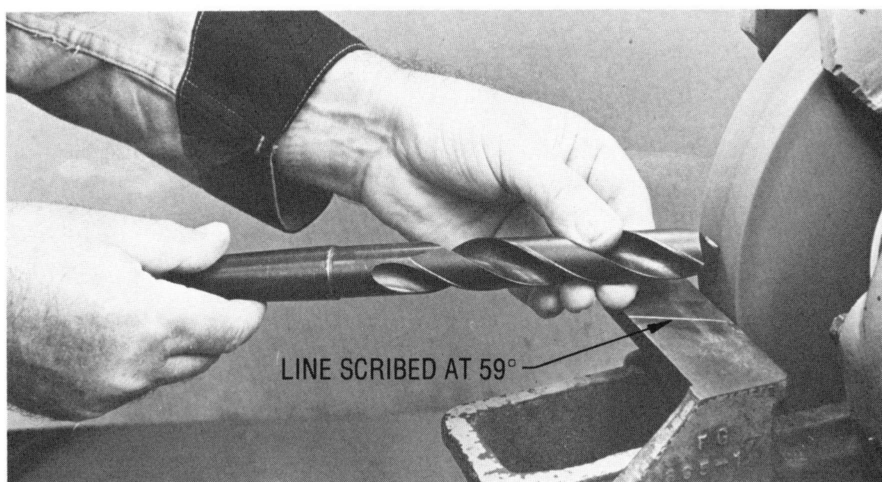


**Fig. 29-9** A line scribed on the tool rest will help to hold the drill at an angle of 59° to the grinding-wheel face. (*Courtesy Kostel Enterprises Ltd.*)

necessary to straighten the face or remove grooves.

**3.** Adjust the tool rest so that it is no farther than 1/16 in. (1.59 mm) from the wheel face. (See N-8.)

**4.** Rest one hand on the tool rest and hold the drill near the point. With the other hand, hold the drill shank slightly lower than the point (Fig. 29-8).

**5.** Move the drill so that it is at an angle of approximately 59° to the grinding wheel face (Fig. 29-9). (A line scribed on the tool rest at an angle of 59° to the wheel face will help to keep the drill at the proper angle.)

**6.** Hold the lip of the drill parallel to the grinder tool rest.

**7.** Bring the lip of the drill against the revolving grinding wheel and slowly lower the drill shank (Fig. 29-8). *Note:* Do not rotate the drill. (See N-9.)

**8.** Without moving the position of the body or the hands, move the drill far enough away from the wheel to turn the drill one-half turn.

**9.** Grind the other lip of the drill in the same manner that was described in step 7.

**10.** Check the angles and lengths of both lips with a drill-point gage (Fig. 29-10).

**11.** Repeat steps 7 to 10 until the cutting lips are sharp and the margins free from wear.

N-8 Keep the tool rest adjusted close to the wheel face to prevent work from being drawn between the rest and wheel.

N-9 Cool the drill in water occasionally; never allow it to overheat.

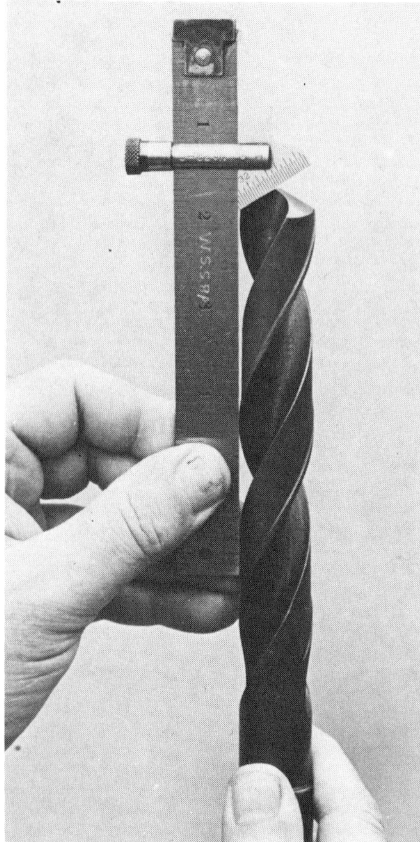

**Fig. 29-10** The angle and length of the lips may be checked with a drill-point gage. (*Courtesy Kostel Enterprises Ltd.*)

N-1 The center head should be used only on work that is true in shape (round, square, or octagonal).

N-2 Always be sure that the scriber point is sharp in order to produce an accurate layout.

## REVIEW QUESTIONS

**1.** Name three things which affect the type of chip produced by a drill.
**2.** What two factors affect the size and condition of a hole?
**3.** What is the effect of
(a) Too little lip clearance
(b) Too much lip clearance
**4.** How much lip clearance should there be for:
(a) General-purpose drills
(b) Hard materials
(c) Very small drills
**5.** What is the recommended point angle for:
(a) General-purpose drills
(b) Drills for plastics and nonferrous metals
(c) Drills for hard, tough material
**6.** Name four characteristics of a correctly ground drill.
**7.** Name seven factors which may indicate that a drill point should be resharpened.
**8.** What seven factors can cause the premature dulling of a drill?
**9.** At what angle should a general-purpose drill be held to the face of a grinding wheel?
**10.** How should the drill cutting edge be held in relation to the grinder tool rest for sharpening?

# UNIT 30

# DRILLING CENTER HOLES

Work that is to be machined between centers on a lathe must have a center hole in each end so that the workpiece can be supported by the lathe centers. This also allows the workpiece to be removed from the lathe any number of times and quickly replaced with the assurance that the machined diameters will run true with the center holes.

## OBJECTIVES

After completing this unit you should be able to:

**1.** Identify three methods of laying out centers in workpieces and know the advantage of each
**2.** State the advantages of the two types of center drills
**3.** Drill center holes to the proper size on a drill press

## LAYING OUT CENTER LOCATIONS

Round workpieces are often centered on a lathe. However, when the work is of such a size or shape that it cannot be held satisfactorily in a lathe chuck for centering, center holes may be drilled on a drill press. Since it is important that the holes be centered as accurately as possible, it is necessary to locate the center of the stock before drilling. To locate the center of the stock, a center head, a hermaphrodite caliper, a surface gage, or a bell center punch can be used.

### Center Head

The center head forms a center square when clamped to the rule. It can be used for locating centers of round, square, and octagonal stock. (See N-1.)

#### To Lay Out the Center of Round Work

**1.** Place the work in a vise and remove the burrs or sharp edges with a file.
**2.** Apply layout dye to both ends of the work.
**3.** Hold the center head firmly against the end of the work (Fig. 30-1) and keep the rule flat against the end.
**4.** With a *sharp* scriber, draw a line along the edge of the rule. Tilt the scriber at a slight angle to keep the point firmly against the edge of the rule. (See N-2.)
**5.** Rotate the center head 90° and scribe another line (Fig. 30-1).
**6.** Repeat the procedure on the other end of the work.
**7.** With a sharp prick punch, lightly punch the intersection of the two lines.
**8.** Using a pair of dividers, check the accuracy of the punch marks (Fig. 30-2).
**9.** Correct the location of the punch marks if necessary and then enlarge them with a center punch.

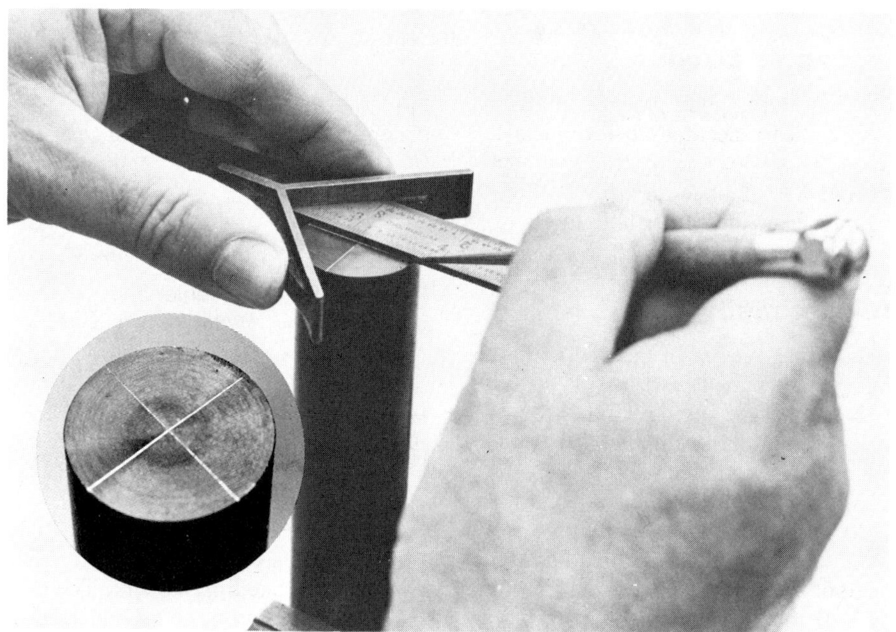

Fig. 30-1 Using the center head to locate the center of round stock.

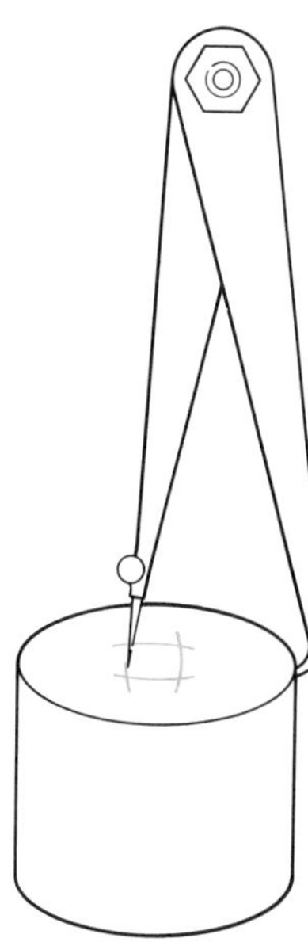

Fig. 30-3 A hermaphrodite caliper may be used to locate the center of rough work.

Fig. 30-2 Checking the accuracy of the center layout with dividers. (*Courtesy Kostel Enterprises Ltd.*)

N-3 Keep the surface plate clean and free from burrs for accurate layout work.

## Hermaphrodite Caliper

The hermaphrodite caliper (Fig. 30-3) is generally used to locate the centers of round work or work which has been cast and is not quite round. It has one bent leg and one straight leg which contains a sharp point used to scribe layout lines. Hermaphrodite calipers may also be used to scribe lines parallel with a machined edge or shoulder. When setting this tool to a size, place the bent leg on the end of a rule and adjust the other leg until the scriber point is at the desired graduation.

### To Locate the Centers of Round Work

**1.** Place the work in a vise and remove the burrs or sharp edges with a file.

**2.** Apply layout dye to both ends.

**3.** Set the hermaphrodite caliper to approximately one-half the work diameter.

**4.** With the thumb of one hand, hold the bent leg just below the edge of the work and scribe an arc.

**5.** Move the bent leg a quarter of a turn and scribe an arc. Repeat until four arcs are scribed on each end of the work (Fig. 30-3).

**6.** Lightly center-punch the center of the four arcs and check the accuracy of the centers.

## Surface-Gage Method

**1.** Remove the burrs and apply layout dye to both ends of the workpiece. (See N-3.)

**2.** Place the work in a V-block on a surface plate.

**3.** Set the surface-gage scriber point to approximately the center of the workpiece.

**4.** Scribe a line across each end of the stock.

**5.** Scribe four lines on each end of the workpiece (Fig. 30-4) by rotating the work about 90° (one-quarter turn) for each line.

**6.** Lightly center-punch the center of the square and check accuracy of the layout.

Fig. 30-4 Locating the center of a round workpiece using a surface gage.

Fig. 30-5 Moving the center punch mark to bring it to center.

N-4 Never start a lathe when testing the accuracy of the center layout.

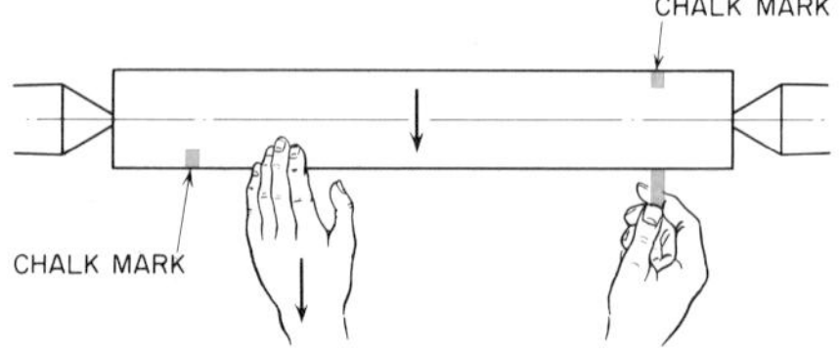


Fig. 30-6 Testing the accuracy of center layout on a lathe.

## Checking the Accuracy of Center Layout

It is considered good practice to check any layout for accuracy before metal is cut. This is especially true with center-hole layout since an error in this layout causes the center hole to be drilled off center.

### Divider Method

1. Place one leg of the divider in the light center punch mark.
2. Adjust the divider so that the other leg is on a line and exactly on the edge of the workpiece (Fig. 30-2).
3. Revolve the divider one-half turn (180°) and check to see that the leg of the divider is in the same relation to the edge of the workpiece.
4. If the divider leg is not in the same relation to the edge of the workpiece at each end of the line, lightly move the center punch mark to correct the error (Fig. 30-5).
5. Repeat this procedure on the other line until the punch mark is exactly in the center of the work.

### Lathe Center Method

1. Deepen the center punch marks at both ends and mount the work between lathe centers.
2. Adjust the lathe center tension so that the work can spin freely.
3. Hold a piece of chalk in one hand which is steadied against any part of the lathe (e.g., compound rest).
4. Spin the work by hand and bring the chalk close to the work surface until it lightly marks the high spot (Fig. 30-6). (See N-4.)
5. Repeat step 4 on the other end of the work.
6. Fasten the work in a vise, tip the center punch, and with a hammer move the center punch mark toward the chalk mark (Fig. 30-5).
7. Retest the workpiece between centers and repeat steps 3 to 6 until the work runs true.

## DRILLING CENTER HOLES

Work that is to be turned between the centers on a lathe must have a center hole drilled in each end so that the work may be supported by the lathe centers. After the center locations have been laid out on both ends of the workpiece, a *combined drill and countersink,* commonly called a *center drill,* is used to drill the center holes.

Two types of center drills are available, the plain or regular type (Fig. 30-7) and the bell type (Fig. 30-8). The regular type has a 60° angle with a small drill located on the end. The bell type is similar but has a secondary bevel near the large diameter which produces a clearance angle near the top of the hole. This clearance angle prevents the top edge of the 60° bearing surface from becoming burred or damaged. Center drills are available in a wide variety of sizes to suit different diameters of work. Table 30-1 lists information regarding regular- and bell-type center drills and the diameters for which each should be used.

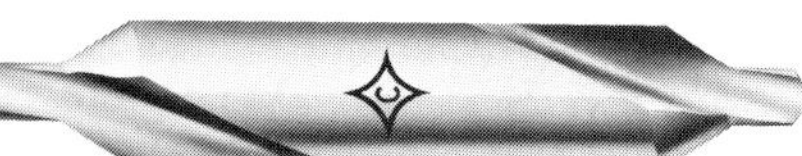

Fig. 30-7 Regular-type center drill. (*Courtesy The Cleveland Twist Drill Co.*)

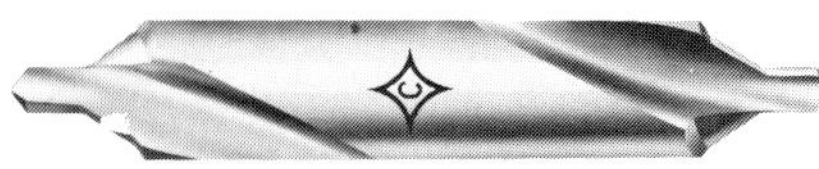

Fig. 30-8 Bell-type center drill. (*Courtesy The Cleveland Twist Drill Co.*)

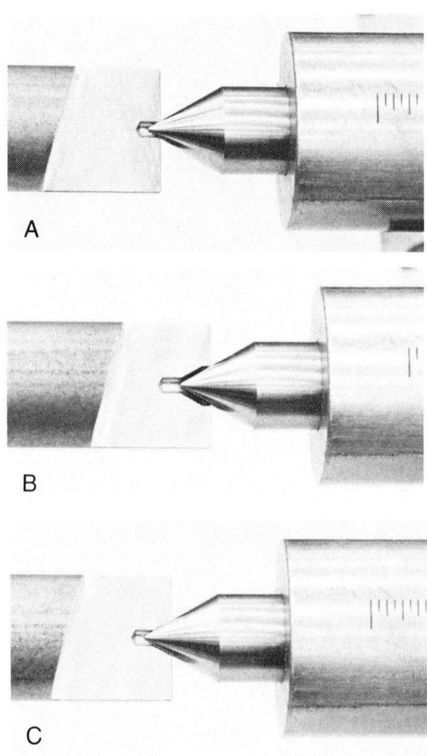


Fig. 30-9 Improperly and properly drilled center holes. (A) Too shallow; (B) too deep; (C) correct size. (*Courtesy Kostel Enterprises Ltd.*)

**TABLE 30-1 CENTER DRILL SIZES**

| Size | | Work diameter | | Diameter of countersink C | Drill-point diameter | Body size |
|---|---|---|---|---|---|---|
| Regular type | Bell type | Inches | Millimeters | | | |
| 1 | 11 | 3/16–5/16 | 3–8 | 3/32 | 3/64 | 1/8 |
| 2 | 12 | 3/8–1/2 | 9.5–12.5 | 9/64 | 5/64 | 3/16 |
| 3 | 13 | 5/8–3/4 | 15–20 | 3/16 | 7/64 | 1/4 |
| 4 | 14 | 1–1½ | 25–40 | 15/64 | 1/8 | 5/16 |
| 5 | 15 | 2–3 | 50–75 | 21/64 | 3/16 | 7/16 |
| 6 | 16 | 3–4 | 75–100 | 3/8 | 7/32 | 1/2 |
| 7 | 17 | 4–5 | 100–125 | 15/32 | 1/4 | 5/8 |
| 8 | 18 | 6 and over | 150 and over | 9/16 | 5/16 | 3/4 |

Fig. 30-10 Drill chucks are used to grip straight-shank cutting tools. (*Courtesy Kostel Enterprises Ltd.*)

N-5 Be sure to wear proper safety glasses.

N-6 Too high a feed pressure may break the center drill point.

In order to provide a good bearing surface for the work on the lathe centers, it is important that the center holes be drilled to the correct size (see Table 30-1). The center hole illustrated in Fig. 30-9A is too shallow and will not provide an adequate bearing surface. The center hole in Fig. 30-9B is too deep and will not allow the taper of the lathe center to contact the taper of the center hole. The center hole illustrated in Fig. 30-9C is drilled to the proper depth which provides a good bearing surface for the lathe centers.

## Drill Chucks

Drill chucks are the most common devices used on a drill press for holding straight-shank cutting tools. Most drill chucks contain three jaws that move simultaneously (all at the same time) when the outer sleeve is turned, or on some types of chucks when the outer collar is raised. The three jaws hold the straight shank of a cutting tool securely and cause it to run accurately. There are generally two common types of drill chucks: the key type and the keyless type (Fig. 30-10).

## To Drill a Lathe Center Hole

Center holes should be as smooth and accurate as possible to provide a good bearing surface and reduce the friction and wear between the work and lathe centers.

**1.** Check the center-hole layout to be sure that it is correct.

**2.** Obtain the correct-size center drill to suit the diameter of the work being drilled (Table 30-1).

**3.** Fasten the center drill in the drill chuck. To prevent breakage, do not have more than ½ in. (12.7 mm) of the center drill extending beyond the drill chuck.

**4.** Set the drill press to the proper speed for the size of the center drill being used. A speed of 1200 to 1500 r/min is suitable for most center drills. (See N-5.)

**5.** Fasten a clamp or table stop on the left side of the table to prevent the vise from swinging during the drilling operation (Fig. 30-11).

**6.** Set the vise on its side on a clean drill-press table.

**7.** Press the work firmly against the bottom of the vise and then tighten it securely.

**8.** With the vise against the table clamp on stop, locate the center punch mark of the work under the drill point.

**9.** Start the drill press and with the hand-feed lever carefully feed the center drill into the work. (See N-6.)

**10.** Frequently raise the drill from the work and apply a few drops of cutting fluid.

**11.** Shut off the machine.

**12.** Measure the diameter at the top of the center hole with a rule (Fig. 30-12).

**13.** Continue drilling until the top of the countersunk hole is the correct size.

**14.** Inspect the hole for smoothness; if scratches or rings are present, apply cutting fluid and lightly bring the center drill into the hole.

**15.** Center-drill both ends to the same diameter.

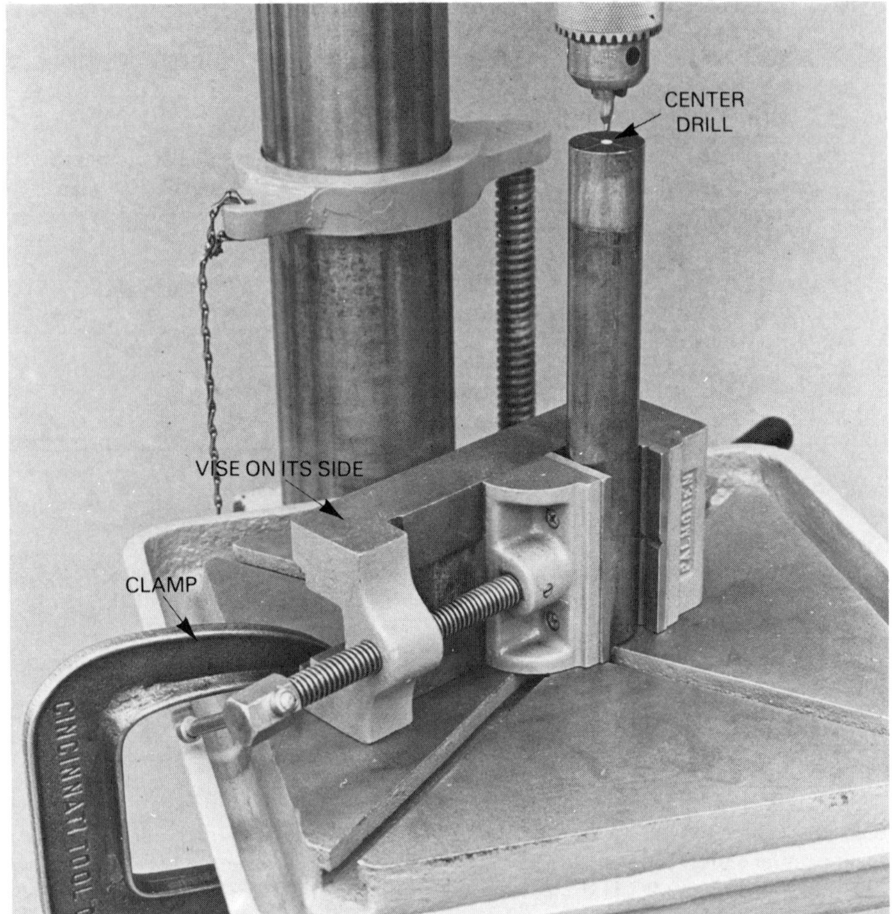


**Fig. 30-11** The work is held in a vise which is set on its side for center drilling. (*Courtesy Kostel Enterprises Ltd.*)

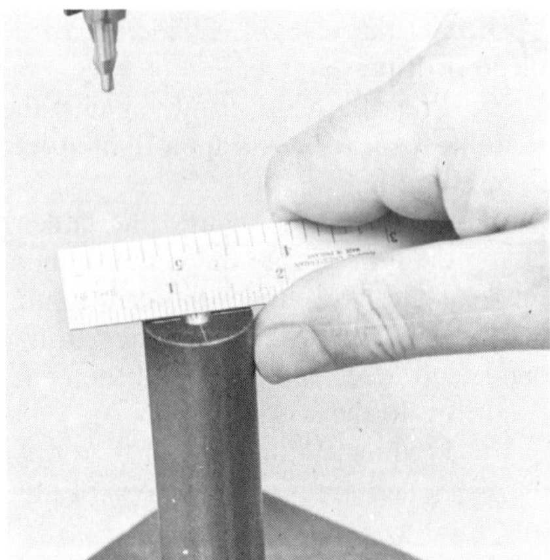

**Fig. 30-12** Measuring the size of the center hole with a rule. (*Courtesy Kostel Enterprises Ltd.*)

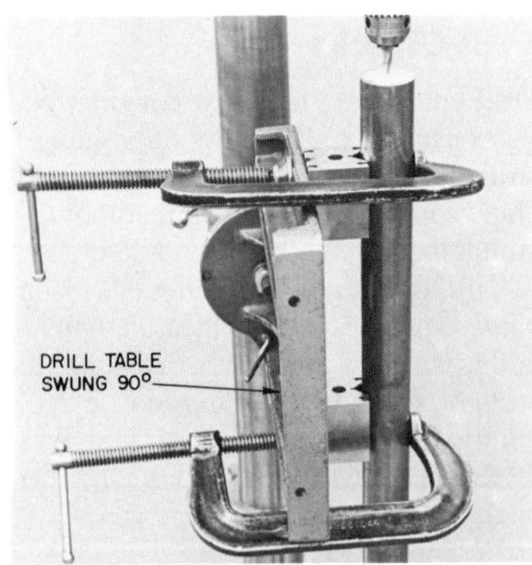


**Fig. 30-13** The drill-press table can be set in a vertical position to accommodate long workpieces for drilling. (*Courtesy Kostel Enterprises Ltd.*)

When the workpiece is too long to be held in a vise, the centers can be hand-drilled with a portable drill, or the workpiece can be set in V-blocks fastened to the table, as shown in Fig. 30-13.

## REVIEW QUESTIONS

**1.** For what type of work can the center head be used to locate centers?

**2.** Why must the scriber be sharp when laying out?

**3.** For what purpose is a hermaphrodite caliper used?

**4.** How many arcs should be scribed when locating centers with a hermaphrodite caliper?

**5.** Briefly describe how the accuracy of the center layout may be tested with dividers.

**6.** What drill is used to produce center holes in a workpiece?

**7.** State the difference between a regular- and a bell-type center drill.

**8.** Why should center holes not be drilled

(a) Too shallow

(b) Too deep

**9.** Why is it important that center holes be drilled as smoothly and accurately as possible?

**10.** Why should a clamp or table stop be fastened to the drill-press table for any drilling operation?

# UNIT 31

# DRILLING HOLES

A wide variety of materials may be drilled in a drill press. Although the size or shape of the drill may vary to suit the work, the procedure for drilling holes is basically the same and must be followed for best results. Use the proper drill speeds and feeds for best cutting-tool results and observe the safety precautions for your own protection.

## OBJECTIVES

After completing this unit you should be able to:

**1.** Lay out hole locations accurately for drilling

**2.** Drill holes in work held in a vise

**3.** Drill holes in round work

**4.** Drill work which is fastened to a drill-press table

**5.** Drill to an accurate layout

**6.** Drill holes in thin material

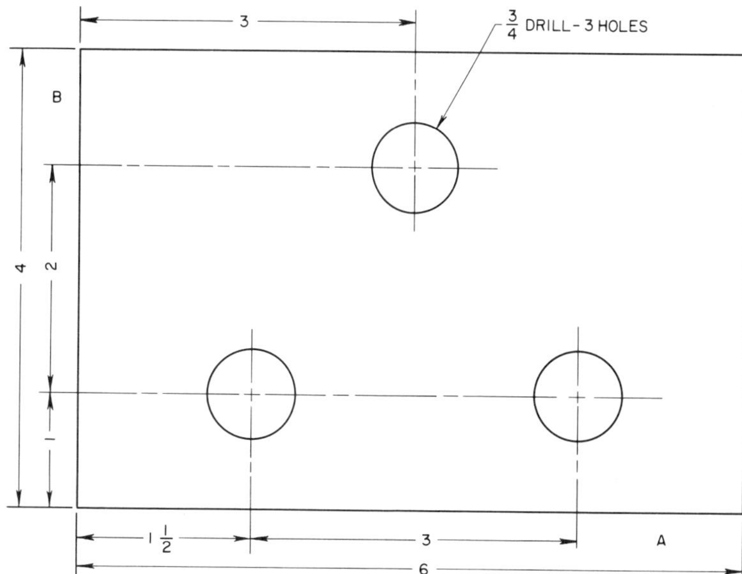


**Fig. 31-1** A sample layout required on a steel plate.

## TO LAY OUT A WORKPIECE FOR DRILLING

Most layouts can be produced with a few basic tools: surface plate, scriber, surface gage, adjustable square, rule, dividers, prick punch, center punch, and hammer.

Certain basic procedures must be followed in making any layout. For example, the steel plate, shown in Fig. 31-1, must be laid out to the dimensions indicated. It is assumed that the plate has been machined square and to the sizes shown on the drawing.

N-1 Burrs on a workpiece cause inaccurate layouts and can cut your hands since they are sharp.

**1.** Remove the burrs from all edges of the plate. (See N-1.)

**2.** Clean the workpiece and apply a thin coating of layout dye.

**3.** Clean the top of the surface plate, place edge *A* of the work on the surface plate, and clamp it to an angle plate (Fig. 31-2).

**4.** Set the surface gage to a height of 1 in. (25.4 mm) and scribe the lower line across the work (Fig. 31-2).

**Fig. 31-2** The workpiece clamped to an angle plate for scribing all horizontal lines. (*Courtesy Kostel Enterprises Ltd.*)

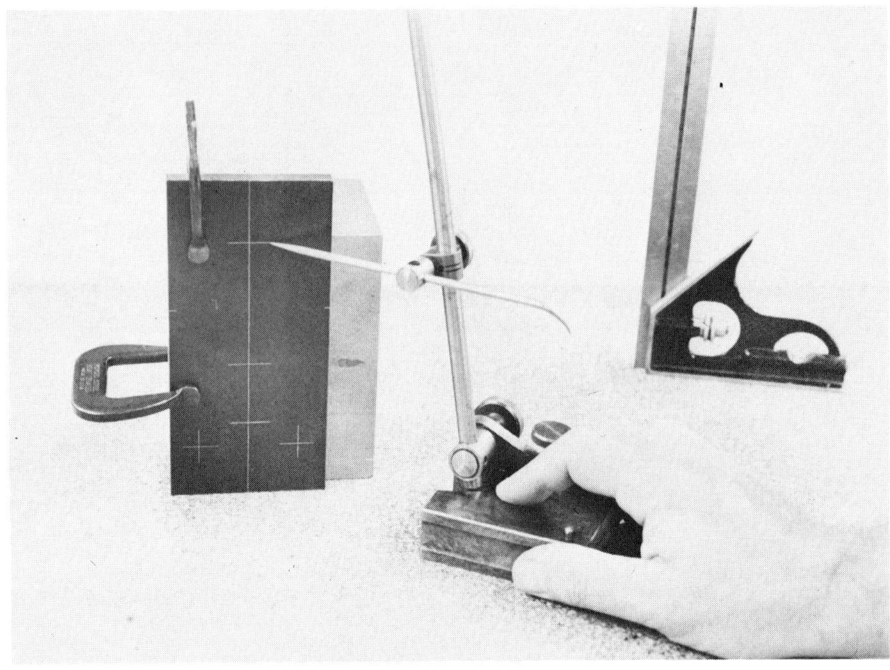

**Fig. 31-3** The angle plate is turned 90° (on its side) for scribing the vertical or intersecting lines. (*Courtesy Kostel Enterprises Ltd.*)

**5.** Set the surface gage to 3 in. (76.2 mm) and scribe the upper line.

**6.** Turn the angle plate 90° (on its side) and scribe the horizontal lines (Fig. 31-3) by setting the surface gage first to a height of 1½ in. (38.1 mm) and then to 4½ in. (114.3 mm).

**7.** Place the workpiece on a bench and lightly prick-punch the intersections of the lines. (See N-2.)

**8.** Using a magnifying glass, check the accuracy of the punch marks and adjust if necessary.

**9.** Set the dividers to ⅜ in. (9.52 mm) and scribe each circle to indicate the diameter of the holes (¾ in. or 19.05 mm) (Fig. 31-5).

**10.** Set the dividers to 5⁄16 in. (7.93 mm) and scribe three ⅝-in.- (15.87-mm-) diameter proof circles using the same centers (Fig. 31-6).

**11.** Prick-punch the circles as shown in Fig. 31-6.

**12.** Using a center punch, enlarge each center hole in order to provide a larger punch mark for the drill to follow.

N-2 Intersecting lines may also be laid out by using a combination square and scriber (Fig. 31-4).

N-3 Do not clamp the work or vise to the drill-press table when spotting holes.

## TO SPOT A HOLE WITH A CENTER DRILL

The chisel edge at the end of the web of a drill is generally larger than the center punch mark on the work and therefore it is difficult to start a drill at the exact location. To prevent a drill from wandering off center, it is considered good practice to first *spot every center punch mark with a center drill.* The small point on the center drill will accurately follow the center punch mark and provide a guide for the larger drill which will be used.

**1.** Mount a small-size center drill in the drill chuck. (See N-3.)

**2.** Mount the work on parallels in a vise or set it on the drill-press table.

**3.** Set the drill press speed to about 1500 r/min.

**4.** Bring the point of the center drill into the center punch mark and allow the work to center itself with the drill point (Fig. 31-7).

**5.** Continue drilling until about one-third of the tapered section of the drill has entered the work.

**6.** Spot all the holes which are to be drilled.

## TO DRILL A HOLE IN WORK HELD IN A VISE

The most common method of holding small workpieces is by means of a vise, which may be held by hand against a table stop or clamped to the table. When drilling holes larger than ½ in. (12.7 mm) in diameter, the vise should be clamped to the table.

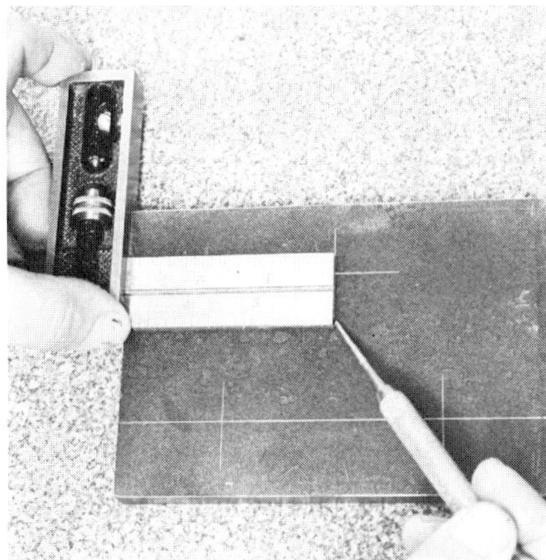
**Fig. 31-4** Intersecting lines may be laid out using a combination square and scriber. (*Courtesy Kostel Enterprises Ltd.*)

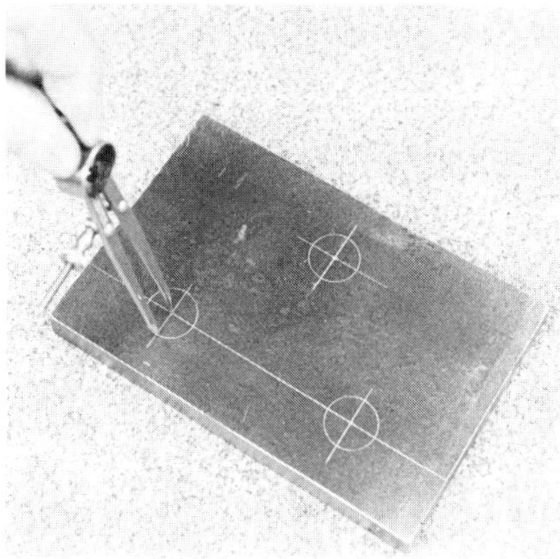
**Fig. 31-5** Scribing the diameter circles with dividers. (*Courtesy Kostel Enterprises Ltd.*)

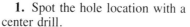

N-4 A table stop or clamp fastened to the table can stop the vise from swinging and prevent injury.

N-5 Ease up on the drilling pressure as the drill starts to break through the work to prevent the drill from breaking or jamming in the work.

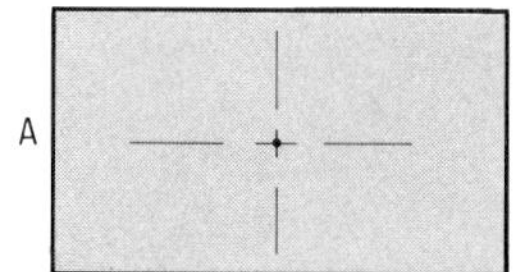

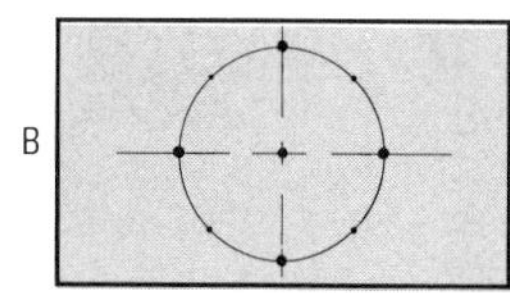

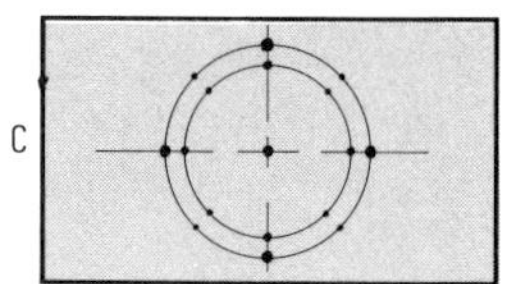

**Fig. 31-6** The layout procedure for a hole to be drilled.

**Fig. 31-7** A center drill is used to spot the location of a hole accurately. (*Courtesy Kostel Enterprises Ltd.*)

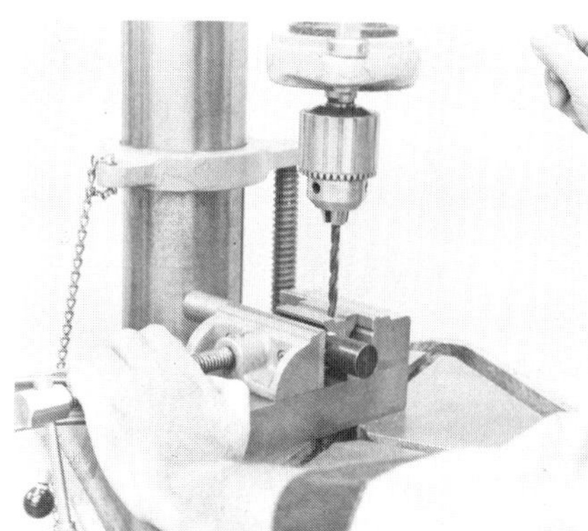
**Fig. 31-8** For small holes, the vise may be hand held, but a stop or clamp should be fastened to the drill press table. (*Courtesy Kostel Enterprises Ltd.*)

**1.** Spot the hole location with a center drill.

**2.** Mount the correct-size drill in the drill chuck.

**3.** Set the drill press to the proper speed for the size of drill and the type of material to be drilled.

**4.** Fasten a clamp or stop on the left side of the table (Fig. 31-8). (See N-4.)

**5.** Mount the work on parallels in a drill vise and tighten it securely.

**6.** With the vise against the table stop, locate the spotted hole under the center of the drill.

**7.** Start the drill-press spindle and begin to drill the hole.

(a) For holes up to ½ in. (12.7 mm) in diameter, hold the vise against the table stop by hand (Fig. 31-8).

*or*

(b) For holes over ½ in. (12.7 mm) in diameter,

- Lightly clamp the vise to the table with another clamp.
- Drill until the full drill point is into the work.
- With the drill revolving, keep the drill point in the work and tighten the clamp holding the vise securely.

**8.** Raise the drill occasionally and apply cutting fluid during the drilling operation.

**9.** Ease up on the drilling pressure as the drill point breaks through the work. This will prevent the drill from being pulled into the work and possibly breaking the drill or damaging the work. (See N-5.)

## TO DRILL ROUND WORK IN A V-BLOCK

V-blocks may be used to hold round work for drilling. Round work is seated in the accurately machined V-groove

**Fig. 31-9** When using a larger drill, the vise should be clamped to the table. (*Courtesy Kostel Enterprises Ltd.*)

**Fig. 31-10A** Using a square and a rule to align the center punch mark on round work. (*Courtesy Kostel Enterprises Ltd.*)

**Fig. 31-10B** Turn the workpiece until the line on the end of the work is in line with the blade of the square. (*Courtesy Kostel Enterprises Ltd.*)

N-6 Always use parallels or a scrap flat block under the work to prevent holes from being drilled in the table.

and small diameters may be held in place with a U-shaped clamp, while larger diameters are fastened with strap clamps.

**1.** Select a V-block to suit the diameter of round work to be drilled. If the work is long, use a pair of V-blocks.

**2.** Mount the work in the V-block and then rotate it until the center punch mark is in the center of the workpiece. Check that the distance from both sides is equal with a rule and square (Fig. 31-10A).

**3.** Tighten the U-clamp securely on the work in the V-block *or* hold the work and V-block in a vise as shown in Fig. 31-10B.

**4.** Spot the hole location with a center drill.

**5.** Mount the proper drill size and set the machine to the correct speed.

**6.** Drill the hole being sure that the drill does not hit the V-block or vise when it breaks through the work.

## DRILL-PRESS TABLE

When work is too large to be held in a vise or other work-holding device, the work may be clamped to the drill-press table. Any types of clamps may be used as long as they do not interfere with the machining operation. If holes must be drilled through the work, it is necessary to place the work on parallels to prevent drilling into the drill-press table. The work can be clamped directly to the table if the holes are not to be drilled through the workpiece. (See N-6.)

The workpiece must be held securely for any drill-press operation. The clamps, bolts, and packing blocks must be located properly and the work clamped securely enough to prevent movement, but not tight enough to spring or distort the work. The correct clamping procedure is illustrated in Fig. 31-11A. Note that the step or packing block is slightly higher than the work and the bolt is located close to the workpiece so that the main pressure is applied to the work. Figure 31-11B illustrates an incorrect clamping procedure in which most of the pressure is applied to the step or packing block. Incorrect clamping occurs whenever the step or packing block is lower than the part being clamped, or when the bolt is closer to the block than to the work.

### Clamping Work to the Table

**1.** Remove any burrs from the workpiece. Clean the workpiece and the drill-press table.

**2.** Mount the workpiece on the drill table. Place a piece of paper between the work and drill-press table to prevent the work from shifting during the drilling operation. If holes are to be drilled through the part, mount the work on parallels or use a subbase under a rough

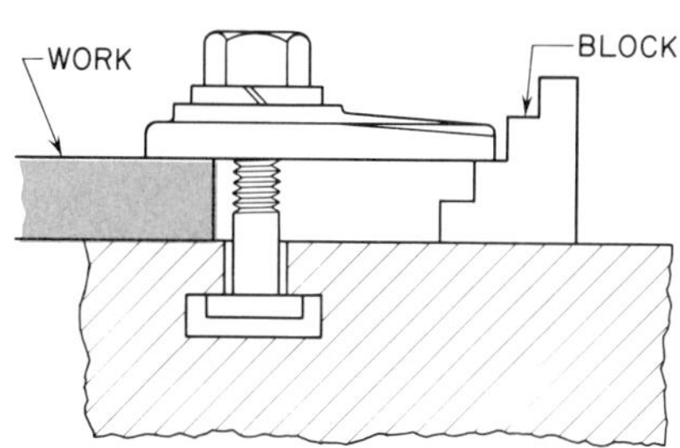


**Fig. 31-11A** The work is correctly clamped because most of the pressure is applied to the part.

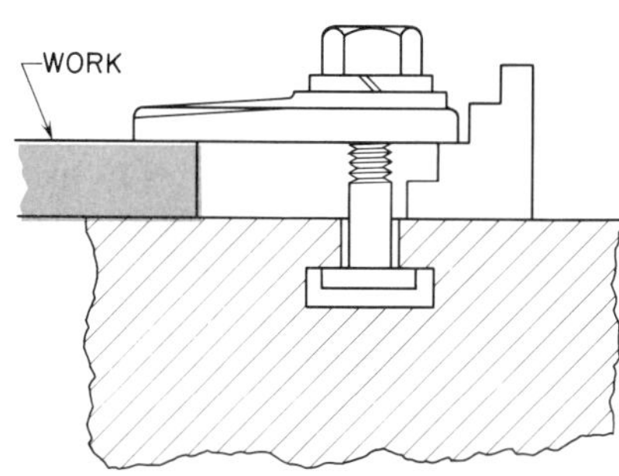


**Fig. 31-11B** The work is incorrectly clamped because most of the pressure is applied to the step block.

N-7 Incorrect clamping can result in broken drills and ruined work.

N-8 Never clamp the work when spotting holes.

N-9 Cutting fluid aids the cutting action and produces more accurate holes.

casting to prevent the marring of the drill-press table.

**3.** If the part does not lie flat on the table, use shims to prevent the work from rocking.

**4.** Locate the work so that the center of the hole to be drilled is aligned with the center of the drill.

**5.** Select straps, bolts, and packing blocks to suit the workpiece.

**6.** Fasten the work securely and properly to the table (Fig. 31-12). Check that the clamps do not interfere with the drilling operation. (See N-7.)

### To Drill Work Fastened to a Drill Table

Because of the size or shape of the workpiece, it is sometimes necessary to clamp the work directly to the table. When work is clamped to the table, always be sure to use parallels between the table and work so that the drill will not cut into the table.

**1.** Spot the hole location with a center drill. (See N-8.)

**2.** Mount the correct size drill and set the drill-press speed and feed.

**3.** Set the work on a suitable set of parallels.

**4.** Locate the spotted hole under the point of the twist drill.

**5.** Select suitable clamps, step blocks, and bolts and position them on the work as shown in Fig. 31-12.

**6.** Lightly tighten each clamp.

**7.** Start the drill revolving and feed the drill until about one-half the drill point has entered the work. If the work is lightly clamped, the spotted hole will align itself with the drill point.

**8.** While the revolving drill is still in contact with the work, tighten both clamps securely.

**9.** Occasionally apply cutting fluid during the drilling operation. (See N-9.)

**10.** Ease up on the drilling pressure as the drill begins to break through the work.

## TO DRILL TO AN ACCURATE LAYOUT

If a hole must be drilled to an exact location, the position of the hole must

**Fig. 31-12** Work properly clamped to a drill-press table. (*Courtesy Kostel Enterprises Ltd.*)

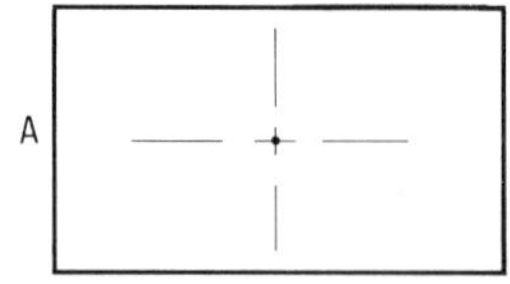

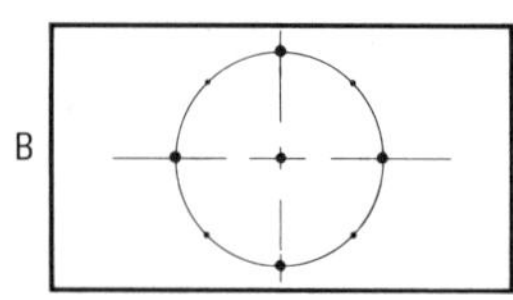

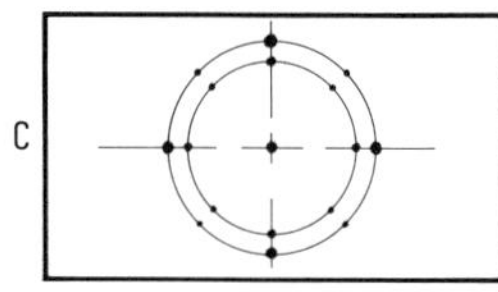


**Fig. 31-13** The procedure for accurately laying out a hole location.

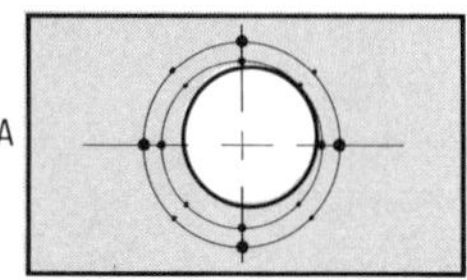

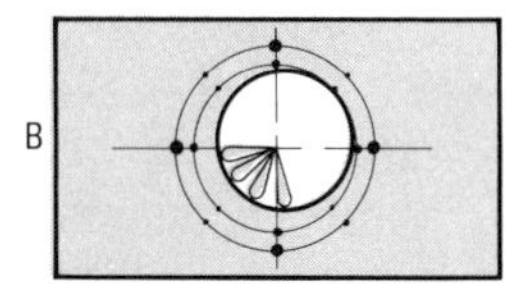

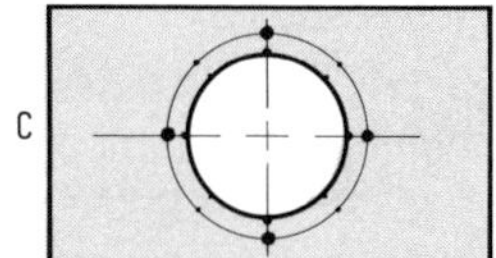


**Fig. 31-14** The procedure for drawing a drill to a layout.

N-10 Always lay out from machined edges for accurate layout work.

N-11 The drill point must be drawn to the center of the scribed circle before the drill cuts to its full diameter.

be accurately laid out, as shown in Fig. 31-13A. During the drilling operation it may be necessary to draw the drill point over so that it is concentric with the layout (Fig. 31-13B).

**1.** Clean and coat the surface of the work with layout dye.

**2.** Locate the position of the hole from two machined edges of the workpiece and scribe the lines as shown in Fig. 31-13A. (See N-10.)

**3.** Lightly prick-punch where the two lines intersect.

**4.** Check the accuracy of the punch mark with a magnifying glass and correct if necessary.

**5.** With a pair of dividers, scribe a circle to indicate the diameter of the hole required.

**6.** Scribe a test circle 1/16 in. (1.5 mm) smaller than the hole size.

**7.** Punch four witness marks on circles up to 3/4 in. (19.05 mm) in diameter, and eight witness marks on circles larger than 3/4 in. (19.05 mm).

**8.** Deepen the center of the hole location with a center punch to provide a larger indentation for the drill to follow.

**9.** Center-drill the work to just beyond the depth of the drill point.

**10.** Mount the proper-size drill in the machine and drill a hole to a depth equal to one-half to two-thirds of the drill diameter.

**11.** Examine the drill-point indentation: it should be concentric with the inner proof circle (Fig. 31-14A).

**12.** If the spotting is off center, cut shallow V-grooves with a cape or diamond point chisel on the side toward which the drill must be moved (Fig. 31-14B).

**13.** Start the drill in the spotted and grooved hole. The drill will be drawn toward the direction of the grooves.

**14.** Continue cutting grooves into the spotted hole until the drill point is drawn to the center of the scribed circles, as shown in Fig. 31-14C. (See N-11.)

**15.** Continue to drill the hole to the desired depth.

## DRILLING A LARGE HOLE

Drills over 1/2 in. (12.7 mm) in diameter generally have taper shanks and are driven by the taper in the drill-press spindle. The size of the tapered hole in

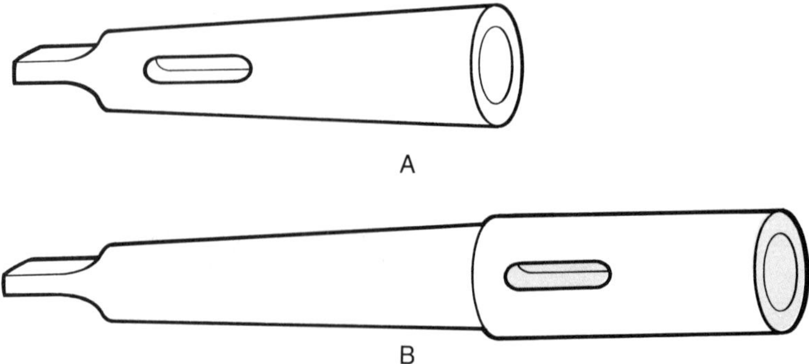


**Fig. 31-15** Drill sleeves and sockets are used to adapt the tapered shank of cutting tools to the drill-press spindle taper. (A) Sleeve; (B) socket.

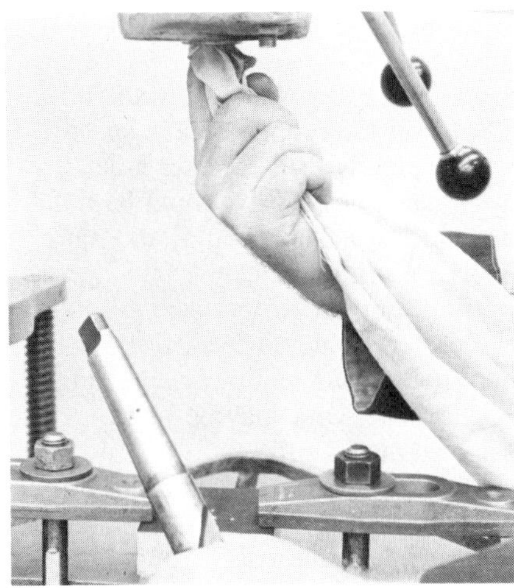

**Fig. 31-16** Cleaning the inside taper of the drill spindle before inserting a tapered-shank cutting tool. (*Courtesy Kostel Enterprises Ltd.*)

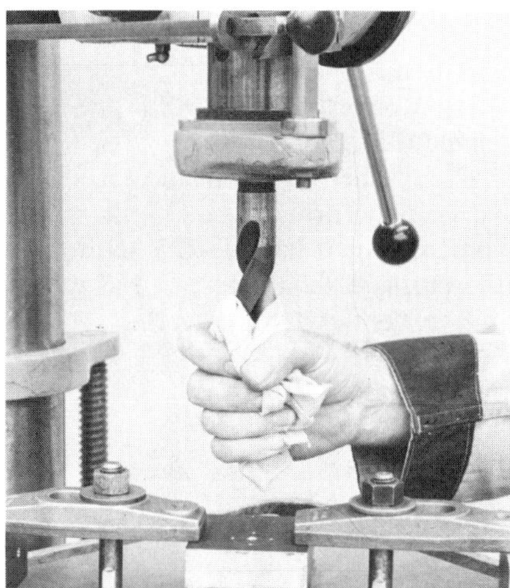

**Fig. 31-17** Using a cloth around the twist drill to prevent hand injury when inserting a drill into the drill-press spindle. (*Courtesy Kostel Enterprises Ltd.*)

N-12 Never make adjustments to any machine tool until the power is shut off at the machine.

N-13 Burrs or dirt can damage the tapers on the cutting tool and in the drill-press spindle and keep the cutting tool from running true.

the drill-press spindle is in proportion to the size of the machine; the larger the machine, the larger the spindle hole. The size of the tapered shank on drill-press cutting tools is also in proportion to the size of the tool. Therefore accessories such as drill sleeves and sockets are necessary to make the size of the cutting-tool shank suit the size of the drill-press spindle taper, and vice versa.

*Drill sleeves* (Fig. 31-15A) are used to adapt the cutting-tool shank to the machine spindle if the taper on the cutting tool is smaller than the tapered hole in the spindle.

A *drill socket* (Fig. 31-15B) is used when the hole in the spindle of the drill press is too small for the taper shank of the drill. The drill is first mounted in the socket and then the socket is inserted into the drill-press spindle. Drill sockets may also be used as extension sockets for extra length.

A *drift* (Fig. 31-18), a wedge-shaped tool, is used to remove a taper shank tool from the drill-press spindle. The drift should be inserted into the spindle slot with its rounded edge up. Place a wooden block on the drill-press table to prevent the drill from marring the table when it is removed. Sharply strike the end of the drift with a hammer to remove the tool from the drill-press spindle.

## Mounting and Removing Taper Shank Tools

Care should be used when mounting and removing taper shank tools in order to preserve their accuracy and ensure that the taper in the drill-press spindle or on the cutting tool will not be damaged.

### Mounting

**1.** Be sure that there is no way that the drill press can be turned on accidentally. *(Turn off the power at the machine.)* (See N-12.)

**2.** Clean the drill-press spindle taper with a cloth (Fig. 31-16).

**3.** Remove any burrs and clean the drill shank thoroughly. (See N-13.)

**4.** Raise the drill-press spindle to its UP position.

**5.** Hold the drill with a rag and turn the drill until the tang fits into the spindle slot (Fig. 31-17).

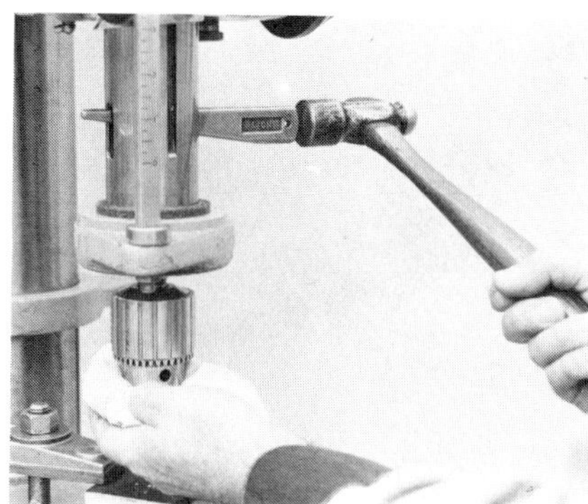

**Fig. 31-18** Removing a drill chuck with a drill drift. (*Courtesy Kostel Enterprises Ltd.*)

**6.** Give the drill a quick upward snap to properly seat the tapered shank.

### Removing

**1.** Be sure that there is no way that the drill press can be turned on accidentally. *(Turn off the power at the machine.)*

**2.** Place a wooden block on the drill-press table under the cutting tool to be removed.

**3.** Turn the drill-press spindle by hand until the slot in the spindle aligns with the slot in the quill.

**4.** Select the proper-size drill drift.

**5.** Insert the drift in the quill slot with its rounded edge up.

**6.** Hold the cutting tool with one hand and tap the drill drift with a hammer to remove the tool from the spindle (Fig. 31-18).

## Pilot Holes for Large Drills

As the size of a drill increases, the thickness of the web of the drill increases to strengthen the drill. The thicker web results in a longer chisel point on the drill. Since the chisel point does not cut, larger drills require more pressure to feed them into the work. To relieve some of this drilling pressure and provide a guide for the larger drill to follow, a pilot hole is first drilled in the work at the location of the hole.

The size of pilot hole drilled should be only slightly larger than the thickness of the web of the drill to be used (Fig. 31-19). If the pilot hole is drilled too large, the following drill may chatter, drill the hole out of round, or

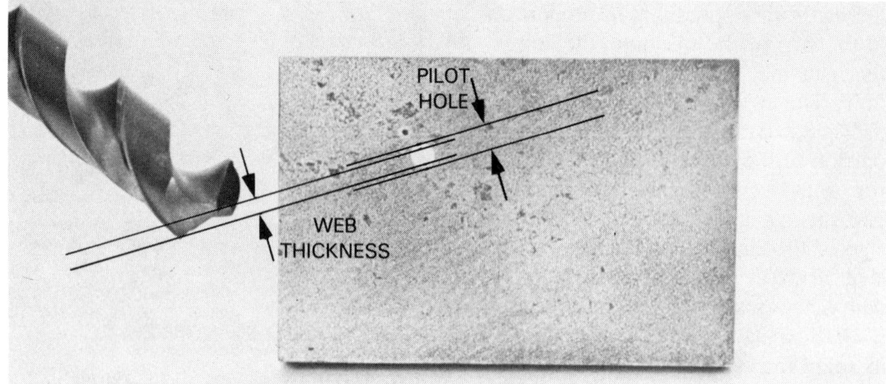


**Fig. 31-19** The size of the pilot drill should be slightly larger than the thickness of the drill web. (*Courtesy Kostel Enterprises Ltd.*)

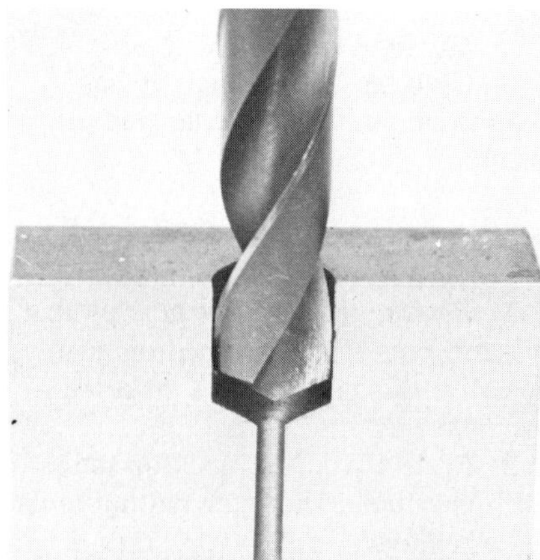

**Fig. 31-20** A pilot hole reduces the drilling pressure and prevents a larger drill from wandering. (*Courtesy Kostel Enterprises Ltd.*)

**N-14** **Do not raise or lower the drill-press table, otherwise the drill spindle and the pilot hole will not be in line.**

damage the top of the hole. Care must be used to drill the pilot hole on center because the larger drill *will follow the pilot hole*. This method may also be used to drill average-size holes when the drill press is small and does not have sufficient power to drive the drill through the solid metal.

## Drilling Large Holes

**1.** Check the print and select the proper drill for the hole required.

**2.** Measure the thickness of the web at the point. Select a drill with a diameter slightly larger than the web thickness (Fig. 31-19).

**3.** Mount the workpiece on the table.

**4.** Adjust the height and position of the table so that the drill chuck can be removed and the larger drill placed in the spindle without having to lower the table after the pilot hole is drilled. Lock the table securely in this position.

**5.** Place a center drill in the drill chuck, set the proper spindle speed, and accurately drill a center hole. *Note:* The center drill should be used first since it is short, rigid, and more likely to follow the center punch mark.

**6.** Using the proper-size pilot drill and correct spindle speed, drill the pilot hole to the required depth (Fig. 31-20). The work may be lightly clamped at this time.

**7.** Shut off the machine, leaving the pilot drill in the hole.

**8.** Clamp the work securely to the table.

**9.** Raise the drill spindle and remove the drill and drill chuck (Fig. 31-18). (See N-14.)

**10.** Clean the taper shank of the drill and the drill-press spindle hole. Remove any burrs on the drill shank with an oilstone.

**11.** Mount the large drill in the spindle (Fig. 31-17).

**12.** Set the proper spindle speed and feed and drill the hole to the required depth. If hand feed is used, the feed pressure should be eased as the drill breaks through the work.

# TO DRILL WORK FASTENED TO AN ANGLE PLATE

An angle plate is a precision tool, machined to an accurate 90° angle, which may be used to hold special-shaped workpieces for drilling (Fig. 31-21).

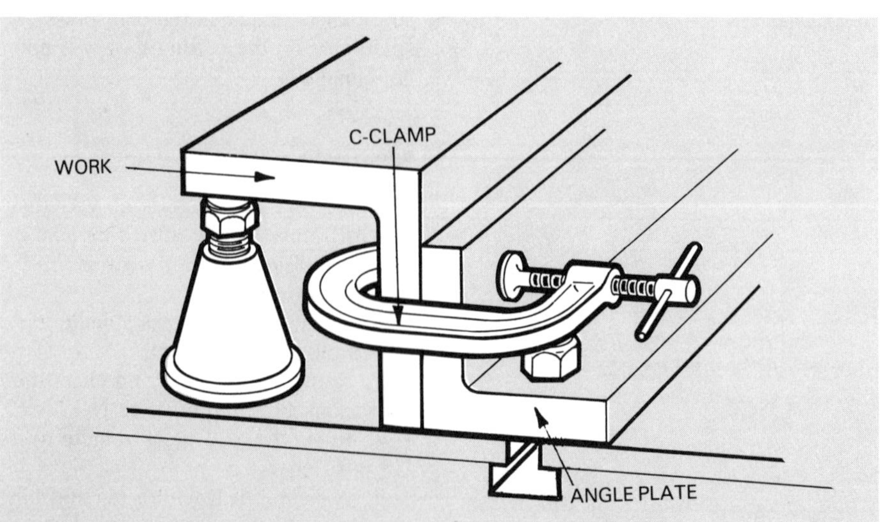


**Fig. 31-21** A workpiece clamped to an angle plate and supported with a screw jack.

1. Select the proper-size angle plate to suit the workpiece.
2. Thoroughly clean the drill-press table and the angle plate.
3. Lightly clamp the angle plate to the drill-press table.
4. Clamp the work lightly to the angle plate.
5. Align or level the work and then securely tighten the clamps.
6. Move the angle plate until the point of the center drill is aligned with the location of the hole.
7. Securely fasten the angle plate to the drill-press table.
8. Use a screw jack to support the edge of the work (Fig. 31-21).
9. Spot the hole location and then drill the hole to the required size.

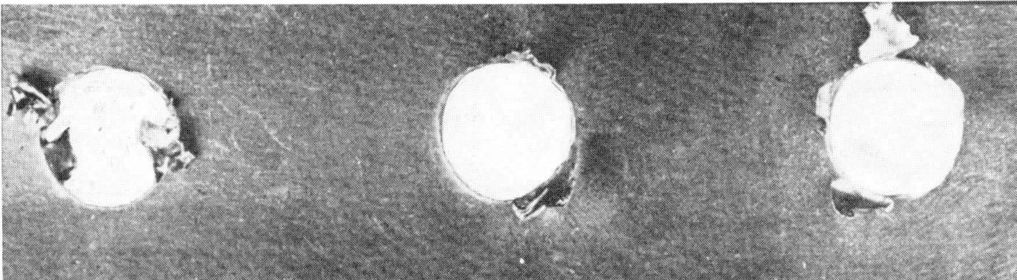

**Fig. 31-22** Regular ground twist drills produce poor holes in thin metal. (*Courtesy Cincinnati Milacron Co.*)

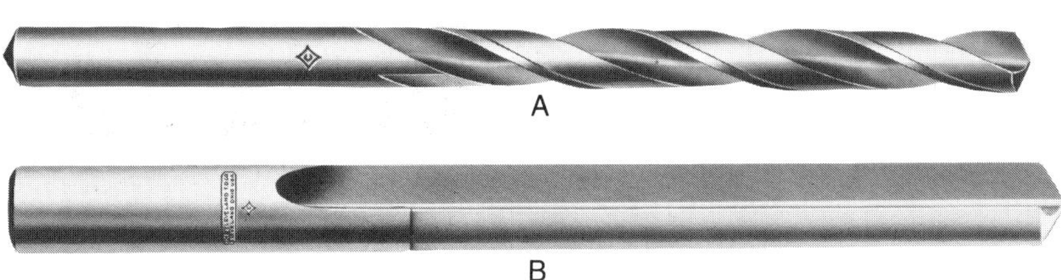

A

B

**Fig. 31-23** Two types of drills used on thin metal. (A) Low helix; (B) straight flute. (*Courtesy The Cleveland Twist Drill Co.*)

**Fig. 31-24** A short flat ground on the lips of a drill makes it suitable for drilling thin materials. (*Courtesy Kostel Enterprises Ltd.*)

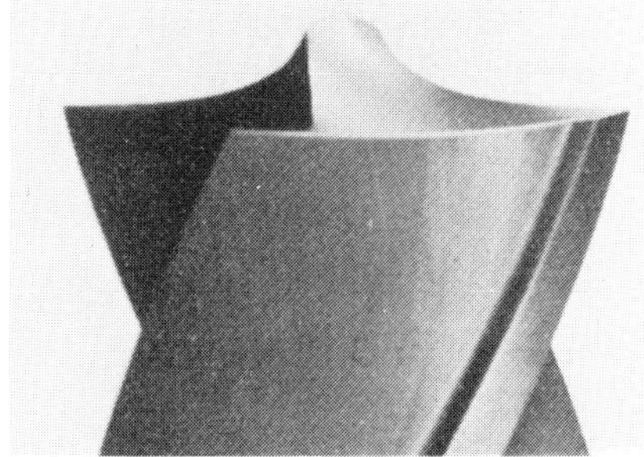

**Fig. 31-25** Best holes are produced in thin material using a drill ground with a center bit and a flat end. (*Courtesy Cincinnati Milacron Co.*)

**Fig. 31-26** Almost burr-free holes are produced with a center bit drill. (*Courtesy Cincinnati Milacron Co.*)

## DRILLING THIN MATERIAL

Drilling holes in sheet metal, particularly with standard twist drills over ½ in. (12.7 mm) in diameter, is generally difficult and often results in damaged work. When drilling thin material, the standard twist drill has a tendency to "hook" into the material. This action is caused by the rake angle created by the helix of the flutes and the drill point. This type of drill does not produce a clean, round hole in the thin material (Fig. 31-22). As a result, special drills and specially ground twist drills should be used when drilling thin materials.

### Drill Types and Requirements

The use of a low helix drill, Fig. 31-23A, or a straight-fluted drill having no rake angle, Fig. 31-23B, will improve the quality of holes drilled in thin material. If a straight-fluted drill and a low helix drill are not available, the standard twist drill can be modified by grinding a short flat on the lip of the drill to remove the rake angle (Fig. 31-24).

Another factor which affects the hole quality is the manner in which the drill point is ground. The kind of drill point which is most suitable for drilling thin material is shown in Fig. 31-25. It is ground with a small point in the center to position the drill in the punch mark. The remainder of the drill point is ground to an angle of 5° from the flat (Fig. 31-25), and a lip clearance of about 12°. As the drill point penetrates the work, the outer edges act as a trepanning tool and produce a round, almost burr-free hole (Fig. 31-26).

The *hole saw* (Fig. 31-27) is another tool used to produce holes in sheet metal. This is particularly useful when large holes in thin material are desired. The hole saw consists of a thin outer heat-treated tool steel shell which has saw teeth ground on the end of the shell. The hole saw is guided into position by a small pilot drill

**Fig. 31-27** A hole saw provides an easy way of producing holes in thin metal. (*Courtesy Kostel Enterprises Ltd.*)

N-15 Clamp thin work securely; it can cause serious cuts if a drill jams and revolves the work.

mounted in the center of the shank (Fig. 31-27). This pilot drill engages in the center punch mark, penetrates the metal first, and acts as a guide for the saw.

### Drilling Thin Material

When drilling holes in thin material, care must be taken to clamp the work as near as possible to the hole being drilled to prevent the material from riding up the drill. The workpiece should be placed on a wooden block or board so that the drill will clear the workpiece and will not cut into the drill press table.

**1.** Mount a properly ground drill or hole saw of the required diameter in the drill chuck.
**2.** Set the spindle speed for the size of drill being used.
**3.** Clean the drill table and place a suitable piece of wood on the table (Fig. 31-27).
**4.** Place the workpiece so that the hole position is over the wood.
**5.** Bring the point of the drill down until the point engages in the center punch hole.
**6.** Clamp the work securely. (See N-15.)
**7.** Drill the hole as required.

## REVIEW QUESTIONS

**1.** Why is it necessary to remove burrs from a workpiece before starting to lay out?
**2.** Why is a center drill suitable for spotting hole locations?
**3.** How deep should hole locations be spotted?
**4.** Name two methods of holding small workpieces for drilling.
**5.** On what side of the table should a clamp or stop be fastened?
**6.** When should a vise be clamped to the table for drilling holes?
**7.** What may occur if drilling pressure is not eased as a drill breaks through the work?
**8.** For what purpose may V-blocks be used?
**9.** Why should parallels be used between the table and workpiece?
**10.** Where should the bolts and step blocks be placed when clamping work to the drill-press table?
**11.** Why should cutting fluid be used when drilling holes?
**12.** From what edges should a layout be made?
**13.** How deep should a hole location be center-drilled?
**14.** How deep should a hole be drilled before the drill-point indentation is compared with the layout?
**15.** Explain the procedure for drawing a hole over toward the layout.
**16.** Why should burrs and dirt be removed from the tapers of cutting tools and the drill-press spindle?
**17.** What is the purpose of a wooden block on the drill-press table when mounting and removing cutting tools?
**18.** State two purposes of a pilot hole.
**19.** What size should the pilot hole be in relation to the web of the larger drill?
**20.** When are angle plates used in drill-press work?
**21.** Name three types of drills which can be used for drilling thin material.

# UNIT 32

# REAMERS

A *reamer* is a rotary cutting tool used to enlarge previously drilled or bored holes to accurate dimensions. Reamers may have two or more straight or helical flutes in the body. These flutes provide for chip clearance during the reaming operation. All straight reamers, as compared with tapered reamers, cut on the chamfer or entering lead only. Straight reamers *do not cut* along the margins on the body diameter. Reamers with helical flutes provide a smooth shear-cutting action and produce a good surface finish.

## OBJECTIVES

The purpose of this unit is to enable you to:

**1.** Identify and state the purpose of the main parts of a reamer
**2.** Recognize and know the purpose of four common machine reamers
**3.** Determine how much material must be left in a hole for reaming

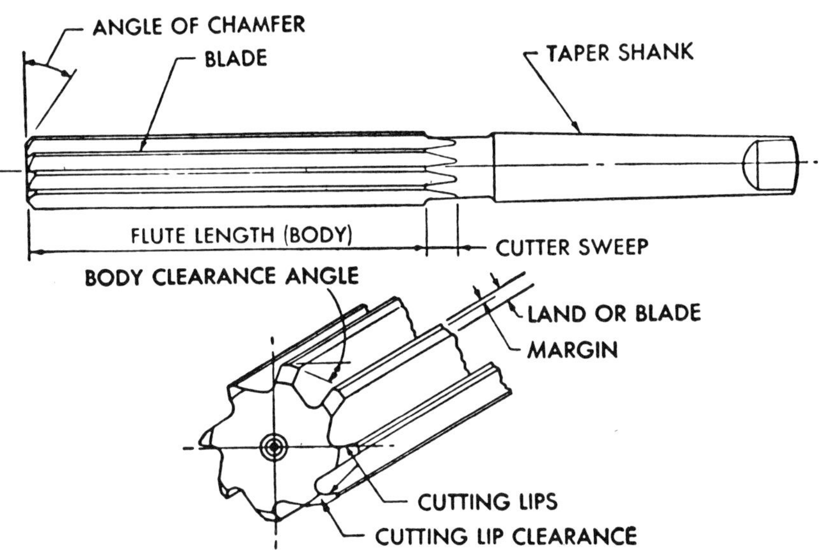


**Fig. 32-1** The main parts of a reamer.

N-1 Helical-fluted reamers produce better surface finishes.

**4.** Machine-ream a hole on a drill press

**5.** Hand-ream a hole on a drill press squarely

## STANDARD REAMER PARTS

There are several different types of reamers; however, the parts of these reamers are similar. The main parts of a reamer are the *body* and the *shank*, (Fig. 32-1).

The *body* of the reamer contains several helical or straight grooves called *flutes*. The teeth, or *lands*, are located between these flutes. The *margin* on the top of each land extends from the *body-clearance angle* to the leading edge of each flute. The margin is ground with *body clearance* to reduce the friction while the reamer is cutting. The *chamfer* ground on the ends of the lands provides the cutting edges for most machine reamers. The angle of the chamfer is from 40 to 50°. *Point clearance* or *chamfer relief* is ground on the end of each tooth so that the reamer will cut as it is advanced into the hole.

The *shank*, which is used to drive a machine reamer, may be straight or tapered. Hand reamers have a square on the end so that a wrench can be used to turn the reamer into the work.

## TYPES OF REAMERS

Reamers are available in a variety of types and sizes to suit various applications. All reamers fall into the two main categories, *machine reamers* and *hand reamers*.

### Machine Reamers

Machine or chucking reamers are available in a wide variety of types and sizes for reaming on drill presses. These reamers are supplied either with straight shanks, which are held and driven by a drill chuck, or with tapered shanks, which fit directly into the taper of the drill-press spindle. Most machine reamers are available with straight or helical flutes. Helical fluted reamers provide a shear-cutting action, clear the chips out of a hole readily, and produce a better surface finish. They are particularly suited for reaming holes having slots or keyways. (See N-1.)

*Rose chucking reamers* (Fig. 32-2A) are designed to cut on the end only and are particularly suited for machine reaming. The end of each tooth has a 45° chamfer which is relieved to provide cutting edges. The lands are ground cylindrically and are not relieved; they serve only to guide the reamer into the hole. Rose reamers are considered to be roughing tools and are used where the hole finish and accuracy are not critical.

*Fluted chucking reamers* (Fig. 32-2B) are used to produce holes to close tolerances and with high finishes. The fluted chucking reamer cuts on the

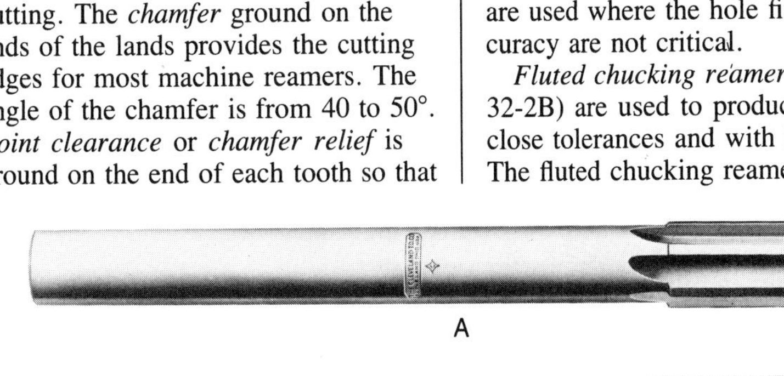

A

B

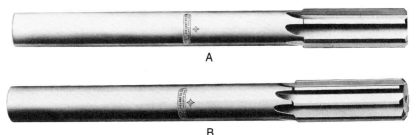

**Fig. 32-2** Types of chucking reamers. (A) Rose; (B) fluted. (*Courtesy the Cleveland Twist Drill Co.*)

**Fig. 32-3** Jobber's reamers have longer teeth than chucking reamers. (*Courtesy Kostel Enterprises Ltd.*)

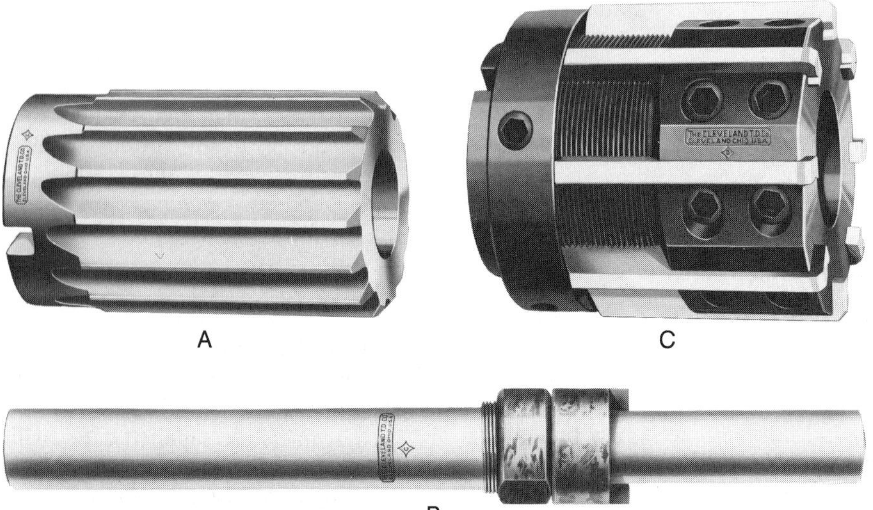

**Fig. 32-4** Shell reamers are available in solid and adjustable types. (A) Solid; (B) arbor; (C) adjustable. (*Courtesy The Cleveland Twist Drill Co.*)

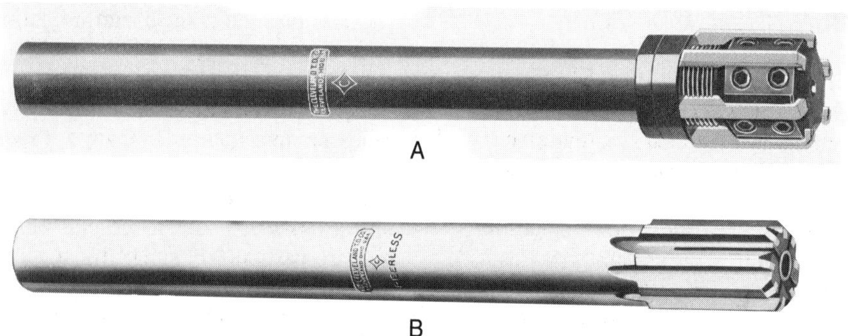

**Fig. 32-5** Expansion reamers can be adjusted to compensate for wear to obtain a specific size. (A) Expansion; (B) adjustable. (*Courtesy The Cleveland Twist Drill Co.*)

N-2 Shell reamers are more economical for larger holes.

N-3 Carbide reamers can be run at two to three times the speed of high-speed steel reamers.

chamfer portion only. The lands are relieved to narrow margins which are provided with very little back taper. Fluted reamers have more teeth than rose reamers for the same diameter and are considered to be finishing tools primarily.

*Jobber's reamers* (Fig. 32-3), a type of fluted reamer, have a longer body than a fluted chucking reamer. The jobber's reamer has a 45° chamfer on the end and is used for machine operations where long cutting edges are required.

*Fluted shell reamers* consist of two parts: the reamer (Fig. 32-4A) and the arbor (Fig. 32-4B). The reamer is held on the arbor by means of a slight taper [⅛-in. (3.17-mm) taper per foot] and is driven by means of two lugs. Shell reamers are used for roughing operations. They are also available in the adjustable type, Fig. 32-4C.

The advantages of shell reamers are:

- They are economical since several sizes of reamers can be used with one arbor. (See N-2.)
- When the reamer becomes worn or damaged, it can be discarded and the arbor used with another reamer.
- They are more economical for reaming larger holes.

The *expansion reamer* (Fig. 32-5A) consists of a slotted high-strength steel body. High-speed steel or carbide blades are brazed into the body. This reamer can be expanded by means of a tapered threaded plug fitted into the end of the reamer. This expansion feature, approximately .005 in. (0.12 mm) for a ½-in. (12.7-mm) reamer, means that the reamer can be expanded and resharpened to its original size several times.

The *adjustable chucking reamer* (Fig. 32-5B) has high-speed steel or carbide blades which may be replaced individually or in sets. The blades are adjustable to approximately .015 in. (0.38 mm) over or under the reamer size. The adjustable feature and the replaceable blade make this an economical reamer.

The *carbide-tipped reamer* (Fig. 32-6) was developed to meet the need for higher production rates. All types of machine reamers are available with carbide-tipped blades. These blades significantly outlast high-speed steel blades. Carbide blades are suited to the machining of castings where sand and scale create a problem for high-speed steel reamers. Because of their abrasion resistance and ability to withstand high cutting temperatures, carbide-tipped reamers can be run at higher speeds and still maintain their sharp cutting edges and size. (See N-3.)

*Machine taper reamers* (Fig. 32-7) are available in a variety of standard tapers and are used to ream tapered holes. To preserve the cutting edges of a finishing reamer, a roughing or old reamer should be used to bring the hole close to the required size, followed with a finishing reamer.

Fig. 32-6 A carbide-tipped chucking reamer.

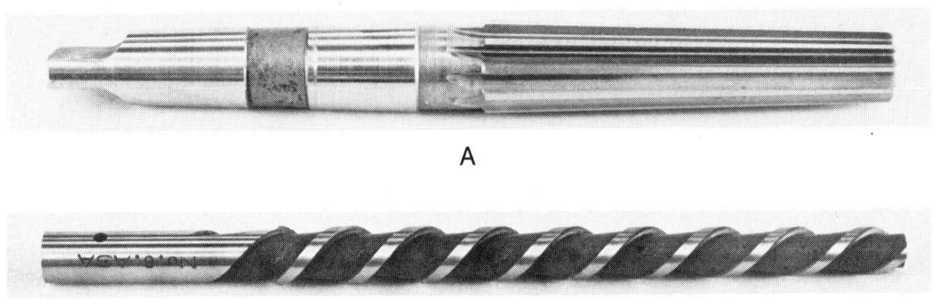


Fig. 32-7 Types of machine taper reamers. (A) Standard taper; (B) taper pin. *(Courtesy Kostel Enterprises Ltd.)*

## Hand Reamers

A hand reamer may be used to finish a hole to a high degree of accuracy and finish. Hand reamers, available in several types, are provided with a square on the shank end for turning the reamer. When hand-reaming in a drill press, always use a stub center in the drill chuck to keep the reamer aligned. Most types of hand reamers are available in high-speed steel or carbon steel and with straight or helical teeth. (See N-4.)

N-4 Never use a hand reamer under power.

*Solid-type hand reamers* (Fig. 32-8) are general-purpose reamers used for most standard hole sizes. The end of the reamer is tapered slightly for a distance equal to its diameter so that the reamer will start easily in the hole. The teeth on straight-fluted hand reamers may be unevenly spaced to reduce chatter and improve the surface finish. Chatter is also reduced by the use of a left-hand helical-fluted reamer. The left-hand helical reamer is also recommended for reaming a hole in which there is a slot or keyway.

N-5 Rotate the reamer in a clockwise direction as it is being pulled out of a hole.

N-6 To eliminate reamer chatter, reduce the speed and increase the feed.

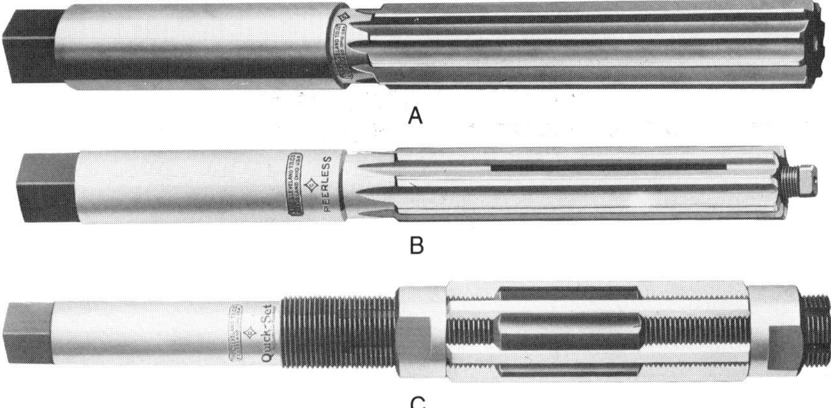


Fig. 32-8 Various types of hand reamers. (A) Straight fluted; (B) expansion; (C) adjustable. *(Courtesy The Cleveland Twist Drill Co.)*

### Reamer Care

The finish of the reamed hole, the accuracy of the hole, and the life of the reamer all depend upon the care the reamer receives. Reamers are expensive and have a relatively short life. Therefore, every precaution should be taken to ensure that they are used and stored properly. The following suggestions will help to prolong the life of a reamer and produce accurate holes.

- Never turn a reamer backward at any time, since this will ruin the cutting edges. (See N-5.)
- Always store reamers in separate compartments. If stored in the same container, they must be separated by some suitable material such as cardboard, plastic, or wood.
- Never roll, drop, or place reamers on the metal surfaces of benches or machines.
- Always use the proper speed when reaming.
- Never feed a reamer too fast. This may cause it to jam in the work.
- Use the proper cutting fluid for the metal being reamed.
- Never permit a reamer to chatter. This will dull the cutting edges quickly. (See N-6.)
- Use helical-fluted reamers for deep holes or holes which have keyways or grooves.

## REAMING FACTORS

When a hole is not required to be accurately finished, it may be drilled to size. However, if the hole must be accurate in size and shape and have a good finish, it should be drilled and then reamed. Speed, feed, and the reaming allowance are the three main factors which affect the accuracy of the reamed hole and the life of the reamer. The following factors also affect the reaming operation:

- The type of work material
- The rigidity of the setup
- The use of the correct cutting fluid
- The depth of the hole
- The type and condition of the reamer
- The condition of the machine
- The type of finish required

**TABLE 32-1 RECOMMENDED STOCK ALLOWANCES FOR REAMING**

| Hole size | | Allowance | |
|---|---|---|---|
| Inch | Millimeter | Inch | Millimeter |
| ¼ | 6.35 | .010 | 0.25 |
| ½ | 12.7 | .015 | 0.38 |
| ¾ | 19.05 | .018 | 0.46 |
| 1 | 25.4 | .020 | 0.51 |
| 1¼ | 31.75 | .022 | 0.56 |
| 1½ | 38.1 | .025 | 0.63 |
| 2 | 50.8 | .030 | 0.76 |
| 3 | 76.2 | .045 | 1.14 |

The speed, feed, and stock allowance are variables which are controlled easily by the operator and only these factors will be discussed.

## Stock Allowance

There is no definite rule for the amount of stock to be left on the hole diameter for reaming because of the many factors which must be considered. If too much material is left, the flutes may clog and the reamer may break; if too little material is left, there is a burnishing (rubbing) action rather than a cutting action which tends to dull the reamer.

The average stock allowances for machine-reaming are listed in Table 32-1. These allowances may be varied slightly to suit various factors which affect the reaming operation.

For hand-reaming, the allowances are less because of the difficulty in forcing the reamer into the hole. The allowance for hand-reaming should never exceed .005 in. (0.12 mm).

## Reaming Speeds

The speed at which the hole is reamed depends mainly on the type of material being machined and the diameter of the reamer. The reaming speeds for most materials are about two-thirds of the speed used for drilling. (See N-7.) For example, if a drill whose diameter is 1 in. (25.4 mm) is revolved at 382 r/min to drill a piece of machine steel, the reamer should be run at two-thirds of 382 or about 255 r/min. Carbide-tipped reamers may be run at two to three times the speeds shown for high-speed steel reamers. Table 32-2

N-7 Too high a speed will ruin a reamer.

**TABLE 32-2 RECOMMENDED REAMING SPEEDS (HIGH-SPEED STEEL REAMERS)**

| Material | Speeds | |
|---|---|---|
| | surface feet per minute | m/min |
| Steel castings | 20–30 | 6–9 |
| Tool steel | 35–40 | 10–12 |
| Cast iron | 50–70 | 15–21 |
| Machine steel | 50–70 | 15–21 |
| Brass and aluminum | 130–200 | 40–60 |

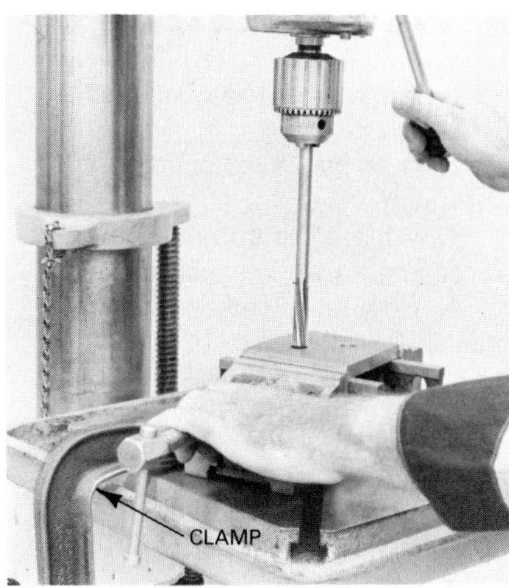


**Fig. 32-9** Reaming holes smaller than ½″ (12.7 mm) in a drill press holding the vise against a stop. (*Courtesy Kostel Enterprises Ltd.*)

N-8 Place the strap clamps and bolts correctly to clamp securely and avoid springing the work.

may be used as a guide in selecting the speed in surface feet per minute for the material to be reamed.

## Feeds

The feeds are generally higher for reaming than for drilling. The amount of feed used will vary with the material to be reamed; but in general, a feed between .001 and .0025 in. (0.02 and 0.06 mm) per flute per revolution provides a good starting point. Too low a feed may result in glazing, excessive wear, or chatter. Too high a feed tends to reduce the accuracy of the hole and produce a poor finish. Use as high a feed as possible to produce the hole finish and accuracy required.

## Reaming a Straight Hole

A hole which is to be finished to size should be reamed immediately after it is drilled and while it is still located under the drill spindle. This will ensure that the reamer follows the same center as the drill.

### Machine Reaming a Straight Hole

**1.** Mount the work on parallels and secure it for drilling (see Unit 31). (See N-8.)

**2.** Drill the hole to the proper size for the reamer to be used, leaving the proper allowance for the reamer (Table 32-1).

**3.** Remove the drill and mount the proper reamer in the spindle. If a taper shank reamer is used, be sure to clean the taper on the shank and in the drill-press spindle. Do not move the work at this time.

**4.** Adjust the spindle speed to suit the reamer and the work material (Table 32-2).

**5.** Set the proper feed rate for the reamer being used.

**6.** Set the depth stop to prevent the reamer from hitting the drill-press table.

**7.** Start the machine and carefully lower the spindle until the chamfer on the reamer just touches the lip of the hole (Fig. 32-10).

**8.** Apply cutting fluid and engage the automatic feed.

**9.** After the reamer reaches the required depth, use the hand-feed lever to raise the reamer until it is clear of the hole.

**10.** Shut off the machine and remove the burr from the edge of the hole.

**11.** Check the hole for size and accuracy.

### Hand Reaming a Straight Hole

When only a small amount of material is left after drilling, the hole may be

**Fig. 32-10** Reaming larger holes than ½″ (12.7 mm), the work should be clamped to the table. (*Courtesy Kostel Enterprises Ltd.*)

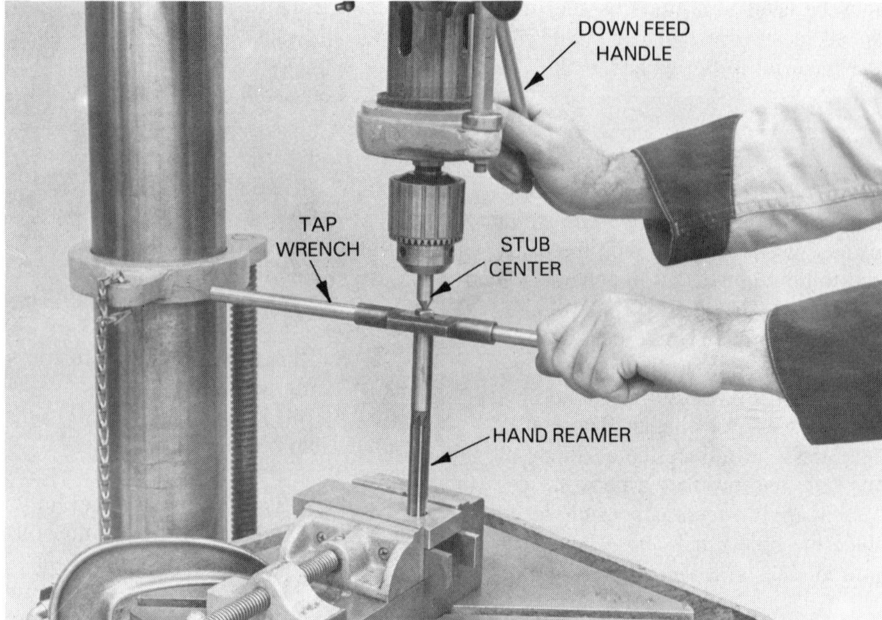


**Fig. 32-11** Hand reaming in a drill press. (*Courtesy Kostel Enterprises Ltd.*)

N-9 Never turn a reamer backward.

reamed by hand before it is removed from the drill-press table.

**1.** Mount the work on parallels and secure it for drilling (Fig. 32-11).

**2.** Drill the hole to the proper size leaving an allowance for the hand reamer to be used. The reaming allowance should be no more than .005 in. (0.12 mm) for a 1-in. (25.4-mm) diameter hole.

**3.** Remove the drill and mount a stub center in the drill-press spindle (Fig. 32-11). Do not move the location of the table or work.

**4.** Place the end of the reamer in the hole and place a tap wrench on the square of the reamer.

**5.** Lower the spindle until the stub center engages in the center hole in the top of the reamer.

**6.** Apply cutting fluid.

**7.** Turn the tap wrench clockwise and at the same time maintain sufficient pressure with the stub center to advance the reamer into the work. Do not force the reamer by advancing the center too rapidly. (See N-9.)

**8.** Ream the hole to the proper depth.

**9.** Continue to turn the reamer clockwise and at the same time pull upward on the reamer until it is clear of the hole.

**10.** Remove the burrs from the hole.

N-1 If a hole must be tapped, do not move the work or table after the hole has been drilled.

## REVIEW QUESTIONS

**1.** For what purpose are reamers used?

**2.** How may hand reamers be identified?

**3.** What is the difference between a rose and a fluted chucking reamer?

**4.** How much can the following reamers be adjusted?
- (a) Expansion
- (b) Adjustable

**5.** State two advantages of carbide reamers over high-speed steel reamers.

**6.** Why should a reamer never be turned backward?

**7.** Why should a reamer never be fed too fast into a hole?

**8.** What type of reamer should be used for reaming holes with keyways or slots?

**9.** Name three main factors which affect the accuracy of a reamed hole and the life of a reamer.

**10.** How much material should be left in a hole for hand reaming?

**11.** At what speed should a reamer be run?

**12.** What is the result of
- (a) Too low feed
- (b) Too high feed

# UNIT 33

# TAPPING A HOLE

The drill press may be used for either hand or machine tapping. When hand tapping, the drill press is used to guide the tap squarely into the hole while the tap is being turned into the work by hand. Hand tapping is generally used when only a few holes are to be tapped. Machine tapping is preferred when many holes must be tapped.

Machine tapping involves the use of a tapping attachment mounted in the drill-press spindle. The tapping operation should be done immediately after the drilling operation to obtain the best accuracy and to avoid duplication of the setup. (See N-1.) This sequence is especially important when tapping by hand.

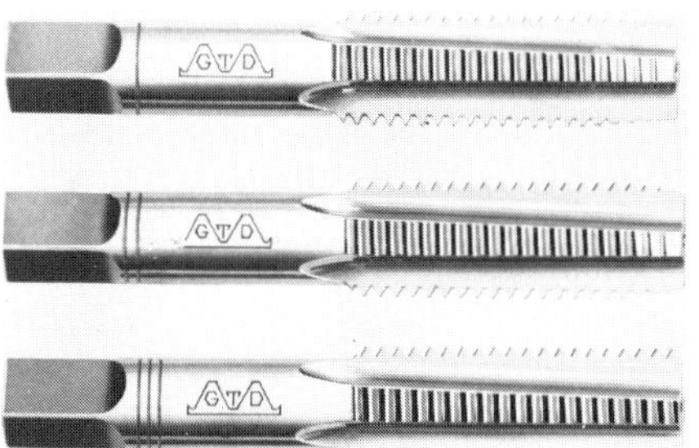

**Fig. 33-1** A set of hand taps. (*Courtesy Greenfield Tap & Die Co.*)

N-2 The stub center keeps the tap in line with the hole.

N-3 The most common machine tap is the gun tap.

N-4 Use helical-fluted taps for threading holes with keyways or slots.

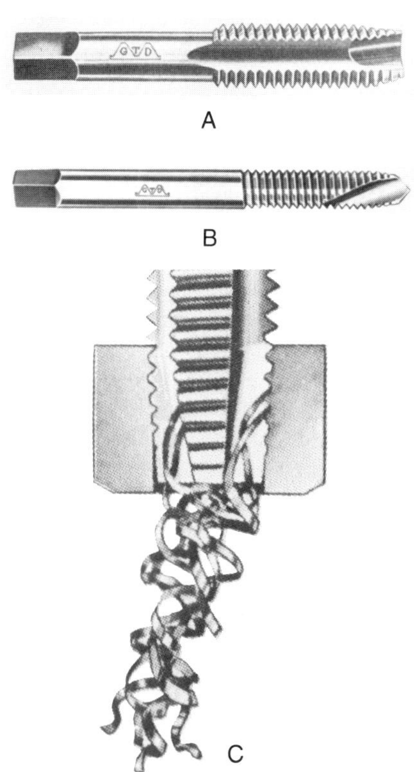


**Fig. 33-2** Types of machine taps (A) Gun tap; (B) stub flute tap; (C) cutting action. (*Courtesy Greenfield Tap & Die Co.*)

# OBJECTIVES

After completing this unit you will be able to:

**1.** Identify and state the purpose of the three hand taps
**2.** Recognize and use the gun and fluted machine taps
**3.** Calculate tap drill sizes for inch and metric taps
**4.** Tap holes by hand
**5.** Set up work and tap holes in a drill press

# TYPES OF TAPS

Taps are hardened and ground precision cutting tools used to produce internal threads. There are many types of taps manufactured and most can be used either for hand or machine operations.

## Hand Taps

Fractional and screw-size machine taps with square ends and standardized linear dimensions are usually called hand taps; however, these taps are used more often in machines than by hand. Hand taps (Fig. 33-1) are available in sets containing three taps: taper, plug, and bottoming. The *taper tap,* which has approximately six threads tapered on its end, is used for easy starting in the hole. The taper and *plug taps* may be used to thread holes which go through the workpiece. The *bottoming tap* is used to thread to the bottom of a blind hole (one that does not go through the workpiece). When using hand taps on a drill press, it is important that the tap be guided by holding it in a drill chuck or with a stub center. (See N-2.)

## Machine Taps

Taps which are used with power must be designed to withstand the torque required to thread the hole and must be able to eliminate the problem of chips jamming in the flutes. Although most taps may be used with power, gun taps, stub-flute taps, spiral-fluted taps, and fluteless taps are designed specifically for machine use.

A *gun tap* (Fig. 33-2A) may be used for machine- or hand-tapping operations. (See N-3.) The gun tap is especially suited for machine tapping because its point angle ''shoots'' the chips out ahead of the tap (Fig. 33-2C). This chip-clearing action eliminates the clogging and loading of the flutes, which is the prime cause of tap breakage. Since the flutes are not required for chip removal, they are shallower and therefore strengthen the tap.

Gun taps may contain two, three, or more flutes and may be used to tap through holes or blind holes which are deep enough to accommodate the chips.

The *stub-flute tap* (Fig. 33-2B) has a short angular flute ground on the chamfered end with the rest of the body left solid. Stub-flute taps are stronger than gun taps and are used to tap shallow through holes of not more than one tap diameter in depth in soft, stringy materials.

Helical-fluted taps, (Fig. 33-3), commonly called *spiral-fluted taps,* have spiral flutes to draw the chips out of a hole (Fig. 33-3C). (See N-4.) They are also used to bridge a keyway or other gap in a hole to be threaded. The helical flutes provide a shear-cutting action and are most often used to draw the chips out of a blind hole in materials that produce stringy chips. Spiral-fluted taps are ideal for

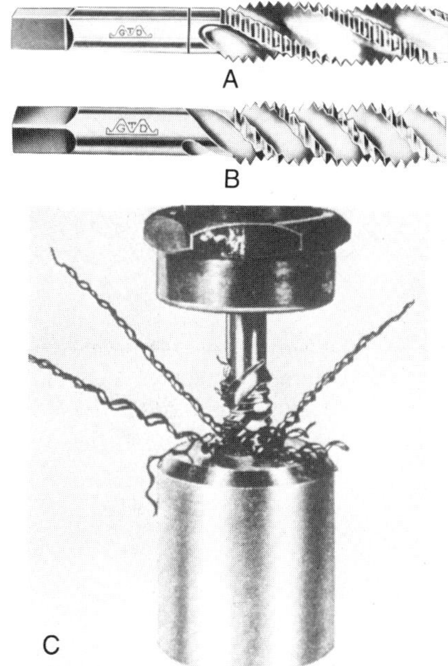


**Fig. 33-3** Types of spiral-fluted taps: (A) Regular spiral; (B) fast spiral; (C) cutting action. (*Courtesy Greenfield Tap & Die Co.*)

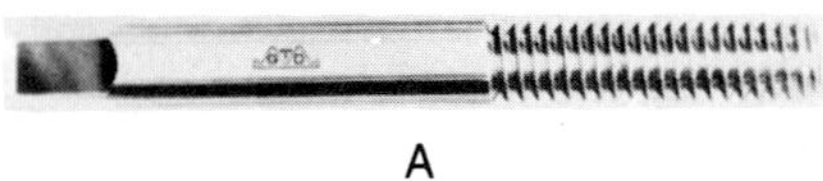


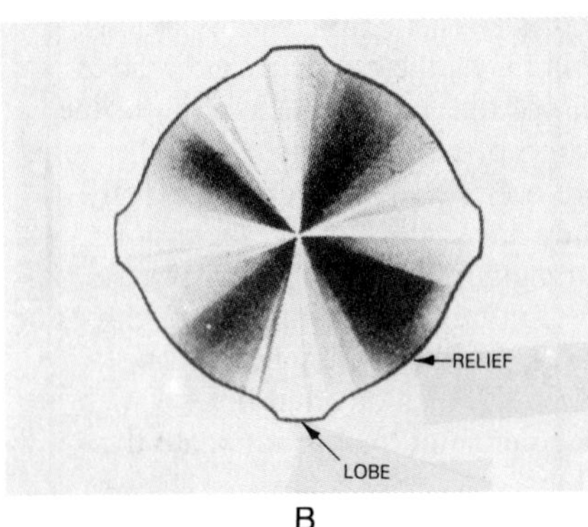


**Fig. 33-4** The characteristics of a fluteless tap: (A) Fluteless tap; (B) lobes of the tap. (*Courtesy Greenfield Tap & Die Co.*)

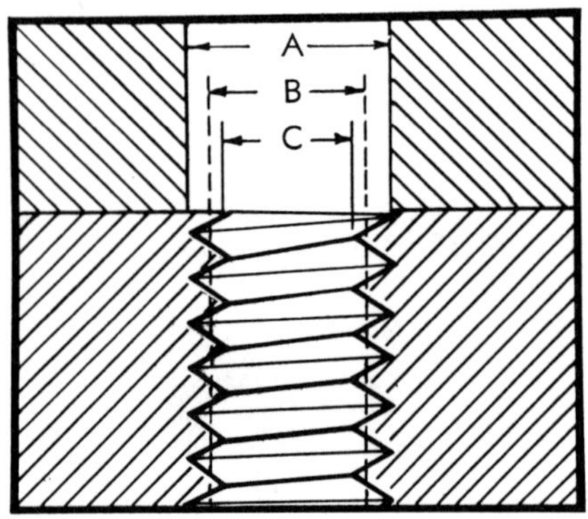


**Fig. 33-5** A cross section of a tapped hole. (*Courtesy Kostel Enterprises Ltd.*)

tapping soft steel, aluminum, magnesium, brass, copper, and die cast metal. The fast spiral taps (Fig. 33-3B) provide for faster clearing of chips; therefore, these taps are used to tap deep holes.

The *fluteless tap* (Fig. 33-4A) is actually a forming tool which produces an internal thread by displacing rather than cutting the metal. The cross section of the fluteless tap, (Fig. 33-4B) shows the lobes which correspond to the cutting edges of a standard tap. The relief forms the lobes and corresponds to the flutes of a standard tap. Since no chips are formed, fluteless taps are valuable where chip removal is a problem. If fluteless taps are applied properly, they provide excellent results in tapping materials such as copper, brass, die castings, aluminum, lead, leaded steels, and similar ductile materials. Improved thread finish and some increase in strength, due to the cold working of the metal, are some of the advantages of producing internal threads with fluted taps.

# TAP DRILL SIZE

## Conventional Taps

Before a tap is used to cut an internal thread, the hole must be drilled in the workpiece to the correct *tap drill size* (Fig. 33-5). This drill size leaves the proper amount of material in the hole for the tap to cut the thread. The tap drill, which is always smaller than the tap diameter, leaves enough material in the hole to produce approximately 75 percent of a full thread.

The tap drill size for a National Form thread (75 percent of a full thread) is calculated using the following formula.

$$\text{Tap drill size} = \text{major diameter} - \frac{1}{\text{threads/inch}}$$

$$= D - \frac{1}{N}$$

**Example**

Calculate the tap drill size required for a 1-in.–8 NC thread.

$$\text{T.D.S.} = D - \frac{1}{N}$$

$$= 1'' - \tfrac{1}{8}$$

$$= \tfrac{7}{8}''$$

## Metric Taps

Metric taps are identified with the letter M followed by the nominal diameter of the thread in millimeters times the pitch in millimeters. A tap with the markings M 2.5 × 0.45 would indicate:

| | |
|---|---|
| M | A metric thread |
| 2.5 | The nominal diameter of the thread in millimeters |
| 0.45 | The pitch of the thread in millimeters |

## Tap Drill Sizes for Metric Threads

The tap drill sizes for metric threads may be calculated by subtracting the pitch from the nominal diameter.

$$\text{T.D.S.} = D - P$$

For example: Calculate the tap drill size for an M12 × 1.75 thread.

$$\text{T.D.S.} = 12 - 1.75$$

$$= 10.25 \text{ mm}$$

Refer to Table 6 in the appendix for tap drill sizes for metric threads.

# TAPPING IN A DRILL PRESS

The drill press can be used to tap a hole by hand or with power. Whenever possible, a hole should be tapped immediately after it is drilled and before the work setup is disturbed. This will ensure that the tapped hole is aligned and square with the workpiece.

## Tapping a Hole by Hand

1. Mount the work on parallels and align the center punch mark with the center of the drill-press spindle.
2. Clamp the work securely to the drill-press table. Be careful not to distort the work during clamping.
3. Adjust the height of the drill-press table to ensure that the drill may be removed from the drill chuck without having to move the table or the work.
4. Spot the hole location with a center drill.

N-5 Be sure that the tap drill is smaller than the tap diameter.

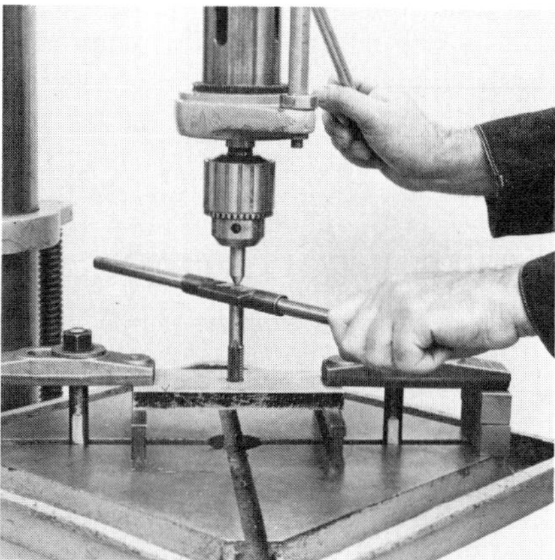

**Fig. 33-6** Guiding a tap in a drill press with a stub center. (*Courtesy Kostel Enterprises Ltd.*)

N-6 Use the proper speed to prolong tap life and produce the best threads.

N-7 Use care when tapping to avoid a broken tap and spoiled work.

**5.** Drill the hole to the correct tap drill size for the size of tap to be used. (See N-5.)

**6.** Without moving the drill-press table or the workpiece, remove the drill and mount a stub center in the drill chuck.

**7.** Place the tap in the drill hole and lower the drill-press spindle until the stub center fits into the center hole in the tap shank (Fig. 33-6).

**8.** Turn the tap wrench clockwise (for right-hand threads) to start the tap in the hole.

**9.** Continue to tap the hole in the usual manner, keeping the stub center in contact with the tap at all times.

## Tapping Speeds

The speed at which a tap operates is very important since it affects the life of the tap and the quality of threads produced. Excessive speed develops a great amount of heat which causes the cutting edges of the tap to break down quickly. It also causes loading or galling, which is the welding of small particles of the material being tapped to the cutting face of the tap. High cutting speeds also prevent adequate lubrication from reaching the cutting zone and therefore aggravate the problem of chip disposal. (See N-6.)

As the cutting edge of the tap dulls because of too much speed, the quality of the thread also deteriorates. A dull tap may produce torn or rough threads, oversize or undersize threads, and burrs. Taps which are loaded produce rough oversize threads.

The following factors must be considered when selecting the speed for a tap.

- The type of work material
- The physical characteristics of the hole
- The type and condition of the fixture, holder, and machine
- The type of cutting fluid to be used

Table 33-1 lists suggested tapping speeds for a variety of materials. These speeds should serve only as a guide or starting point. It may be necessary to raise or lower the listed speed to arrive at the best tapping speed for each job.

## Tapping Problems

Tapping problems are divided into two categories: (1) tap failure and (2) poor quality or incorrect size of the thread. These two problems usually are related to each other since the condition that causes oversize or torn threads and the chipping of tap teeth also results in a broken tap. Some of the more common causes of tap failure and eventual breakage are illustrated in Fig. 33-7. (See N-7.)

Chips packed at the bottom of the hole can cause the tap to chip and break (Fig. 33-7A). This condition can be corrected by cleaning out the chips frequently, drilling a deeper hole, or using a shallower thread depth. If the flutes become clogged with chips (Fig. 33-7B) more tapping torque (turning power) is required. Clogged flutes may cause oversize threads, chipping, or fracturing of the tap. This can be corrected by using a larger tap drill, a tap with more chip space, or a spiral-fluted tap.

Figure 33-7C and D illustrates taps which are not aligned correctly with the holes. This usually results in a broken tap and can be avoided by ensuring that the hole is aligned properly with the tapping spindle.

## Tapping a Hole Using Power

When a large number of holes must be threaded, they are generally tapped by power since it is faster and more accu-

**TABLE 33-1 RECOMMENDED TAPPING SPEEDS FOR HIGH-SPEED STEEL TAPS**

| Material | Surface feet per minute | m/min | Material | Surface feet per minute | m/min |
|---|---|---|---|---|---|
| Aluminum | 90–100 | 27–30 | Steel—cast | 20–30 | 6–9 |
| Brass | 90–100 | 27–30 | —machine | 40–60 | 12–18 |
| Bronze | 40–60 | 12–18 | —nickel | 25–35 | 8–11 |
| Iron, cast | 70–80 | 21–24 | —stainless | 15–25 | 5–8 |
| Iron, malleable | 35–60 | 11–18 | —tool | 25–35 | 8–11 |

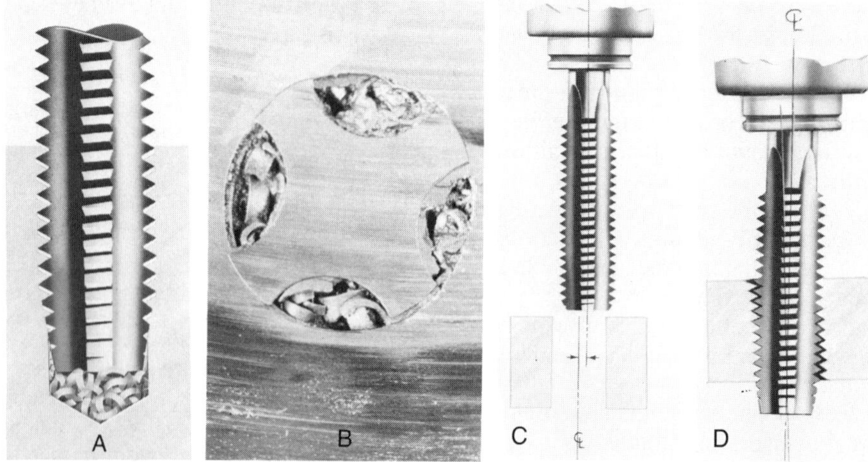


**Fig. 33-7** Common causes of broken taps. (A) Chips packed at the bottom of a hole; (B) flutes clogged; (C) tap not aligned; (D) tap axis not aligned. (*Courtesy The Cleveland Twist Drill Co.*)

**Fig. 33-8** A tapping attachment mounted in a drill-press spindle. (*Courtesy The Ridge Tool Co.*)

rate than hand tapping. A tapping attachment (Fig. 33-8) is mounted in the drill-press spindle for power tapping. This attachment contains a friction clutch which provides a steady, even drive to the tap. If too much pressure is applied, the friction clutch will slip and the tap will stop turning before breakage occurs. As a result, both the life of the tap and its cutting edges are prolonged.

The tapping attachment is constructed so that when the drill-press spindle is fed downward, the tap revolves clockwise. Releasing the down pressure causes the tap to stop. Up pressure (raising) on the downfeed handle engages the reversing mechanism of the tapping attachment and causes the tap to turn counterclockwise and back out of the hole.

Two- or three-fluted machine taps are used for power tapping because of their ability to clear chips out of the hole. The tapping speed for most materials ranges from 20 to 60 surface feet per minute (6 to 18 m/min), but it may be necessary to adjust this speed to suit the tap size and the material being tapped.

**1.** Mount the tapping attachment in the drill-press spindle (Fig. 33-8).

**2.** Fasten a machine or gun tap of the proper size in the tapping attachment.

**3.** Set the spindle speed for the tap size and the material to be tapped.

**4.** Make sure that the drill-press spindle is free to slide in the sleeve.

**5.** Align the hole to be threaded with the tap and clamp the work in this position.

**6.** Start the drill-press spindle and apply cutting fluid to the tap.

**7.** Feed the tap into the hole. If the tap stops revolving, raise the downfeed handle to bring the tap out of the hole and clear the chips.

**8.** When the tap is to the required depth, apply upward pressure on the downfeed handle to reverse the tap and bring it out of the hole.

## REVIEW QUESTIONS

**1.** Name the three hand taps in a set.

**2.** What taps can be used for tapping through holes?

**3.** Name four taps which can be used under power.

**4.** Which taps "shoot" the chips out ahead of the tap?

**5.** What tap is generally used for tapping shallow holes in soft, stringy material?

**6.** Why must the tap drill always be smaller than the tap diameter?

**7.** Calculate the tap drill size for:
(a) $3/8$–16 N.C.
(b) $1/2$–13 N.C.

**8.** What is the purpose of the stub center when tapping a hole by hand on the drill press?

**9.** What effect does too high a speed have when tapping by power?

**10.** Name four factors that affect the speed at which a tap should be run.

**11.** List four reasons why taps break.

**12.** Briefly explain how a tapping attachment causes a tap to turn.

# UNIT 34

# DRILL-PRESS OPERATIONS

Although the drill press is primarily used for drilling and reaming holes, it can perform many other operations. The variety of cutting and finishing tools which can be used in a drill press allows operations such as countersinking, counterboring, spot-facing, buffing, and spot-finishing to be performed.

## OBJECTIVES

The aim of this unit is to enable you to:

1. Recognize and know the purpose of various countersinks
2. Countersink holes for flat head machine screws
3. Counterbore and spot-face holes
4. Finish metal surfaces by buffing and spot-finishing

## COUNTERSINKING

*Countersinking* is the process of making a cone-shaped opening at the top of a drilled hole for a flat or an oval head machine screw. Countersinks may also be used to remove burrs from the top of a drilled hole. The countersink or cutting tool used for this operation is available with included angles of 60 and 82°. An 82° countersink is used to produce the tapered hole for flat head bolts or machine screws; a 60° countersink is used to produce lathe center holes.

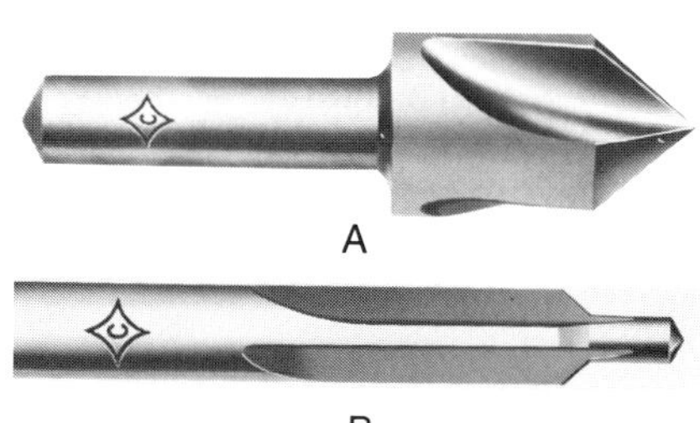


**Fig. 34-1** Two standard types of countersinks: (A) Self-centering; (B) pilot type. (*Courtesy The Cleveland Twist Drill Co.*)

Countersinks are available as a self-centering type, Fig. 34-1A, or with a pilot, Fig. 34-1B.

A different style of countersinking and deburring tool is shown in Fig. 34-2. These tools are chatter-free since they have only one cutting edge. The countersinks shown in Fig. 34-2A and B are available in a variety of sizes and with included angles of 82 and 100°. They are sharpened easily by grinding the inside of the hole at the cutting edge with a small mounted grinding wheel. Special grinding fixtures are available to sharpen these tools.

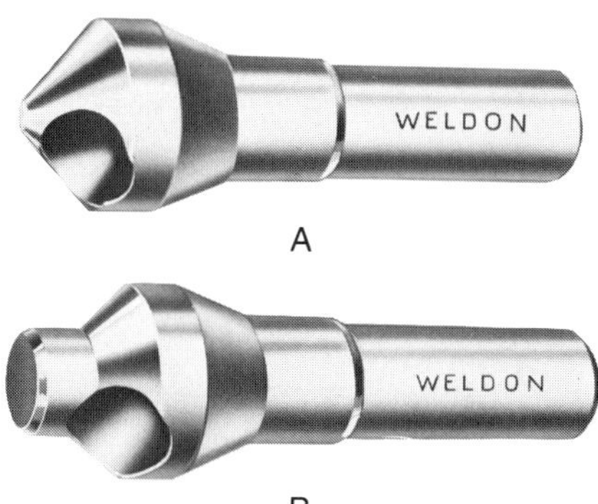


**Fig. 34-2A and B** Countersinks and deburring tools. (*Courtesy Weldon Tool Company.*)

### To Countersink a Hole for a Machine Screw

1. Mount an 82° countersink in the drill chuck.
2. Adjust the spindle speed to about one-quarter that used for drilling. (See N-1.)
3. Place the workpiece on the drill table.
4. With the spindle stopped, lower the countersink into the hole. Clamp the work if necessary. If a pilot-type countersink is used, the pilot should be a slip fit in the drilled hole. The pilot will center the cutting tool and the work.

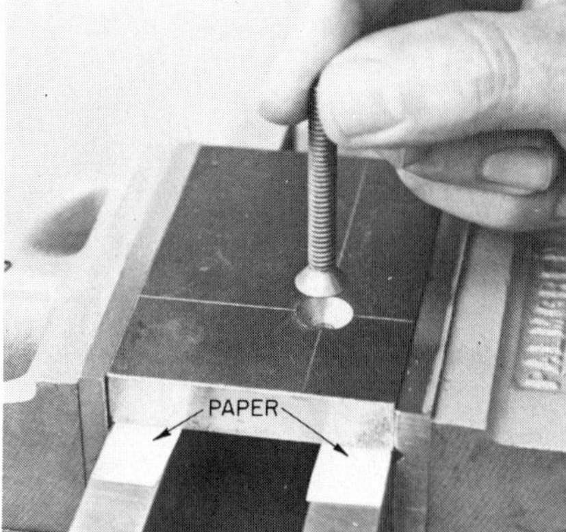


**Fig. 34-3** The diameter of the countersunk hole may be checked with the head of a flat head machine screw. (*Courtesy Kostel Enterprises Ltd.*)

5. Raise the countersink slightly, start the machine, and feed the countersink by hand until the proper depth is reached. The diameter may be checked by placing an inverted screw in the countersunk hole (Fig. 34-3). (See N-2.)
6. If several holes are to be countersunk, set the depth stop so that all the holes will be the same depth. The gage should be set when the spindle is stationary and the countersink is in the hole.
7. Countersink all of the holes to the depth set on the gage.

## COUNTERBORING

*Counterboring* is the operation of enlarging the top of a drilled hole to a given depth (Fig. 34-4). The enlarged

**Fig. 34-4** A counterbore is used to enlarge the end of a hole. (*Courtesy Kostel Enterprises Ltd.*)

N-1 Use a slower speed to prevent the countersink from chattering and produce a clean, smooth hole.

N-2 The diameter of the top of the countersunk hole should be slightly larger than the head of the machine screw.

**Fig. 34-5** A set of counterbores with solid pilots.

N-3 Be sure that the pilot is a slip fit in the drilled hole.

N-4 Set the depth gage on the drill press if several holes of the same size must be counterbored.

N-5 It is important that all marks or scratches be removed from the surface before buffing.

N-6 Do not apply too much compound; the excess will only be thrown off the pad and wasted.

N-7 Do not hold the workpiece with a cloth; it can become caught in the revolving pad and cause serious injury.

hole creates a square shoulder which will accommodate a bolt or cap screw head or the shoulder of a pin.

*Counterbores* (Fig. 34-5) are available in a variety of styles and sizes. All counterbores have a pilot, either solid or interchangeable, to center the counterbore with the drilled hole. The pilot should be a slip fit for the drilled hole if the best results are to be obtained when counterboring.

### Counterboring a Hole

**1.** Position the work and fasten it securely to the table.

**2.** Drill a hole in the work to suit the body of the bolt or pin. (See N-3.)

**3.** Mount the proper-size counterbore in the drill-press spindle.

**4.** Set the spindle speed to about one-quarter of the regular drilling speed.

**5.** Bring the counterbore close to the work surface and align the pilot properly with the drilled hole.

**6.** Start the machine, apply cutting fluid, and counterbore to the required depth. (See N-4.)

## SPOT-FACING

*Spot-facing* is the operation of machining a smooth, flat surface around a hole to provide a square seat for a washer, bolt head, or nut. Spot-facing is often done on the raised surface or lug of a machine casting to provide a better surface for fastening the machine to the floor.

A counterbore of the right diameter with a pilot to fit the drilled hole may be used as a spot-facing tool (Fig. 34-6A). A boring bar fitted with a double-edged cutting tool and with a pilot on the end to fit the drilled hole (Fig. 34-6B), may also be used to spot-face. The procedure for spot-facing generally is the same as that for counterboring.

## BUFFING

Buffing is the operation of producing a highly polished finish on the surface of a workpiece. Abrasive cloth is used first to remove machine marks (Fig. 34-7A). The surface is then brought into contact with a revolving buffing wheel which has been coated with jeweler's rouge or a polishing compound (Fig. 34-7B).

### Buffing a Round Workpiece

**1.** Carefully polish the surface of the workpiece with fine abrasive cloth.

**2.** Mount a buffing pad on a suitable arbor and place it in the drill chuck. The edge of the buffing pad must be free of any foreign particles. (See N-5.)

**3.** Set the drill press to the top spindle speed.

**4.** Start the machine and hold the polishing compound against the edge of the revolving pad until the periphery of the pad is coated. (See N-6.)

**5.** Hold the workpiece firmly in both hands and press it against the edge of the buffing pad (Fig. 34-8). (See N-7.)

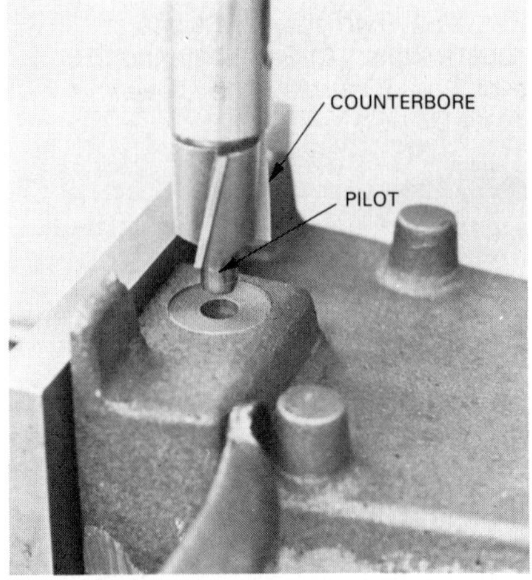


**Fig. 34-6A** A large counterbore with a pilot may be used for spot-facing. (*Courtesy Kostel Enterprises Ltd.*)

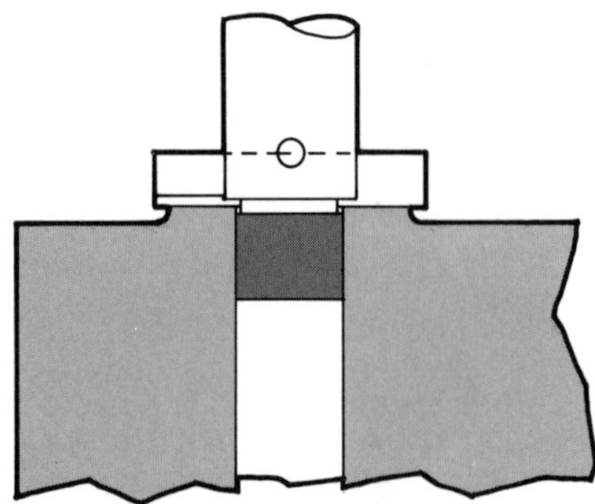

**Fig. 34-6B** A boring bar with a pilot and a double-edged cutting tool may be used for spot-facing.

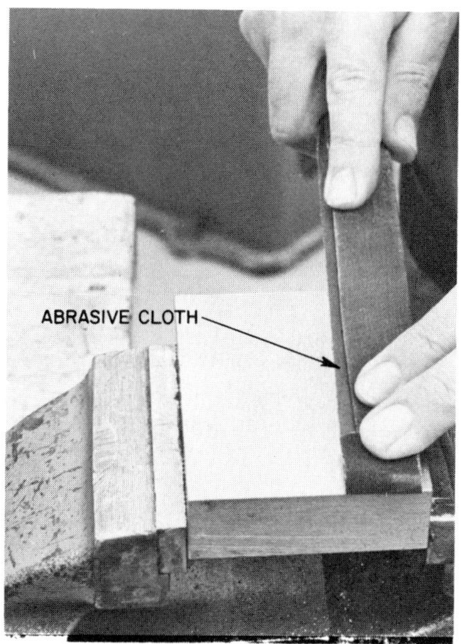

**Fig. 34-7A** A workpiece polished with fine abrasive. (*Courtesy Kostel Enterprises Ltd.*)

**Fig. 34-7B** A buffed workpiece is much shinier. (*Courtesy Kostel Enterprises Ltd.*)

**6.** Slowly rotate the work while keeping it in contact with the revolving buffing wheel.

**7.** Continue steps 5 and 6 until the entire work surface is buffed.

**8.** Stop the machine.

**9.** Wipe the surface of the work with a clean dry cloth to remove any buffing compound.

## SPOT-FINISHING

*Spot-finishing* (Fig. 34-9) is the process of producing circular spots (mottled finish) on the surface of a flat workpiece. Spot-finishing is performed by revolving a brass or hardwood rod in a drill chuck and bringing it into contact with the metal surface which has been coated with an abrasive paste.

### Spot-finishing a Metal Surface

**1.** Polish the surface of the workpiece with abrasive cloth.

**2.** Secure a wooden dowel or brass rod approximately 3 in. (76.2 mm) long. The size of the rod should be the diameter of the spots required.

**3.** Face one end square in a lathe.

**4.** Mount the dowel or rod in the drill chuck with the faced end down.

**5.** Set the spindle speed to about 150 r/min.

**6.** Align one edge of the workpiece with the edge of the spot-finishing tool and clamp a straightedge to the drill-press table (Fig. 34-9).

**Fig. 34-8** Buffing a round piece. (*Courtesy Kostel Enterprises Ltd.*)

**7.** Spread a thin coating of abrasive compound over the surface to be finished.

**8.** Place the work against the straightedge with the corner under the spot-finishing tool.

**9.** Start the machine, hold the work against the straightedge, and lower the spindle until the revolving dowel contacts the work.

**10.** Apply sufficient pressure until a satisfactory circular pattern is produced on the surface.

**11.** Move the work a distance of about two-thirds the diameter of the dowel. Hold the work against the straightedge.

**12.** Make the second spot and continue this procedure until the first row of spots is complete.

**13.** Set the straightedge back a distance equal to about two-thirds the diameter of the dowel and continue to spot the workpiece.

**Fig. 34-9** A straightedge is used to keep the spot-finishing pattern parallel to the edge of the work. (*Courtesy Kostel Enterprises Ltd.*)

**14.** Repeat these procedures until the work surface is spot-finished.

**15.** Remove the workpiece, clean the surface, and coat it with oil.

## REVIEW QUESTIONS

**1.** What is the purpose of countersinking?

**2.** What countersink is used for:
(a) Flat head machine screws
(b) Lathe center holes

**3.** What speed should be used when countersinking?

**4.** State the purpose of counterboring.

**5.** What is the purpose of spot-facing?

**6.** Name two tools which can be used for spot-facing.

**7.** How should a workpiece be prepared before finishing a surface with a buffing wheel?

**8.** What speed should be used when buffing?

**9.** Name two materials which can be used for spot-finishing rods.

**10.** How can the spot-finishing pattern be kept parallel to the edge of the work?

# SECTION

# THE ENGINE LATHE

The lathe, probably one of the earliest machine tools developed, is one of the most versatile and widely used machines. Because a larger percentage of the material cut in a machine shop is cylindrical, the basic lathe has led to the development of turret lathes, screw machines, boring mills, numerically controlled lathes, and turning centers. The progress in the design of the basic engine lathe and its related machines has been responsible for the development and production of thousands of products we use every day.

Some of the common operations performed on a lathe are: facing, taper turning, parallel turning, thread cutting, knurling, boring, drilling, and reaming. The engine lathe is generally used for machining individual parts to the required specifications. It is also used when a small number of similar parts are required (in short-production runs). It is the backbone of a machine shop, so a thorough knowledge of it is essential for any machinist.

# UNIT 35

# LATHE TYPES AND CONSTRUCTION

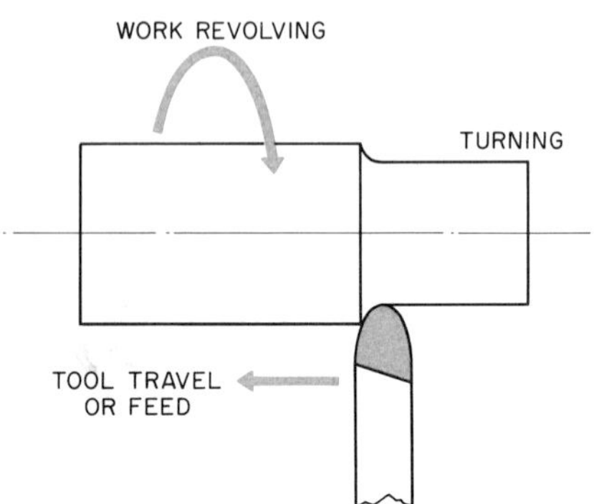


**Fig. 35-1A** The workpiece held for rough turning. (*Courtesy South Bend Lathe, Inc.*)

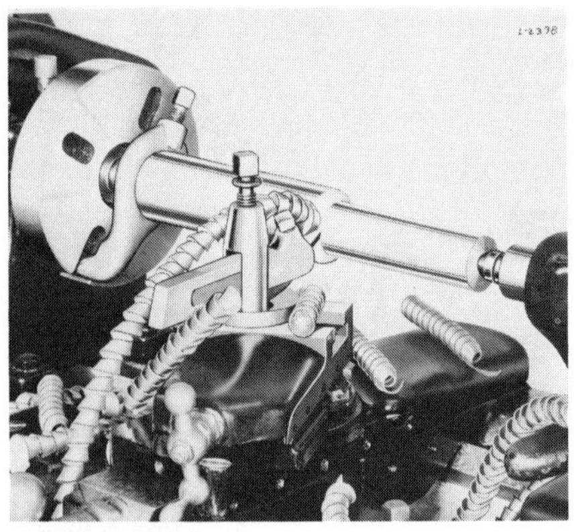

**Fig. 35-1B** The cutting action of a lathe.

**Fig. 35-2** A bench lathe mounted on a cabinet. (*Courtesy Logan Engineering Co.*)

The engine lathe is the most common lathe used in a machine shop. It is one of the oldest and still one of the most important machine tools in use today. Starting with the spring pole or treadle-foot-operated wood-turning lathe of the thirteenth century, it has undergone continual refinements and improvements until today it is a high-precision machine tool capable of machining round parts to very close tolerances.

The main function of the engine lathe is to turn cylindrical shapes and workpieces, Fig. 35-1A. This is done by rotating the metal held in a workholding device while a cutting tool is forced against its circumference. Figure 35-1B shows the cutting action of a tool on work being machined in a lathe.

## OBJECTIVES

Upon completion of this unit, you should be able to:

1. Identify and state the purpose of four types of engine lathes
2. Know the purpose of manufacturing and production lathes
3. Identify and know the purpose of a lathe's main operative parts
4. Observe the safety precautions required when operating a lathe

## TYPES OF LATHES

Lathes are generally classified under three categories: engine lathes, manufacturing lathes, and production lathes.

### Engine Lathes

*Engine lathes,* which include bench, speed, precision, toolroom, and gap-bed lathes, are available in various sizes.

A *bench lathe* (Fig. 35-2) is a small engine lathe which can be mounted on a bench or a metal cabinet. Bench lathes are generally small in size and used for light machining on small workpieces.

A *speed lathe* (Fig. 35-3) which can be mounted on a bench or cabinet, is noted for the fast setup and changeover of work, ease of operation, and low maintenance. Speed lathes are used for light machining operations, turning, polishing, and finishing on small precision work.

The *toolroom lathe* (Fig. 35-4) is equipped with special attachments and accessories to allow a variety of precision operations to be performed. It is generally used to produce tools and gages which are used in tool and die work.

The *gap-bed lathe* (Fig. 35-5) has a section of the bed, below the faceplate, which can be removed to increase the maximum work diameter that can be revolved.

### Manufacturing Lathes

*Manufacturing lathes* (Fig. 35-6A and B) are basically engine lathes which have been modified by the addition of a tracer attachment or a digital readout system. Tracer lathes are used to duplicate parts which may be too difficult or costly to produce on other types of lathes. Figure 35-6A shows a hydraulic tracer attachment on a lathe and the part being produced. Lathes equipped with digital readout systems are used to speed the production of parts normally produced on an engine lathe (Fig. 35-6B).

### Production Lathes

*Production lathes* are generally used when a large number of duplicate parts must be produced. Turret lathes, single-spindle automatic lathes, and the numerically controlled lathes are the common machines in this group.

The *turret lathe* (Fig. 35-7) is used to produce a large number of duplicate parts which may require operations such as turning, drilling, boring, reaming, facing, and threading. On some turret lathes, as many as 20 different tools can be mounted on a ram- or saddle-type turret, and each tool may be rotated into position quickly and accurately. Once the tools have been set, each part is quickly and accurately produced.

A *single-spindle automatic lathe* (Fig. 35-8) is designed to automatically mass-produce parts which require primarily turning and facing oper-

**Fig. 35-3** A speed lathe used for light machining and finishing operations. (*Courtesy of Harding Bros. Inc.*)

**Fig. 35-4** A toolroom lathe. (*Courtesy Monarch Machine Tool Co.*)

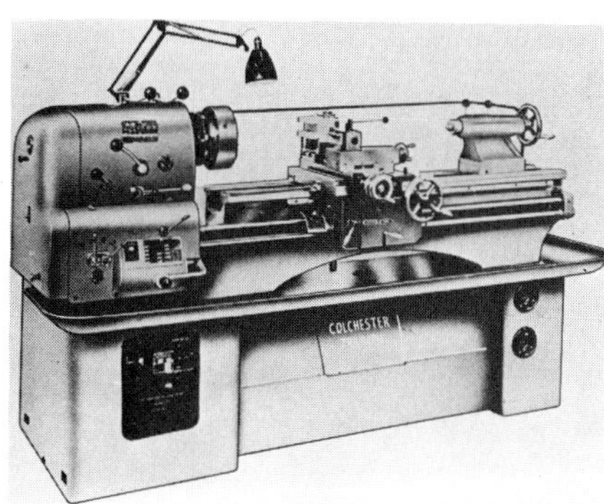

**Fig. 35-5** A gap-bed lathe. (*Courtesy Colchester Lathe & Tool Co.*)

**Fig. 35-6A** A hydraulic tracer attachment mounted on a lathe. (*Courtesy Cincinnati Milacron Co.*)

**Fig. 35-6B** A lathe equipped with a digital readout system. (*Courtesy Bendix Corporation*)

ations. Automatic lathes generally have two toolslides mounted on the carriage. The front-slide tooling is used for turning and boring operations. The rear-slide tooling is used for facing, undercutting, chamfering, and necking operations.

The *numerically controlled lathe* or turning center (Fig. 35-9) is one of the latest modifications of the basic engine lathe. This lathe, controlled by numerical tape, is used primarily for turning operations and can economically and automatically produce shafts of almost any shape. On the model illustrated in Fig. 35-9, a circular or crown turret which can be indexed holds eight carbide tools rigidly for performing various turning operations. The numerically controlled lathe can outperform

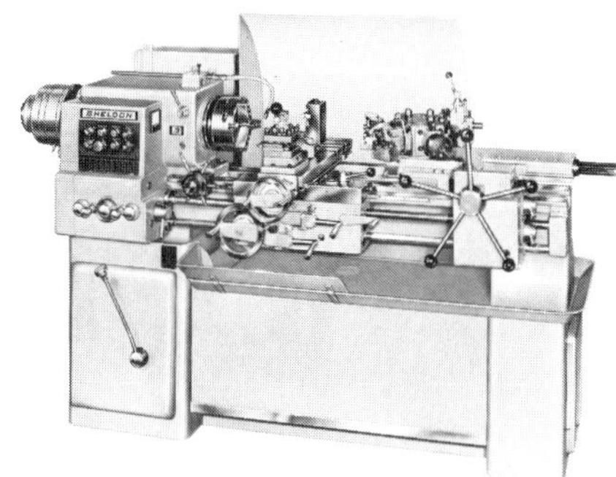

**Fig. 35-7** A turret lathe used to mass-produce parts. (*Courtesy Sheldon Machine Co.*)

**Fig. 35-8** A single-spindle automatic lathe. (*Courtesy Jones & Lamson Co.*)

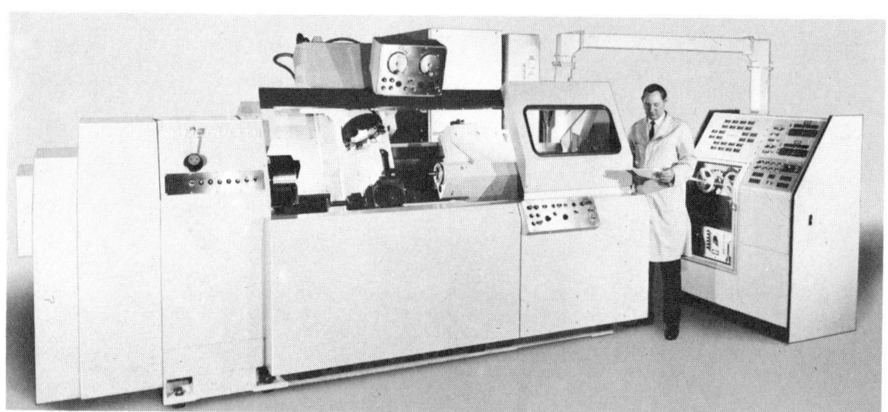

**Fig. 35-9** A numerically controlled lathe. (*Courtesy Cincinnati Milacron Co.*)

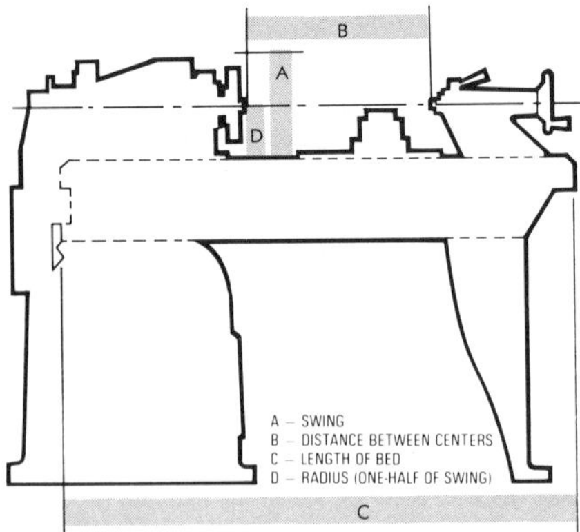


**Fig. 35-10** The size of a lathe is determined by the swing and length of the bed.

most types of lathes and provides savings in tooling, setup, and cycle time.

## SIZE OF THE ENGINE LATHE

The size of an engine lathe is determined by the maximum diameter of work which may be revolved or swung over the bed (Fig. 35-10). The length of a lathe is stated by the length of the bed. The most common inch lathes have a 9- to 30-in. swing, with a capacity of 16 in. to 12 ft between centers. A typical inch lathe may have a 13-in. swing, a 6-ft-long bed, and a capacity to turn work 36 in. long between centers. The average metric lathe used in school shops may have a 230- to 330-mm swing and have a bed length of from 500 to 3000 mm in length.

## LATHE PARTS

The main function of a lathe is to provide a means of rotating a workpiece against a cutting tool, thereby removing metal. All lathes, regardless of design or size, are basically the same (Fig. 35-11). They provide:

**1.** A support for the lathe accessories or the workpiece
**2.** A way of holding and revolving the workpiece
**3.** A means of holding and moving the cutting tool

### Bed

The *bed* (Fig. 35-12) is a heavy rugged casting made to support the working parts of the lathe. Machined ways on its top section guide and align the major parts of the lathe. Many lathes are made with flame-hardened and ground ways to reduce wear and maintain accuracy.

### Headstock

The *headstock* (Fig. 35-13) is clamped on the left-hand side of the bed. The *headstock spindle,* a hollow cylindrical shaft supported by bearings, provides a drive from the motor to work-holding devices. A live center and sleeve, a face plate, or a chuck can be fitted to

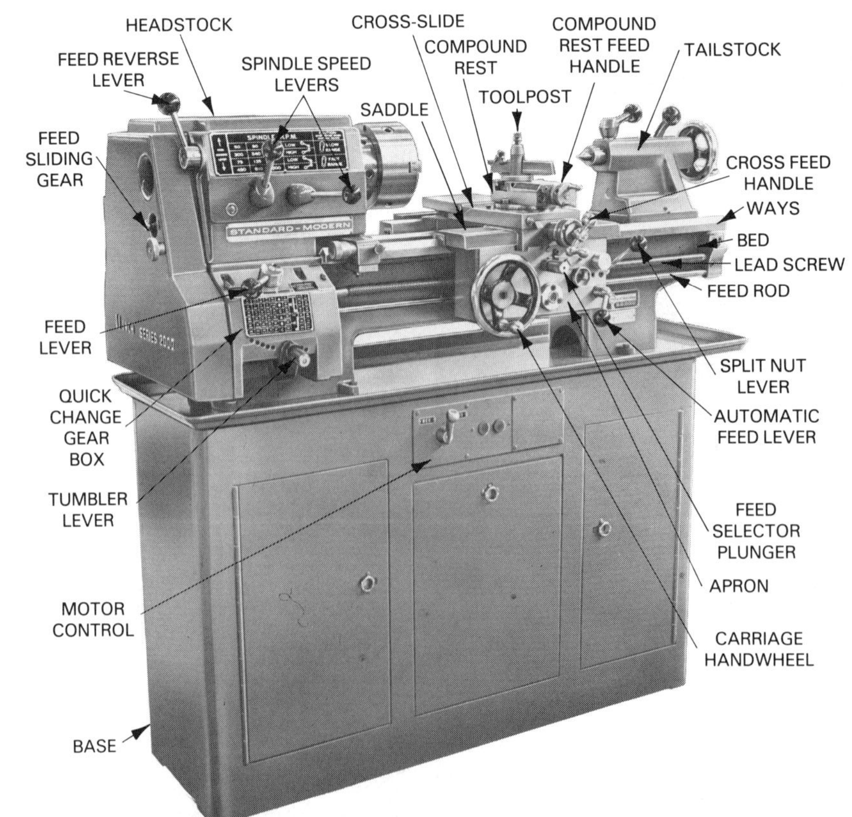


**Fig. 35-11** The main parts of an engine lathe. (*Courtesy Standard-Modern Tool Co.*)

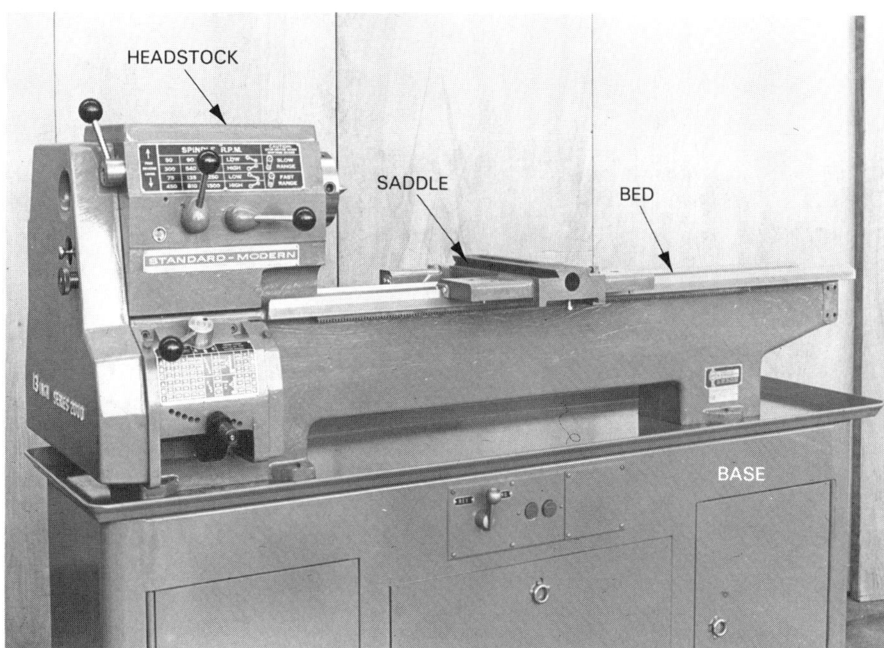


**Fig. 35-12** The lathe bed is the foundation of a lathe. (*Clive Peacock Photography Limited.*)

the spindle nose to hold and drive the work. The live center has a 60° point which provides a bearing surface for the work to turn between centers. (See N-1.)

N-1 Lathe centers should be made of high-speed steel.

Headstock spindles are driven by a set of gears in the headstock; however, some models may still be driven by a cone pulley and belt. The lathe with a cone pulley drive is generally called a belt-driven lathe; the gear-driven lathe is referred to as a geared-head lathe.

**Fig. 35-13** A gear-drive headstock. (*Courtesy The Colchester Lathe Co.*)

The *feed reverse lever* can be placed in three positions. The top position provides a forward direction to the *feed rod* and *lead screw,* the center position is neutral, and the bottom position reverses the direction of the feed rod and lead screw.

## Quick-Change Gearbox

The *quick-change gearbox* (Fig. 35-14), which contains a number of different-sized gears, provides the feed rod and lead screw with various speeds for turning and thread-cutting operations. The feed rod and lead screw provide a drive for the carriage when either the *automatic feed lever* or the *split-nut lever* is engaged. (See N-2.)

N-2 The feed rod moves the carriage automatically.

## Carriage

The carriage (Fig. 35-15) supports the cutting tool and is used to move it along the bed of the lathe for turning operations. The carriage consists of three main parts—the saddle, the apron, and the cross-slide. (See N-3.)

N-3 The automatic-feed lever engages the feed rod to move the carriage.

The *saddle,* an H-shaped casting mounted on the top of the lathe ways, supports the *cross-slide,* which provides a cross movement for the cutting tool. The *compound rest,* used to support the cutting tool, can be swiveled at any angle for taper-turning operations. The cross-slide and the compound rest are moved by feed screws. Each of these is provided with a graduated collar to make accurate settings possible for the cutting tools.

The *apron* is fastened to the saddle and houses the feeding mechanisms which provide an automatic feed to the carriage. The automatic feed lever is used to engage feeds to the carriage. The *apron handwheel* can be turned by hand to move the carriage along the bed of the lathe. This handwheel is connected to a gear which meshes in a rack fastened to the bed of the lathe. The *feed directional plunger* can be shifted into three positions. The in position engages the longitudinal feed for the carriage. The center or neutral position is used in thread-cutting to allow the split-nut lever to be engaged. The out position is used when an automatic crossfeed is required.

## Tailstock

The tailstock (Fig. 35-16) is made up of two units. The top half can be adjusted on the base by two adjusting screws for aligning the dead and live centers for parallel turning. These screws can also be used for offsetting

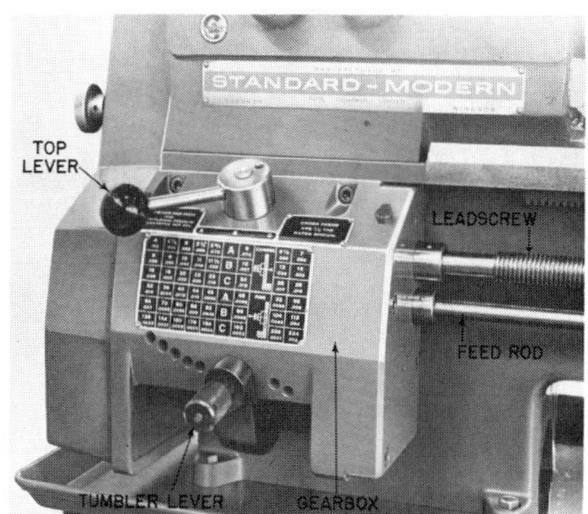


**Fig. 35-14** A quick-change gearbox enables fast settings of feeds and speeds. (*Courtesy Standard-Modern Tool Co.*)

**Fig. 35-15** The main parts of a lathe carriage are the saddle, apron and cross-slide. (*Courtesy Standard-Modern Tool Co.*)

N-4 Only offset the tailstock for taper turning when absolutely necessary.

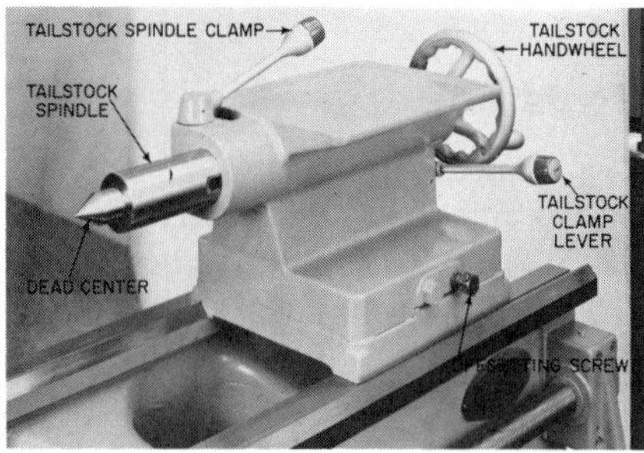


**Fig. 35-16** The tailstock of a lathe. (*Courtesy Cincinnati Milacron Co.*)

the tailstock for taper turning between centers. (See N-4.) The tailstock can be locked in any position along the lathe bed by tightening the *clamp lever,* or *nut*. One end of the dead center is tapered to fit into the *tailstock spindle;* the other end has a 60° point to provide a bearing support for work turned between centers. Standard tapered tools, such as reamers and drills, can be held in the tailstock spindle. A *spindle-binding lever,* or lock handle, is used to hold the tailstock spindle in a fixed position. The *tailstock handwheel* moves the tailstock spindle in or out of the tailstock casting. It can also be used to provide a hand feed for drilling and reaming operations.

## SAFETY PRECAUTIONS

A lathe can be very dangerous if not handled properly, even though it is equipped with various safety guards and features. It is up to the operator to observe various safety precautions and prevent accidents. He must realize that a clean and orderly area around a machine will go a long way in preventing accidents. (See N-5.)

N-5 Follow safe practices to prevent accidents to yourself and others.

The following are some of the more important safety regulations which should be observed when operating a lathe:

**Fig. 35-17** Loose clothing can be caught in a machine.

1. Always wear approved safety glasses when operating any machine.
2. Never attempt to run a lathe until you are familiar with its operation.
3. Never wear loose clothing, rings, or watches when operating a lathe. These could be caught by the revolving parts of a machine and cause a serious accident (Fig. 35-17).
4. Always stop the lathe before taking measurements of any kind.
5. Always use a brush to remove chips. *Do not* handle them by hand because they are very sharp (Fig. 35-18).
6. Before mounting or removing accessories, always shut off the power supply to the motor.
7. Do not take heavy cuts on long slender pieces. This could cause the work to bend and fly out of the machine (Fig. 35-19).

**Fig. 35-18** A paint brush is safe for removing chips and dirt from a lathe.

**Fig. 35-19** Heavy cuts on slender work may cause it to be thrown out of the machine.

Fig. 35-20 Oil or grease on the floor around a machine is dangerous.

**8.** Do not lean on the machine, but stand erect, keeping your face and eyes away from flying chips.

**9.** Keep the floor around a machine clean and free of grease and oil, which could cause dangerous falls (Fig. 35-20).

## REVIEW QUESTIONS

**1.** Explain the cutting action of a lathe.

**2.** What is the main function of a lathe?

**3.** List six operations which may be performed on a lathe.

**4.** Name four types of engine lathes.

**5.** Name three types of production lathes.

**6.** State four parts of the head-stock.

**7.** What is the purpose of the quick-change gearbox?

**8.** State one purpose for each apron part:

(a) Apron handwheel
(b) Automatic feed lever
(c) Feed directional plunger
(d) Compound-rest feed screw

**9.** What part of the tailstock is used to move the tailstock spindle?

**10.** List four important safety precautions that should be observed when operating a lathe.

# UNIT 36

# CUTTING TOOLS AND TOOLHOLDERS

To machine metal in a lathe, a cutting tool called a *toolbit* is used. Toolbits are usually made of high-speed steel in various sizes. The smaller sizes are used on small lathes, and the larger toolbits on lathes of greater capacity.

## OBJECTIVES

Upon completion of this unit you should be able to:

**1.** Identify and know the purpose of the various toolbit angles and clearances

**2.** Prepare a grinder for sharpening a general-purpose lathe toolbit

**3.** Grind and sharpen a general-purpose toolbit

**4.** Identify and state the purpose of the common lathe toolholders

## CUTTING TOOLS

In order to cut various shapes, specially shaped toolbits are used (Fig. 36-1). Left-hand toolbits, which have their cutting edge on the right-hand side, are used for turning work toward the tailstock. Right-hand cutting tools, which have the cutting edge on the left-hand side, are used for cutting toward the headstock. A lathe cutting tool is generally known by the operation it performs. For example, a roughing tool is used to rough-turn work, a threading tool is used for threadcutting, etc.

To produce various surfaces, faces, and forms, the cutting edge of the toolbit must be precisely ground. The cutting edge must be shaped to the proper form and then relieved with clearance angles to allow the edge to cut into the metal. The end-relief (clearance) angle is ground on the end of the toolbit to allow the point to be fed into the work. The side relief (clearance) angle is ground below the

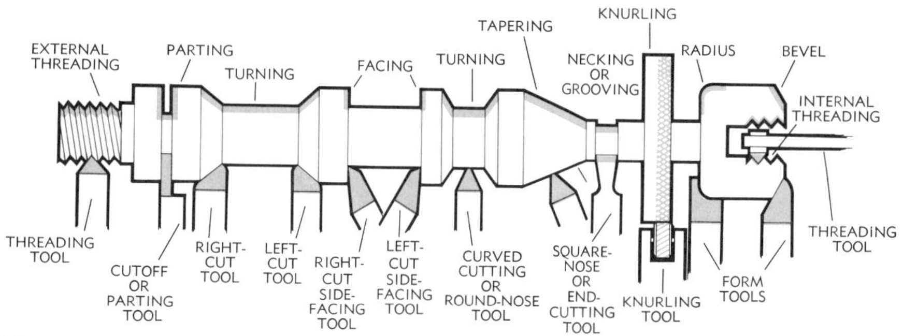


Fig. 36-1 Common cuts made by various cutting tools.

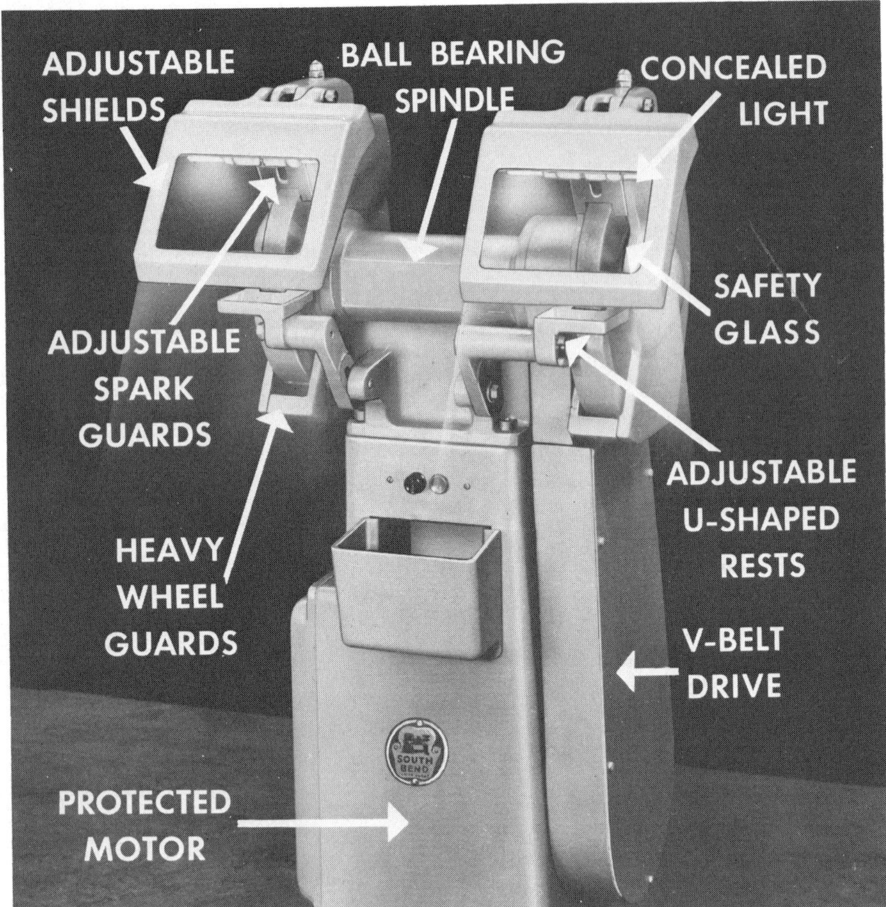


**Fig. 36-2** The main parts of a pedestal grinder.

**Fig. 36-3** The face of a grinding wheel severely loaded.

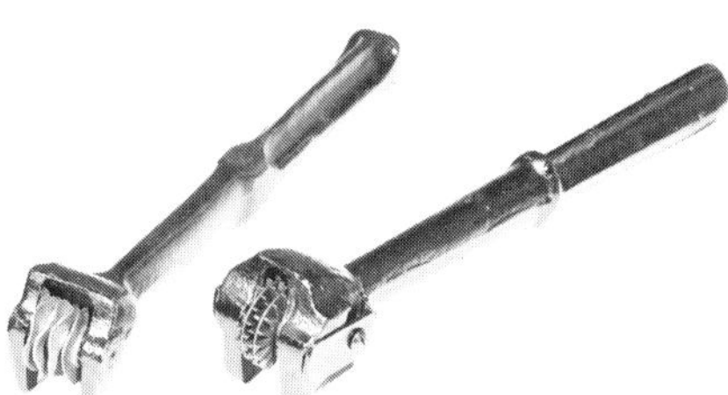

**Fig. 36-4** Types of mechanical wheel dressers. (*Courtesy The Norton Company.*)

cutting edge to allow the toolbit to be fed lengthwise along the work.

## BENCH OR PEDESTAL GRINDER

Since most lathe toolbits are ground by hand, it is important to be able to set up the grinder and use it safely in order to produce the best results. The bench or pedestal grinder is used for sharpening single-pointed cutting tools and rough grinding metal. Because the work is usually held in the hand, this type of grinding is called *offhand grinding*. The pedestal grinder (Fig. 36-2) consists of an electric motor with a grinding wheel mounted on each end of the spindle. One wheel is usually coarse-grained for the fast removal of metal; the other is fine-grained for finish grinding. The U-*shaped work rests* provide a rest for either the work or the hands while grinding.

*Aluminum oxide* and *silicon carbide* are the two types of grinding wheels generally used on bench or pedestal grinders. These manufactured abrasives are superior to the natural abrasives such as emery, sandstone, corundum, and quartz, because they contain no impurities and have a more uniform grain structure. Grinding wheels made from aluminum oxide are used to grind carbon and high-speed steel toolbits. Grinding wheels made of silicon carbide are used to grind single-pointed cemented-carbide cutting tools.

### Grinder Safety

Because grinding wheels operate at very high speeds and the grinding particles are very fine, it is important to observe the following safety precautions.

**1.** Always wear approved safety glasses when operating a grinder. (See N-1.)

**2.** Always stand to one side of the wheel when starting a grinder. Never stand in line with a grinding wheel.

**3.** Allow a new wheel to run for about one minute before using. If a wheel is going to break, it will break in the first minute.

N-1 The safety glass eyeshields on a grinder provide eye protection, but it is good safety practice to also wear safety glasses while grinding.

4. Always use a wheel guard which covers at least one-half of the grinding wheel.

5. Never run a grinding wheel faster than the speed recommended on its blotter.

6. Do not grind on the side of a wheel unless it is designed for this purpose.

7. Never force a grinding wheel by jamming work into it.

8. Always remove burrs produced by grinding with a file or a hone.

N-2 Always wear approved safety glasses to avoid eye injury.

N-3 Keep the work rests adjusted to within 1/16 in. (1.58 mm) of the wheel to prevent work from being jammed between the rest and the wheel.

N-4 A properly ground toolbit will cut efficiently and produce a good surface finish.

## Dressing and Truing a Wheel

When a grinding wheel is used, several things happen to it:

1. Small metal particles imbed themselves in the wheel and cause it to become loaded or clogged (Fig. 36-3).

2. The abrasive grains become worn smooth and the wheel loses its cutting action.

3. Grooves become worn in the face of the wheel.

Any one of these reasons requires dressing or truing a wheel. *Dressing* is the process of reconditioning the wheel to make it cut better. *Truing* refers to shaping a wheel to a desired shape and to make its grinding surface run true with its axis. Both truing and dressing may be done at the same time with a mechanical wheel dresser or an abrasive stick dresser.

A mechanical wheel dresser (Fig. 36-4), commonly called a star dresser, is generally used to dress an offhand grinding wheel. It consists of a number of hardened, pointed disks mounted loosely in a handle.

Fig. 36-5 Dressing a grinder wheel with a mechanical wheel dresser. (*Courtesy The Norton Company.*)

### To Dress and True a Wheel

1. Adjust the grinder work rests so that when the lugs of the mechanical dresser are against its edge, the dresser rolls just touch the face of the wheel (Fig. 36-5).

2. Wear an approved pair of safety glasses. (See N-2.)

3. Stand to one side of the wheel and then start the grinder.

4. Hold the dresser down firmly with its lugs against the edge of the work rest.

5. Move the dresser across the face of the wheel in a steady motion for a trial pass.

6. After each pass, tilt the holder up slightly to advance the dresser disks into the wheel.

7. When the wheel is dressed, stop the grinder and adjust the work rests to within 1/16 in. (1.58 mm) of the wheel face and tighten securely. (See N-3.)

## Grinding a Lathe Toolbit

Grinding lathe toolbits is an operation with which the lathe operator must become familiar. The angles at which a toolbit cuts best should be duplicated on cutting tools of the same type. A lathe tool cuts better if it is provided with the *proper* amount of end and side relief. Too much relief angle (side and end), although it provides a keener cutting edge, weakens a toolbit. Too little side or end relief prevents a lathe tool from cutting. (See N-4.)

Table 36-1 lists the recommended rake and relief angles for high-speed steel and cemented-carbide lathe toolbits. It may be necessary to vary slightly from these recommendations to compensate for factors such as:

- The size and condition of the lathe
- The shape and hardness of the workpiece
- The type of cutting operation being performed

**TABLE 36-1 RAKE AND RELIEF ANGLES IN DEGREES FOR LATHE TOOLBITS**

| Material | Side relief | | End relief | | Side rake | | Back rake | |
|---|---|---|---|---|---|---|---|---|
| | High-speed steel | Carbide | High-speed steel | Carbide | High-speed steel | Carbide | High-speed steel | Carbide |
| Aluminum | 12 | 6 to 10 | 8 | 6 to 10 | 15 | 10 to 20 | 35 | 0 to 10 |
| Brass | 10 | 6 to 8 | 8 | 6 to 8 | 5 to −4 | 8 to −5 | 0 | 0 to −5 |
| Bronze | 10 | 6 to 8 | 8 | 6 to 8 | 5 to −4 | 8 to −5 | 0 | 0 to −5 |
| Cast iron | 10 | 5 to 8 | 8 | 5 to 8 | 12 | 6 to −7 | 5 | 0 to −7 |
| Machine steel | 10 to 12 | 5 to 10 | 8 | 5 to 10 | 12 to 18 | 6 to −7 | 8 to 15 | 0 to −7 |
| Stainless steel | 10 | 5 to 8 | 8 | 5 to 8 | 15 to 20 | 6 to −7 | 8 | 0 to −7 |
| Tool steel | 10 | 5 to 8 | 8 | 5 to 8 | 12 | 6 to −7 | 8 | 0 to −7 |

N-5 Cool the toolbit frequently to prevent overheating, which damages the tool. Never quench cemented-carbide cutting tools.

N-6 While grinding, move the toolbit back and forth across the face of the wheel. This helps to grind faster and prevents grooving the wheel.

## To Grind a Lathe Toolbit

A general-purpose toolbit (Fig. 36-6) can be used for a variety of lathe operations such as rough turning, finish turning, shoulder turning, and facing. This can result in a saving of time and tool setup which is required when a number of different toolbits are used.

**1.** Hold the toolbit firmly while supporting the hands on the grinder tool rest.

**2.** Tilt the bottom of the toolbit in toward the wheel and grind the 10° side-relief angle and form required on the left side of the toolbit (Fig. 36-7A).

**3.** Grind until the side cutting edge is about ½ in. (12.7 mm) long and the point is over about one-fourth the width of the toolbit (Fig. 36-7B).

**4.** High-speed steel toolbits must be cooled frequently. (See N-5.)

**5.** Hold the back end of the toolbit lower than the point and grind the 15° end-relief angle on the right side. At the same time the end cutting edge should form an angle of from 70 to 80° with the side cutting edge (Fig. 36-8A and B).

**6.** Hold the toolbit about 45° to the axis of the wheel (Fig 36-9A and B), tilt the bottom of the toolbit in, and grind the 14° side rake on the top of the toolbit. (See N-6.)

**7.** Grind the side rake the entire length of the side cutting edge but do not grind the top of the cutting edge below the top of the toolbit.

**8.** Grind a slight radius on the point being sure to keep the same end and side-relief angles.

**9.** Use an oilstone to hone the point and cutting edge of the toolbit to remove sharp edges, produce a keener cutting edge, and improve its cutting action.

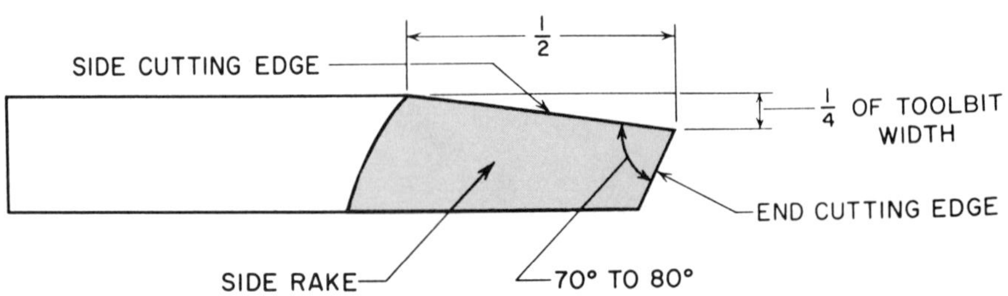


**Fig. 36-6** The top view of a general-purpose tool bit showing its shape and dimensions. (*Courtesy Kostel Enterprises Ltd.*)

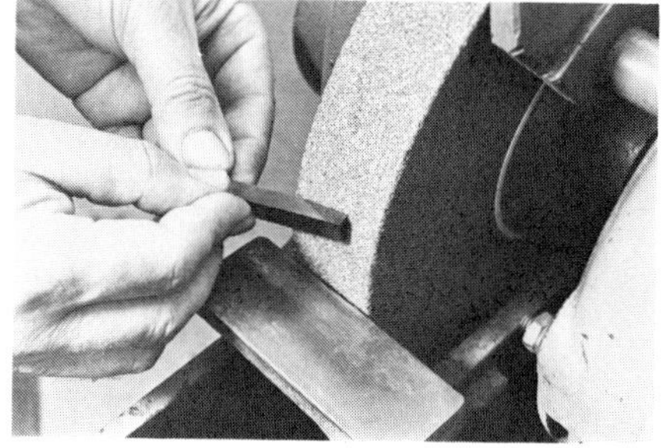

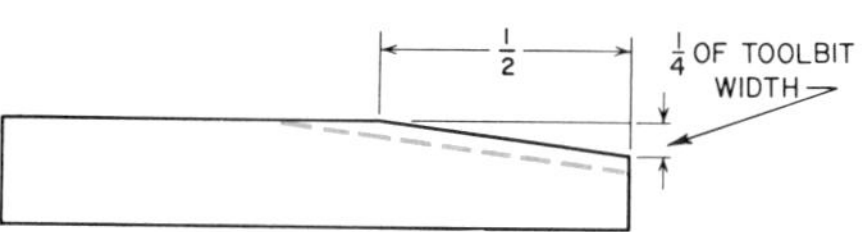


Fig. 36-7 Grinding the side cutting edge and the side-relief angle on a toolbit. (*Courtesy Kostel Enterprises Ltd.*)

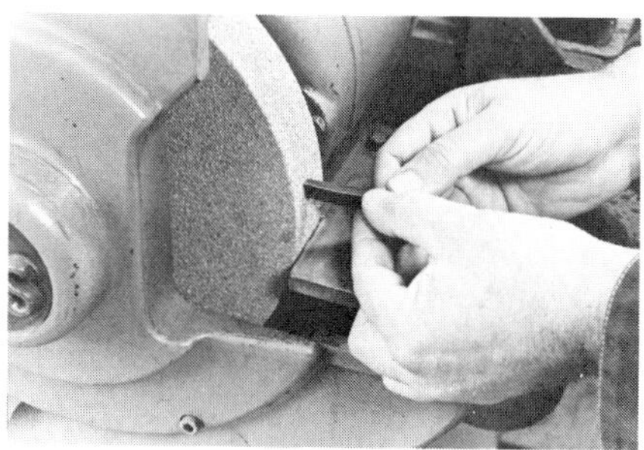

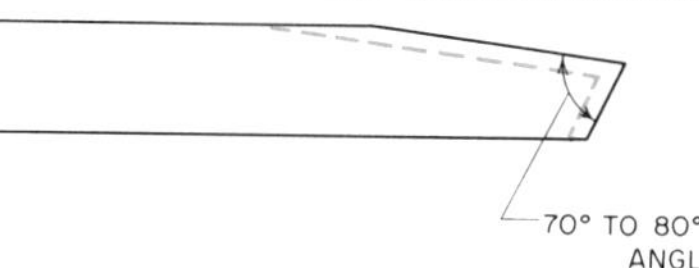


Fig. 36-8 Shaping the point and grinding the end-relief angle. (*Courtesy Kostel Enterprises Ltd.*)

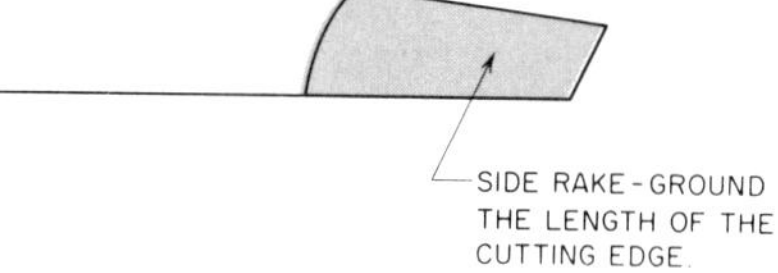


Fig. 36-9 Grinding the side rake on the top of the toolbit. (*Courtesy Kostel Enterprises Ltd.*)

### To Sharpen a General-Purpose Toolbit

A general-purpose lathe toolbit, ground to the shape and dimensions shown in Fig. 36-6, can be quickly sharpened by grinding only the *end cutting edge*. It is important to maintain the same shape and end-relief angle when sharpening a toolbit.

After the worn portion is removed, grind a slight radius on the point and hone the cutting edge. When the side-cutting edge becomes too short, after repeated sharpenings, regrind the whole toolbit to the original shape and dimensions.

## LATHE TOOLHOLDERS AND TOOLPOSTS

Lathe cutting tools are generally held by two methods:

- In *toolholders,* which provide a means of rigidly holding the cutting tool. The most common toolholders are those used for turning, cutting-off, threading, and boring operations.
- In *toolposts,* which provide a means of holding either a toolholder or a cutting tool. The most common are the standard (round), turret, heavy-duty (open-side), and quick-change toolposts.

## Toolholders

### Toolholders for High-Speed Steel Toolbits

Toolholders for high-speed toolbits are manufactured in three styles:

The *left-hand offset toolholder* (Fig. 36-11A) is used for machining work close to the headstock, facing operations, and cutting from right to left. It is identified by the letter L.

The *straight toolholder* (Fig. 36-11B) is used for taking cuts in either direction. It is a general-purpose toolholder identified by the letter S.

The *right-hand offset toolholder* (Fig. 36-11C) is used for machining work close to the tailstock, facing operations, and cutting from left to right. It is identified by the letter R.

All toolholders of this type accommodate square toolbits, ranging from 3/16- to 3/4-in. square, which are held in place by a set-screw on the top of the toolholder. The square hole which accommodates the toolbit is at an angle of 15 to 20° to the base of the toolholder. This provides the proper back rake for high-speed steel toolbits when machining on a lathe.

### Toolholders for Brazed Carbide-Tipped Toolbits

Toolholders of this type are similar to those used for high-speed steel toolbits. They are available in five sizes to accept toolbit shanks from 1/4- to 5/8-in. square. Since carbide cutting tools require little or no back rake, the hole in the toolholder is parallel to the base (Fig. 36-12).

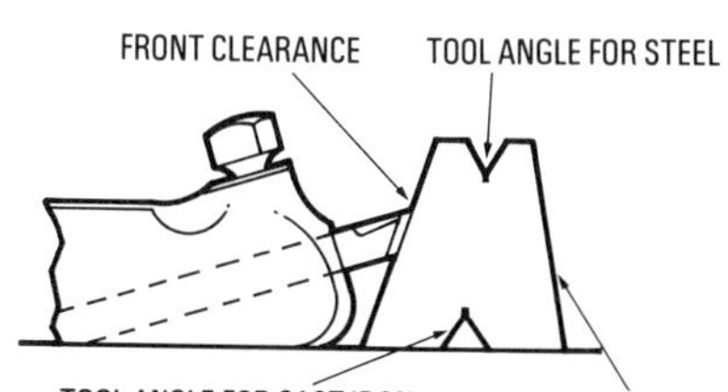


Fig. 36-10 Checking the end-relief angle with a toolbit grinding gage.

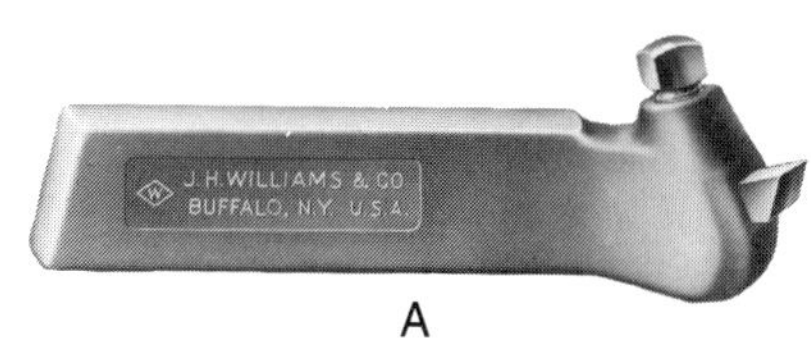


A

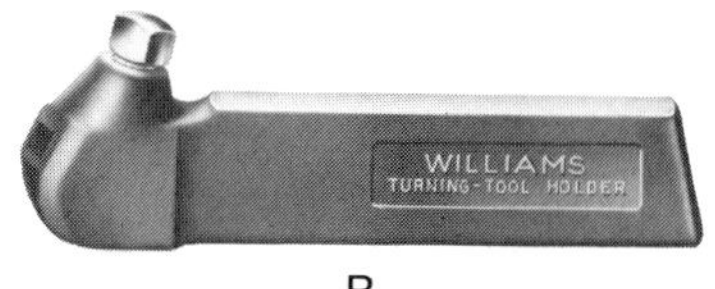


B

C

Fig. 36-11 The most common lathe toolholders: (A) Left-hand; (B) straight; (C) right-hand (*Courtesy J. H. Williams & Co.*).

Fig. 36-12 Carbide toolholders hold the toolbit parallel to the base of the holder. (*Courtesy J. H. Williams & Co.*)

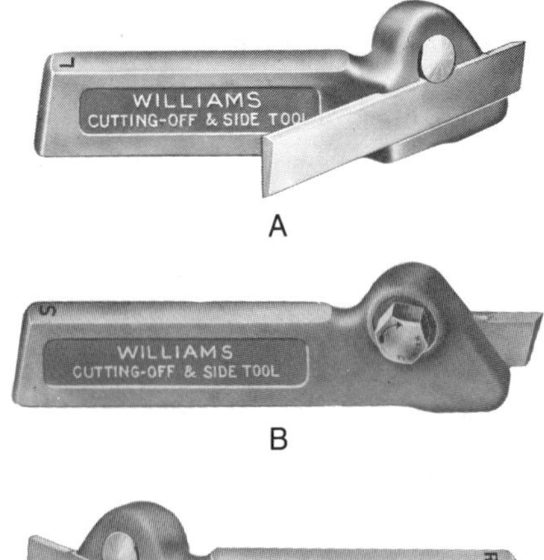


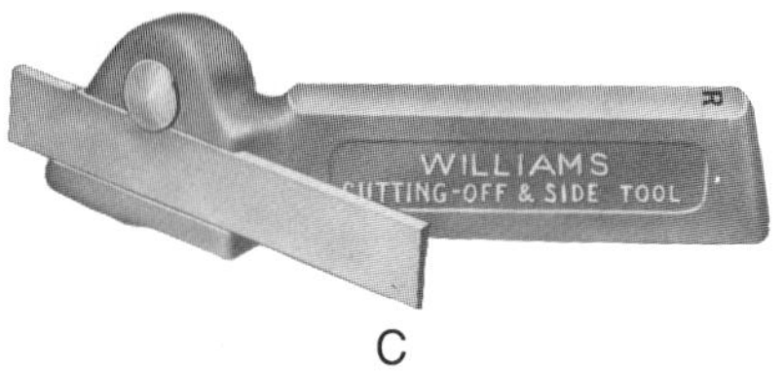


Fig. 36-13 Types of solid cutting-off toolholders: (A) Left-hand; (B) straight; (C) right-hand (*Courtesy J. H. Williams & Co.*).

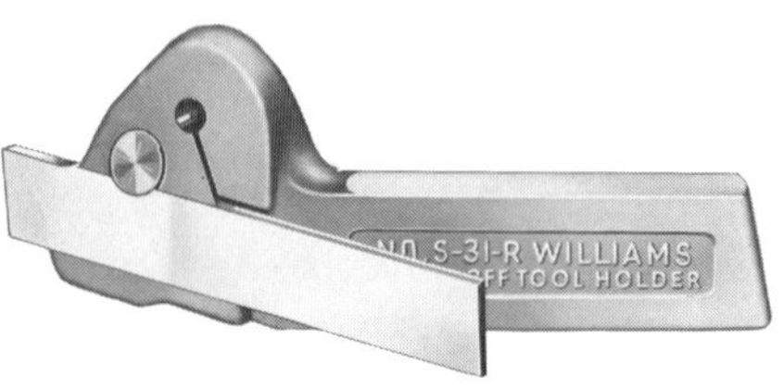


Fig. 36-14 Spring-type toolholders reduce the chances of broken tools and damaged work during cutting-off operations. (*Courtesy J. H. Williams & Co.*)

Fig. 36-15 A preformed thread-cutting toolholder. (*Courtesy J. H. Williams & Co.*)

Carbide turning toolholders are available in straight, right-hand, and left-hand offset types.

## Cutting-off Toolholders

A cutting-off or parting tool is generally used when work is to be parted off, grooved, or undercut. Parting tools are long, thin blades with suitable side clearance to prevent them from binding when in use. They are held tightly in place by means of a cam or wedging action provided in the toolholder.

Cutting-off toolholders may be of the solid type or the spring type.

*Solid-type toolholders* (Fig. 36-13), are available in straight, right-hand offset, and left-hand offset shanks.

*Spring-type toolholders* (Fig. 36-14), are designed to relieve excess pressure on the cutting blade. The gooseneck design of the holder reduces cutting-tool chatter and minimizes tool breakage caused when work "climbs" onto the cutting tool. Spring-type toolholders are made with straight and right-hand shanks.

## Threading Toolholders

Threading in a lathe is generally performed by holding a toolbit, ground to the desired thread form, in a toolholder. A special toolholder (Fig. 36-15) with a cam-shaped cutter ground to 60° is available for thread cutting. The cutter is form-relieved and should be ground only on the top face when sharpening is required. The tool height is adjustable; after the height has been set, the tool is then locked into position.

## Boring Toolholders

Boring toolholders are made in several styles to accommodate the wide range of internal machining which can be performed on a lathe.

*Light boring toolholders* (Fig. 36-16A), held in the toolpost, are used for small holes and light cuts. The boring tools are generally made of round stock with forged ends. They may be ground for boring or internal threading as required.

A boring tool suitable for heavier work (Fig. 36-16B) is held in the toolpost. This consists of a boring bar mounted in a toolholder and a toolbit fitted into the end of the boring bar. Boring bars generally allow toolbits

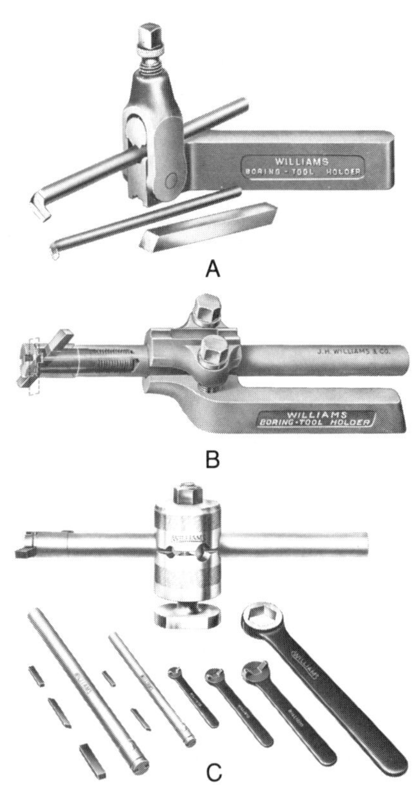


Fig. 36-16 Types of boring toolholders: (A) Light; (B) regular; (C) Heavy-duty (*Courtesy J. H. Williams & Co.*).

to be held at 45 or 90° to the axis of the bar.

The *heavy-duty boring bar holder* (Fig. 36-16C) is mounted on the compound rest. This toolholder can accommodate three bars of different diameters, which permits the use of the largest bar possible for the job. A tool bit is held in the end of the bar, either at 45 or 90° to the axis of the bar.

# Toolposts

## Standard (Round) Toolpost

The toolpost usually supplied with an engine lathe is the standard or round type (Fig. 36-17). This toolpost, which fits into the T-slot of the compound rest, provides a means of holding and adjusting a toolholder or a cutting tool. A concave ring and wedge (rocker) provide a means of adjusting the cutting tool height.

## Turret-Type Toolpost

Turret-type toolposts (Fig. 36-18) are designed to hold four cutting tools, which can be easily indexed for use as required. Several operations, such as

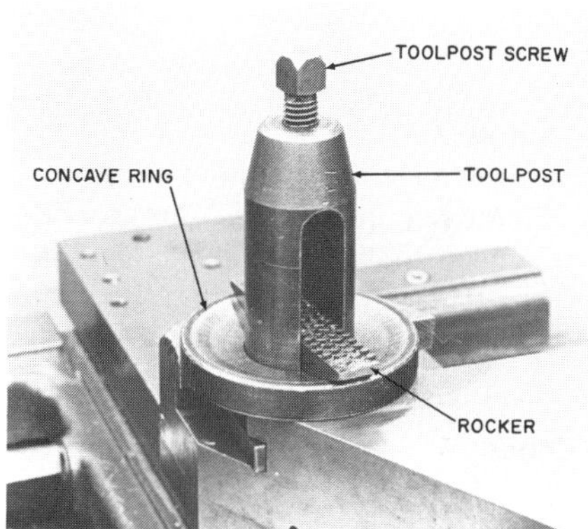


**Fig. 36-17** A standard toolpost.

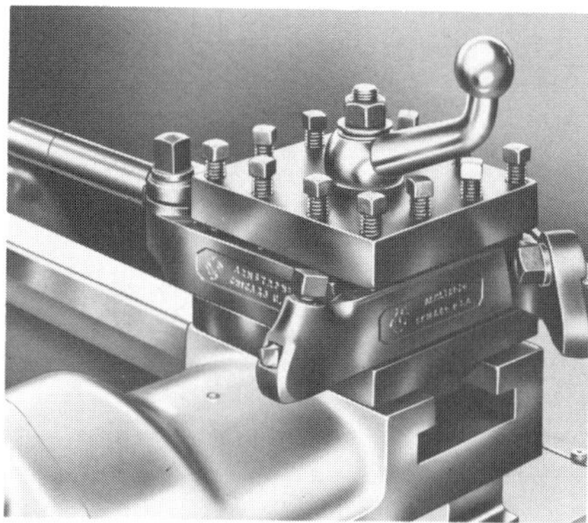

**Fig. 36-18** The turret toolpost can hold up to four different cutting tools and be indexed quickly for performing various operations on a workpiece. (*Courtesy Cincinnati Milacron Co.*)

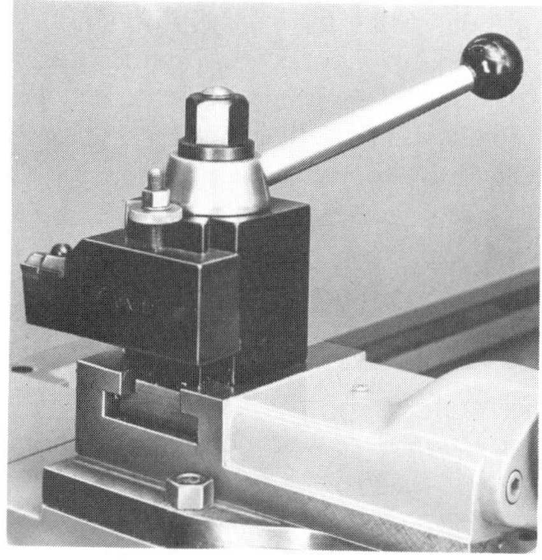

**Fig. 36-19** The dovetailed toolpost allows various cutting tools to be changed rapidly and accurately. (*Courtesy Cincinnati Milacron Co.*)

N-1 The correct lathe speed will give longer cutting tool life and also machine the work in the shortest time.

turning, grooving, threading, and parting, may be performed on a workpiece by loosening the locking handle and rotating the holder until the desired tool bit is in the cutting position. This reduces the setup time for various toolbits, thereby increasing production.

The turret-type toolpost, which fits into the slot on the compound rest, is designed for use with solid-type tools and toolholders for throwaway carbide inserts.

### Quick-Change Toolpost and Holders

Quick-change toolholders are made in different styles to accommodate different types of cutting tools. Each holder is dovetailed and fits on a dovetailed toolpost (Fig. 36-19), which is mounted on the compound rest.

The tool is held in position by setscrews and is generally sharpened in the holder, after which it is preset. After a tool becomes dull, the unit (the holder and tool) may be replaced with another preset unit. This is useful where many parts of one size are being machined since the cutting point on the toolbit, having been preset in the toolroom, is in exactly the same position as the tool it replaces. Each toolholder (Fig. 36-20) fits onto the dovetail on the toolpost and is locked in position by means of a clamp. A knurled nut on each holder provides for vertical adjustment of the unit.

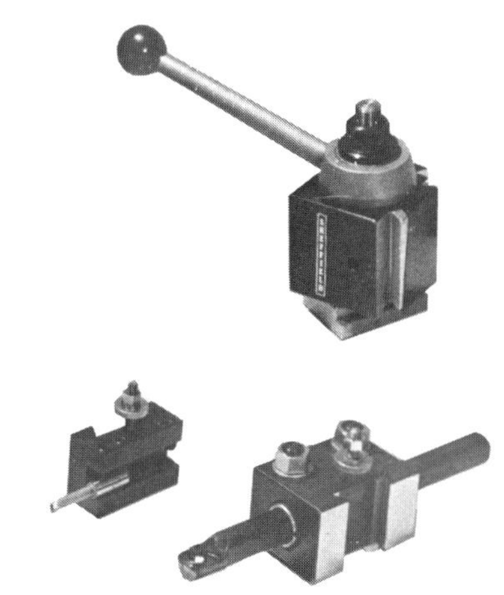

**Fig. 36-20** A quick-change boring bar holder.

## REVIEW QUESTIONS

1. What is the purpose of the end- and side-relief angles on a general-purpose toolbit?
2. What type of grinding wheel should be used to grind:
   (a) High-speed steel toolbits
   (b) Cemented-carbide cutting tools
3. List five important safety precautions which should be observed when using a grinder.
4. Name three reasons for dressing a grinding wheel.
5. How close should the work rests be set to the wheel? Explain why.
6. List the suggested angles and clearances for a high-speed steel toolbit for cutting machine steel.
   (a) Side relief
   (b) End relief
   (c) Side rake
   (d) Back rake
7. How long should the side cutting edge be on a general-purpose toolbit?
8. Why should the cutting edge be honed after grinding?
9. Name the tool used to check the angles and clearances of a lathe toolbit.
10. Name three toolholders and state the purpose of each.
11. What type of cutting-off toolholder reduces chatter and tool breakage?
12. List the advantages of:
    (a) Turret-type toolpost
    (b) Quick-change toolpost

# UNIT 37

# CUTTING SPEED AND FEEDS

The speed at which work revolves in a lathe is a very important factor which can affect the production rate and also the life of a cutting tool. Too slow a lathe speed will result in the loss of valuable time; too fast a speed will cause the cutting tool to break down quickly, resulting in time being wasted in resharpening the cutting tool. It is important that the correct speed and feed be used for each work material and the type of cutting tool being used. (See N-1.)

## OBJECTIVES

After completing the unit, you will be able to:

**1.** Calculate and set the correct speeds for the material to be cut
**2.** Set the lathe to the proper feed for each machining operation
**3.** Calculate the time required for making a cut on a workpiece

## CUTTING SPEED

*Cutting speed* for lathe work may be defined as the rate at which a point on the circumference of the work passes the cutting tool in a minute. Cutting speed may be expressed in feet per minute (ft/min) or meters per minute (m/min). For example, if machine steel has a cutting speed of 100 ft/min (30 m/min), the lathe speed should be set so that 100 ft (30 m) of the work circumference passes the cutting tool in 1 minute. The recommended cutting speed (CS) for various material is listed in Table 37-1. These cutting speeds have been determined by the metal and cutting tool manufacturers as being the best for cutting tool life and production rates. (See N-2.)

N-2 The cutting speed varies for each type of work material.

### r/min Calculations

To be able to calculate the number of revolutions per minute at which to set a lathe, the diameter of the work and the cutting speed of the material must be known. The revolutions per minute at which the lathe should be set can be found by applying one of the following formulas. (See N-3.)

N-3 The diameter and type of material to be cut affect the revolutions per minute set on a lathe.

### Inch Calculations

$$\text{r/min} = \frac{\text{CS (ft)} \times 12}{\pi \times \text{work dia. (in.)}}$$

$$= \frac{\text{CS} \times 12}{3.1416 \times D}$$

Since most lathes are provided with only a limited number of speed settings, a simpler formula can be used to calculate the r/min. (See N-4.) The 3.1416 of the formula will divide into 12 approximately four times. This results in the following formula, which is close enough for most lathe operations.

N-4 When calculating revolutions per minute for inch work, be sure that the diameter of the work is in inches.

$$\text{r/min} = \frac{\text{CS} \times 4}{D \text{ (in.)}}$$

**Example:**
To calculate the revolutions per minute required to finish-turn a 2-in.-diameter piece of machine steel (Cutting speed for machine steel is 100.):

$$\text{r/min} = \frac{\text{CS} \times 4}{D}$$

$$= \frac{100 \times 4}{2}$$

$$= 200$$

### Metric Calculations

The revolutions per minute at which the lathe should be set when using metric measurements is as follows:

$$\text{r/min} = \frac{\text{CS (m)} \times 1000}{\pi \times \text{work dia. (mm)}}$$

$$= \frac{\text{CS} \times 1000}{3.1416 \times D}$$

Since only a few machines are equipped with variable-speed drives which allow them to be set to the exact calculated speed, a simplified formula can be used to calculate r/min. (See N-5.) The $\pi$ (3.1416) on the bottom line of the formula will divide into 1000 of the top line approximately 320 times. This results in a simplified formula which is close enough for most lathe operations.

N-5 When calculating revolutions per minute for metric work, be sure that the diameter of the work is in millimeters.

$$\text{r/min} = \frac{\text{CS} \times 320}{D \text{ (mm)}}$$

**Example:**
Calculate the revolutions per minute required to finish-turn a 45-mm-diameter piece of machine steel. (CS of machine steel = 30 m/min.)

$$\text{r/min} = \frac{\text{CS} \times 320}{D}$$

$$= \frac{30 \times 320}{45}$$

$$= \frac{9600}{45}$$

$$= 213.3$$

**TABLE 37-1 LATHE CUTTING SPEEDS IN FEET PER MINUTE AND METERS PER MINUTE USING A HIGH-SPEED TOOLBIT**

| | Facing, Turning, Boring | | | | | |
|---|---|---|---|---|---|---|
| | Rough cut | | Finish cut | | Threading | |
| Material | ft/min | m/min | ft/min | m/min | ft/min | m/min |
| Machine steel | 90 | 27 | 100 | 30 | 35 | 11 |
| Tool steel | 70 | 21 | 90 | 27 | 30 | 9 |
| Cast iron | 60 | 18 | 80 | 24 | 25 | 8 |
| Bronze | 90 | 27 | 100 | 30 | 25 | 8 |
| Aluminum | 200 | 61 | 300 | 93 | 60 | 18 |

**Fig. 37-1** The speed of a geared-head lathe is regulated by the position of the headstock levers. (*Courtesy Standard-Modern Tool Co.*)

N-6 Never change speeds on a geared-head or belt-driven lathe while it is running.

N-7 Set the proper feed in order to machine the work in the shortest time and still produce the desired surface finish.

## Setting Lathe Speeds

Engine lathes are designed to operate at various spindle speeds, for machining different-sized diameters and types of material. These speeds are measured in revolutions per minute and are changed by means of gear levers or a variable-speed adjustment and by cone pulleys on older-model lathes. When setting the spindle speed, always set the machine as close as possible to the calculated speed but never higher. If the cutting action is satisfactory, the speed may be increased slightly; however, if the cutting action is not satisfactory or the work vibrates or chatters, reduce the speed and increase the feed. (See N-6.)

On a belt-driven lathe, the various speeds are obtained by moving the belt to various-sized pulleys. On the geared-head lathe (Fig. 37-1) speeds are changed by moving the speed levers into their proper positions according to the revolutions-per-minute chart that is fastened to the headstock. While shifting the lever positions, place one hand on the drive plate or chuck and turn it slowly by hand. This will enable the levers to engage the gear teeth without clashing. (See N-6.)

Some lathes are equipped with a variable-speed headstock and any speed within the range of the machine can be set. The spindle speed is set while the lathe is running by turning a speed-control knob until the desired speed is indicated on the speed dial.

## Lathe Feed

The feed of a lathe is defined as the distance the cutting tool advances along the length of the work for every revolution of the spindle. For example, if

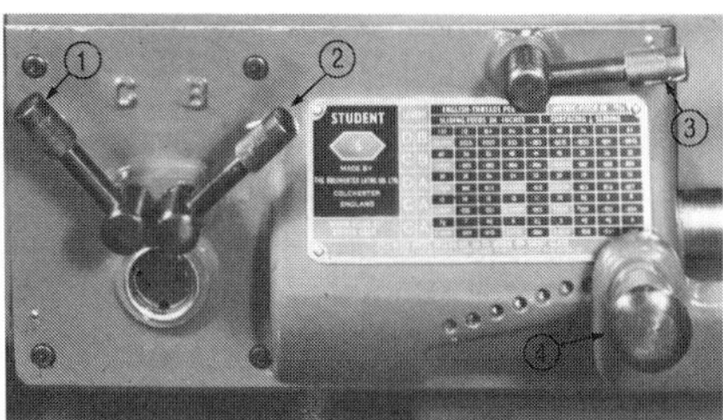


**Fig. 37-2** The quick-change gearbox controls the lathe feeds. (*Courtesy The Colchester Lathe Co.*)

the lathe is set for an .008-in. (0.20-mm) feed, the cutting tool will travel along the length of the work .008 in. (0.20 mm) for every complete turn that the work makes. The feed of an engine lathe is dependent on the speed of the lead screw or feed rod. This is controlled by the change gears in the *quick-change gearbox* (Fig. 37-2).

Whenever possible, only two cuts should be taken to bring a diameter to size: a roughing cut and a finishing cut. Since the purpose of a roughing cut is to remove excess material quickly and since surface finish is not too important, a coarse feed should be used. The finishing cut is used to bring the diameter to size and produce a good surface finish; therefore a fine feed should be used. For general-purpose machining a .010- to .015-in. (0.25- to 0.40-mm) feed for rough turning and a .003- to .005-in. (0.07- to 0.12-mm) feed for finish turning is recommended. Table 37-2 lists the recommended feeds for cutting various materials when using a high-speed steel cutting tool. (See N-7.)

### To Set the Lathe Feed

**1.** From the quick-change gearbox plate, select the amount of feed required.

**TABLE 37-2 FEEDS FOR VARIOUS MATERIALS (USING A HIGH-SPEED CUTTING TOOL)**

| Material | Rough cuts | | Finish cuts | |
|---|---|---|---|---|
| | Inches | Millimeters | Inches | Millimeters |
| Machine steel | .010–.020 | 0.25–0.50 | .003–.010 | 0.07–0.25 |
| Tool steel | .010–.020 | 0.25–0.50 | .003–.010 | 0.07–0.25 |
| Cast iron | .015–.025 | 0.40–0.65 | .005–.012 | 0.13–0.30 |
| Bronze | .015–.025 | 0.40–0.65 | .003–.010 | 0.07–0.25 |
| Aluminum | .015–.030 | 0.40–0.75 | .005–.010 | 0.13–0.25 |

N-8 Before turning on the lathe, be sure that all levers are fully engaged by turning the headstock spindle by hand. If all the levers are correctly engaged, the feed rod should turn as the spindle is revolved.

**2.** Move the tumbler lever 4 (Fig. 37-2) into the hole directly below the row in which the selected amount of feed is found.

**3.** Follow the row in which the selected feed is found to the left and set the *feed-change levers* (1 and 2) to the letters indicated on the feed chart. (See N-8.)

**4.** Set lever 3 (lead screw engaging lever) to the down position.

## CALCULATING MACHINING TIME

To calculate the time required to machine any workpiece, factors such as speed, feed, and depth of cut must be considered. By applying the following formula, the time required to take a cut can be readily calculated.

$$\text{Time required} = \frac{\text{length of cut}}{\text{feed} \times \text{r/min}}$$

**Example:**

Calculate the time required to take a roughing cut (.015 feed) off an 18-in.-long piece of 2-in.-diameter machine steel.

$$\text{r/min} = \frac{\text{CS} \times 4}{D} = \frac{90 \times 4}{2} = 180$$

$$\text{Cutting time} = \frac{\text{length of cut}}{\text{feed} \times \text{r/min}} = \frac{18}{.015 \times 180} = 6.6 \text{ min}$$

**Example:**

Calculate the time required to take a finish cut (0.10-mm feed) off a 250-mm-long piece of 30-mm-diameter machine steel.

$$\text{r/min} = \frac{\text{CS} \times 320}{D} = \frac{30 \times 320}{30} = 320$$

$$\text{Cutting time} = \frac{\text{length of cut}}{\text{feed} \times \text{r/min}} = \frac{250}{0.10 \times 320} = 7.8 \text{ min}$$

## REVIEW QUESTIONS

**1.** What will happen if lathe speed is:
 (a) Too slow
 (b) Too fast

**2.** Define cutting speed.

**3.** See Table 37-1 and calculate the revolutions per minute required to rough turn the following:
 (a) ¾-in.-diameter machine steel
 (b) 25-mm-diameter bronze

**4.** Name three methods of changing speed on lathes.

**5.** Define lathe speed.

**6.** What is the purpose of:
 (a) The rough cut
 (b) The finishing cut

**7.** What feed is recommended for finish turning
 (a) Machine steel
 (b) Aluminum

**8.** What part of the lathe controls the rate of feed to the carriage?

9. Calculate the time required to take a roughing cut from
   (a) 1-in.-diameter piece of tool steel 12 in. long (CS 70 ft/min, .015 feed).
   (b) 25-mm-diameter piece of aluminum 300 mm long (CS 61 m/min, 0.50-mm feed).

# UNIT 38

# CENTERING WORK IN A LATHE

Center holes in round workpieces can be quickly and accurately drilled in a lathe providing that the workpiece can be held in a chuck. Before center drilling, it is important that the operator be able to mount and remove lathe spindle accessories and to face the ends of work squarely.

**Fig. 38-1** The type L tapered spindle nose.

## OBJECTIVES

After completing this unit, you will be able to:

1. Mount and remove accessories from three types of lathe spindle noses
2. Face work held in a lathe chuck
3. Drill correct-size center holes in work held in a lathe chuck

## MOUNTING AND REMOVING CHUCKS

Lathe accessories such as chucks and drive plates are fitted to the headstock of a lathe. The proper procedure for mounting and removing these accessories must be followed in order not to damage the lathe spindle and/or accessories, and to preserve the accuracy of the machine.

### Types of Lathe Spindles

There are two common types of lathe spindle noses upon which accessories such as drive plates and chucks are mounted. They are the tapered spindle nose and the cam-lock spindle nose. The threaded spindle nose may be still found on older lathes.

The *tapered spindle nose* (Fig. 38-1) has a $3\frac{1}{2}$-in. taper per foot (291 mm/m). The chuck is fitted on the taper and positioned by the drive key. The threaded lock ring, fitted at the left end of the spindle nose, threads onto the chuck or drive plate and locks it in position. The *cam-lock spindle nose* (Fig. 38-2) has a very short taper [3-in. taper per foot (250 mm/m)]. The chuck or drive plate is located by the taper on the spindle and held in position by three cam-lock devices.

The *spring collet chuck* (Fig. 38-3) consists of a collet held in a collet

**Fig. 38-2** The type D-1 cam-lock spindle nose.

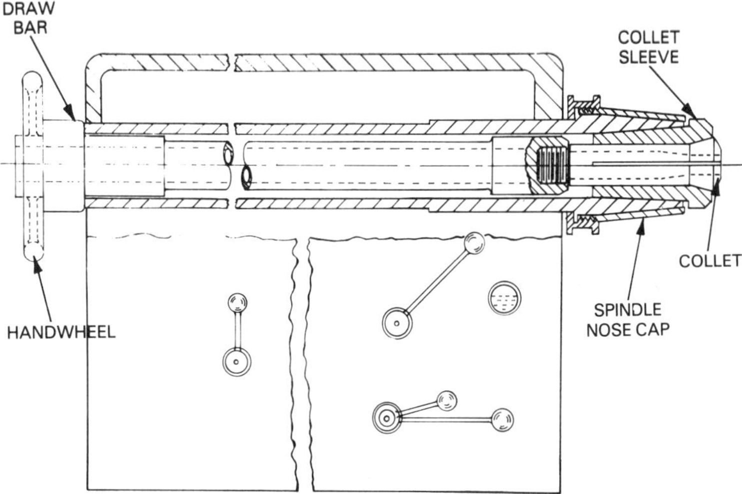


**Fig. 38-3** A cross-sectional view of a headstock showing how a collet chuck is held.

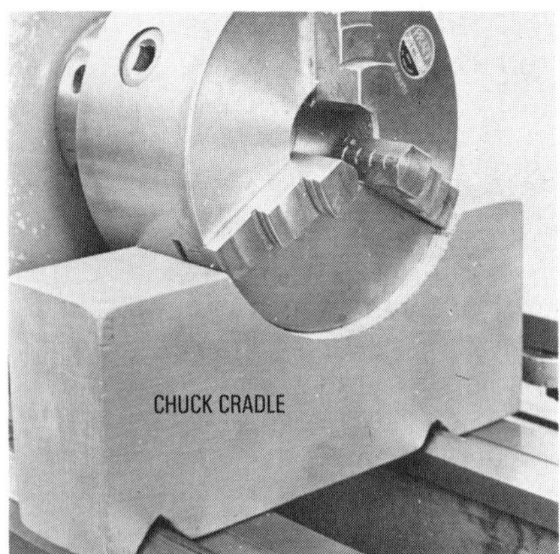


**Fig. 38-4** A chuck cradle prevents damaging the lathe bed when mounting and removing accessories. (*Courtesy Kostel Enterprises Ltd.*)

N-1 Never use power when removing lathe-spindle accessories.

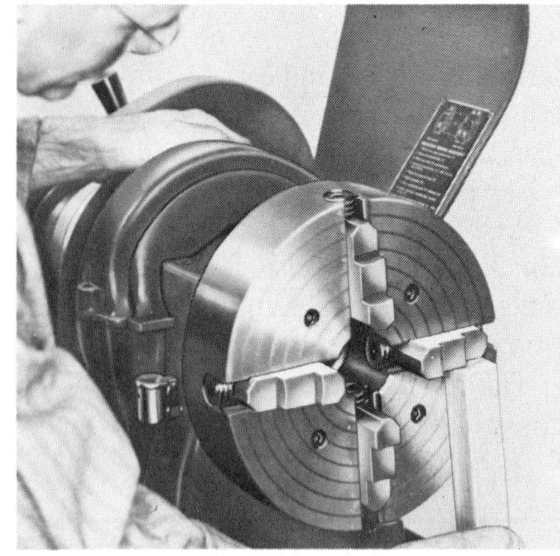

**Fig. 38-5** A hardwood block should be used when removing a chuck from a threaded spindle nose. (*Courtesy South Bend Lathe Works.*)

N-2 Fill the chuck spindle hole with a cloth or paper towel to keep chips and dirt out of the threads.

sleeve which is fitted into the lathe spindle. As the handwheel (and hollow draw-in bar) is tightened, the collet is drawn into the sleeve and tightens on the workpiece. There are collets for round, square, and hexagonal workpieces. Since these collets have an adjustment range of only a few thousandths of an inch, only true and sized work may be held in them.

A *rubber-flex collet* attachment which mounts on the spindle nose of the lathe has a much wider range than the spring collet and can hold much larger work. Each rubber-flex collet, which has a set of tapered steel bars mounted in rubber, has a range of about ⅛ in. (3.2 mm).

## To Remove a Chuck

**1.** Set the lathe in the slowest speed. *Shut off the electric switch.*

**2.** Place a chuck cradle under the chuck (Fig. 38-4).

**3.** Remove the chuck or accessory by following the steps outlined below, depending on the type of lathe spindle nose. (See N-1.)

### Threaded Spindle Nose

(a) Turn the lathe spindle by hand until a chuck-wrench socket is in the top position.
(b) Insert the chuck wrench into the hole and pull it sharply counterclockwise (toward you).

*or*

(a) Place a block or short stick under the chuck jaw, as shown in Fig. 38-5.
(b) Revolve the lathe spindle by hand in a clockwise direction until the chuck is loosened on the spindle.
(c) Remove the chuck from the spindle and store it with the jaws in the up position where it will not be damaged. (See N-2.)

### Taper Spindle Nose

(a) Secure the proper C-spanner wrench.
(b) Place it around the front of the lock ring of the spindle with the handle in an upright position (Fig. 38-6).
(c) Place one hand on the curve of the spanner wrench to prevent it from slipping off the lock ring.

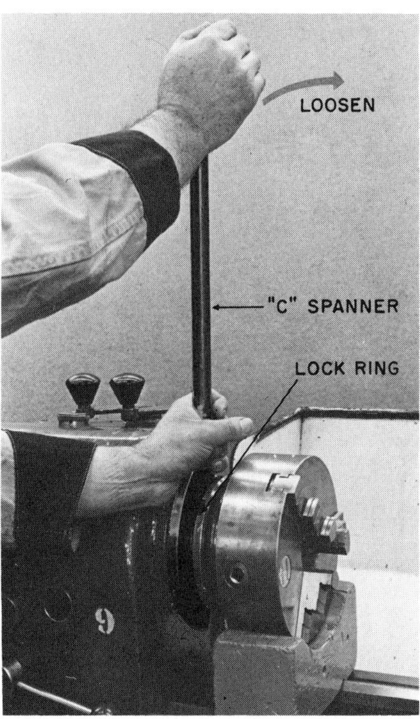


**Fig. 38-6** A C-spanner wrench is used to loosen the lock ring on a taper spindle nose.

**Fig. 38-7** Plug the tapered hold with a cloth to prevent dirt and chips from entering. (*Courtesy Kostel Enterprises Ltd.*)

(d) With the palm of the other hand, sharply strike the handle of the wrench in a clockwise direction.
(e) Hold the chuck with one hand while turning the lock ring clockwise with the other hand.
(f) If the lock ring becomes tight, use the spanner wrench to break the taper contact between the spindle and chuck.
(g) Plug the hole in the chuck (Fig. 38-7) and store it with the jaws in the up position.

Fig. 38-8 Turn the chuck wrench counterclockwise to align the registration lines of each cam lock with the lines on the spindle. (*Courtesy Kostel Enterprises Ltd.*)

Fig. 38-9 Breaking the taper contact between the lathe spindle and chuck. (*Courtesy Kostel Enterprises Ltd.*)

N-3 Dirt and chips cause chucks to run out of true and permanently damage the spindle surfaces.

N-4 Never use power when mounting lathe-spindle accessories.

N-5 Jamming a chuck too tightly makes it difficult to remove and may damage the threads.

### Cam-Lock Spindle Nose

(a) With the proper-size wrench turn each cam-lock counterclockwise until its registration line matches the registration line on the spindle nose, or is at the 12 o'clock position (Fig. 38-8).
(b) With one hand sharply strike the chuck to remove it from the spindle (Fig. 38-9).
(c) Remove and store the chuck properly.

## To Mount a Chuck

The following procedures apply to any accessories mounted *on* a lathe spindle nose.

1. Set the lathe to the slowest speed and *shut off the electric switch*.
2. Clean all surfaces of the spindle nose and the mating parts of the chuck. (See N-3.)
3. Place a cradle block on the lathe bed in front of the spindle and place the chuck on the cradle (Fig. 38-4).
4. Slide the cradle close to the lathe spindle nose and mount the chuck.
5. Mount the chuck or accessory by following the steps outlined below depending on the type of lathe spindle nose. (See N-4.)

### Threaded Spindle Nose

(a) Revolve the lathe spindle by hand in a counterclockwise direction and bring the chuck up to the spindle.
(b) If the chuck and spindle are clean and correctly aligned, the chuck should easily thread onto the lathe spindle.
(c) When the chuck adaptor plate is within 1/16 in. (1.5 mm) of the spindle shoulder, give the chuck a quick turn to seat it against the spindle shoulder. (See N-5.)

### Taper Spindle Nose

(a) Revolve the lathe spindle by hand until the key on the spindle nose aligns with the keyway in the tapered hole of the chuck (Fig. 38-10).
(b) Slide the chuck onto the lathe spindle and at the same time turn the lock ring in a counterclockwise direction (Fig. 38-11).
(c) Tighten the lock ring securely with a spanner wrench by striking it sharply downward (Fig. 38-12) when standing at the front of the machine.

Fig. 38-10 Before mounting, align the key in the spindle with the keyway in the chuck. (*Courtesy Kostel Enterprises Ltd.*)

Fig. 38-11 Hold the chuck on the spindle with one hand and tighten the lock ring in a downward direction with the other hand. (*Courtesy Kostel Enterprises Ltd.*)

### Cam-Lock Spindle Nose

(a) Align the registration line of each cam lock with the registration line on the lathe spindle nose (Fig. 38-13).
(b) Revolve the lathe spindle by hand until the clearance holes in the spindle align with the cam-lock studs of the chuck (Fig. 38-13).
(c) Slide the chuck onto the spindle.
(d) Securely tighten each cam lock in a clockwise direction (Fig. 38-14).

Fig. 38-12 Tightening the lock ring on a taper nose spindle chuck. (*Courtesy Kostel Enterprises Ltd.*)

Fig. 38-13 Mounting a chuck on a cam-lock spindle nose.

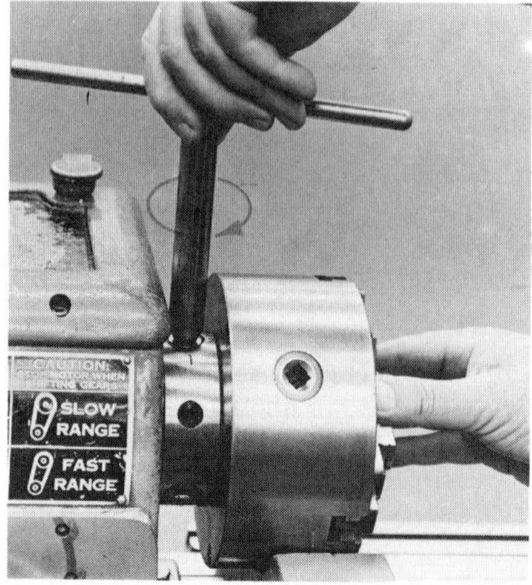


Fig. 38-14 Hold the chuck with one hand and turn each cam lock clockwise until all are tight. (*Courtesy Kostel Enterprises Ltd.*)

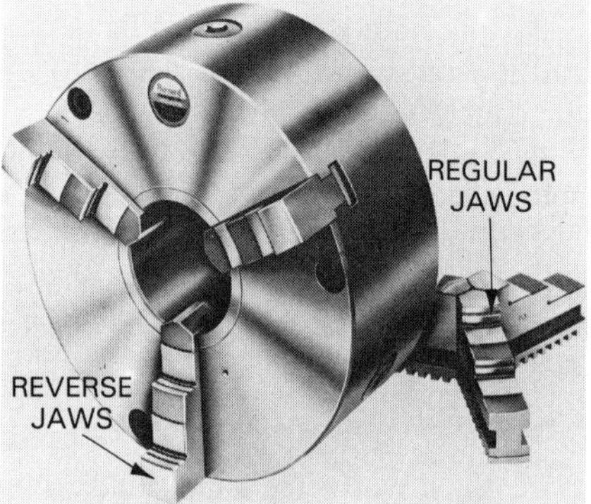


Fig. 38-15 Three-jaw chucks are equipped with a regular and a reversed set of jaws. (*Courtesy The Clausing Corporation.*)

## THREE-JAW CHUCK WORK

The three-jaw universal chuck is used to hold round or hexagonal work for machining. With proper care, this chuck should be able to hold work to within .002 in. (0.05 mm) of concentricity even after long use. Three-jaw chucks are supplied with two sets of jaws, a regular set and a reversed set (Fig. 38-15). The *regular set* is used to grip outside diameters and also inside diameters of large work. The *reversed set* is used to grip the outside of large-diameter work. All chuck jaws are stamped with the same serial number as the chuck and also numbered 1, 2, or 3 to match the slot in which they fit (Fig. 38-16). They have been fitted and ground true for that chuck and must never be used on another chuck.

### To Face Work in a Chuck

The purpose of facing work in a chuck is to obtain a true flat surface and cut the work to length.

**1.** Set the work in the chuck so that no more than three times its diameter extends beyond the chuck jaws (Fig. 38-17).

**2.** Swivel the compound rest 30° to the right if only one surface on the work must be faced (Fig. 38-18) *or* swivel the compound rest 90° to the cross slide if a series of steps or shoul-

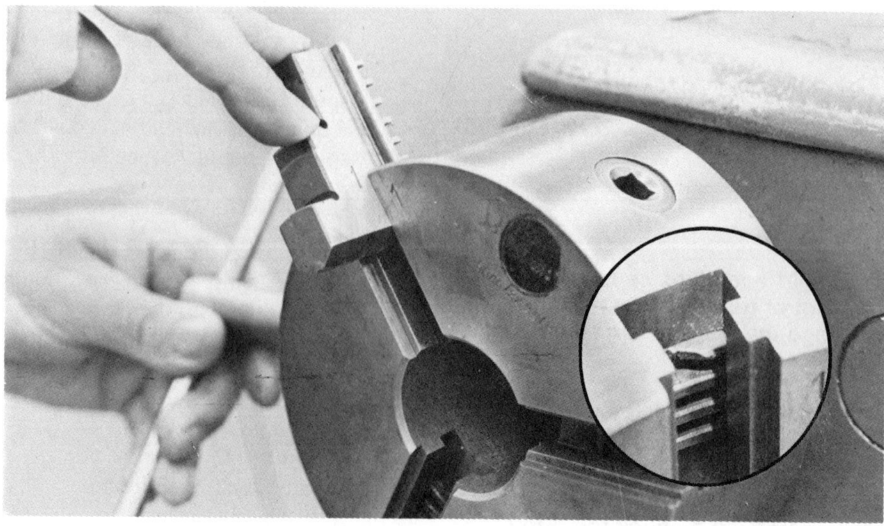

Fig. 38-16 Never insert jaws into a chuck unless they have the same serial number as the chuck. (*Courtesy Kostel Enterprises Ltd.*)

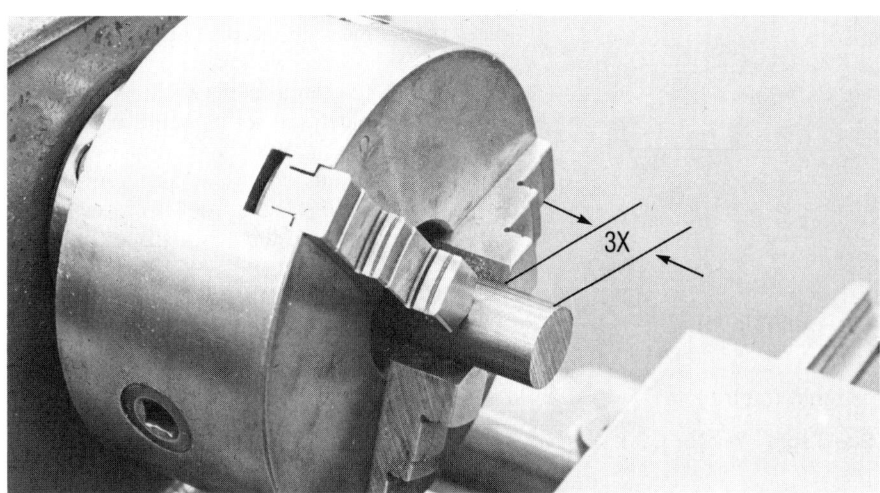


**Fig. 38-17** The workpiece should not extend more than three times the diameter beyond the chuck jaws. (*Courtesy Kostel Enterprises Ltd.*)

**Fig. 38-18** Set the compound rest at 30° to the right and place the toolpost to the left side of the compound rest. (*Courtesy Kostel Enterprises Ltd.*)

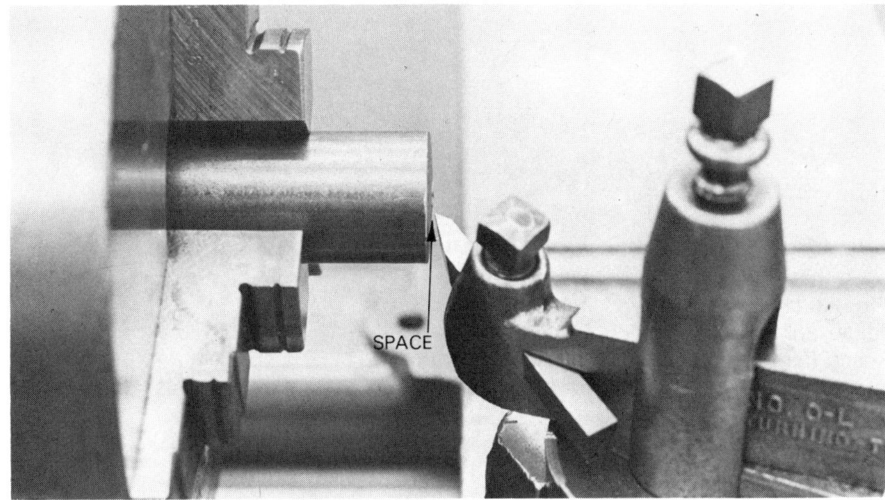


**Fig. 38-19** A toolbit set up for facing the end of a workpiece. (*Courtesy Kostel Enterprises Ltd.*)

ders must be faced to accurate length on the same workpiece.

**3.** Fasten a facing toolbit in the toolholder and set its point to center height.

**4.** Adjust the toolholder until the point of the facing tool is closest to the work and there is a space left along the side (Fig. 38-19).

**5.** Set the lathe speed for facing. (See N-6.)

N-6 $\text{r/min (inch)} = \frac{CS \times 4}{D}$

$\text{r/min (metric)} = \frac{CS \times 320}{D}$

(See Table 37-1 for CS.)

**6.** Move the carriage until the toolbit starts a light cut at the center of the surface to be faced.

**7.** Lock the carriage in position and set the depth of cut with the compound rest handle (Fig. 38-20).

(a) Twice the amount to be removed if the compound rest is set at 30°.

(b) The same as the amount to be removed if the compound rest is set at 90° to the cross slide.

**8.** Face the work to length.

## To Drill Center Holes

Center holes can be drilled on a lathe in round or hexagonal work without

**Fig. 38-20** Lock the carriage in position by tightening the carriage lock using two-finger pressure on the wrench. (*Courtesy Kostel Enterprises Ltd.*)

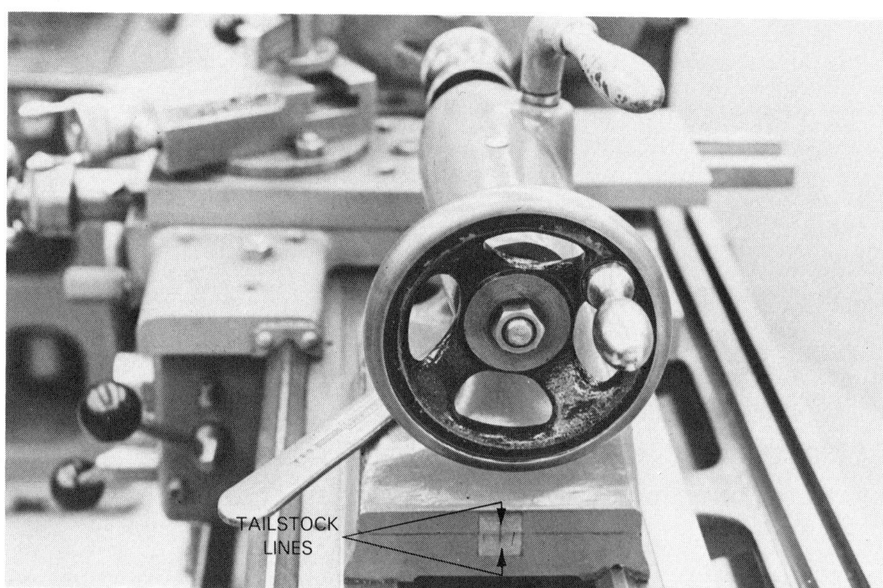


**Fig. 38-22** Check to see that the lines on the top and bottom half of the tailstock are in line. (*Courtesy Kostel Enterprises Ltd.*)

**Fig. 38-21** Fasten the proper-size center drill in the drill chuck. (*Courtesy Kostel Enterprises Ltd.*)

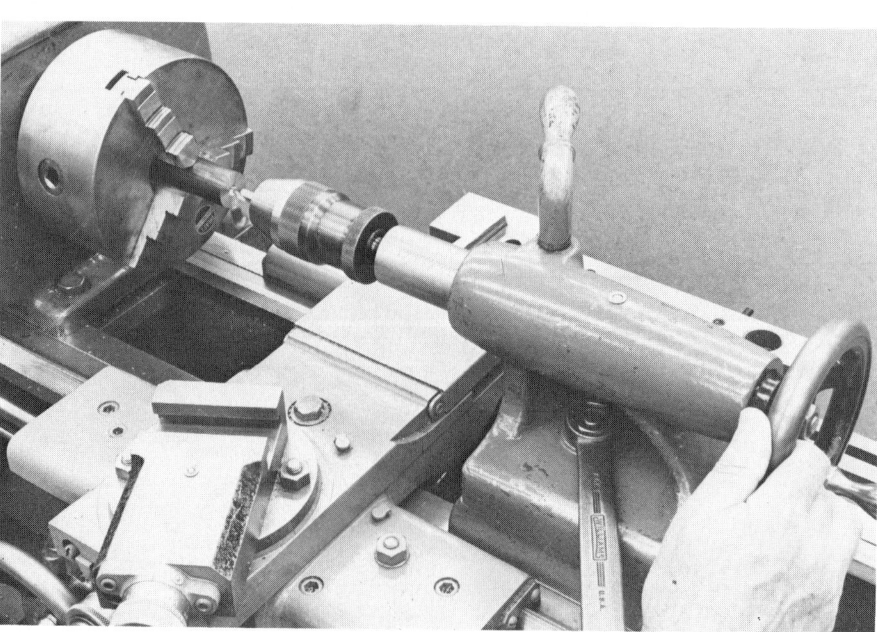

**Fig. 38-23** Feeding the center drill into the work by turning the tailstock handwheel. (*Courtesy Kostel Enterprises Ltd.*)

having to lay out the location of the center.

**1.** Grip the work short in a three-jaw chuck. No more than three times the diameter should extend beyond the chuck jaws.
**2.** Square the end of the work by facing.
**3.** Mount a drill chuck in the tailstock spindle. (See N-7.)
**4.** Select the proper center drill to suit the work diameter and fasten it in the drill chuck (Fig. 38-21). See Table 30-1.
**5.** Check the lines on the back of the tailstock to see that they are aligned. Correct if necessary (Fig. 38-22). (See N-8.)
**6.** Set the lathe speed to approximately 1200 to 1500 r/min.
**7.** Move the tailstock until the center drill is close to the work and then lock the tailstock clamp nut.
**8.** Start the lathe spindle and turn the tailstock handwheel to feed the center drill into the work (Fig. 38-23).
**9.** Frequently apply cutting fluid and drill the center hole until the top of the hole is to the correct diameter [about ¼ in. for 1-in.-diameter work (6.35 mm for 25.4-mm-diameter work)].

N-7 Clean the tailstock spindle taper and the shank of the drill chuck before assembling.

N-8 If the tailstock is off center, the center drill can be broken.

## REVIEW QUESTIONS

**1.** Name three types of spindles found on lathes.
**2.** How is the chuck fastened in position on the following?
(a) Tapered spindle nose
(b) Cam-lock spindle nose
**3.** Why should a chuck cradle be used when mounting or removing a chuck?
**4.** Briefly explain how a chuck can be removed from a tapered spindle nose.
**5.** Why is it important that dirt and chips be cleaned from mating parts of a chuck and spindle?
**6.** Briefly explain how to mount a chuck on a
(a) Tapered spindle nose
(b) Cam-lock spindle nose
**7.** How far should the work extend past the jaws when facing work in a lathe chuck?
**8.** How should the toolbit be set for facing?
**9.** Why must the tailstock be in line when drilling center holes?
**10.** How deep should a center hole be drilled?

# UNIT 39

# SETUP FOR MACHINING BETWEEN CENTERS

A large percentage of lathe work is machined between centers and it is important that the lathe be set up properly in order to produce accurate work in the shortest time possible. The alignment of lathe centers and the cutting tool setup both can affect the accuracy of the work produced, and therefore the operator should follow both of these procedures carefully for best results.

## OBJECTIVES

Upon completion of this unit you should be able to:

**1.** Mount and remove headstock and tailstock centers
**2.** Align lathe centers by the tailstock-lines, trial-cut, and dial-indicator method
**3.** Set up the cutting tool for machining between centers
**4.** Set up the workpiece for machining between centers

## MOUNTING AND REMOVING LATHE CENTERS

In order to machine parallel work that is concentric with the center holes in the work, it is very important that the centers be mounted correctly in a lathe. Dirt, burrs, or chips on the tapered shanks of the lathe centers or in the lathe spindle holes can prevent the center from seating properly in the spindle and result in spoiled work. (See N-1.)

N-1 Dirt or chips on mating parts can damage both parts and cause inaccurate work.

### Mounting a Center in the Headstock

**1.** Clean the tapered hole in the spindle with a cloth wrapped around a stick (Fig. 39-1).

**Fig. 39-1** Cleaning the lathe spindle taper with a cloth wrapped around a stick. (*Courtesy Kostel Enterprises Ltd.*)

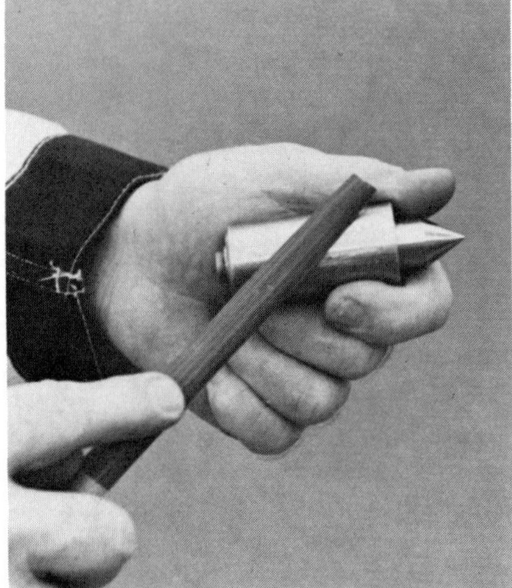

**Fig. 39-2** Removing burrs on the live center sleeve with a hand file. (*Courtesy Kostel Enterprises Ltd.*)

N-2 If the center is not inserted with a sharp snap, the pressure of a cut will force the center into the headstock and the workpiece will become loose between centers.

N-3 Holding the lathe center with a cloth will prevent hand injury from the sharp point of the center.

N-4 The work will be tapered if the lathe centers are not in line.

N-5 Do not move the crossfeed handle until the graduated collar reading has been noted.

**2.** Remove all burrs from the center sleeve and lathe center with a file or hone (Fig. 39-2).

**3.** Clean the internal taper of the center sleeve and the taper of the lathe center.

**4.** Seat the center into the center sleeve with a sharp snap. (See N-2.)

**5.** Insert the center sleeve into the lathe spindle for a short distance.

**6.** With a sharp snap, seat it into the headstock spindle.

### Mounting a Center in the Tailstock

**1.** Clean the tapered hole in the tailstock spindle and the taper on the lathe center.

**2.** Extend the tailstock spindle about 2 in. (50.8 mm) beyond the tailstock.

**3.** Insert the center into the tailstock for a short distance.

**4.** With a sharp snap, seat it into the tailstock spindle.

### Removing a Headstock Center

**1.** Be sure that the lathe is stopped.

**2.** Wrap a cloth around the headstock center and hold it with one hand. (See N-3.)

**3.** Insert a knockout bar in the headstock spindle.

**4.** Tap the back of the center with the knockout bar until it is removed from the headstock spindle.

**5.** Clean and store the center where it will not be damaged.

### Removing a Tailstock Center

**1.** Wrap a cloth around the tailstock center and hold it with one hand.

**2.** Turn the tailstock handwheel counterclockwise to draw the spindle into the tailstock.

**3.** When the end of the center contacts the knockout screw it will be removed from the tailstock spindle.

**4.** Clean and store the center where it will not be damaged.

## ALIGNMENT OF LATHE CENTERS

To produce a parallel diameter when work is machined between centers, it is important that the headstock and tailstock centers be in line. If the lathe centers are not aligned with each other, the diameter cut will be tapered (larger at one end than the other end). (See N-4.) There are three common methods used to check the alignment of lathe centers:

**1.** By having the lines on the back of the tailstock in line (which should always be done as a first step in aligning centers)

**2.** By taking a light trial cut near each end of the work and measuring the diameters with a micrometer

**3.** By using a dial indicator and a parallel test bar (which achieves greatest accuracy)

The lathe tailstock consists of two halves, the *baseplate* and the *tailstock body* (Fig. 39-3). The tailstock body can be adjusted either toward or away from the cutting tool by means of the two adjusting screws (G and F). This allows the tailstock center to be aligned with the headstock center.

### To Align Centers by the Tailstock Graduations

**1.** Loosen the tailstock clamp nut or lever (Fig. 39-4).

**2.** Loosen one adjusting screw and tighten the opposite one depending on the direction the tailstock must be moved.

**3.** Continue adjusting the screws until the line on the tailstock body matches the line on the baseplate (Fig. 39-3).

**4.** Tighten the loose adjusting screw to hold the tailstock body in position.

**5.** Recheck the tailstock graduations alignment and then tighten the tailstock clamp nut or lever.

### To Align Centers With the Trial-Cut Method

**1.** Take a trial cut from the work at the tailstock end (section A) deep enough to produce a true diameter (Fig. 39-5). This cut should be about ¼ in. (6.35 mm) long.

**2.** Disengage the automatic feed and note the reading on the graduated collar of the crossfeed screw. (See N-5.)

**3.** Turn the crossfeed handle counterclockwise to bring the cutting tool away from the work.

**Fig. 39-3** The tailstock body can be adjusted in order to align lathe centers. (*Courtesy Kostel Enterprises Ltd.*)

**Fig. 39-4** The tailstock can be locked in any position on the bed by tightening the clamp lever or nut. (*Courtesy Kostel Enterprises Ltd.*)

**Fig. 39-5** Checking the alignment of the lathe centers with the trial-cut method. (*Courtesy Kostel Enterprises Ltd.*)

**4.** Move the carriage until the cutting tool is about 1 in. (25.4 mm) from the lathe dog.

**5.** Turn the crossfeed handle clockwise until it is at the same graduated collar setting as it was at section A.

**6.** Cut section B about ½ in. (12.7 mm) long (Fig. 39-5).

**7.** Stop the lathe and measure both diameters with a micrometer (Fig. 39-6).

**8.** If both diameters are the same, the lathe centers are in line. If the diameters are different, loosen and then adjust the tailstock one-half the difference between the two diameters:

(a) Away from the cutting tool if the diameter at the tailstock end is smaller (Fig. 39-7A).

(b) Toward the cutting tool if the diameter at the tailstock end is larger (Fig. 39-7B).

**9.** If necessary, readjust the tailstock and continue to take trial cuts until both diameters are the same.

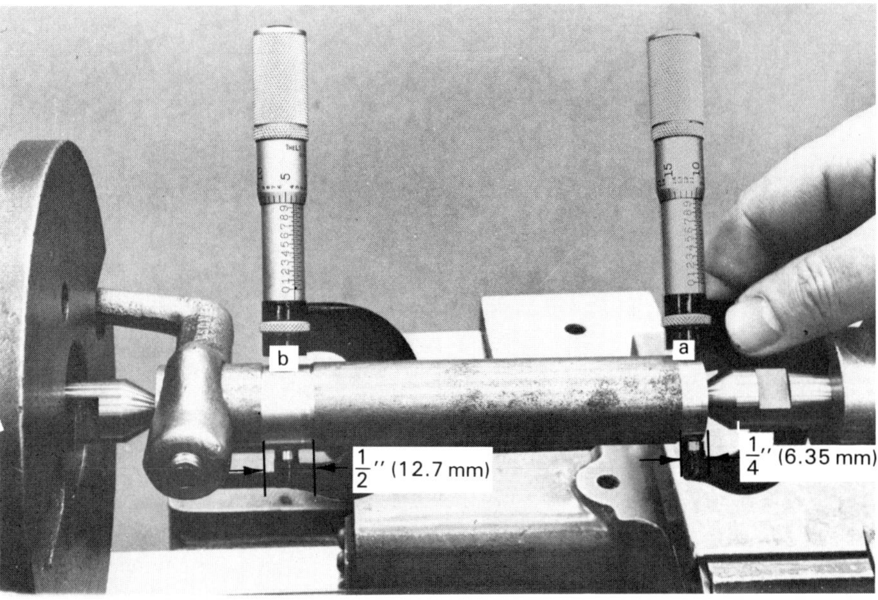


**Fig. 39-6** Checking the turned surfaces for size with a micrometer. (*Courtesy Kostel Enterprises Ltd.*)

**Fig. 39-7A** If the diameter at the tailstock end is *smaller,* move the tailstock body (top half) *away* from the cutting tool. (*Courtesy Kostel Enterprises Ltd.*)

**Fig. 39-7B** If the diameter at the tailstock end is *larger,* move the tailstock body (top half) *toward* the cutting tool. (*Courtesy Kostel Enterprises Ltd.*)

## To Align Centers with a Test Bar and Dial Indicator

**1.** Clean the lathe centers and the center holes in the test bar.

**2.** Adjust the test bar snugly between centers and then tighten the tailstock spindle clamp.

**3.** Mount a dial indicator, with the indicator plunger on center and in a horizontal position, in the toolpost or on the lathe carriage.

**4.** Turn the crossfeed handle until the indicator registers approximately one-quarter of a revolution on section A (Fig. 39-8). (See N-6.)

**5.** Note the indicator reading at section A.

**6.** Turn the carriage handwheel to bring the indicator into contact with section B (Fig. 39-9).

**7.** Compare the two indicator readings and if they are not the same, re-

**Fig. 39-8** Checking the lathe center alignment with a test bar and dial indicator. (*Courtesy Kostel Enterprises Ltd.*)

N-6 Do not move the position of the crossfeed handle.

**Fig. 39-9** If both readings on the test bar are the same, the centers are in line. (*Courtesy Kostel Enterprises Ltd.*)

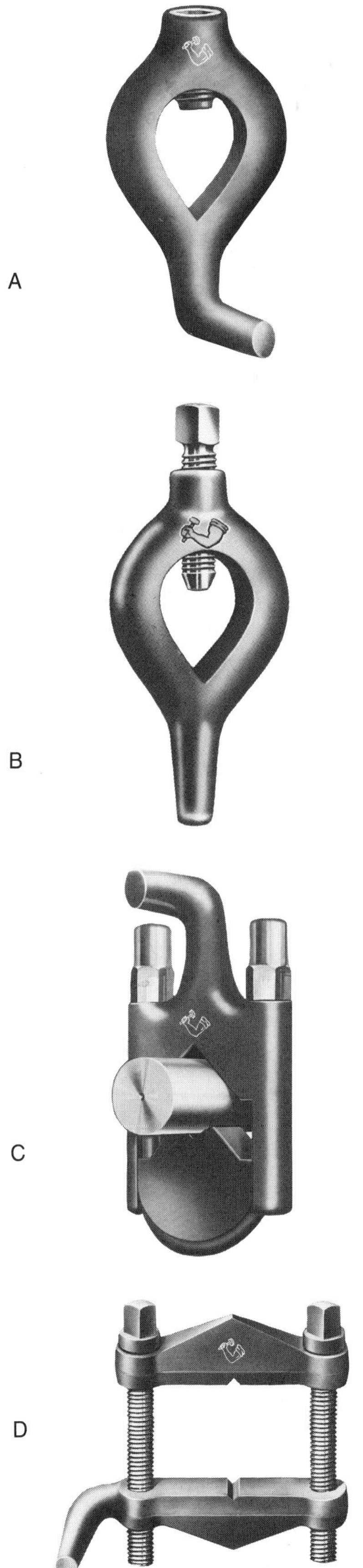


Fig. 39-10 Common types of lathe dogs. (*Courtesy Armstrong Bros. Tool Co.*)

turn the carriage until the indicator again registers on section A.

**8.** Loosen the tailstock clamp nut and by means of the tailstock adjusting screws (Fig. 39-3) move the tailstock in the proper direction the difference between the readings at sections A and B.

**9.** Tighten the tailstock clamp nut and repeat steps 4 to 7 to recheck the center alignment.

## LATHE DOGS

Work machined between lathe centers is driven by a lathe dog. The lathe dog is provided with an opening to receive and clamp the workpiece and a bent tail. The tail of the dog fits into a slot in the lathe driveplate to drive the workpiece.

Lathe dogs are made in a variety of sizes and styles to suit various operations and workpieces.

The *standard bent-tail lathe dog* (Fig. 39-10A) is most commonly used to drive round workpieces on a lathe. They are available in sizes to accommodate diameters ranging from ⅜ to 6 in. (9.52 to 152.4 mm). Standard lathe dogs are available with square-head or headless setscrews to clamp the workpiece. The headless screw is safer than the square-head setscrew since it has no protruding head.

The *straight-tail dog* (Fig. 39-10B) may be used for the same purpose as the bent-tail type; however, it must be driven by a stud attached to the driveplate. The straight tail and the locking screw end are approximately the same weight, thus providing a balanced lathe dog. Straight-tail dogs are often used in precision work where centrifugal force may cause inaccuracies in the workpiece.

The *safety clamp lathe dog* (Fig. 39-10C) is especially valuable for use on finished work which may be damaged by the setscrew of a lathe dog. The sliding jaw gives the safety clamp lathe dog a wide range of adjustment and allows it to be mounted on a workpiece while it is between the lathe centers.

The *clamp lathe dog* (Fig. 39-10D) can be used to hold round, square, rectangular, hexagonal, and odd-shaped workpieces. The two setscrews allow this dog to hold a wide range of workpiece sizes.

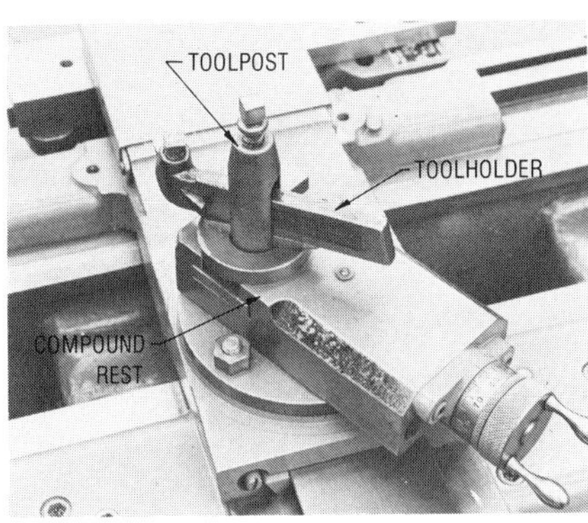


Fig. 39-11 Set the toolpost to the left-hand side of the compound rest with the toolholder held short. (*Courtesy Kostel Enterprises Ltd.*)

Fig. 39-12 The toolholder is gripped short and the toolbit is set to center. (*Courtesy Kostel Enterprises Ltd.*)

### To Set Up the Cutting Tool

The setup of the toolholder and the cutting tool is very important, since it can affect the accuracy and efficiency of the metal-removal process. If the setup is not correct, the cutting tool may not cut well, or it may dig in and ruin the workpiece.

**1.** Move the toolpost to the left-hand side of the compound rest (Fig. 39-11).

**2.** Mount a toolholder so that its setscrew is close to the toolpost. For the best rigidity, have about the width of a thumb between the toolholder screw and the toolpost (Fig. 39-12).

**3.** Set the toolholder so that it is at right angles to the work or pointing slightly toward the tailstock.

**4.** Insert the desired toolbit in the toolholder so that it extends only about ½ in. (12.7 mm).

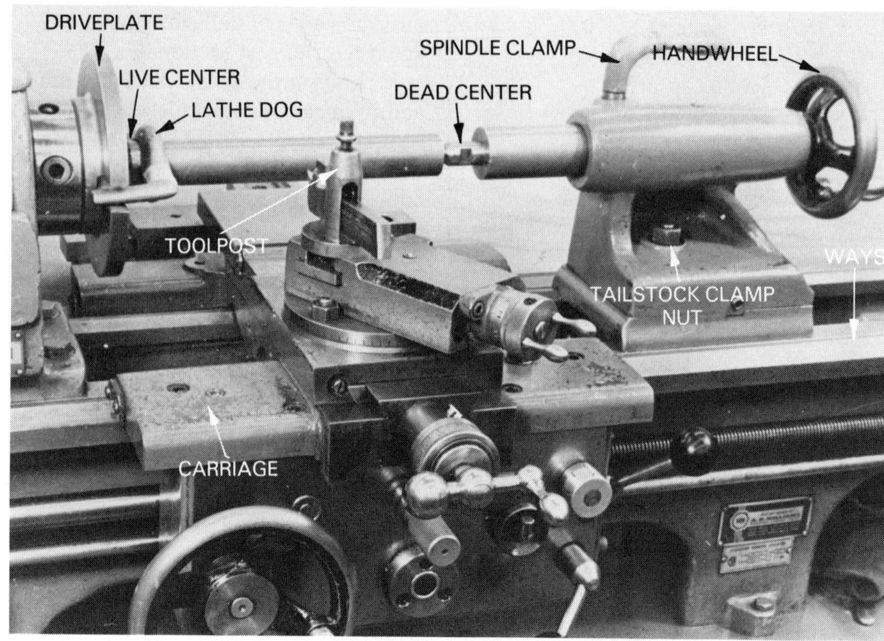


**Fig. 39-13** Work mounted between lathe centers for machining. (*Courtesy Kostel Enterprises Ltd.*)

N-7 Tightening the toolholder setscrew too tightly may break the toolbit.

N-8 If the live center is not running true, the machined diameters will not be concentric with the center holes.

N-9 Center lubricant helps to reduce the heat and friction caused by the revolving work.

N-10 If the lathe dog binds in the drive plate slot, it prevents the work from seating on the lathe center and causes inaccurate work.

**5.** Use two-finger pressure on the wrench to tighten the toolholder setscrew. (See N-7.)

**6.** Adjust the toolholder until the point of the cutting tool is even with the lathe center point (Fig. 39-12).

**7.** Tighten the toolpost screw securely to prevent the toolholder from moving under the pressure of the cut.

## To Mount Work between Centers (Fig. 39-13)

**1.** Check the live center by holding a piece of chalk close to it while it is revolving. If the live center is not running true, the chalk will mark only the high spot. (See N-8.)

**2.** If this happens, remove the live center from the headstock and clean the tapers on the center and the headstock spindle.

**3.** Replace the live center and check again for trueness.

**4.** Check the lathe center alignment.

**5.** Adjust the tailstock spindle until it extends about 2½ to 3 in. (63.5 to 76.2 mm) beyond the tailstock.

**6.** Loosen the tailstock clamp nut or lever.

**7.** Place the lathe dog on the end of the work with the bent tail pointing to the left.

**8.** Lubricate the dead center with a suitable center lubricant (Fig. 39-14). (See N-9.)

**9.** Place the end of the work with the lathe dog on the live center and slide the tailstock toward the headstock until the dead center supports the other end of the work.

**10.** Tighten the tailstock clamp nut or lever (Fig. 39-15).

**11.** Adjust the tail of the dog in the slot of the drive plate and tighten the lathe dog screw. (See N-10.)

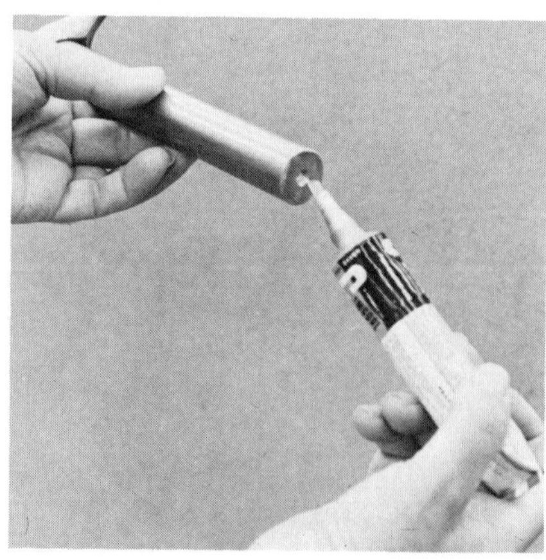

**Fig. 39-14** Applying a lubricant to the center hole. (*Courtesy Kostel Enterprises Ltd.*)

**Fig. 39-15** Using two-finger pressure to tighten the tailstock clamping nut. (*Courtesy Kostel Enterprises Ltd.*)

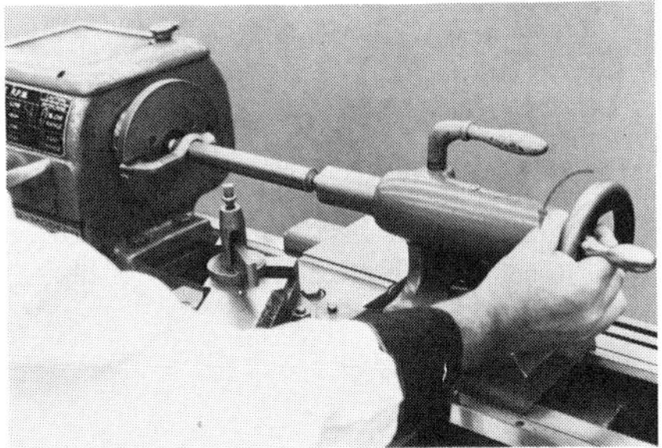

**Fig. 39-16** Turning the tailstock handwheel backward to to adjust the center tension. (*Courtesy Kostel Enterprises Ltd.*)

**Fig. 39-17** Checking the center adjustment of the workpiece for freeness and end play. (*Courtesy Kostel Enterprises Ltd.*)

**12.** Turn the drive plate by hand until the slot and lathe dog are parallel with the bed of the lathe.

**13.** Hold the tail of the dog *up* in the slot and tighten the tailstock handwheel only enough to hold the lathe dog in the *up position.*

**14.** Turn the tailstock handwheel backward only until the lathe dog drops in the drive plate slot (Fig. 39-16).

**15.** Hold the tailstock handwheel in this position and tighten the tailstock spindle clamp with the other hand.

**16.** Check the tension of the work between centers. The tail of the dog should drop of its own weight and there should be no end play between the lathe centers and the work (Fig. 39-17).

**17.** Move the carriage to the furthest position (left-hand end) of cut and revolve the lathe spindle by hand to see that the dog does not hit the compound rest.

## REVIEW QUESTIONS

**1.** Why is it important that dirt, burrs, and chips be cleaned from lathe centers before they are mounted?

**2.** What may happen if a center is not seated properly in the headstock spindle?

**3.** Why should a cloth be held over a headstock center when it is being removed?

**4.** Why is it important that the headstock and tailstock centers be in line?

**5.** Name three methods of aligning lathe centers.

**6.** In which direction should the tailstock body be moved if:

(a) The work diameter is larger at the tailstock end

(b) The work diameter is smaller at the tailstock end

**7.** Briefly explain how to align lathe centers using a test bar and dial indicator.

**8.** Name four types of lathe dogs.

**9.** How far should the toolholder extend beyond the toolpost?

**10.** How far should the tailstock spindle extend beyond the tailstock?

**11.** Why should the dead center be lubricated?

**12.** What is the correct tension on work mounted between lathe centers?

# UNIT 40

# MACHINING BETWEEN CENTERS

Much of the work machined on a lathe is held between the tailstock and headstock centers. Since both centers are in a fixed position, work can be machined, removed from the lathe, and replaced for additional machining with the assurance that the machined diameters will be true (concentric) with the center holes. In training programs, where it is often necessary to remove work a number of times before it is completely machined, it saves much valuable time which would be necessary to set up the work accurately with other work-holding methods. The most common operations performed on work mounted between centers are facing, parallel turning, and shoulder turning.

## OBJECTIVES

The aim of this unit is to enable you to:

**1.** Set up the lathe and face work squarely and to accurate lengths

**2.** Machine parallel diameters to within ±.002 in. (0.05 mm) accuracy for parallelism and size

**3.** Machine square, filleted, and beveled shoulders to within 1/64 in. (0.39 mm) accuracy

## FACING

Facing is a squaring operation that is performed on the ends of work after it has been cut off by a saw. In order to produce a flat surface when facing between centers, the lathe centers must be in line. The purposes of facing are to:

**1.** Provide a true flat surface, square with the axis of the work

**2.** Make a smooth surface from which to take measurements

**3.** Cut work to a required length

**Fig. 40-1** The toolholder is gripped short and the point of the toolbit is set even with the center. (*Courtesy Kostel Enterprises Ltd.*)

N-1 The toolbit, which is hard and brittle, may break if the toolholder setscrew is too tight.

N-2 $\text{r/min (inch)} = \dfrac{CS \times 4}{D}$

$\text{r/min (metric)} = \dfrac{CS \times 320}{D}$

N-3 Locking the carriage will ensure that a flat surface will be cut.

N-4 When facing, all cuts should begin at the center and feed to the outside to produce a flat, smooth surface.

## To Face Work between Centers

**1.** Set the toolholder to the left-hand side of the compound rest.

**2.** Mount a facing tool in the toolholder having it extending only about ½ in. (12.7 mm) and tighten the toolholder setscrew using only two-finger pressure. (See N-1.)

**3.** Adjust the toolholder until the point of the facing tool is at the same height as the lathe center point, then tighten the toolpost lightly (Fig. 40-1).

**4.** Tap the toolholder until the point of the cutting tool is closest to the work and there is a space along the side (Fig. 40-2).

**5.** Tighten the toolpost securely to keep the toolholder in position.

**6.** If the entire end must be faced, insert a half center in the tailstock spindle (Fig. 40-2).

**7.** Bring the toolbit to the end of the work by moving the carriage with the apron handwheel.

**8.** Set the lathe to the correct speed for the material being cut. (See N-2.)

**9.** Move the toolbit in as close to the center of the work as possible and set the depth of cut by using the apron handwheel.

**10.** Tighten the carriage lock with two-finger pressure on the wrench (Fig. 40-3). (See N-3.)

**11.** Feed the tool out by turning the crossfeed screw handle slowly and steadily with two hands, thus cutting from the center outward. (See N-4.)

**12.** Loosen the carriage lock before setting the depth of other cuts.

## Facing with the Compound Rest Set at 30°

Work can be accurately faced to length by using the graduated collar of the compound rest.

**1.** Swivel the compound rest at 30° to the cross-slide (Fig. 40-4).

**2.** Move the toolpost to the left-hand side of the compound rest.

**3.** Set the facing tool to center. Check it against the lathe center point (Fig. 40-1).

**4.** Bring the toolbit close to the center of the surface to be faced.

**5.** Lock the carriage in position to stop it from moving during a cut (Fig. 40-3).

**Fig. 40-2** The toolholder is swiveled until the point of the facing tool is close to the work with a space along the side of the cutting edge.

**Fig. 40-3** The carriage is locked in position by tightening the carriage lock nut. (*Courtesy Kostel Enterprises Ltd.*)

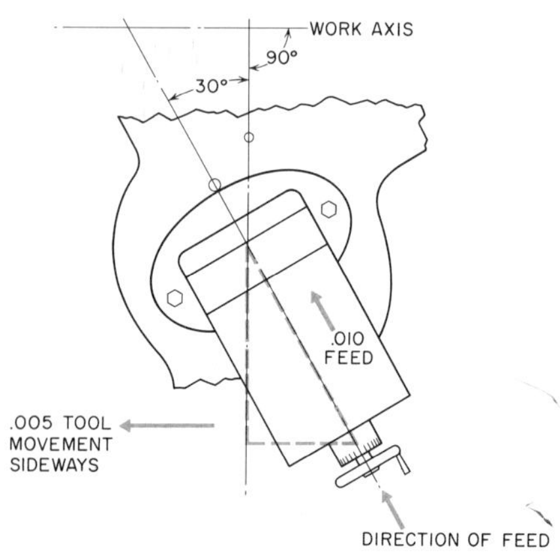


**Fig. 40-4** Accurate facing is possible when the compound rest is set at 30°.

N-5 With the compound rest at 30°, the amount of side movement of the cutting tool is always *one-half the amount fed.*

**Fig. 40-5** Accurate facing of steps or shoulders is possible when the compound rest is set at 90°.

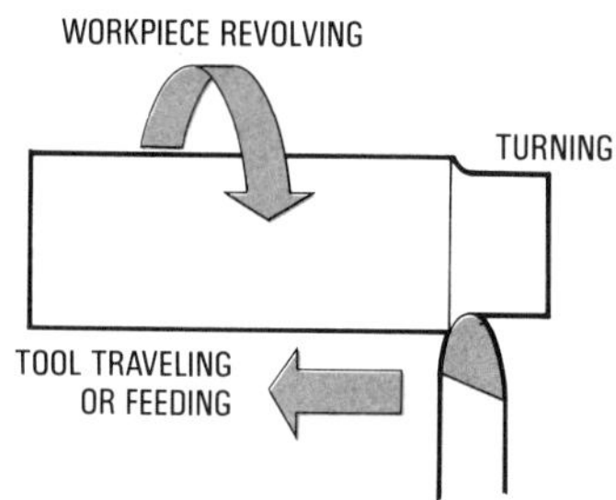


**Fig. 40-6** The principle of machining a diameter on a lathe.

**6.** Feed the compound rest screw until a light cut is started and face the surface.

**7.** Measure the work and calculate the amount of material to be removed.

**8.** Feed the compound rest screw double the amount of material to be removed. For example, if .010 in. (0.25 mm) must be removed, feed the compound rest .020 in. (0.50 mm). (See N-5.)

### Facing with the Compound Rest at 90°

A series of short steps or shoulders can be accurately machined along the length of a piece of work when the compound rest is set at 90° to the cross slide (Fig. 40-5). The amount of toolbit travel is the same as the reading on the graduated collar of the compound rest feed screw.

**1.** Swivel the compound rest at 90° to the cross slide (Fig. 40-5).

**2.** Set the cutting tool to center.

**3.** Rough turn all shoulders to within 1/32 in. (0.79 mm) of the required length.

**4.** Set the cutting tool for facing the shoulders (Fig. 40-2).

**5.** Bring the top slide of the compound rest back as far as possible to ensure sufficient travel to cut all shoulders.

**6.** Set the cutting tool to take a light cut from the end of the workpiece (or a shoulder), and then lock the carriage in position (Fig. 40-3).

**7.** Machine each shoulder to length by feeding the compound rest the required distance.

## PARALLEL TURNING

Work is generally machined on a lathe for two reasons: to cut it to size and to produce a true diameter. Work that must be cut to size and also be the same diameter along the entire length of the workpiece involves the operation of parallel turning (Fig. 40-6). In order to produce a parallel diameter, the headstock and tailstock centers must be in line. Many factors determine the amount of material which can be removed on a lathe at one time. However, whenever it is possible, work should be cut to size in two cuts: a roughing cut and a finishing cut.

## GRADUATED MICROMETER COLLARS

*Graduated micrometer collars* are sleeves or bushings that are mounted on the compound rest and crossfeed screws (Fig. 40-7). They assist the operator to set the cutting tool accurately to remove the required amount of material from the workpiece. The micrometer collars on lathes using the inch system of measurement are usually graduated in thousandths of an inch (.001 in.). The collars on lathes using the metric system of measurement are usually graduated in fiftieths of a millimeter (0.02 mm).

The graduated collar indicates only the distance that the cutting tool has been moved toward the work. On any machines where the work revolves (lathes, cylindrical grinders, boring mills, etc.) the cutting tool should therefore be set in only half the amount of metal to be removed, because material is removed from the circumference of the workpiece.

### Inch Lathes

The circumference of the crossfeed and compound rest screw collars on lathes using the inch system of measurement is usually divided into 100 or 125 equal divisions, each having a value of .001 in. Therefore, if the crossfeed screw is turned clockwise 20 graduations, the cutting tool will be moved

**Fig. 40-7** Micrometer collars assist the operator in setting the correct depth of cut. (*Courtesy South Bend Lathe Inc.*)

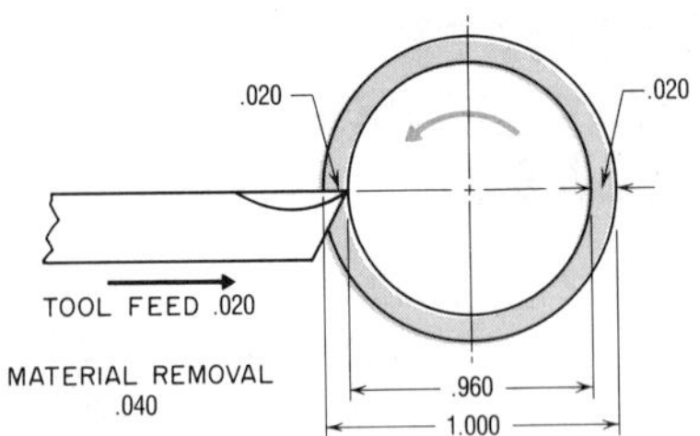


**Fig. 40-8** The work in the lathe revolves, so a .020-in. depth of cut reduces the diameter by .040 in.

**Fig. 40-9** A light cut must be made around the entire circumference of the work before setting a depth of cut. (*Courtesy Kostel Enterprises Ltd.*)

**N-6** The trial cut produces a true diameter and sets the graduated collar and the cutting tool to the diameter.

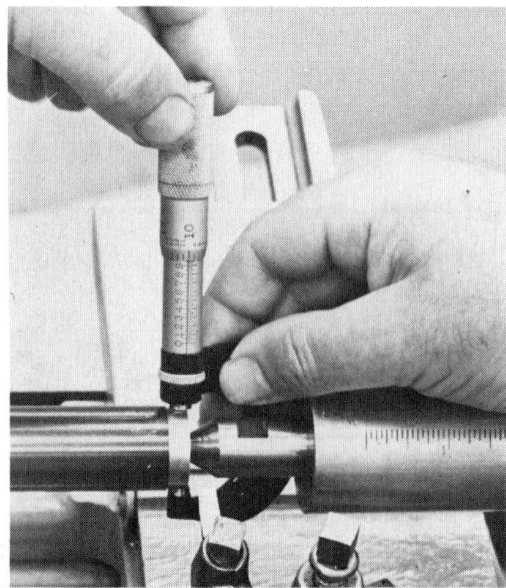

**Fig. 40-10** Measuring the diameter of a trial cut with a micrometer. (*Courtesy Kostel Enterprises Ltd.*)

.020 in. toward the work. Because the work in a lathe revolves, a .020-in. depth of cut will be taken from the entire work circumference, thereby reducing the diameter .040 in. (2 × .020 in.) (Fig. 40-8).

## Metric Lathes

The circumference of the crossfeed and compound rest screw collars on lathes using the metric system of measurement is usually divided into 200 or 250 equal divisions, each having a value of 0.02 mm. Therefore if the crossfeed screw is turned clockwise 30 graduations, the cutting tool will be moved 30 × 0.02 mm or 0.60 mm toward the work. Because the work in a lathe revolves, a 0.60-mm depth of cut will remove 1.20 mm from the diameter of a workpiece.

### To Set an Accurate Depth of Cut

**1.** Move the toolpost to the left-hand side of the compound rest.

**2.** Set the cutting tool to center and tighten the toolpost screw *securely*.

**3.** Start the lathe and move the carriage until the toolbit overlaps the right-hand end of the workpiece by approximately 1⁄16 in. (1.58 mm).

**4.** Feed the toolbit in with the crossfeed handle until a light cut is made around the entire circumference of the work (Fig. 40-9).

**5.** Turn the carriage handwheel until the toolbit just clears the right-hand end of the work.

**6.** Turn the crossfeed handle clockwise .005 in. (0.12 mm) and take a trial cut about 1⁄4 in. (6.35 mm) long. (See N-6.)

**7.** Stop the lathe and be sure that the crossfeed handle setting is not moved.

**8.** Turn the carriage handwheel until the toolbit clears the right-hand end of the work.

**9.** Measure the diameter of the trial cut with a micrometer and then calculate the amount of metal yet to be removed (Fig. 40-10).

**10.** Turn the crossfeed handle clockwise until the graduated collar moves one-half the amount of material to be removed. For example, if .012 in. (0.30 mm) must be removed, the crossfeed handle should be set in .006 in. (0.15 mm).

**11.** Take another trial cut 1⁄4 in. (6.35 mm) long and measure the diameter with a micrometer.

**12.** Adjust the crossfeed handle setting if necessary.

## Rough Turning

Rough turning is used to remove most of the excess material as quickly as possible and to true the work diameter. The roughing cut should be taken to within 1⁄32 in. (0.79 mm) of the finished size of the workpiece. Generally one roughing cut should be taken if up to 1⁄2 in. (12.7 mm) is to be removed from the diameter (Fig. 40-11).

### To Rough Turn a Diameter

**1.** Mount a rough turning or general-purpose cutting tool in the toolholder.

**2.** Set the toolpost on the left side of the compound rest.

**3.** Have the toolholder extend as little as possible beyond the toolpost and set the point of the cutting tool

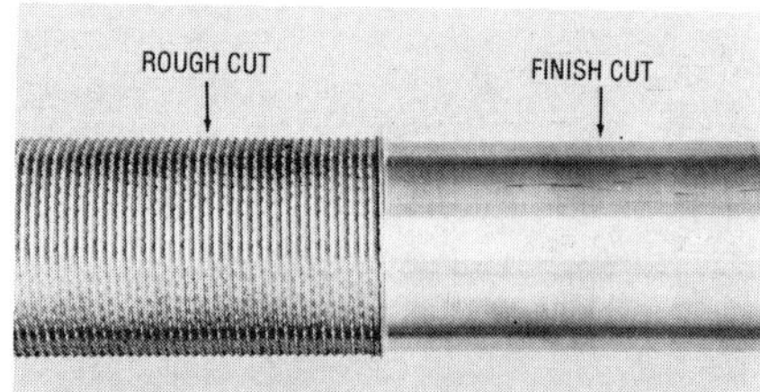


**Fig. 40-11** The surface finish produced by rough and finish turning. (*Courtesy Kostel Enterprises Ltd.*)

**Fig. 40-12** The toolholder should be set as close as possible (about the width of a thumb) to the toolpost and the point of the toolbit even with the lathe center.

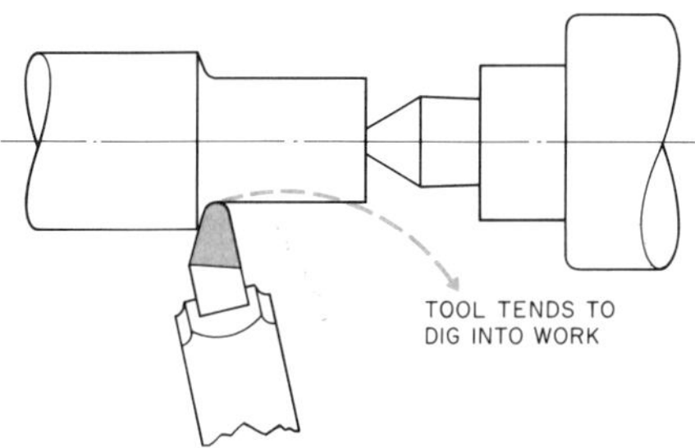


**Fig. 40-13A** If an incorrectly set toolholder moves under pressure of a cut, the work will be cut undersize.

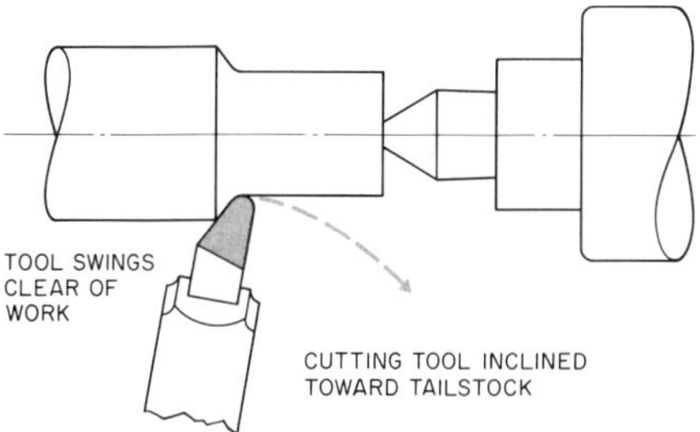


**Fig. 40-13B** If a correctly set toolholder moves under the pressure of a cut, the cutting tool will swing away from the turned work.

N-7 A rough-cut feed of .010 to .015 in. (0.25 to 0.40 mm) is suitable for most work.

N-8 A finish-cut feed of .003 to .005 in. (0.07 to 0.12 mm) is suitable for most work.

even with the lathe center point (Fig. 40-12).

**4.** Adjust the toolholder so that it is pointing slightly toward the tailstock (Fig. 40-13B).

**5.** Tighten the toolpost screw securely.

**6.** Set the lathe speed for the type and size of material being cut (Table 37-1).

**7.** Set the quick-change gearbox for the rough-cut feed (Fig. 40-14). (See N-7.)

**8.** Take a light trial cut about ¼ in. (6.35 mm) long at the right-hand end of the work.

**9.** Stop the lathe but *do not move the crossfeed handle setting or the graduated collar.*

**10.** Turn the carriage handwheel until the cutting tool clears the right-hand end of the work.

**11.** Measure the diameter of the trial cut and calculate how much material must be removed.

**12.** Turn the crossfeed handle clockwise one-half the calculated amount and take another trial cut.

**13.** Measure the diameter with a micrometer and reset the depth of cut if necessary.

**14.** Turn the rough cut to the required length.

## Finish Turning

The purpose of finish turning is to bring the workpiece to the required size and to produce a good surface finish. Generally only one finish cut is required since no more than .030 to .050 in. (0.76 to 1.27 mm) should be left on the diameter for the finish cut. The toolbit should have a slight radius on the point and the lathe should be set for a .003- to .005-in. (0.07- to 0.12-mm) feed. Be sure that the lathe centers are aligned exactly, otherwise a taper will be cut on the workpiece.

**Fig. 40-14** Move the levers as indicated on the quick-change gearbox chart for rough and finish feeds. (*Courtesy Kostel Enterprises Ltd.*)

**Fig. 40-15** Always take a trial cut for about ¼ in. (6.35 mm) along the workpiece. (*Courtesy Kostel Enterprises Ltd.*)

### To Finish Turn a Diameter

**1.** Check the alignment of the lathe centers.

**2.** Set the lathe to the correct speed for finish turning (Table 37-1).

**3.** Set the quick-change gearbox for the finish-cut feed (Table 37-2). (See N-8.)

**4.** Take a *light trial cut* ¼ in. (6.35 mm) long from the diameter at the tailstock end.

**5.** Disengage the automatic feed and stop the lathe but *do not move the crossfeed handle setting.*

**6.** Turn the carriage handwheel until the cutting tool clears the right-hand end of the work.

**7.** Measure the diameter with a micrometer and calculate the amount of material which must still be removed (Fig. 40-10).

**8.** Turn the crossfeed handle *clockwise* one-half the calculated amount (the difference between the trial cut size and the finished diameter) and take a trial cut ¼ in. (6.35 mm) long (Fig. 40-15).

**9.** Stop the lathe and measure the diameter.

**10.** Reset the crossfeed handle setting if necessary and turn the diameter to the required length.

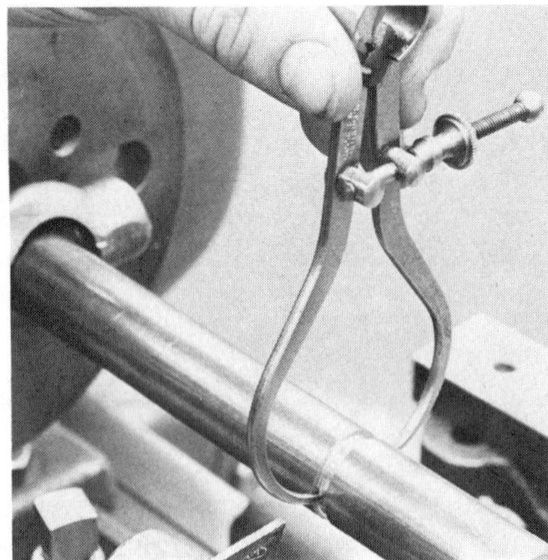

Fig. 40-16 Checking the diameter of a groove with an outside caliper. (*Courtesy Kostel Enterprises Ltd.*)

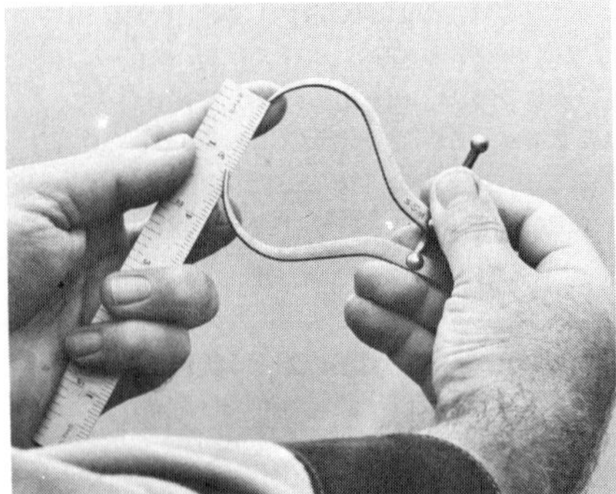

Fig. 40-17 Setting an outside caliper to a rule. (*Courtesy Kostel Enterprises Ltd.*)

**N-9 Inaccurate measurements result when a caliper is forced over a diameter.**

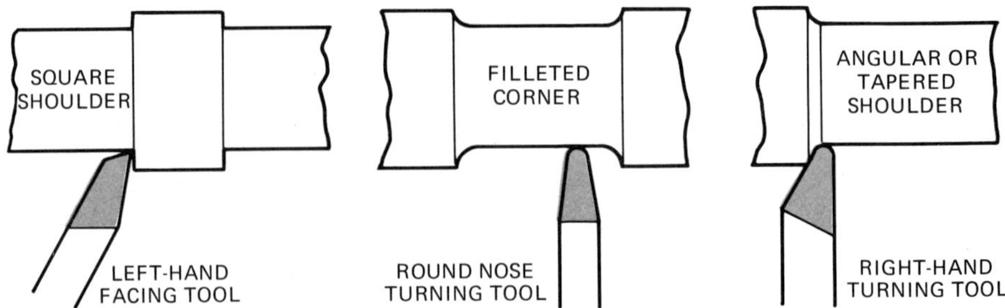


Fig. 40-18 Three common types of shoulders and the turning tools used for each.

## To Check a Diameter with an Outside Caliper

Whenever possible, a diameter should be measured with a micrometer; there may be times, however, when a micrometer is not available or it is necessary to measure narrow grooves. Diameters may be measured within a reasonable degree of accuracy with outside calipers if they are properly set and used.

**1.** Stop the machine. Never measure work that is revolving—this practice is dangerous and the measurement taken will be inaccurate.

**2.** Turn the adjusting screw until the legs of the caliper lightly contact the center of the diameter.

**3.** Check the setting by holding the spring of the caliper between the thumb and index finger and allowing the legs to drop over the diameter (Fig. 40-16). If it is correct, the caliper should just drop over the diameter, pulled by gravity. (See N-9.)

**4.** Adjust the caliper until just a *slight drag* is felt as the caliper legs pass over the diameter.

**5.** Use a rule to measure the distance between the two caliper legs (Fig. 40-17).

## Shoulder Turning

Whenever more than one diameter is machined on a shaft, the section joining each diameter is called a shoulder or step. The square, filleted, and chamfered shoulders are most commonly used in machine shop work (Fig. 40-18).

### To Machine a Square Shoulder

**1.** Lay out the length of the shoulder with a center punch mark or cut a light groove at this point with a sharp toolbit (Fig. 40-19A and B).

**2.** Rough and finish turn the small diameter to within 1/32 in. (0.79 mm) of the required length.

**3.** Mount a facing tool and set it for the facing operation (Fig. 40-2).

**4.** Start the lathe and feed the cutting tool in until it lightly marks the small diameter near the shoulder.

**5.** Note the reading on the crossfeed graduated collar (Fig. 40-20).

**6.** Turn the carriage handwheel to start a light cut.

Fig. 40-19 Marking the length of a shoulder. (A) With a center punch mark; (B) by cutting a light groove on the workpiece.

**Fig. 40-20** The cutting tool can be returned to the same position any number of times if the crossfeed graduated collar setting is noted. (*Courtesy Kostel Enterprises Ltd.*)

N-10 Apply cutting fluid to avoid chatter and to produce a good finish.

N-11 Lay out the length of the small diameter for beveled shoulders.

**7.** Face the shoulder by turning the crossfeed handle *counterclockwise*.

**8.** Return the crossfeed handle to the original graduated collar setting.

**9.** Repeat steps 6 and 7 until the shoulder is to the correct length.

## To Machine a Filleted Shoulder

**1.** Lay out the length of the shoulder with a center punch mark or by cutting a light groove at this point (Fig. 40-19A and B).

**2.** Rough and finish turn the small diameter to the correct length minus the radius to be cut. For example, a 3-in. (76.2-mm) length with a 1/8-in. (3.17-mm) radius should be turned 2 7/8 in. (73 mm) long.

**3.** Mount the correct radius toolbit and set it to center (Fig. 40-21).

**4.** Set the lathe for one-half the turning speed.

**5.** Coat the small diameter, near the shoulder, with chalk or layout dye.

**6.** Start the lathe and feed the cutting tool in until it lightly marks the small diameter near the shoulder.

**7.** Slowly feed the cutting tool sideways with the carriage handwheel until the shoulder is cut to the correct length. (See N-10.)

## Machining a Beveled (Angular) Shoulder

Beveled or angular shoulders are used to eliminate sharp corners and edges, to make parts easier to handle, and to improve the appearance of the part. They are sometimes used to strengthen a part by eliminating the sharp corner found on a square shoulder. Shoulders are beveled at angles ranging from 30 to 60°; however, the most common is the 45° bevel.

**1.** Lay out the position of the shoulder along the length of the workpiece. (See N-11.)

**2.** Rough and finish turn the small diameter to size.

**3.** Mount a side cutting tool in the toolholder and set it to center.

**4.** Use a protractor and set the side cutting edge of the toolbit to the desired angle (Fig. 40-22).

**5.** Apply chalk or layout dye to the small diameter as close as possible to the shoulder location.

**6.** Set the lathe spindle to approximately one-half the turning speed.

**Fig. 40-21** A radius tool is used to produce a filleted shoulder. (*Courtesy Kostel Enterprises Ltd.*)

**Fig. 40-22** Using a protractor to set the toolbit side cutting edge to 45°.

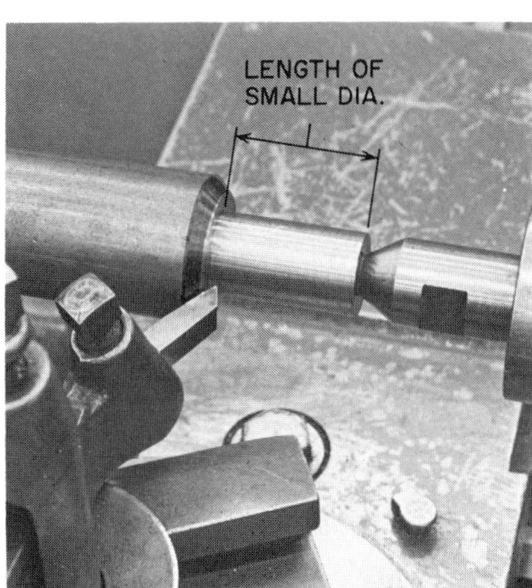


**Fig. 40-23** Machining the beveled shoulder with the side of a toolbit.

**Fig. 40-24** The compound rest swiveled to cut a large beveled shoulder.

N-12 Note the crossfeed graduated collar reading in case other cuts must be taken.

N-13 To reduce the possibility of cutting into the small diameter, take all cuts from the center outward.

**7.** Bring the point of the toolbit in until it just removes the chalk or layout dye.

**8.** Turn the carriage handwheel by hand to feed the cutting tool into the shoulder (Fig. 40-23). (See N-12.)

**9.** Apply cutting fluid to assist the cutting action and to produce a good surface finish.

**10.** Machine the beveled shoulder until it is the required size.

If the size of the shoulder is large, and chatter occurs when cutting with the side of the toolbit, it may be necessary to cut the beveled shoulder using the compound rest (Fig. 40-24).

**1.** Set the compound rest to the desired angle.

**2.** Adjust the toolbit so that only the point does the cutting.

**3.** Machine the bevel by feeding the compound rest by hand. (See N-13.)

## REVIEW QUESTIONS

**1.** Name three reasons for facing work.

**2.** Explain how the toolbit should be set for facing.

**3.** Why are half-centers used when facing?

**4.** In which direction should facing cuts be taken?

**5.** With the compound rest set at 30°, how much will be removed if the compound feedscrew is moved .008 in. (0.20 mm)?

**6.** How should the compound rest be set for machining a series of short steps?

**7.** What is the value of each graduated collar division on

(a) Inch lathes

(b) Metric lathes

**8.** How far should the graduated collar be set to turn

(a) A 1.120-in. diameter to .980 in.

(b) A 22.34-mm diameter to 21.80 mm

**9.** What is the purpose of the small trial cut when turning a diameter?

**10.** What is the purpose of rough turning?

**11.** How should the toolholder be set for rough turning to prevent it from digging in?

**12.** What speed and feed should be used to rough turn a 1-in.- (25.4-mm-) diameter piece of machine steel?

**13.** What is the purpose of finish turning?

**14.** How much material should be left on a diameter for finish turning?

**15.** What finish cut feed is recommended for most work?

**16.** Name three common shoulders used in machine shop work.

**17.** What are two methods used to lay out the length of shoulders?

**18.** Why should cutting fluid be used when cutting a filleted shoulder?

**19.** Name two methods of machining angular shoulders.

# UNIT 41

# FILING, POLISHING, AND KNURLING

Operations such as filing, polishing, and knurling are performed on round work produced in a lathe to improve either the surface finish or the appearance. Filing and polishing are used to bring the work to size and to improve the surface finish. Knurling is used to provide a hand grip and to improve the appearance of the work.

## OBJECTIVES

After completing this unit you should be able to:

**1.** Select and properly use the correct file to finish a diameter to within ±.001 of size

**2.** Select and use abrasive cloth for polishing various types of work material

**3.** Set up and use a knurling tool to produce good patterns on work diameters

## FILING IN A LATHE

Filing in a lathe is used to remove burrs, tool marks, and sharp corners. It is not considered good practice to file a diameter to size, for too much filing will tend to produce a diameter which is out-of-round. Whenever filing

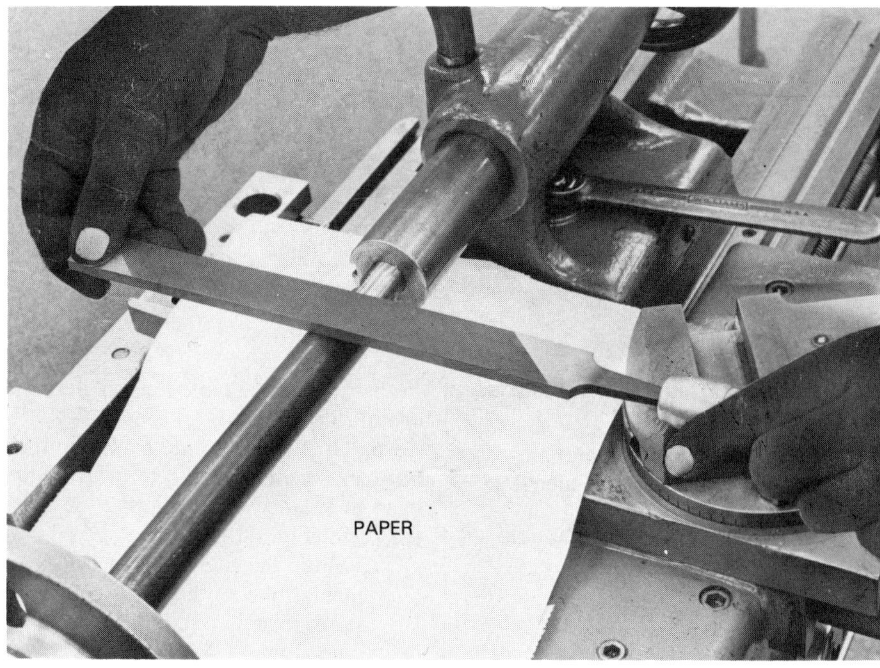


Fig. 41-1 The left-hand method of filing in a lathe is recommended for safety reasons.

is used to finish work to size, no more than .002 to .003 in. (0.05 to 0.07 mm) should be left on the diameter. The National Safety Council recommends filing with the left hand so that arms and hands can be kept clear of the headstock (Fig. 41-1).

When filing or polishing in a lathe, it is good practice to cover the lathe bed with a piece of paper to prevent filings from getting into the slides and causing wear and damage to the lathe. (See N-1.)

N-1 It is dangerous to cover the lathe bed with a cloth since it can become caught in the revolving work.

## Filing Suggestions

Whenever it is necessary to file on a lathe, the *following suggestions should be observed* so that the operation is done safely and efficiently.

**1.** Never use a file without a handle. The file tang is sharp and could cause serious hand injury.

**2.** File left-handed to keep your arms and hands away from the revolving driveplate or chuck.

**3.** Apply just enough pressure on the forward stroke to make the file cut. Pressure on the return stroke causes the file to dull quickly.

**4.** Use long, slow strokes when filing in a lathe. 30 to 40 strokes per minute are recommended. (See N-2.)

N-2 Fast, short strokes will produce a work diameter which is out of round.

**5.** Keep oil off a file or a surface to be filed. Oil clogs the file teeth and prevents them from cutting.

**6.** Clean a file frequently with a file card to keep the teeth clean and free of chips. (See N-3.)

N-3 Prolong the life of a file by cleaning the teeth whenever it is necessary.

**7.** Use a revolving tailstock center when filing in a lathe. If a dead center is used, check the *center tension* occasionally, since the speed of the revolving work causes friction.

## To File Work in a Lathe

**1.** Cover the lathe bed with paper (Fig. 41-1).

**2.** Set the lathe at twice the speed used for turning.

**3.** Adjust the work tension freely between centers. (Use a revolving dead center if one is available.) (Fig. 41-2.)

**4.** Disengage the lead screw by placing the reverse lever in a neutral position.

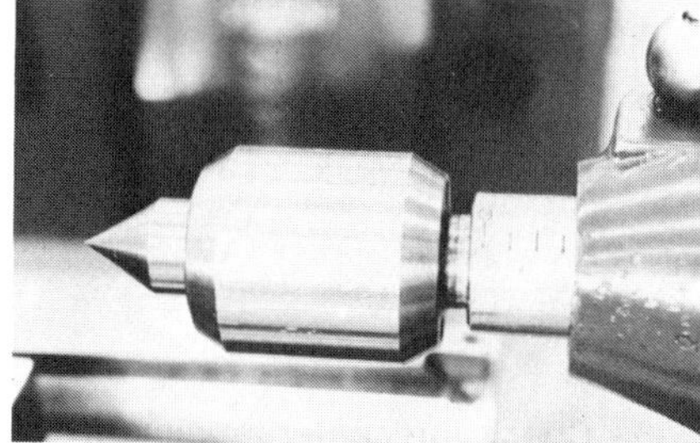

Fig. 41-2 A revolving dead center eliminates friction between the work and lathe center. (*Courtesy Kostel Enterprises Ltd.*)

**Fig. 41-3A** A long-angle lathe file clears chips rapidly. (*Courtesy Nicholson File Co.*)

**Fig. 41-3B** A mill file is used for general-purpose filing. (*Courtesy Nicholson File Co.*)

N-4 Before starting to file, it is wise to observe the following:

1. Remove rings and watches.
2. Roll up coat and shirt sleeves above the elbows.
3. Be sure that there is a handle on the file.

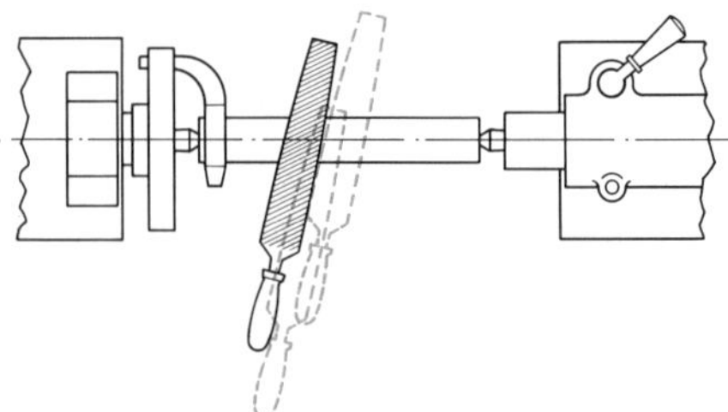

**Fig. 41-4** Each file stroke should overlap the previous stroke by one-half the file width.

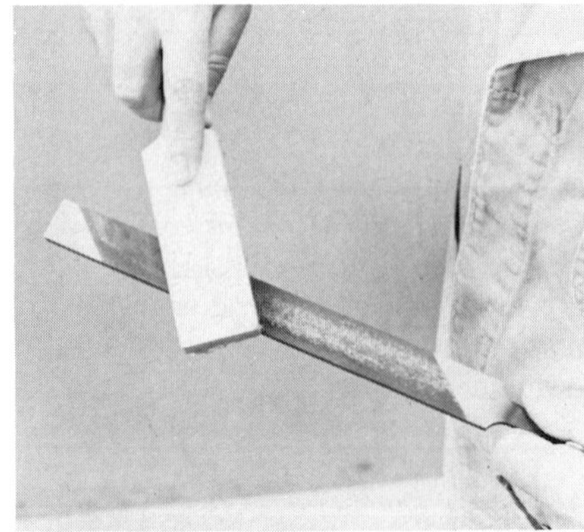

**Fig. 41-5** A file brush will clean the chips from the teeth of a file to prevent scratches on the work.

N-5 Check the center tension frequently because the high speed used for polishing causes the work to heat and expand quickly.

**5.** Select a suitable long angle lathe or mill file (Fig. 41-3). (See N-4.)

**6.** Grip the lathe file handle in the left hand, using the fingers of the right hand to balance and guide the file at the point (Fig. 41-1).

**7.** Move the file along the work after each stroke so that each cut overlaps approximately one-half the width of the file (Fig. 41-4).

**8.** Use long strokes and apply pressure only on the forward stroke.

**9.** Use approximately 30 to 40 strokes per minute.

**10.** If the file loads up with cuttings, clean it with a file brush and rub a little chalk on the file teeth (Fig. 41-5).

## POLISHING

Polishing is a finishing operation which generally follows filing to improve the surface finish on the work. The finish obtained on the diameter is directly related to the coarseness and type of the abrasive cloth used. A fine-grit abrasive cloth produces the best surface finish. Aluminum oxide abrasive cloth should be used for polishing most ferrous metals and silicon carbide abrasive cloth on nonferrous metals.

### To Polish Work in a Lathe

**1.** Be sure that all loose clothing is tucked in to prevent it from becoming caught by the revolving work.

**2.** Cover the lathe bed with paper (Fig. 41-6).

**3.** Set the lathe at a high speed and disengage the lead screw and feed rod.

**4.** Mount work between centers freely with a little end play, or use a rotating dead center. (See N-5.)

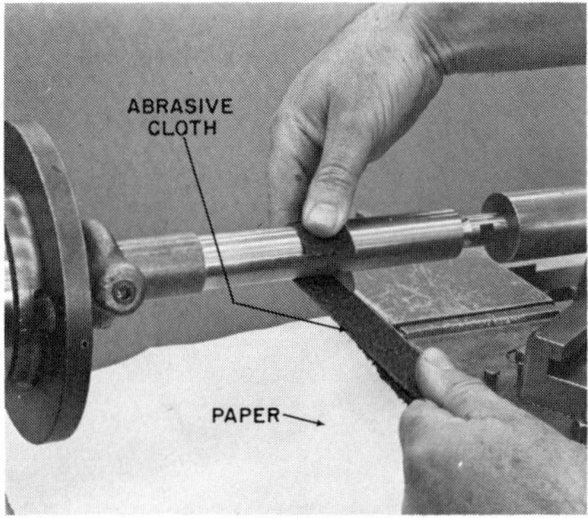


**Fig. 41-6** A high surface finish can be produced with abrasive cloth.

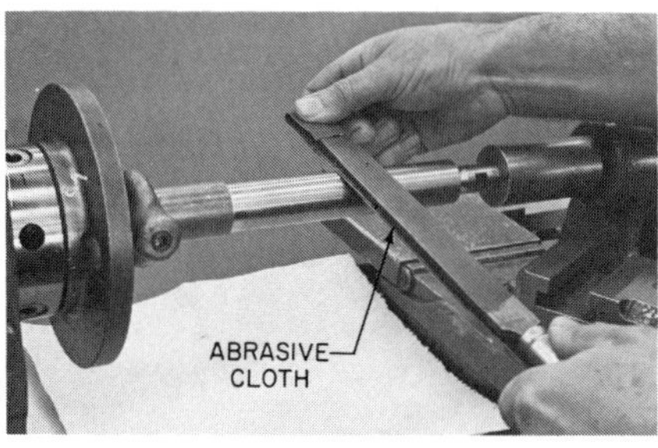


**Fig. 41-7** More pressure can be applied during polishing by holding a piece of abrasive cloth under the file.

**5.** Use a piece of 80 to 100 grit abrasive cloth about 1 in. (25.4 mm) wide for rough polishing.

**6.** Hold the abrasive cloth as shown in Fig. 41-6 to prevent the top end of the abrasive cloth from wrapping around the work and injuring the fingers.

**7.** Hold the long end of the abrasive cloth securely with one hand while the fingers of the other hand press the cloth against the diameter (Fig. 41-6).

**8.** Slowly move the abrasive cloth back and forth along the diameter to be polished.

**9.** Use a piece of 120 to 180 grit abrasive cloth for finish polishing.

**10.** Apply a few drops of oil to the abrasive cloth for the final passes along the diameter to improve the surface finish.

### Polishing with Abrasive Cloth Under a File

Another method of polishing is by holding a piece of abrasive cloth under a file and using long strokes similar to filing (Fig. 41-7). The abrasive cloth should be slightly wider than the file

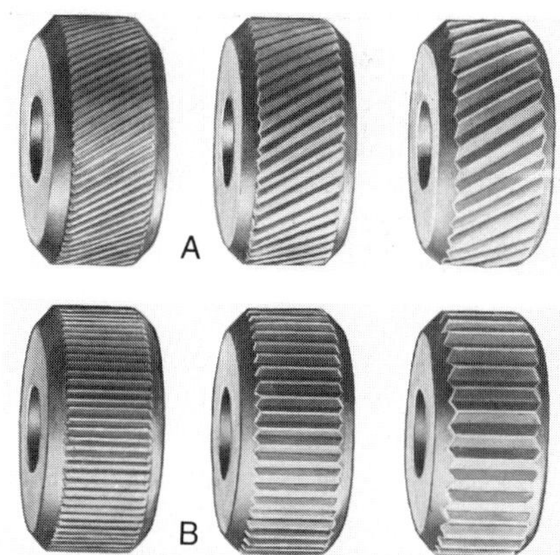


Fig. 41-8 Diamond- and straight-pattern fine, medium, and coarse knurling rolls. (*Courtesy J. H. Williams & Co.*)

N-6 Knurling is a forming operation.

**Fig. 41-9A** A knurling tool with one set of rolls in a self-centering head. (*Courtesy J. H. Williams & Co.*)

**Fig. 41-9B** A knurling tool with three sets of rolls in a revolving head. (*Courtesy J. H. Williams & Co.*)

N-7 Use a heavy-duty revolving tailstock center whenever possible because the center tension cannot be checked while the knurling rolls are in contact with the work.

to ensure that the file does not contact the work surface. This method allows more pressure to be applied during polishing, ensures parallel diameters, and allows polishing to be performed in a specific area.

## KNURLING

Knurling is a process of impressing diamond-shaped or straight indentations on the surface of work. The purposes of knurling are to improve the appearance of the work and to provide a better grip. It is done by forcing a knurling tool containing a set of hardened cylindrical patterned rolls against the surface of revolving work. Diamond and straight pattern rolls in three styles—fine, medium, and coarse—are illustrated in Fig. 41-8. (See N-6.)

Knurling tools have a heat-treated body held in the toolpost and a set of hardened rolls mounted in a movable head. The knurling tool shown in Fig. 41-9A contains one set of rolls mounted in a self-centering head. The knurling tool in Fig. 41-9B contains three sets of rolls (fine, medium, and coarse) mounted in a revolving head which pivots on a hardened steel pin.

### To Knurl a Diamond Pattern

**1.** Mount the work between centers with the required length of knurled section marked on the work. (See N-7.)

**2.** Set the lathe at one-fourth the speed used for turning.

**3.** Set the quick-change gearbox for a feed of .010 to .020 in. (0.25 to 0.50 mm).

**4.** Set the center of the floating head of the knurling tool even with the dead center point of the lathe (Fig. 41-10).

**5.** Adjust the knurling tool so that it is at right angles to the work (Fig. 41-11).

**6.** Tighten the toolpost screw securely so that the knurling tool will not move during the knurling operation.

**7.** Set the knurling tool near the end of the work so that only one-half to three-fourths of the width of the knurling roll is on the work (Fig. 41-12). This generally results in easier starting and a better knurling pattern.

**Fig. 41-10** Aligning the center of the floating head with the lathe center point. (*Courtesy Kostel Enterprises Ltd.*)

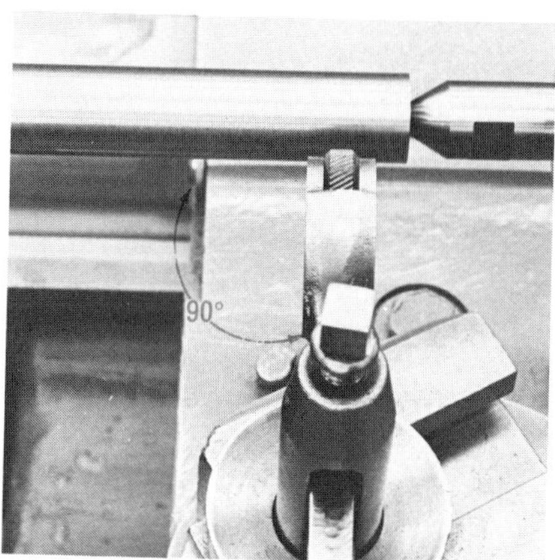


**Fig. 41-11** The knurling tool should be set at 90° or slightly pointing toward the tailstock. (*Courtesy Kostel Enterprises Ltd.*)

**Fig. 41-12** Force the knurling rolls into the work about .025 in. (0.63 mm) to obtain a good knurl pattern.

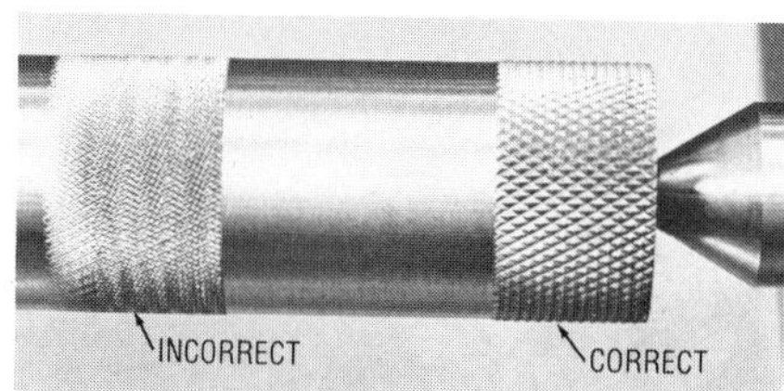


**Fig. 41-13** Correct and incorrect knurling patterns. (*Courtesy Kostel Enterprises Ltd.*)

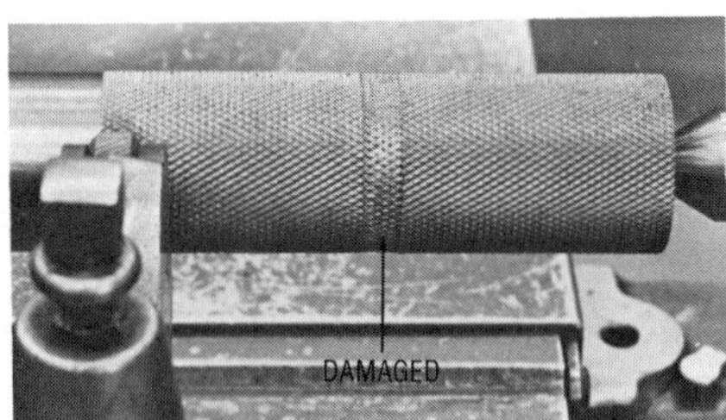


**Fig. 41-14** Disengaging the feed will damage the knurling pattern. (*Courtesy Kostel Enterprises Ltd.*)

**8.** Force the knurling tool into the work approximately .025 in. (0.65 mm) and start the lathe

*or*

Start the lathe and then force the knurling tool into the work until the diamond pattern comes to a point.

**9.** Stop the lathe and examine the pattern. If necessary, reset the knurling tool.

(a) If the pattern is incorrect (Fig. 41-13), it is usually because the knurling tool is not set on center.

(b) If the knurling tool is on center and the pattern is not correct, it is generally due to worn knurling rolls. In this case it will be necessary to set the knurling tool off square slightly so that the corner of the knurling rolls can start the pattern.

**10.** Once the pattern is correct, engage the automatic carriage feed and apply cutting fluid to the knurling rolls. (See N-8.)

N-8 Never disengage the automatic feed until the full length is knurled.

**11.** Knurl to the proper length.

**12.** If the knurling pattern is not to a point, reverse the lathe feed and take another pass across the work.

## REVIEW QUESTIONS

**1.** What is the purpose of filing and polishing work in a lathe?

**2.** Why should the lathe bed be covered with paper before filing and polishing?

**3.** Why is filing left-handed recommended?

**4.** How can a file be kept clean and free of chips?

**5.** Why is a revolving tailstock center recommended when filing and polishing?

**6.** What type of abrasive cloth should be used for polishing

(a) Ferrous metals

(b) Nonferrous metals

**7.** What is the advantage of polishing with an abrasive cloth under a file?

**8.** Name two purposes of knurling.

**9.** Explain how the knurling tool should be set up for knurling.

**10.** How should the knurling tool be started for performing the knurling operation?

**11.** If the knurling pattern is not correct, how can it be corrected?

**12.** Why should the lathe feed not be disengaged during knurling?

# UNIT 42

# TAPERS

A taper may be defined as a uniform increase or decrease in the diameter of a piece of work measured along its length. Tapers in the inch system are expressed in taper per foot or taper per inch. Metric tapers are expressed as a ratio of 1 mm per unit of length; for example 1:20 taper would have a 1-mm change in diameter in 20 mm of length. A taper provides a rapid and accurate method of aligning machine parts and an easy method of holding tools such as twist drills, lathe centers, and reamers. The American Standards Association classifies tapers used on machines and tools as self-holding tapers and self-releasing, or steep tapers. (See N-1.)

N-1 Tapers provide a fast and accurate means of aligning machine parts and holding cutting tools.

## OBJECTIVES

The purpose of this unit is for you to:

**1.** Identify and know the purpose of self-holding and self-releasing inch and metric tapers

**2.** Calculate and cut short tapers with the compound rest

**3.** Calculate and machine tapers by offsetting the tailstock

**4.** Calculate and cut tapers using the taper attachment

*Self-holding tapers* (Fig. 42-1) are those which remain in position due to the wedging action of the taper. The inch-designed tapers of this series are composed of the Morse; Brown and Sharpe; and ¾-in. taper per foot series (Table 42-1). *Steep tapers* (Fig. 42-2), such as those used on milling machine arbors and accessories, are held in the machine by a draw bolt and are driven by lugs or keys.

## INCH TAPERS

Some of the tapers included in Table 42-1 are taken from the Morse, and Brown and Sharpe series. The following describes these and others used in machine shop work.

**1.** *Morse taper,* approximately ⅝-in. taper/ft, is a standard taper used for

Fig. 42-1 The tang of a self-holding taper prevents slippage. (*Courtesy The Cleveland Twist Drill Co.*)

Fig. 42-2 A steep taper with a draw bolt to hold the arbor in the spindle.

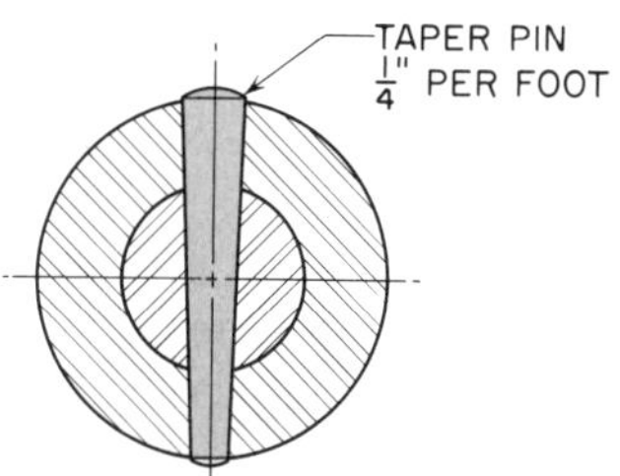


Fig. 42-3 Tapered pins allow machine parts to be assembled quickly and accurately.

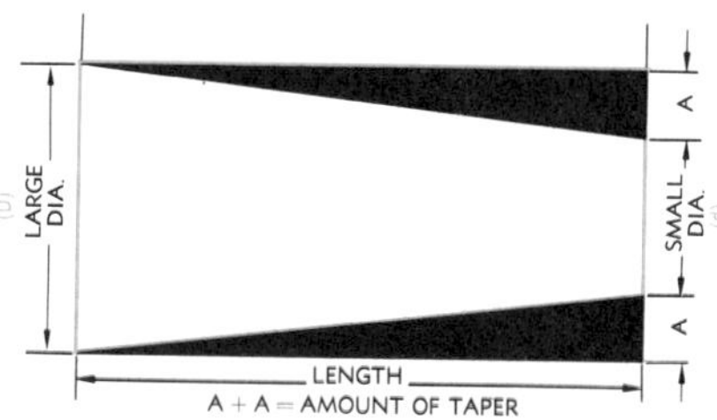


Fig. 42-4 The main parts of an inch taper.

N-2 Inch tapers may be expressed in taper per inch, taper per foot, and degrees.

N-3
$D$ = diameter at the large end of the taper
$d$ = diameter at the small end of the taper
T.L. = total length of the tapered section

**TABLE 42-1 BASIC DIMENSIONS OF SELF-HOLDING TAPERS**

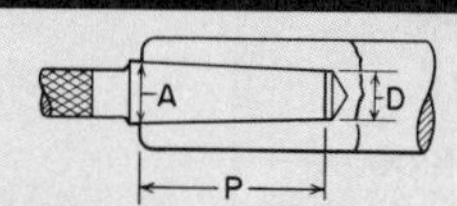


| Number of taper | Taper per foot | Diameter at gage line *A* | Diameter at small end *D* | Length *P* | Series origin |
|---|---|---|---|---|---|
| .239 | .502 | .2392 | .200 | 15/16 | Brown and Sharpe taper series |
| .299 | .502 | .2997 | .250 | 1 3/16 | |
| .375 | .502 | .3752 | .3125 | 1 1/2 | |
| *0 | .624 | .3561 | .252 | 2 | Morse taper series |
| 1 | .5986 | .475 | .369 | 2 1/8 | |
| 2 | .5994 | .700 | .572 | 2 9/16 | |
| 3 | .6023 | .938 | .778 | 3 3/16 | |
| 4 | .6233 | 1.231 | 1.020 | 4 1/16 | |
| 4 1/2 | .624 | 1.500 | 1.266 | 4 1/2 | |
| 5 | .6315 | 1.748 | 1.475 | 5 3/16 | |
| 6 | .6256 | 2.494 | 2.116 | 7 1/4 | |
| 7 | .624 | 3.270 | 2.750 | 10 | |
| 200 | .750 | 2.000 | 1.703 | 4 3/4 | 3/4-in. taper per foot series |
| 250 | .750 | 2.500 | 2.156 | 5 1/2 | |
| 300 | .750 | 3.000 | 2.609 | 6 1/4 | |
| 350 | .750 | 3.500 | 3.063 | 7 | |
| 400 | .750 | 4.000 | 3.516 | 7 3/4 | |
| 450 | .750 | 4.500 | 3.969 | 8 1/2 | |
| 500 | .750 | 5.000 | 4.422 | 9 1/4 | |
| 600 | .750 | 6.000 | 5.328 | 10 3/4 | |
| 800 | .750 | 8.000 | 7.141 | 13 3/4 | |
| 1000 | .750 | 10.000 | 8.953 | 16 3/4 | |
| 1200 | .750 | 12.000 | 10.766 | 19 3/4 | |

*Taper 0 is not a part of the self-holding taper series. It has been added to complete the Morse taper series.

twist drills, reamers, end mills, and lathe center shanks. The Morse taper has eight standard sizes from 0 to 7.

**2.** *Brown and Sharpe taper,* 1/2-in. taper/ft, is a standard taper used in all Brown and Sharpe machines, cutters, and drive shanks.

**3.** *Jarno taper,* .600-in. taper/ft, is used for some machine spindles.

**4.** *Standard taper pin,* 1/4-in. taper/ft, is a standard for all tapered pins used in the fabrication of machinery. They are listed by numbers from 0 to 10 (Fig. 42-3).

**5.** *Standard milling machine taper,* 3 1/2-in. taper/ft. This is a self-releasing taper used exclusively on milling-machine spindles and equipment.

## Inch-Taper Calculations

Most inch tapers are expressed in taper per foot or degrees. If this information is not supplied, it is generally necessary to calculate the taper per foot of the workpiece. Taper per foot is the amount of difference between the large and small diameter of the taper in 12 in. of length. For example, if the tapered section on a piece of work is 12 in. long and the large diameter is 1 in. and the small diameter is 1/2 in., the taper per foot would be the difference between the large and small diameters, or 1/2 in. The main parts of an inch taper are: the amount of taper, the length of the tapered part, the large diameter, and the small diameter (Fig. 42-4). (See N-2.)

Since not all tapers are 12 in. long, if the small diameter, large diameter, and length of the tapered section are known, then the taper per foot can be calculated by applying the following formula (See N-3):

$$\text{taper/ft} = \frac{(D - d) \times 12}{\text{T.L.}}$$

**Example:**

To calculate the taper per foot of the workpiece shown in Fig. 42-5:

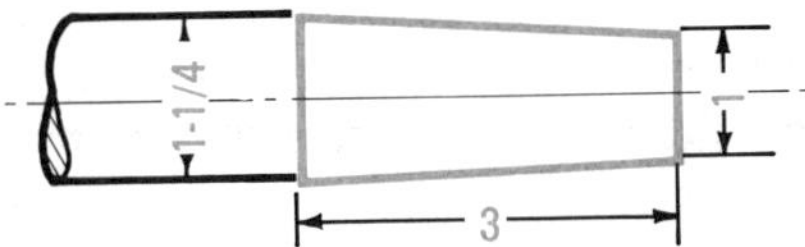

Fig. 42-5 The dimensions of a tapered workpiece.

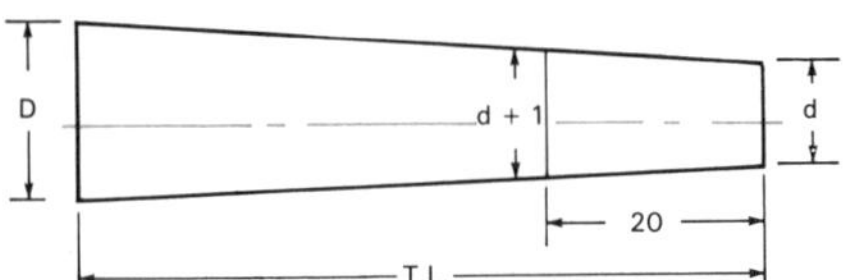

Fig. 42-6 The characteristics of a metric taper.

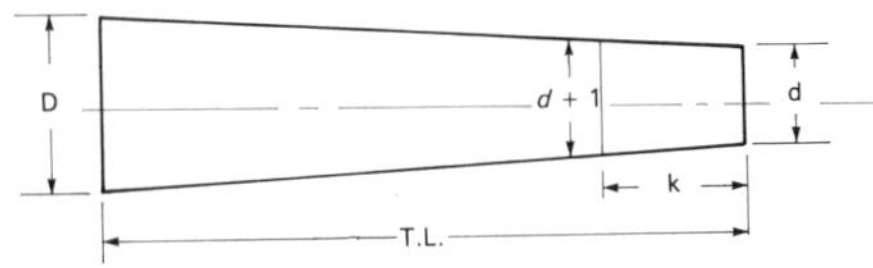

Fig. 42-7 Dimensions of a metric taper.

N-4 Parts of a metric taper:
$D$ = large diameter
$d$ = small diameter
T.L. = total length of taper
O.L. = total length of work
$k$ = taper per unit length

N-5
Taper/ft = taper per foot
O.L. = overall length of work
12 = inches per foot
2 = offset is taken from the centerline of the work

N-6
O.L. = overall length of work
T.L. = length of the tapered section
$D$ = diameter at the large end of the taper
$d$ = diameter at the small end of the taper

$$\text{taper/ft} = \frac{(1^1/_4 - 1) \times 12}{3} = \frac{^1/_4 \times 12}{3} = 1 \text{ in.}$$

If taper per inch is required, divide taper per foot by 12. For example, the 1-in. taper/ft of the previous problem would have .083 taper/in. (1 in. ÷ 12 in.).

After the taper per foot has been calculated, no further calculations are necessary if the taper is to be cut with a taper attachment. If the taper is to be cut by the tailstock offset method, the amount to offset the tailstock must be calculated.

## METRIC TAPERS

Metric tapers are expressed as a ratio of one millimeter per unit of length. In Fig. 42-6 the work would taper 1 mm in a distance of 20 mm. This taper would be expressed as a ratio of 1:20 and would be indicated on a drawing as taper = 1:20.

Since the work tapers 1 mm in 20 mm of length, the diameter at a point 20 mm from the small diameter ($d$) will be 1 mm larger ($d + 1$).

Some common metric tapers are:

| | |
|---|---|
| Milling machine spindle | 1:3.429 |
| Morse taper shank | approximately 1:20 |
| Tapered pins and pipe threads | 1:50 |

### Metric-Taper Calculations

If small diameter $d$, unit length of taper $k$, and total length of taper T.L. are known, large diameter $D$ may be calculated.

In Fig. 42-7, the large diameter $D$ will be equal to the small diameter plus the amount of taper. The amount of taper for the unit length $k$ is $(d + 1) - (d)$, or 1 mm. Therefore the amount of taper per millimeter of unit length = $1/k$.

The total amount of taper will be the taper per millimeter $1/k$ multiplied by the total length of taper (T.L.). (See N-4.)

$$\text{Total taper} = \frac{1}{k} \times \text{T.L.} \quad \text{or} \quad \frac{\text{T.L.}}{k}$$

$$D = d + \text{total amount of taper}$$

$$D = d + \frac{\text{T.L.}}{k}$$

**Example:**

Calculate large-diameter $D$ for a 1:30 taper having a small diameter of 10 mm and a length of 60 mm.

Since taper is 1:30,

$k = 30$

$$D = d + \frac{\text{T.L.}}{k} = 10 + \frac{60}{30} = 10 + 2 = 12 \text{ mm}$$

## TAILSTOCK OFFSETTING

The tailstock offset method is often used to produce tapers in a lathe on work turned between centers when a taper attachment is not available. To produce a taper, the amount of tailstock offset must first be calculated by applying one of the following formulas.

### Inch-Tailstock Offset Calculations

$$\text{Tailstock offset} = \frac{\text{taper/ft} \times \text{O.L.}}{12 \times 2}$$

(See N-5.) To calculate the tailstock offset for a 10-in.-long piece of work which has a ¾-in. taper/ft:

$$\text{Tailstock offset} = \frac{\frac{3}{4} \times 10}{24} = \frac{3}{4} \times \frac{1}{24} \times 10 = \frac{5}{16} \text{ in.}$$

In cases where it is not necessary to find the taper per foot, a simplified formula can be used to calculate the amount of tailstock offset (see N-6):

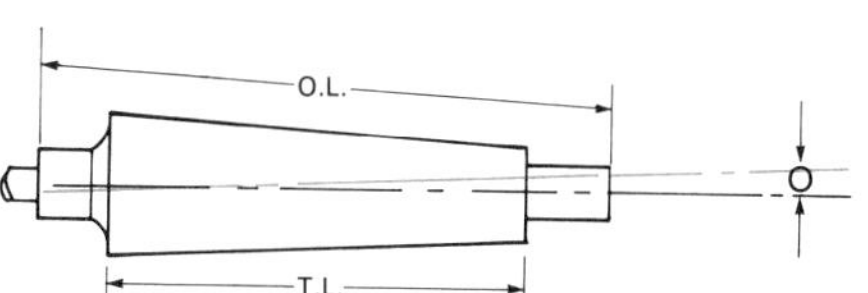


Fig. 42-8 Metric taper turning by the tailstock offset method.

N-7 In order to calculate the tailstock offset for metric tapers, the large-end diameter must be known.

$$\text{Tailstock offset} = \frac{\text{O.L.}}{\text{T.L.}} \times \frac{(D - d)}{2}$$

**Example:**

Using the simplified formula to find the tailstock offset for the following piece of work: large diameter is 1 in., small diameter is 23⁄32 in., the length of the taper is 6 in., and the overall length of the work is 18 in.

$$\text{Tailstock offset} = \frac{18}{6} \times \frac{\left(1 - \frac{23}{32}\right)}{2}$$

$$= \frac{18}{6} \times \frac{9}{64}$$

$$= \frac{27}{64} \text{ in.}$$

## Metric-Tailstock-Offset Calculations

If the taper is to be turned by offsetting the tailstock, the amount of offset is calculated as follows. (See N-7.) See Fig. 42-8.

$$\text{Tailstock offset} = \frac{(D - d)}{2 \times \text{T.L.}} \times \text{O.L.}$$

where $D$ = large diameter
$d$ = small diameter
T.L. = length of taper
O.L. = total length of work

**Example:**

Calculate the tailstock offset required to turn a 1:30 taper 60 mm long on a workpiece 300 mm long. The small diameter of the tapered section is 20 mm.

Large diameter of taper $D$ $= d + \frac{\text{T.L.}}{k}$

$$= 20 + \frac{60}{30}$$

$$= 20 + 2$$

$$= 22 \text{ mm}$$

$$\text{Tailstock offset} = \frac{(D - d)}{2 \times \text{T.L.}} \times \text{O.L.}$$

$$= \frac{(22 - 20)}{2 \times 60} \times 300$$

$$= \frac{2}{120} \times 300$$

$$= 5 \text{ mm}$$

## Taper Attachments

Turning a taper using the taper attachment provides many advantages in producing both internal and external tapers. The most important are:

**1.** Setup is simple. The taper attachment is easy to connect and disconnect.

**2.** Live and dead centers are not adjusted, so center alignment is not disturbed.

**3.** Greater accuracy can be achieved since one end of the guide bar is graduated in degrees and the other end in inches of taper per foot or in a ratio of 1 mm per unit of length.

**4.** The taper can be produced between centers or on projecting work from any holding device such as a chuck or a collet, regardless of the length of the work.

**5.** Internal tapers can be produced with the same taper setup as for external tapers.

**6.** A great range of tapers can be produced, and this is of special advantage when production is a factor and various tapers are required on a unit.

There are two common types of taper attachments in use:

**1.** The plain taper attachment (Fig. 42-9).

**2.** The telescopic taper attachment (Fig. 42-10).

To use the plain taper attachment, the crossfeed screw nut must be disengaged from the cross slide. When a telescopic taper attachment is used, the crossfeed screw is not disengaged, and the depth of cut can be set by the crossfeed handle.

## Inch-Taper-Attachment Offset Calculations

Most tapers cut on a lathe with the taper attachment are expressed in taper per foot. If the taper per foot of the taper on the workpiece is not given, it may be calculated by using the following formula:

$$\text{Taper/ft} = \frac{(D - d) \times 12}{\text{T.L.}}$$

**Example:**

Calculate the taper per foot for a taper with the following dimensions: large-diameter $D$, 1⅜ in.; small-di-

**Fig. 42-9** A plain taper attachment. (*Courtesy Kostel Enterprises Ltd.*)

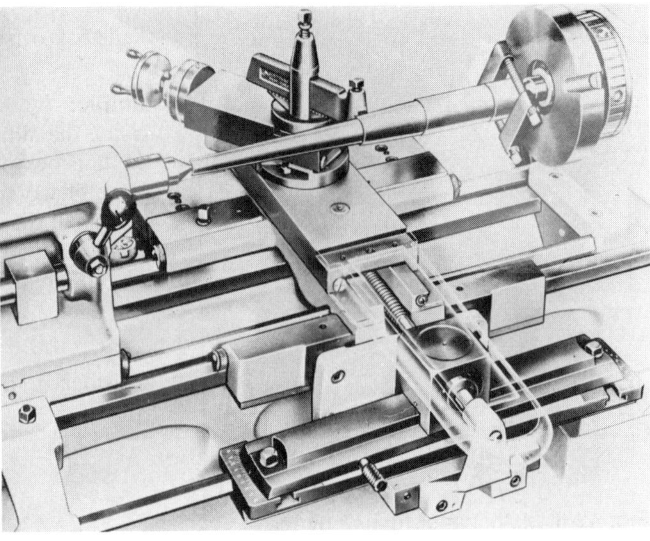

**Fig. 42-10** A telescopic taper attachment.

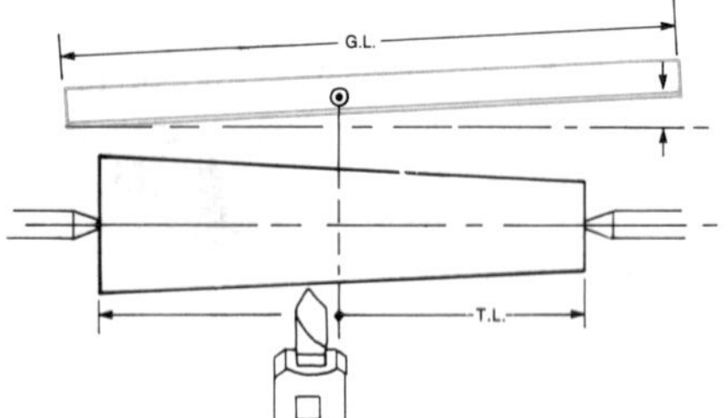


**Fig. 42-11** Principle of a taper attachment.

ameter $d$, $^{15}\!/_{16}$ in.; length of tapered section (T.L.), 7 in.

$$\text{Taper/ft} = \frac{\left(1\frac{3}{8} - \frac{15}{16}\right) \times 12}{7}$$

$$= \frac{\frac{7}{16} \times 12}{7}$$

$$= \frac{3}{4} \text{ in.}$$

## Metric-Taper Attachment Offset Calculations

When the taper attachment is used to turn a taper, the amount the guide bar is set over may be determined as follows:

**1.** If the angle of taper is given on the drawing, set the guide bar to one-half the included angle (Fig. 42-11).

**2.** If the angle of taper is not given on the drawing, use the following formula to find the amount of guide bar setover.

$$\text{Guide bar setover} = \frac{(D - d)}{2} \times \frac{\text{G.L.}}{\text{T.L.}}$$

where $D$ = large diameter of taper

$d$ = small diameter of taper

T.L. = length of taper

G.L. = length of taper attachment guide bar

**Example:**

Calculate the amount of setover for a 500-mm-long guide bar to turn a 1:50 × 250-mm-long taper on a workpiece. The small diameter of the taper is 25 mm.

$$\text{Large diameter of taper} = d + \frac{\text{T.L.}}{k}$$

$$= 25 + \frac{250}{50}$$

$$= 30 \text{ mm}$$

$$\text{Guide bar setover} = \frac{(D - d)}{2} \times \frac{\text{G.L.}}{\text{T.L.}}$$

$$= \frac{(30 - 25)}{2} \times \frac{500}{250}$$

$$= \frac{5}{2} \times 2$$

$$= 5 \text{ mm}$$

# TAPER TURNING

Tapers can be cut on a lathe by using the taper attachment, offsetting the tailstock, and setting the compound rest to the angle of the taper.

## To Cut a Taper Using a Taper Attachment

The procedure for machining a taper using either a plain or telescopic taper attachment, on work mounted between centers or in a chuck, is basically the same with only minor adjustments required. The procedure for setting the

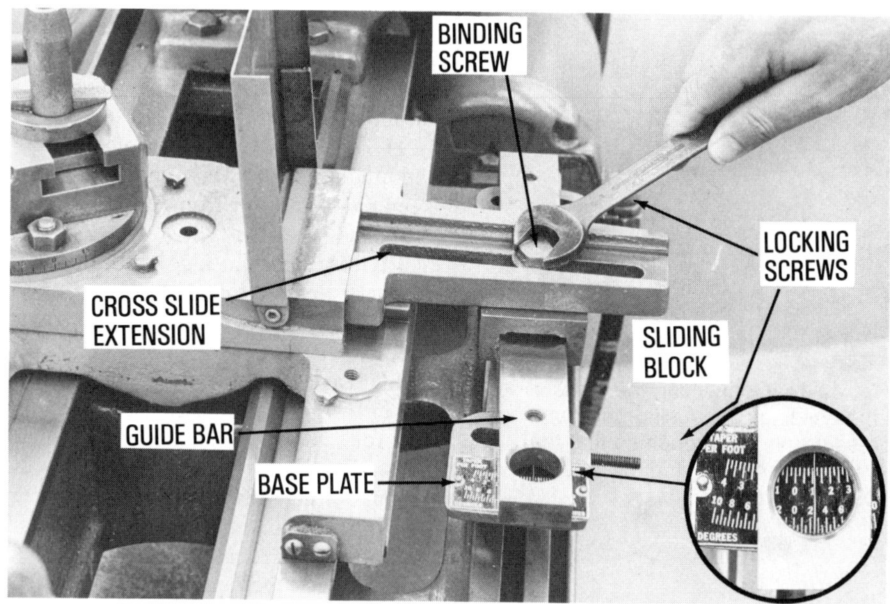


**Fig. 42-12** Setting the taper attachment to the required taper. (*Courtesy Kostel Enterprises Ltd.*)

plain taper attachment and cutting a taper are outlined as follows:

**1.** Clean and oil the *guide bar*.

**2.** Loosen the guide-bar lock nuts so that it is free to move on the *base plate*.

**3.** By adjusting the *locking screws*, offset the end of the guide bar the required amount; or for inch tapers, set the taper attachment to the required taper per foot (Fig. 42-12).

**4.** Tighten the guide-bar lock nuts.

**5.** Swivel the compound rest so that it is at about 30° to the cross-slide.

**6.** Set the cutting tool to center and tighten the toolpost securely.

**7.** Mount the work in the lathe and mark the length to be tapered. (See N-8.)

N-8 A light groove cut into the work to indicate the length to be cut will be visible when the work is revolving.

**8.** Feed the cutting tool in until it is about ¼ in. (6.35 mm) from the diameter of the work.

**9.** Remove the *binding screw* which connects the cross slide and the crossfeed screw nut (Fig. 42-13).

**Fig. 42-13** Removing the binding screw which connects the cross slide to the crossfeed screw nut. (*Courtesy Kostel Enterprises Ltd.*)

**Fig. 42-14** Connecting the cross slide to the sliding block using only two-finger pressure on the wrench so that the threads will not be stripped. (*Courtesy Kostel Enterprises Ltd.*)

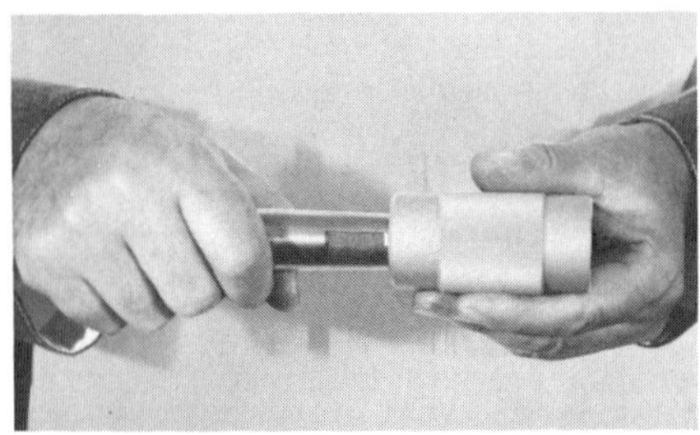

**Fig. 42-15** Using a taper ring gage to check the accuracy of a taper.

N-9 Before starting a new cut, move the carriage until the cutting tool clears the end of the work by ½ in. (12.7 mm) to remove the play (backlash) in the taper attachment.

N-10 When using the plain taper attachment, the depth of cut must be set by the compound rest handle.

N-11 $T.O. = \frac{tpf \times O.L.}{24}$

**10.** Use the binding screw to connect the cross-slide extension to the sliding block using two-finger pressure on the wrench (Fig. 42-14).

**11.** Insert a plug in the hole where the binding screw was removed to keep chips and dirt from damaging the crossfeed screw.

**12.** Move the carriage until the cutting tool clears the right-hand end of the work by about ½ in. (12.7 mm). (See N-9.)

**13.** Take a light trial cut for about ¹⁄₁₆ in. (1.58 mm) and check the diameter for size.

**14.** Set the depth of the roughing cut [about ¹⁄₁₆ in. (1.58 mm) larger than the finish size] and rough cut the taper to the required length. (See N-10.)

**15.** Check the taper for fit (Fig. 42-15). See page 215 on fitting a taper to a gage.

**16.** Readjust the taper attachment setting if necessary and take a light trial cut from the taper.

**17.** When the taper fit is correct, cut the taper to size.

## Tailstock Offset Method

The tailstock offset method of cutting tapers is generally used when a lathe is not equipped with a taper attachment and the work is mounted between centers. The tailstock center must be moved out of line with the headstock center enough to produce the desired taper (Fig. 42-16). Since the tailstock can be offset only a certain amount, the range of tapers which can be cut is limited.

### To Offset the Tailstock

**1.** Calculate the amount the tailstock must be offset to cut the desired taper on the work. (See N-11.)

**2.** Loosen the tailstock clamp nut.

**3.** Loosen one tailstock adjusting screw and tighten the opposite one until the tailstock offset is correct (Fig. 42-17).

**4.** Tighten the adjusting screw that was loosened and recheck the offset with a rule.

**5.** Correct the setting if necessary and then tighten the tailstock clamp nut.

**6.** Mount the work between centers and cut the taper to size by following operations 12 to 15 of "To Cut a Taper Using a Taper Attachment" (p. 212).

## Compound Rest Method

The compound rest is used to cut short, steep tapers, that are given in degrees, on work mounted in a chuck or between centers. The compound rest must be set to the required angle and then the cutting tool is advanced along the taper using the compound rest feed handle.

### To Cut a Taper Using the Compound Rest

**1.** Check the blueprint for the angle of the taper in degrees.

**2.** Loosen the compound rest lock nuts.

**3.** Swivel the compound rest to the required angle (Fig. 42-18).

(a) If the included angle is given as in Fig. 42-18A, set the compound rest to one-half the included angle.

(b) If the angle is given on one side only, as in Fig. 42-18B, set the compound rest to that angle.

**4.** Tighten the compound rest lock nuts using only two-finger pressure on the wrench to avoid stripping the thread on the compound rest studs (Fig. 42-19).

**5.** Set the toolbit on center and then swivel the toolholder so that it is at 90° to the compound rest (Fig. 42-20).

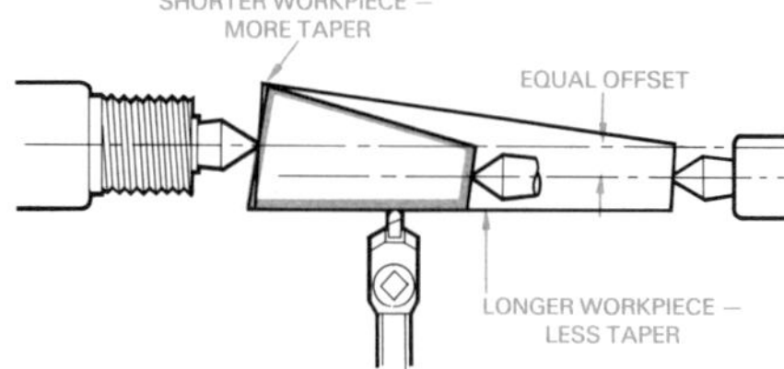


**Fig. 42-16** The amount of taper cut when using the tailstock offset method varies with the length of the workpiece.

**Fig. 42-17** Offsetting the tailstock for taper turning with the adjusting screws. (*Courtesy Kostel Enterprises Ltd.*)

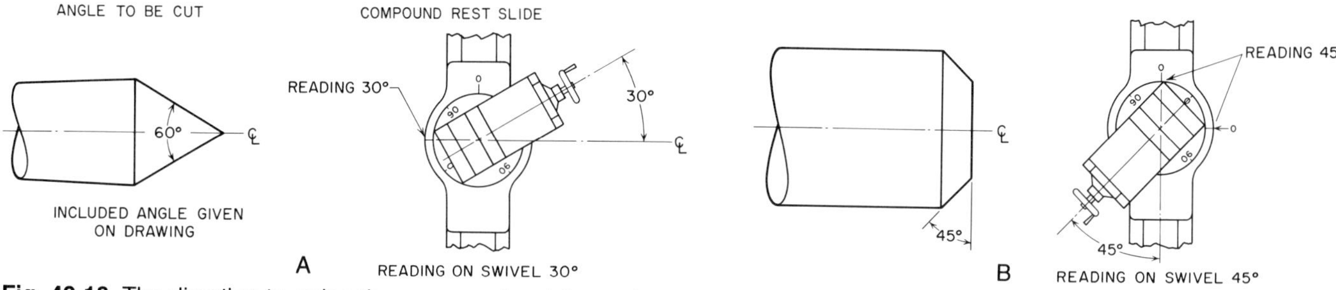


**Fig. 42-18** The direction to swing the compound rest for cutting various angles.

**Fig. 42-19** When tightening the compound rest lock nuts, use only two-finger pressure on the wrench. (*Courtesy Kostel Enterprises Ltd.*)

**Fig. 42-20** Cutting a taper by feeding the compound rest. (*Courtesy Kostel Enterprises Ltd.*)

N-12 Taper/in. $= \frac{\text{tpf}}{12}$

**6.** Bring the toolbit close to the diameter to be cut using the carriage handwheel and crossfeed handle.

**7.** Cut the taper by turning the *compound rest feed screw.*

**8.** Check the taper for size and angle.

## CHECKING A TAPER

External tapers can be checked for accuracy of size or fit by using a taper ring gage, a standard micrometer, or a special taper micrometer.

### To Check a Taper with a Ring Gage

**1.** Draw three equally spaced light lines with chalk or mechanics' blue along the length of the taper (Fig. 42-15).

**2.** Insert the taper into the gage and turn counterclockwise one-half turn, then remove it for inspection.

**3.** If the chalk is rubbed from the whole length of the taper, the taper is correct.

**4.** If the chalk lines are rubbed from only one end, the taper is incorrect.

**5.** By making slight adjustments to the taper attachment and taking trial cuts, machine the taper until the fit is correct.

### To Check an Inch Taper with a Standard Micrometer

**1.** Calculate the amount of taper per inch of the taper. (See N-12.)

**2.** Clean the tapered section of the work and apply layout dye as shown in Fig. 42-21.

**3.** Lay out two lines exactly 1 in. apart (Fig. 42-21).

**4.** Measure the taper with a micrometer at both lines so that the left edge of the micrometer anvil and spindle just touch the line.

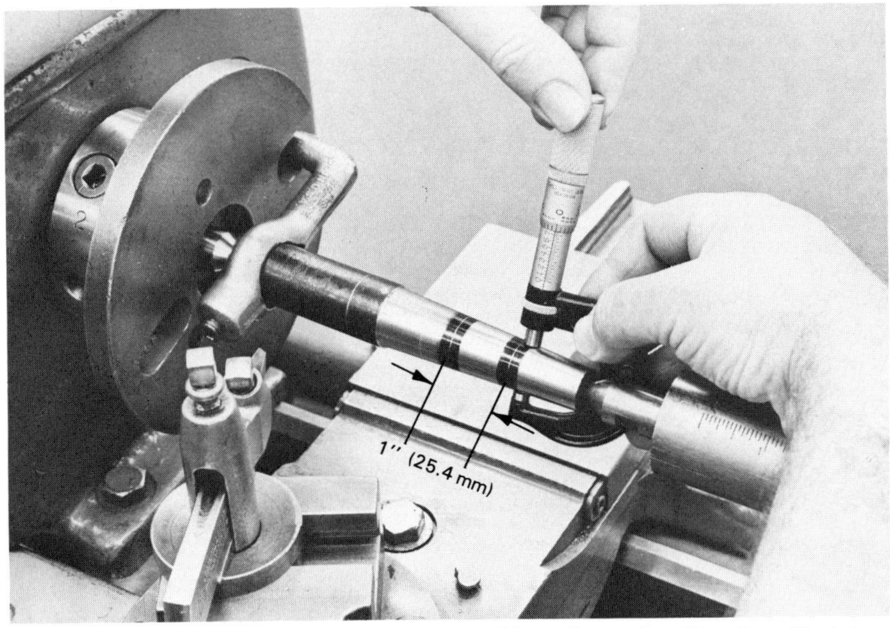


Fig. 42-21 Checking the accuracy of a taper with a micrometer. (*Courtesy Kostel Enterprises Ltd.*)

**5.** Subtract the difference between the two readings and compare the answer with the required taper per inch.

**6.** If necessary, adjust the taper attachment setting to correct the taper.

### To Check a Metric Taper with a Metric Micrometer

**1.** Check the drawing for the taper required.

**2.** Clean the tapered section of the work and apply layout dye.

**3.** Lay out two lines on the taper which are the same distance apart as the second number in the taper ratio. Example: If the taper is 1:20, the lines would be 20 mm apart.

**4.** Measure the diameters carefully with a metric micrometer at the two lines. The difference between these two diameters should be 1 mm for each unit of length. (See N-13.)

**5.** If necessary, adjust the taper attachment setting to correct the taper.

N-13 If the work is long enough, lay out the lines at double or triple the length of the tapered section and increase the difference in diameters by the appropriate amount. For instance, on a 1:20 taper, the lines may be laid out 60 mm apart or three times the unit length of the taper. Therefore the difference in diameters would then be 3 × 1, or 3 mm. This will give a more accurate check of the taper.

## REVIEW QUESTIONS

**1.** Define a taper and state its purpose.

**2.** Name three ways in which inch tapers are expressed.

**3.** How are metric tapers expressed?

**4.** How much taper per foot do the following have?

(a) Morse tapers

(b) Brown and Sharpe tapers

(c) Jarno tapers

(d) Standard taper pin

**5.** Calculate the taper per foot for the following workpiece: large diameter: 1.625; small diameter: 1.425; taper length: 3 in.

**6.** How is a taper which increases 3 mm in 120 mm of length expressed on a drawing?

**7.** Calculate the large diameter for a 1:50 taper having a small diameter of 6 mm and a length of 75 mm.

**8.** Calculate the tailstock offset required to cut .625 taper per foot on a piece of work 15 in. long.

**9.** Calculate the tailstock offset required to cut a 1:40 taper 120 mm long on a workpiece 240 mm long. The small diameter of the taper is 50 mm.

**10.** List six advantages of cutting tapers with a taper attachment.

**11.** Calculate the amount of setover required for a 500-mm-long guide bar to cut a 1:30 × 180-mm-long taper on a workpiece. The small diameter of the taper is 30 mm.

**12.** At what angle should the compound rest be set when using a plain taper attachment?

**13.** What types of tapers are generally cut with the compound rest?

**14.** How should tapers be checked with a taper ring gage?

**15.** List the steps required to check a taper with a standard micrometer.

# UNIT 43

# GROOVING AND FORM TURNING

Operations such as grooving and form turning can be performed on work mounted between centers or work held in a chuck. Grooves are used to provide a relief at the end of a thread or a seat for snap or O-rings. Form turning is used to produce a concave or convex form on the external or internal surface of the workpiece.

## OBJECTIVES

Upon completion of this unit you should be able to:

1. Recognize and know the purpose of three common grooves
2. Cut various grooves in work held in a chuck or between centers
3. Machine concave and convex forms freehand

## GROOVING

*Grooving,* sometimes called *necking, recessing,* or *undercutting,* is the process of cutting a grooved form on a cylinder. The shape of the cutting tool and the depth to which it is fed determines the shape of the groove. Square, round, or V-shaped grooves are the most common in machine shop work.

## TYPES OF GROOVES

*Square grooves* (Fig. 43-1A) are cut at the end of the section to be threaded in order to provide a channel into which the threading tool may run. If the groove is cut against a shoulder, it allows the mating part to fit squarely against the shoulder. When a diameter is to be finished to size by grinding, a groove is generally cut against the shoulder to provide clearance for the grinding wheel and to ensure a square corner. Square grooves may be cut with a cutoff (parting) tool or a toolbit (Fig. 43-1A) ground to the size required.

*Round grooves* (Fig. 43-1B) serve the same purpose as square grooves and are generally used on parts subject to strain. The round groove eliminates the sharpness of a square corner and strengthens the part at the point where it may fracture. A round-nose (radius) toolbit (Fig. 43-1B), ground to the desired radius, is used to cut round grooves.

*V-shaped grooves* (Fig. 43-1C) are most commonly found on pulleys driven by V-belts. The V-shaped groove eliminates much of the slippage which occurs in other belt drives. The V-groove can also be cut at the end of a thread to provide a channel into which the threading tool may run. A toolbit ground to the desired angle (Fig. 43-1C) can be used to cut shallow V-grooves. Larger V-grooves, such as those found on pulleys, should be cut with the lathe compound rest set to form each side individually.

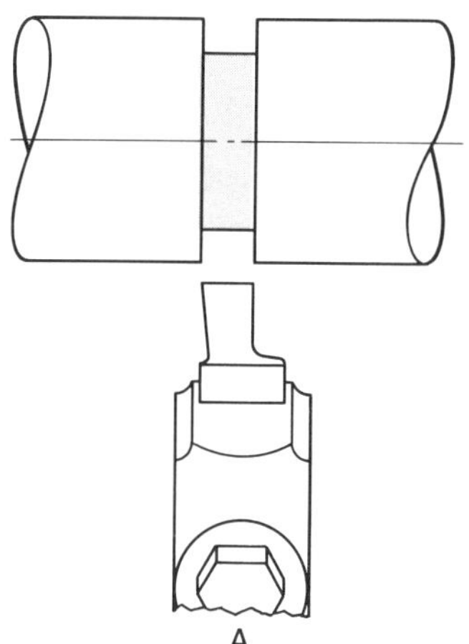


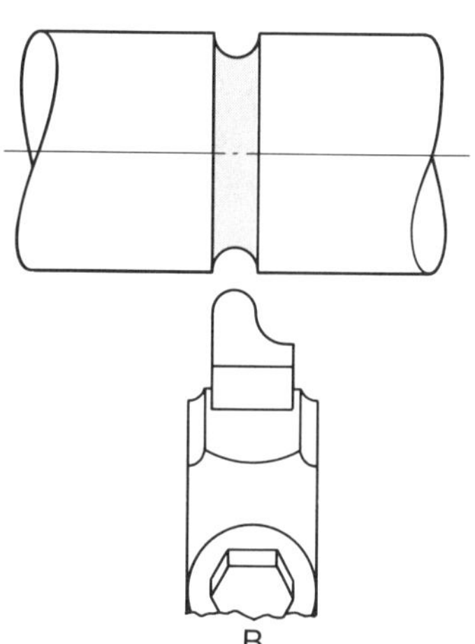


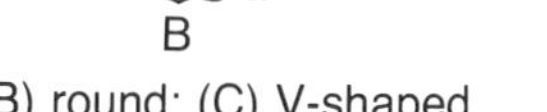

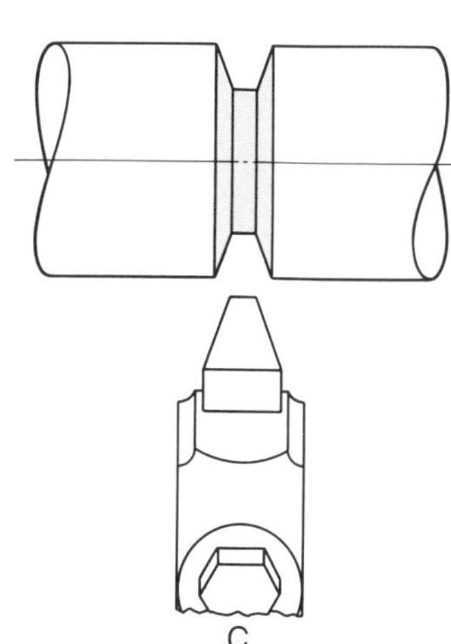


**Fig. 43-1** Three common types of grooves: (A) Square; (B) round; (C) V-shaped.

Fig. 43-2 Use care when grooving between centers to avoid bending the work.

Fig. 43-3 Setting the point of a grooving tool even with the lathe center point. (*Courtesy Kostel Enterprises Ltd.*)

N-1 Use a steady rest to support slender work when grooving between centers.

N-2 Use a steady feed when grooving.
1. *Too-fast feed* will cause the tool to jam or bend the work.
2. *Too-slow feed* will cause the tool to rub and chatter.

## Cutting a Groove

The operation of grooving reduces the diameter and tends to weaken the workpiece at that point; care should therefore be exercised during the grooving operation. This is especially true with work held between centers, since the workpiece is weakened and may bend at the groove, ruining the workpiece and damaging the cutting tool (Fig. 43-2). (See N-1.)

Since the shape of the groove is generally governed by the cutting tool shape, the cutting of various grooves is similar and only a general operation of grooving is outlined.

### To Cut a Groove

1. Lay out the location of the groove, using a center punch and layout tools.
2. Set the lathe at one-half the turning speed.
3. Mount the proper-shaped toolbit in the toolholder and set the cutting tool to center (Fig. 43-3).
4. Locate the toolbit on the work at the position where the groove is to be cut.
5. Start the lathe and feed the cutting tool toward the work using the crossfeed handle until the toolbit lightly marks the work (Fig. 43-4).
6. Hold the crossfeed handle in position and then set the graduated collar to zero (Fig. 43-4).
7. Calculate how far the crossfeed screw must be turned to cut the groove to the proper depth.
8. Apply cutting fluid frequently and groove the work to the proper depth. (See N-2.)
9. Stop the lathe and check the depth of the groove with outside calipers.
10. It is best to move the carriage by hand a little to the right and left while grooving to overcome chatter.

## FREEHAND FORM TURNING

Freehand turning of curved forms involves the operation of the carriage and crossfeed controls to produce the desired shape on the workpiece. Freehand turning is generally used when it is uneconomical to make a template and set up a tracer attachment if only a few parts are required. The beginning lathe operator generally finds this a difficult operation since it involves the coordination of both hands. However, by practicing and following the basic steps, an operator can soon produce curved forms with reasonable accuracy.

Fig. 43-4 Feed the parting tool until it marks the work and then set the crossfeed screw collar to zero. (*Courtesy Kostel Enterprises Ltd.*)

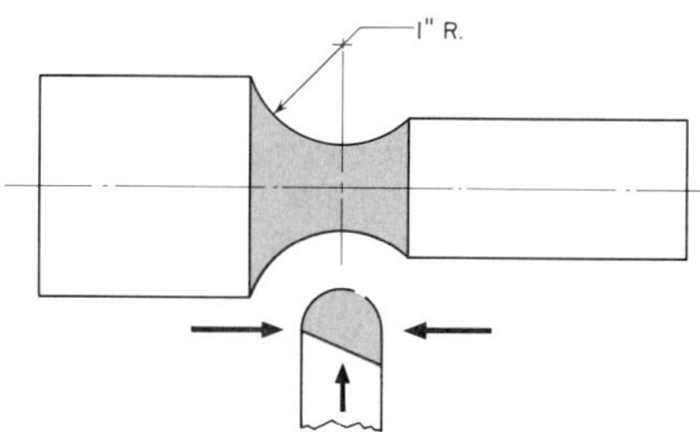


**Fig. 43-5** Machining a concave form on a round workpiece.

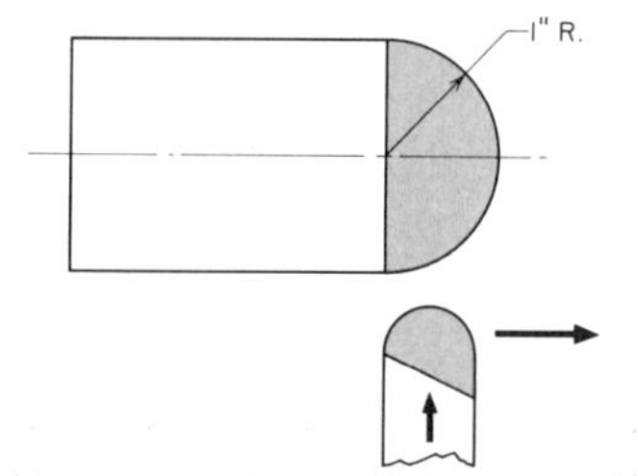


**Fig. 43-6** Machining a convex form on the end of a workpiece.

N-3 If chatter occurs during form turning, reduce the lathe speed until the chatter stops.

N-4 To eliminate chatter, move the lathe carriage *slightly* to the left and right (longitudinally) while feeding the toolbit.

N-5 Take light cuts until the proper coordination between the carriage and crossfeed is attained.

N-6 If a radius gage is not available, use the circumference of a piece of steel having the correct diameter to check the accuracy of the radius. For example: A 2-in. (50.8-mm) -diameter equals a 1-in. (25.4-mm) -radius.

N-7 If a 1-in. (25.4-mm) -radius gage is not available, drill or bore a 2-in. (50.8-mm) -hole in a thin piece of metal and then cut the piece through the center. This produces two 1-in. (25.4-mm) -radius gages.

## Form-Turning Hints

**1.** The lathe speed should be approximately one-half the turning speed. This is necessary since so much of the toolbit cutting edge is in contact with the workpiece at one time. (See N-3.)

**2.** A gooseneck tool is recommended when turning a large radius since it tends to reduce chatter and produce a smoother surface.

**3.** Apply cutting fluid to produce a good cutting action and eliminate chatter.

**4.** Feed the toolbit into the work slowly while applying cutting fluid. (See N-4.)

**5.** Form turning tends to cause the work to spring, so support the workpiece with a steady rest wherever possible.

**6.** Abrasive cloth should be used to improve the surface finish of a contour form. To avoid altering the shape of the contour, filing should be kept to a minimum.

## Turning a Concave Form Freehand

**1.** Mount the workpiece on the lathe.

**2.** With a sharp, pointed toolbit, cut a light groove at each end and also in the center of the concave form.

**3.** Mount a round-nose (radius) toolbit on center.

**4.** Set the lathe speed to approximately one-half the turning speed.

**5.** Move the carriage to bring the edge of the radius toolbit close to the start of the concave form.

**6.** Place one hand on the carriage handwheel *(not the handle)* and the other hand on the crossfeed handle.

**7.** Slowly turn the carriage handwheel, feeding the toolbit toward the center of the form, while the other hand turns the crossfeed handle clockwise, moving the tool into the work.

**8.** Take successive cuts from each side (Fig. 43-5), always starting at the large end of the form and working toward the center. (See N-5.)

**9.** Check the small diameter of the form with a caliper and the accuracy of the form with a radius gage. (See N-6.)

**10.** Cut the form as close to the size and contour as possible.

**11.** Use a half-round file to remove the machine marks and bring the radius to the required shape.

## Turning a Convex Radius Freehand

The same procedure used to machine a concave form can basically be used to produce a convex form. The following steps outline the procedure for machining the 1-in. (25.4-mm) radius on the work illustrated in Fig. 43-6.

**1.** With a pointed toolbit, cut a light groove 1 in. (25.4 mm) from the end.

**2.** Using a round-nose toolbit, take cuts starting at the 1-in. (25.4-mm) line and feed toward the end of the work (Fig. 43-6).

**3.** Take successive cuts, *always* starting at the outside diameter and cutting toward the end.

**4.** Check the accuracy of the form with a radius gage. (See N-7.)

**5.** Machine the radius as close to size and shape as possible.

**6.** With a file, remove the machine marks and finish the form.

# HYDRAULIC TRACER ATTACHMENT

When many duplicate parts, having several radii or contours which may be difficult to produce, are required, they may be easily made on a hydraulic tracer lathe or on a lathe equipped with a hydraulic tracer attachment (Fig. 43-7).

Hydraulic tracer lathes incorporate a means of moving the cross slide by controlled oil pressure. A flat template of the desired contour of the finished piece, or a circular template identical to the finished piece, is mounted in an attachment on the lathe. Automatic control of the tool slide, and duplication of the part, is achieved by a stylus which bears against the template surface. As the carriage is fed along automatically, the stylus follows the contour of the template. The stylus arm actuates a control valve regulating the flow of oil into a cylinder incorporated in the tool slide base. A piston connected to the tool slide is moved in or out by the flow of oil to the cylinder. This movement causes the tool slide

**Fig. 43-7** Machining an intricate form with a hydraulic tracer attachment mounted on the cross slide of the lathe.

(and toolbit) to move in or out as the carriage moves along, duplicating the profile of the template on the workpiece.

### Advantages of a Tracer Attachment

**1.** Intricate forms, difficult to produce by other means, can be readily produced.
**2.** Various forms, tapers, and shoulders can be produced in one cut.
**3.** Duplicate parts can be produced rapidly and accurately.
**4.** Accuracy and finish of the part does not depend on the skill of the operator.

## REVIEW QUESTIONS

**1.** Name and state the purpose of three common types of grooves.
**2.** How are larger V-grooves cut?
**3.** Why should care be exercised when grooving work mounted between centers?
**4.** What effect do the following have when grooving:
(a) Too-fast feed
(b) Too-slow feed
**5.** How can chatter be overcome while grooving?
**6.** When is freehand form turning generally used?
**7.** Why should cutting fluid be used when form turning?
**8.** In which direction should cuts be taken when machining a concave form?
**9.** If a radius gage of the correct size is not available, how can the following be checked?
(a) Concave radius
(b) Convex radius
**10.** List four advantages of a hydraulic tracer attachment for producing intricate forms and contours.

# UNIT 44

# THREADS AND THREAD CUTTING

Threads have been used throughout history as fastening devices to hold parts together. The early Romans used screws, which were filed by hand, in wine presses and parts of suits of armor were held together with screws in the fifteenth century. Over the years, the art of producing threads has continually improved and today threads are mass-produced to exact specifications by taps, die chasers, thread milling, thread rolling, and grinding. However, many threads are still cut on an engine lathe, especially if a special thread form or size is required.

## OBJECTIVES

Upon completion of this unit, you should be able to:

**1.** Recognize common thread forms and know the purpose of each
**2.** Calculate the thread dimensions necessary to cut threads
**3.** Set up a lathe and cut external unified threads
**4.** Set up an inch lathe to cut metric threads
**5.** Set up a lathe and cut an external Acme thread
**6.** Set up a lathe and cut internal threads

## STANDARD THREAD FORMS

A thread is a helical ridge of uniform section formed on the inside or outside of a cylinder or a cone. Threads are used for several purposes.

**1.** For fastening devices such as screws, bolts, studs, and nuts.
**2.** To provide accurate measurement as in a micrometer.
**3.** To transmit motion. The threaded lead screw on the lathe causes the carriage to move along when threading.
**4.** To increase torque. Heavy work can be raised with a screw jack.

Some of the common types of thread forms are shown in Fig. 44-1.

**1.** *American National Thread* (Fig. 44-1A) is listed under three main divisions: National Coarse, National Fine, National Special. This thread is commonly known as a locking thread form in America.
**2.** *Unified Screw Thread* (Fig. 44-1B) was the result of a need for a common system for use in Canada, United States, and England. This thread incorporates the features of the American National Form and the British Standard Whitworth threads. Threads in the Unified series are interchangeable with American National and Whitworth threads of the same pitch and diameter.
**3.** *International Metric Thread* (Fig. 44-1C) is a standard thread currently used throughout Europe. It is presently used in North America mainly on instruments and spark plugs.
**4.** *American National Acme Thread* (Fig. 44-1D) is generally classified as a power-transmission type.
**5.** *Square Thread* (Fig. 44-1E) is used for maximum transmission power Because of its shape, friction between its matching threads is kept to a minimum.

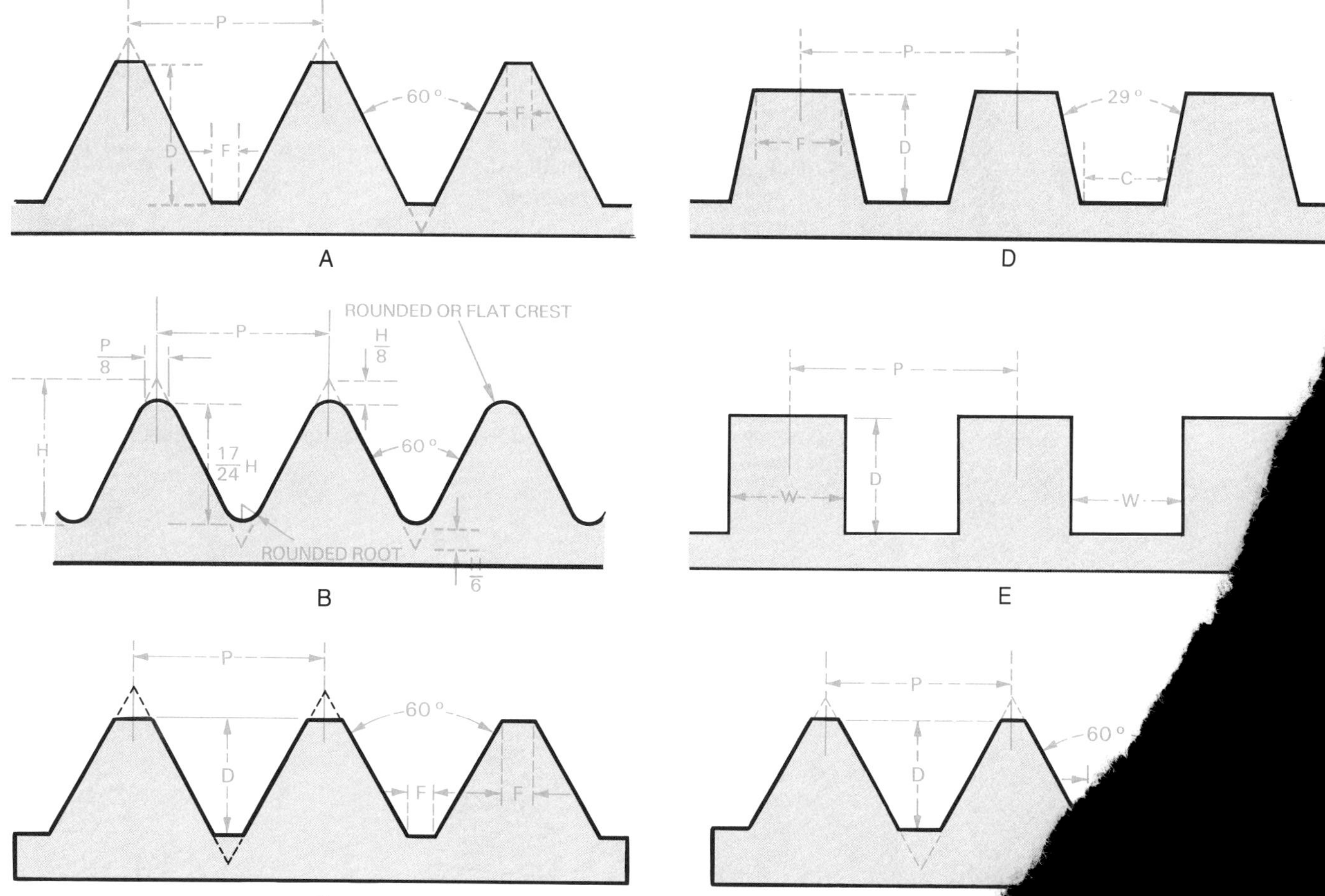


**Fig. 44-1** Common standard thread forms: (A) American National Form Thread; (B) Unified S
Thread; (D) American National Acme Thread; (E) Square Thread; (F) ISO Metric Threa

N-1 ISO thread identification
M8 × 1.25
8 mm (major diameter)
1.25 mm (pitch)

6. *ISO Metric Threads*. One of the world's major industrial problems has been the lack of an international thread standard whereby the thread used in any country could be interchanged with that of another country. In April 1975, the Industrial Organization for Standardization (ISO) drew up an agreement covering a standard metric thread profile, the sizes and pitches for the various threads in the new ISO Metric Thread Standard. (See N-1.) The new series has only 25 thread sizes ranging in diameter from 1.6 mm to 100 mm. See Table 44-1 for this series.

The ISO metric thread (Fig. 44-1F) has a 60° included angle and a crest equal to 0.125 times the pitch, which is similar to the National Form thread. The main difference, however, is the depth of the external thread, which is 0.6134 times the pitch. Because of these dimensions, the root of the thread is larger than that of the National Form thread. The root of the ISO metric thread is one-fourth the pitch (0.250*P*).

The new ISO series will not only simplify thread design but will generally produce stronger threads for a given diameter and pitch and will reduce the large inventory of fasteners now required by industry.

## THREAD TERMS AND CALCULATIONS

Screw threads form a very important part of every component that is made, from a tiny wrist watch to a large earth mover. To understand thread theory and screw cutting, you must know certain thread terminology. All threads have common thread terms (Fig. 44-2).

**Angle of thread**—The angle included between the sides of the thread, e.g., the thread angle of the new ISO Metric Thread and that of the American National Form is 60°.

**Major diameter**—The largest diameter of the thread on the screw or nut.

**Minor diameter**—The smallest diameter of an external or internal screw thread.

**Number of threads**—The number of roots or crests per inch of the threaded length. This term does not apply to metric threads.

**Pitch**—The distance from a point on one thread to the corresponding point on the next thread measured parallel to the axis.

**Lead**—The distance a screw thread advances axially for one complete revolution.

**Crest**—The top surface joining the two sides of a thread.

**Root**—The bottom surface joining the sides of two adjacent threads.

**Side**—The surface of the thread which connects the crest with the root.

**Depth of thread**—The distance between the crest and the root of a thread measured perpendicular to the axis.

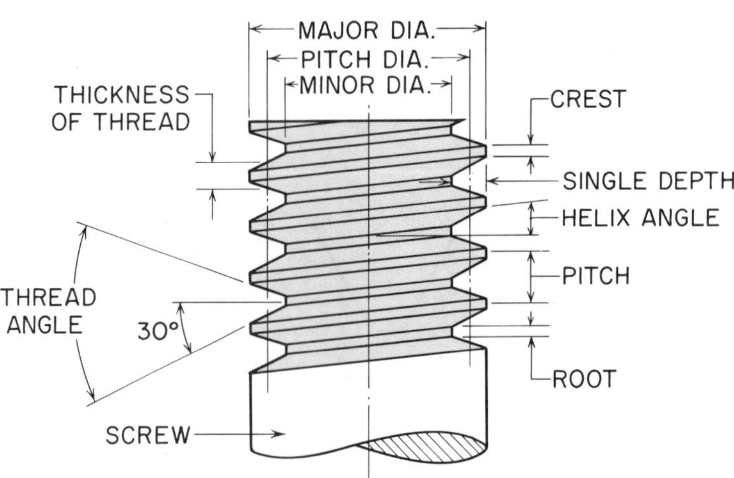


Fig. 44-2 The main parts of a screw thread.

**TABLE 44-1 ISO METRIC PITCH-DIAMETER COMBINATIONS**

| Nominal dia. (mm) | Thread pitch (mm) | Nominal dia. (mm) | Thread pitch (mm) |
|---|---|---|---|
| 1.6 | 0.35 | 20 | 2.5 |
| 2.0 | 0.40 | 24 | 3.0 |
| 2.5 | 0.45 | 30 | 3.5 |
| 3.0 | 0.50 | 36 | 4.0 |
| 3.5 | 0.60 | 42 | 4.5 |
| 4.0 | 0.70 | 48 | 5.0 |
| 5.0 | 0.80 | 56 | 5.5 |
| 6.0 | 1.00 | 64 | 6.0 |
| 8.0 | 1.25 | 72 | 6.0 |
| 10.0 | 1.50 | 80 | 6.0 |
| 12.0 | 1.75 | 90 | 6.0 |
| 14.0 | 2.00 | 100 | 6.0 |
| 16.0 | 2.00 | | |

N-2 The leadscrew should revolve downward for right-hand threads and upward for left-hand threads.

**Fig. 44-3** The quick-change gearbox controls the speed of the feed rod and the leadscrew.

## Calculations for American National Form Thread

$P$ = Pitch of thread

$= \dfrac{1}{\text{Number of threads per inch}}$

$D$ = Depth of thread

$= .6495 \times \text{pitch}$

$= \dfrac{.6495}{\text{Number of threads per inch}}$

$F$ = Width of flat on crest or root

$= \dfrac{P}{8}$

$= \dfrac{1}{8 \times \text{Number of threads per inch}}$

$N$ = Number of threads per inch

**Example:**

Find the pitch, depth, and minor diameter of a 1-in.—8 N.C. thread.

Major diameter $= 1.000$ in.

Pitch $= \dfrac{1}{N}$

$= \dfrac{1}{8}$ in.

Depth of thread $= .6495 \times P$

$= .6495 \times \dfrac{1}{8}$

$= .081$ in.

Minor diameter $=$ Major dia. $- (D + D)$

$= 1.000 - (.081 + .081)$

$= .838$ in.

## Calculations for ISO Metric Threads

$P$ = Pitch in millimeters

$D$ = Depth of external thread

$= 0.6134 \times \text{pitch}$

$FC$ = Width of flat at the crest

$= 0.125 \times \text{pitch}$

$FR$ = Width of flat at the root

$= 0.250 \times \text{pitch}$

**Example:**

What is the pitch, depth, minor diameter, width of crest, and width of root for a M 14 × 2 thread?

Pitch $= 2$ mm

Depth $= 0.6134 \times 2$

$= 1.226$ mm

Minor diameter $=$ Major diameter $- (D + D)$

$= 14 - (1.226 + 1.226)$

$= 11.54$ mm

Width of crest $= 0.125 \times \text{pitch}$

$= 0.125 \times 2$

$= 0.25$ mm

Width of root $= 0.250 \times \text{pitch}$

$= 0.250 \times 2$

$= 0.50$ mm

# THREAD CUTTING

Thread cutting on a lathe is a process of producing a ridge of uniform section by cutting a continuous groove around a cylinder. This is done by taking successive light cuts with a threading toolbit the same shape as the thread form. In order to produce an accurate thread, it is important that the lathe, the cutting tool, and the work be set up properly.

## To Set the Quick-Change Gearbox for Threading

The quick-change gearbox is designed to speed up changing gears for thread cutting. This unit has gear changes which transmit the proper ratio between the spindle and the lead screw.

**1.** Check the blueprint for the number of threads per inch required.

**2.** On the quick-change gearbox chart find the whole number which represents the number of threads per inch.

**3.** Engage the tumbler lever in the hole at the bottom of the vertical column in which the number is located (Fig. 44-3).

**4.** Set the top lever in the proper position as indicated on the chart for the thread required.

**5.** Engage sliding gear in or out as required.

**6.** Turn the driveplate or chuck by hand and make sure that the lead screw revolves. (See N-2.)

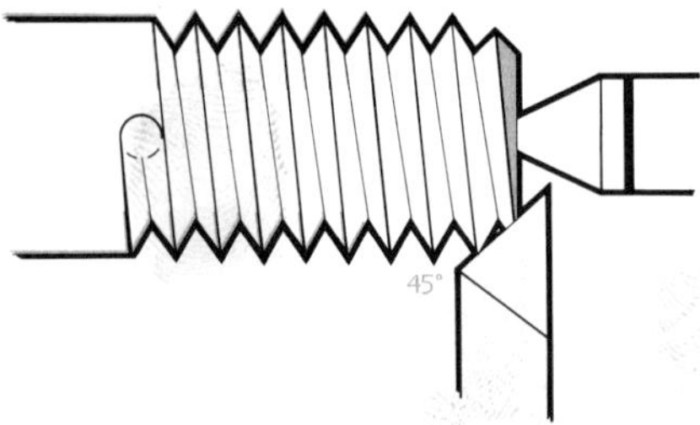


**Fig. 44-4** The ends of all threads should be chamfered smaller than the minor diameter to protect the thread.

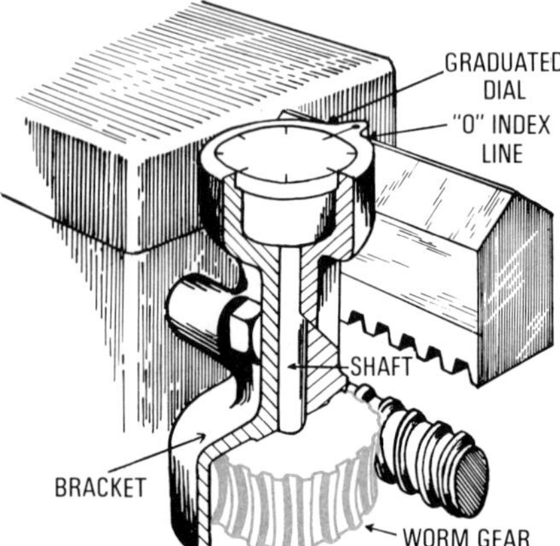


**Fig. 44-5** The thread chasing dial indicates when the split-nut should be engaged for taking successive cuts on a thread.

N-3 Revolve the lathe spindle by hand when setting the speed levers to be sure they are engaged properly.

**7.** Recheck the complete setup before thread cutting.

## To Chamfer a Diameter for Threading

The purpose of chamfering before threading is to allow the nut or part to start on easily, to improve appearance, to remove burrs, and to protect the start of the thread. It is preferable to chamfer work before threading, but this operation may also be done after threading (Fig. 44-4).

**1.** Set the toolbit to center.
**2.** Swivel the toolholder until the left-hand cutting edge is at an angle of 30 to 45° (Fig. 44-4).
**3.** Tighten the toolpost securely.
**4.** Set the lathe at three-quarters the turning speed.
**5.** Feed the toolbit in by hand using the crossfeed handle and apron handwheel.
**6.** For large chamfers, the lathe should be run slower and cutting fluid is recommended.
**7.** The chamfer should be slightly deeper than the minor diameter of the thread.

## Thread Chasing Dial

The thread chasing dial is an indicator with a revolving dial which can be either fastened to the carriage or built into it (Fig. 44-5). The chasing dial shows the operator when to engage the split-nut lever in order to take successive cuts in the same groove or thread. It also indicates the relationship between the ratio of the number of turns of the work and the leadscrew with respect to the position of the cutting tool and the thread groove.

The thread chasing dial is connected to a worm gear which meshes with the threads of the leadscrew. The dial is graduated into eight divisions, four numbered ones and four unnumbered ones, and it revolves as the leadscrew turns. Figure 44-6 indicates when the split-nut lever should be engaged for cutting various numbers of threads per inch.

## To Set Up a Lathe for Threading (60° Thread)

**1.** Set the lathe speed to about one-quarter the speed used for turning. (See N-3.)

| THREADS PER INCH TO BE CUT | WHEN TO ENGAGE SPLIT NUT | READING ON DIAL |
|---|---|---|
| EVEN NUMBER OF THREADS | ENGAGE AT ANY GRADUATION ON THE DIAL — 1, 1 ½, 2, 2 ½, 3, 3 ½, 4, 4 ½ | 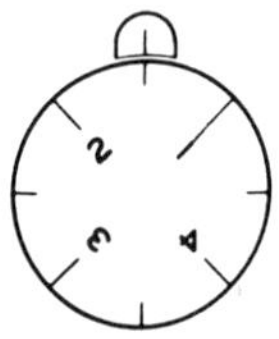 |
| ODD NUMBER OF THREADS | ENGAGE AT ANY MAIN DIVISION — 1, 2, 3, 4 | 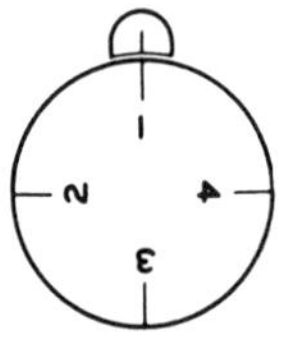 |
| FRACTIONAL NUMBER OF THREADS | 1/2 THREADS, E.G. 11 1/2 ENGAGE AT EVERY OTHER MAIN DIVISION 1 & 3, OR 2 & 4 OTHER FRACTIONAL THREADS ENGAGE AT SAME DIVISION EVERY TIME | 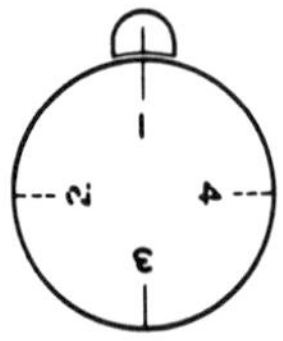 |
| THREADS WHICH ARE A MULTIPLE OF THE NUMBER OF THREADS PER INCH IN THE LEAD SCREW | ENGAGE AT ANY TIME THAT SPLIT NUT MESHES | USE OF DIAL UNNECESSARY |

**Fig. 44-6** Split-nut engagement rules for inch thread cutting.

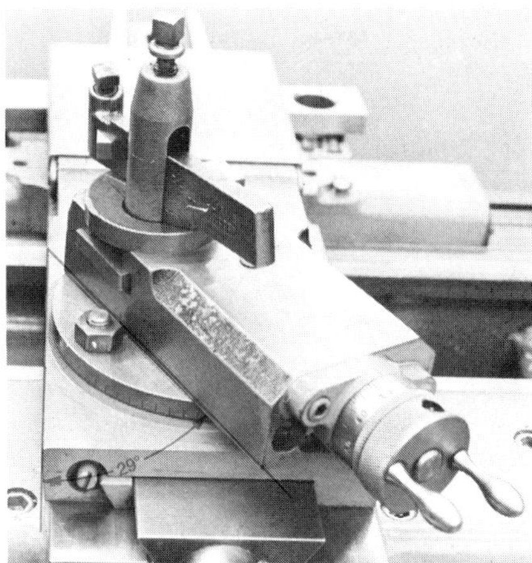

**Fig. 44-7** The compound rest is swiveled to the right when cutting a right-hand thread. (*Courtesy Kostel Enterprises Ltd.*)

**Fig. 44-8** Adjust the toolholder until the point of the threading tool is even with the center. (*Courtesy Kostel Enterprises Ltd.*)

N-4 The driving slot is marked in case the work is removed from the lathe, it can be replaced in the same position so that the threading tool will cut in the same groove.

N-5 Split-nut engagement:
Even threads — Any line
Odd threads — Numbered lines

**2.** Set the quick-change gearbox for the required number of threads per inch (Fig. 44-3).

**3.** Engage the lead screw.

**4.** Secure a 60° threading toolbit and check the angle using a thread center gage.

**5.** Set the compound rest at 29° to the right (to the left for a left-hand thread), as in Fig. 44-7.

**6.** Set the point of the toolbit even with the dead center point (Fig. 44-8).

**7.** Set the toolbit at right angles to the centerline of the work using a thread center gage (Fig. 44-9).

**8.** Arrange the apron feed lever in the neutral position and check the engagement of the split-nut lever.

## To Cut a Thread

Thread cutting is a lathe operation that requires a great deal of attention and skill. It involves manipulation of the lathe parts, correlation of the hands, and strict attention to the operation being performed. Before proceeding to cut the thread, it is wise to take several trial passes without cutting, in order to get the feel of the machine.

**1.** Mount the work in the lathe and check that the diameter to be threaded is .002 in. (0.05 mm) undersize.

**2.** With chalk, mark the drive plate slot that is driving the lathe dog. (See N-4.)

**3.** Mark the length to be threaded by cutting a light groove at this point with the threading tool while the lathe is revolving.

**4.** Chamfer the end of the work with the side of the threading tool.

**5.** Move the carriage until the point of the threading tool is near the right-hand end of the work.

**6.** Turn the crossfeed handle until the threading tool is close to the diameter, but stop when the handle is at the 3 o'clock position (Fig. 44-10).

**7.** Hold the crossfeed handle in this position and set the graduated collar to zero.

**8.** Start the lathe, and turn the compound rest handle until the threading tool lightly marks the work.

**9.** Move the carriage to the right until the toolbit clears the end of the work.

**10.** Feed the compound rest clockwise about .003 in. (0.08 mm).

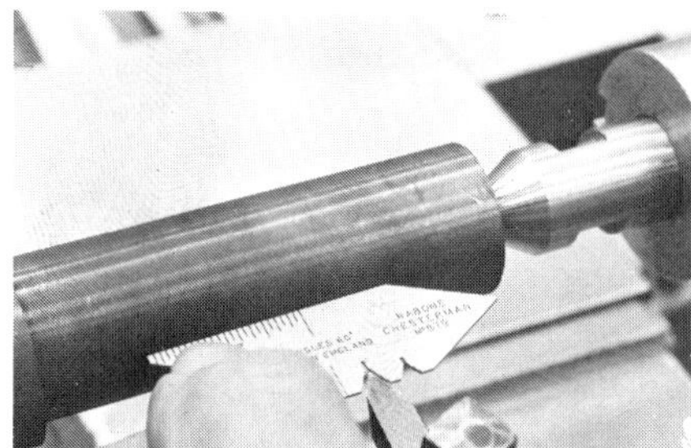

**Fig. 44-9** Setting a threading tool square with the work using a thread center gage. (*Courtesy Kostel Enterprises Ltd.*)

**Fig. 44-10** Thread cutting is made easier when the crossfeed handle is at the 3 o'clock position. (*Courtesy Kostel Enterprises Ltd.*)

**Fig. 44-11** Withdraw the toolbit, then disengage the split-nut lever. (*Courtesy Kostel Enterprises Ltd.*)

**11.** Engage the split-nut lever on the correct line of the thread chasing dial and take a trial cut along the length to be threaded. (See N-5.)

**12.** At the end of the cut, turn the crossfeed handle counterclockwise to move the toolbit away from the work and disengage the split-nut lever (Fig. 44-11).

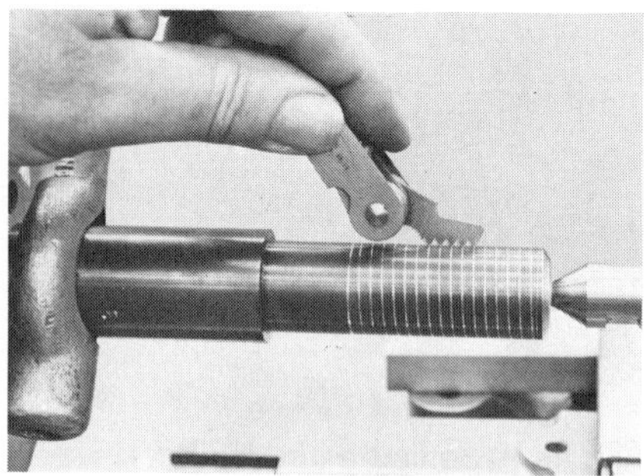

Fig. 44-12 Checking the number of threads per inch with a thread pitch gage. (*Courtesy Kostel Enterprises Ltd.*)

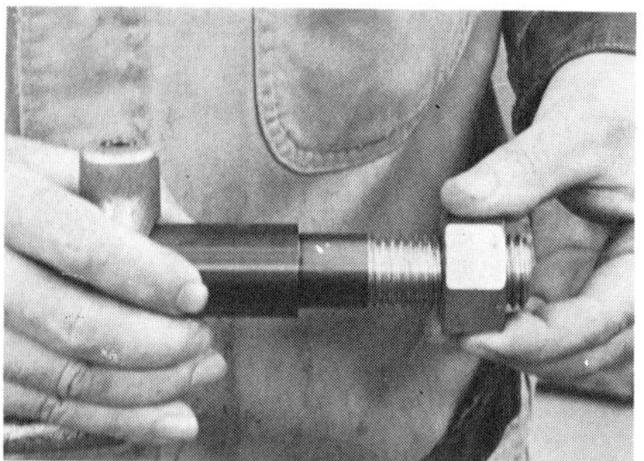

Fig. 44-13 Checking the accuracy of a thread with a master nut.

N-6 When using Tables 44-2A and B, be sure that the flat on the threading tool point is the proper width, otherwise the thread width will not be correct.

**TABLE 44-2A DEPTH SETTINGS WHEN CUTTING 60° NATIONAL FORM THREADS**

| tpi | Compound rest setting 0° | 30° | 29° |
|---|---|---|---|
| 24 | .027 | .031 | .0308 |
| 20 | .0325 | .0375 | .037 |
| 18 | .036 | .0417 | .041 |
| 16 | .0405 | .0468 | .046 |
| 14 | .0465 | .0537 | .0525 |
| 13 | .050 | .0577 | .057 |
| 11 | .059 | .068 | .0674 |
| 10 | .065 | .075 | .074 |
| 9 | .072 | .083 | .082 |
| 8 | .081 | .0935 | .092 |
| 7 | .093 | .1074 | .106 |
| 6 | .108 | .1247 | .1235 |
| 4 | .1625 | .1876 | .1858 |

**13.** Stop the lathe and check the number of threads per inch with a thread pitch gage (Fig. 44-12). If the number of threads per inch produced by the trial cut is not correct, recheck the quick-change gearbox setting.

**14.** After each cut, turn the carriage handwheel to bring the toolbit to the start of the thread and return the cross-feed handle to zero.

**15.** Set the depth of all threading cuts with the compound rest handle. For National Form threads, use Table 44-2A; for ISO metric threads, see Table 44-2B. (See N-6.)

**16.** Apply cutting fluid and take successive cuts until the top (crest) and the bottom (root) of the thread are the same width.

**17.** Remove the burrs from the top of the thread with a file.

**18.** Check the thread with a master nut and take further cuts, if necessary, until the nut fits the thread freely with no end play (Fig. 44-13).

## To Convert an Inch Lathe to Metric Threading

Metric threads may be cut on a standard quick-change gear lathe by using a pair of change gears having 50 and 127 teeth, respectively. Since the leadscrew has inch dimensions and is designed to cut threads per inch, it is necessary to convert the pitch in millimeters into threads per inch. To do this, it is first necessary to understand the relationship of the inch and the metric systems of measurement. 1 in. = 2.54 cm. Therefore the ratio of inches to centimeters is 1:2.54, or 1/2.54.

**TABLE 44-2B DEPTH SETTINGS WHEN CUTTING 60° ISO METRIC THREADS**

| Pitch, mm | Compound rest setting (mm) 0° | 30° | 29° |
|---|---|---|---|
| 0.35 | 0.19 | 0.21 | 0.21 |
| 0.4 | 0.21 | 0.25 | 0.24 |
| 0.45 | 0.24 | 0.28 | 0.27 |
| 0.5 | 0.27 | 0.31 | 0.30 |
| 0.6 | 0.32 | 0.37 | 0.36 |
| 0.7 | 0.37 | 0.43 | 0.42 |
| 0.8 | 0.43 | 0.50 | 0.49 |
| 1.0 | 0.54 | 0.62 | 0.62 |
| 1.25 | 0.67 | 0.78 | 0.77 |
| 1.5 | 0.81 | 0.93 | 0.92 |
| 1.75 | 0.94 | 1.09 | 1.08 |
| 2.0 | 1.08 | 1.25 | 1.24 |
| 2.5 | 1.35 | 1.56 | 1.55 |
| 3.0 | 1.62 | 1.87 | 1.85 |
| 3.5 | 1.89 | 2.19 | 2.16 |
| 4.0 | 2.16 | 2.50 | 2.47 |
| 4.5 | 2.44 | 2.81 | 2.78 |
| 5.0 | 2.71 | 3.13 | 3.09 |
| 5.5 | 2.98 | 3.44 | 3.40 |
| 6.0 | 3.25 | 3.75 | 3.71 |

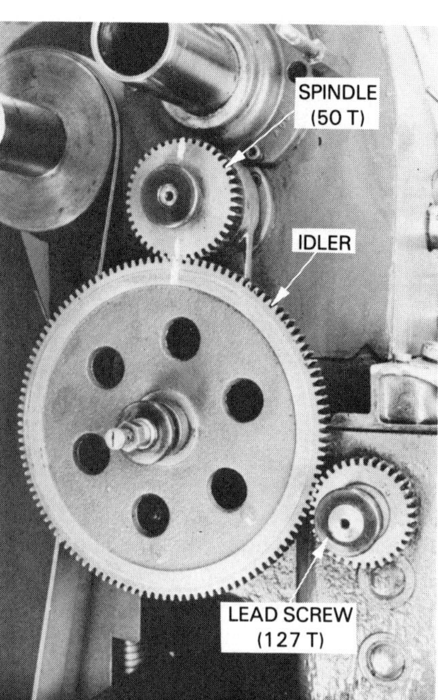


Fig. 44-14 The gears required for converting an inch lathe to cut metric threads.

To cut a metric thread on a lathe, it is necessary to install certain gears in the gear train which will produce a ratio of 1:2.54. These gears are:

$$\frac{1}{2.54} \times \frac{50}{50} = \frac{50 \text{ teeth}}{127 \text{ teeth}}$$

In order to cut metric threads, two gears having 50 and 127 teeth must be placed in the gear train of the lathe. The 50 tooth gear is used as the spindle or drive gear and the 127 tooth gear is placed on the leadscrew (Fig. 44-14).

## To Cut a 2.5-mm Metric Thread on a Standard Quick-Change Gear Lathe

**1.** Mount the 127 tooth gear on the leadscrew.

**2.** Mount the 50 tooth gear on the spindle.

N-7 Never disengage the split nut until the thread has been cut to depth, because metric threads have no definite relation to the thread chasing dial.

N-8 If the split nut is disengaged before the tool is aligned in the thread groove, the tool will not be in the correct position when the split-nut lever is engaged and may ruin the thread.

**3.** Convert the 2.5-mm pitch to threads per centimeter.

$$10 \text{ mm} = 1 \text{ cm}$$

$$\text{Pitch} = \frac{10}{2.5} = 4 \text{ threads/cm}$$

**4.** Set the quick-change gearbox to 4 threads/in. By means of the 50 and 127 tooth gears, the lathe will now cut 4 threads/cm, or 2.5-mm pitch.

**5.** Set up the lathe for thread cutting. See the section above entitled "To Set Up a Lathe for Threading (60° Thread)" (p. 224).

**6.** Take a light trial cut. At the end of the cut back out the cutting tool and stop the machine but *do not disengage the split nut*. (See N-7.)

**7.** Reverse the spindle rotation until the cutting tool has just cleared the end of the threaded section.

**8.** Check the thread with a metric screw pitch gage.

**9.** Cut the thread to the required depth.

## To Reset a Threading Tool

A threading tool must be reset whenever it is necessary to remove partly threaded work and finish it at a later time, if the threading tool has had to be removed for regrinding, or if the dog slips on the work.

**1.** Mount the work and set up the lathe for thread cutting.

**2.** With the threading toolbit clear of the work, start the lathe and engage the split-nut lever on the correct line for the number of threads per inch being cut.

**3.** Allow the carriage to travel until the toolbit is opposite any portion of the unfinished thread.

**4.** Stop the lathe, but be sure to *leave the split-nut lever engaged*. (See N-8.)

**5.** Feed the toolbit into the thread groove using *only* the compound rest and crossfeed handles until the right-hand side of the toolbit touches the right-hand side of the thread (left side for left-hand threads) (Fig. 44-15).

**6.** Set the crossfeed graduated collar to zero without moving the crossfeed handle.

**7.** Back out the threading tool using the crossfeed handle, disengage the split-nut lever, and move the carriage until the toolbit clears the start of the thread.

**8.** Set the crossfeed handle back to zero and take a trial cut without setting the compound rest.

**9.** Set the depth of cut using the compound rest handle and take successive cuts to finish the thread.

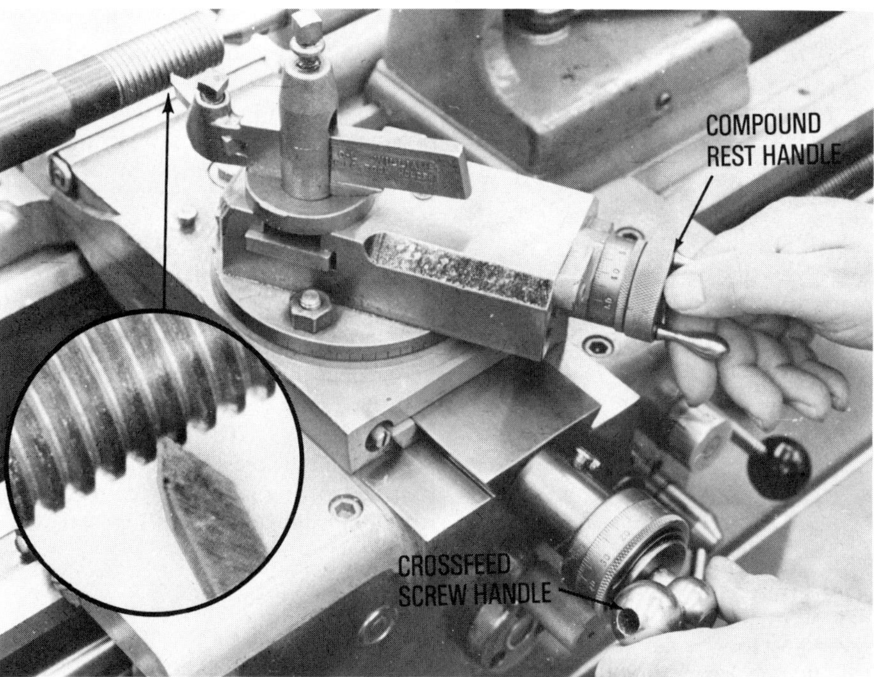


**Fig. 44-15** Use only the crossfeed and the compound rest handles to reset a threading tool into a partially cut thread. (*Courtesy Kostel Enterprises Ltd.*)

**Fig. 44-16** A groove is cut to the minor diameter of the thread to allow space for the threading toolbit to start cutting a left-hand thread.

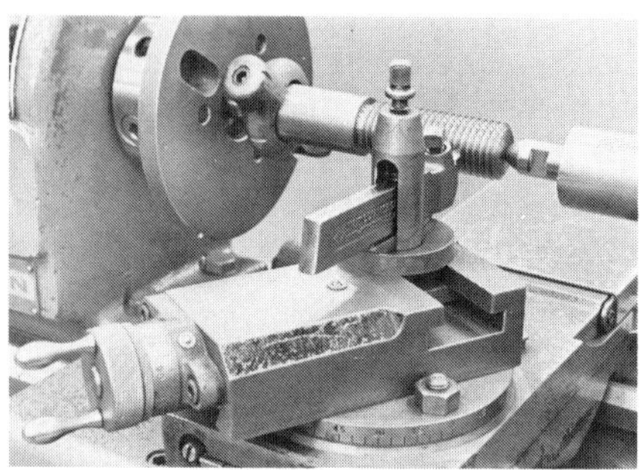

**Fig. 44-17** The compound rest is set at 29° to the left when cutting 60° left-hand threads. (*Courtesy Kostel Enterprises Ltd.*)

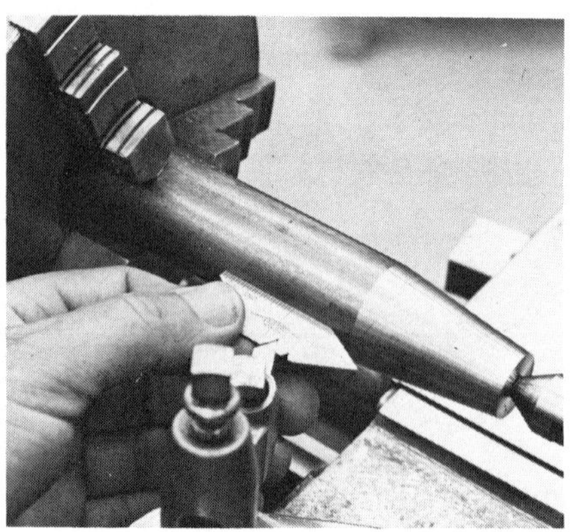

**Fig. 44-18** The toolbit must be set square with the axis of the work when cutting threads on a tapered section.

N-9 It is most important that the threading tool be set at 90° to the axis of the work and not square with the tapered section.

### Cutting a Left-Hand 60° Thread

Left-hand threads are used where a right-hand thread may loosen due to the rotation of a spindle. In this case, the nut on the left-hand thread would tighten. The pedestal grinder is an example of right- and left-hand thread applications on the same spindle. The thread on the right-hand end of the spindle (when standing in front of the grinder) has a right-hand thread. The thread on the left end is left-hand. If these were reversed, the grinding wheels would loosen in use.

Although the procedure for cutting right- and left-hand threads is the same, there are a few changes in the machine setup.

**1.** Check the diameter of the part to be threaded with the drawing and determine the number of threads per inch to be cut.

**2.** Set the spindle speed and the quick-change gearbox as for right-hand threads.

**3.** Engage the leadscrew to revolve in the opposite direction as for a right-hand thread.

**4.** Set the compound rest at 29° to the left (Fig. 44-17).

Note: Be sure that the groove at the left-hand end of the thread is cut to the minor diameter.

**5.** Set up the threading tool and square it with the work using a thread center gage.

**6.** When cutting the left-hand thread, the threading tool is set to the left end of the section to be threaded.

**7.** Proceed to cut the thread to the same dimensions as for a right-hand thread.

### Cutting a Thread on a Tapered Section

When a thread such as a pipe thread is required on a workpiece, either the taper attachment or the offset tailstock method may be used for cutting the taper. (See N-9.) The threads are then cut on the tapered section, using the same tapering setup and the regular setup for thread cutting (Fig. 44-18).

## ACME THREAD

The *Acme thread* is gradually replacing the square thread because it is stronger and easier to cut with taps and dies. It is used extensively for leadscrews because the 29° angle formed by its sides allows the split-nut to be engaged easily during thread cutting.

The Acme thread is provided with .010-in. clearance for both the crest and root. The hole for an internal Acme thread is cut .020 in. larger than the minor diameter of the screw, and the major diameter of a tap or internal thread is .020 in. larger than the major diameter of the screw. This provides .010-in. clearance between the screw and nut on both the top and bottom.

### To Cut an Acme Thread

**1.** Grind a toolbit to fit the end of the Acme thread gage (Fig. 44-19). Be sure to provide sufficient side clearance so that the tool will not rub while cutting the thread.

**2.** Grind the point of the tool flat until it fits into the slot of the gage indicating the number of threads per inch to be cut. *Note:* If a gage is not available, the width of the toolbit point may be calculated:

$$\text{Width of point} = \frac{.3707}{N} - .0052$$

**3.** Set the quick-change gearbox to the required number of threads per inch.

**4.** Set the compound rest 14½° to the right (half the included thread angle).

**5.** Set the Acme threading tool on center and square it with the work using the gage shown in Fig. 44-20.

**6.** At the right-hand end of the work, cut a section ¹⁄₁₆ in. (1.59 mm) long to the minor diameter. This will indicate when the thread is to the full depth.

**7.** Cut the thread to the proper depth by feeding the cutting tool using the compound rest.

## THREAD MEASUREMENT

There are several methods of checking threads for depth, angle, and accuracy. Most commonly used are thread gages, thread micrometers, and a finished master hexagon nut.

A finished master hexagon nut can be used for checking all general-pur-

**Fig. 44-19** An Acme thread gage is used to check the angle of the threading tool.

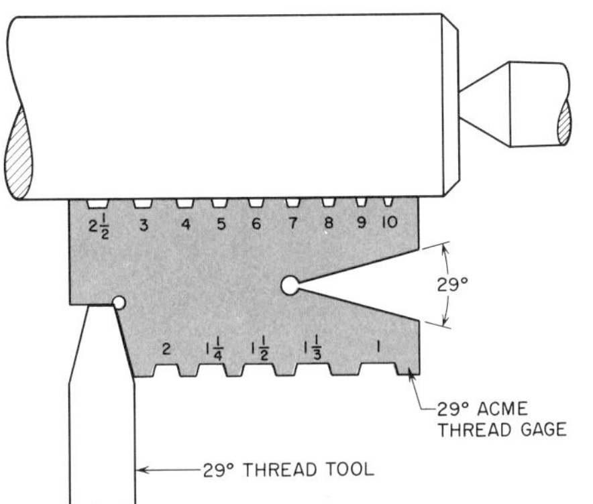


**Fig. 44-20** An Acme thread gage being used to set up and align the cutting tool with the workpiece.

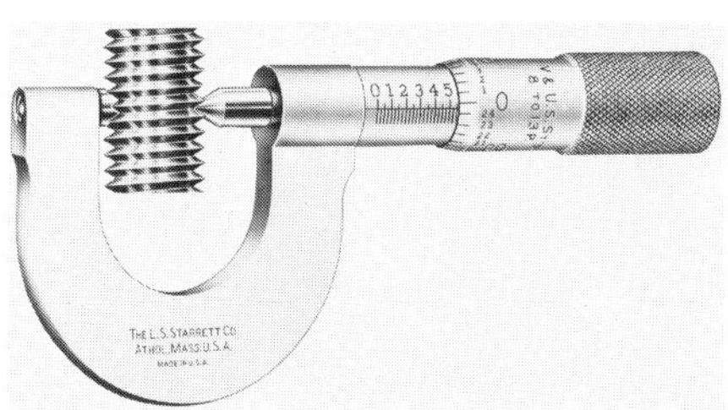

**Fig. 44-21** A thread micrometer measures the pitch diameter of a thread. (*Courtesy The L.S. Starrett Company.*)

N-10 The burrs must be removed and the diameter must be the correct size before using the one-wire method.

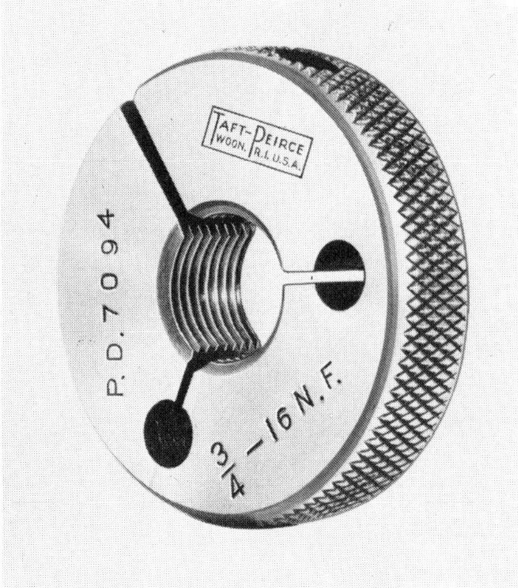


**Fig. 44-22** A thread ring gage is used for checking threads on production work. (*Courtesy Taft-Peirce Mfg. Co.*)

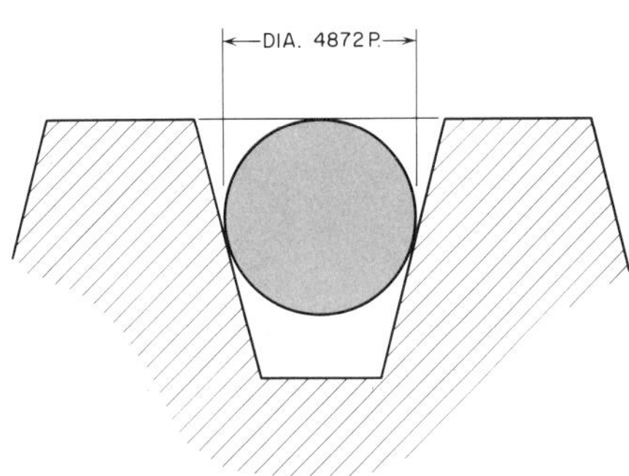


**Fig. 44-23** Using one wire to measure the accuracy of an Acme thread.

pose threads. The thread should be cut deep enough to allow the nut to turn on freely with no end play. Metric and inch thread micrometers (Fig. 44-21) can be used to check the pitch diameter of threads to an accuracy of .001 in. (0.02 mm). These micrometers are made to measure certain ranges of threads. Metric screw thread micrometers are available for checking thread diameters from 0 to 100 mm.

A thread ring gage (Fig. 44-22) is often used in production work for testing threads. It is a hardened standard ring gage that can be adjusted to compensate for wear or tolerance.

### Measuring Acme Threads

For most purposes, the *one-wire method* of measuring Acme threads is accurate enough. A single wire or pin of the correct diameter is placed in the thread groove (Fig. 44-23) and measured with a micrometer. The thread is the correct size when the micrometer reading over the wire is the same as the major diameter of the thread and the wire is tight in the thread. (See N-10.)

The diameter of the wire to be used can be calculated as follows:

$$\text{Wire diameter} = .4872 \times \text{pitch}$$

For example, if 6 threads per inch are being cut, the wire diameter should be:

$$\text{Wire size} = .4872 \times \frac{1}{6}$$
$$= .081 \text{ in.}$$

## CUTTING INTERNAL THREADS

Most internal threads are cut with taps; however, there are times when a tap of a specific size is not available and the thread must be cut on a lathe. Internal threading, or cutting threads in a hole, is an operation performed on work held in a chuck, collet, or mounted on a faceplate. The threading tool is similar to a boring toolbit, except the shape is ground to the form of the thread to be cut.

### Types of Taps

Hand taps (Fig. 44-24) are available in sets containing three taps: taper, plug, and bottoming. When using hand taps

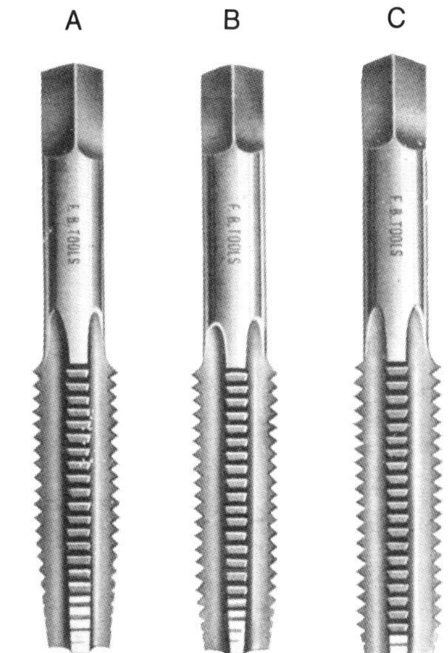


**Fig. 44-24** A set of hand taps: (A) Taper; (B) plug; (C) bottoming.

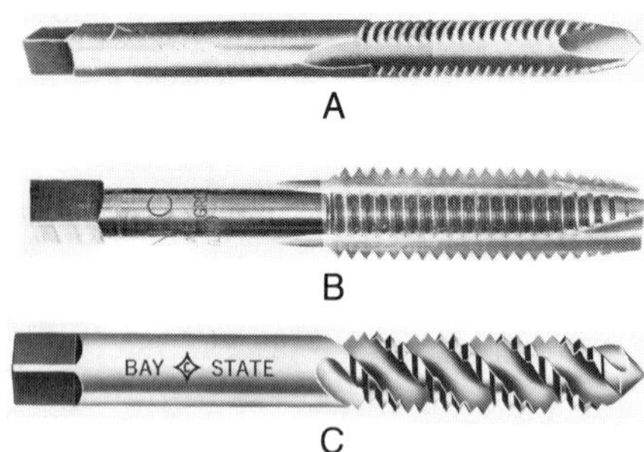


Fig. 44-25 Special-purpose taps: (A) Two-fluted spiral point; (B) three-fluted spiral point; (C) spiral-fluted. *(Courtesy Kostel Enterprises Ltd.)*

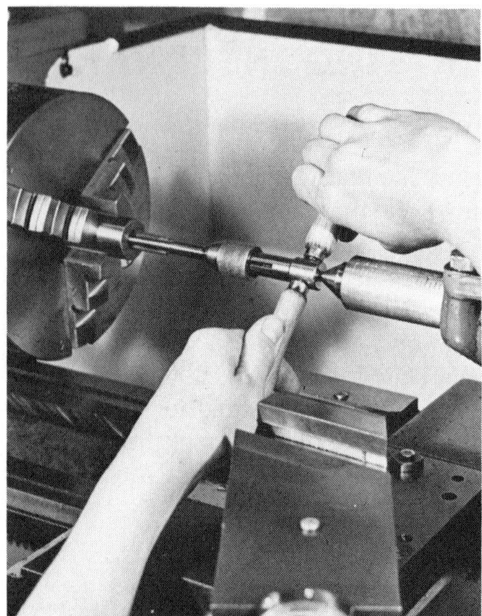

Fig. 44-26 Cutting an internal thread on a lathe with a tap.

N-11 Never use hand taps under power.

N-12

$$\text{T.D.S. (inch)} = D - \frac{1}{N}$$

$$\text{T.D.S. (metric)} = D - P$$

N-13 It is good practice to remove the tap from the hole occasionally to clear the chips.

N-14
$$\text{T.D.S.} = D - \frac{1}{N} = 1.375 - \frac{1}{6} = 1.209$$

N-15 The boring bar should be as large as possible and held short to prevent chatter.

to cut internal threads on a lathe, it is important that the tap be rotated into the work by hand and not by power. (See N-11.)

*Special-purpose taps* (Fig. 44-25) are available in a variety of types and sizes to suit various metals or operations. Most of these taps can be used by hand or in a machine for power tapping. The shape of the initial cutting edge on spiral point taps causes the chips to advance ahead of the tap.

## Tap Drill Size

Before a tap is used to cut an internal thread, a hole must be drilled in the workpiece to the correct *tap drill size*. This drill size leaves the proper amount of material in the hole for a tap to cut the thread. The tap drill, *which is always smaller than the tap diameter,* leaves enough material in the hole to produce approximately 75 percent of a full thread. (See N-12.)

## To Tap a Hole in a Lathe

Internal threads may be cut on a lathe by using the proper-size tap. A standard tap may be used for this operation; however, a gun tap is preferred because the chips are cleared ahead of the tap. The tap is aligned by placing the point of the dead center in the shank end of the tap to guide it while the tap is turned by means of a tap wrench. The lathe spindle is locked and the tap is turned by hand (Fig. 44-26).

**1.** Mount the work in the chuck and face the end.

**2.** Center drill a hole so that the top edge of the hole is slightly larger than the tap diameter.

**3.** Select the proper tap drill size for the tap to be used. See Table 5 at the back of the book.

**4.** Set the lathe to the proper speed for the diameter of the drill being used.

**5.** Drill the tap drill hole to the required depth using cutting fluid if required.

**6.** Stop the lathe and lock the spindle or put the lathe in its slowest speed. *Do not start the lathe.*

**7.** Start a taper tap, turning it by hand.

**8.** Apply cutting fluid frequently while tapping the hole (Fig. 44-26).

**9.** Keep the dead center snug into the shank of the tap during this operation by turning the tailstock handwheel while turning the tap with the other hand.

**10.** Back off the tap every half turn to break the chip. (See N-13.)

**11.** Remove the taper tap and complete tapping the hole with a plug or bottoming tap.

Many internal threads are finished with a tap; however, if a certain size of tap is not available, or when it is important that the thread is concentric with the diameter, it must be cut on the lathe. The workpiece to be threaded may be held in a collet, chuck, on a faceplate, or supported in a steady rest. The internal threading operation is very similar to cutting external threads on a lathe. However, extra care should be taken due to the spring of the boring bar and the inability of the operator to see the operation as easily as when cutting an external thread.

## To Cut a 1⅜-in.—6 NC Internal Thread

**1.** Calculate the tap drill size of the thread. (See N-14.)

**2.** Mount the work to be threaded in a chuck, collet, or on a faceplate.

**3.** Drill a hole, approximately 1/16 in. smaller than the tap drill size in the workpiece. For this thread it would be $1.209 - .062 = 1.147$, or a 1 5/32-in. hole.

**4.** Mount a boring tool in the lathe and bore the hole to the tap drill size (1.209). (See N-15.)

**5.** Recess the start of the hole to the major diameter of the thread (1.375) 1/16 in. (1.5 mm) for length. During the thread-cutting operation, this will indicate when the thread is cut to depth.

**6.** If the thread does not go through the workpiece, a recess should be cut at the end of the thread to the major diameter (Fig. 44-27). This recess should be wide enough to allow the threading tool to "run out" and permit time to disengage the split-nut lever.

**7.** Set the compound rest at 29° to the left (Fig. 44-28); to the right for left-hand threads.

**8.** Mount a threading toolbit into the boring bar and set it to center.

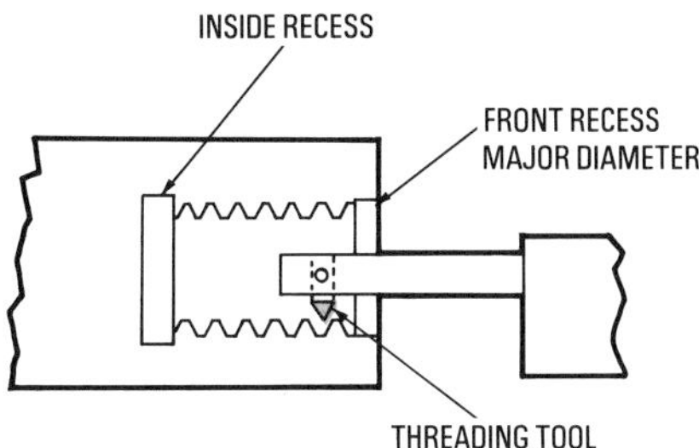


**Fig. 44-27** A recess at the front and back of a hole is desirable when cutting internal threads.

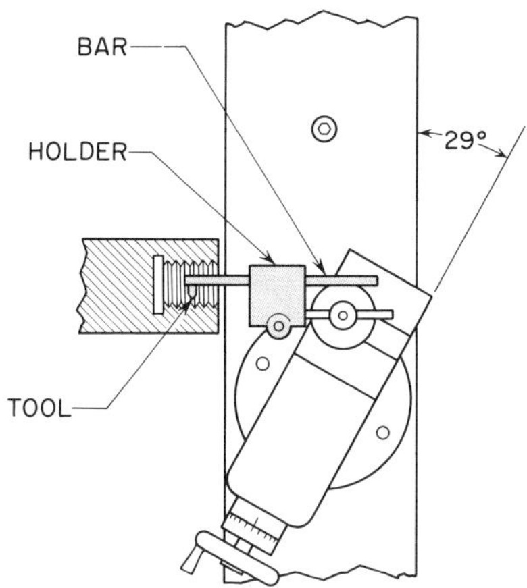


**Fig. 44-28** The compound rest is set at 29° to the left for cutting a right-hand internal thread.

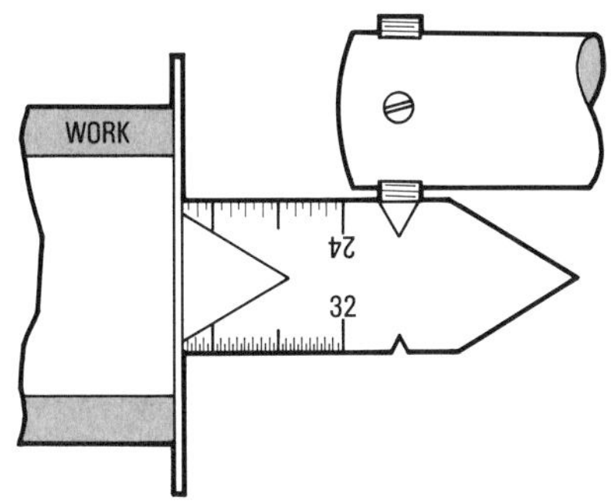


**Fig. 44-29** Using a thread center gage to square the threading tool with the work.

N-16 See Tables 44-2A and B for the amount of compound rest feed required to cut various threads.

N-17 The last few cuts should only be .001 in. (0.02 mm) deep in order to eliminate any spring in the boring bar.

**9.** Square the threading tool with a thread center gage (Fig. 44-29).

**10.** Place a mark on the boring bar, measuring from the threading tool, to indicate the length of hole to be threaded. This will show when the split-nut lever should be disengaged.

**11.** Start the lathe and turn the crossfeed handle *out* until the threading tool just scratches the internal diameter.

**12.** Set the crossfeed and compound rest graduated collars to zero and clear the cutting tool from the hole with the carriage handwheel.

**13.** Turn the compound rest handle counterclockwise to move the toolbit out .003 to .005 in. (0.07 to 0.12 mm), and take a trial cut.

**14.** At the end of each cut (when the mark on the boring bar is even with the edge of the workpiece), disengage the split-nut lever and turn the crossfeed handle clockwise to clear the toolbit from the thread.

**15.** Move the carriage to the right until the toolbit is clear of the work.

**16.** Check the pitch of the thread with a screw pitch gage.

**17.** Return the crossfeed handle to zero and set the depth of cut by feeding the compound rest counterclockwise about .015 in. (0.40 mm).

**18.** Take successive cuts, decreasing depths until the thread is to the proper depth. (See N-16.) If the point of the threading tool is the correct width, the amount of compound rest feed can be calculated by applying the following formula:

$$\text{Compound rest feed} = \frac{.750}{N} = \frac{.750}{7} = .107$$

**19.** As the thread becomes deeper, it is necessary to decrease the amount of compound rest feed to decrease the spring of the boring bar. (See N-17.)

**20.** Check the thread for fit with a thread plug gage or a bolt.

# REVIEW QUESTIONS

## Standard Thread Forms

**1.** Define a thread.

**2.** State four purposes of threads.

**3.** Name six thread forms and state the included angle for each.

**4.** Briefly define:
(a) Minor diameter
(b) Pitch
(c) Lead

**5.** For a ¾-in.—10 UNC thread calculate:
(a) Depth of thread
(b) Minor diameter
(c) Width of root

**6.** For a M16 × 3 metric thread calculate:
(a) Depth of thread
(b) Minor diameter
(c) Width of root

## Thread Cutting

**7.** In which direction should the lead screw revolve for cutting?
(a) Right-hand threads
(b) Left-hand threads

**8.** How deep should the chamfer be cut on work to be threaded?

**9.** What purpose does the thread chasing dial serve?

**10.** When, on the thread chasing dial, should the split-nut lever be engaged for cutting?
(a) Even threads
(b) Odd threads

**11.** At what angle should the compound rest be set for cutting a 60° right-hand thread?

**12.** How is the threading toolbit set square with the work?

**13.** Why should the driveplate slot be marked with chalk when thread cutting?

**14.** What is the purpose of first taking a light trial cut before cutting a thread?

**15.** What lathe part is used to set the depth of each thread cut?

**16.** How should an inch lathe be set up to cut an M8 × 1.25 metric thread?

**17.** What parts of a lathe are used to reset a threading tool into a partially cut thread?

**18.** How does the lathe setup for a left-hand thread differ from a right-hand thread?

**19.** At what angle should the compound rest be set to cut an Acme thread?

## Thread Measurement

**20.** List three methods of checking the accuracy of a thread.

**21.** What instrument can be used to measure the pitch diameter of a thread?

**22.** What diameter wire should be used to check an 8-pitch Acme thread?

### Cutting Internal Threads

**23.** Name three types of taps which can be used to produce internal threads.

**24.** How can a tap be kept in alignment when tapping a hole in a lathe?

**25.** Calculate the tap drill size of a 1¼-in.—7 UNC internal thread.

**26.** When cutting internal threads, what is the purpose of:

(a) A recess at the front of the hole

(b) A recess at the end of a hole

# UNIT 45

# STEADY RESTS, FOLLOWER RESTS, AND MANDRELS

The use of lathe accessories such as steady rests, follower rests, and mandrels increases the variety of work that can be machined on a lathe. Steady and follower rests are used to provide a support for long, slender workpieces to prevent them from springing and bending. Mandrels provide a means of holding and driving parts such as gears, flanges, and pulleys.

## OBJECTIVES

After completing this unit, you should be able to:

**1.** Set up and use a steady rest to machine a long, slender shaft

**2.** Set up and use a follower rest to support a shaft immediately next to the cutting tool

**3.** Mount and machine work on various types of mandrels

## STEADY REST

A steady rest (Fig. 45-1) is used to support long, slender work and prevent it from springing while being machined between centers. A steady rest may also be used when it is necessary to perform a machining operation on the end of a workpiece which is held in a chuck. The steady rest is fastened to the lathe bed and its three jaws are adjusted to the surface of the work to provide a supporting bearing. The jaws on a steady rest are generally made of soft material, such as fiber or brass, to prevent damaging the work surface.

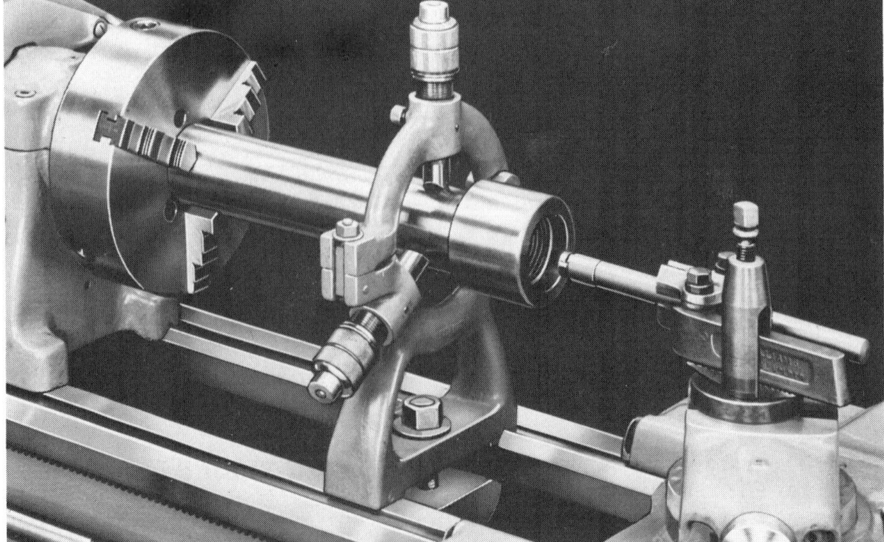

**Fig. 45-1** Using a steady rest to support the end of long work held in a chuck.

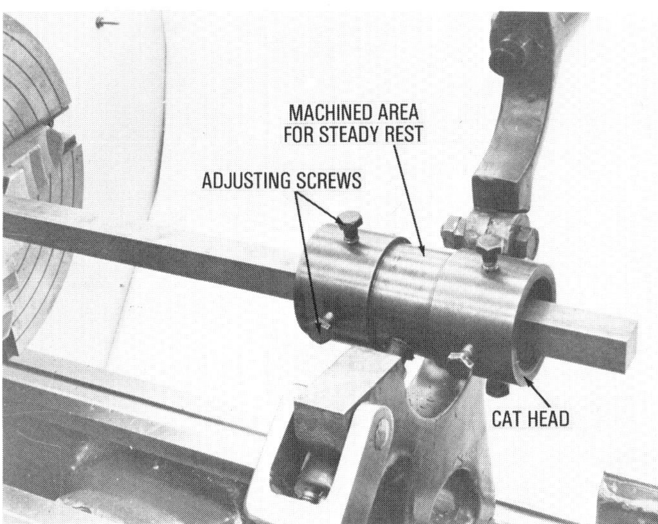


**Fig. 45-2** A cathead can be adjusted to provide a true surface for the steady rest even if square work is being turned. (*Courtesy Kostel Enterprises Ltd.*)

**Fig. 45-3** Mounting a steady rest to prevent long, small-diameter work from springing during the machining operation. (*Courtesy Kostel Enterprises Ltd.*)

N-1 The white or red lead will smear as the chuck jaws contact the work.

N-2 If the work is not aligned properly, a taper will be cut.

**Fig. 45-4** The follower rest is used to support long slender work between centers while thread cutting.

Some steady rests have rollers attached to the jaws to provide good support for the work.

### To Set Up a Steady Rest

**1.** Mount the work between centers; or set up and true the work in a chuck.

(a) If the work diameter is not round, turn a true spot on the diameter (slightly wider than the steady rest jaws) at the point where the steady rest will be supporting the work. Long work in a chuck should be first supported by the tailstock center. If the diameter is rough, turn a section for the steady rest and one near the chuck to the same diameter.

(b) If it is impossible to turn a true diameter (due to the shape of the workpiece), mount and adjust a cathead (Fig. 45-2) on the work.

**2.** Move the carriage to the tailstock end of the lathe.

**3.** Place the steady rest on the lathe bed at the desired position. If work diameter is turned and held in a chuck, slide the steady rest up close to the chuck.

**4.** Adjust the lower two jaws to the work diameter, using a paper feeler to provide clearance between the jaws and the work.

**5.** Slide the steady rest to the desired position and fasten it in place (Fig. 45-3).

**6.** Close the top section of the steady rest and adjust the top jaw, using a paper feeler.

**7.** Apply white or red lead to the diameter at the steady rest jaws. (See N-1.)

**8.** Start the lathe and carefully adjust each jaw until it just touches the diameter.

**9.** Tighten the lock screw on each jaw and then apply a suitable lubricant.

**10.** Before machining, it is wise to indicate the top and front of the turned diameter at the chuck and at the steady rest, to check for alignment. If the indicator reading varies, adjust the steady rest until it is correct. (See N-2.)

## FOLLOWER REST

A follower rest, mounted on the saddle, moves along with the carriage to prevent work from springing up and away from the cutting tool. The follower rest, positioned immediately behind the cutting tool, can be used to support long work for successive operations such as thread cutting (Fig. 45-4).

### To Set Up a Follower Rest

**1.** Mount the work between centers.

**2.** Fasten the follower rest to the saddle of the lathe.

**3.** Position the cutting tool in the toolpost so that it is just to the left of the follower rest jaws.

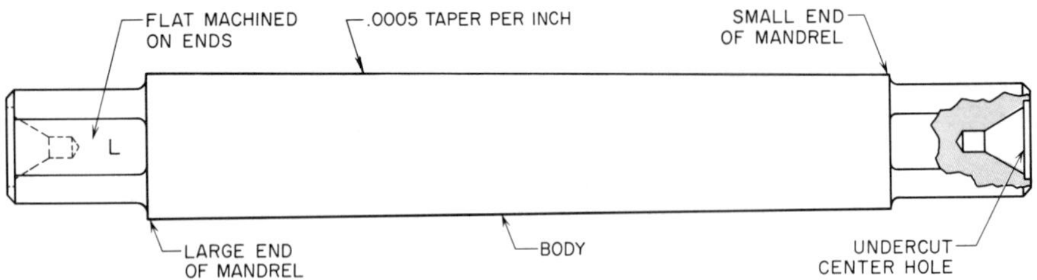


**Fig. 45-5** Characteristics of a standard solid mandrel.

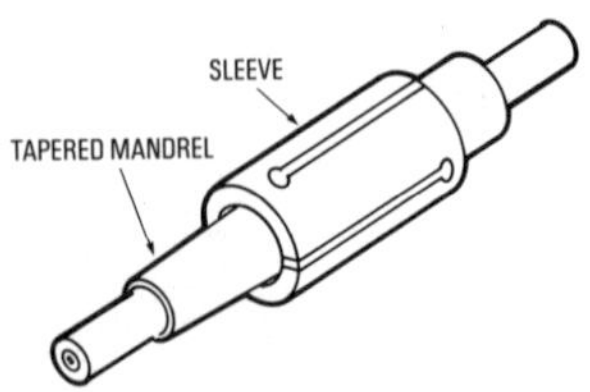


**Fig. 45-6A** Expansion mandrels are used to hold workpieces with odd-sized holes.

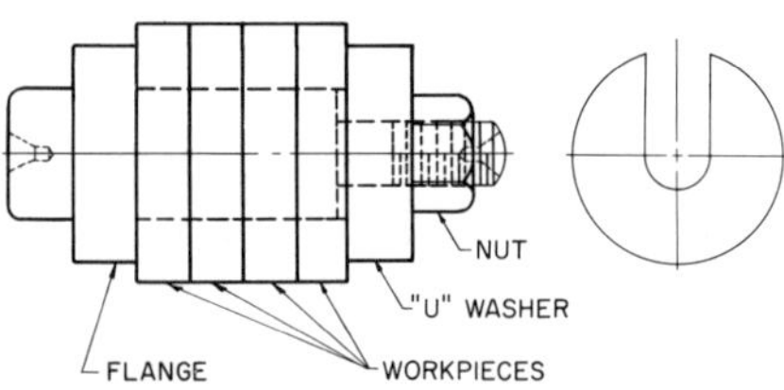


**Fig. 45-6B** A gang mandrel is used to hold a number of identical pieces for machining. (*Courtesy Cincinnati Milacron Co.*)

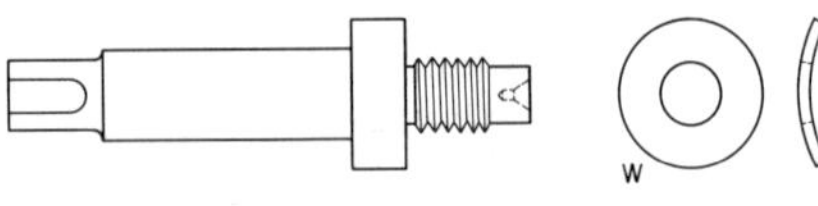


**Fig. 45-6C** A threaded mandrel is used to hold a part with a threaded hole.

**Fig. 45-6D** Taper shank mandrels are mounted in the lathe headstock spindle.

N-3 Use paper feelers or red lead when setting the jaws to the work surface.

N-4 A mandrel is used to hold workpieces; an arbor is used to hold cutting tools.

N-5 Oil will prevent scoring as the mandrel is pressed into the workpiece.

**4.** Turn the work diameter, for approximately 1½ in. (38.1 mm) long, to the desired size.

**5.** Adjust both jaws of the follower rest until they lightly contact the turned diameter. (See N-3.)

**6.** Tighten the lock screw on each jaw.

**7.** Lubricate the work and the follower rest jaws to prevent marring the finished diameter.

**8.** If successive cuts are required, readjust the follower rest jaws as in steps 4 and 5.

# MANDRELS

A *mandrel* (Fig. 45-5) is a precision tool which, when pressed into the hole of a workpiece, provides centers for a machining operation. They are especially valuable for thin work such as flanges, pulleys, gears, etc., where the outside diameter must run true with the inside diameter and it would be difficult to hold the work in a chuck.

## Characteristics of a Standard Mandrel

**1.** Mandrels are usually hardened and ground, and tapered .006 to .008 in. (0.15 to 0.20 mm) per foot.

**2.** The nominal size is near the middle and the small end is usually .001 in. (0.02 mm) under; the large end is .004 in. (0.10 mm) over the nominal size.

**3.** Both ends are turned smaller than the body and provided with a flat so that the lathe dog does not damage the mandrel.

**4.** The size of the mandrel is stamped on the large end.

**5.** The center holes, which are recessed slightly, are large enough to provide a good bearing surface and to withstand the strain caused by machining.

## Types of Mandrels

Many types of mandrels are used to suit various types of workpieces or machining operations. Descriptions and purposes of some of the more common types follow. (See N-4.)

The *solid mandrel* (Fig. 45-5) is available for most of the standard hole sizes. It is a general-purpose mandrel which may be used for a variety of workpieces.

The *expansion mandrel* (Fig. 45-6A) consists of a sleeve, with four or more grooves cut lengthwise, fitted over a solid mandrel. A taper pin fits into the sleeve to expand it to hold work that does not have a standard-size hole. Another form of expansion mandrel has a slotted bushing fitting over a tapered mandrel. Various-size bushings can be used with this mandrel, increasing its range.

The *gang mandrel* (Fig. 45-6B) is used to hold a number of identical parts for a machining operation. The body of the mandrel is parallel (no taper) and has a shoulder or flange on one end. The other end is threaded for a locking nut.

The *threaded mandrel* (Fig. 45-6C) is used for holding workpieces having a threaded hole. An undercut at the shoulder ensures that the workpiece will seat squarely and is not canted on the threads.

The *taper-shank mandrel* (Fig. 45-6D) may be fitted to the tapered hole in the headstock spindle. The projecting portion may be machined to any desired form to suit the workpiece. This type of mandrel is often used for small workpieces or those which have blind holes.

## To Mount Work on a Mandrel

**1.** Secure a mandrel to fit the hole in the workpiece.

**2.** Thoroughly clean the mandrel and apply a thin film of oil on the diameter (Fig. 45-7). (See N-5.)

**3.** Clean and remove any burrs from the hole in the workpiece.

**4.** Start the small end of the mandrel into the hole by hand. (The large end has the size stamped on it.)

**5.** Place the work on an arbor press with a machined surface down so that the hole is at right angles (Fig. 45-8).

**6.** Press the mandrel firmly into the workpiece.

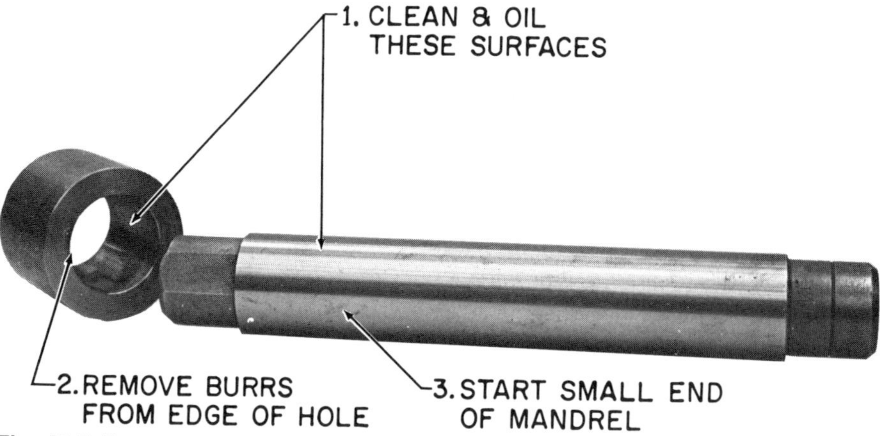


**Fig. 45-7** Preparing the mandrel and workpiece before assembling.

**Fig. 45-8** Using an arbor press to force a mandrel into a workpiece.

**Fig. 45-9** Checking the trueness of a live center with a dial indicator.

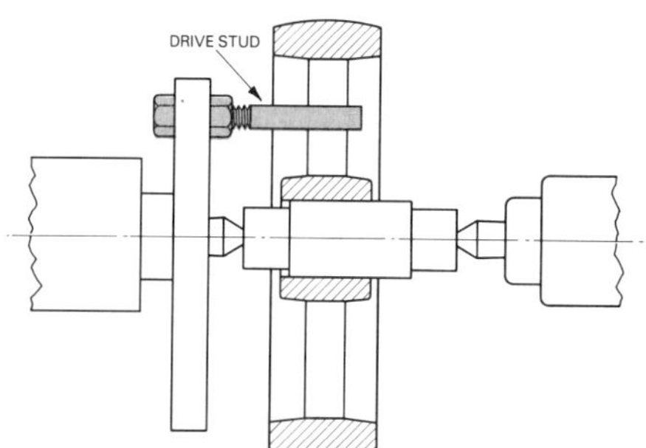


**Fig. 45-10** Large work turned on a small mandrel should be driven directly from the driveplate by means of a suitable drive stud.

## Machining Work on a Mandrel

The workpiece is pressed on a solid mandrel and is held securely only by friction. Therefore, whenever possible, take all cuts toward the large end of the mandrel. This tends to keep the workpiece tight on the mandrel. If it is necessary to cut toward the small end, only light cuts should be taken. The following precautions should be observed when mounting a mandrel on a lathe, and also during the machining operation.

**1.** Use an indicator to check that the lathe headstock center is running true (Fig. 45-9). (See N-6.)

**2.** Thoroughly clean the lathe and mandrel centers.

**3.** Check the alignment of the lathe centers with a test bar and dial indicator to ensure that parallel work is turned.

**4.** Mount a lathe dog on the large end of the mandrel (the size is stamped on this end).

**5.** Mount the mandrel and workpiece on the lathe and carefully adjust the lathe center tension. (See N-7.)

**6.** When machining the outside diameter of a part, always cut toward the large end of the mandrel so that the part is forced on and not off the mandrel. (See N-8.)

**7.** When machining small parts, use cutting fluid whenever possible to keep the heat of the workpiece to a minimum. If the part is overheated, it may expand and slip on the mandrel.

**8.** Light cuts should be taken on large-diameter work mounted on a small mandrel to eliminate slippage, overcome chattering and springing, or bending the mandrel. If possible, drive the workpiece directly from the driveplate by means of a suitable stud (Fig. 45-10). This prevents the workpiece from slipping or turning on the mandrel.

N-6 If the lathe center is not true, the mandrel will not run true and the machined diameter will not be concentric with the hole.

N-7 Use a revolving tailstock center whenever possible to avoid overheating and ruining the lathe center and the mandrel.

N-8 Take all cuts toward the lathe dog which should be fastened on the large end of the mandrel.

## REVIEW QUESTIONS

**1.** What purpose does a steady rest serve?
**2.** How should the steady rest jaws be adjusted to the work diameter?
**3.** Why should the top of the turned diameter be checked for alignment before machining?
**4.** For what purpose are follower rests used?
**5.** On what part of the lathe is the follower rest fastened?
**6.** Why is it necessary to lubricate the jaws of a follower rest?
**7.** What purpose do mandrels serve?
**8.** State the difference between a mandrel and an arbor.
**9.** How is the large end of a mandrel marked?
**10.** Name five mandrels and state the purpose of each.
**11.** How should the work and mandrel be prepared before assembly?
**12.** Why should the headstock center be running true?
**13.** In which direction should cuts be taken on work mounted on a mandrel? Why?

**Fig. 46-1** All jaws of a universal chuck move in or out simultaneously. *(Courtesy The Clausing Corporation.)*

# UNIT 46

# MACHINING IN A CHUCK

Lathe accessories such as chucks, driveplate, and face plates are fitted to and driven by the headstock spindle. Chucks are used to hold workpieces that would be difficult or impossible to hold by any other method. Since the chuck jaws are adjustable, it is possible to hold almost any shaped workpiece for machining. The most common lathe chucks are the three-jaw, the four-jaw, and the spring-collet chuck.

## OBJECTIVES

The aim of this unit is to enable you to:

**1.** Set up and machine work in a three-jaw chuck
**2.** Set up and machine work in a four-jaw chuck
**3.** Set up and machine work in collet chucks
**4.** Face work in a chuck to accurate lengths
**5.** Groove and cut off work held in a chuck

## MOUNTING WORK IN A CHUCK

The proper procedure for mounting and removing chucks must be followed in order not to damage the lathe spindle and/or accessory, and to preserve the accuracy of the lathe. There are three types of lathe spindle noses: the threaded spindle nose, the tapered spindle nose, and the cam-lock spindle nose. See Unit 38 for the correct procedure for mounting and removing chucks from each type of spindle nose.

Although there are several types of lathe chucks and the method of mounting work in each may vary, it is important that the workpiece and chuck jaws be cleaned before work is mounted. This not only ensures greater accuracy in machining but also helps preserve the accuracy of the chuck.

### Three-Jaw Chuck Work

Round and hexagonal work can be quickly and conveniently mounted in a three-jaw universal chuck (Fig. 46-1)., Since all jaws move simultaneously, the work is automatically centered to within about .003 in. (0.07 mm). As the scroll plate becomes worn, the chuck may lose this accuracy. When this occurs, work should be mounted in a four-jaw independent chuck to maintain the desired accuracy.

To preserve the accuracy of a universal chuck, work which has an outer scale should not be gripped without the protection of soft shims under the chuck jaws.

#### Mounting Work in a Three-Jaw Chuck

**1.** Clean the surface of the workpiece and the jaws of the chuck.
**2.** Using the proper chuck wrench, open the jaws slightly more than the size of the work. (See N-1.)
**3.** Insert the work into the chuck, leaving the proper amount protruding

N-1 Be sure that the chuck wrench fits the chuck sockets; otherwise the chuck sockets and the wrench will be damaged.

N-2 Never leave a chuck wrench in a chuck.

N-3 Be sure that the start of the scroll thread does not protrude into the jaw slide.

(Fig. 46-2). This amount depends on the operation to be performed.

**4.** With the left hand, tighten the chuck wrench while the right hand slowly rotates the workpiece. This causes the work to center properly in the chuck.

**5.** Tighten the chuck jaws securely by using the chuck wrench only. (See N-2.)

## Assembling the Jaws in a Three-Jaw Universal Chuck

When chuck jaws are changed in a universal chuck, care must be taken to assemble them in the proper order, otherwise the chuck jaws will not run true. All chucks and the jaws supplied with each are marked with the same serial number. It is important to match the serial numbers of the chuck and the jaws being inserted in it.

**1.** Thoroughly clean the jaws and the jaw slides in the chuck.

**2.** Turn the chuck wrench clockwise until the start of the scroll thread is almost showing at the back edge of slide 1. (See N-3.)

**3.** Insert jaw 1 (in slot 1) and press down with one hand while turning the chuck wrench clockwise with the other (Fig. 46-3A).

**4.** After the scroll thread has engaged in the jaw, continue turning the chuck wrench clockwise until the start of the scroll thread is near the back edge of groove 2.

**5.** Insert jaw 2 and repeat steps 3 and 4 (Fig. 46-3B).

**6.** Insert the third jaw in the same manner (Fig. 46-3C).

**Fig. 46-2** The work should not extend past the jaws more than three times its diameter. (*Courtesy Kostel Enterprises Ltd.*)

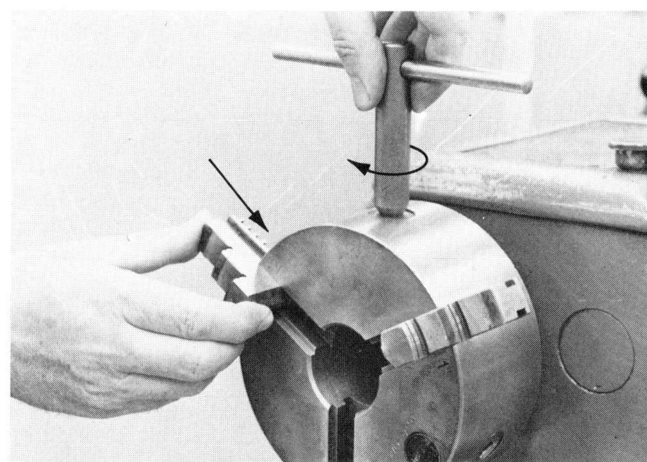

**Fig. 46-3B** Placing jaw 2 into the universal chuck. (*Courtesy Kostel Enterprises Ltd.*)

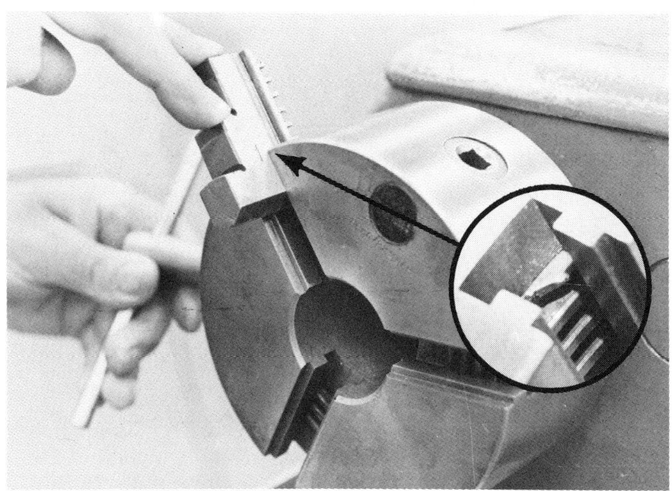

**Fig. 46-3A** The scroll plate should engage jaw 1 first.

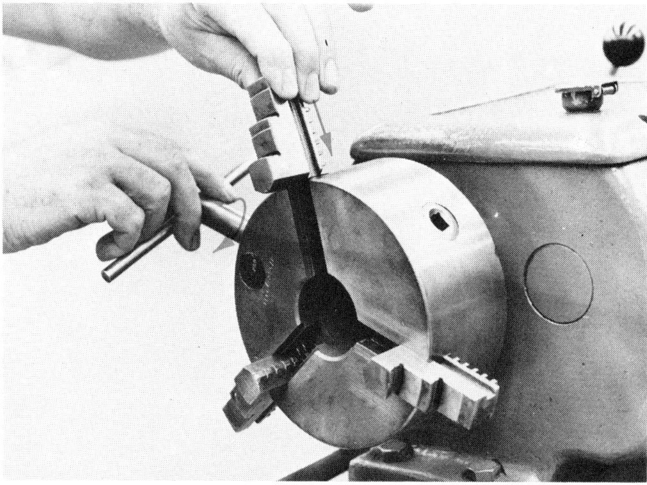

**Fig. 46-3C** Assembling jaw 3 into the chuck. (*Courtesy Kostel Enterprises Ltd.*)

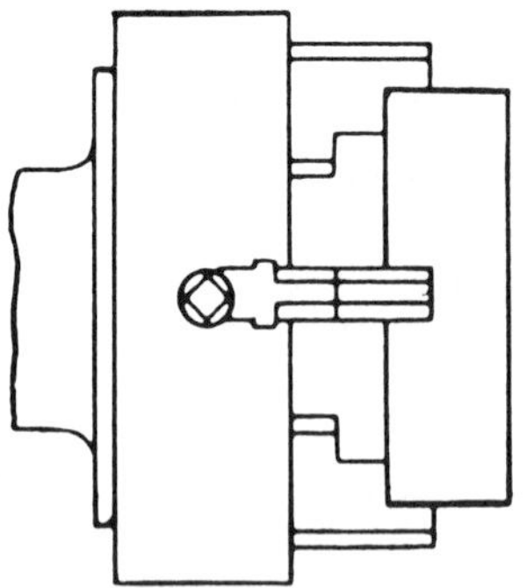
Fig. 46-4A The chuck jaws are reversed to hold large diameter work.

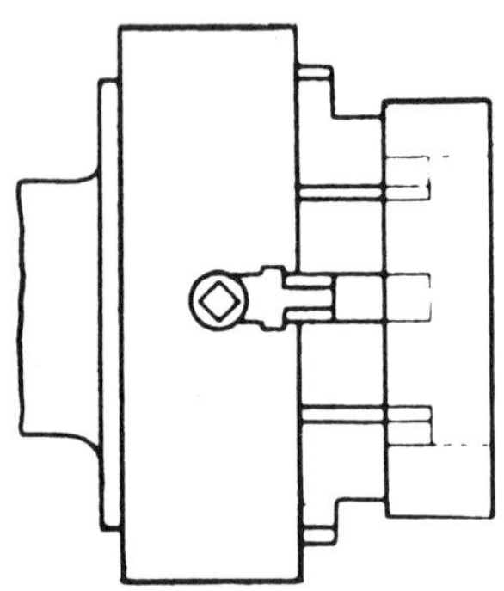
Fig. 46-4B The jaws in the normal position are holding the inside diameter of a workpiece.

N-4 Support your hand on the compound rest so that the chalk will be held steady.

## Four-Jaw Independent Chucks

When it is necessary for a workpiece to run absolutely true, it is generally mounted in a four-jaw independent chuck. Since each jaw can be adjusted independently, work can be trued to within .001 in. (0.02 mm) accuracy or less. The jaws of this chuck are reversible and allow a wide range of work to be gripped either externally or internally (Fig. 46-4A and B).

Round, square, octagonal, hexagonal, and irregularly shaped workpieces can be held in a four-jaw independent chuck. Work can be adjusted to run concentric or off center as required. The face of the chuck has a number of evenly spaced concentric grooves which permit quick and approximate positioning of the chuck jaws.

### Truing Work in a Four-Jaw Chuck (Approximate Methods)

**1.** Measure the diameter of the workpiece.
**2.** Adjust the chuck jaws to the approximate size of the workpiece by using the rings on the face of the chuck.
**3.** Place the work in the chuck and tighten the jaws lightly on the work.
**4.** Adjust all jaws so that they are in the same relation to one ring in the chuck face (Fig. 46-5).
**5.** True the workpiece, using any of the following methods.

(a) *Chalk method*

i. Start the lathe and slowly bring a piece of chalk toward the revolving workpiece until it lightly marks the surface (Fig. 46-6). (See N-4.)
ii. Stop the lathe and see if the chalk mark is uniform around the work circumference. A uniform mark is an indicator that the work is true.
iii. If the chalk mark appears only on a part of the circumference, adjust the workpiece by loosening the jaw opposite the chalk mark and tightening the jaw by the chalk.
iv. Repeat this procedure until the chalk mark appears lightly around the whole circumference.

Fig. 46-5 The rings on the chuck face can be used to adjust the jaws to approximately the size of the work. (*Courtesy Kostel Enterprises Ltd.*)

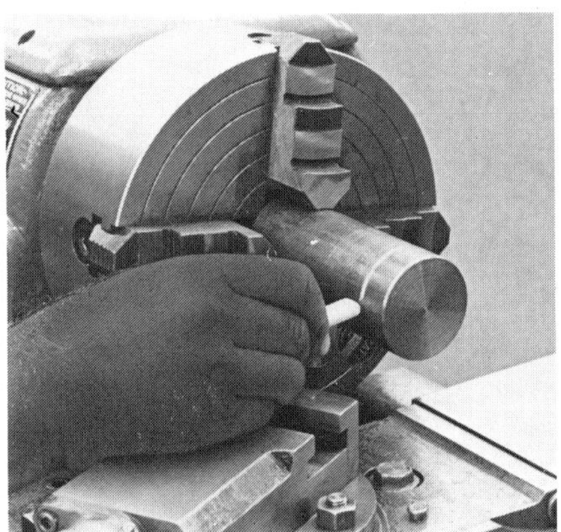
Fig. 46-6 Aligning work in a four-jaw chuck using the chalk method.

Fig. 46-7 Using a surface gage to align work in a four-jaw chuck. (*Courtesy Kostel Enterprises Ltd.*)

**Fig. 46-8** Using the back of a toolholder to align work in a four-jaw chuck.

N-5 Adjust only the two opposite jaws at any time; otherwise it will be very difficult to true the workpiece in a four-jaw chuck.

N-6 Do not use an extension on the chuck wrench handle or the chuck may be damaged.

(b) *Surface Gage method*

i. Place a surface gage on the lathe carriage and adjust the scriber point until it is on center and close to the work circumference (Fig. 46-7).

ii. Revolve the lathe by hand to find the low spot on the work.

iii. Loosen the jaw *nearest* the low spot and tighten the jaw *opposite* the low spot to adjust the work closer to center.

iv. Repeat steps 2 and 3 until the work is running true.

(c) *Toolholder method:* The back of a lathe toolholder is another method of approximately truing a workpiece. Once the toolholder is set as in Fig. 46-8, follow steps 2 to 4 of the surface gage method to true the work in the chuck.

**6.** Tighten all chuck jaws uniformly so that the work is held securely.

**7.** When long pieces of work are mounted in the chuck, it is necessary to check to see that both ends of the work run true. When the work runs reasonably true near the chuck, check and true the other end by tapping the high spot with a soft-faced hammer. Repeat this procedure until both ends of the work run true.

## Truing Work in a Four-Jaw Independent Chuck Using a Dial Indicator

When work must be set up accurately in a chuck, a dial indicator should be used to check the concentricity of the work (Fig. 46-9).

**1.** Mount the work and true it approximately by means of the chalk, surface gage, or toolholder method.

**2.** Mount a dial indicator in the toolpost of the lathe (Fig. 46-9) with its plunger in a horizontal position.

**3.** Adjust the crossfeed handle until the indicator needle registers about .020 in. (0.50 mm).

**4.** Take the lathe "out of gear" and revolve the spindle slowly by hand.

**5.** Note the highest and lowest reading on the indicator. (See N-5.)

**6.** Loosen the jaw at the low reading no more than one-eighth of a turn and tighten the opposite jaw (at the high reading) one-half the difference between the two readings (Fig. 46-10).

**7.** Continue to adjust *only* these two jaws until the indicator reading is the same for both of them.

**8.** Adjust the other two jaws in the same manner until the indicator needle shows no movement at any point around the circumference.

**9.** Tighten all jaws evenly and recheck the indicator reading on the workpiece. (See N-6.)

**Fig. 46-9** Truing work in a four-jaw chuck using a dial indicator. (*Courtesy Kostel Enterprises Ltd.*)

Fig. 46-10 The work revolved one-half turn to get the high reading. (*Courtesy Kostel Enterprises Ltd.*)

## Collet Chucks

Collet chucks are used on lathes for holding and machining small parts, since they allow the parts to be mounted quickly and accurately. Because it is easy to impair the accuracy of the collets, the following precautions should be observed when using a draw-in or spring collet chuck.

**1.** Be sure the surface of the work is true and free of scale and burrs.

**2.** The size of work held in a spring collet should not be more than .002 to .003 in. (0.05 to 0.07 mm) over or under the collet size. (See N-7.)

N-7 Never hold rough work or work that is too small so that the collet is not damaged.

**3.** Never tighten a collet unless there is a workpiece in it. If pressure is left on the collet, it may spring out of shape.

**4.** Indicate the first workpiece in each lot to check for runout and taper. Should either of these occur, remove the collet chuck and thoroughly clean

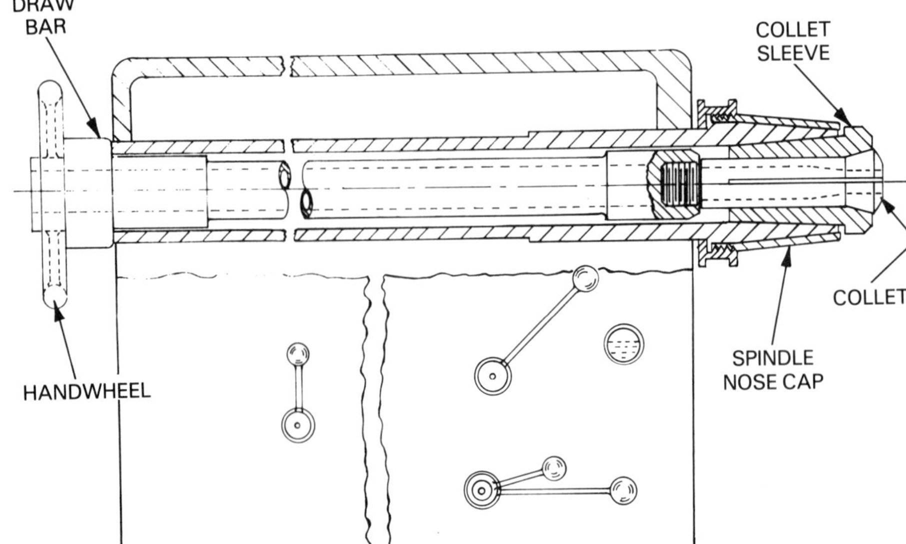


**Fig. 46-11** A draw-in collet chuck assembly mounted in the headstock spindle.

Fig. 46-12 A spring collet chuck mounts on a lathe spindle. (*Courtesy Cushman Industries Inc.*)

Fig. 46-13 The Jacobs spindle nose chuck with rubber-flex collets.

N-8 Long work can be held in a collet provided that the other end is supported by the tailstock center or a steady rest.

N-9 If a short workpiece (less than ¾ in. or 19.05 mm long) must be held in a collet, place a plug of the same diameter as the work in the back of the collet. This will ensure that the work will be held securely in the collet.

N-10 If it is not possible to support longer work, take light cuts and use extreme care when machining.

and remove any burrs from the collet, chuck, and the lathe spindle.

### Use of Collet Chucks

*Draw-in Collet Chucks*

**1.** Clean the spindle nose and the lathe spindle hole.

**2.** Clean the bearing surface of the collet chuck spindle adapter.

**3.** Mount the collet chuck on the spindle nose and insert the drawbar in the back of the lathe spindle (Fig. 46-11).

**4.** Insert the proper collet into the chuck.

**5.** Start the thread of the drawbar onto the thread of the collet.

**6.** Insert the work into the collet, leaving a maximum of three times the diameter protruding. (See N-8.)

**7.** Turn the handwheel to tighten the collet onto the work.

**8.** Proceed with the machining operations.

*Spring Collet Chucks*

Spring collet chucks (Fig. 46-12) are mounted on the spindle in the same manner as a chuck or driveplate. Since they are not held in place by a draw bar, spring collet chucks can hold larger-size stock than the draw-in collet chuck. The principle and application of spring collet chucks are the same as the draw-in collet chuck.

*The Jacobs Spindle Nose Collet Chuck*

The Jacobs collet chuck (Fig. 46-13) has a wider range than the spring collet chucks. The rubberflex collet consists of a series of metal inserts molded into rubber. Since these collets are very flexible, each collet has a range of approximately ⅛ in. (3.17 mm) [1/16 in. (1.58 mm) over and under the nominal size]. This allows a much wider range of workpieces to be accommodated with fewer collets.

*Mounting Work in a Spindle Nose Collet Chuck*

**1.** Clean the lathe spindle nose and the taper in the chuck adapter plate.

**2.** Set the speed-change levers at a low spindle speed to lock the spindle.

**3.** Mount the chuck on the spindle nose.

**4.** Clean the body cone and the inside of the nose.

**5.** Clean the internal and external surfaces of the collet.

**6.** Place the proper collet in the body cone.

**7.** Place the nose collar over the collet and thread the nose into the chuck by rotating the handwheel clockwise.

**8.** Insert the work into the collet and rotate the handwheel clockwise to tighten the collet jaws onto the workpiece. (See N-9.)

**9.** Press the lock ring in toward the spindle nose and turn the handwheel slightly until the lock ring snaps into place. This locks the chuck in position.

**10.** Set the proper spindle speed and machine the workpiece.

## MACHINING OPERATIONS

The operations for machining work in a chuck are basically the same as for machining between centers. The work machined in a chuck is generally short in length and usually does not require additional support. However, if a length of more than three times the diameter extends beyond the chuck jaws, the work should be supported by a steady rest or tailstock center. Use a revolving tailstock center whenever possible, since the expansion of the work during the machining operation will not affect this center. (See N-10.) Although any type of machining operation can be performed on work held in a chuck, the most common operations are facing, turning to size, and cutting off.

### Hints on Chuck Work

All diameters should be machined on a workpiece before it is removed from a three-jaw chuck. If the work is removed and then replaced in the chuck for further machining, it will probably not run true.

**1.** Remove the headstock center and the spindle sleeve before mounting a chuck.

**2.** Clean the lathe spindle and the chuck adapter before mounting the chuck.

**3.** Tighten the chuck jaws around the most rigid part of the work to prevent distortion of the workpiece.

**4.** If the work projects more than three times the diameter of the stock, it should be supported by a revolving tailstock center or steady rest.

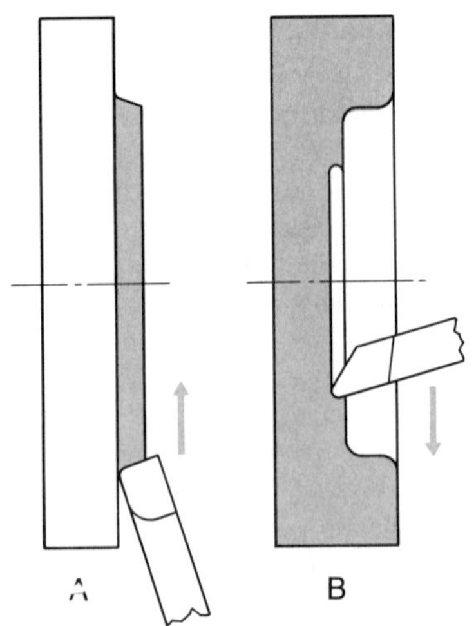


Fig. 46-14 The cutting tool is fed radially to face a surface flat and square with the axis of the work.

N-11 If the work slides into the chuck and comes off the dead center, it may bend.

N-12 The toolholder should point only to the left for finishing cuts.

N-13 With the compound rest set at 30°, a .020-in. feed on the graduated collar moves the cutting tool sideways .010 in. or half the amount fed.

**5.** Never grip the work on a diameter smaller than the diameter to be machined unless absolutely necessary, otherwise the work may be bent.

**6.** Tighten the chuck securely so that the workpiece is not moved into the chuck by the pressure of the cut. (See N-11.)

**7.** Always set the toolbit point on center. If the toolbit is set too low, the work may be bent.

**8.** Position the toolpost on the left side of the compound rest. This prevents the jaws from striking the compound rest when cutting close to the chuck.

**9.** Set the toolholder at 90° to the work or slightly to the right to prevent the tool from "digging in" on a heavy cut. (See N-12.)

**10.** Move the carriage until the toolbit is at the extreme left end of travel and the toolbit is within ⅛ in. (3.17 mm) of the work surface. Rotate the chuck one turn by hand to see that the jaws do not strike the compound rest.

**11.** Never use an air hose to clean a chuck.

**12.** Oil the chuck sparingly.

**13.** Store chucks in suitable compartments when not in use.

**14.** Never leave a chuck wrench in a chuck.

# FACING

Facing is the process of machining the end or face to produce a smooth, flat surface which is square with the axis of the work. This operation is similar to facing between the centers and is performed for the following reasons:

- To square the end surface
- To have an accurate surface from which to measure
- To cut work to length

Work faced in a chuck generally consists of shoulders and larger surfaces than those faced between centers. Work can be faced more conveniently in a chuck since the lathe centers do not interfere with the machining operation. The workpiece can be held in either a three- or four-jaw chuck which allows both external and internal surfaces to be faced (Fig. 46-14A and B).

## Facing in a Chuck

**1.** Mount and true the work in the chuck.

**2.** Set the compound rest 30° to the right if only one surface is to be faced (Fig. 46-15) (see N-14), or set the compound rest at 90° to the cross slide if a series of short steps or shoulders

Fig. 46-15 Accurate facing is possible when the compound rest is set at 30°. (*Courtesy Kostel Enterprises Ltd.*)

Fig. 46-16 Set the compound rest at 90° if a series of short steps or shoulders must be faced accurately. (*Courtesy Kostel Enterprises Ltd.*)

Fig. 46-17 In some cases, one toolbit can be used for both rough and finish facing cuts.

N-14 With the compound rest set at 90°, a .020-in. feed on the graduated collar will move the cutting tool .020 in. or the amount fed.

N-15 Feed the toolbit from the outside to the center for the roughing cut; from the center outward for the finish cut.

N-16 Be sure to remove the parallels before starting the lathe.

must be faced to accurate lengths (Fig. 46-16).

**3.** Fasten a facing tool in the lathe toolholder (Fig. 46-17). The toolbit shown in Fig. 46-17 may be used for both roughing and finishing cuts when one surface is being cut.

**4.** Adjust the toolholder so that the side clears the surface to be faced (Fig. 46-18).

**5.** Set the point of the facing tool exactly on center.

**6.** Set the lathe speed to suit the diameter and type of material being cut.

**7.** Start the lathe and move the carriage so that the point of the facing tool is just touching the surface to be faced.

**8.** Turn the crossfeed handle and bring the toolbit out to clear the work diameter.

**9.** Lock the carriage in position and set the depth of the roughing cut to within .015 in. (0.39 mm) of the finished size with the compound rest handle. The roughing cut should be as large as possible; if necessary, take more than one roughing cut. (See N-15.)

**10.** Take the roughing cut feeding the toolbit in toward the center with the crossfeed handle or automatic crossfeed (Fig. 46-17).

**11.** Measure the work to determine the amount yet to be removed.

**12.** Set the depth of the finish cut with the compound-rest handle.

**13.** Take the finish cut, feeding from the center to the outside diameter using the crossfeed handle or automatic feed.

**14.** Remove the burrs from the edge of the workpiece with a file.

**15.** If the other end must be faced parallel, reverse the work in the jaws and tap the piece snugly against the chuck face; or if the work is shorter than the jaws, place parallels between the work and chuck face. Tap the work against the parallels and tighten the chuck jaws. (See N-16.)

**16.** Face the work to length.

## TURNING DIAMETERS

Much lathe work is machined between centers because the work setup is quick and simple. However, there is a lot of work which cannot be held between centers and this type of work is generally held in a three- or four-jaw chuck for machining. Since turning in a chuck is similar to turning between centers, the operations of rough and finish turning will only be briefly reviewed. For a more detailed explanation of these operations refer to Unit 40. Whenever possible, machine the work to size in two cuts: one roughing cut and one finish cut.

Fig. 46-18 A toolbit set for facing the end of a workpiece. (*Courtesy Kostel Enterprises Ltd.*)

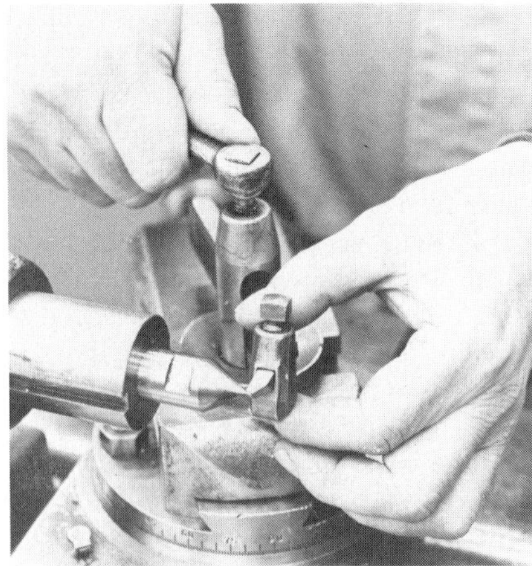

**Fig. 46-19** The toolholder should be gripped short and the cutting tool set to the lathe center point.

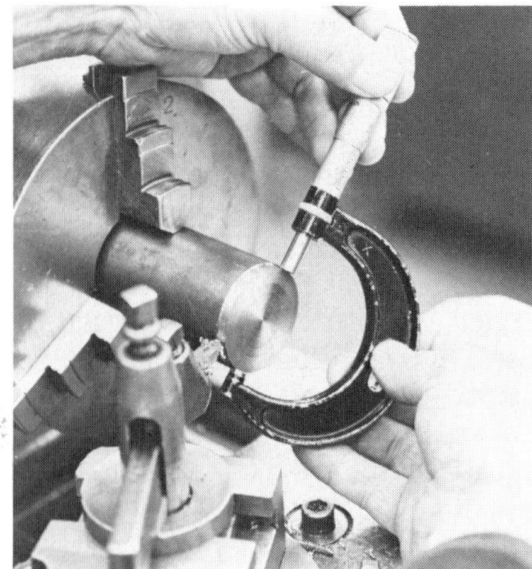

**Fig. 46-20** Measuring the diameter of the trial cut.

N-17 Longer work must be supported by a revolving tailstock center or steady rest.

N-18 Take the rough cut to within $\frac{1}{32}$ in. (0.79 mm) of the finish size.

N-19 Never grip work on a small diameter and try to cut off or groove the large diameter. This can result in bent work.

## Rough and Finish Turning

**1.** Mount the work in the chuck having no more than three times the diameter extending beyond the chuck jaws. (See N-17.)

**2.** Mount a general-purpose toolbit in a straight or left-hand toolholder.

**3.** Move the toolpost to the left side of the compound rest and grip the toolholder short.

**4.** Set the cutting tool point to center and tighten the toolpost screw securely (Fig. 46-19).

**5.** Set the lathe speed and feed for rough turning.

**6.** Take a trial cut, to a true diameter, $\frac{1}{4}$ in. (6.35 mm) long at the end of the work.

**7.** Stop the lathe and measure the diameter of the trial cut (Fig. 46-20).

**8.** Set the crossfeed graduated collar in one-half the amount of metal to be removed for the rough cut.

**9.** Take a trial cut $\frac{1}{4}$ in. (6.35 mm) long and measure the diameter again (Fig. 46-20). (See N-18.)

**10.** Rough turn the diameter to the required length.

**11.** Hone the cutting tool or replace it with a finish turning toolbit.

**12.** Set the speeds and feeds for finish turning.

**13.** Take a light trial cut $\frac{1}{4}$ in. (6.35 mm) long and calculate the amount to be removed.

**14.** Set the crossfeed handle in one-half the amount to be removed and take another trial cut.

**15.** Finish turn the diameter to the required length.

# CUTTING OFF WORK IN A CHUCK

When several identical, short parts are required, they are often made from bar stock inserted through the lathe spindle and held in a chuck or collet. After each part is machined to the correct size and shape, it is cut from the longer bar with a cutting-off or parting tool mounted in the toolpost. Cutoff tools may also be used for grooving and undercutting operations.

Most work is cut off with an inserted-blade cutting-off tool (Fig. 46-21); however, a parting tool ground from a square toolbit is often more satisfactory for small work.

## Hints on Cutting Off

Since cutting-off tools are thin and more fragile than most lathe tools, it is important that certain precautions be observed in their use. The following suggestions should help eliminate most of the problems encountered during cutoff operations.

**1.** Work must be held securely in the chuck or collet.

**2.** Extend the workpiece only enough to allow the cut to be made as close to the chuck jaws as possible.

**3.** Work with more than one diameter should be held on the larger diameter when cutting off. (See N-19.)

**4.** Extend the cutoff blade beyond the toolholder only enough to cut off the part; one-half the diameter plus $\frac{1}{8}$ in. or 3.17 mm for clearance (Fig. 46-22).

A

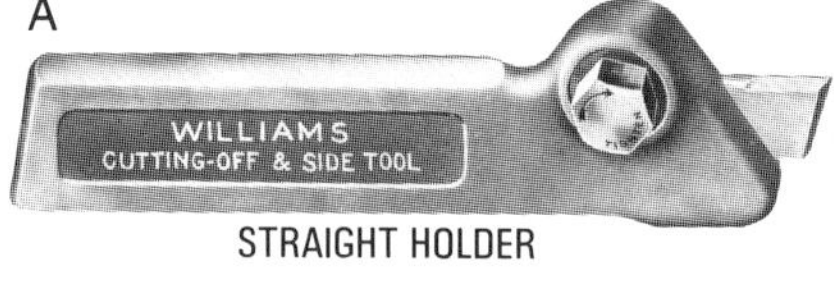


B

C

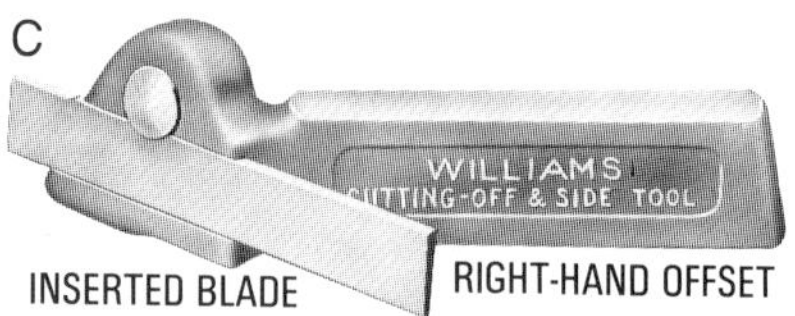


**Fig. 46-21** Types of inserted-blade cutoff tools.

**Fig. 46-22** Grip the cutoff toolholder short and have the blade extending only enough to cut off the work. *(Courtesy Kostel Enterprises Ltd.)*

**Fig. 46-23** Position the cutting tool at 90° to the work. (*Courtesy Kostel Enterprises Ltd.*)

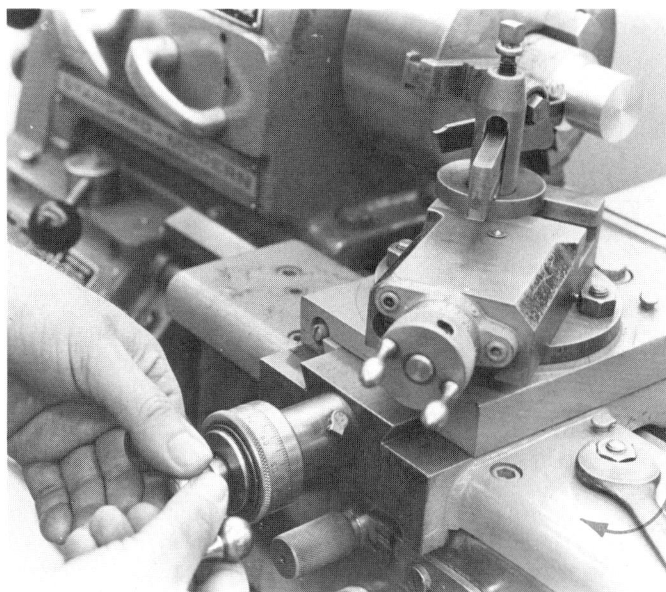

**Fig. 46-24** Using both hands to feed the cutting tool steadily into the workpiece. (*Courtesy Kostel Enterprises Ltd.*)

N-20 Too slow a feed dulls the cutting edge; too fast a feed causes jamming and tool breakage.

N-21 Apply cutting fluid to assist the cutting action and to prevent the tool from jamming.

N-22 Moving the carriage slightly is advisable on all parting operations, especially when grooves are deeper than ¼ in. (6.35 mm).

**5.** Grip the toolholder as short as possible with the blade set on center.

**6.** Feed the tool steadily during grooving or parting operations. (See N-20.)

**7.** Always wear safety glasses when using a cutting-off tool.

**8.** Use cutting fluid on steel. Cut brass and cast iron dry.

## To Cut Off Work Held in a Chuck

**1.** Mount the work in the chuck with the part to be cut off as close to the jaws as possible.

**2.** Mount a right-hand offset cutoff tool on the left side of the compound rest (Fig. 46-22).

**3.** Set the cutting-off toolholder as close to the toolpost as possible to prevent vibration and chatter.

**4.** Have the cutting blade extending beyond the holder half the diameter to be cut off plus ⅛ in. (3.17 mm) for clearance.

**5.** Set the cutting tool to center and at 90° to the work and tighten the toolpost screw securely (Fig. 46-23).

**6.** Set the lathe to one-half the turning speed.

**7.** Move the cutting tool to the position of the cut.

**8.** Start the lathe and steadily feed the cutting tool into the work with the crossfeed handle (Fig. 46-24). (See N-21.)

**9.** While feeding the tool into the work, move the cutting tool back and forth a few thousandths with carriage handwheel to prevent the tool from binding in the groove. (See N-22.)

**10.** Remove the burrs from both sides of the groove with a file before the part is cut off.

**11.** Feed the tool slowly when the part is almost cut off.

## REVIEW QUESTIONS

**1.** Name three common chucks which can be mounted on the headstock spindle.

**2.** How does a three-jaw chuck operate?

**3.** List the procedure for assembling the jaws into a three-jaw chuck.

**4.** What is the purpose of the rings or grooves in the face of a four-jaw chuck?

**5.** List three methods of approximately aligning work in a four-jaw chuck.

**6.** Explain how to true work in a four-jaw chuck to within .001 in. (0.02 mm) of trueness.

**7.** Name three types of collet chucks used on lathes.

**8.** List three precautions which should be observed with spring collet chucks.

**9.** How much range do Jacobs rubberflex collets have?

## Machining Operations

**10.** If work projects more than three times its diameter beyond the chuck jaws, how should it be supported?
**11.** What can be done to prevent the chuck jaws from striking the compound rest?
**12.** List three purposes of facing.
**13.** If the compound rest is fed in .030 in. (0.76 mm), how much would be removed from the face of work if:
(a) The compound rest is set at 30°
(b) The compound rest is set at 90°
**14.** How can the carriage be prevented from moving during the facing operation?
**15.** In which direction should the cutting tool be fed for rough and finish facing?
**16.** What is the purpose of the trial cut when turning diameters?
**17.** How close to finish size should the rough cut be taken?
**18.** Name three common parting toolholders.
**19.** How far should the blade extend beyond the cutoff toolholder?
**20.** While grooving or parting, what is the effect of
(a) Too-slow feed
(b) Too-fast feed
**21.** Why should the cutting tool be moved back and forth sideways slightly when cutting off?

# UNIT 47

# DRILLING, BORING, AND REAMING

The variety of tools available greatly increases the number of operations which can be performed on work held in a lathe. Some of the cutting tools are mounted in the tailstock spindle while others are fastened in the toolpost. On work held in a chuck or fastened to a faceplate, operations such as drilling, boring, and reaming can be performed on the end of a workpiece.

## OBJECTIVES

After completing this unit you should be able to:

**1.** Accurately spot and drill various-size holes with a drill held in a drill chuck or the tailstock spindle
**2.** Select and use various boring bars to machine a hole to within ±.001 in. (0.02 mm) accuracy
**3.** Ream a hole by power in a lathe
**4.** Ream a hole by hand in a lathe

## DRILLING

Work held in a chuck can be drilled quickly and accurately in a lathe. The drill, held in a drill chuck or in the tailstock spindle, is brought against the revolving work by turning the tailstock handwheel. Drilling generally precedes lathe operations such as boring, reaming, and tapping.

Various methods can be used to hold drills in a lathe, depending on the size of drill being used. The most common methods are:

- Straight shank drills—In a drill chuck.
- Taper shank drills—In the tailstock spindle taper.
- Large taper shank drills—In a drill holder.

### Drilling with the Drill Mounted in a Drill Chuck

**1.** Mount and true the work in a chuck.
**2.** Face the end of the workpiece square and to size.
**3.** Set the lathe to the proper speed for the size and type of material to be drilled.
**4.** Check the tailstock and make sure that it is aligned with the headstock center.
**5.** With a center drill, spot the hole until about three-quarters of the tapered portion of the center drill enters the work (Fig. 47-1).
**6.** Mount the correct size drill in a drill chuck (Fig. 47-2).

In some cases where the drill is sufficiently rigid, it is possible to start the drill without the use of a center hole. The drill should be supported by the end of the toolholder (Fig. 47-2) to

Fig. 47-1 A center hole provides a guide for the drill and keeps it from wandering. (*Courtesy Kostel Enterprises Ltd.*)

Fig. 47-2 The end of the toolholder may be used to prevent the drill from wobbling when starting a hole.

Fig. 47-3 Using a rule to measure the depth of a hole being drilled.

N-1 If the drill flutes become clogged with chips, the drill may jam in the hole and break.

N-2 The pilot hole should be a little larger than the thickness of the drill web.

prevent the drill from wobbling as it starts into the workpiece.

**7.** Start the lathe and turn the tailstock handwheel to feed the drill into the work.

**8.** Apply cutting fluid frequently and drill the hole to depth.

**9.** Back the drill out of the hole occasionally to clear the chips from the drill flutes. (See N-1.)

**10.** Check the depth of the hole using a rule or the graduations on the tailstock spindle (Fig. 47-3).

**11.** Always *ease up* the drill pressure as a drill starts to break through the work.

### Drilling with the Drill Mounted in the Tailstock Spindle

If a hole larger than the capacity of the drill chuck is required, it is necessary to mount a tapered shank drill in the tailstock spindle (Fig. 47-4), or in a drill holder (Fig. 47-5).

Because of the wide web on a large drill, there is a tendency for the drill to wander, particularly at the start of the hole. To overcome this, a pilot hole, slightly larger than the web of the drill, should be first drilled in the metal. The use of a pilot hole reduces the tendency for the drill to wander since the point of the drill does not contact the metal and cannot pivot off center. By using a pilot hole, a drill cuts fairly close to size, provided that it is correctly ground. (See N-2.)

Fig. 47-4 A taper shank drill mounted in the spindle is prevented from turning by a lathe dog.

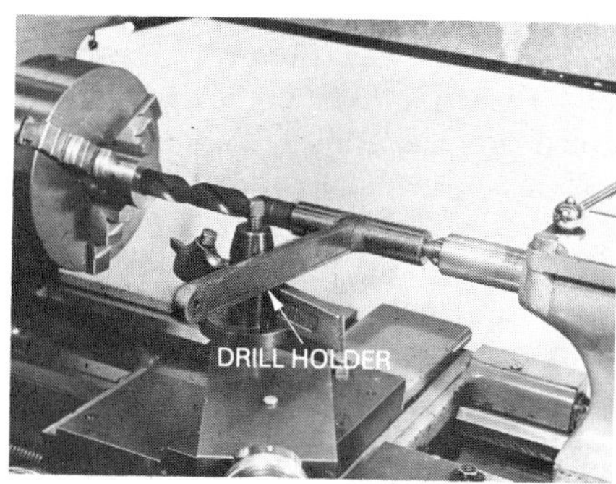


Fig. 47-5 Drilling a hole with a drill mounted in a drill holder and supported on the dead center.

## BORING

Boring is the process of enlarging and truing a drilled or cored hole with a single-point cutting tool. When a hole is drilled in a workpiece in a lathe, the hole may not be concentric with the outside diameter. This occurs if the drill becomes dull or wanders due to hard spots or sand holes in the metal. If the hole location and size must be accurate, it is necessary to bore the hole so that the hole is concentric with the diameter of the workpiece. Boring is also used to finish an off-size hole for which no drill or reamer is available (Fig. 47-6).

### Boring Tools and Holders

Since boring is a turning operation performed on the inside of a hole, the boring tool is similar to a turning tool. Boring tools may be of two types, the solid forged type and the boring bar type with a toolbit mounted in the end of the bar.

Fig. 47-6 Boring may be used to true the inside diameter of a hole. (*Courtesy R. K. LeBlond Machine Tool Co.*)

**Fig. 47-7** Forged boring bars are used for small holes and light cuts.

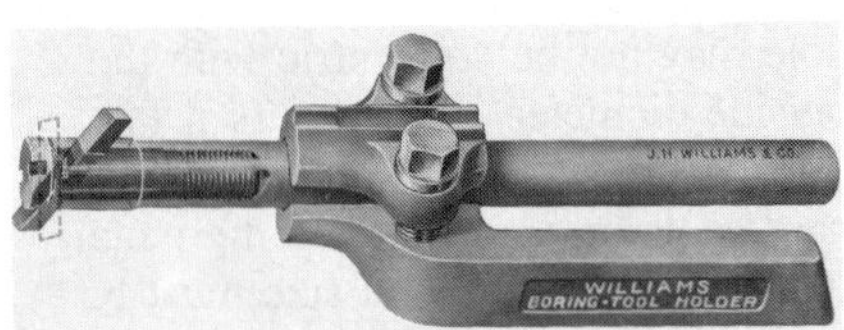


**Fig. 47-8** A boring bar and toolholder used for general-purpose boring. (*Courtesy J. H. Williams & Co.*)

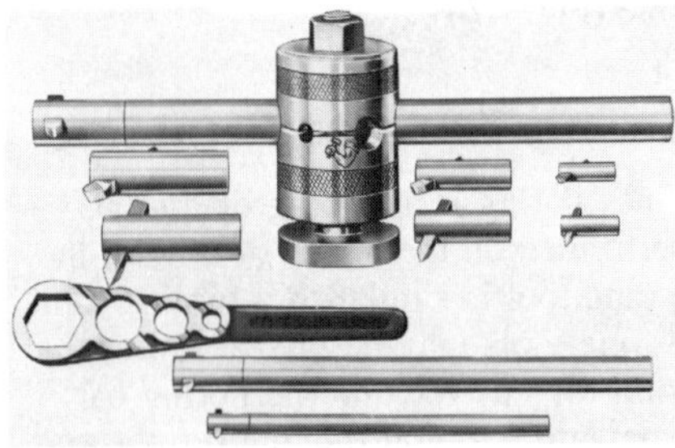

**Fig. 47-9** A heavy-duty boring bar set. (*Courtesy Armstrong Bros. Tool Co.*)

N-3 The boring bar should be level, parallel to the centerline of the lathe and held as short as possible to minimize chatter.

N-4 The speed for boring is the same as for turning and is calculated on the hole diameter.

N-5 Leave the hole .010 to .020 in. (0.25 to 0.51 mm) undersize for the finish cut.

N-6 If chatter occurs, reduce the speed and increase the feed.

The *solid forged boring tool* (Fig. 47-7) is generally made from high-speed steel and then ground to resemble a left-hand turning tool. Boring tools of this type are generally used for light-duty machining and are held in a special toolholder which is mounted in the toolpost.

A *boring bar toolholder* (Fig. 47-8) is mounted in the toolpost and is used for heavier cuts than the forged boring tool and for general-purpose boring. Various-size bars, having broached holes to accommodate square toolbits can be mounted in the toolholder.

A *heavy-duty boring bar set* (Fig. 47-9) is mounted in the compound rest and can accommodate three different sizes of boring bars. The construction of this unit permits easy changing of the boring bars to suit the size of hole. This allows the operator to use the largest bar for each job which permits the use of greater speeds and feeds, thereby increasing production.

## Boring

**1.** Mount the workpiece in a chuck.

**2.** Face and center drill the workpiece.

**3.** Set the lathe to the proper speed for drilling.

**4.** Drill the hole to the required depth and to within 1⁄16 in. (1.59 mm) of the finished diameter.

**5.** Mount the boring bar holder on the left side of the compound rest (Fig. 47-6).

**6.** Mount the largest boring bar which can be accommodated in the drilled hole. (See N-3.)

**7.** Set the point of the toolbit just slightly above center because there is a tendency for the tool to spring down when cutting.

**8.** Set the lathe to the proper speed for the material to be bored and select a medium feed .008 to .010 in. (0.20 to 0.25 mm) for rough boring. (See N-4.)

**9.** Start the lathe and turn the crossfeed handle *counterclockwise* until the boring tool touches the inside diameter of the hole.

**10.** Take a light trial cut about .005 in. (0.12 mm) deep and about 1⁄4 in. (6.35 mm) long at the right-hand end of the work.

**11.** Stop the lathe and measure the diameter of the bored section, using a telescopic gage or an inside micrometer (Fig. 47-10A and B).

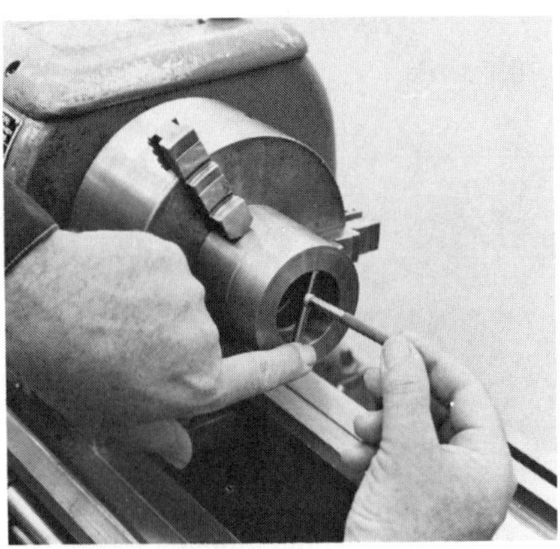

**Fig. 47-10A** Measuring a small hole using a telescopic gage. (*Courtesy Kostel Enterprises Ltd.*)

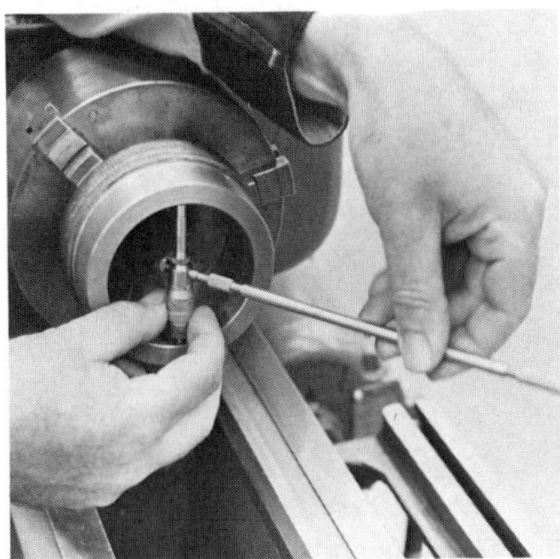

**Fig. 47-10B** Measuring a larger hole with an inside micrometer. (*Courtesy Kostel Enterprises Ltd.*)

**12.** Calculate the amount of material to be removed from the hole for the roughing cut. (See N-5.)

**13.** Feed the cutting tool *out* (counterclockwise), using the crossfeed handle *one-half* the amount of material to be removed.

**14.** Take a trial cut 1⁄4 in. (6.35 mm) long and recheck the diameter.

**15.** Take the roughing cut for the required length. (See N-6.)

**16.** Stop the lathe and move the carriage to the right until the boring tool clears the hole. *Do not move the crossfeed handle*.

**17.** Set a fine feed .005 in. (0.12 mm) for the finish cut.

**18.** Start the lathe, take a light trial cut about 1⁄4 in. (6.35 mm) long, and check the diameter for size.

**19.** When the size is correct, take the finishing cut.

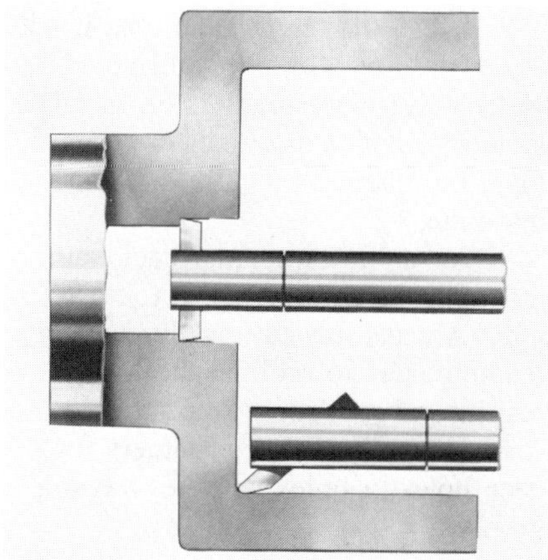

**Fig. 47-11** A boring bar may be used for finish boring a hole and squaring the shoulder of a counterbore. (*Courtesy Armstrong Bros. Tool Co.*)

N-7 Set the compound rest at 90° and lock the carriage to face a shoulder to an accurate depth.

N-8 Nominal size is the size stamped or marked on the reamer.

## COUNTERBORING

*Counterboring* is the process of enlarging a hole for a definite depth. Holes are counterbored to provide recesses for bolts heads, nuts, and differences in shaft diameters, and to act as a guide when internal threading. A counterbored hole has parallel sides and the shoulder between the bored and the counterbored hole is square (Fig. 47-11).

The procedure for counterboring is the same as for boring a hole except that it goes only part way through an existing hole. After the hole is bored to size, the shoulder is generally faced square with the hole. (See N-7.)

## REAMING

*Reaming* is the process of enlarging a drilled or bored hole to produce an accurately sized and shaped hole with a good surface finish. This operation is performed with a tool called a reamer. Reamers are divided into two classes: machine reamers and hand reamers.

### Machine Reamers

Machine reamers may be divided into two classes:

- Roughing reamer—Cuts on the ends only
- Finishing reamer—Cuts on the ends and along the length of the tooth

### Roughing Reamers

Roughing reamers are designed to produce holes quickly and to within a few thousandths under the nominal size of the hole.

The *rose reamer* (Fig. 47-12) is the most commonly used machine reamer. The ends of the teeth are chamfered to 45° and are provided with clearance so that all the cutting is done on the end of the reamer. The lands have no clearance and are used to guide the reamer in the hole. Rose reamers remove material quickly, and are particularly suited to the machining of cored holes. They are often made about .003 to .005 in. (0.07–0.12 mm) under the nominal size so that the hole can be finished to size with a finishing reamer. (See N-8.)

### Finishing Reamers

Finishing reamers are used to finish a hole which has been bored or rough-reamed to within about .005 in. (0.12 mm) of the finish size.

*Fluted chucking reamers* (Fig. 47-13) have more teeth than a rose reamer for any given diameter and may have straight or tapered shanks. They are designed to cut on the full length of the land to produce a smooth and accurately sized hole. The ends of the teeth are slightly chamfered for end cutting.

The *jobber's reamer* (Fig. 47-14) is available with straight or helical flutes. This reamer cuts along the full length of the lands, which are longer than

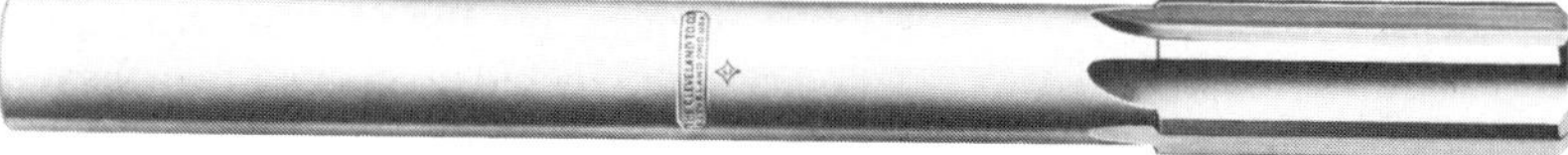

**Fig. 47-12** A rose reamer cuts on the end angle only.

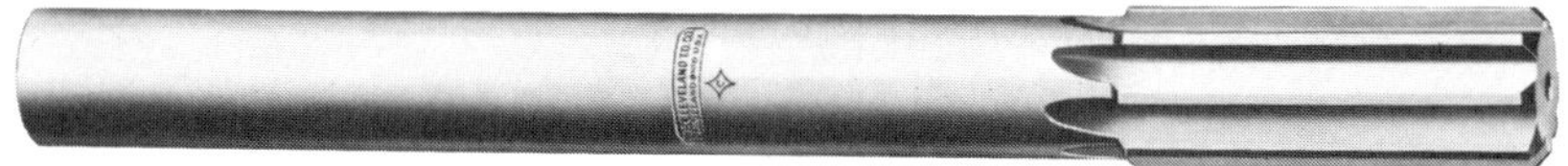

**Fig. 47-13** A fluted chucking reamer cuts on the ends as well as on the full length of the land.

**Fig. 47-14** A jobber's reamer has longer flutes than a chucking reamer and a tapered shank.

**Fig. 47-15** A chucking expansion reamer is adjusted to diameter by a small set screw in the end.

**Fig. 47-16** The adjustable reamer has inserted blades.

**Fig. 47-17** Taper machine reamers.

N-9 The reamer cutting edges can be nicked or damaged through improper storing.

N-10 Never turn the lathe spindle or reamer backward; this will damage the reamer.

N-11 Too high a speed will dull the cutting edges of a reamer.

those of a fluted chucking reamer. It is used as a precision reamer and is designed to remove only a few thousandths from the hole diameter.

*Expansion reamers* are sometimes used as finishing reamers and can be adjusted slightly to compensate for wear. The most common expansion reamers are the chucking expansion (Fig. 47-15) and the adjustable blade reamer (Fig. 47-16).

### Taper Reamers

*Taper machine reamers* (Fig. 47-17) are manufactured in all standard tapers and with tapered shanks so that they can be mounted directly in the spindle of the machine. The hole is generally step-drilled first and followed by the roughing and finishing taper reamers.

## Reamer Care

Proper care of a reamer is important if it is to produce an accurate and smoothly finished hole. The following points should be observed if the reamer is to give top performance.

**1.** Store reamers in separate containers or compartments. Plastic or cardboard tubes provide good reamer storage. (See N-9.)

**2.** Clean the reamer and coat it with a light film of oil before storing.

**3.** Never place reamers on metal surfaces such as machines or bench tops. Lay them on a cloth or a piece of masonite.

**4.** *Never* turn a reamer backward; this dulls the cutting edges. (See N-10.)

**5.** Always use cutting fluid when reaming steel to prolong the life of the reamer and produce a smooth hole.

**6.** Use helical fluted reamers for deep holes or holes with keyways or grooves.

## Reaming Allowances

How a hole has been produced determines the amount of material to be left for reaming. Rough-drilled or punched holes should be rough-reamed before a finishing reamer is used. The amount of material left for reaming purposes depends on several factors such as the condition of the machine, the type of material, the finish desired, the speed and feed of the reamer, and the use of cutting fluid. Table 47-1 shows suggested stock allowances for machine reaming.

## Reaming Speeds and Feeds

Reaming speed is about one-half the drilling speed for similar materials, however the speed is also affected by factors such as:

- The rigidity of the setup
- The type of work material
- The surface finish and tolerance required

Higher reaming speeds may be used on rigid setups. Holes which require a fine finish and close tolerance should be reamed at slower speeds. Cutting fluid permits higher speeds and produces a better surface finish in the hole. (See N-11.)

Table 47-2 gives suggested reaming speeds for high-speed steel reamers. Carbide-tipped reamers can be run at about twice the speed shown. If chatter occurs while reaming, reduce the speed until the chatter stops.

Reaming feeds are generally about twice the feeds used for drilling. Too slow a feed will produce a glazed surface, cause the reamer to dull, and chatter. Too fast a feed reduces the ac-

**TABLE 47-1 MACHINE REAMING ALLOWANCES**

| Hole diameter | | Allowance | |
|---|---|---|---|
| Inch | mm | Inch | mm |
| ¼ | 6.35 | .010 | 0.25 |
| ½ | 12.70 | .015 | 0.38 |
| 1 | 25.40 | .020 | 0.50 |
| 1½ | 38.10 | .025 | 0.63 |
| 2 | 50.80 | .030 | 0.76 |
| 3 | 76.20 | .045 | 1.14 |

**TABLE 47-2 RECOMMENDED REAMING SPEEDS (HIGH-SPEED STEEL)**

| Material | Speed | |
|---|---|---|
| | ft/min | m/min |
| Aluminum | 140–200 | 43–61 |
| Brass | 130–180 | 40–55 |
| Bronze | 50–100 | 15–30 |
| Cast iron | 50–70 | 15–21 |
| Machine steel | 50–70 | 15–21 |
| Stainless steel | 40–50 | 12–15 |
| Magnesium | 170–250 | 52–76 |

**Fig. 47-18** Setup for reaming in a lathe.

N-12 Feed the drill slowly as it breaks through the work to prevent drill breakage.

N-13 The taper on the reamer shank or in the tailstock spindle can be damaged if the reamer turns.

curacy of the hole and produces a poor surface finish. Feed a taper reamer slowly since it is cutting along its entire length.

## Reaming a Straight Hole

Although the reaming of straight and tapered holes is similar, different procedures are used, especially with machine and hand reamers.

**1.** Check that the tailstock center is in line.

**2.** Mount and machine the work in a chuck.

**3.** Face and spot the hole location with a center drill.

**4.** Select the proper-size drill which will leave enough material for reaming.

(a) 1/64 in. (0.39 mm) smaller for holes up to ½ in. (12.7 mm) in diameter.

(b) 1/32 in. (0.79 mm) smaller for holes over ½ in. (12.7 mm) in diameter.

**5.** Mount the drill in the drill chuck.

**6.** Apply cutting fluid and drill the hole to the required depth. (See N-12.)

**7.** Remove the drill and mount a reamer in the drill chuck or tailstock spindle (Fig. 47-18). For reamers over ⅝-in. (15.87-mm) diameter, fasten a lathe dog near the reamer shank and support the tail on the compound rest to prevent it from turning. (See N-13.)

**8.** Slide the tailstock until the reamer is close to the work and lock the tailstock in position.

**9.** Set the lathe to one-half the drilling speed (Table 47-2).

**10.** Apply cutting fluid and slowly feed the reamer into the hole with the tailstock handwheel.

**11.** Occasionally bring the reamer out of the hole to clear the chips from the flutes and add cutting fluid.

**12.** When the hole has been reamed, stop the lathe and remove the reamer from the hole.

**13.** Clean the reamer and store it where it will not be nicked or damaged.

**14.** Remove the burr from the edge of the reamed hole with a scraper.

## Hand Reaming

If a hole is produced slightly undersize by a machine reamer, it can be brought to size by a hand reamer (Fig. 47-19).

## Reaming with a Hand Reamer

**1.** Drill or bore the hole to within .005 in. (0.12 mm) of the finished size.

**2.** Lock the headstock spindle or set the lathe to the slowest speed.

**3.** Mount a center in the tailstock spindle.

**4.** Place the cutting end of the reamer into the hole and slide the tailstock until the lathe center enters the center hole of the reamer shank.

**5.** Lock the tailstock in position.

**6.** Apply cutting fluid to the reamer.

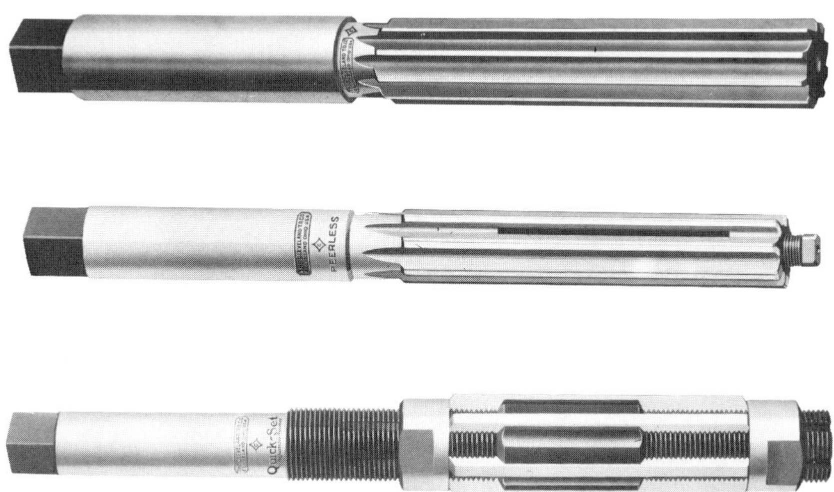


**Fig. 47-19** Straight-fluted hand reamers.

N-14 Never turn a reamer backward.

**7.** Using a wrench, turn the reamer clockwise with the left hand. At the same time, slowly turn the tailstock handwheel to support and advance the reamer into the hole. (See N-14.)

**8.** Remove the reamer occasionally by turning it clockwise, clear the chips from the flutes, and apply cutting fluid.

**9.** Ream the hole to the required depth.

**10.** Remove any burrs from the edge of the hole with a scraper.

# REVIEW QUESTIONS

## Drilling

**1.** Name three methods of holding a drill in a lathe.

**2.** What is the purpose of center drilling the workpiece before drilling a hole?

**3.** How may a large drill be started when a center hole is not used?

**4.** What is the purpose of a pilot hole?

## Boring

**5.** Define the operation of boring.

**6.** Name three types of boring bars and state the purpose of each.

**7.** How close to the finish size should a hole be drilled?

**8.** How can chatter be overcome while boring?

**9.** What is the purpose of counterboring?

## Reaming

**10.** What is the purpose of reaming?

**11.** What is the difference between a rose and a fluted reamer?

**12.** List five important points which should be observed with reamers.

**13.** Name three factors which affect the reaming speed.

**14.** What is the effect of

(a) Too-slow feed

(b) Too-fast feed

**15.** How much material should be left for machine reaming

(a) Holes up to ½ in. (12.7 mm) diameter?

(b) Holes over ½ in. (12.7 mm) diameter?

**16.** How much material should be left in a hole for hand reaming?

# SECTION 

# MILLING MACHINES

illing machines are machine tools used to accurately produce one or more machined surfaces on the workpiece by means of one or more rotary milling cutters having single or multiple cutting edges. The workpiece is held securely on the *worktable* of the machine or in a holding device clamped to the table. It is then brought into contact with a revolving cutter.

The milling machine is a versatile machine tool which can handle a variety of operations such as milling flat and irregular-shaped surface, cutting gear and thread, drilling, boring, reaming, and slotting. Its versatility makes it one of the most important machine tools used in machine shop work.

# UNIT 48

# HORIZONTAL MILLING MACHINES AND ATTACHMENTS

Because of the great variety of parts required by industry, a wide range of milling machines and accessories has been developed to produce these parts more efficiently than could be possible with one type of milling machine.

## OBJECTIVES

This unit will enable you to know:

1. The types of horizontal milling machines used in industry and their application
2. The main parts of the universal horizontal milling machine
3. The purpose and use of the operational parts of the horizontal milling machine
4. The purpose and use of the various attachments and accessories used on the horizontal milling machine

## TYPES OF HORIZONTAL MILLING MACHINES

Milling machines are classified under the following headings:

1. *Manufacturing type,* in which the cutter height is controlled by vertical movement of the headstock
2. *Knee-and-column type,* in which the relationship between the cutter height and the work is controlled by vertical movement of the table

## MANUFACTURING-TYPE MILLING MACHINES

Manufacturing-type milling machines are used primarily for quantity production of identical parts. This type of machine may be either semi-automatic or fully automatic and is of simple but sturdy construction. Fixtures clamped to the table hold the workpiece for a variety of milling operations, depending on the type of cutters or special spindle attachments used. After this machine has been set up, the operator is required only to load and unload the machine and start the automatic cycle, which is controlled by *cams* and preset *trip dogs*.

Some of the more common manufacturing-type milling machines are discussed as follows:

*The plain manufacturing type* (Fig. 48-1) has one horizontal spindle and one headstock. It is sometimes equipped with a reciprocating table cycle which permits feeding and rapid traversing in both directions. On this type of machine, *reciprocal milling* is possible by means of two identical fixtures mounted on opposite ends of the table. While work is being machined at one end, a new piece is being loaded into the fixture at the other end.

*The duplex manufacturing type* (Fig. 48-2) is similar to the plain type, except that it has two horizontal spindles mounted in two independently adjustable headstocks. It can be used to perform two identical or two different milling operations on one or more pieces at the same time.

*The small plain automatic knee-and-column type* (Fig. 48-3) is similar to the plain horizontal mill. It is used for production milling of small- or medium-sized parts. The table is operated by power and controlled automatically by trip dogs mounted on the front of the table.

*The unit-type automatic fixed-bed milling machine* (Fig. 48-4) consists of a group of small-sized manufacturing milling machine units mounted on a common base. As many as four units can be mounted on one base for producing small parts at a high rate of speed.

*The small plain automatic fixed-bed type* (Fig. 48-5) has complete automatic cycle control of the table by a cycle selector and trip dogs. The machine may also be equipped with *automatic rise and fall* of the spindle carrier, permitting quick and economical milling of surfaces on different levels or between obstructions.

*The tracer-controlled milling machine* (Fig. 48-6) has a hydraulic or electric circuit designed to automati-

**Fig. 48-1** A plain manufacturing-type milling machine. (*Courtesy Cincinnati Milacron Co.*)

**Fig. 48-3** The small plain automatic knee-and-column-type milling machine is used for small-lot production. (*Courtesy Cincinnati Milacron Co.*)

**Fig. 48-2** The duplex manufacturing-type milling machine has two milling heads. (*Courtesy Cincinnati Milacron Co.*)

**Fig. 48-4** The unit-type automatic feed-bed milling machine. (*Courtesy Cincinnati Milacron Co.*)

cally control the relative positions of the cutter and the workpiece by a tracer stylus riding on a cam, template, or model. This machine is used for efficient accurate reproduction of curved or irregular surfaces. If the tracer is disengaged, it can be used for standard milling operations.

*The numerical controlled milling machine* (Fig. 48-7) is one of the most efficient developments in the machine industry. Milling machines are particularly suited to numerical control, as a result of which, a large percentage of milling machines produced are of this type.

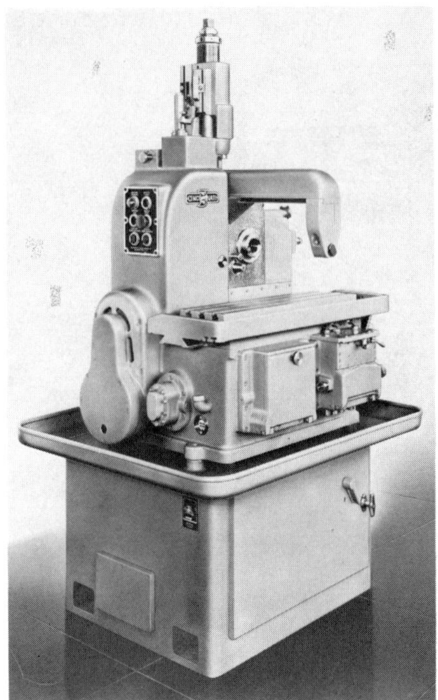

**Fig. 48-5** The small plain automatic fixed-bed machine has complete automatic control. (*Courtesy Cincinnati Milacron Co.*)

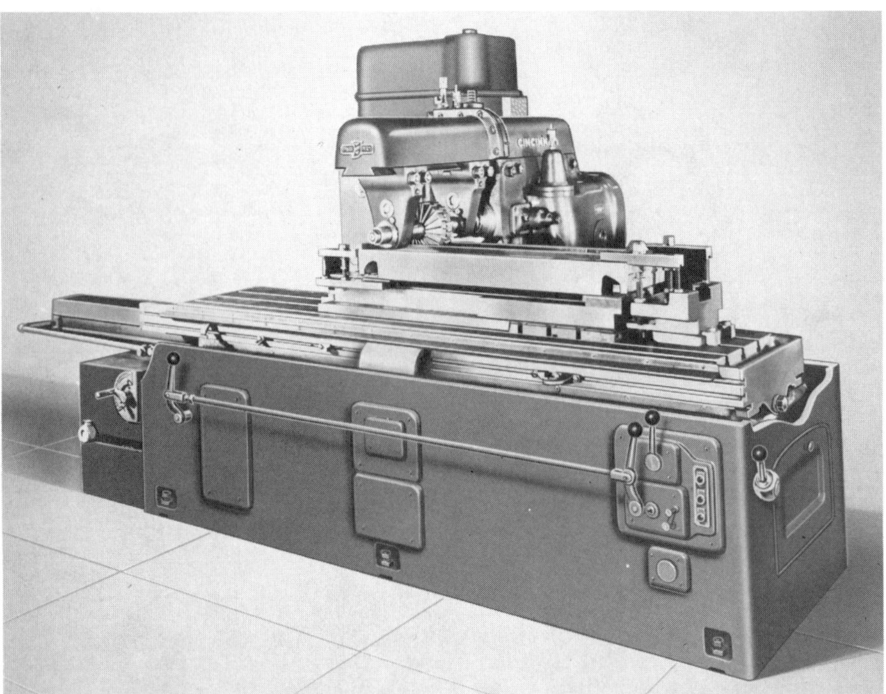

**Fig. 48-6** Tracer-controlled milling machines are used for accurate reproduction of curved and irregular surfaces. (*Courtesy Cincinnati Milacron Co.*)

Numerical control is the control of the (milling) machine by programmed numerical data, usually stored on tape or cards. These data (or instructions) are fed into a unit which decodes the information and directs the machine to perform the operation in a given sequence in order to produce the part rapidly and accurately. Numerical control is especially suited to the production of identical parts which require close tolerances.

**Fig. 48-7** Numerical controlled milling machines can produce duplicate parts efficiently and accurately. (*Courtesy Cincinnati Milacron Co.*)

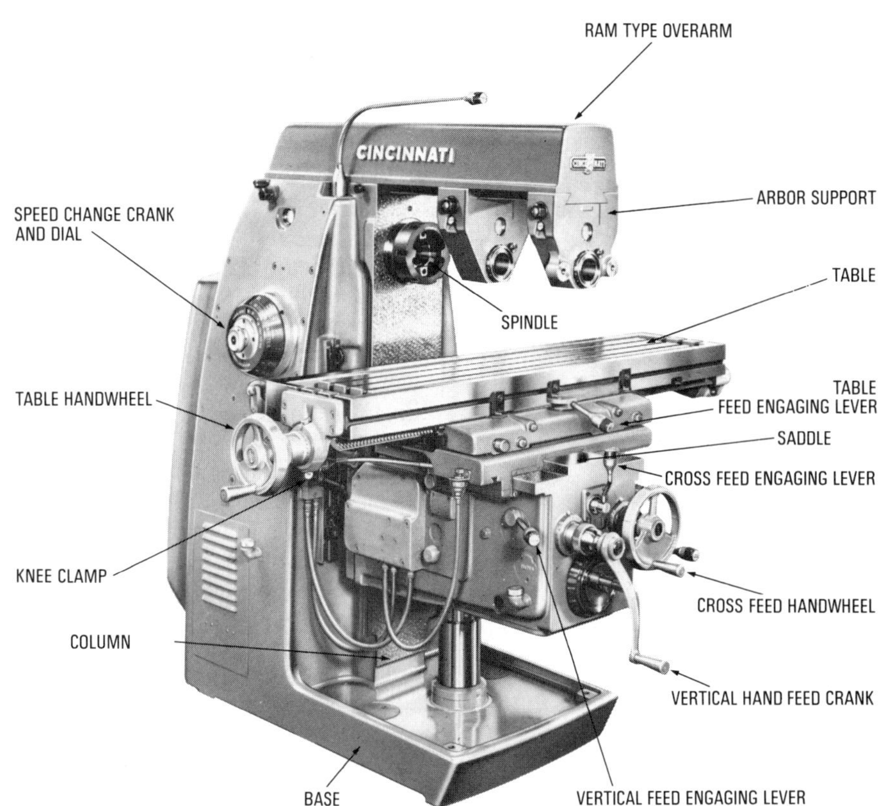


**Fig. 48-8** The universal horizontal milling machine. (*Courtesy Cincinnati Milacron Co.*)

## KNEE-AND-COLUMN-TYPE HORIZONTAL MILLING MACHINES

Since the knee-and-column milling machine is the most commonly used in school shops, jobbing shops, and tool rooms, only this type will be discussed in detail.

In order to be able to operate any machine properly, the student should know the parts of this machine and the purpose of these parts.

Horizontal milling machines fall into two categories:

1. Plain horizontal milling machines
2. Universal horizontal milling machines

These two milling machines are quite similar in appearance. The only difference is the addition of a *table swivel housing* between the table and the saddle of the universal machine. (See N-1.) This permits the table to be swung up to 45° in either direction for angular and helical milling operations. cause of this, the universal horizontal milling machine (Fig. 48-8) is more versatile. It is used extensively in machine shops. Both these machines can be fitted with various attachments such as a dividing head, rotary table, slotting and rack cutting attachments, as well as various special fixtures.

N-1 The table on a universal milling machine can be swiveled.

N-2 Use care whenever power feeds are used.

### Universal Horizontal Milling Machine Parts

The *base* gives support and rigidity to the machine and also acts as a reservoir for cutting fluid.

The *column* is the large vertical section of the machine, mounted on the base. It provides support for the knee and table. The motor and spindle speed gears are generally mounted in the column.

The *knee,* which supports the table assembly, is attached to the column face and is moved vertically on the column by means of the *knee elevating screw*. It is moved manually by the *vertical handfeed crank* and automatically by the *vertical feed engaging lever*. (See N-2.)

The *saddle* is fitted on top the knee and may be moved in or out manually

**Fig. 48-9** A vertical milling attachment. *(Courtesy Cincinnati Milacron Co.)*

**Fig. 48-10** Cutting an internal keyway with a slotting attachment. (*Courtesy Cincinnati Milacron Co.*)

N-3 The arbor support should be located as close to the cutter as possible.

by means of the crossfeed handwheel or automatically by the crossfeed engaging lever. The *swivel table housing,* fastened to the saddle on a universal milling machine, enables the table to be swiveled 45° to either side of the centerline.

The *table* rests on guideways in the saddle and travels longitudinally in a horizontal plane. It supports the vise and the work. The *crossfeed handwheel* is used to move the table manually toward or away from the column. The *crossfeed engaging lever* is used to move the table toward or away from the column by power feed.

The *table handwheel* is used to move the table to the right or left manually. The *table feed engaging lever* is used to move the table to the right or left automatically by power. The *feed dial* is used to regulate the table feeds.

The *spindle* provides the drive for arbors, cutters, and attachments used on a milling machine. Most milling machines are equipped with two or more *arbors* on which many different-shaped cutters may be mounted.

The *spindle speed dial* is set by a crank that is turned to regulate the spindle speed. On some milling machines, the spindle speed changes are made by means of two levers.

The *overarm* provides for correct alignment and support of the arbor and various attachments. It can be adjusted and locked in various positions depending on the length of the arbor and the position of the cutter.

The *arbor support* is fitted to the overarm and can be clamped at any location on the overarm. Its purpose is to align and support various arbors and attachments. (See N-3.)

# MILLING MACHINE ATTACHMENTS AND ACCESSORIES

A wide variety of accessories are available to increase the versatility and productivity of a milling machine. There are three classes of accessories:

**1.** Those which increase the versatility of the milling machine such as the vertical milling attachment and the slotting attachment

**2.** Those which hold the cutters such as arbors, collets, and adapters

**3.** Those which hold the workpiece for machining such as vises, fixtures, and clamps

## Milling Machine Attachments

### Vertical Milling Attachment

The *vertical milling attachment* (Fig. 48-9), which may be mounted on the face of the column or the overarm, enables a plain or universal milling machine to be used as a vertical milling machine. Angular surfaces may be machined by swinging the head, parallel to the face of the column, to any angle up to 45° on either side of the vertical position. On some models the head may be swung as much as 90° to either side. Vertical attachments enable the horizontal milling machine to be used for such operations as face milling, end milling, drilling, boring, and T-slot milling.

A modification of the vertical milling attachment is the *universal milling attachment* which may be swung in two planes, parallel to the column and at right angles to it, for the cutting of compound angles. The vertical and universal attachments are also manufactured in *high-speed* models which permit the efficient use of small- and medium-sized end mills and cutters for such operations as die sinking and key seating.

### Slotting Attachment

The *slotting attachment* (Fig. 48-10) converts the rotary motion of the spindle into reciprocating motion for cutting keyways, splines, templates, and irregular-shaped surfaces. The length of the stroke is controlled by an adjustable crank. The toolslide may be swung to any angle in a plane parallel to the face of the column, making the slotting attachment especially valuable in die work.

## Cutter-holding Accessories

### Arbors, Collets, and Adapters

*Arbors* (Fig. 48-11), used for mounting the milling cutter, are inserted and held in the main spindle by a drawbolt

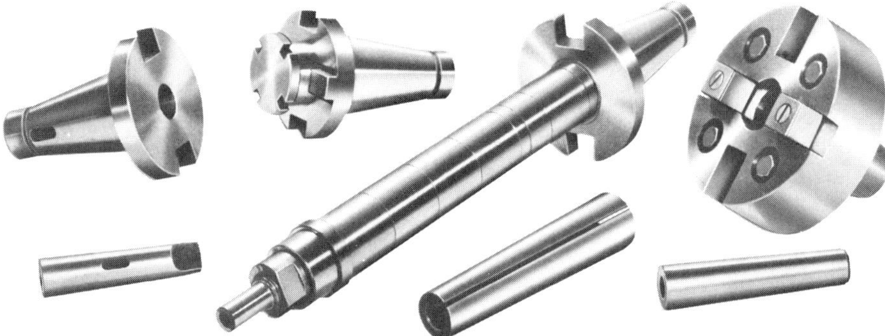

**Fig. 48-11** Arbors, collets, and adapters. (*Courtesy Cincinnati Milacron Co.*)

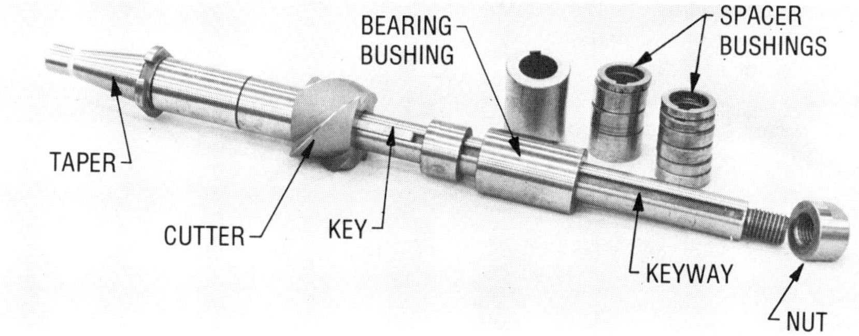


**Fig. 48-12** The milling machine arbor holds and drives the cutter. (*Courtesy Kostel Enterprises Ltd.*)

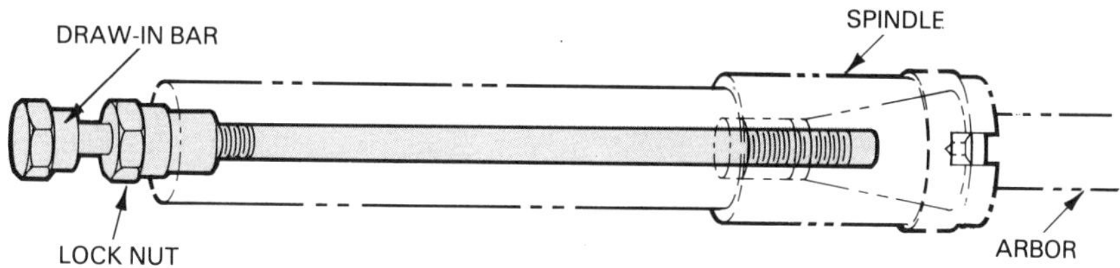


**Fig. 48-13** The milling machine arbor is held securely in the spindle by the draw-in bar and lock nut.

N-4 A collet permits the easy and rapid change of cutting tools.

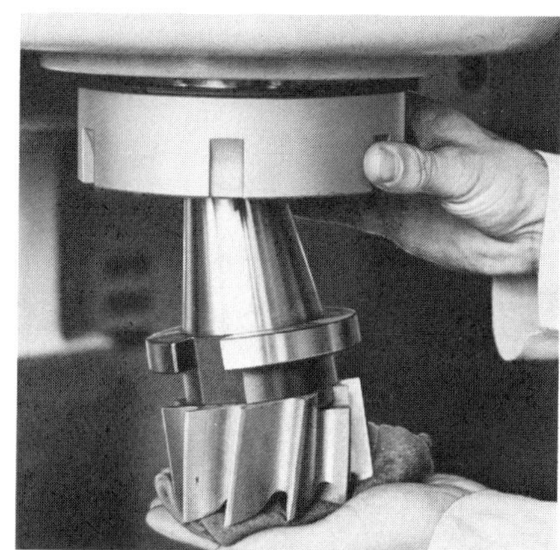

**Fig. 48-14** A quick-change adapter and collets permit easy changing of cutting tools. (*Courtesy Cincinnati Milacron Co.*)

or a special quick-change adapter. One end of the arbor is held in the milling machine spindle while the outer end is supported by one or more arbor supports. The cutter is positioned on the arbor by collars and is prevented from turning on the arbor by a square key (Fig. 48-12).

On all modern milling machines, the various arbors and adapters are mounted in the spindle by means of the American Standard Milling Machine Taper (3½ in./ft). While the taper permits the arbor to center itself accurately with the machine spindle, the arbor is held in the spindle by a drawbar (Fig. 48-13). The arbor is pulled securely into the spindle by tightening the lock nut. Two drive keys on the spindle, which fit into matching keyways on the arbor flange, prevent the arbor from turning when a heavy cut is being taken.

The *collet adapter* is held in the spindle in the same manner as the arbor. It provides a means of mounting tools with tapered shanks, such as drills, in the spindle of the milling machine.

*Quick-change adapters* and collets (Fig. 48-14) used for holding a wide range of cutting tools have made it possible to perform several operations such as milling, drilling, and boring without changing the setup of the workpiece. (See N-4.)

## Work-holding Devices

There are several devices used in industry for holding work to be milled. The most commonly used are: vise, V-blocks, angle plates, fixtures, dividing head, and clamps.

### Vises

Milling machine vises are the most widely used work-holding devices for milling. They can be used for holding square, round, and rectangular pieces for the cutting of keyways, grooves, flat surfaces, angles, gear racks, and T-slots. Milling vises are manufactured in three styles.

The *plain vise* (Fig. 48-15) may be bolted to the table so that its jaws are parallel or at right angles to the axis of the spindle. The vise is positioned quickly and accurately by keys on the bottom which fit into the T-slots on the table.

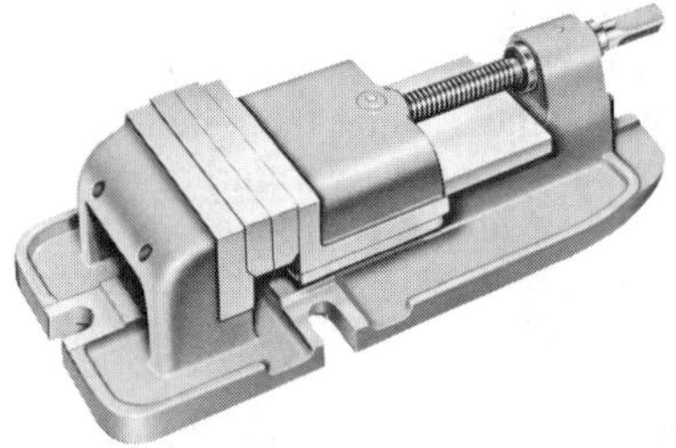

Fig. 48-15 A plain milling machine vise. (*Courtesy Cincinnati Milacron Co.*)

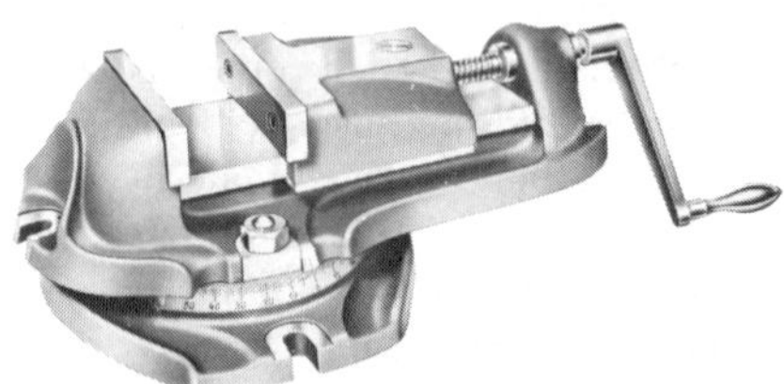

Fig. 48-16 A swivel-base vise permits angular cuts to be made. (*Courtesy Pedersen Milling Machine Co.*)

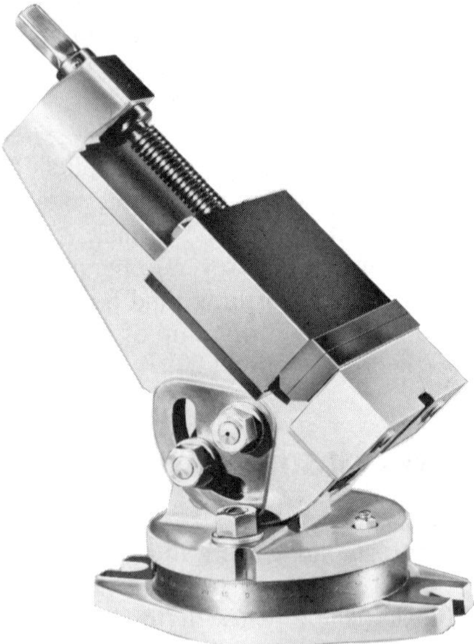

Fig. 48-17 A universal vise permits compound angles to be machined. (*Courtesy Cincinnati Milacron Co.*)

N-5 A compound angle is a surface which slopes in more than one plane.

N-6 The shape of the workpiece determines the type of clamp to be used.

The *swivel-base vise* (Fig. 48-16) is similar to the plain vise, except that it has a swivel base which enables the vise to be swiveled through 360° in a horizontal plane.

The *universal vise* (Fig. 48-17) may be swiveled through 360° in a horizontal plane and may be tilted from 0 to 90° in a vertical plane. It is used chiefly by toolmakers, mouldmakers, and diemakers, since it permits the setting of compound angles for milling. (See N-5.)

*V-blocks* usually have a 90° V-shaped groove and a tongue which fits into the table slot to allow proper alignment for milling special shapes, flat surfaces, or keyways in round work.

*Angle plates* are used for holding large work or special shapes when machining one surface square with another.

Fig. 48-18 Milling a workpiece held in a fixture. (*Courtesy Cincinnati Milacron Co.*)

Fig. 48-19A Finger clamp. (*Courtesy J. H. Williams & Co.*)

Fig. 48-19B U-clamp. (*Courtesy J. H. Williams & Co.*)

Fig. 48-19C Straight clamp. (*Courtesy J. H. Williams & Co.*)

A *fixture* (Fig. 48-18) is a special holding device made to hold a particular workpiece for one or more milling operations on a production basis. It provides an easy setup method but is limited to the job for which it is made.

## Indexing or Dividing Head

This very useful accessory permits the cutting of bolt heads, gear teeth, and ratchets, etc. When connected to the leadscrew of the milling machine, it will revolve the work as required to cut helical gears and flutes in drills and reamers, etc. The dividing head will be fully discussed later in this section.

## Clamps

When work cannot be held in a vise or fixture, it is often clamped to the table. Three common clamps are shown in Fig. 48-19A to C. Other modifications of these clamps are the *double-finger* and the *gooseneck* types.

Clamps are held in position by a T-bolt that fits into the T-slot of the milling machine table. One end of the clamp is supported by the work and the other end by a step block or a suitable packing block. The height of the step block or packing block should be *slightly higher* than the workpiece. (See N-6.)

# REVIEW QUESTIONS

**1.** Name six operations that can be performed on a horizontal milling machine.

**2.** Name six manufacturing-type milling machines and state the main purpose of each.

**3.** What is the difference between a plain and a universal horizontal milling machine?

**4.** Where is the reservoir for the cutting fluid located?

**5.** What purpose does the column serve?

**6.** Name the two controls which may be used to raise or lower the knee.

**7.** Name two methods of moving the saddle toward or away from the column.

**8.** What two parts provide support and alignment to the arbor?

A

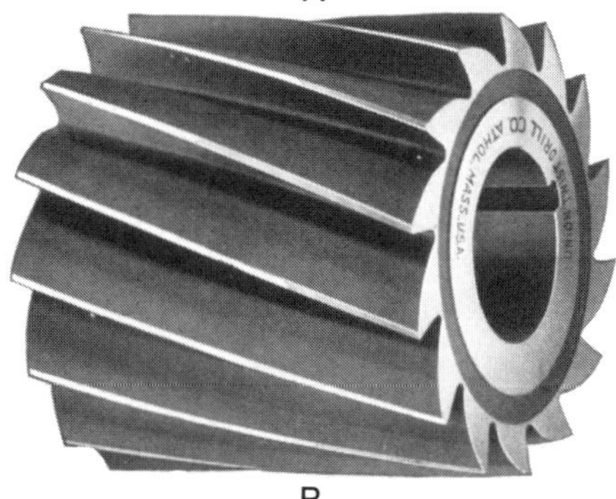
B

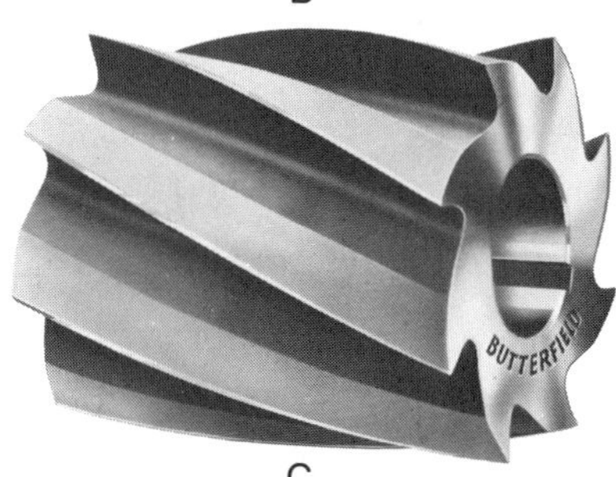
C

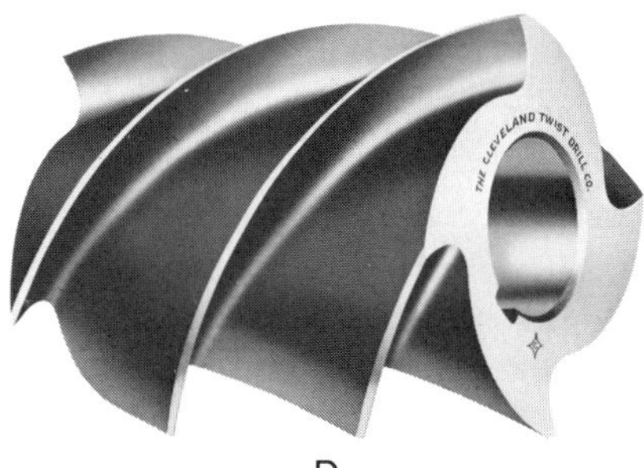
D

**Fig. 49-1** Plain milling cutters. (A) Light-duty plain (*Courtesy The Cleveland Twist Drill Co.*); (B) light-duty (helical teeth); (C) heavy-duty plain (*Courtesy The Butterfield Division, Union Twist Drill Co.*); (D) high-helix plain (*Courtesy The Cleveland Twist Drill Co.*).

### Milling Attachments and Accessories

**9.** What is the purpose of the vertical milling attachment?

**10** State four uses for the slotting attachment.

**11.** How is the cutter positioned on the arbor?

**12.** What is the taper per foot for the Standard American Milling Machine taper?

**13.** How is the arbor held securely in the spindle?

**14.** How is the arbor prevented from turning in the spindle?

**15.** What is the advantage of a quick-change adapter?

**16.** Name three types of milling machine vises.

**17.** Name four other methods of holding work for milling.

**18.** How high should the step or packing block be in relation to the work clamped?

# UNIT 49

# MILLING CUTTERS AND PROCESSES

A milling cutter is a rotary cutting tool having a cylindrical body and equally spaced teeth around the periphery. As the cutter rotates on its axis, the teeth engage with the workpiece and remove metal in the form of chips.

## OBJECTIVES

After completing this unit, you should be able to:

**1.** Identify six types of standard milling cutters and state the purpose of each

**2.** Identify four types of special-purpose cutters and state the use of each

**3.** State the purpose of eight milling processes

**4.** Understand the advantages and disadvantages of climb and conventional milling

## MILLING CUTTERS

Milling cutters generally fall into two categories: solid cutters and inserted-tooth cutters.

The *solid cutters* are those in which the teeth have been cut into the body of the cutter. The teeth may be straight (parallel) or helical (at an angle) to the axis of the cutter. Solid-type cutters are generally made of high-speed steel.

*Inserted-tooth cutters* have removable and replaceable teeth which are fastened or locked into the body of the cutter. These blades may be made of high-speed steel or cemented carbides. The inserted-tooth construction is generally used on large cutters and the blades can be quickly replaced when they become dull.

Since it would be impossible to cover all types and varieties of milling cutters, only some of the more common types will be covered in this unit.

### Plain Milling Cutters

Probably the most widely used milling cutter is the plain milling cutter which is used to produce a *flat surface parallel to its axis*. These cutters may be of several types as shown in Fig. 49-1.

*Light-duty plain milling cutters* (Fig. 49-1A), which are less than ¾ in. (19.05 mm) wide will usually have straight teeth; those over ¾ in. (19.05 mm) wide have a helix angle of about 25° (Fig. 49-1B). This type of cutter is used only for light milling operations since it has too many teeth to permit the chip clearance required for heavier cuts.

*Heavy-duty plain milling cutters* (Fig. 49-1C) have fewer teeth than the light-duty type, which provide for better chip clearance. The helix angle varies up to 45°. This greater helix angle on the teeth produces a smoother surface due to the shearing action and reduces chatter. Less power is required with this cutter than with straight-tooth and small-helix angle cutters.

*High-helix plain milling cutters* (Fig. 49-1D) have helix angles from 45 to over 60°. They are particularly suited to the milling of wide and intermittent surfaces in contour and profile milling. Although this type of cutter is usually mounted on the milling machine arbor, it is sometimes shank-mounted with a pilot on the end and used for milling elongated slots.

A

B

Fig. 49-2 Side milling cutters. (A) Straight teeth; (B) staggered teeth. (*Courtesy The Cleveland Twist Drill Co.*)

Fig. 49-3 Face milling cutter (*Courtesy The Butterfield Division, Union Twist Drill Co.*)

## Side Milling Cutters

*Side milling cutters* are comparatively narrow cylindrical milling cutters with teeth on each side as well as on the periphery. They are used for cutting slots and vertical faces.

Side milling cutters may have straight teeth (Fig. 49-2A) or staggered teeth (Fig. 49-2B). Staggered-tooth cutters have each tooth offset alternately to the right and left with an alternately opposite helix angle on the periphery. These cutters have free cutting action at high speeds and feeds. They are particularly suited for milling deep, narrow slots.

## Face Milling Cutters

*Face milling cutters* (Fig. 49-3) are generally over 6 in. (152.4 mm) in diameter and have *inserted teeth* held in place by a wedging device. The corners of this type of cutter are beveled; most of the cutting action occurs at these bevels and on the periphery of the cutter.

Face milling cutters under 6 in. are called *shell end mills* (Fig. 49-4). They are solid, multiple-tooth cutters with teeth on the face and the periphery. They are usually held on a stub arbor which may be threaded or employ a key in the shank to drive the cutter. Shell end mills are more economical than large solid end mills because they are cheaper to replace when broken or worn out.

## Angular Cutters

*Angular cutters* have teeth that are neither parallel nor perpendicular to the cutting axis. They are used for milling angular surfaces, such as grooves, serrations, chamfers, reamer teeth, etc. They may be divided into two groups:

*Single-angle milling cutters* have teeth on the angular surface and may or may not have teeth on the flat side. The included angle between the flat face and the angular face designates the cutters, such as 45 or 60° angular cutter.

*Double-angle milling cutters* (Fig. 49-5) have two intersecting angular surfaces with cutting teeth on both. When these cutters have equal angles on both sides of the line at a right angle to the axis (symmetrical), they are designated by the size of the included angle. When the angles formed are not the same (unsymmetrical), the cutters are designated by specifying the angle on both sides of the plane or line.

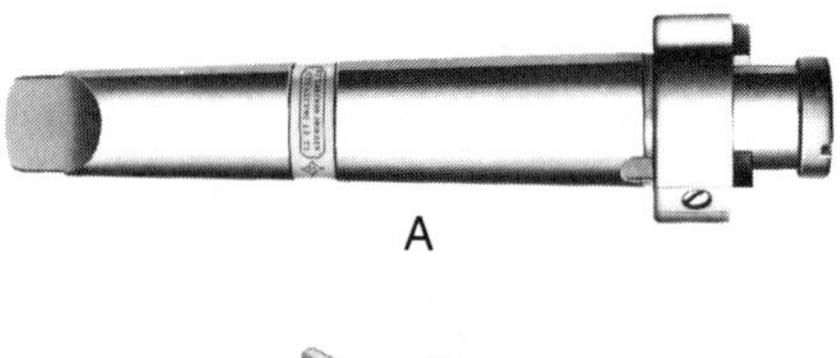

A

B

Fig. 49-4 (A) Arbor; (B) shell end mill. (*Courtesy The Cleveland Twist Drill Co.*)

Fig. 49-5 Double-angle milling cutter. (*Courtesy The Cleveland Twist Drill Co.*)

## Formed Cutters

*Formed cutters* (Fig. 49-6) incorporate the exact shape of the part to be produced, permitting exact duplication of irregular-shaped parts more economically than most other means. Formed

A

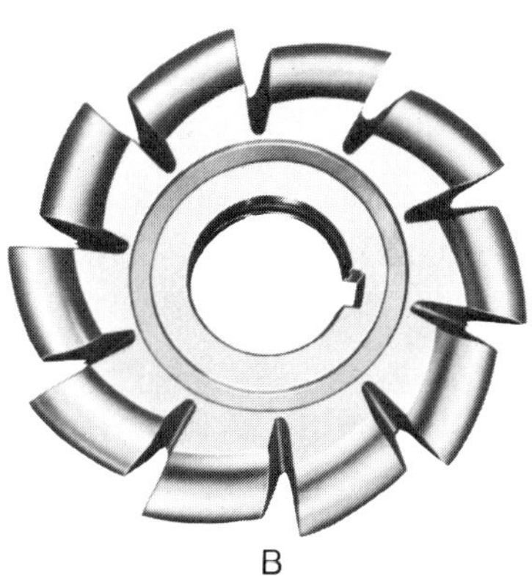

B

**Fig. 49-6** Formed cutters: (A) Convex; (B) gear tooth.

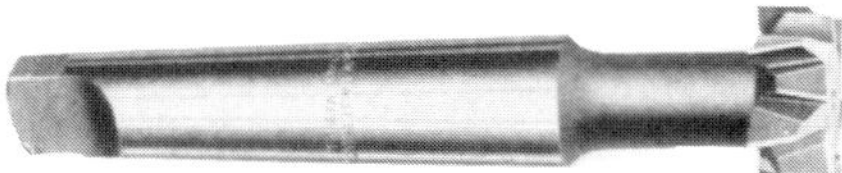

**Fig. 49-7** T-slot cutter. (*Courtesy The Butterfield Division, Union Twist Drill Co.*)

**Fig. 49-8** Woodruff keyseat cutter. (*Courtesy The Butterfield Division, Union Twist Drill Co.*)

cutters are particularly useful for the production of small parts. Examples of *form relieved cutters* are concave, convex, and gear cutters.

Formed cutters are sharpened by grinding the tooth face. Tooth faces are radial and may have positive, zero, or negative rake, depending on the cutter application. It is imperative that the original rake on the tooth be maintained so that the profiles of the tooth and of the work are not changed. If the tooth face rake is maintained exactly, the cutter may be resharpened until the teeth are too thin for use.

## T-Slot Cutter

The *T-slot cutter* (Fig. 49-7) is used to cut the wide horizontal groove at the bottom of a T-slot after the narrow vertical groove has been machined with an end mill or a side milling cutter. It consists of a small side milling cutter with teeth on both sides and an integral shank for mounting, similar to an end mill.

## Woodruff Keyseat Cutters

The *Woodruff keyseat cutter* (Fig. 49-8) is similar in design to plain and side milling cutters. Smaller sizes up to approximately 2 in. (50.8 mm) in diameter are made with a solid shank and straight teeth; larger sizes are mounted on an arbor and have staggered teeth on both the sides and the periphery. They are used for milling semicylindrical keyseats in shafts.

## Flycutters

A *flycutter* (Fig. 49-9) consists of an adapter which may accommodate one or more single-pointed cutting tools. It is usually fitted in the spindle of the vertical milling attachment and used for face milling operations. Roughing and finishing facing cuts may be performed in one pass if the flycutter has two or more cutting tools mounted in it. One tool is set out slightly farther from the center and slightly higher than the opposite cutting tool. This permits the higher tool to make a roughing cut and the lower tool, the finish cut.

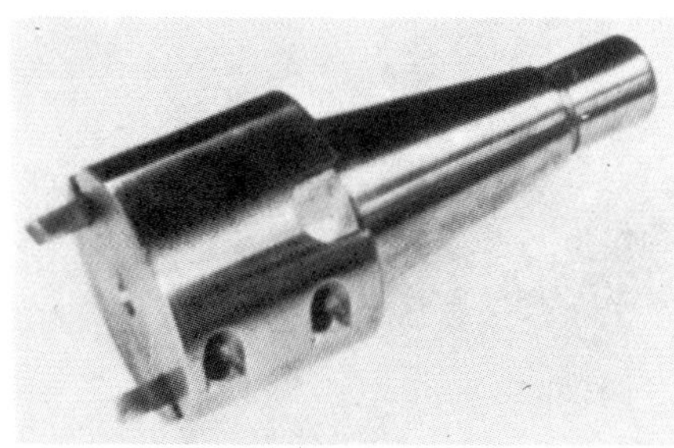

**Fig. 49-9** Flycutter.

**Fig. 49-10** Metal slitting saw. (*Courtesy Union Twist Drill Co.*)

## Metal Saws

*Metal slitting saws* (Fig. 49-10) are basically thin plain milling cutters. Some have sides relieved, or "dished," to prevent rubbing or binding when in use and others have side cutting teeth. Slitting saws are made in widths from 1/32 to 3/16 in. (0.8 to 5 mm). Because of their thin cross section, they should be operated at approximately one-quarter to one-eighth of the feed per tooth used for other cutters. For nonferrous metals, their speed can be increased. Unless a special driving flange is used for slitting saws, it is *not* advisable to key the saw to the milling arbor. The arbor nut should be pulled up as tightly as possible *by hand only*. Since slitting saws are so easily broken, some operators find it desirable to "climb" or "down" mill when sawing. However, to overcome the play between the leadscrew and nut, the backlash eliminator should be engaged.

## End Mills

*End mills* have cutting teeth on the end as well as on the periphery and are fitted to the spindle by a suitable adapter. They are of two types: the *solid end mill* in which the shank and

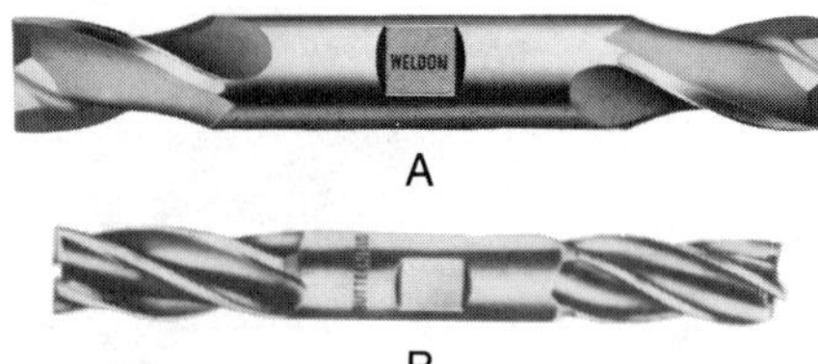


**Fig. 49-11** End mills. (A) Two flute (*Courtesy Weldon Tool Company*); (B) four flute (*Courtesy The Butterfield Division, Union Twist Drill Co.*).

the cutter are integral; and the *shell end mill* which, as previously stated, employs a separate shank.

Solid end mills, generally smaller than shell end mills, may have either straight or helical flutes. They are manufactured with straight and taper shanks and with two or more flutes. The two-flute type, sometimes called *slot drills* (Fig. 49-11A), have flutes which meet at the cutting end, forming two cutting lips across the bottom. These lips are of different lengths, one extending beyond the center axis of the cutter. This permits the two-flute end mill to be used in a milling machine for drilling a hole to start a slot that does not extend to the edge of the metal. When a slot is being cut with a two-flute end mill, the depth of cut should not exceed one-half the diameter of the cutter. When the four-flute end mill (Fig. 49-11B) is used for slot cutting, it is usually started at the edge of the metal or in a previously drilled hole.

## MILLING PROCESSES

Although the majority of operations performed on a knee-and-column-type machine are either plain milling or side milling, several other operations or combinations of operations may be performed.

*Plain* or *slab milling* (Fig. 49-12) is the production of a horizontal flat surface parallel to the axis of the milling machine arbor. The workpiece may be held in a vise or fixture or fastened directly to the table.

*Side milling* (Fig. 49-13) is the process of machining a vertical flat surface perpendicular to the axis of the milling machine arbor. This operation is performed by the combined action of the peripheral and side teeth on a side milling cutter.

*Face milling* (Fig. 49-14) is used to produce a flat surface parallel to the column of the machine. This is done by means of a face milling cutter mounted in the milling-machine spindle. Face milling may also be done using a vertical milling attachment to produce horizontal flat surfaces. Both the periphery and the end of the teeth do the cutting.

*End milling* is an operation similar to face milling but using a much smaller cutter. Cutting is done on the end of the cutter as well as on the periphery. This operation is used for facing small surfaces, milling slots or grooves, producing internal recesses, and truing the edges of a workpiece.

*Slotting* is the process of cutting grooves or slots in the workpiece. A staggered-tooth side milling cutter or an end mill can be used for this operation.

**Fig. 49-12** Milling the surface of a workpiece held in a fixture. (*Courtesy Cincinnati Milacron Co.*)

**Fig. 49-13** Side milling a vertical surface. (*Courtesy Cincinnati Milacron Co.*)

Fig. 49-14 Face milling a flat surface with a shell end mill.

*Sawing* may be performed on the milling machine with a thin metal slitting saw. Since these saws are easily broken, sawing should be done only on machines equipped with a backlash eliminator and using the climb-milling process.

*Straddle milling* (Fig. 49-15) involves the use of two side milling cutters to machine the opposite sides of a workpiece parallel in one cut. The cutters are separated on the arbor by a spacer or spacers of the required length so that the distance between the inside faces of the cutters is equal to the desired size. Applications of straddle milling are the milling of square and hexagonal heads on bolts.

*Gang milling* (Fig. 49-16), a fast method of milling since several surfaces can be machined in one pass, is used a great deal in production work. Two or more cutters are used on the arbor to produce the desired shape. The cutters may be a combination of plain and side milling cutters.

## Conventional and Climb Milling

The direction in which the workpiece is fed into the cutter indicates whether *conventional* (up) milling or *climb* (down) milling is being used.

The most commonly used method of feeding is to bring the work against the rotation direction of the cutter *(conventional, or up milling)* (Fig. 49-17). However, if the machine is equipped with a backlash eliminator (Fig. 49-19), certain types of work can best be milled by *climb milling* (Fig. 49-18).

Although climb milling is not as widely used as conventional milling, it has certain advantages and disadvantages.

### Advantages

**1.** It is particularly suited to the machining of thin and hard-to-hold parts, since the workpiece is forced against the table or holding device by the cutter.

**2.** Work need not be clamped as tightly.

**3.** Consistent parallelism and size may be maintained, particularly on thin parts.

**4.** It may be used where breakout at the edge of the workpiece could not be tolerated.

**5.** It requires up to 20 percent less power to cut by this method.

**6.** It may be used when cutting off stock or when milling deep, thin slots.

### Disadvantages

**1.** It cannot be used unless the machine has a backlash eliminator and the table gibs have been tightened.

Fig. 49-15 Straddle milling two sides of a hexagon in one pass.

Fig. 49-16 Gang milling four surfaces in one pass.

Fig. 49-17 Conventional or up milling. Note that the work is fed against the cutter rotation direction. (*Courtesy Kostel Enterprises Ltd.*)

Fig. 49-18 Climb or down milling. The work is fed into a cutter which revolves down into the work. (*Courtesy Kostel Enterprises Ltd.*)

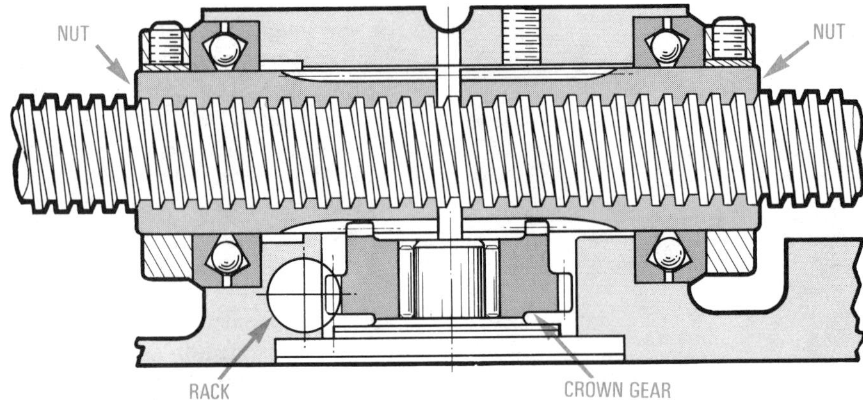


Fig. 49-19 The backlash eliminator permits the operation of climb milling. (*Courtesy Cincinnati Milacron Co.*)

**2.** It cannot be used for machining castings or hot rolled steel, since the hard outer scale will damage the cutter.

### Backlash Eliminator

This device, when engaged, eliminates the backlash (play) between the nut and the table leadscrew, permitting the operation of *climb* (down) milling. It allows the workpiece to be fed into the cutter at an even rate rather than being quickly drawn into the cutter. It prevents damage to the cutter and the workpiece.

The backlash eliminator (Fig. 49-19) works as follows. Two independent nuts are mounted on the leadscrew. These nuts engage a common crown gear which in turn meshes with a rack. Axial movement of the rack is controlled by the backlash eliminator engaging knob located on the front of the saddle. By turning the knob ''in,'' the nuts are forced to move along the leadscrew in opposite directions, removing all backlash.

## REVIEW QUESTIONS

**1.** What type of surface is produced by a plain milling cutter and a side milling cutter?

**2.** Name two cutters that may be used for face milling operations.

**3.** How does cutting take place with a face milling cutter?

**4.** For what purpose are formed cutters used?

**5.** What precaution should be taken when using a metal slitting saw?

**6.** State four uses for an end mill.

**7.** Describe the cutter setup to machine a bolt head 1 in. across the flats.

**8.** What is the advantage of gang milling?

**9.** By means of two labeled diagrams show the difference between climb milling and conventional milling.

**10.** Name three purposes of the backlash eliminator.

# UNIT 50

# MOUNTING AND REMOVING ARBORS AND CUTTERS

Any hand or machine tool can be dangerous if improperly used. A person learning to operate the milling machine must first learn the safety regulations governing the use of this machine and the work setups required to perform the job safely.

## OBJECTIVES

When you have completed this unit you should be able to:

**1.** Follow the safety regulations and operate a milling machine properly

**2.** Mount and remove a milling machine arbor

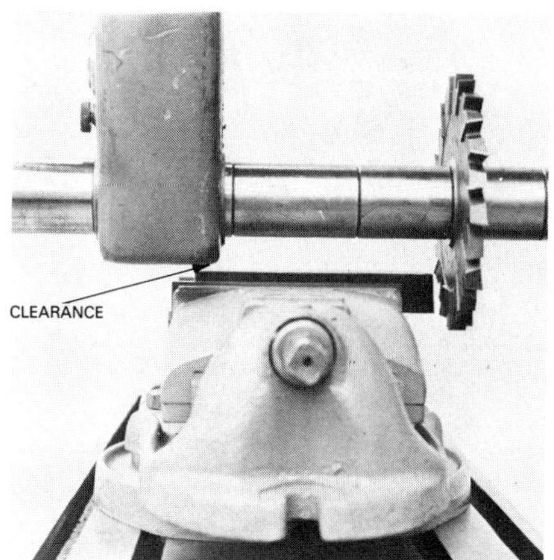


**Fig. 50-1** Check that the arbor support and the cutter is clear of the work. (*Courtesy Kostel Enterprises Ltd.*)

N-1 Milling cutters have many sharp teeth; handle them with care.

N-2 Never remove milling chips by hand.

N-3 Good housekeeping is important in the machine shop.

**3.** Mount and remove a milling cutter

## MILLING MACHINE SAFETY

The milling machine, like any other machine, demands the total attention of the operator and a thorough understanding of the hazards involved in its operation. The following rules should be observed when operating the milling machine.

**1.** Be sure that the work and the cutter are mounted securely before taking a cut.

**2.** Always wear safety glasses.

**3.** When mounting or removing milling cutters, always hold them with a cloth to avoid being cut. (See N-1.)

**4.** While setting up work, move the table as far as possible from the cutter to avoid cutting your hands.

**5.** Be sure that the cutter and machine parts clear the work (Fig. 50-1).

**6.** *Never* attempt to mount, measure, or adjust work until the cutter is *completely* stopped.

**7.** Keep hands, brushes, and rags away from a revolving milling cutter *at all times*.

**8.** When using milling cutters, do not use an excessively heavy cut or feed. This can cause a cutter to break and the flying pieces could cause serious injury.

**9.** Always use a *brush,* not a rag, to remove the cuttings after the cutter has stopped revolving. (See N-2.)

**10.** Never reach over or near a revolving cutter; keep hands at least 12 in. or 300 mm from a revolving cutter.

**11.** Keep the floor around the machine free of chips, oil, and cutting fluid. (See N-3.)

## MOUNTING AND REMOVING A MILLING MACHINE ARBOR

The milling arbor is used to hold the cutter during the machining operation. When mounting or removing an arbor, the proper procedure should be followed in order to preserve the accuracy of the machine.

### The Arbor Assembly

The milling cutter is driven by a key which fits into the keyways on the arbor and cutter (Fig. 50-2). This prevents the cutter from turning on the arbor. Spacer and bearing bushings hold the cutter in position on the arbor after the nut has been tightened. The tapered end of the arbor is held securely

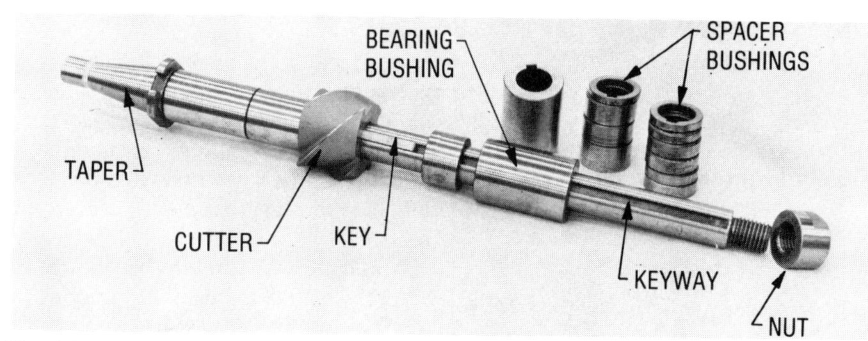


**Fig. 50-2** A milling arbor assembly.

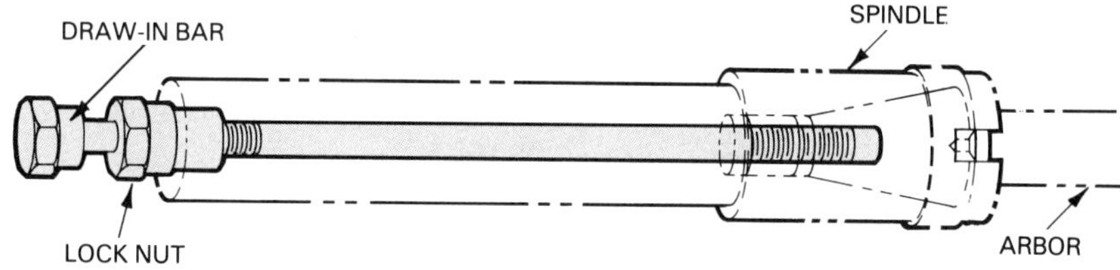


**Fig. 50-3** A draw-in bar holds the arbor firmly in place in the spindle.

Fig. 50-4 The arbor is secured in the spindle with a draw-in bar.

Fig. 50-6 To break the taper fit, tap the end of the draw-in bar with a soft hammer.

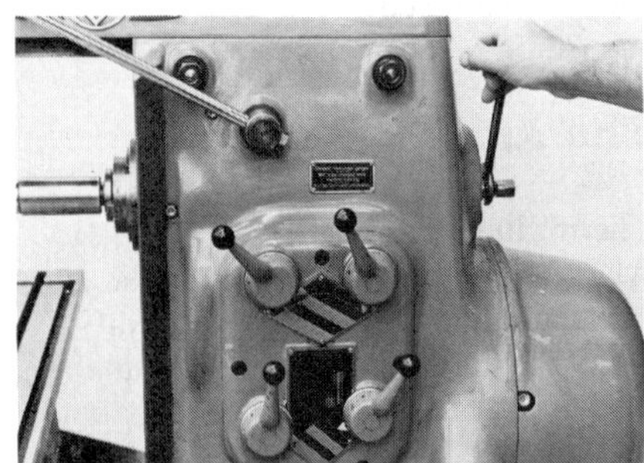

Fig. 50-5 Tighten the draw-in bar lock nut to hold the arbor securely. (*Courtesy Kostel Enterprises Ltd.*)

N-4 Always tighten the nut only hand tight with the wrench.

N-5 Never use a hard-faced hammer or tool to remove the arbor.

N-6 If the cutter teeth are pointing in the wrong direction, their cutting edges will be ruined as soon as they contact the workpiece.

in the machine spindle by a draw-in bar (Fig. 50-3). The outer end of the arbor assembly is supported by the bearing bushing and the arbor support.

### To Mount an Arbor

**1.** Clean the tapered hole in the spindle and the taper on the arbor, using a clean cloth.

**2.** Check that there are no cuttings or burrs in the taper which would prevent the arbor from running true.

**3.** Check the bearing bushing and remove any burrs using a honing stone.

**4.** Place the tapered end of the arbor in the spindle and align the spindle-driving lugs with the arbor slots.

**5.** Place the right hand on the draw-in bar (Fig. 50-4) and turn the thread into the arbor approximately 1 in. (25.4 mm).

**6.** Tighten the draw-in bar lock nut securely against the back of the spindle (Fig. 50-5). (See N-4.)

### To Remove an Arbor

**1.** Remove the milling machine cutter.

**2.** Loosen the lock nut on the draw-in bar approximately two turns (Fig. 50-3).

**3.** With a soft-faced hammer, hit the end of the draw-in bar until the arbor taper is free in the spindle (Fig. 50-6). (See N-5.)

**4.** With one hand hold the arbor, and unscrew the draw-in bar from the arbor with the other hand (Fig. 50-4).

**5.** Carefully remove the arbor from the tapered spindle so as not to damage the spindle or arbor tapers, and store it in its proper place.

**6.** Leave the draw-in bar in the spindle for further use.

## Milling Cutters

Milling cutters must be changed frequently so it is important that the following sequence be followed in order not to damage the cutter, the machine, or the arbor.

### To Mount a Milling Cutter

**1.** Remove the arbor nut and collar and place them on a piece of masonite (Fig. 50-7).

**2.** Clean all surfaces of cuttings and burrs.

**3.** Check the direction of the arbor rotation.

**4.** Slide the spacing collars on the arbor to the position desired for the cutter (Fig. 50-7).

**5.** Fit a key into the arbor keyway where the cutter is to be located.

**6.** Hold the cutter with a cloth and mount it on the arbor being sure that the cutter teeth point in the direction of the arbor rotation (Fig. 50-8). (See N-6.)

**Fig. 50-7** A piece of masonite is used to protect the milling machine table when changing an arbor or cutter. (*Courtesy Kostel Enterprises Ltd.*)

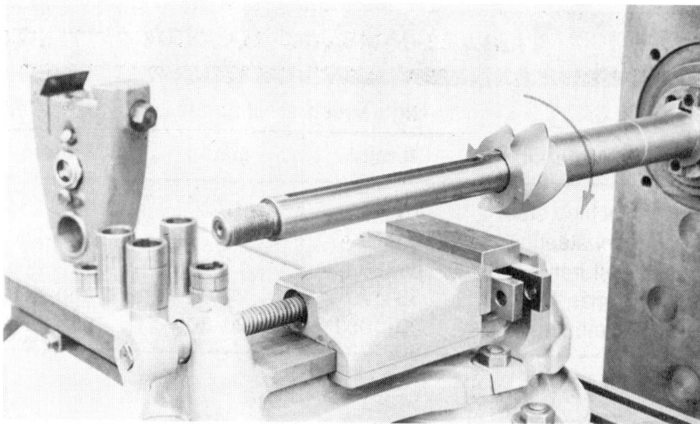

**Fig. 50-8** The cutter teeth must point in the direction that the spindle is rotating. (*Courtesy Kostel Enterprises Ltd.*)

N-7 The shorter the distance between the arbor support and the column, the more rigid will be the setup.

**7.** Slide the arbor support in place and be sure that it is on a bearing bushing on the arbor. (See N-7.)

**8.** Put on additional spacers leaving room for the arbor nut. Tighten the nut by hand.

**9.** Lock the overarm in position.

**10.** Tighten the arbor nut firmly with a wrench using only hand pressure.

**11.** Lubricate the bearing collar in the arbor support.

**12.** Make sure that the arbor and arbor support will clear the work (Fig. 50-1).

### Removing a Milling Cutter

N-8 Tightening or loosening the arbor nut without the arbor support in place may bend the arbor.

**1.** Be sure that the arbor support is in place and supporting the arbor on a bearing bushing before using a wrench on the arbor nut. This will prevent bending the arbor. (See N-8.)

**2.** Clean all cuttings from the arbor and cutter.

**3.** Loosen the arbor nut, using a wrench.

**4.** Loosen the arbor support and remove it from the overarm.

**5.** Remove the nut, spacers, and cutter. Place them on a board (Fig. 50-7) and not on the table surface.

**6.** Clean the spacer and nut surfaces and replace them on the arbor. *Do not use a wrench to tighten the arbor nut at this time.*

**7.** Place the cutter in the proper storage.

## REVIEW QUESTIONS

**1.** Name two things that must be checked before a cut is taken.

**2.** Why should milling cutters always be handled with a cloth?

**3.** What can happen if an excessively heavy cut is taken or too fast a feed is used?

**4.** How should chips be removed?

**5.** How far should hands be kept from a revolving cutter?

**6.** What is the purpose of the milling machine arbor?

**7.** What prevents the cutter from slipping on the arbor?

**8.** How is the arbor held securely in the spindle?

**9.** How is the outer end of the arbor supported?

**10.** Why should collars and cutters be placed on masonite rather than the milling machine table?

**11.** How tight must the arbor nut be when using a wrench?

**12.** What precaution must be taken before the arbor nut is loosened?

# UNIT 51

# CUTTING SPEEDS, FEEDS, AND DEPTH OF CUT

The most important factors affecting the efficiency of a milling operation are cutter speed, feed, and depth of cut. If the cutter is run too slowly, valuable time will be wasted, while excessive speed results in loss of time

**TABLE 51-1 MILLING MACHINE CUTTING SPEEDS**

| Material | High-speed steel cutter ft/min | High-speed steel cutter m/min | Carbide cutter ft/min | Carbide cutter m/min |
|---|---|---|---|---|
| Machine steel | 70–100 | 21–30 | 150–250 | 45–75 |
| Tool steel | 60–70 | 18–20 | 125–200 | 40–60 |
| Cast iron | 50–80 | 15–25 | 125–200 | 40–60 |
| Bronze | 65–120 | 20–35 | 200–400 | 60–120 |
| Aluminum | 500–1000 | 150–300 | 1000–2000 | 150–300 |

in replacing and regrinding cutters. Somewhere between these two extremes is the efficient *cutting speed* for the material being machined.

The rate at which the work is fed into the revolving cutter is important. If the work is fed too slowly, time will be wasted and cutter chatter may occur which shortens the *tool life* of the cutter. (See N-1.) If work is fed too fast, the cutter teeth can be broken. Much time will be wasted if several shallow cuts are taken instead of one deep or roughing cut. Therefore *speed, feed,* and *depth of cut* are three important factors in any milling operation. (See N-2.)

N-1 Tool life refers to the number of pieces produced by a cutting tool between sharpenings. A cutter with short tool life will produce fewer pieces than a cutter with long tool life.

N-2 It is good practice to take only one roughing and one finishing cut.

## OBJECTIVES

After completing this unit you should be able to:

**1.** Select cutting speeds and calculate the revolutions per minute for various cutters and materials

**2.** Select and calculate the proper feeds for various cutters and materials

**3.** Follow the correct procedure for taking roughing and finishing cuts

## CUTTING SPEEDS

The cutting speed for a milling cutter is the speed, either in feet per minute or in meters per minute, that the periphery of the cutter should travel when machining a certain metal. The speeds used for milling cutters are much the same as those used for any cutting tool. Several factors must be considered when determining the proper revolutions per minute at which to machine a metal. The most important are:

- Type of work material
- Cutter material
- Diameter of the cutter
- Surface finish required
- Depth of cut being taken
- Rigidity of the machine and work setup

Since different types of metals vary in hardness, structure, and machinability, different cutting speeds must be used for each type of metal and for various cutter materials. The cutting speeds for the more common metals are shown in Table 51-1.

To get optimum use from a cutter, the proper speed at which the cutter should be revolved must be determined. (See N-3.) When machining machine steel, a high-speed steel cutter would have to achieve a surface speed of about 90 ft/min (27.43 m/min). Since the diameter of the cutter affects this speed, it is necessary to consider its diameter in the calculation. The following examples illustrate how the formula is developed.

N-3 Too fast a speed will shorten the cutter tool life; too slow a speed will waste time.

### Inch Calculations

**Example:**

Calculate the speed required to revolve a 3-in. high-speed steel milling cutter when cutting machine steel (90 surface feet per minute).

**1.** Determine the circumference of the cutter or the distance a point on the cutter would travel in one revolution.

$$\text{Circ. of cutter} = 3 \times 3.1416$$

*or*

$$\text{Cutter circ.} = 3 \times 3.1416$$

**2.** To determine the proper cutter speed, or revolutions per minute, divide the cutting speed by the circumference of the cutter.

$$\text{r/min} = \frac{\text{cutting speed (ft)}}{\text{circumference (in.)}}$$

$$= \frac{90}{3 \times 3.1416}$$

Since the numerator is in feet and the denominator in inches, the numerator must be changed to inches. Therefore:

$$\text{r/min} = \frac{12 \times \text{cutting speed}}{3 \times 3.1416}$$

N-4 CS = Cutting speed of material

D = Diameter of cutter

N-5 r/min (inch) $= \frac{4 \times CS}{D}$

N-6 r/min (metric) $= \frac{CS \times 320}{D}$

Because it is usually impossible to set a machine to the exact revolutions per minute, it is permissible to consider that 3.1416 will divide into 12 approximately four times. The formula now becomes (See N-4):

$$\text{r/min} = \frac{4 \times CS}{D}$$

Using this formula and Table 51-1, it is possible to calculate the proper cutter or spindle speed for any material and cutter diameter.

**Example:**

At what speed should a 2-in.-diameter carbide cutter revolve to mill a piece of cast iron [cutting speed (CS) = 150]? (See N-5.)

$$\text{r/min} = \frac{4 \times 150}{2}$$
$$= 300$$

## Metric Calculations

The revolutions per minute at which the milling machine should be set when using metric measurements is as follows:

$$\text{r/min} = \frac{\text{CS (m)} \times 1000}{\pi \times \text{Cutter diam. (mm)}}$$
$$= \frac{CS \times 1000}{3.1416 \times D}$$

Since only a few machines are equipped with variable-speed drives which allow them to be set to the exact calculated speed, a simplified formula can be used to calculate revolutions per minute. The $\pi$ (3.1416) on the bottom line of the formula will divide into the 1000 of the top line approximately 320 times. This results in a simplified formula which is close enough for most milling operations.

$$\text{r/min} = \frac{\text{CS (m)} \times 320}{D \text{ (mm)}}$$

**Example:**

Calculate the revolutions per minute required for a 75-mm-diameter high-speed steel milling cutter when cutting machine steel (CS = 30 m/min). (See N-6.)

$$\text{r/min} = \frac{30 \times 320}{75}$$
$$= \frac{9600}{75}$$
$$= 128$$

# MILLING FEEDS AND DEPTH OF CUT

The two other factors which affect the efficiency of a milling operation are the *milling feed* or the rate at which the work is fed into thc milling cutter and the *depth of cut* taken on each pass.

## Feeds

*Feed* is the rate at which the work moves into the revolving cutter; it is measured in either inches per minute or millimeters per minute. The *milling feed* is determined by multiplying the chip size (chip per tooth) desired, the number of teeth in the cutter, and the revolutions per minute of the cutter.

*Chip per tooth* is the amount of material which should be removed by each tooth of the cutter as it revolves and advances into the work. See Table

**TABLE 51-2 RECOMMENDED FEED PER TOOTH (HIGH-SPEED STEEL CUTTERS)**

| | Side mills | | End mills | | Plain helical mills | | Saws | |
|---|---|---|---|---|---|---|---|---|
| Material | Inches | Millimeters | Inches | Millimeters | Inches | Millimeters | Inches | Millimeters |
| Machine steel | .007 | 0.18 | .006 | 0.15 | .010 | 0.25 | .002 | 0.05 |
| Tool steel | .005 | 0.13 | .004 | 0.10 | .007 | 0.18 | .002 | 0.05 |
| Cast iron | .007 | 0.18 | .007 | 0.18 | .010 | 0.18 | .002 | 0.05 |
| Bronze | .008 | 0.20 | .009 | 0.23 | .011 | 0.28 | .003 | 0.08 |
| Aluminum | .013 | 0.33 | .011 | 0.28 | .018 | 0.46 | .005 | 0.13 |

51-2 for the recommended chip per tooth for some of the more common metals.

The feed rate on a milling machine depends on a variety of factors such as:

- Depth and width of cut
- Design or type of cutter
- Sharpness of the cutter
- Workpiece material
- Strength and uniformity of the workpiece
- Type of finish and accuracy required
- Power and rigidity of the machine

### Inch Calculations

The formula used to find the work feed in inches per minute is

$$\text{Feed in./min} = N \times \text{c.p.t.} \times \text{r/min}$$

where $N$ = number of teeth in milling cutter
c.p.t. = chip per tooth for a particular cutter and metal, as given in Table 51-2 (see N-7)
r/min = revolutions per minute of the milling cutter

N-7 The spindle speed (revolutions per minute) must always be calculated before the feed rate can be calculated.

N-8 $\text{r/min (inch)} = \frac{4 \times CS}{D}$

**Example:**

Find the feed in inches per minute using a 3.5-in.-diameter 12-tooth helical cutter to cut machine steel (CS = 80). It would first be necessary to calculate the proper revolutions per minute for the cutter. (See N-8.)

$$\text{r/min} = \frac{4 \times 80}{3.5}$$
$$= 91$$

$$\text{Feed (in./min)} = N \times \text{c.p.t.} \times \text{r/min}$$
$$= 12 \times .010 \times 91$$
$$= 10.9 \text{ or } 11$$

N-9 To change from millimeters to meters, divide by 1000.

N-10 $\text{r/min (metric)} = \frac{CS \times 320}{D}$

N-11 $\text{Feed (mm/min)} = N \times \text{c.p.t.} \times \text{r/min.}$

N-12 Too heavy a cut may damage the work and cutter; too light a cut may dull the cutter.

### Metric Calculations

The formula used to find work feed in millimeters per minute is the same as the formula used to find the feed in inches per minute, except that mm/min is substituted for in./min. (See N-9.)

**Example:**

Calculate the feed in millimeters per minute for a 75-mm-diameter, six-tooth helical carbide milling cutter when machining a cast-iron workpiece (CS = 60).

First calculate the revolutions per minute of the cutter. (See N-10 and N-11.)

$$\text{r/min} = \frac{60 \times 320}{75}$$
$$= \frac{19{,}200}{75}$$
$$= 256$$

$$\text{Feed (mm/min)} = N \times \text{c.p.t.} \times \text{r/min}$$
$$= 6 \times 0.18 \times 256$$
$$= 276.4$$
$$= 276 \text{ mm/min}$$

## DEPTH OF CUT

Where a smooth accurate finish is desired, it is considered good milling practice to take a roughing and finishing cut. Roughing cuts should be deep, with a feed as heavy as the work and the machine will permit. Heavier cuts may be taken with helical cutters having fewer teeth than with those having many teeth. Cutters with fewer teeth are stronger and have a greater chip clearance than cutters with more teeth.

Finishing cuts should be light, with a finer feed than is used for roughing cuts. (See N-12.) The depth of the cut should be at least 1/64 in. (0.39 mm). Lighter cuts and extremely fine feeds are not advisable, since the chip taken by each tooth will be thin and the cutter will often rub the surface of the work, rather than bite into it, thus dulling the cutter. When a fine finish is required, the feed should be reduced rather than the cutter speeded up; more cutters are dulled by high speeds than by high feeds.

## REVIEW QUESTIONS

**1.** What will happen if a milling cutter is run
  (a) Too slow
  (b) Too fast

**2.** Define cutting speed.

**3.** List six factors which must be considered when determining the proper revolutions per minute to machine a workpiece.

4. Calculate the proper spindle speed for a 5-in.-diameter carbide cutter when machining cast iron (CS = 125).

5. At what revolutions per minute should the spindle be set for a 100-mm-diameter milling cutter to cut aluminum (CS = 150 m)?

6. Define milling feed.

7. What is chip per tooth?

8. List five factors which will affect the feed rate.

9. Calculate the feed in inches per minute using a ¾-in.-diameter carbide end mill having four teeth to cut bronze (CS = 20).

10. What is the minimum amount of material that should be left for a finish cut?

# UNIT 52

# MACHINING A FLAT SURFACE

Before a workpiece is machined, it is most important that the machine and the workpiece be set up properly, and that the operator understands the machine controls and how to proceed with the operation.

## OBJECTIVES

After completing this unit, you should be able to:

1. Align the milling machine table to within ±.001 in. (0.02 mm) accuracy of the column
2. Align the milling machine vise parallel and at 90° to the table travel
3. Set up the workpiece and machine a flat surface with a helical milling cutter

## ALIGNING THE UNIVERSAL MILLING MACHINE TABLE

If the workpiece is to be machined accurately to a layout or have cuts made square or parallel to a surface, it is always good practice to align the table of a universal milling machine and the vise or fixture used to hold the workpiece.

### Procedure

1. Clean the table and the face of the column thoroughly.

2. Mount a dial indicator on the table by means of a magnetic base or any suitable mounting device (Fig. 52-1).

3. Move the table toward the column until the dial indicator registers approximately one-half a revolution and set the dial to zero. (See N-1.)

N-1 Always indicate from the table to the face of the column, never from the column to the table.

Fig. 52-1 Aligning a universal milling table parallel to the column face.

4. Using the table feed handwheel, move the table along the width of the column. Note the reading on the dial indicator.

5. If there is any movement of the indicator hand, loosen the locking devices on the swivel-table housing, adjust for half the difference of the needle movement, and lock the table housing.

6. Recheck the table for alignment and adjust if necessary.

## ALIGNING THE MILLING MACHINE VISE

Whenever accuracy is required on the workpiece, it is necessary to align the device which holds the workpiece. This may be a vise, angle plate, or special fixture. Since most work is held in a vise, the alignment of this accessory will be outlined. Although the base of the vise is graduated in degrees, it can be set accurately only by means of a dial indicator as outlined in the following. (See N-2.)

N-2 When both the table and the vise must be aligned, the table must always be aligned first.

### To Align the Vise Parallel to the Table Travel

1. Clean the surface of the table and the bottom of the vise.

2. Mount and fasten the vise to the table.

3. Swivel the vise until the solid jaw is approximately parallel with the table slots. (See N-3.)

N-3 Be sure that the solid jaw is clean and free of burrs.

4. Mount a dial indicator to the arbor or to a cutter (Fig. 52-2).

Fig. 52-2 An indicator assembly fastened to the arbor for aligning the vise. (*Courtesy Kostel Enterprises Ltd.*)

5. Tighten a clean parallel in the vise with about 1 in. (25.4 mm) sticking above the vise jaws.

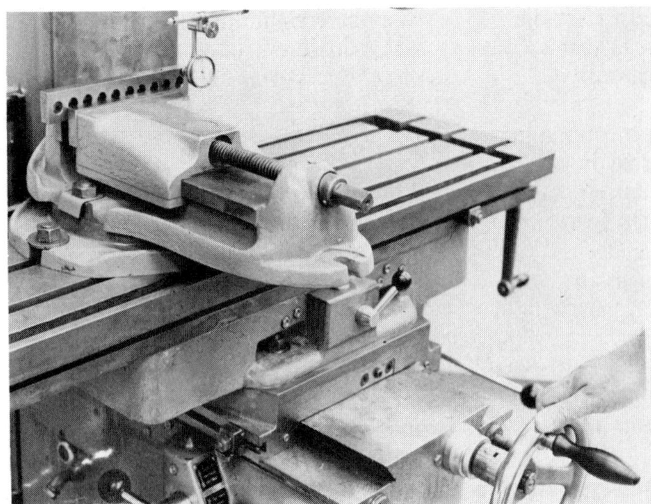

**Fig. 52-3** Setting the indicator to a parallel held in a vise. (*Courtesy Kostel Enterprises Ltd.*)

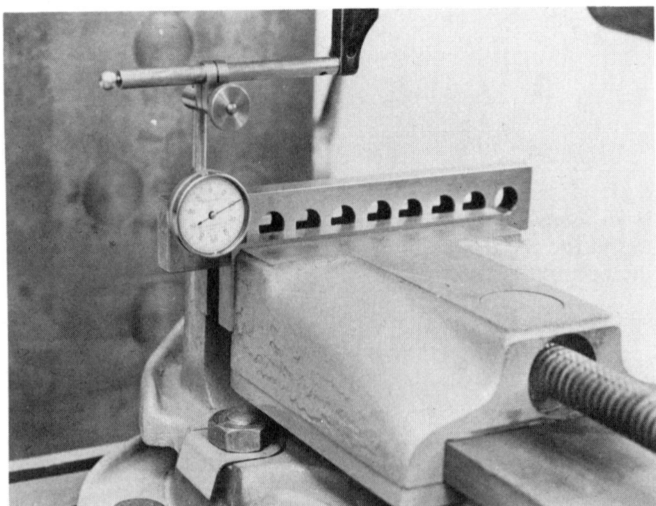

**Fig. 52-4** Checking the indicator reading at the left end of the parallel. (*Courtesy Kostel Enterprises Ltd.*)

N-4 Always tap the vise so that the parallel *moves away* from the indicator.

N-5 Be sure that the table has been aligned parallel to the face of the column before the vise is aligned.

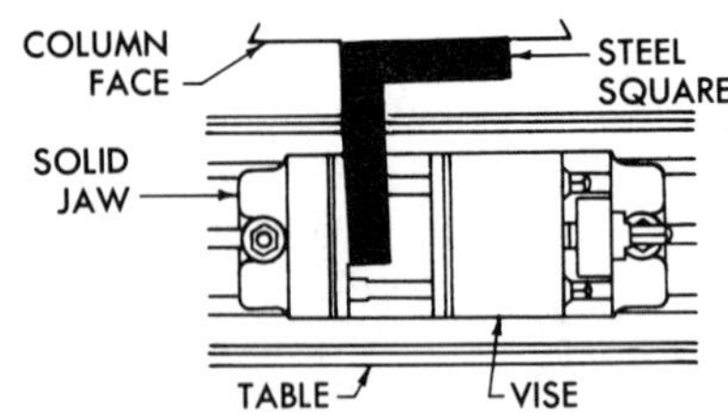


**Fig. 52-5** Using a square to align the vise at right angles to the table.

**Fig. 52-6** Removing the burrs from a workpiece. (*Courtesy Kostel Enterprises Ltd.*)

**6.** Raise the table until the indicator plunger is about ⅜ in. (9.52 mm) below the top of the parallel at the right-hand end (Fig. 52-3).

**7.** Adjust the table with the crossfeed handwheel until the indicator registers about one-half a revolution against the parallel.

**8.** Set the bezel to zero.

**9.** Move the table along the length of the parallel and note the reading at the left-hand end (Fig. 52-4).

**10.** If there is any difference in the two indicator readings, loosen the nuts on the upper or swivel part of the vise.

**11.** Tap the vise with a soft-faced hammer until the indicator needle registers half the difference between the two readings. (See N-4.)

**12.** Tighten the vise lock nuts.

**13.** Recheck the indicator reading at both ends of the parallel and readjust if necessary.

### To Align the Vise at 90° to the Table Travel

Whenever it is necessary to machine steps, grooves, or surfaces at 90° to another edge, the solid jaw of the vise must be aligned. If the work does not require great accuracy, the vise jaw can be aligned at 90° to the table by placing a square against the column face and the solid jaw of the vise (Fig. 52-5). If great accuracy is required, the dial indicator method should be used to align the solid jaw of the vise.

**1.** Clean the surface of the table and the bottom of the universal vise.

**2.** Mount and fasten the vise to the table.

**3.** Swivel the vise until the solid jaw is approximately at 90° to the table (Fig. 52-5).

**4.** Fasten a dial indicator to the arbor or milling cutter (Fig. 52-2).

**5.** Adjust the table until the indicator registers about half a revolution against the solid jaw of the vise or a parallel held in a vise. (See N-5.)

**6.** Set the bezel to zero.

**7.** Move the table toward the column with the crossfeed handwheel and note the reading of the indicator at the other end of the vise jaw or parallel.

**8.** Loosen the nuts on the upper, or swivel, part of the vise.

**9.** Adjust the vise one-half the difference in the indicator reading by tapping the vise by hand or with a soft-faced hammer.

**10.** Tighten the vise and re-indicate along the solid jaw.

**11.** Repeat steps 8 to 10 until the indicator reads the same at both ends of the solid jaw.

## MILLING A FLAT SURFACE

One of the most common operations performed on a milling machine is that of machining a flat surface. Flat surfaces are generally machined on a workpiece with a helical milling cutter. The work may be held in a vise or clamped to a table.

**1.** Remove all burrs from all edges of the work with a mill file (Fig. 52-6).

**2.** Clean the vise and workpiece. (See N-6.)

**3.** Align the vise to the column face of the milling machine using a dial indicator.

**4.** Set the work in the vise using parallels and paper feelers under each corner to make sure that the work is seated on the parallels.

**5.** Tap the work lightly with a soft hammer until the paper feeler is tight between the work and the parallels (Fig. 52-7). (See N-7.)

**6.** Select a plain helical cutter wider than the work to be machined.

**7.** Mount the cutter on the arbor for conventional milling (Fig. 52-8).

**8.** Set the proper speed for the size of cutter and material.

**9.** Set the feed to approximately .003 to .005 in. (0.08 to 0.13 mm) chip per tooth.

**10.** Start the cutter and raise the work, using a paper feeler between the cutter and the work (Fig. 52-9).

**11.** Stop the spindle when the cutter just cuts the paper. (See N-8.)

**12.** Set the graduated collar on the elevating screw handwheel to zero (Fig. 52-10).

**13.** Move the work clear of the cutter and set the depth of cut using the graduated collar.

N-6 Burrs will cause inaccurate work.

N-7 Remove the vise handle before machining so that it does not jam against the arbor.

N-8 Check that there is clearance between the top of the vise and the arbor support.

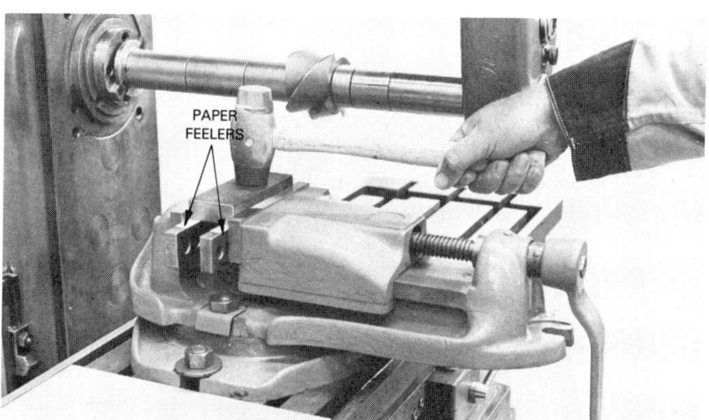


**Fig. 52-7** Seating the workpiece on parallels using paper feelers and a lead hammer. (*Courtesy Kostel Enterprises Ltd.*)

**Fig. 52-8** Work and milling cutter set up for conventional or up milling. (*Courtesy Kostel Enterprises Ltd.*)

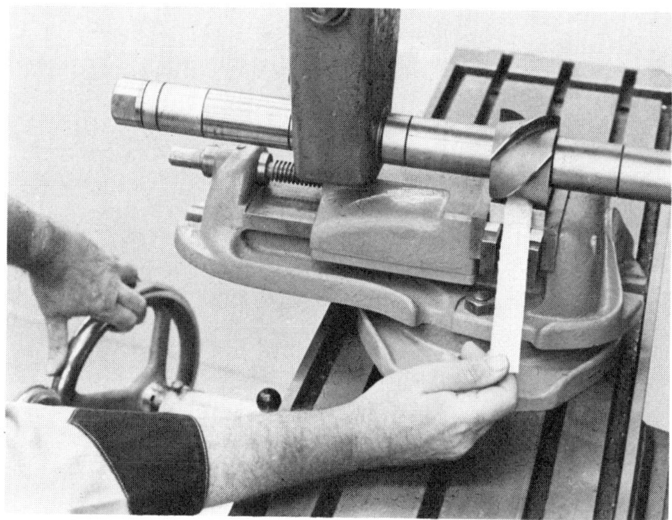

**Fig. 52-9** Raising the workpiece to the cutter. *Caution:* Grasp the paper *lightly* between the fingers. (*Courtesy Kostel Enterprises Ltd.*)

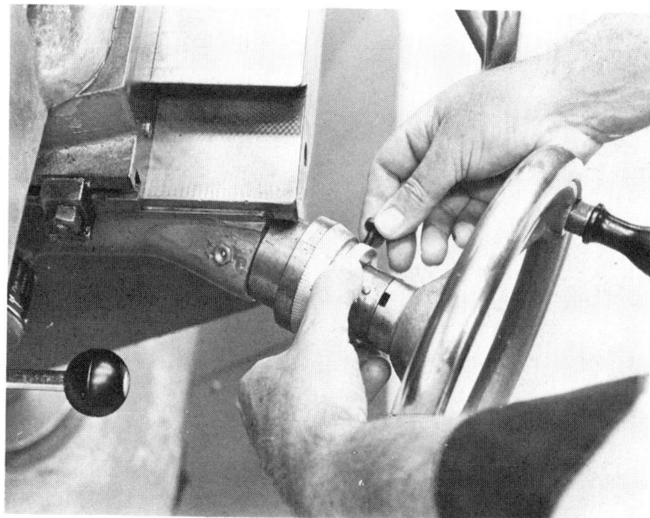

**Fig. 52-10** Setting the vertical feed graduated collar to zero. (*Courtesy Kostel Enterprises Ltd.*)

N-9 Cuts of less than .010 in. (0.25 mm) will dull the cutter.

**14.** For roughing cuts use a depth of not less than ⅛ in. (3.17 mm) and .010 to .025 in. (0.25 to 0.63 mm) for finish cuts. (See N-9.)
**15.** Set the table dogs for the length of cut.
**16.** Engage the feed and cut side 1.
**17.** Set up and cut the remaining sides as required.

## REVIEW QUESTIONS

**1.** Where must the indicator be mounted to align the table parallel to the face of the column?
**2.** Briefly outline how the table is adjusted if there is a difference in the indicator readings on the column.
**3.** What handwheel is used to move the table when aligning the vise
(a) Parallel to the table travel
(b) 90° to the table
**4.** In what direction should the vise be tapped when aligning it with a dial indicator?
**5.** Name two methods used to align the vise at 90° to the table.
**6.** Why are paper feelers used when setting up the workpiece?
**7.** Describe the operation of setting the workpiece in the vise for milling.
**8.** What is the purpose of a paper feeler between the cutter and the workpiece?
**9.** What depth of cut should be used for
(a) Roughing cut
(b) Finishing cut

# UNIT 53

# THE INDEXING OR DIVIDING HEAD

The *indexing or dividing head* is one of the most important attachments for the milling machine. It is used to divide the circumference of a workpiece into equally spaced divisions when milling such items as gears, splines, squares, hexagons, etc. It may also be used to rotate the workpiece at a predetermined ratio to the table feed rate to produce cams and helical grooves on gears, drills, reamers, etc.

## OBJECTIVES

After completing this unit you will be able to:

**1.** Calculate the indexing required for simple and direct indexing.
**2.** Set up and use the dividing head to machine a square and a hexagon using simple and direct indexing.

## INDEXING THEORY

The universal dividing head set consists of the *headstock* with *index plates,* headstock *change gears* and *quadrant, universal chuck, footstock,* and the *center rest.* (Fig. 53-1.)

The index head consists of a 40-tooth *worm wheel* fastened to the *index-head spindle* engaging with a *single-threaded worm* attached to the *index crank* (Fig. 53-2). Since there are 40 teeth in the worm wheel, one complete turn of the index crank will cause the spindle to rotate $\frac{1}{40}$ of a turn. Therefore, 40 turns of the index crank revolve the spindle 1 complete turn, thus making a ratio of 40:1.

Work may be indexed by either *simple* or *direct* indexing. The formula for calculating the number of turns for simple indexing is:

$$\frac{40}{\text{Number of divisions to be cut}} = \text{Number of turns of index crank}$$

**Example:**
For four divisions,

$$\frac{40}{4} = \text{10 full turns of the index crank}$$

If it is required to cut a reamer with eight equally spaced flutes, the indexing would be

$$\frac{40}{8} = \text{5 full turns}$$

For six divisions it would be

$$\frac{40}{6} = 6\tfrac{2}{3} \text{ turns}$$

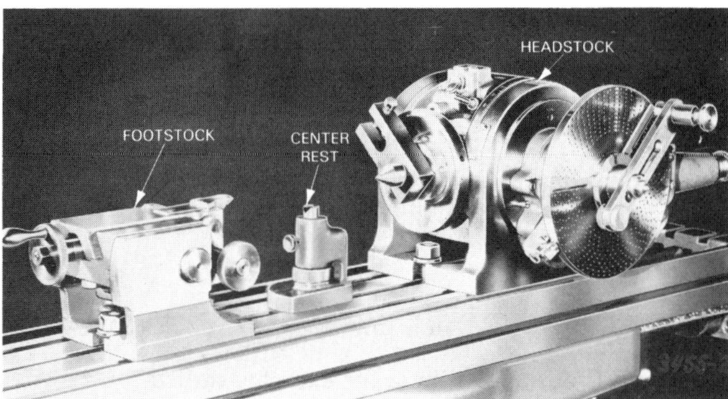

**Fig. 53-1** A universal dividing head set. (*Courtesy Cincinnati Milacron Co.*)

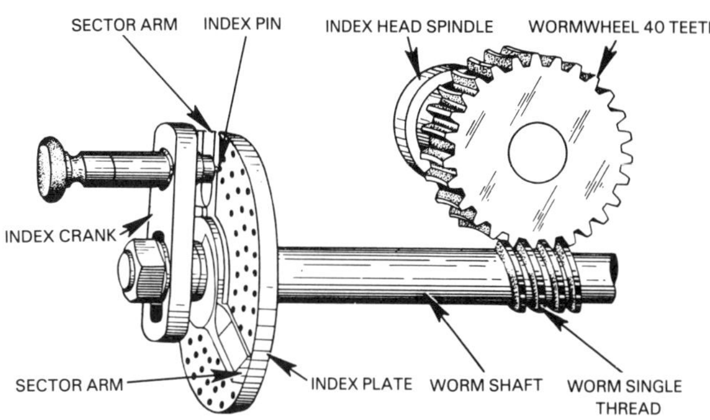

**Fig. 53-2** Section showing the internal mechanism of a dividing head.

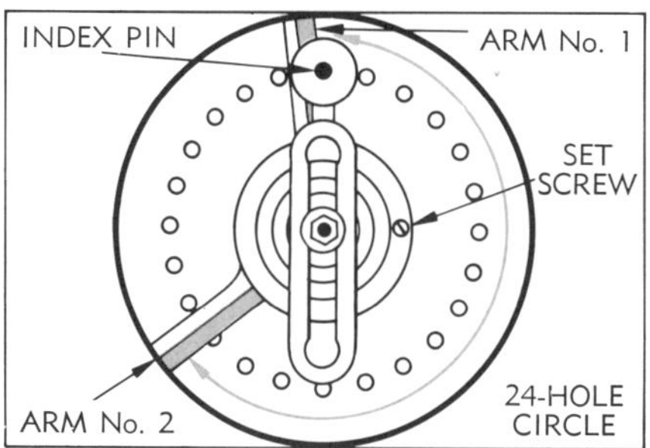

**Fig. 52-3A** Sector arms set for 16 holes on a 24-hole circle (two thirds turn of crank).

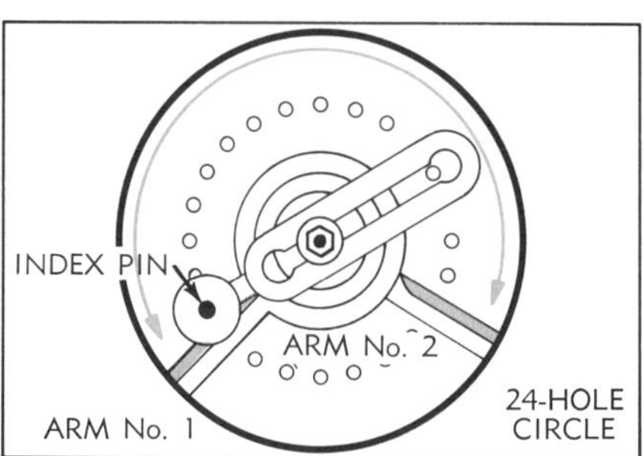

**Fig. 53-3B** Position of sector arms after 6⅔ turns of the crank.

N-1 If the pin is turned past the required hole, an error will result in the spacing. The pin must be turned back at least one-half turn to eliminate backlash, and then brought back to the proper hole.

N-2 Do not count the hole the pin is in, otherwise the indexing will be incorrect.

The six turns are easily made, but what of the two-thirds of a turn? This two-thirds of a turn involves the use of the index plate and sector arms.

## INDEX PLATE

The index plate is a circular plate provided with a series of circles of equally spaced holes into which the index crank pin engages.

To get two-thirds of a turn, choose any circle with a number of holes that is evenly divisible by 3, such as 24 (Table 53-1); then take ⅔ of 24 = 16 holes on a 24-hole circle. For the six divisions we thus have 6 full turns plus 16 holes on a 24-hole circle.

### To Set the Indexing Crank and Sector Arms

To set the dividing head for six turns and 16 holes on a 24-hole circle, proceed as follows:

**1.** Loosen the index crank nut, set the index pin into a hole on the 24-hole circle, and then retighten the nut.

**2.** Place the beveled edge of a sector arm against the index crank pin, then count 16 holes on the 24-hole circle, *not including the hole in which the index crank is engaged* (Fig. 53-3A).

**3.** Move the other sector arm just beyond the sixteenth hole and tighten the lock screws.

**4.** After the first cut has been made and the cutter returned for the next cut, withdraw the index crank pin, make six full turns clockwise, plus the 16 holes between the sector arms.

**5.** Stop between the last two holes, gently tap the index crank, and allow the pin to snap into place.

**6.** Move the sector arms around against the pin, which is ready for the next division (Fig. 53-3B). (See N-1.)

### To Mill a Hexagon Using Simple Indexing

A hexagon to be 1⅝ in. across flats is to be milled on a 2-in.-diameter shaft (Fig. 53-4). To cut the hexagon, check the alignment of the index centers and then mount the work.

$$\text{Indexing for 6 divisions} = \frac{40}{6} \text{ or } 6\tfrac{2}{3} \text{ turns}$$

**1.** Select a hole circle that the denominator 3 will divide into, such as 24. (See N-2.)

**TABLE 53-1 INDEX PLATE HOLE CIRCLES**

| Brown & Sharpe | | Cincinnati Standard Plate | |
|---|---|---|---|
| Plate 1 | 15-16-17-18-19-20 | One side | 24-25-28-30-34-37-38-39-41-42-43 |
| Plate 2 | 21-23-27-29-31-33 | | |
| Plate 3 | 37-39-41-43-47-49 | Other side | 46-47-49-51-53-54-57-58-59-62-66 |

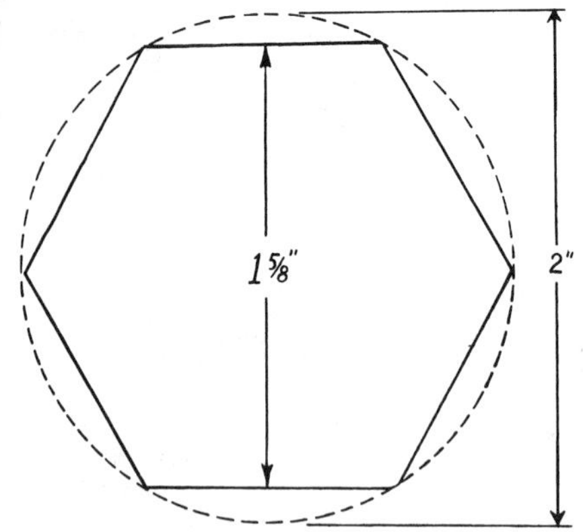


Fig. 53-4 A 1⅝-in. hexagon cut on a 2-in.-diameter shaft.

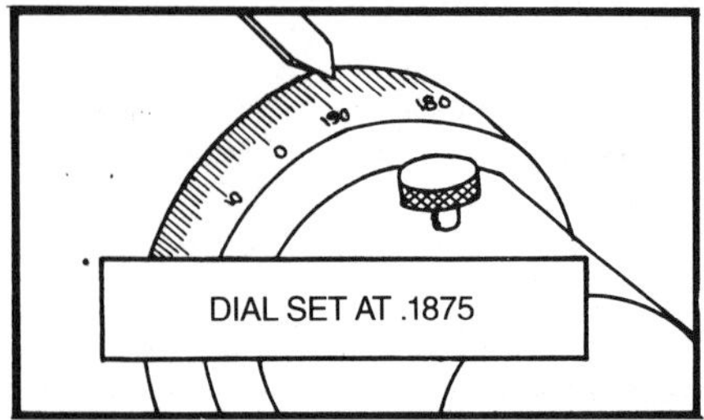


Fig. 53-5 Elevating screw dial set at .1875.

N-3 Use a long piece of paper to keep the hands as far away from the revolving cutter as possible.

**2.** Set the sector arms to 16 holes on a 24-hole circle (Fig. 53-3A). *Do not count the hole the pin is in.*

**3.** From the hole the pin is in, count 16 holes and adjust the other sector arm to just beyond the sixteenth hole (Fig. 53-3B).

**4.** Start the machine and set the cutter to the top of the work by using a paper feeler. (See N-3.)

**5.** Set the graduated dial to zero.

**6.** Calculate the depth of cut (Fig. 53-4). 2 in. − 1⅝ in. = ⅜ in. (.375 in.). As only one-half the material is to be removed from each side,

$$\text{Depth of cut} = \frac{.375 \text{ in.}}{2} = .1875 \text{ in.}$$

**7.** Move work clear of cutter.

**8.** Raise the table .1875 in. (Fig. 53-5).

**9.** Mill one side.

**10.** Index 20 turns (3 × 6⅔) and mill the opposite side (Fig. 53-6).

**11.** With a micrometer check the distance across the flats and adjust the depth setting and recut if necessary (Fig. 53-7).

**12.** Continue indexing 6⅔ turns until all six sides are cut.

## PLAIN OR DIRECT INDEXING

Direct indexing is the simplest form of indexing but can be used only for milling divisions that divide evenly into 24, 30, or 36. The common divisions that can be obtained are listed in Table 53-2.

**Example:**

What direct or plain indexing is required to mill four flats on a round shaft?

$$\text{Indexing} = \frac{24}{4} = 6 \text{ holes in 24-hole}$$

circle for each flat milled.

or

$$\frac{36}{4} = 9 \text{ holes in 36-hole circle for each flat milled}$$

### To Mill a Square by Direct Indexing

**1.** Disengage the worm and worm shaft by turning the worm disengaging shaft lever.

**2.** Adjust the plunger behind the index plate into the 24-hole circle (Fig. 53-8).

**3.** Mount the workpiece in the dividing head chuck or between centers.

**4.** Adjust the cutter height and cut the first side.

**5.** Remove the plunger pin by turning the plunger-pin lever (Fig. 53-8).

**6.** Turn the plate, attached to the dividing head spindle, one-half turn (12 holes) and engage the plunger pin).

**7.** Take the second cut.

Fig. 53-6 Two opposite sides are milled first so that the size may be checked. (*Courtesy Kostel Enterprises Ltd.*)

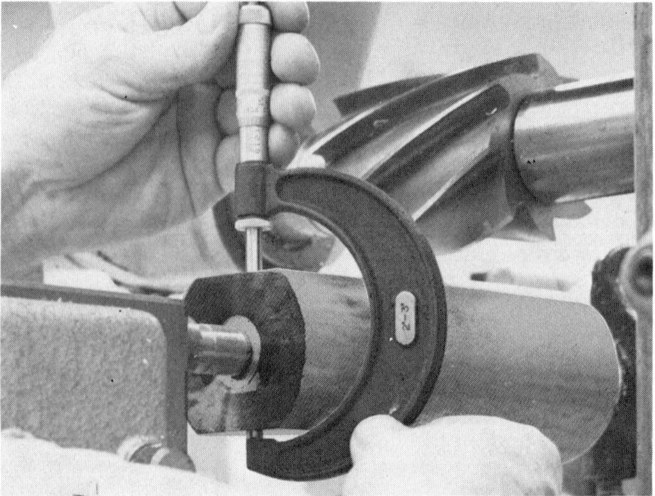

Fig. 53-7 Measuring the distance across the flats. (*Courtesy Kostel Enterprises Ltd.*)

**TABLE 53-2**

| Plate hole circles | Divisions for which plate can be used | | | | | | | | | | | | | |
|---|---|---|---|---|---|---|---|---|---|---|---|---|---|---|
| 24 | 2 | 3 | 4 | ... | 6 | 8 | ... | ... | 12 | ... | ... | 24 | | |
| 30 | 2 | 3 | ... | 5 | 6 | ... | ... | 10 | ... | 15 | ... | ... | 30 | |
| 36 | 2 | 3 | 4 | ... | 6 | ... | 9 | ... | 12 | ... | 18 | ... | ... | 36 |

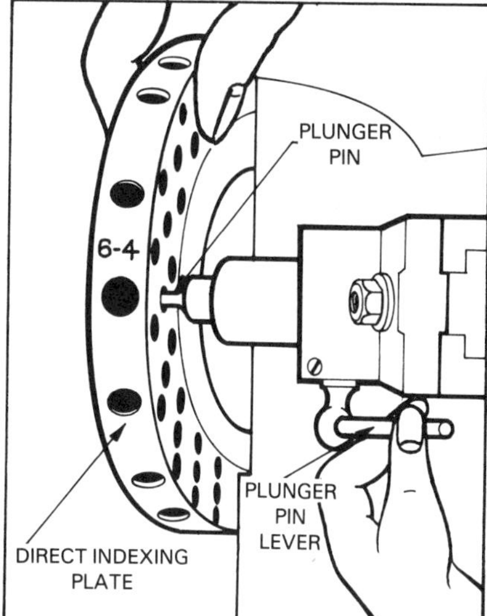


**Fig. 53-8** Limited indexing is achieved by using the direct indexing plate and plunger pin.

**8.** Measure the work across the flats and adjust the work height if required.

**9.** Cut the remaining sides by indexing 6 holes for the third side and 12 holes for the fourth side.

## ANGULAR INDEXING

When the angular distance between divisions is given, instead of the number of divisions, the setup for simple indexing may be used; however, the method of calculating the indexing is changed.

One complete turn of the index crank turns the work $\frac{1}{40}$ of a turn, or $\frac{1}{40}$ of 360°, which equals 9°.

When the angular dimension is given in degrees, the indexing is then calculated as follows:

$$\text{Indexing in degrees} = \frac{\text{degrees required}}{9}$$

**Example 1:**
Calculate the indexing for 45°.

$$\text{Indexing} = \frac{45}{9}$$
$$= 5 \text{ complete turns}$$

**Example 2:**
Calculate the indexing for 60°.

$$\text{Indexing} = \frac{60}{9}$$
$$= 6\tfrac{2}{3}$$
$$= 6 \text{ complete turns, 12 holes on an 18-hole circle}$$

If the dimensions are given in *degrees* and *minutes,* it will be necessary to convert the degrees into minutes (number of degrees × 60 minutes) and add the answer to the minutes required.

The indexing in minutes is calculated as follows.

$$\text{Indexing in minutes} = \frac{\text{minutes required}}{540}$$

## REVIEW QUESTIONS

**1.** What is the purpose of a dividing or indexing head?

**2.** What two types of indexing may be done on a dividing head?

**3.** What is the formula for calculating simple indexing?

**4.** Calculate the simple indexing for 24, 29, and 72 divisions using the Brown & Sharpe dividing head.

**5.** List the procedure for setting the sector arms for 10 holes on a 27-hole circle.

**6.** What will be the depth of cut required to mill a $1\frac{1}{2}$-in. square on the end of a $2\frac{1}{8}$-in. shaft?

**7.** After one side of a hexagon is machined, what side should be machined next? Explain why.

**8.** What direct indexing is required to cut a
(a) Pentagon (5 sides)
(b) Octagon (8 sides)

**9.** Calculate the indexing required for 36° and 75°.

# UNIT 54

# GEAR CUTTING

When it is required to transmit motion from one shaft to another, several methods may be used, such as belts, pulleys, and gears. If the shafts are parallel to each other and quite a distance apart, a flat belt and large pulleys may be used to drive the second shaft, the speed of which may be controlled by the size of the pulleys.

When the shafts are closer together, a V-belt, which tends to reduce the excessive slippage of a flat belt, may be used. With both these methods, the speed of the driven shaft may not be accurate due to slippage between the

Fig. 54-1 A spur gear and pinion are used for slower speeds.

Fig. 54-2 Internal gears provide speed reductions with a minimum space requirement.

N-1 Torque is the turning power of a shaft.

driving and driven members (belts and pulleys). In order to eliminate slippage and produce a positive drive, gears are used.

## OBJECTIVES

After completing this unit you should be able to:

1. Identify the different types of gears used in industry
2. Apply the formula for calculating gear parts
3. Select the proper cutter and cut a spur gear

## GEARS AND GEARING

*Gears* are used to transmit power positively from one shaft to another by means of successively engaging teeth (in two gears). They are used when exact speed ratios and power transmission must be maintained. Gears may also be used to increase or decrease the speed of the driven shaft, thus decreasing or increasing the *torque* of the driven member. (See N-1.)

Shafts in a gear drive or train are generally parallel. They may, however, be driven at any angle by means of suitably designed gears.

## TYPES OF GEARS

*Spur gears* (Fig. 54-1) are generally used to transmit power between two parallel shafts. The teeth on these gears are straight and parallel to the shafts to which they are attached. When two gears of different sizes are in mesh, the larger one is called the *gear* and the smaller one is called the *pinion*. Spur gears are used where slow to moderate speed drives are required.

*Internal gears* (Fig. 54-2) are used where the shafts are parallel and the centers must be closer together than could be achieved with spur or helical gearing. This arrangement provides for a stronger drive since there is a greater area of contact than with the conventional gear drive. It also provides speed reductions with a minimum of space requirement. Internal gears are used on heavy-duty tractors where much torque is required.

Fig. 54-3 Helical gears for shafts which are parallel.

*Helical gears* may be used to connect parallel shafts (Fig. 54-3) or shafts which are at an angle. Because of the progressive rather than intermittent action of the teeth, helical gears run more smoothly and quietly than spur gears. Since there is more than one tooth in engagement at any one time, helical gears are stronger than spur gears of the same size and pitch. However, special bearings (thrust bearings) are often required on shafts to overcome the end thrust produced by these gears as they turn.

On most installations where it is necessary to overcome end thrust, *herringbone gears* (Fig. 54-4) are used. This type of gear resembles two helical gears placed side by side, with one-half having a left-hand helix and the other half a right-hand helix. These gears have a smooth continuous action and eliminate the need for thrust bearings.

When two shafts are located at an angle with their axial lines intersecting at 90°, power is generally transmitted by means of *bevel gears* (Fig. 54-5A). When the shafts are at right angles and the gears are of the same size, they are called *miter gears* (Fig. 54-5B). However, it is not necessary that the shafts be only at right angles in order to transmit power. If the axes of the shafts intersect at any angle other than 90°, the gears are known as *angular bevel gears* (Fig. 54-5C). Bevel gears

**Fig. 54-4** Herringbone gears eliminate end thrust on shafts.

**Fig. 54-5A** Bevel gears transmit power at 90°.

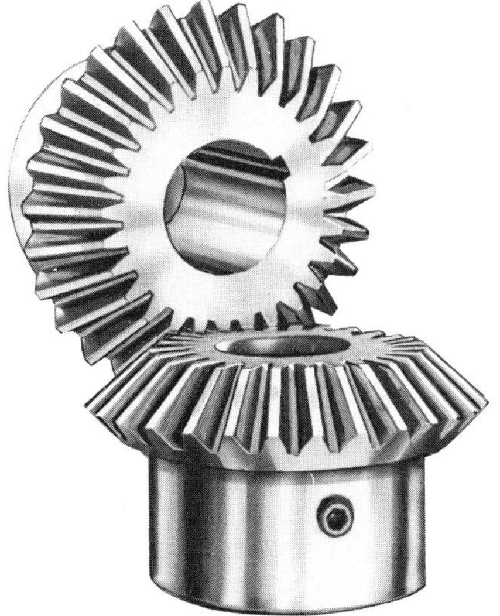

**Fig. 54-5B** Miter gears transmit power at 90° and the driver and driven gears are the same size.

**Fig. 54-5C** Angular bevel gears are used for shafts which are not at right angles.

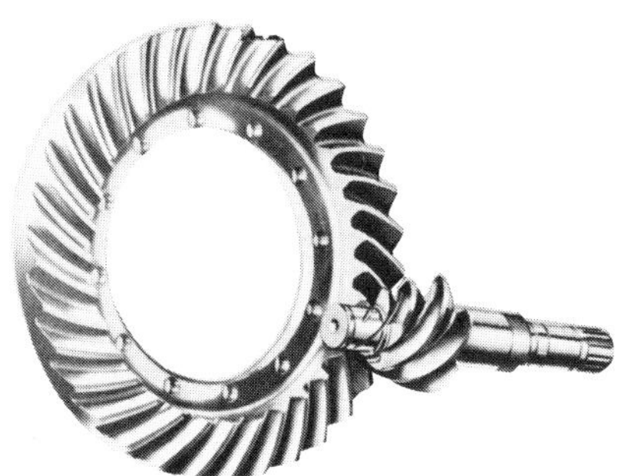

**Fig. 54-5D** Hypoid gears are used in automotive drives.

have straight teeth very similar to spur gears.

A modification of bevel gears having helical teeth are known as *hypoid gears* (Fig. 54-5D). The shafts of these gears, although at right angles, are not in the same plane and therefore do not intersect. Hypoid gears are used in automobile drives.

**Fig. 54-6** A worm and worm gear are used for speed reduction.

**Fig. 54-7** A rack and pinion converts rotary motion to linear motion.

When shafts are at right angles and considerable reduction in speed is required, a *worm and worm gear* (Fig. 54-6) may be used. The worm which meshes with the worm gear may be a single- or multiple-start thread. A worm with a double-start thread will revolve the worm gear twice as fast as a worm with a single-start thread and the same pitch.

When it is necessary to convert rotary motion to linear motion, a *rack and pinion* (Fig. 54-7) may be used. The rack, which is actually a straight or flat gear, may have straight teeth to mesh with a spur gear, or angular teeth to mesh with a helical gear.

## GEAR TERMINOLOGY

A knowledge of the more common gear terms is desirable to understand

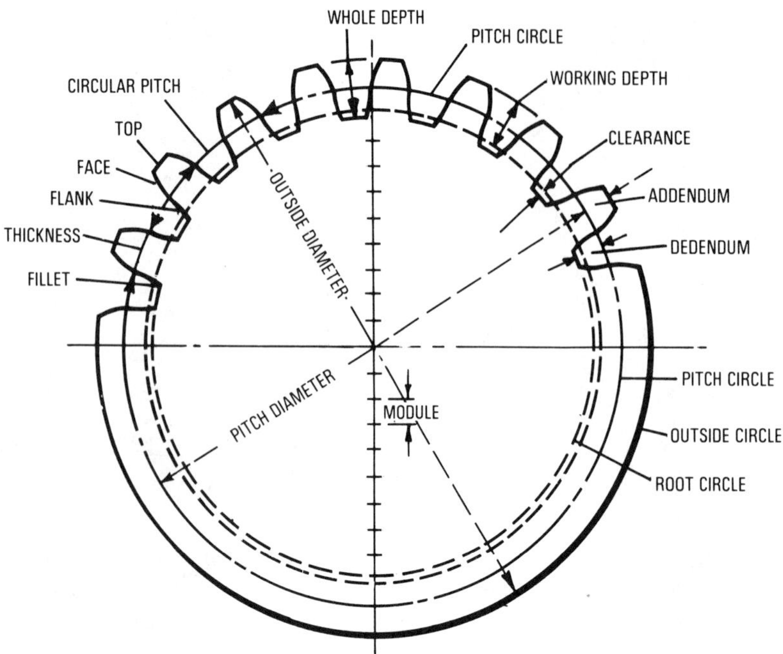


**Fig. 54-8** Parts of a gear.

N-2 Circular pitch calculations are used when cutting a rack.

N-3 The involute is the shape of the sides of most gear teeth.

N-4 A rack, which converts rotary motion to linear motion, consists of a series of gear teeth cut on a flat bar or surface.

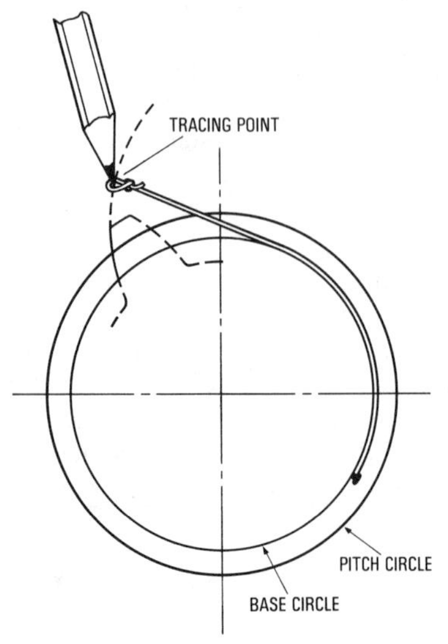


**Fig. 54-9** Method of generating an involute.

gearing and make the calculations necessary to cut a gear. Most of these terms are applicable to either inch or metric gearing, although the method of calculating dimensions may differ. These methods are explained as applicable to the inch and metric gear-cutting sections.

- *Addendum*— the height of the tooth above the pitch circle
- *Center distance*—the shortest distance between the axes of two mating gears
- *Chordal thickness*—the thickness of the tooth measured at the pitch circle
- *Circular pitch*—the distance from a point on one tooth to a corresponding point on the next tooth measured on the pitch circle (see N-2)
- *Clearance*—the distance between the top of one tooth and the bottom of the mating tooth space
- *Dedendum*—the distance from the pitch circle to the bottom of the tooth
- *Diametral pitch*—the ratio of the number of teeth for each inch of pitch diameter of the gear
- *Involute*—the curved line produced by a point of a stretched string when it is unwrapped from a given cylinder (Fig. 54-9) (see N-3)
- *Module* (metric gears)—the pitch diameter of a gear divided by the number of teeth; an actual dimension, unlike diametral pitch, which is a ratio of the number of teeth to the pitch diameter
- *Linear pitch*—the distance from a point on one tooth to the corresponding point on the next tooth of a gear rack (see N-4)
- *Pitch diameter*—the diameter of the pitch circle
- *Tooth thickness*—the thickness of the tooth measured on the pitch circle
- *Whole depth*—the full depth of the tooth

## Spur Gear Cutters

Involute gear cutters are used to cut most spur and helical gears.

Gear cutters are available in many sizes ranging from 1 to 48 diametral pitch. Comparative sizes of teeth ranging from 4 DP to 16 DP are shown in Fig. 54-10.

These cutters are generally made in sets of eight and are numbered from 1 to 8 (Fig. 54-11). Note the gradual change in shape from the cutter 1, which has almost straight sides to the much more curved sides of the cutter 8. As shown in Table 54-1, the cutter 1 is used for cutting any number of teeth in a gear from 135 teeth to a rack, while the cutter 8 will cut only 12 and 13 teeth. It should be noted that in order for gears to mesh, they must be of the same diametral pitch; the number of cutter permits *only* a more accurate meshing of the teeth.

## Cutting a Spur Gear

In order to cut a spur gear on a milling machine, the operator must first check that all sizes and dimensions on the

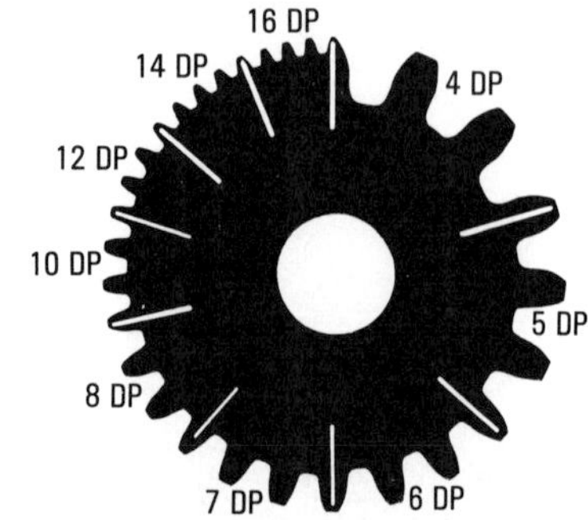


**Fig. 54-10** Comparative tooth sizes 4 to 16 DP.

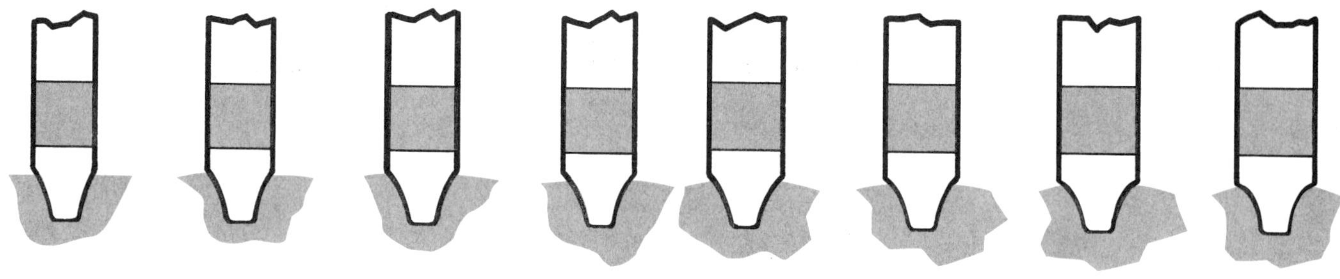


**Fig. 54-11** Involute profiles for a set of eight gear cutters 10 DP.

**TABLE 54-1 INVOLUTE GEAR CUTTERS**

| Cutter number | Range of teeth |
|---|---|
| 1 | 135 to a rack |
| 2 | 55–134 |
| 3 | 35–54 |
| 4 | 26–34 |
| 5 | 21–25 |
| 6 | 17–20 |
| 7 | 14–16 |
| 8 | 12–13 |

N-5 $\text{O.D.} = \frac{N + 2}{DP}$

N-6 $\text{W.D.} = \frac{2.157}{DP}$

N-7 $\text{Indexing} = \frac{40}{N}$

N-8 If centers are not aligned, tapered gear teeth will result.

N-9 Do not count the hole in which the pin is engaged.

print are correct. He should also pay particular attention to the proper cutter selection and machine setup.

The procedure for machining a spur gear is outlined in the following example. (See N-5, N-6, and N-7.)

**Example:**

A 52-tooth gear with an 8 diametral pitch is required.

**1.** Calculate all the necessary gear data. (See Table 54-2.)

$$\text{OD} = \frac{54}{8} = 6.750$$

$$\text{Whole depth of tooth} = \frac{2.157}{8} = .2697$$

Cutter number = 3 (35 to 54 teeth)

Indexing (using Cincinnati standard plate)

$$= \frac{40}{52} = \frac{10}{13} \times \frac{3}{3}$$

= 30 holes on the 39 hole circle

**2.** Turn the gear blank to the proper dimensions.

**3.** Press the gear blank firmly onto the mandrel.

**4.** Align the table of the milling machine parallel to the column.

**5.** Mount the index head and footstock, and check the height alignment of the index centers (Fig. 54-12). (See N-8.)

**6.** Set the dividing head so that the index pin fits into a hole on the 39-hole circle and the sector arms are set for 30 holes. (See N-9.)

**7.** Mount the mandrel (and workpiece), with the large end toward the indexing head, between the index centers.

(a) The footstock center should be adjusted up tightly into the mandrel and locked in position.

(b) The dog should be tightened properly on the mandrel and the tail of the dog should not bind in the slot.

(c) The tail of the dog should then be locked in the driving fork of the dividing head by means of the set screws. This ensures that there will be no play between the dividing head and the mandrel.

(d) The dog should be far enough from the gear blank to ensure that the cutter will not hit the dog when the gear is being cut.

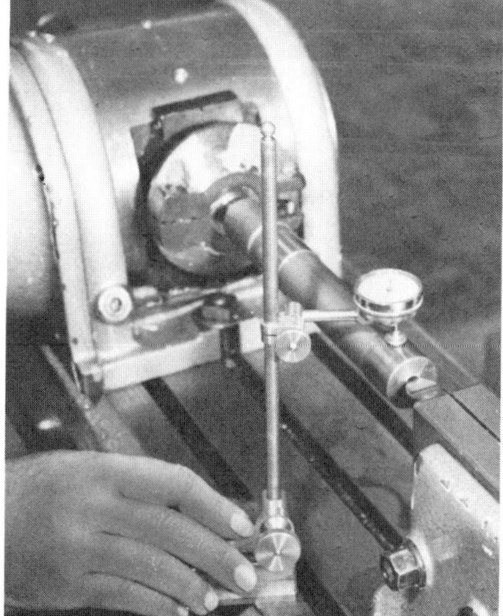

**Fig. 54-12** Checking the height alignment of the index centers with a dial indicator.

**TABLE 54-2 RULES AND FORMULAS FOR SPUR GEARS**

| To obtain | Knowing | Rule | Formula |
|---|---|---|---|
| Addendum | Diametral pitch | Divide 1 by the diametral pitch | $A = \frac{1}{DP}$ |
| Center distance | Diametral pitch | Divide the total number of teeth in both gears by twice the diametral pitch | $CD = \frac{N + n}{2 \times DP}$ |
| Circular pitch | Diametral pitch | Divide 3.1416 by the diametral pitch | $CP = \frac{3.1416}{DP}$ |
| Clearance | Diametral pitch | Divide .157 by the diametral pitch | $Cl = \frac{.157}{DP}$ |
| Dedendum | Diametral pitch | Divide 1.157 by the diametral pitch | $D = \frac{1.157}{DP}$ |
| Diametral pitch | Number of teeth and pitch diameter | Divide the number of teeth by the pitch diameter | $DP = \frac{N}{PD}$ |
| Diametral pitch | Number of teeth and outside diameter | Add 2 to the number of teeth and divide the sum by the outside diameter | $DP = \frac{N + 2}{OD}$ |
| Number of teeth | Outside diameter and diametral pitch | Multiply the outside diameter by the diametral pitch and subtract 2 | $N = OD \times DP - 2$ |
| Number of teeth | Pitch diameter and diametral pitch | Multiply the pitch diameter by the diametral pitch | $N = PD \times DP$ |
| Pitch diameter | Number of teeth and diametral pitch | Divide the number of teeth by the diametral pitch | $PD = \frac{N}{DP}$ |
| Pitch diameter | Outside diameter and number of teeth | Multiply the number of teeth by the outside diameter, and divide the product by the number of teeth plus 2 | $PD = \frac{N \times OD}{N + 2}$ |
| Outside diameter | Number of teeth and diametral pitch | Add 2 to the number of teeth and divide the sum by the diametral pitch | $OD = \frac{N + 2}{DP}$ |
| Tooth thickness | Diametral pitch | Divide 1.5708 by the diametral pitch | $T = \frac{1.5708}{DP}$ |
| Whole depth | Diametral pitch | Divide 2.157 by the diametral pitch | $WD = \frac{2.157}{DP}$ |

**8.** Move the table close to the column to keep the setup as rigid as possible.

**9.** Mount an 8 DP— 3 cutter on the milling machine arbor over the approximate center of the gear. Be sure to have the cutter rotating toward the indexing head.

**10.** Center the gear blank with the cutter by either of the following methods:

(a) Place a square against the outside diameter of the gear (Fig. 54-13). With a pair of inside calipers or a rule, check the distance between the square and the side of the cutter. Adjust the table until the distances from both sides of the gear blank to the sides of the cutter are the same.

(b) A more accurate method of centralizing the cutter is to use gage blocks in lieu of the inside calipers or rule.

**11.** *Lock the cross slide.*

**12.** Start the milling cutter and run the work under the cutter. (See N-10.)

**13.** Raise the table until the cutter *just* touches the work. This can be done by using a chalk mark on the gear blank or by placing a piece of paper

N-10 Be sure that the cutter revolves so that the teeth point in the direction of the arbor rotation.

**Fig. 54-13** Centering the gear cutter and the workpiece.

**Fig. 54-14** Setting the micrometer collar on the vertical feed screw to zero. (*Courtesy Kostel Enterprises Ltd.*)

N-11 The center of the cutter must clear the edge of the work before the trip dog is set.

N-12 After each tooth has been cut, the cutter should be stopped before the table is returned to prevent marring the finish on the gear teeth.

between the gear blank and the cutter to indicate when the cutter is just touching the work.

**14.** Set the graduated micrometer collar on the vertical feed to zero (Fig. 54-14).

**15.** Move the work clear of the cutter by means of the longitudinal feed handle and raise the table to about two-thirds the depth of the tooth (.180); then lock the knee clamp.

**16.** Slightly notch all gear teeth to check for correct indexing.

**17.** Rough out the first tooth and set the automatic feed trip dog after the cutter is clear of the work. (See N-11.)

**18.** Return the table to the starting position.

**19.** Cut the remaining teeth and return the table to the starting position.

**20.** Loosen the knee clamp, raise the table to the proper depth of .270 in., and *lock the knee clamp*. (See N-12.)

**21.** Finish cut all teeth.

## REVIEW QUESTIONS

**1.** What is the advantage of a gear drive over a belt drive?

**2.** State three uses of gears.

**3.** Where may the following gear drives be used?

(a) Spur
(b) Internal
(c) Herringbone
(d) Hypoid

**4.** State two advantages of helical gears over spur gears.

**5.** What is the purpose of a worm and worm-gear arrangement?

**6.** What is the purpose of a rack and pinion?

**7.** Define the following gear terms.

(a) Pitch diameter
(b) Diametral pitch
(c) Addendum
(d) Dedendum

**8.** What two factors must be considered when selecting a gear cutter?

**9.** What is the formula for each of the following gear parts?

(a) Outside diameter
(b) Whole depth of tooth

**10.** What is the formula for determining the proper indexing for the gear teeth?

**11.** How should the mandrel be facing when setting it between the index centers?

**12.** Name two methods of centering the gear blank with the cutter.

**13.** What should be the depth of the first (roughing) cut when cutting a gear?

**14.** What precaution should be taken so that the finish of a tooth will not be marred?

# UNIT 55

# VERTICAL MILLING MACHINES

Much of the work on a horizontal milling machine is best done with a vertical milling attachment. The time involved in setting up a vertical attachment prevents the machine from being used for other milling operations at the same time; therefore, the vertical milling machine has become popular in industry. Some of the operations which can be carried out on these machines are: face milling, end milling, keyway cutting, dovetail cutting, T-slot and circular-slot cutting, gear cutting, drilling, boring, and jig boring. Because of the vertical spindle, many of the facing operations can be done with a fly-cutter, which reduces the cost of cutters considerably. Also, since most cutters are much smaller than for the horizontal mill, the cost of the cutters for the same job is usually much less for a vertical milling machine.

## OBJECTIVES

After completing this unit, you should be able to:

**1.** Recognize and state the purpose of four types of vertical milling machines

**2.** Identify and state the purpose of the main operative parts of a vertical mill

**3.** Identify and know the purpose of various cutters used on a vertical mill

**Fig. 55-1** Standard vertical milling machine. (*Courtesy Cincinnati Milacron Co.*)

## TYPES OF VERTICAL MILLING MACHINES

The standard milling machine (Fig. 55-1) has all the construction features of a plain horizontal milling machine, except that the cutter spindle is mounted in a vertical position. The spindle head on most vertical milling machines may be swiveled, which permits the machining of angular surfaces. The spindles on most vertical milling machines have a short travel, which facilitates step milling and the drilling and boring of holes.

The cutters used are of the end mill or shell end mill type. This type of machine is particularly suited to the use of the rotary table, permitting the machining of circular grooves and the positioning of holes which have been laid out with angular measurements.

The *ram-type* vertical milling machine (Fig. 55-2) is a lighter type than the standard vertical machine. Because of its simplicity and the ease of setup, it has become increasingly popular. Machines of this type are generally used for lighter types of milling machine work.

*The numerically controlled vertical milling machine* (Fig. 55-3) is designed with rigid features for production operations. It can control the machine to move the table longitudinally (*X* axis), move the table crosswise (*Y* axis), and control the vertical movement of the spindle (*Z* axis). The magnetic or punched tape, which contains the machining information and procedures, activates various machine controls and servomechanisms to automatically produce a part.

**Fig. 55-2** Ram-type vertical milling machines. (*Courtesy Bridgeport Machine Division of Textron Inc.*)

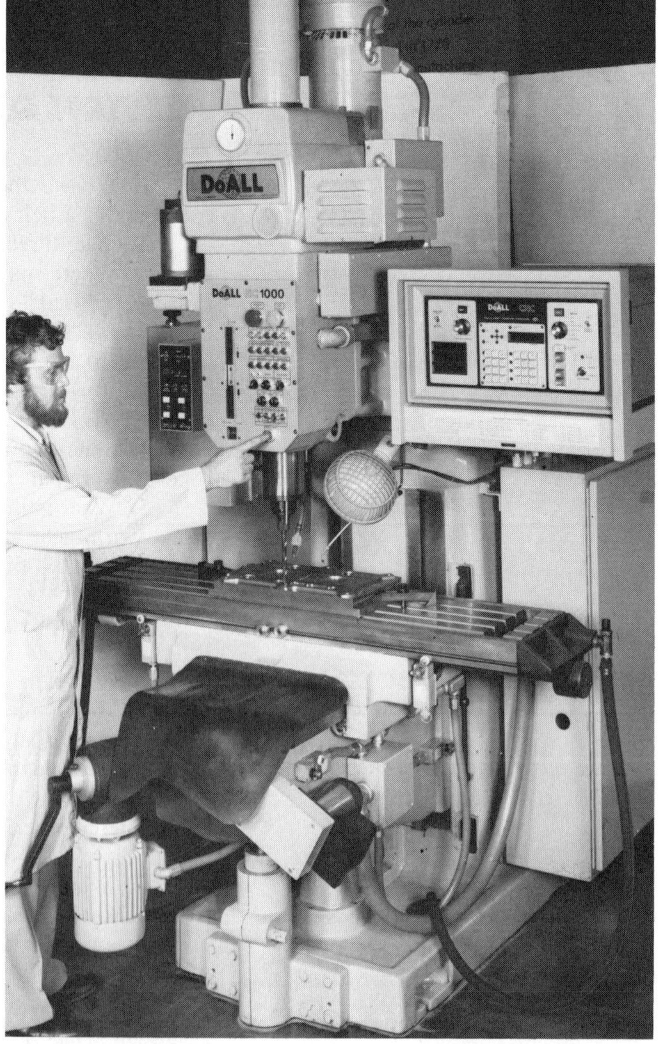


**Fig. 55-3** Rigid ram numerically controlled vertical milling machine. (*Courtesy The DoAll Company.*)

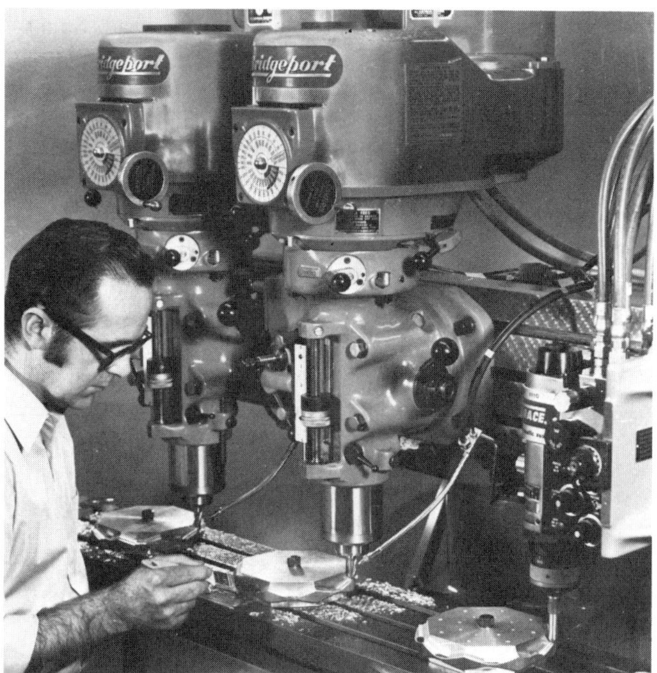

**Fig. 55-4** A tracer-controlled vertical milling machine. (*Courtesy Bridgeport Machine Division of Textron Inc.*)

**Fig. 55-5** An optical scanner controlling the machine movements by following the lines on a drawing. (*Courtesy Bridgeport Milling Machines, a Division of Textron Inc.*)

Most milling machines can be equipped with numerical control units. Once the machining information and procedures are programmed on punched or magnetic tape, the tape reader on the unit automatically activates machine controls to follow the instructions and machine the part accordingly.

The *tracer-controlled milling machine* (Fig. 55-4) duplicates the shape of a master or template on a workpiece. The hydraulic tracer unit is activated by a stylus or tracer finger which contacts the master or template. Only a light pressure on the stylus when it is in contact with the master will cause the cutter to move up, down, or sideways to coincide with the stylus movement. As the master moves under the stylus, the workpiece assumes the identical form of the master.

The *optical-scanner-controlled milling machine* (Fig. 55-5) can duplicate flat parts directly from sharp black lines on a drawing, sketch, or print. Electric signals, generated by the scanner, pick up the lines and feed these signals to servomechanisms which control the table movement. The optical scanner reproduces parts accurately since it follows the lines on a drawing or print to within ±.001 in. (0.02 mm) accuracy.

## VERTICAL MILL PARTS

A vertical milling machine (Fig. 55-6) has basically the same parts as a horizontal plain milling machine. Instead of the cutter fitting into a horizontal spindle, it fits into a vertical spindle. On most machines the head can be swiveled 90° to either side of the centerline for the cutting of angular surfaces. The vertical milling machine is especially useful for face and end milling operations (Fig. 55-7A and B).

The *base* (Fig. 55-6) is made of ribbed cast iron. It may contain a coolant reservoir.

The *column* is cast in one piece with the base. The machined face of the column provides the ways for the vertical movement of the knee. The upper part of the column is machined to receive a *turret* on which the *overarm* is mounted.

The *overarm* may be round or of the more common *ram* type. It may be adjusted toward or away from the column to increase the capacity of the machine.

The *head* is attached to the end of the ram (or overarm) (Fig. 55-6). On universal-type machines, the ram may be swiveled in two planes. The motor which provides the drive to the *spindle*

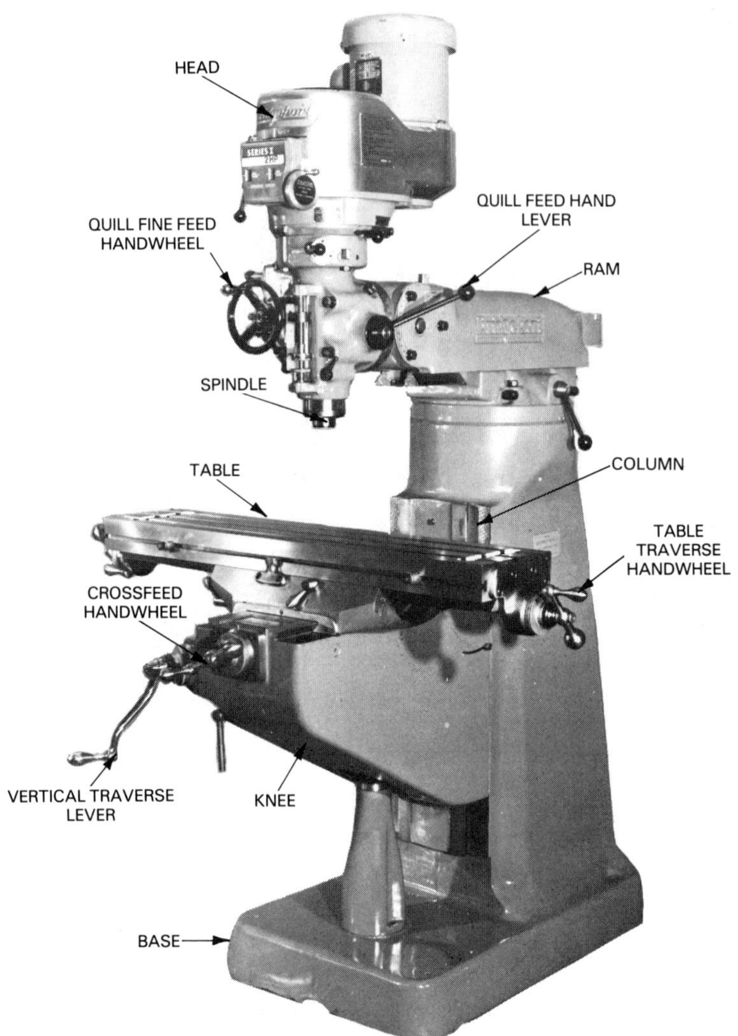


**Fig. 55-6** A vertical milling machine is useful for face and end milling operations. (*Courtesy Bridgeport Machine Division of Textron Inc.*)

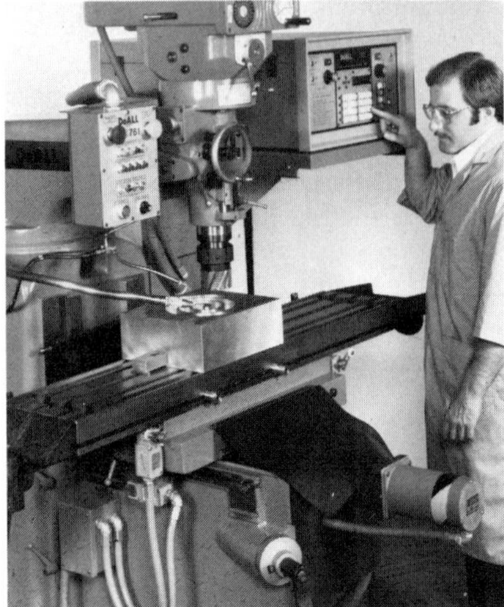

**Fig. 55-7A** End milling the top surface and inside of a casting fastened to an angle plate. (*Courtesy The DoAll Company.*)

**Fig. 55-7B** Milling a surface with the head tilted to the required angle. (*Courtesy Bridgeport Machine Division of Textron Inc.*)

is mounted on top of the head. Spindle-speed changes may be made by means of gears and V-belts or variable-speed pulleys on some models. The spindle mounted in the *quill* may be fed by means of the *quill-feed hand lever,* the *quill-fine-feed handwheel,* or by automatic-power feed.

The *knee* moves up and down the face of the column and supports the *saddle* and *table*. On most machines, all table movements are controlled manually. Automatic-feed-control units for any table movement are usually added as accessories.

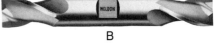

A

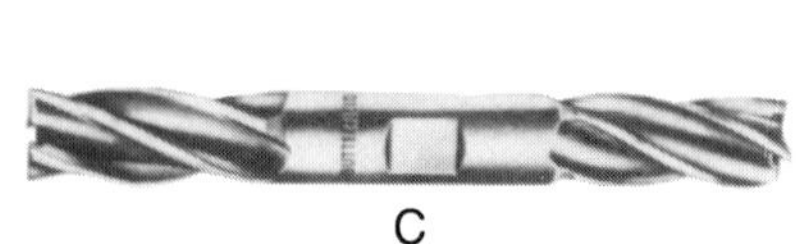


B

C

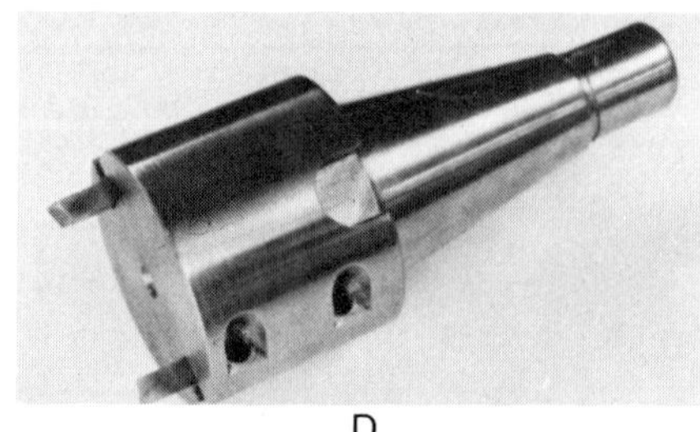

D

**Fig. 55-8** Types of vertical milling cutters. (A) Shell end mill and arbor; (B) two-fluted end mill (*Courtesy Weldon Tool Company*); (C) four-fluted end mill; (D) flycutter.

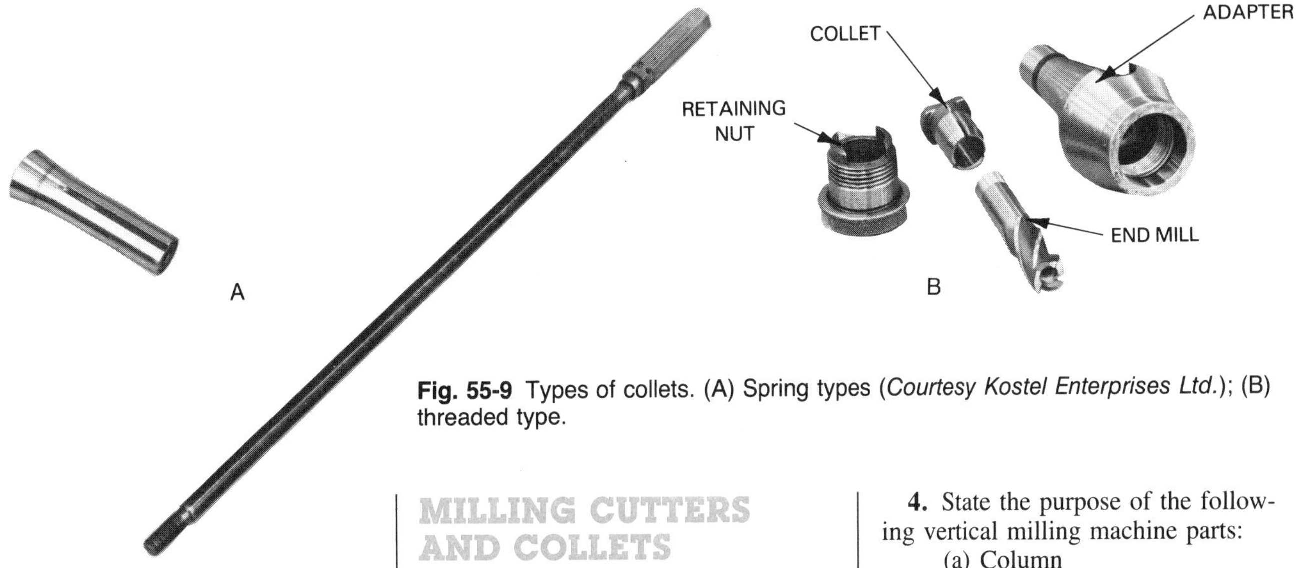

Fig. 55-9 Types of collets. (A) Spring types (*Courtesy Kostel Enterprises Ltd.*); (B) threaded type.

## MILLING CUTTERS AND COLLETS

Most machining on the vertical mill is done with either an end mill, a shell end mill, or a flycutter.

*End mills* have cutting teeth on the end as well as on the periphery and are fitted to the spindle by a suitable adaptor. They may be of two types: the *solid end mill* or the *shell end mill* (Fig. 55-8A), which is fitted to a separate shank.

The solid end mills may have two or more flutes. Two-fluted end mills (Fig. 55-8B) have different length lips on the end and may be used for drilling shallow holes. The four-fluted type (Fig. 55-8C) requires a starting hole before milling a slot in the center of a workpiece.

Larger work surfaces may be machined by means of a *flycutter* (Fig. 55-8D), which holds two or more single-pointed cutting tools. Flycutters provide an economical means of machining work surfaces.

*Collets* provide a means of holding solid end mills in the spindle. They may be of the spring type (Fig. 55-9A) or the threaded type (Fig. 55-9B).

## REVIEW QUESTIONS

**1.** List seven operations which can be performed on a vertical milling machine.

**2.** Name two common types of vertical milling machines.

**3.** What is the purpose of the following milling machines?
- (a) Numerically controlled
- (b) Tracer-controlled
- (c) Optical-scanner-controlled

**4.** State the purpose of the following vertical milling machine parts:
- (a) Column
- (b) Head
- (c) Spindle
- (d) Knee

**5.** Name two types of end mills used in a vertical mill.

**6.** What type of end mill can be used for drilling a shallow hole?

**7.** For what purpose are flycutters used?

**8.** Name two types of collets used to hold end mills.

# UNIT 56

# SETTING UP THE VERTICAL MILL

In order to produce accurate work on a vertical milling machine, it is important that the vertical head and vise be aligned properly. To prolong the life of the cutter, be sure that the cutter speed and feed are correct for the type and size of the cutter and the work material being cut.

## OBJECTIVES

After completing this unit, you should be able to:

**1.** Align the toolhead at 90° to the table

**2.** Align the vise to within ±.001 in. (0.02 mm) of parallel to the table travel (See N-1.)

N-1 The head must be aligned square (at 90°) to the table to produce accurate work.

**Fig. 56-1** The indicator dial should be set at zero at the front of the table. (*Courtesy Kostel Enterprises Ltd.*)

**Fig. 56-2** The difference between the back and front reading is the amount the head is out of alignment. (*Courtesy Kostel Enterprises Ltd.*)

**3.** Calculate the proper cutting speed and feed to be used with various cutting tools and work material

## ALIGNING THE VERTICAL HEAD

Proper alignment of the head is very important when machining holes or when face milling. If the head is not at an angle of 90° to the table, the holes will not be square with the work surface when the cutting tool is fed by hand or by automatic feeds. When face milling, the machined surface will be stepped if the head is not square with the table. Although all heads are graduated in degrees and some have vernier devices used for setting the head, it is well to check the spindle alignment before machining work.

N-2 Always set the indicator to the high reading on the table first to prevent the indicator plunger from being caught in the T-slots of the table.

**1.** Mount a dial indicator on a suitable rod held in the spindle or a drill chuck (Fig. 56-1).

**2.** Position the indicator over the front of the table. (See N-2.)

**3.** Carefully lower the spindle until the indicator button touches the table and the dial indicator registers about one-half revolution (Fig. 56-1).

**4.** Set the bezel at zero.

**5.** Carefully rotate the spindle 180° by hand until the button bears on the back of the table. Compare the readings (Fig. 56-2).

**6.** If there is any difference in the readings, loosen the locking nuts on the swivel mounting.

**7.** Adjust the head until the indicator registers one-half the difference between the two readings. Tighten the locking nuts (Fig. 56-3).

**8.** Recheck the accuracy of the head and adjust if necessary.

**9.** Rotate the vertical mill spindle 90° and set the dial indicator at one end of the table (Fig. 56-4A).

**10.** Rotate the machine spindle 180° and check the reading for the other end of the table (Fig. 56-4B).

**11.** If the two readings do not coincide, repeat steps 6 and 7 until the readings are the same.

**12.** Tighten the locking nuts on the swivel mount.

**13.** Recheck the readings and adjust if necessary.

**Fig. 56-3** Tightening the lock screws on the head. (*Courtesy Kostel Enterprises Ltd.*)

Fig. 56-4A Setting the indicator dial to zero. (*Courtesy Kostel Enterprises Ltd.*)

Fig. 56-4B Checking the difference between the right- and left-hand table indicator readings. (*Courtesy Kostel Enterprises Ltd.*)

## ALIGNING THE MILLING MACHINE VISE

Whenever accuracy is required on the workpiece, it is necessary to align the device which holds the workpiece. This may be a vise, angle plate, or special fixture. Since most work is held in a vise, the alignment of this accessory will be outlined.

### To Align the Vise Parallel to the Table Travel

**1.** Clean the surface of the table and the bottom of the vise.

**2.** Mount and fasten the vise.

**3.** Swivel the vise until the solid jaw is approximately parallel with the table slots.

**4.** Mount a dial indicator to a rod held in a chuck or spindle.

**5.** Adjust the table until the indicator registers about half a revolution against a parallel held in a vise.

**6.** Set the bezel to zero (Fig. 56-5A).

**7.** Move the table along for the length of the jaw and note the reading of the indicator (Fig. 56-5B).

**8.** Loosen the nuts on the upper or swivel part of the vise.

**9.** Adjust the vise to half the movement of the needle by tapping it

Fig. 56-5A Setting the indicator dial to zero at the right-hand end of the parallel.

Fig. 56-5B Checking the difference between the indicator readings at the right- and left-hand end of the parallel.

**Fig. 56-6** When indicator readings at both ends of the parallel are the same, tighten the vise locknuts.(*Courtesy Kostel Enterprises Ltd.*)

N-3 Always tap the vise so that the parallel moves away from and *not toward* the indicator.

N-4 Too high a speed will dull the cutter quickly; too low a speed may cause cutter breakage.

N-5 $\text{r/min (inch)} = \frac{CS \times 4}{D}$

by hand or with a soft hammer in the appropriate direction. (See N-3.)

**10.** Tighten the vise at this setting (Fig. 56-6).

**11.** Recheck and adjust if necessary.

## CUTTING SPEEDS

The cutting speed for a milling cutter is the speed, either in feet per minute or in meters per minute, that the periphery of the cutter should travel when machining a certain metal. The speeds used for milling machine cutters are much the same as those used for any cutting tool. Several factors must be considered when setting the revolutions per minute to machine a surface. The most important are the:

- Material to be machined
- Type of cutter
- Finish required
- Depth of cut
- Rigidity of the machine and the workpiece.

It is good practice to start from the calculated revolutions per minute and then progress until maximum tool life and economy is accomplished. The speed of a milling machine cutter is calculated as for a twist drill or a lathe workpiece. (See N-4.)

### Inch Cutting Speeds

**Example:**
Find the revolutions per minute to mill a keyway in a machine steel shaft using a ¼-in. high-speed steel end mill (CS = 80).

$$\text{r/min} = \frac{CS \times 4}{D} = \frac{80 \times 4}{1/4} = 1280$$

**Example:**
Calculate the revolutions per minute required to end mill a ¾-in.-wide groove in aluminum with a high-speed cutter (CS = 600). (See N-5.)

$$\text{r/min} = \frac{600 \times 4}{3/4} = 3200$$

### Metric Cutting Speeds

The revolutions per minute at which the milling machine should be set when using metric measurements is as follows:

$$\text{r/min} = \frac{(CS,\ m) \times 1000}{(\pi \times D,\ mm)} = \frac{CS \times 1000}{3.1416 \times D}$$

**TABLE 56-1 MILLING MACHINE CUTTING SPEEDS**

| | High-speed steel cutter | | Carbide cutter | |
|---|---|---|---|---|
| **Material** | ft/min | m/min | ft/min | m/min |
| Machine steel | 70–100 | 21–30 | 150–250 | 45–75 |
| Tool steel | 60–70 | 18–20 | 125–200 | 40–60 |
| Cast iron | 50–80 | 15–25 | 125–200 | 40–60 |
| Bronze | 65–120 | 20–35 | 200–400 | 60–120 |
| Aluminum | 500–1000 | 150–300 | 1000–2000 | 150–300 |

**TABLE 56-2 RECOMMENDED FEED (CHIP) PER TOOTH (HIGH-SPEED STEEL END MILLS)**

| Cutter diameter | | Machine steel | | Tool steel | | Cast iron | | Bronze | | Aluminum | |
|---|---|---|---|---|---|---|---|---|---|---|---|
| Inches | Millimeters | Inches | Millimeters | Inches | Millimeters | Inches | Millimeters | Inches | Millimeters | Inches | Millimeters |
| 1/8 | 3.17 | .0005 | 0.01 | .0005 | 0.01 | .0005 | 0.01 | .0005 | 0.01 | .002 | 0.05 |
| 1/4 | 6.35 | .001 | 0.02 | .001 | 0.02 | .001 | 0.02 | .001 | 0.02 | .003 | 0.07 |
| 3/8 | 9.52 | .002 | 0.05 | .002 | 0.05 | .002 | 0.05 | .002 | 0.05 | .003 | 0.07 |
| 1/2 | 12.7 | .003 | 0.07 | .002 | 0.05 | .003 | 0.07 | .003 | 0.07 | .005 | 0.12 |
| 3/4 | 19.05 | .004 | 0.10 | .003 | 0.07 | .003 | 0.07 | .003 | 0.07 | .006 | 0.15 |
| 1 | 25.4 | .005 | 0.12 | .003 | 0.07 | .004 | 0.10 | .004 | 0.10 | .007 | 0.17 |

Note: For flycutters and shell end mills use the feeds listed in Table 51-2.

N-6 $\text{r/min (metric)} = \dfrac{CS \times 320}{D}$

N-7 Too fast a feed will crowd the cutter and not allow the chips to escape; too slow a feed tends to dull a cutter.

N-8 $\text{r/min (inch)} = \dfrac{CS \times 4}{D}$

By dividing the $\pi$ (3.1416) into 1000, a simplified revolutions per minute formula is arrived at which is close enough for most milling operations.

$$\text{r/min} = \frac{CS \times 320}{D}$$

**Example:**

At what revolutions per minute should a 75-mm carbide flycutter revolve to cut a piece of machine steel (CS 60 m)? (See N-6.)

$$\text{r/min} = \frac{60 \times 320}{75} = \frac{19{,}200}{75} = 256$$

**Example:**

Calculate the revolutions per minute required to mill a groove 25 mm wide in a piece of aluminum using a high-speed end mill (CS = 200 m).

$$\text{r/min} = \frac{200 \times 320}{25} = \frac{64{,}000}{25} = 2560$$

## MILLING FEEDS

*Feed* is the rate at which the work moves into the revolving cutter, and it is measured either in inches per minute or in millimeters per minute. *Chip per tooth* is the amount of material which should be removed by each tooth of the cutter as it revolves and advances into the work. (See N-7.) The *milling feed* is determined by multiplying the chip size (chip per tooth) desired, the number of teeth in the cutter, and the revolutions per minute of the cutter.

### Inch Calculations

The formula used to find the work feed in inches per minute is the same as the formula used to find the feed in millimeters per minute, except that inches per minute is substituted for millimeters per minute:

in./min = $N$ × chip per tooth × r/min

where N = number of teeth in milling cutter

c.p.t. = chip per tooth for a particular cutter and metal, as given in Table 56-2

r/min = number of revolutions per minute of milling cutter

**Example:**

Find the feed in inches per minute using a 3/4-in.-diameter four-fluted end mill to cut machine steel (CS = 80). It would first be necessary to calculate the proper revolutions per minute for the cutter. (See N-8.)

$$\text{r/min} = \frac{4 \times 80}{3/4} = 426.6, \text{ or } 427$$

$$\text{Feed (in./min)} = N \times \text{c.p.t.} \times \text{r/min} = 4 \times .004 \times 427 = 6.8$$

N-9 r/min (metric) $= \frac{CS \times 320}{D}$

N-10 Be sure that the levers are properly engaged before starting the motor.

## Metric Calculations

The formula used to find work feed in millimeters per minute is:

$$\text{mm/min} = N \times \text{c.p.t.} \times \text{r/min}$$

**Example:**

Calculate the feed in millimeters per minute for a 25-mm-diameter (see N-9), four-flute high-speed steel end mill when machining a cast-iron workpiece (CS = 20 m)

$$\text{r/min} = \frac{20 \times 320}{25}$$
$$= \frac{6400}{25}$$
$$= 256$$

$$\text{Feed (mm/min)} = N \times \text{c.p.t.} \times \text{r/min}$$
$$= 4 \times 0.12 \times 256$$
$$= 122.8$$

The calculated feeds given for both examples would be possible only under ideal conditions. Under average operating conditions and especially when taking finishing cuts, it is suggested that the feed rate be set to approximately one-half the amount calculated. The feed can then be gradually increased to the capacity of the machine and the finish desired.

## Setting Speeds and Feeds

Since the setting of speeds and feeds on vertical milling machines varies with the manufacturer, it is almost impossible to set down the procedure for each machine. However, the following general rules apply to each machine:

**1.** Calculate the revolutions per minute for the size and type of cutter, and the work material to be cut.

**2.** Be sure that the power to the motor is shut off.

**3.** Turn the spindle by hand slowly when setting speed levers. (See N-10.)

**4.** On a variable-speed head, start the spindle revolving before setting the proper speed.

**5.** Calculate the feed rate for the size and type of cutter, and the work material to be cut.

**6.** Set feed levers when the spindle is stopped or revolving slowly.

## REVIEW QUESTIONS

**1.** Why is it necessary to align the vertical head square with the table?

**2.** At what four points on the table should the indicator reading be checked?

**3.** How far should the head be adjusted in relation to the indicator readings at both ends of the table?

**4.** In which direction should the vise be tapped when it is being aligned?

**5.** Define cutting speed.

**6.** List five factors which must be considered when setting the spindle speed.

**7.** Calculate the revolutions per minute for a 5/8-in.-diameter carbide cutter used to cut tool steel (CS = 160).

**8.** Calculate the revolutions per minute for a 25-mm-diameter high-speed steel cutter used to cut cast iron (CS = 20 m).

**9.** Calculate the feed required for a 1/2-in.-diameter four-fluted high-speed steel end mill for cutting tool steel (CS = 65).

**10.** Calculate the feed required for a 50-mm six-tooth carbide flycutter to cut machine steel (CS = 60 m), (c.p.t. = 0.25 mm).

**11.** Name two precautions which should be taken when setting the speed on a vertical mill.

# UNIT 57

# MACHINING A FLAT SURFACE

In order to produce a true, flat surface on a vertical milling machine, it is important that the head be aligned at 90° to the table. If possible, the cutting tool should be large enough to just overlap the edges of the work so that only one pass is required to machine the surface. An end mill can be used for narrow surfaces, while a shell end mill or a flycutter should be used for wide surfaces.

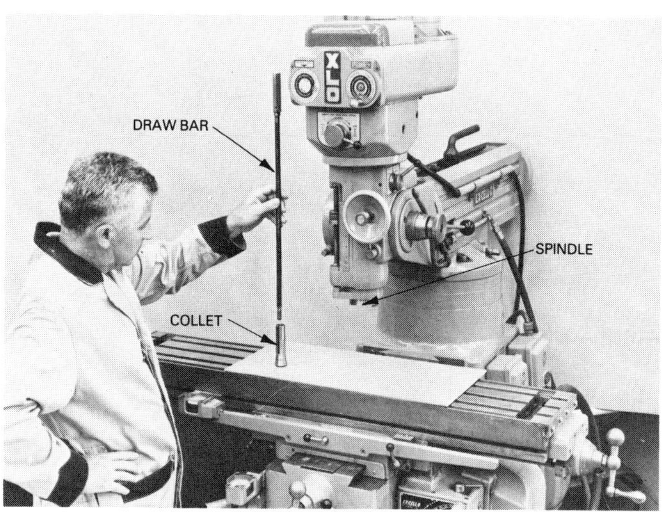


**Fig. 57-1** Place tools and accessories on a piece of masonite to protect the table surface. (*Courtesy Kostel Enterprises Ltd.*)

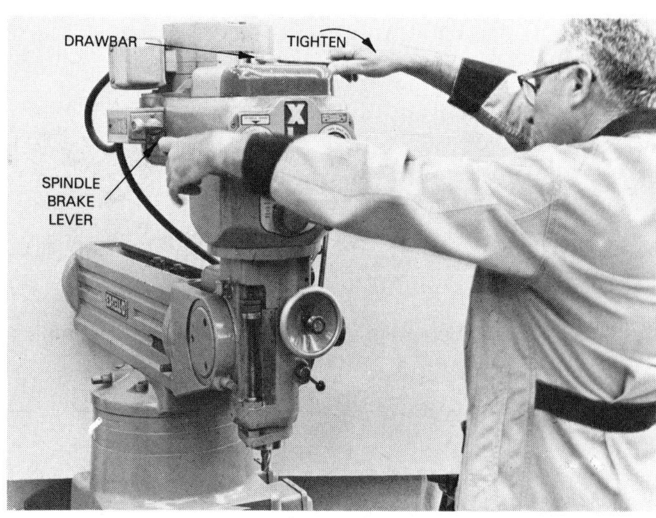


**Fig. 57-2** Tighten the drawbar securely, but only hand tight, otherwise the threads may be damaged. (*Courtesy Kostel Enterprises Ltd.*)

N-1 Protect the table from nicks or burrs which can cause inaccuracy.

N-2 Hold the cutting tool as short as possible to prevent breakage.

N-3 Never use a hammer or extra leverage when tightening the drawbar, otherwise the threads in the collet or on the drawbar may be damaged.

N-4 The sharp tap will break the taper contact between the spindle and collet.

# OBJECTIVES

After completing this unit, you should be able to:

**1.** Mount and remove cutting tools from the vertical mill spindle

**2.** Machine a flat surface with a fly-cutter or end mill

**3.** Machine a block square and parallel to within $\pm .002$ in. (0.05 mm) accuracy

**4.** Machine the ends of a workpiece square with the sides

# MOUNTING AND REMOVING CUTTERS

A wide variety of cutting tools can be held in a vertical milling machine spindle for performing various machining operations. The cutting tools can be held in the spindle by spring collets, collets and adapters, or solid collets.

## Mounting a Cutting Tool in a Collet

**1.** Shut off the electric power to the machine. (See N-1.)

**2.** Place a piece of masonite on the machine table (Fig. 57-1).

**3.** Clean the inside of the spindle taper.

**4.** Slide the drawbar through the top of the spindle.

**5.** Select the proper-size collet to suit the diameter of the cutting tool shank.

**6.** Clean the taper and keyway on the collet.

**7.** Insert the collet into the spindle and hold it up while turning it slowly until the keyway fits into the key of the spindle.

**8.** Hold the collet up and at the same time thread the drawbar into the collet approximately four turns.

**9.** Hold the cutting tool with a rag and insert it into the collet for the full length of the shank. (See N-2.)

**10.** Tighten the drawbar (clockwise) into the collet by hand.

**11.** Hold the spindle-brake lever and tighten the drawbar securely with a wrench (Fig. 57-2). (See N-3.)

## Removing a Cutting Tool from a Collet

**1.** Shut off the electric power to the machine.

**2.** Hold the spindle-brake lever and loosen the drawbar (counterclockwise) with a wrench (Fig. 57-2).

**3.** Loosen the drawbar approximately three full turns.

**4.** Hold the cutting tool with a cloth.

**5.** With a soft-faced hammer, strike down sharply on the head of the drawbar. (See N-4.)

**6.** Remove the cutting tool from the collet and clean and store it where it will not be damaged.

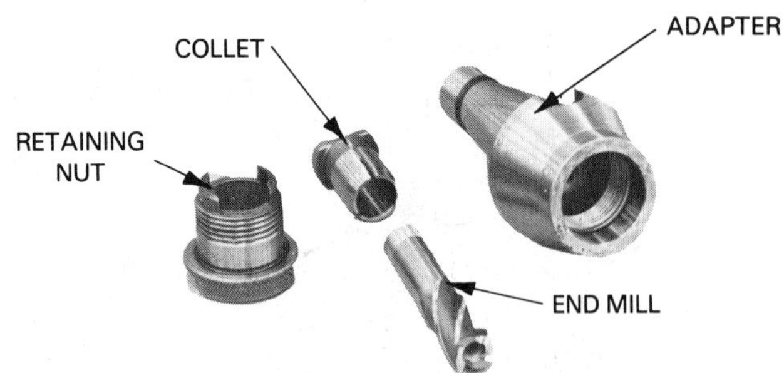


**Fig. 57-3A** The components of a collet and adapter assembly (*Courtesy Kostel Enterprises Ltd.*)

**Fig. 57-3B** Aligning the slots of the adapter with the drive keys on the spindle. (*Courtesy Kostel Enterprises Ltd.*)

N-5 Burrs or dirt will prevent parts from seating properly and will result in the cutter not running true.

N-6 Always use a rag when handling cutters to prevent painful cuts.

## Mounting a Cutting Tool into an Adapter

**1.** Remove all burrs from the adapter and shank of the cutting tool (Fig. 57-3A). (See N-5.)

**2.** Clean the internal taper of the spindle.

**3.** Align the slots of the tapered adapter with the drive keys of the spindle (Fig. 57-3B).

**4.** Hold the adapter up in the spindle and thread the drawbar clockwise into the adapter.

**5.** Hold the brake lever and tighten the drawbar securely with a wrench.

**6.** Clean the inside of the collet and thread the end mill into the collet two or three turns (Fig. 57-4).

**7.** Clean the outside of the collet and slide the retaining nut on the collet.

**8.** Fit the assembled end mill into the adapter.

**9.** Thread the retaining nut all the way into the adapter.

**10.** Hold the retainer nut with one hand and then use a rag to thread the end mill into the collet by hand (Fig. 57-5). (See N-6.)

**11.** Hold the spindle-brake lever and tighten the retainer nut securely.

# TO MACHINE A FLAT SURFACE

To machine a true, flat surface, the vertical head must be aligned at 90° with the surface of the table. The cutting tool used can be either an end mill or a flycutter depending on the width of the workpiece.

**1.** Check that the head is at right angles (90°) to the table.

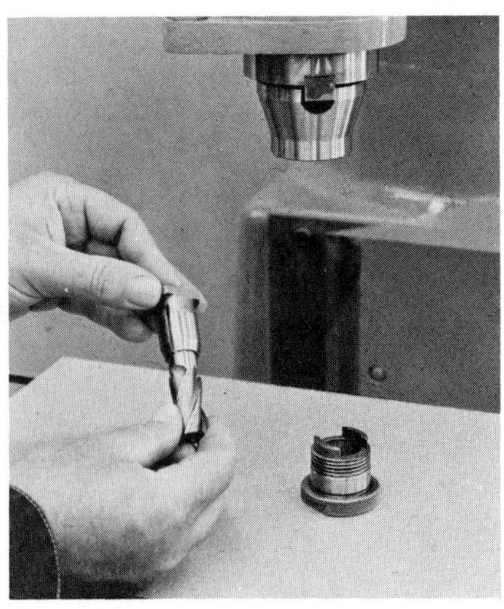

**Fig. 57-4** Threading the end mill into the collet by hand. (*Courtesy Kostel Enterprises Ltd.*)

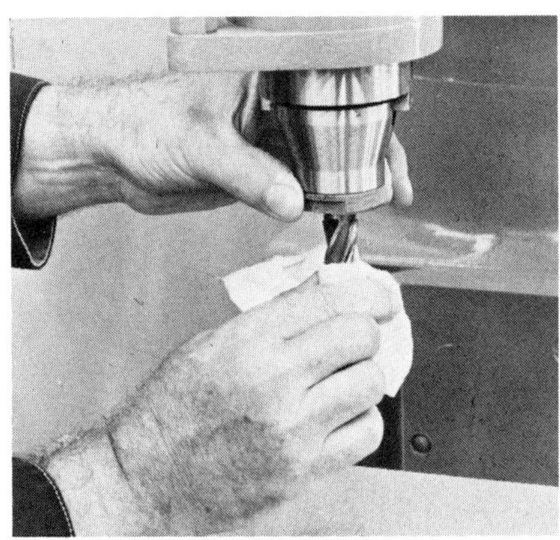

**Fig. 57-5** Using a rag to protect your hand while threading the end mill into the adapter. (*Courtesy Kostel Enterprises Ltd.*)

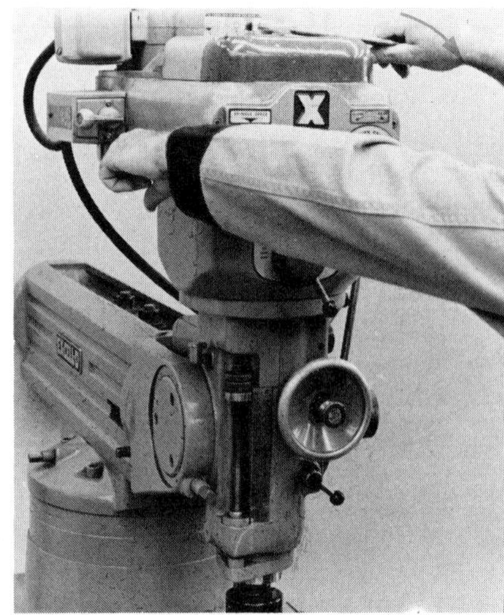

**Fig. 57-6** Tightening the drawbar to hold a flycutter in the spindle. (*Courtesy Kostel Enterprises Ltd.*)

**Fig. 57-7** Mounting the work in a vise with the largest surface facing up. Note the paper feelers between the work and parallels. (*Courtesy Kostel Enterprises Ltd.*)

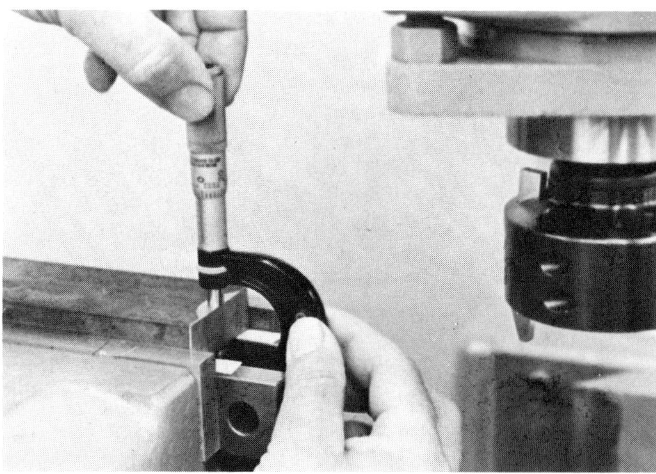

**Fig. 57-8** Measuring the thickness of the work after taking a short trial cut. (*Courtesy Kostel Enterprises Ltd.*)

N-7 Never use too wide a facing cutter since it may jam and throw the head out of alignment.

N-8 $r/min\ (inch) = \frac{CS \times 4}{D}$

$r/min\ (metric) = \frac{CS \times 320}{D}$

N-9 Be sure that the cutter is revolving in the correct direction.

N-10 Lock the knee clamp before taking a cut; unlock it before setting the depth of cut.

N-11 Always feed the work *into the rotation* of the cutter.

**2.** Swivel the vise so that the jaws are parallel with the table slots.

**3.** Mount a suitable end mill or flycutter in the machine spindle (Fig. 57-6). (See N-7.)

**4.** Set the machine speed for the size of the cutter and the material to be cut. (See N-8.)

**5.** Mount the workpiece into the vise. Use parallels if necessary (Fig. 57-7).

**6.** Adjust the table to align the centerline of the work and the centerline of the cutter.

**7.** Move the table until the cutter is over the right-hand end of the work by about ¼ in. (6.35 mm).

**8.** Start the machine and raise the table until the cutter just touches the surface of the work. (See N-9.)

**9.** Move the table so that the work clears the cutter.

**10.** Set the graduated collar to zero without moving the elevating screw.

**11.** Raise the table .010 in. (0.25 mm) and take a trial cut approximately ¼ in. (6.35 mm) long. (See N-10.)

**12.** Stop the machine, clear the cutter, and measure the work thickness (Fig. 57-8).

**13.** Calculate the amount to be removed and raise the table the desired amount.

**14.** Machine the surface to size.

## MACHINING A BLOCK SQUARE AND PARALLEL

In order to mill the four sides of a piece of work so that its sides are square and parallel, it is important that each side be machined in a definite order. It is very important that dirt and burrs be removed from the work, vise, and parallels, since they can cause inaccurate work.

### Machining Side 1

**1.** Clean the vise thoroughly and remove all burrs from the workpiece.

**2.** Set the work on parallels in the center of the vise with the largest surface (side 1) facing up (Fig. 57-9).

**3.** Place short paper feelers under each corner between the parallels and the work.

**4.** Tighten the vise securely.

**5.** With a soft-faced hammer, tap the workpiece down until all paper feelers are tight.

**6.** Mount a flycutter in the milling machine spindle.

**7.** Set the machine for the proper speed for the size of the cutter and the material to be machined. (See Table 56-1).

**8.** Start the machine and raise the table until the cutter just touches near the right-hand end of side 1. (See N-11.)

**9.** Move work clear of the cutter.

**10.** Raise the table about .030 in. (0.76 mm) and machine side 1 using a steady feed rate (Fig. 57-9).

**11.** Take the work out of the vise and remove all burrs from the edges with a file.

### Machining Side 2

**12.** Clean the vise, work, and parallels thoroughly.

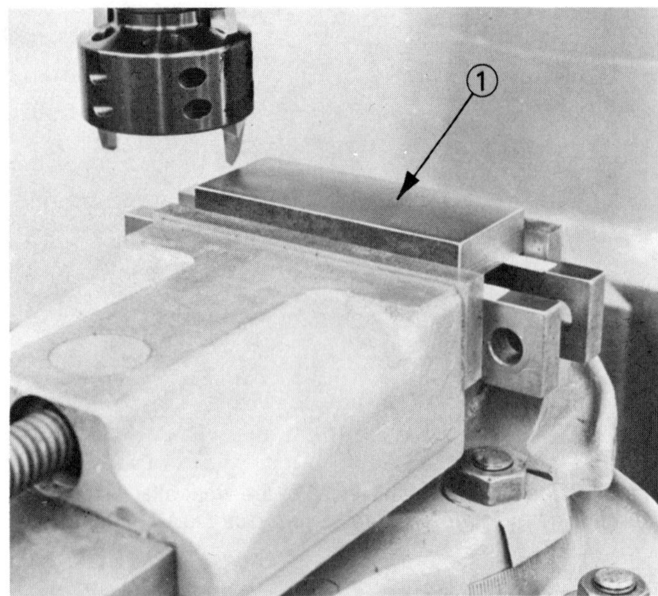


Fig. 57-9 Side 1 or the surface with the largest area should be facing up. (*Courtesy Kostel Enterprises Ltd.*)

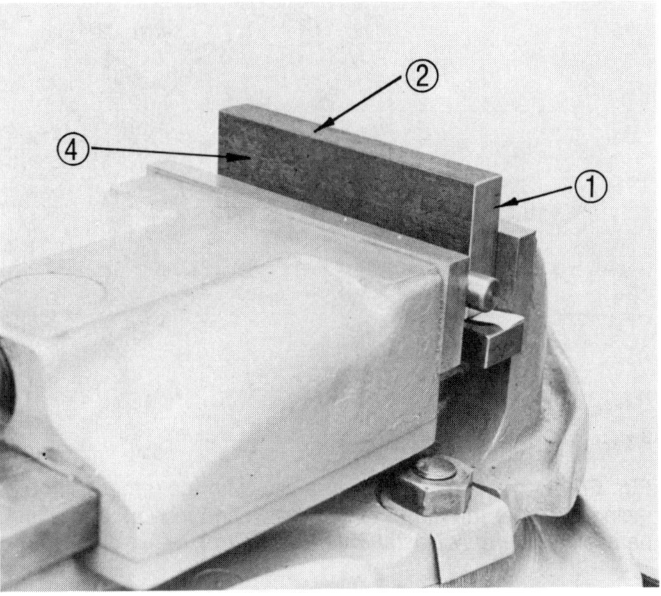


Fig. 57-10 The finished side (1) should be placed against the cleaned solid jaw for machining side 2. (*Courtesy Kostel Enterprises Ltd.*)

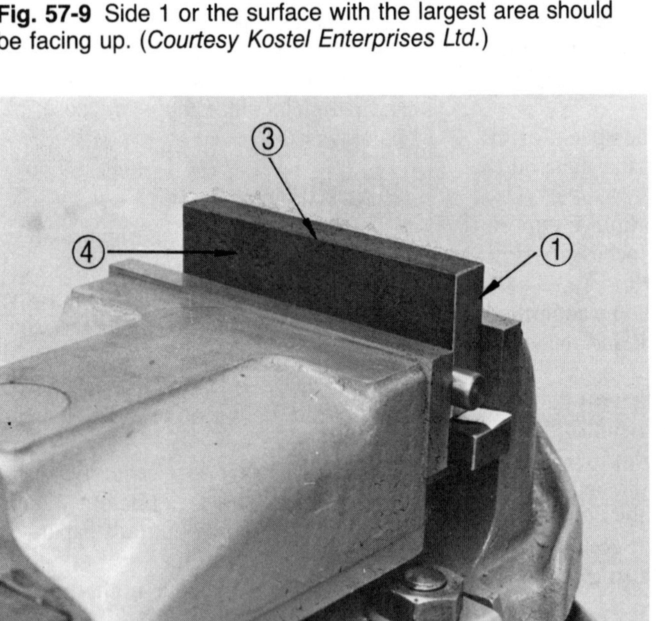


Fig. 57-11 Place side 1 against the solid jaw, with side 2 down for machining side 3. (*Courtesy Kostel Enterprises Ltd.*)

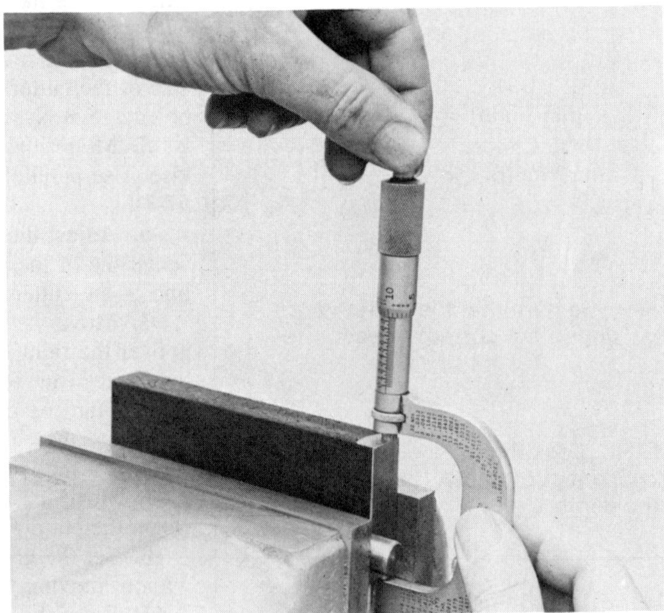

Fig. 57-12 Measuring the trial cut on side 3. (*Courtesy Kostel Enterprises Ltd.*)

N-12 The round bar must be in the *center* of the amount of work being held inside the vise jaws.

**13.** Place the work on parallels, if necessary, with side 1 against the solid jaw and side 2 up (Fig. 57-10).

**14.** Place short paper feelers under each corner between the parallels and the work.

**15.** Place a round bar between side 4 and the movable jaw. (See N-12.)

**16.** Tighten the vise securely and tap the work down until the paper feelers are tight.

**17.** Follow steps 8 to 11 and machine side 2.

## Machining Side 3

**18.** Clean the vise, work, and parallels thoroughly.

**19.** Place side 1 against the solid vise jaw with side 2 resting on parallels if necessary (Fig. 57-11).

**20.** Push the parallel to the left so that the right edge of the work extends about ¼ in. (6.35 mm) beyond the parallel.

**21.** Place short paper feelers under each corner between the parallels and

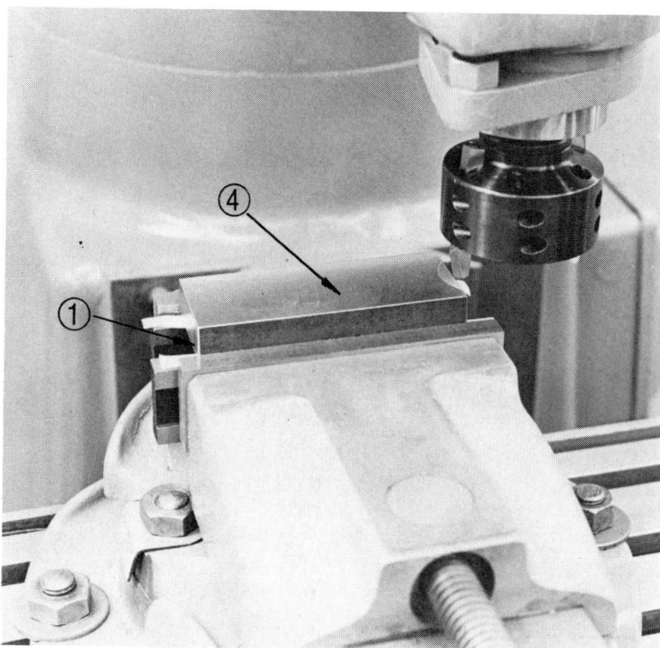


**Fig. 57-13** The work set up properly for machining side 4. Note the paper feelers between the parallel and the work. (*Courtesy Kostel Enterprises Ltd.*)

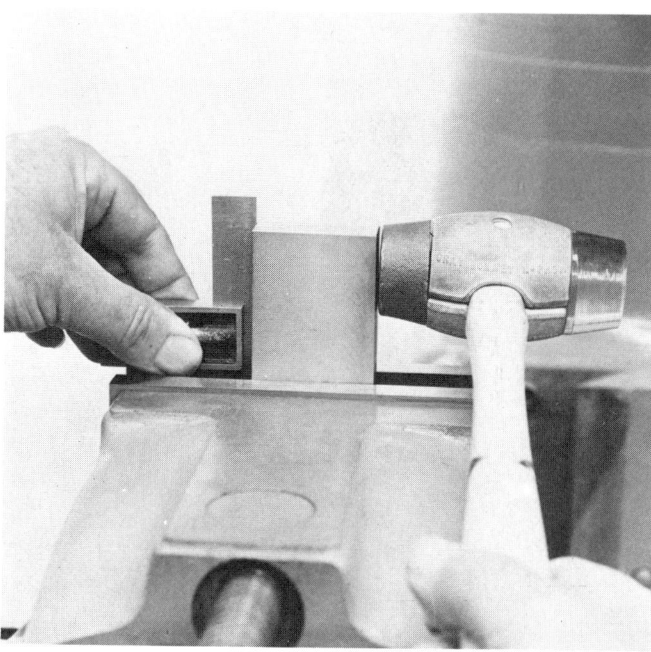

**Fig. 57-14** Setting the edge of the work square for machining the ends. (*Courtesy Kostel Enterprises Ltd.*)

N-13 With three finished surfaces, the round bar is not required when machining side 4.

N-14 Burrs on work will cause inaccurate setups.

the work.

**22.** Place a round bar between side 4 and the movable jaw being sure that the round bar is in the center of the amount of work held *in the vise*.

**23.** Tighten the vise securely and tap the work down until the paper feelers are tight.

**24.** Start the machine and raise the table until the cutter just touches near the right-hand end of side 3.

**25.** Move the work clear of the cutter and raise the table about .010 in. (0.25 mm).

**26.** Take a trial cut about ¼ in. (6.35 mm) long, stop the machine, and measure the width of the work (Fig. 57-12).

**27.** Raise the table the required amount and machine side 3 to the correct width.

**28.** Remove the work and file off all burrs.

### Machining Side 4

**29.** Clean the vise, work, and parallels thoroughly.

**30.** Place side 1 down on the parallels with side 4 up and tighten the vise securely (Fig. 57-13). (See N-13.)

**31.** Place short paper feelers under each corner between the parallels and work (Fig. 57-13).

**32.** Tighten the vise securely.

**33.** Tap the work down until the paper feelers are tight.

**34.** Follow steps 24 to 27 and machine side 4 to the correct thickness.

## MACHINING THE ENDS SQUARE

Two common methods are used to square the ends of workpieces in a vertical mill. Short pieces are generally held vertically in the vise and are machined with an end mill or flycutter (Fig. 57-15). Long pieces are generally held flat in the vise with one end extending past the end of the vise. The end surface is cut square with the body of an end mill (Fig. 57-17).

### Procedure for Short Work

**1.** Set the work in the center of the vise with one of the ends up and tighten the vise lightly.

**2.** Hold a square down firmly on top of the vise jaws and bring the blade into light contact with the side of the work.

**3.** Tap the work until its edge is aligned with the blade of the square (Fig. 57-14).

**4.** Tighten the vise securely and recheck the squareness of the side.

**5.** Take about a .030-in.- (0.76-mm-) deep cut and machine the end square (Fig. 57-15).

**6.** Remove the burrs from the end of the machined surface. (See N-14.)

**Fig. 57-15** One end machined square with the sides of the workpiece. (*Courtesy Kostel Enterprises Ltd.*)

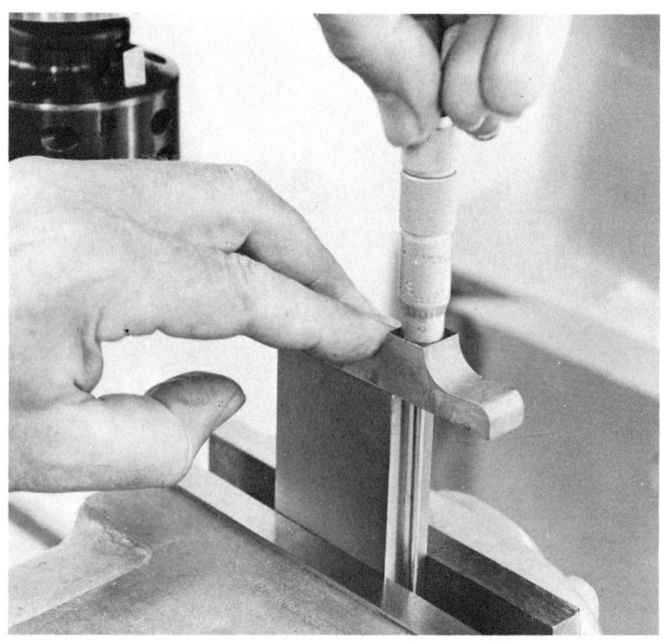

**Fig. 57-16** Measuring the length of the workpiece with a depth micrometer. (*Courtesy Kostel Enterprises Ltd.*)

**7.** Clean the vise and set the machined end on paper feelers in the bottom of the vise.

**8.** Tighten the vise securely and tap the work down until the paper feelers are tight.

**9.** Take a trial cut from the end until the surface cleans up.

**10.** Measure the length of the workpiece with a depth micrometer (Fig. 57-16).

**11.** Raise the table the required amount and machine the work to length.

## Procedure for Long Work

**1.** Align the vise jaws parallel with the table slots.

**2.** Remove all burrs and clean the work, parallels, and vise.

**3.** Set the work on parallels and paper feelers (Fig. 57-17).

**4.** Position the work so the end projects past the vise jaws and parallels by about ¼ in. (6 mm).

**5.** Tighten the vise securely and tap the work down until all paper feelers are tight.

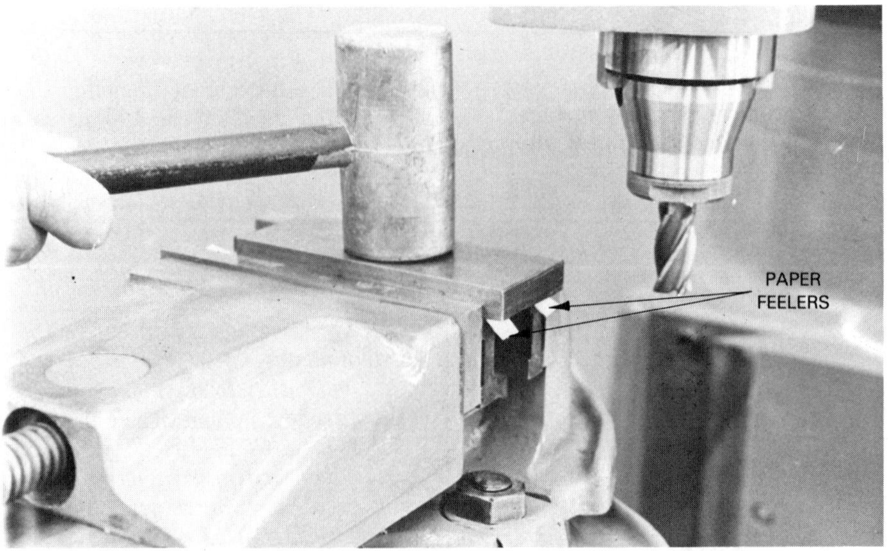


**Fig. 57-17** Long work set flat in the vise for squaring the ends. (*Courtesy Kostel Enterprises Ltd.*)

**Fig. 57-18** One end of the work machined square with the sides. (*Courtesy Kostel Enterprises Ltd.*)

**Fig. 57-19** Measuring the length of the work with an outside micrometer. (*Courtesy Kostel Enterprises Ltd.*)

N-15 Be sure that the cutting edge on the body of the end mill is a little longer than the thickness of work to be cut.

**6.** Mount a three- or four-fluted end mill close up to the collet. (See N-15.)

**7.** Set the spindle speed.

**8.** Carefully move the table until the side of the revolving end mill touches the end surface of the work.

**9.** Set a .030 in. (0.76 mm) depth of cut using the table traverse handle.

**10.** Lock the knee and table lock lever.

**11.** Machine the end square by feeding the work with the crossfeed handle (Fig. 57-18).

**12.** Remove the work and file the burrs off the machined end.

**13.** Set the work on parallels with the other end projecting about ¼ in. (6.35 mm) beyond the vise jaws.

**14.** Take a cut until the end just cleans up.

**15.** Measure the length of the work with a micrometer or vernier caliper (Fig. 57-19).

**16.** Set the depth of cut and machine the work to length.

## REVIEW QUESTIONS

**1.** Why should tools and machine accessories be placed on a piece of masonite instead of on the machine table?

**2.** In which direction should the collet drawbar be tightened?

**3.** Why is it important that a hammer or extra leverage not be used when tightening the drawbar?

**4.** How far should the drawbar be loosened before tapping it to release the collet?

**5.** Why is it important that the head is at 90° to the table when machining surfaces?

**6.** Why should paper feelers be used between the work and parallels?

**7.** How should the work be set in a vise to machine side 1?

**8.** How should the work be set in a vise to machine side 2?

**9.** Draw a sketch of the work setup required for machining side 3 and side 4.

**10.** How are the following generally held in a vise and machined when squaring the ends?

(a) Short workpieces

(b) Long workpieces

**11.** List the procedure for setting up short work for squaring the ends.

**12.** List the procedure for setting up long work for squaring the ends.

# UNIT 58

# VERTICAL MILLING OPERATIONS

The vertical milling machine is one of the most versatile machines in a shop. The many attachments which are available permit operations such as drilling, boring, reaming, cutting slots, keyways, and dovetails, and radius milling

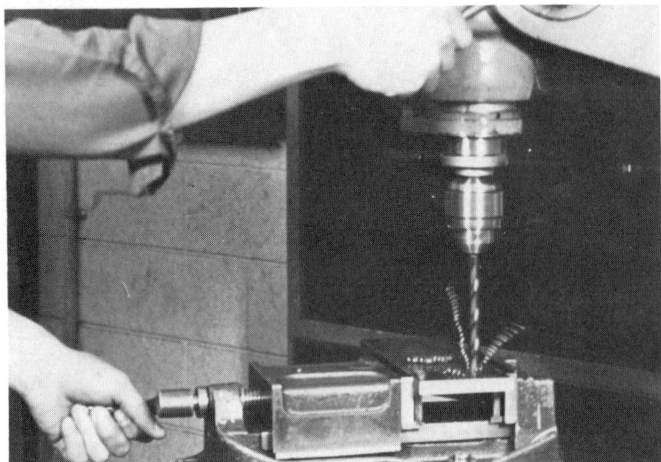

**Fig. 58-1** Drilling a hole to accurate location in a vertical mill.

**Fig. 58-2** Holes should be reamed, after drilling and boring, before the table location is changed.

N-1 Do not move the table location when changing drills, otherwise the hole location will be lost.

N-2 Too high a speed will quickly dull the cutting edges of a reamer.

N-3 Turning a reamer backward will ruin the reamer cutting edges.

to be performed with ease. In many shops, the vertical milling machine has replaced the shaper since it can do all the operations more easily and quickly.

## OBJECTIVES

After completing this unit you should be able to:

**1.** Drill, bore, and ream a hole on a vertical milling machine
**2.** Mill angular surfaces by two methods
**3.** Machine slots and keyways
**4.** Use the rotary table to machine circular slots

## DRILLING, REAMING, AND BORING

### To Drill on a Vertical Mill

Small twist drills are usually held in key-type or keyless drill chucks. Large drills are held in collets or tapered adapters that fit into the spindle. Holes drilled in a vertical milling machine are generally more accurate for size and location than those drilled on a drill press.

**1.** Mount a drill chuck or collet in the spindle.
**2.** Locate the hole center in line with the center of the machine spindle by either of the following:
(a) Using a wiggler mounted in the machine spindle and picking up the center punch mark
(b) Moving the table the required distance from two machined edges of the work
**3.** Tighten the table and saddle clamps when the hole location is aligned with the center of the machine spindle.
**4.** Place a center drill in the chuck and slightly center-drill the hole.
**5.** Mount the proper-size drill in the drill chuck. (See N-1.)
**6.** Set the machine to the proper speed and feed.
**7.** Engage the quill feed (Fig. 58-1) and drill the hole to the required depth.

### To Ream on a Vertical Mill

Reamers are usually held in a drill chuck or an adapter. Two types of reamers—the rose, or fluted; and the precision end-cutting—are used for bringing a hole to size quickly. These are used following the drilling or boring operations. Reaming provides a true, sized, finished hole faster than boring.

**1.** Mount the reamer in the spindle or drill chuck (Fig. 58-2).
**2.** Set the speed and feed for reaming (approximately one-quarter the drilling speed). (See N-2.)
**3.** Apply cutting fluid as the reamer is fed into the hole.
**4.** Stop the machine.
**5.** Remove the reamer from the hole. (Do not turn reamer backward.) (See N-3.)

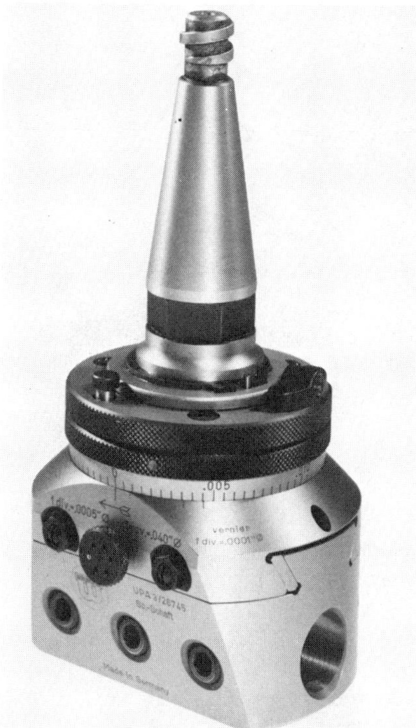

Fig. 58-3 The dovetail offset boring chuck allows accurate settings to be made. (*Courtesy Moore Special Tool Co. Inc.*)

N-4 Do not move the table location when changing tools in the spindle.

### To Bore on a Vertical Mill

Boring is the operation of enlarging and truing a drilled or cored hole with a single-point cutting tool. Many holes are bored on a milling machine to bring them to accurate size and location. The offset boring chuck is especially useful in that it allows accurate settings to be made for removing material from a hole (Fig. 58-3).

**1.** Align the vertical head square (at 90°) to the table.

**2.** Set up and align the work parallel to the table travel.

**3.** Align the center of the milling machine spindle with the reference point or edges of the work.

**4.** Set the graduated dials on the crossfeed and table screws to zero.

**5.** Calculate the coordinate location of all holes.

**6.** Spot all holes with a center drill or spotting tool.

**7.** Drill all holes under ½ in. (12.7 mm) diameter to within 1/64 in. (0.39 mm) of finish size; drill holes over ½ in. (12.7 mm) diameter to within 1/32 in. (0.79 mm) of size. (See N-4.)

**8.** Mount the boring chuck using the largest boring bar or tool as possible (Fig. 58-4A and B).

**9.** Rough bore all holes to within .005 to .007 in. (0.12 to 0.17 mm) of finish size.

**10.** Finish bore all holes to the required size (Fig. 58-5).

## MACHINING ANGULAR SURFACES

Angular surfaces can be produced on a vertical milling machine by:

**1.** Swiveling the toolhead to the required angle and machining with the body of an end mill

**2.** Setting the work on an angle and machining with a flycutter or shell end mill

**3.** Setting the work flat in a vise, swiveling the toolhead, and cutting short angular surfaces on the top of a workpiece with a flycutter

**4.** Setting the work flat in a vise, swiveling the vise to the required angle, and cutting long angular surfaces with an end mill.

Since the most common methods of cutting angular surfaces on a vertical

A

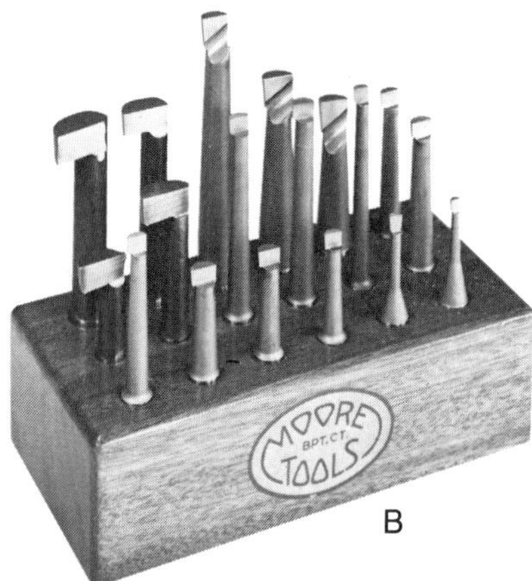



B

Fig. 58-4 Types of boring tools: (A) Solid boring bars; (B) single-point tools. (*Courtesy Moore Special Tool Co. Inc.*)

Fig. 58-5 Boring a hole with an offset boring chuck. (*Courtesy Cincinnati Milacron Co.*)

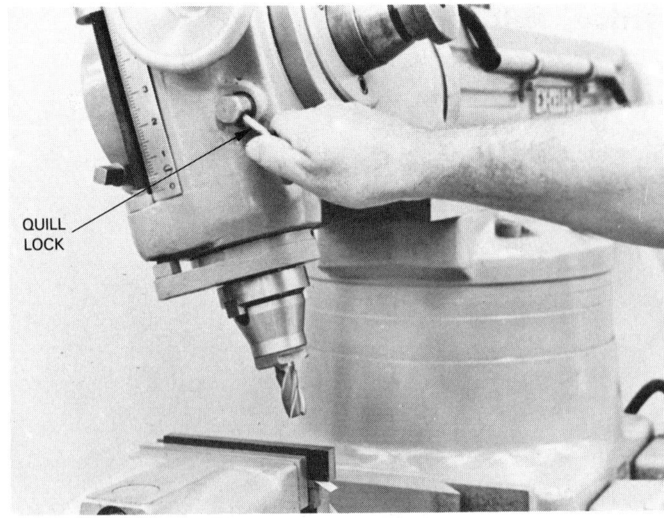


Fig. 58-6 The vertical mill head swiveled to the required angle. (*Courtesy Kostel Enterprises Ltd.*)

Fig. 58-7 Checking the angle before machining to the layout line. (*Courtesy Kostel Enterprises Ltd.*)

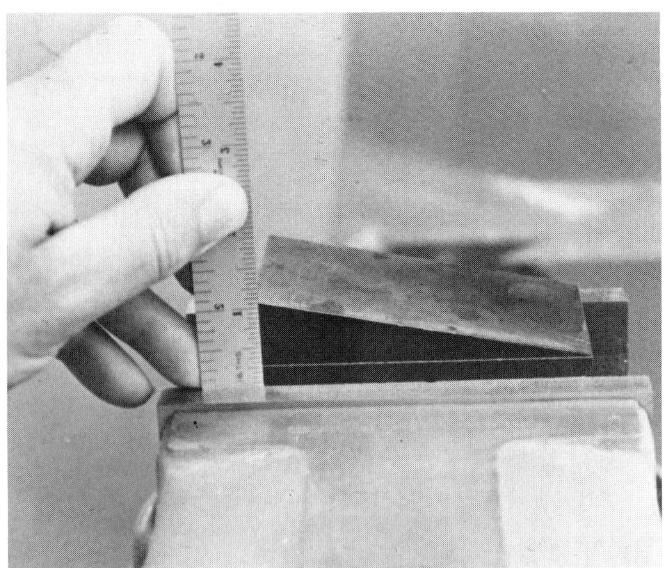

Fig. 58-8 Checking from the top of the vise jaw to the layout line at both ends with a rule. (*Courtesy Kostel Enterprises Ltd.*)

N-5 If the vise is not parallel, the angle cut on the work will not be correct.

N-6 Always check the accuracy of the angle before the cut is made to the layout line.

milling machine are by swiveling the toolhead to the angle or setting the work on an angle, only these two methods will be covered in detail.

## Machining Angular Surfaces by Swiveling the Toolhead

**1.** Lay out the angular surface on the workpiece.

**2.** Clean the vise.

**3.** Align the vise parallel to the table travel. (See N-5.)

**4.** Mount the work on parallels in the vise.

**5.** Swivel the vertical head to the required angle (Fig. 58-6).

**6.** Tighten the quill clamp with the quill in the up position.

**7.** Start the machine and raise the table until the side of the cutter touches the work. Carefully raise the table until the cut is to the desired depth.

**8.** Take a trial cut so that there is a tapered surface of about ½ in. (12.7 mm) long on the workpiece.

**9.** Check the angle with a protractor and recheck the work or toolhead setting if necessary (Fig. 58-7). (See N-6.)

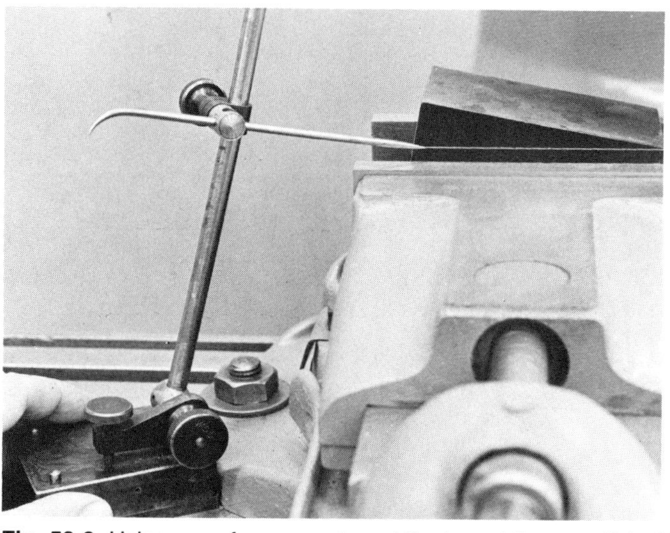

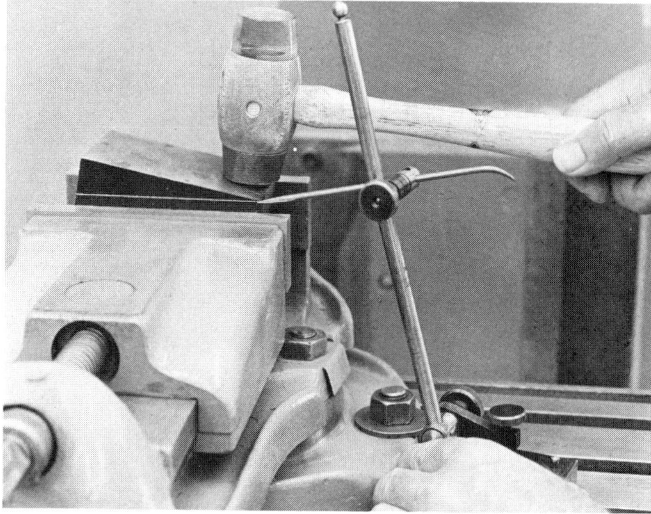

Fig. 58-9 Using a surface gage to set the layout line parallel with the top of the vise jaw. (*Courtesy Kostel Enterprises Ltd.*)

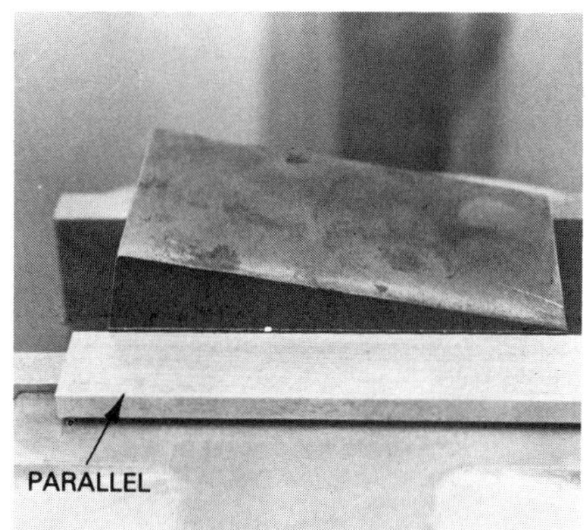


Fig. 58-10 Using a parallel to set the layout line parallel to the top of the vise jaw. (*Courtesy Kostel Enterprises Ltd.*)

N-7 Gripping the quill short will prevent vibration of the spindle.

N-8 If the cut is not parallel with the layout line, readjust the work in the vise.

N-9 Set a square on the machine table to align the layout lines.

**10.** If the angle is correct, cut the tapered surface to the layout line.

### Machining Angular Surfaces by Setting the Work on an Angle

**1.** Lay out the angular surface on the workpiece.

**2.** Check that the vertical head is square with the table.

**3.** Clean the vise.

**4.** Lock the quill clamp with the quill in the up position. (See N-7.)

**5.** Set the workpiece in the vise with the layout line parallel to the top and about ¼ in. (6.35 mm) above the vise jaws. Use any of the following methods to set the layout line parallel to the top of the vise.

(a) Check both ends of the layout line with a rule (Fig. 58-8).

(b) Check both ends of the layout line with a surface gage (Fig. 58-9).

(c) Use a ¼ in. (6.35 mm) thick parallel to set the line parallel (Fig. 58-10).

**6.** Adjust the work under the cutter so that the cut will start at the narrow side of the taper and progress into the thicker metal.

**7.** Take successive cuts of about .125 to .150 in. (3.17 to 3.81 mm) deep, or until the cut is about 1/32 in. (0.79 mm) above the layout line.

**8.** Check to see that the cut and the layout line are parallel to each other. (See N-8.)

**9.** Raise the table until the cutter just touches the layout line.

**10.** Clamp the knee at this setting.

**11.** Take the finishing cut to the layout line.

## CUTTING KEYSEATS, SLOTS, AND DOVETAILS

Keyseats, slots, and dovetails can be easily cut on a vertical mill by using the appropriate cutter in the machine spindle. Keyseats are generally cut with a two- or three-fluted end mill. T-slots are cut with a T-slot cutter; dovetails are cut with a dovetail cutter.

### Cutting Keyseats

**1.** Lay out the position of the keyseat on the shaft (Fig. 58-11).

**2.** Set the work in a vise on a parallel. If the shaft is long, it may be clamped on the table by placing it in one of the table slots or in V-blocks.

**3.** Using the layout lines on the end of the shaft, set up the shafts so that the keyseat layout is in the proper position on top of the shaft. (See N-9.)

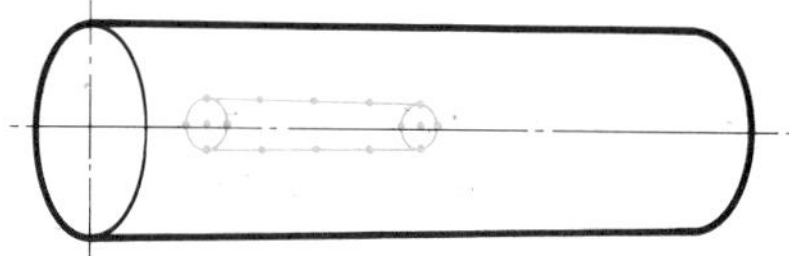

Fig. 58-11 Layout of a keyseat on a shaft.

**Fig. 58-12** Setting the cutter to the side of the work.

**Fig. 58-13** The cutter aligned with the layout and the center of the shaft.

N-10 Always use a two-fluted end mill for cutting keyseats with two blind ends.

**4.** Mount a two- or three-fluted end mill of a diameter equal to the width of the keyseat, in the milling machine spindle.

**5.** Center the workpiece with the machine spindle by carefully touching the revolving end mill up to one side of the shaft (Fig. 58-12). This may also be done by placing a piece of thin paper between the shaft and the end mill.

**6.** Lower the table until the end mill clears the workpiece.

**7.** Move the table over an amount equal to half the diameter of the shaft plus half the diameter of the end mill plus the thickness of the paper (Fig. 58-13). For example: If a ¼-in. slot is required in a 2-in. shaft and the thickness of the paper used is .002 in., the table would be moved over

$$1.000 + .125 + .002 = 1.127$$

Or if a 6-mm slot is required in a 50-mm-diameter shaft and the thickness of the paper used is 0.06 mm, the table should be moved over

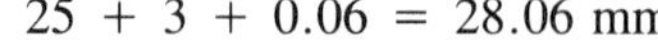

$$25 + 3 + 0.06 = 28.06 \text{ mm}$$

**8.** If the keyseat being cut has two blind ends, adjust the work until one end of the keyseat is aligned with the edge of the end mill. (See N-10.)

**9.** Feed the end mill down (or the table up) until it *just* cuts to its full diameter. If the keyseat has one blind end only, the work is adjusted so that this cut is taken at the end of the work. The work would now be moved clear of the end mill.

**10.** Adjust the depth of cut to one-half the thickness of the key, and machine the keyseat to the proper length.

## Milling T-Slots

T-slots are machined in the tops of machine tables and accessories to receive bolts for clamping workpieces. They are machined in two operations:

**1.** The vertical slot is cut to depth and width with an end mill.

**2.** The T-slot is cut with a T-slot cutter of the proper diameter and thickness (Fig. 58-14).

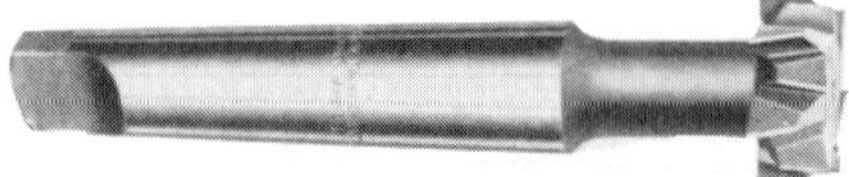

**Fig. 58-14** A T-slot milling cutter.

**Fig. 58-15** A dovetail being machined on a vertical mill.

### Milling Dovetails

Dovetails are used to permit reciprocating motion between two elements of a machine. They are composed of a male and female part and are adjusted by means of a gib. Dovetails may be machined on a vertical milling machine or on a horizontal mill equipped with a vertical milling attachment. A dovetail cutter (Fig. 58-15) is a special single-angle end-mill-type cutter ground to the angle of the dovetail required.

## THE ROTARY TABLE

The *rotary table,* or the *circular milling attachment,* can be used on plain, universal, and vertical milling machines. It provides rotary motion to the workpiece and makes it possible to cut radii, circular grooves, and circular sections not possible by other means. The drilling and boring of holes which have been dimensioned by angular measurements, as well as other indexing operations, are easily accomplished with this accessory.

### Hand-Feed Rotary Table Construction

The rotary table unit (Fig. 58-16) consists of a *base* which is bolted to the milling machine table. Fitting into the base is the *rotary table* on the bottom of which is mounted a *worm gear.* A *worm shaft* mounted in the base meshes with and drives the worm gear. The worm shaft may be quickly disengaged when rapid rotation of the table is required, as when setting work concentric with the table. A *handwheel* is mounted on the outer end of the worm shaft. The bottom edge of the table is graduated in half degrees. On most rotary table units, there is a *vernier scale* on the handwheel collar which permits setting to within 2 minutes of a degree. The table has T-slots cut into the top surface to permit the clamping of work.

A hole in the center of the table accommodates test plugs to permit easy centering of the table with the machine spindle. Work may be centered with the table by means of test plugs or arbors.

Some rotary tables use an *indexing attachment* in place of a handwheel (Fig. 58-17). This attachment is often supplied as an accessory to the standard rotary table. It not only serves the same purpose as a handwheel, but also permits the indexing of work with dividing head accuracy. The worm and worm wheel ratio of rotary tables need not be 40:1 as on most dividing heads. The ratio may be 72:1, 80:1, 90:1, 120:1, or any other ratio. Larger ratios are usually found on larger tables. The method of calculating the indexing is the same as for the dividing head except that the number of teeth in the worm wheel is used rather than 40 as in the dividing head calculations.

**Example:**

Calculate the indexing for five equally spaced holes on a circular

**Fig. 58-16** A hand-feed rotary table. (*Courtesy Cincinnati Milacron Co.*)

**Fig. 58-17** A rotary table with an indexing attachment. (*Courtesy Cincinnati Milacron Co.*)

**Fig. 58-18** Milling a circular slot with work mounted on a rotary table.

N-11 Clean the base of the rotary table and the milling machine table before mounting the rotary table.

N-12 Set the crossfeed and longitudinal feed collars to zero without moving the table setting.

N-13 Tapping the work toward the indicator can damage the indicator.

N-14 Do not move the table when setting the crossfeed and longitudinal dials to zero.

plate using a rotary table with an 80:1 ratio

$$\text{Indexing} = \frac{\text{Number of teeth in worm wheel}}{\text{Number of divisions required}} = \frac{80}{5} = 16 \text{ turns}$$

Accurately spaced holes on circles and segments (such as clutch teeth and teeth in gears too large to be held between index centers) are some of the applications of this type of rotary table.

### To Center the Rotary Table with the Vertical Mill Spindle

**1.** Square the vertical head with the machine table.

**2.** Mount the rotary table on the milling machine. (See N-11.)

**3.** Place a test plug in the center hole of the rotary table.

**4.** Mount an indicator with a grasshopper leg in the machine spindle.

**5.** With the indicator just clearing the top of the test plug, rotate the machine spindle by hand and approximately align the plug with the spindle.

**6.** Bring the indicator into contact with the diameter of the plug, and rotate the spindle by hand.

**7.** Adjust the machine table by the longitudinal and crossfeed handles until the dial indicator registers no movement.

**8.** Lock the machine table and saddle, and recheck the alignment. (See N-12.)

**9.** Readjust if necessary.

### To Center a Workpiece with the Rotary Table

Often it is necessary to perform a rotary table operation on several identical workpieces, each having a machined hole in the center. To quickly align each workpiece, a special plug can be made to fit the center hole of the workpiece and the hole in the rotary table. If there are only a few pieces, which would not justify the manufacture of a special plug, or if the workpiece does not have a hole through its center, the following method can be used to center the workpiece on the rotary table.

**1.** Align the rotary table with the vertical-head spindle.

**2.** Lightly clamp the workpiece to the rotary table in the approximate center.

**3.** Disengage the rotary table worm mechanism.

**4.** Mount an indicator in the machine spindle or on the milling machine table, depending on the workpiece.

**5.** Bring the indicator into contact with the surface to be indicated, and revolve the rotary table by hand.

**6.** With a soft metal bar, tap the work (away from the indicator movement) until no movement is registered on the indicator in a complete revolution of the rotary table. (See N-13.)

**7.** Clamp the workpiece tightly, and recheck the accuracy of the setup.

## RADIUS MILLING

When it is required to mill the ends on a workpiece to a certain radius or to machine circular slots having a definite radius, a sequence should be followed. (Fig. 58-18 illustrates a typical setup.)

### Procedure

**1.** Align the vertical milling machine spindle at 90° to the table.

**2.** Mount a rotary table on the milling machine.

**3.** Center the rotary table with the machine spindle using a test plug in the table and a dial indicator on the spindle.

**4.** Set the longitudinal feed dials and the crossfeed dial to zero. (See N-14.)

**5.** Mount the work on the rotary table, aligning the center of the radial cuts with the center of the table.

**6.** Move either the crossfeed or the longitudinal feed (whichever is more convenient) an amount equal to the radius required.

**7.** *Lock both the table and the saddle*, and remove the handles if convenient.

**8.** Mount a two-fluted end mill of the proper diameter in the spindle.

**9.** Rotate the work, using the rotary table feed handwheel, to the starting point of the cut.

**10.** Set the depth of cut and machine the slot to the size indicated on the drawing, using hand or power feed.

## REVIEW QUESTIONS

**1.** Why are holes drilled in a vertical milling machine?

**2.** How can the center of a hole be located for drilling?

**3.** Name two types of reamers used in a vertical mill.

**4.** Why are reamers used?

**5.** Why should reamers never be turned backward?

**6.** Define the operation of boring.

**7.** To what size should the following holes be drilled before boring?

(a) Holes under ½ in. (12.7 mm)

(b) Holes over ½ in. (12.7 mm)

**8.** Name two common methods of cutting angular surfaces on a vertical mill.

**9.** List three methods which can be used to set work on an angle in a vise.

**10.** What type of cutter should be used to cut keyseats?

**11.** How can an end mill be set to the center of a shaft?

**12.** How are T-slots generally cut?

**13.** What is the purpose of dovetails?

**14.** Why are rotary tables used in machine shop work?

**15.** What is the purpose of the hole in the center of the rotary table?

**16.** Calculate the indexing (using 80:1 ratio) for equally spacing 10 holes on a bolt circle.

**17.** List the main steps required to center a workpiece on a rotary table.

# SECTION

## SHAPER

The metal shaper was developed for the purpose of removing metal to produce a flat surface. This flat surface may be machined horizontally, on an angle, or in a vertical plane. The work is held in a vise or fastened to the table. A cutting tool is driven back and forth by a ram which travels in a horizontal plane with a reciprocating motion. The tool peels a chip off the work on each forward stroke only.

Because of the versatility of the vertical milling machine, many operations previously performed on a shaper are now done on a vertical mill more quickly and easily. Therefore, only the basic shaper controls and machining operations will be covered in this section, since the work setups in both machines are basically the same.

# UNIT 59

# SETTING UP THE SHAPER

Although the vertical milling machine has replaced the shaper for many operations, the shaper is still a useful machine and many are still found in school shops and industry. Since it is possible that a person may encounter this machine, it is important that the basic setups and operations be understood. (See Fig. 59-1.)

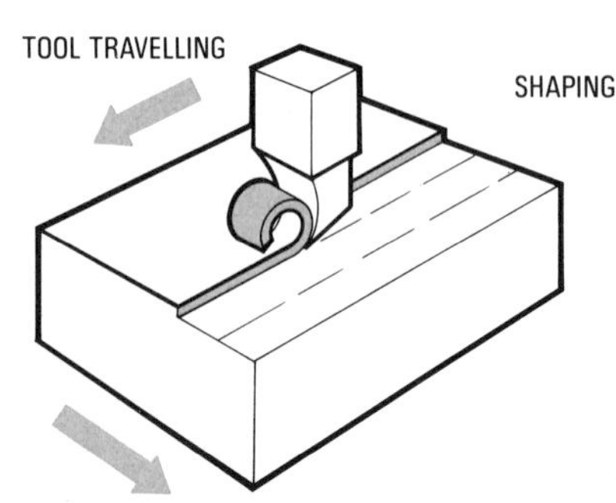


**Fig. 59-1** The cutting action of a shaper.

## OBJECTIVES

After completing this unit, you should be able to:

**1.** Identify and know the purpose of the main shaper parts
**2.** Calculate and set the proper speeds and feeds for various materials
**3.** Set the length and position of the shaper stroke to suit various workpieces

N-1 Hydraulic shapers are best suited for using carbide cutting tools because of the constant feed rate.

## TYPES OF SHAPERS

The three types of shapers commonly found in a machine shop are the *crank, gear,* and *hydraulic-type* shapers. The size of a shaper is determined by the largest cube which it can machine. For example, a 14-in. shaper can machine a block 14 in. × 14 in. × 14 in.; a 300-mm shaper can machine a block 300 mm × 300 mm × 300 mm.

### Crank Shaper

The crank shaper (Fig. 59-2) is the type most commonly used. Its ram is given its reciprocating motion by means of a rocker arm operated by a crank pin from the main driving gear, or "bull wheel." The forward stroke of the shaper occurs when the crank pin is in the upper portion of the rocker arm or 220° of the cycle. The return stroke occurs when the pin is in the lower portion or 140° of the cycle. This action causes the return stroke to be faster than the forward stroke on a 3:2 ratio.

### Gear Shaper

The gear shaper obtains the drive for the ram from a gear and a rack connected to the ram. This shaper also has a mechanism that provides a quick return for the ram.

### Hydraulic Shaper

The hydraulic shaper is driven by the movement of a piston in a cylinder of oil and is controlled by a valve mechanism connected with the oil pump. Oil pressure provided by the pump forces the operating cylinder, connected to the ram, forward. A reversing valve causes the cylinder to reverse and move the ram backward. (See N-1.)

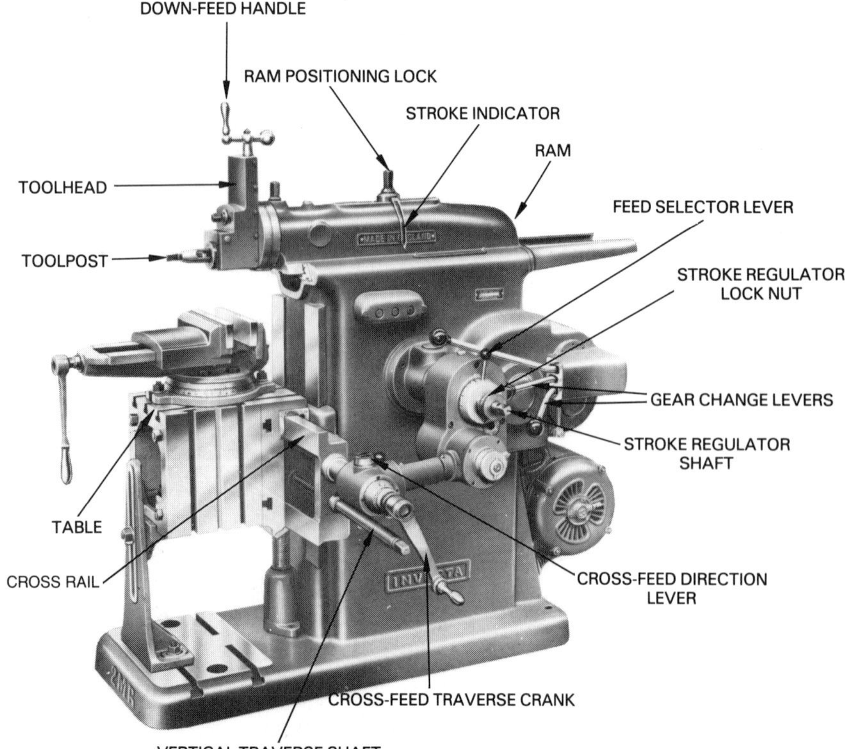


**Fig. 59-2** The main parts of a crank shaper. (*Courtesy Elliott Machine Tools.*)

## SHAPER PARTS

To operate a shaper successfully, a knowledge of the main operative parts and their function is necessary.

The *ram* is a semi-cylindrical form of heavy construction that provides the forward and return strokes to the cutting tool. It contains the *ram-positioning mechanism* and the toolhead. The *ram-adjusting shaft* is used to change

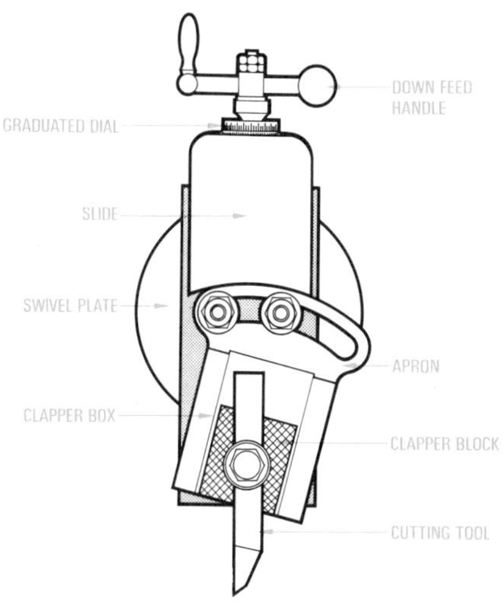


**Fig. 59-3** The parts of a shaper toolhead. (*Courtesy Cincinnati Shaper.*)

the position of the stroke and the *ram-positioning lock* or *clamp* locks the ram in a fixed position.

The *toolhead* (Fig. 59-3) is fastened to the ram to hold the toolholder and the *clapper box* which allows the cutting tool to raise slightly on the return stroke. The *downfeed handle* provides a means of feeding the cutting tool to any given dimension or depth of cut in either thousandths of an inch or hundredths of a millimeter, as measured on its graduated collar.

The *crossfeed traverse crank* is used to provide a horizontal movement to the table in either thousandths of an inch or hundredths of a millimeter with the use of the graduated collar. This screw is used to move the table longitudinally under the cutting tool. The *crossfeed direction lever* may be engaged in a ratchet to provide an automatic feed to the table. The *vertical traverse shaft* is used to lower or raise the table.

The *table* is fastened to the *cross-rail;* it provides a support for the work to be machined. It can be raised or lowered by the elevating screw and moved away from or toward the operator by hand with the crossfeed screw, or by power with the automatic feed mechanism.

The *stroke-regulator shaft* adjusts the length of stroke required and the *stroke-regulator lock nut* is used to lock the mechanism in a fixed position. The *stroke indicator* provides a guide when adjusting the shaper to the proper length of stroke required.

## SHAPER SPEEDS

Speeds and feeds are dependent on many factors that control the efficiency of the shaper use. Factors such as depth of cut, amount of feed, material to be machined, and type of cutting tool control the speeds and feeds for machining. Table 59-1 gives recommended shaper speeds and feeds. (See N-2.)

N-2 Cutting speeds are expressed in either feet per minute (ft/min) or meters per minute (m/min).

### Inch Speed Calculations

To obtain number of strokes per minute,

$$N = \frac{\text{CS} \times 7}{L}$$

where $N$ = number of strokes per minute
$L$ = length of work plus 1 in. for tool clearance
CS = cutting speed of metal in feet per minute
7 = constant for shaper speeds

**Example:**

How many strokes per minute should the ram be set to shape a piece of machine steel 9 in. long?

$$N = \frac{\text{CS} \times 7}{L}$$

$$= \frac{80 \times 7}{10}$$

$$= 56 \text{ strokes/min}$$

### Metric Speed Calculations

When shaping a workpiece with metric dimensions, set the length of stroke to 25 mm longer than the workpiece, to allow for clearance at each end of the stroke. The number of strokes per minute is calculated as follows:

$$\text{Strokes per minute } (N) = \frac{\text{CS (m)}}{L \text{ (m)}} \times \frac{3}{5} \text{ or } 0.6$$

$$3/5 = \text{ratio of forward to return stroke}$$

$$N = \frac{\text{CS}}{L} \times 0.6$$

**TABLE 59-1 SHAPER SPEEDS AND FEEDS**

| Cutting tool | Machine steel | | | | Tool steel | | | | Cast iron | | | | Brass | | | |
|---|---|---|---|---|---|---|---|---|---|---|---|---|---|---|---|---|
| | Speed per minute | | Feed | | Speed per minute | | Feed | | Speed per minute | | Feed | | Speed per minute | | Feed | |
| | ft | m | in. | mm | ft | m | in. | mm | ft. | m | in. | mm | ft | m | in. | mm |
| H.S.S. | 80 | 24 | .010 | 0.25 | 50 | 15 | .015 | 0.38 | 60 | 18 | .020 | 0.51 | 160 | 48 | .010 | 0.25 |
| Carbide | 150 | 46 | .010 | 0.25 | 150 | 46 | .012 | 0.30 | 100 | 30 | .012 | 0.30 | 300 | 92 | .015 | 0.38 |

N-3 $N \text{ (metric)} = \frac{CS}{L} \times 0.6$

N-4 The shaper cuts only on the forward stroke.

N-5 The feed should operate only on the return stroke of the ram.

N-6 Never make any adjustments or take measurements until the shaper is stopped.

The length of the work is usually given in millimeters; therefore divide the length plus the tool clearance by 1000 to bring it to meters.

**Example:**

How many strokes per minute are required to machine a 330-mm-long piece of tool steel (CS 15)? (See N-3.)

$$\text{Length of stroke } L = 330 + 25 \text{ mm clearance}$$

$$\text{Convert } L \text{ to meters} = \frac{355}{1000}$$

$$= 0.36 \text{ m}$$

$$N = \frac{15}{0.36} \times 0.6$$

$$= 25 \text{ strokes/min}$$

## SHAPER FEED MECHANISM

When horizontal surfaces are being machined, the table upon which the work is mounted feeds the work toward the cutting tool automatically on the return stroke of the ram. (See N-4.) When vertical feeds are required, the feed is controlled by turning the downfeed screw handle by hand or automatically if the shaper is equipped with an automatic downfeed mechanism, on each return stroke of the ram.

### To Engage or Disengage the Table Feed

**1.** Move the cutting tool clear of the work.

**2.** Turn the pawl knob until the pawl engages in the sprocket wheel, as in Fig. 59-4.

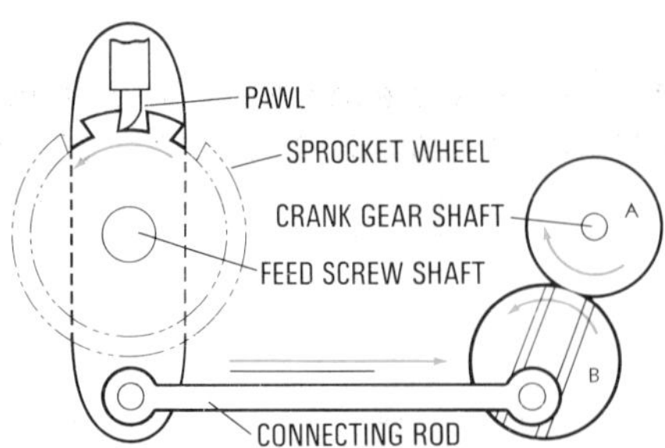


**Fig. 59-4** The shaper table feed mechanism.

**3.** Start the shaper.

**4.** Check the position of the connecting rod in the T-slot in gear B to see that the feed operates on the return stroke of the ram. (See N-5.)

**5.** To set the amount of feed, adjust the connecting rod off center at B so that the pawl will move (index) two or more teeth on the sprocket wheel when the ram is on its return stroke.

**6.** To disengage the feed, raise the pawl knob and turn the pawl 90° to prevent the pawl from entering the teeth in the sprocket wheel.

On hydraulic shapers, the feed direction and the amount of feed are regulated by control and reversing valves. The feed is driven by a gear and a rack which are controlled by a hydraulic system. Feeds can be increased or decreased while the machine is running by adjustment of the valves.

## SAFETY PRECAUTIONS

It is very important that the operator be familiar with the various areas of a shaper which can be hazardous. By following a few basic safety precautions and using common sense, most accidents can be avoided.

**1.** Always wear safety glasses to protect your eyes from flying chips (Fig. 59-5).

**2.** Never operate any machine until you have been properly instructed.

**3.** *Before starting* a shaper, make sure that the work, vise, tool, and ram are securely fastened.

**4.** Check that the tool and toolhead will clear the work and the column on the return stroke. (See N-6.)

**5.** Always prevent chips from flying by using a metal shield or wire screen. This can prevent injury to others.

**6.** *Always* stand parallel to the cutting stroke and *not* in front of it.

**7.** Never attempt to remove chips or reach across the table while the ram is in motion.

**8.** Keep the area around a machine neat and tidy.

**9.** Clean oil and grease from the floor immediately to prevent dangerous falls (Fig. 59-6).

**10.** Never attempt to adjust a machine while it is in motion.

Fig. 59-5 Always wear safety glasses to protect your eyes in a machine shop. (*Courtesy The Clausing Corporation.*)

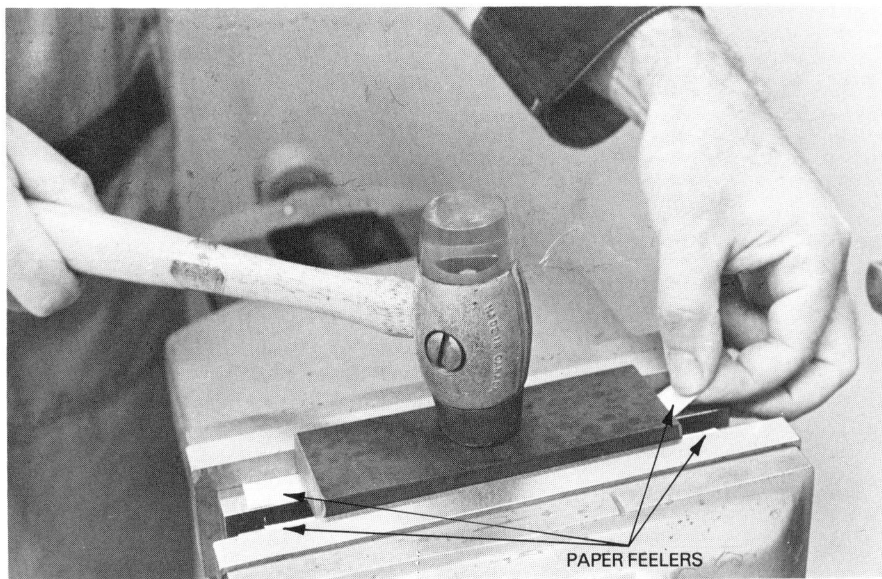


Fig. 59-7 Use paper feelers between the work and parallels when setting work in a vise. (*Courtesy Kostel Enterprises Ltd.*)

Fig. 59-6 Grease and oil on floors can cause dangerous falls.

## SETTING THE SHAPER STROKE

The shaper stroke must be set to suit the length of each workpiece, and also the position where it is held in a vise, on the table, or other work-holding device. If the shaper stroke is longer than necessary, valuable time will be wasted since it will take longer to complete the job.

Since the setting up of work in a shaper vise is exactly the same as that for a milling machine vise, refer to Unit 52, "Milling a Flat Surface," and Unit 57, "Machining a Block Square and Parallel," for the procedure for setting a workpiece in a vise.

### To Set the Length of the Shaper Stroke

**1.** Mount the work in the shaper vise (Fig. 59-7).

**2.** Measure the length of the work and add 1 in. (25.4 mm) for tool clearance (½ in. or 12.7 mm at each end of the stroke).

**3.** Use the start-stop or inching button to jog the motor until the ram is at the extreme back end of the stroke

*or*

Place the crank on the stroke regulator shaft and turn it until the ram is at the back end of the stroke (Fig. 59-8).

**4.** Loosen the stroke-regulator lock nut about one-half turn (Fig. 59-9).

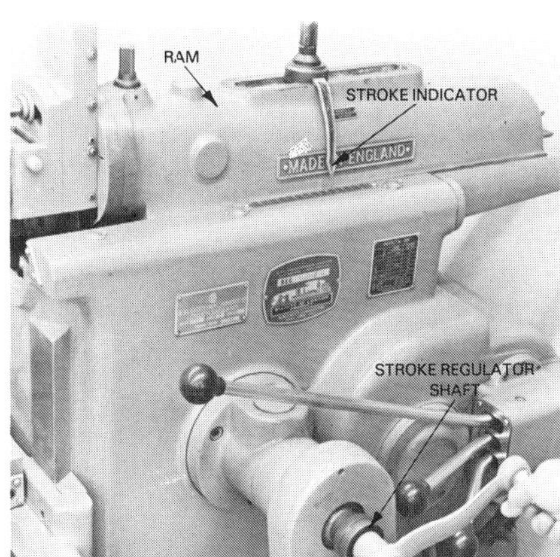


Fig. 59-8 Setting the ram to the back end of the stroke. (*Courtesy Kostel Enterprises Ltd.*)

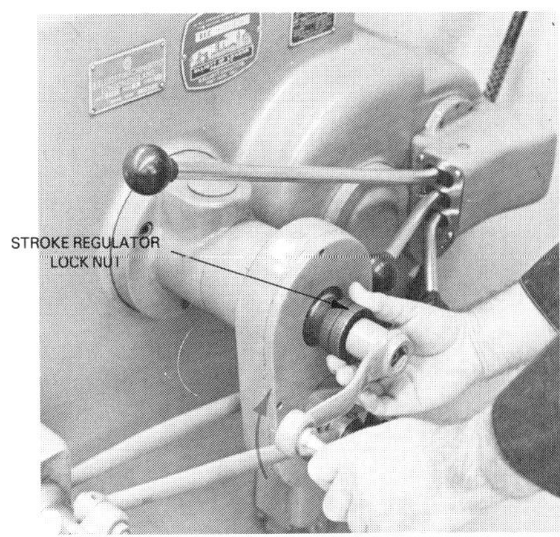


Fig. 59-9 To loosen the locknut, hold it firmly and turn the crank clockwise. (*Courtesy Kostel Enterprises Ltd.*)

**Fig. 59-10** To loosen the ram lock, turn it counterclockwise. (*Courtesy Kostel Enterprises Ltd.*)

**Fig. 59-11** The ram may be moved by hand to bring the toolbit within ½ inch (12.7 mm) of the workpiece.

N-7 Always be sure that the ram is at the *back end* of the stroke before setting the length or position.

N-1 Always adjust the vise away from and not toward the indicator plunger.

**5.** Turn the stroke regulator shaft until the stroke indicator is at the desired length of stroke.

**6.** Tighten the stroke-regulator lock nut.

## To Set the Position of the Shaper Stroke

Before setting the position of the shaper stroke, the length of stroke must be first set as in the previous operation. The ram must then be adjusted so that the cutting tool will machine the full length of the piece. (See N-7.)

**1.** Set the length of the stroke as in the previous operation.

**2.** Be sure that the ram is at the extreme back end of the stroke.

**3.** Loosen the ram lock or clamp (Fig. 59-10).

**4.** Pull the toolhead and ram forward until the toolbit is within ½ in. (12.7 mm) of the work (Fig. 59-11). Or if the ram is equipped with an adjusting screw, turn it until the toolbit is within ½ in. (12.7 mm) of the work.

**5.** Tighten the ram lock or clamp.

**6.** With the cutting tool clear of the work, start the machine and see that the toolbit clears each end of the work by about ½ in. (12.7 mm).

## REVIEW QUESTIONS

**1.** Name three types of shapers.

**2.** How is the size of a shaper determined?

**3.** List four parts which are fastened to the ram.

**4.** What is the purpose of the following parts?

(a) Clapper box
(b) Downfeed handle
(c) Crossfeed traverse crank
(d) Stroke regulator shaft

**5.** Name four factors which affect the shaper speeds and feeds.

**6.** Calculate the number of strokes per minute required to:

(a) Cut a 11-in.-long piece of cast iron (CS 60).
(b) Cut a 155-mm-long piece of machine steel (CS 24).

**7.** On what stroke should the automatic feed move the table?

**8.** What safety precaution should be observed before starting a shaper?

**9.** Why is it unwise to have a shaper stroke longer than necessary?

**10.** How much longer than the workpiece should the shaper stroke be set?

**11.** In what position must the ram be set before setting the length and position of the stroke?

**12.** How close to the end of the work should the cutting tool be positioned?

# UNIT 60

# MACHINING IN A SHAPER

Most work machined in a shaper is held in a vise bolted to the table. Since the setup of work in a shaper vise and a vertical mill vise is exactly the same, this information will not be repeated in this unit. Whenever this information is required, reference will be made to the vertical milling machine units which contain detailed information.

## OBJECTIVES

After completing this unit, you should be able to:

**1.** Set up the work and vise for machining flat surfaces

**2.** Machine a block square and parallel

**3.** Set up the shaper to machine a vertical surface

## ALIGNING THE SHAPER VISE

The body of the vise can be swiveled on a base plate to any angular setting required. When the work must be machined parallel or at right angles to the direction of the cut, the vise jaw must be aligned. To accurately align the vise, a dial indicator is attached to the toolhead of the shaper and passed over the entire length of the solid jaw. With this method, a vise can be aligned to within .001 in. (0.02 mm) accuracy. (See N-1.)

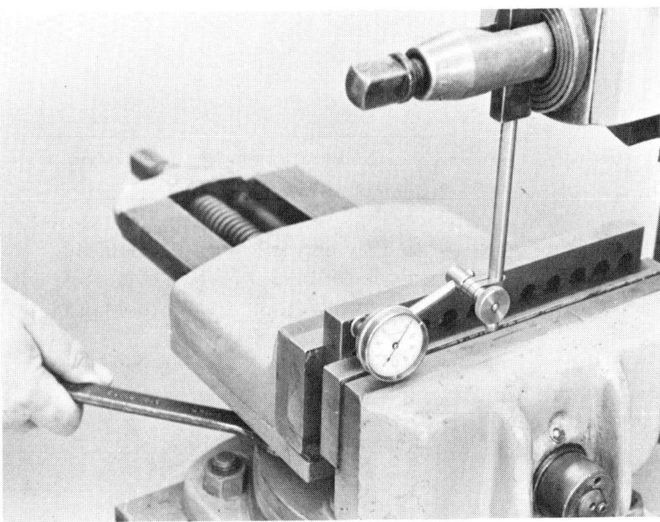

**Fig. 60-1** Aligning the vise parallel to the ram travel. (*Courtesy Kostel Enterprises Ltd.*)

**Fig. 60-2** Aligning the vise at 90° to the ram travel. (*Courtesy Kostel Enterprises Ltd.*)

**Fig. 60-3** Checking the seat of the vise with an indicator and parallels. (*Courtesy Kostel Enterprises Ltd.*)

**Fig. 60-4** The difference between the readings on both parallels is the amount the vise is out of parallelism. (*Courtesy Kostel Enterprises Ltd.*)

N-2 Clean the vise and remove all burrs before testing a vise for parallelism.

To align the shaper vise parallel (Fig. 60-1) or at 90° to the ram travel (Fig. 60-2), see Unit 52.

### To Test the Work Seat for Parallelism

**1.** Set up the indicator as shown in Fig. 60-3. (See N-2.)

**2.** Place a set of parallels on the bottom of the vise.

**3.** Adjust the toolhead until the dial indicator registers approximately one-half revolution on one parallel.

**4.** Move the ram forward by hand for the length of the parallel and note the indicator reading.

**5.** With the crossfeed screw handle, move the vise to obtain a reading across the second parallel (Fig. 60-4).

**6.** To correct any errors in alignment, remove the vise from the table, clean it, and remove burrs from all clamping surfaces on the vise and table.

**7.** Remount the vise on the table and test for accuracy.

## SHAPING A FLAT SURFACE

Most work machined in a shaper is held in a vise. After the work is prop-

**Fig. 60-5** Remove all burrs from the workpiece before setting work in a vise. (*Courtesy Kostel Enterprises Ltd.*)

N-3 Always grip work in the center to prevent springing or twisting the vise.

N-4 Always stop the shaper before adjusting the length or position of the stroke.

N-5 $S \text{ (inch)} = \dfrac{CS \times 7}{L}$

$S \text{ (metric)} = \dfrac{CS \text{ (m)}}{L \text{ (m)}} \times 0.6$

erly set up in the vise, the cutting tool must be set to the surface of the work and the surface is then machined.

### To Mount Work in a Shaper Vise

**1.** Remove all burrs from the workpiece (Fig. 60-5).

**2.** Clean the work and the vise thoroughly.

**3.** Select a set of parallels large enough to raise the work about ¼ in. (6.35 mm) above the vise jaws.

**4.** Set the work on the parallels in the center of the vise. (See N-3.)

**5.** Place a short paper feeler between each corner of the work and the parallels (Fig. 60-6).

**6.** Tighten the vise securely. If the vise is not tightened securely, when the work is tapped down, it will hit the parallels and bounce up.

**7.** Tap the work down onto the parallels with a soft-faced hammer alternating the hammer blows between each end of the work.

**8.** Tap each corner of the work down until all the paper feelers are tight.

**9.** *Do not* retighten the vise after the work has been tapped down because the work will be forced up and off the parallels during the tightening process.

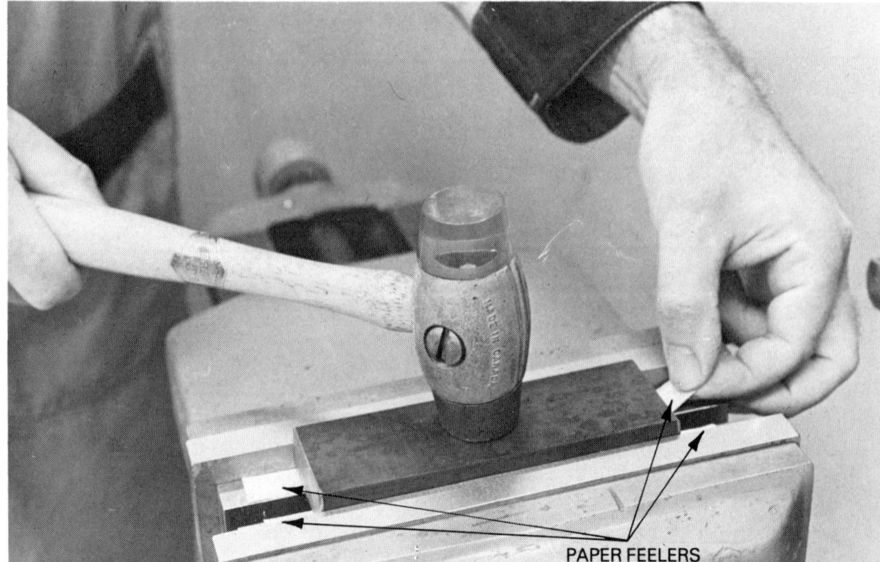


**Fig. 60-6** Mounting work in a vise. Note the paper feelers between the parallels and the work. (*Courtesy Kostel Enterprises Ltd.*)

### To Set the Shaper for Machining

**1.** Mount the work in the vise (Fig. 60-6).

**2.** Set the length and position of the shaper stroke. (See N-4.)

**3.** Loosen the apron lock screw and swing the top of the toolhead apron to the left (Fig. 60-7). This will prevent the toolbit from rubbing on the return stroke.

**4.** Tighten the apron lock screw and set the toolholder vertically.

### To Set the Toolbit to the Work

**5.** Start the shaper and stop it when the toolbit is over the work.

**6.** Hold a piece of paper between the toolbit and the surface of the work.

**7.** Turn the downfeed handle until a slight drag is felt on the paper (Fig. 60-8).

**8.** Loosen the graduated collar lock screw and set the collar to zero.

**9.** Tighten the graduated collar lock screw.

**10.** Move the table until the cutting tool clears the edge of the work by about ¼ in. (6.35 mm) (Fig. 60-9).

### To Machine the Surface

**11.** Set the depth of cut by turning the downfeed handle clockwise the required amount (about .060 in. or 1.52 mm to clean up a surface).

**12.** Set the shaper speed and feed for the material being cut. (See Table 59-1.) (See N-5.)

**13.** Start the shaper and engage the automatic feed being sure that the feed operates on the return stroke of the ram (Fig. 60-10).

**14.** Machine the surface and then remove all burrs from the edges of the workpiece with a file.

## MACHINING A BLOCK SQUARE AND PARALLEL

In order to machine the four sides of a piece of work so that its sides are square and parallel, it is important that each side be machined in a definite order (Fig. 60-11). It is very important

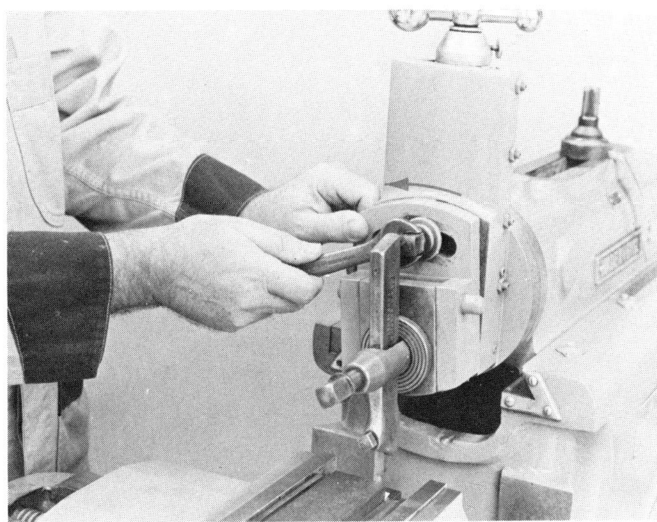

Fig. 60-7 Swivel the toolhead apron to the left to allow the cutting tool to lift on the return stroke. (*Courtesy Kostel Enterprises Ltd.*)

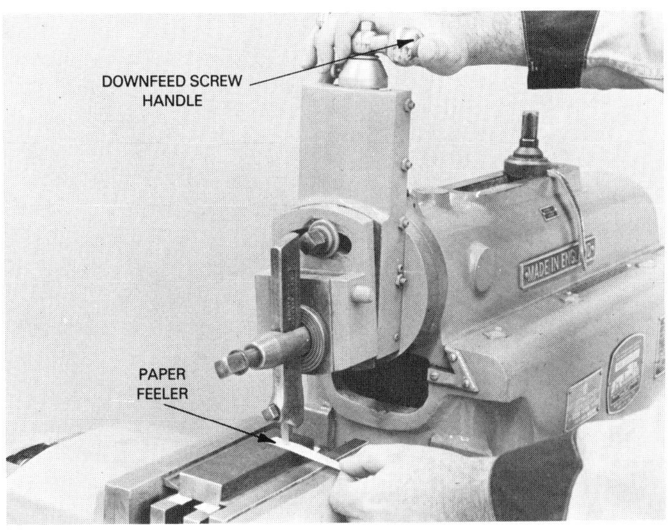


Fig. 60-8 Using a paper feeler to set the cutting tool to the work surface. (*Courtesy Kostel Enterprises Ltd.*)

Fig. 60-9 The cutting tool clearing the edge of the work by 1/4 inch (6.35 mm). (*Courtesy Kostel Enterprises Ltd.*)

Fig. 60-10 The feed should operate on the return stroke of the ram. (*Courtesy Kostel Enterprises Ltd.*)

Fig. 60-11 The sequence for machining the sides of a rectangular piece square and parallel. (*Courtesy Kostel Enterprises Ltd.*)

N-6 With the work past the end of the vise, the vertical surface can be machined without hitting the vise or parallels.

that dirt and burrs be removed from the work, vise, and parallels, since they can cause inaccurate work. Since the setting up of work for machining square and parallel in a shaper vise is exactly the same as in a vertical milling machine vise, refer to Unit 57 for a detailed explanation of each setup.

## MACHINING A VERTICAL SURFACE

A vertical surface can be machined on flat work by feeding the toolhead with the downfeed handle. In order for the machined edge to be square, the toolhead must be set at 90° to the top of the shaper table.

### To Shape a Vertical Surface

1. Align the toolhead with a square (Fig. 60-12).
2. Set the shaper vise at 90° to the ram.
3. Place a suitable pair of parallels in the vise.
4. Adjust the work so that about 1/2 in. (12.7 mm) extends past the end of the vise jaws and parallels (Fig. 60-13). (See N-6.)
5. Tighten the vise securely and then tap the work down until the paper feelers are tight.

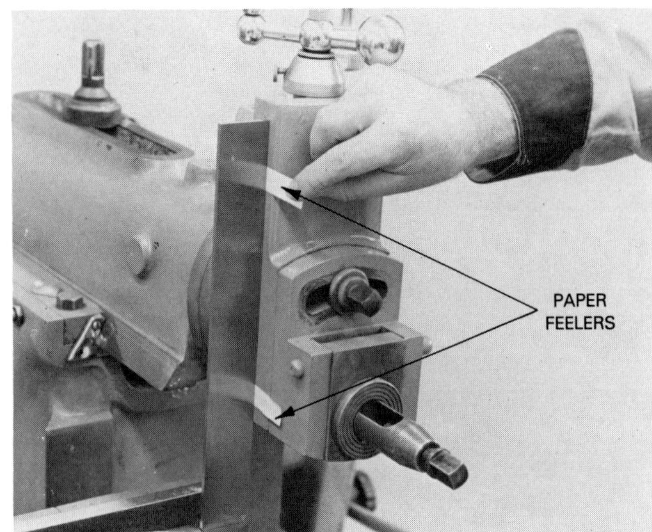


**Fig. 60-12** Aligning the toolhead with a square. (*Courtesy Kostel Enterprises Ltd.*)

**Fig. 60-13** The top of the apron should always be swung away from the vertical surface to be cut. (*Courtesy Kostel Enterprises Ltd.*)

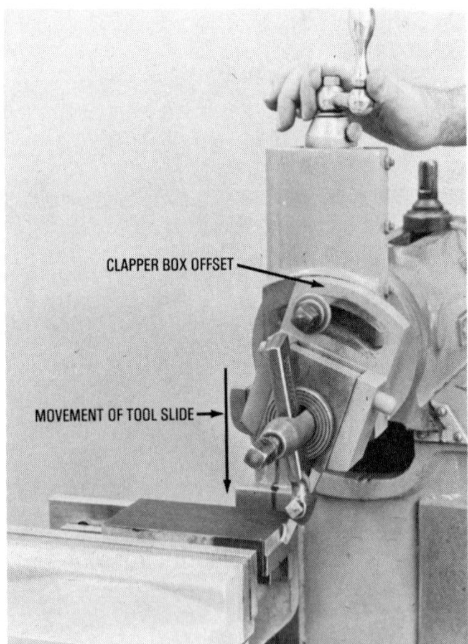


**Fig. 60-14** Feed the toolbit down by the downfeed handle on each return stroke of the ram. (*Courtesy Kostel Enterprises Ltd.*)

N-7 Before starting a shaper for machining a vertical or angular surface, be sure that the toolhead clears the frame on the return stroke.

**6.** Adjust the length and position of the shaper stroke.

**7.** Swing the top of the apron away from the surface to be cut (Fig. 60-13).

**8.** Fasten a left-hand toolholder in the toolpost.

**9.** Mount a facing tool in the toolholder.

**10.** Swivel the toolholder so that it will clear the side of the surface to be cut.

**11.** Raise the toolhead so that the toolbit clears the top of the work by about ¼ in. (6.35 mm).

**12.** Set the depth of cut by turning the crossfeed traverse crank.

**13.** Set the shaper for the proper speed and then start the shaper. (See N-7.)

**14.** Feed the toolbit down by feeding the downfeed handle clockwise about .010 in. (0.25 mm) on each return stroke of the ram (Fig. 60-14).

**15.** If necessary, take additional cuts to bring the work to size.

## MACHINING AN ANGULAR SURFACE

Angular surfaces may be cut on a shaper by three methods:

**1.** Setting the workpiece on the desired angle

**2.** Swiveling the vise to the required angle

**3.** Setting the toolhead to an angle

See Unit 58 for a detailed explanation of how angular work can be set up and machined.

## REVIEW QUESTIONS

**1.** Why is it important that the vise be aligned before machining in a shaper?

**2.** List the main steps to test the work seat for parallelism.

**3.** Why is it important that burrs be removed from the work before making a setup?

**4.** Why is it important that paper feelers be used between the work and parallels?

**5.** What will occur if the vise is retightened after the work has been tapped down?

**6.** How long should the stroke be for machining an 8-in.- or 200-mm-long workpiece?

**7.** Why should the top of the toolhead apron be swung to the left when machining a flat surface?

**8.** On what ram stroke should the feed operate?

**9.** Name one method of aligning the toolhead.

**10.** In which direction should the top of the apron be swung when machining a vertical surface?

# SECTION

## GRINDING MACHINES

The ability to sharpen tools by rubbing them against certain sands or stone was discovered by primitive human beings many centuries ago. Natural abrasives were used to produce the first grinding stone in the Iron Age, which in turn added more precision to the sharpening process. In the early 1900s, grinding took a giant step forward with the ability of individuals to produce abrasives such as silicon carbide and aluminum oxide. Since human beings could not control the quality and hardness of the abrasives, more development was expanded on better machine tools to use this product more effectively. Today surface, cylindrical, centerless, cutter and tool, and abrasive grinders can remove metal quickly and to a high degree of accuracy. In modern industry, the use of abrasives has contributed more to mass production than any other single factor. Only by means of abrasives and precision tools has it been possible to produce the close tolerances and surface finishes required by industry.

# UNIT 61

# ABRASIVES

An abrasive grain is a hard, tough substance containing many sharp projecting cutting edges or points. It is used as a cutting tool to penetrate and cut away material that is softer than itself. Abrasive products such as grinding wheels, coated abrasives, sharpening stones, and abrasive sticks are composed of many abrasive grains which have been bonded together.

**Fig. 61-1** Fused aluminum oxide before crushing. *(Courtesy General Abrasive Co.)*

## OBJECTIVES

After completing this unit, you should be able to:

1. Know about the manufacture and use of aluminum oxide and silicon carbide
2. Understand the types and uses of coated abrasives
3. Know the elements which go into the manufacture of a grinding wheel
4. Interpret the specifications of a grinding wheel from the markings found on the blotter

**Fig. 61-2** Fused silicon carbide before crushing. *(Courtesy General Abrasive Co.)*

## TYPES OF ABRASIVES

Abrasives may be divided into two classes: *natural* and *artificial*.

*Natural abrasives,* such as sandstone, emery, quartz, and corundum, were used extensively prior to the early part of the twentieth century. (See N-1.) However, they have been almost totally replaced by manufactured abrasives with their inherent advantages. One of the best natural abrasives is diamond, but because of the high cost of industrial diamonds (bort), its use in the past was limited mainly to grinding cemented carbides and glass, and sawing concrete, marble, limestone, and granite. However, due to the introduction of synthetic or manufactured diamonds, industrial natural diamonds will become cheaper in cost and will be used on many more grinding applications.

*Manufactured abrasives* are used extensively because their grain size, shape, and purity can be closely controlled. This uniformity of grain size and shape, which ensures that each grain does its share of work, is not possible with natural abrasives. (See N-2.) There are several types of manufactured abrasives: *aluminum oxide, silicon carbide, boron carbide, cubic boron nitride,* and *manufactured diamond.*

### Aluminum Oxide

Aluminum oxide (Fig. 61-1) is probably the most important abrasive since about 75 percent of the grinding wheels manufactured are made from it. It is generally used for high-tensile-strength materials including all ferrous metals except cast iron. (See N-3.)

### Silicon Carbide

Silicon carbide (Fig. 61-2) is suited for grinding materials which have a low-tensile strength and high density, such as cemented carbides, stone, and ceramics. It is also used for cast iron and most nonferrous and nonmetallic materials. It is harder and tougher than aluminum oxide. Silicon carbide may vary in color from green to black. Green silicon carbide is used mainly for grinding cemented carbides and other hard materials. Black silicon carbide is used for grinding cast iron and soft nonferrous metals, such as aluminum, brass, and copper. It is also suitable for grinding ceramics. (See N-4.)

### Boron Carbide

One of the newer abrasives is boron carbide. It is harder than silicon carbide and, next to the diamond, it is the hardest material manufactured. Boron carbide is not suitable for use in grinding wheels and is used only as a loose abrasive and a relatively cheap substitute for diamond dust. Because of its extreme hardness, it is used in the manufacture of precision gages and sand-blast nozzles. Crush dressing rolls made of boron carbide have proven superior to tungsten carbide rolls for the dressing of grinding wheels on multi-form grinders. Boron carbide is also widely accepted as an abrasive used in ultrasonic machining applications.

N-1 Natural abrasives contain impurities and the grain quality and hardness vary.

N-2 Manufactured abrasives contain no impurities and the quality of the grain can be closely controlled.

N-3 Aluminum oxide abrasive is used for most ferrous metals.

N-4 Silicon carbide abrasive is used for nonferrous metals.

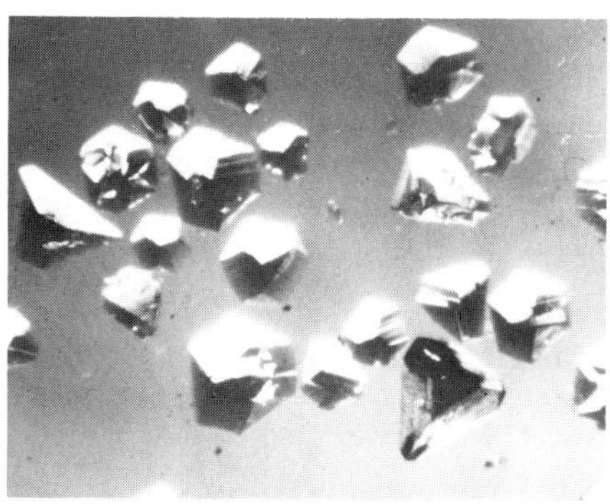

**Fig. 61-3** Crystals of cubic boron nitride. (*Courtesy The General Electric Company*)

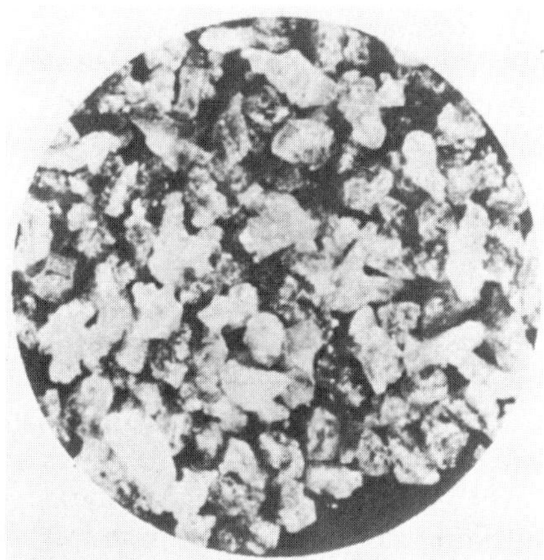

**Fig. 61-4A** Type RVG is used to grind ultrahard materials. (*Courtesy The General Electric Company*)

**Fig. 61-4B** Type MBG-11 is a tough crystal used in metal bond grinding wheels. (*Courtesy The General Electric Company.*)

N-5 Cubic boron nitride is used for grinding very hard metals while generating very little heat.

N-6 Coated abrasives are used to grind or polish metals by hand or machine.

## Cubic Boron Nitride

One of the more recent developments in the abrasive field has been the introduction of cubic boron nitride (Fig. 61-3). This manufactured abrasive has hardness properties between silicon carbide and diamond. The crystal known as Borazon Ⓣ CBN (cubic boron nitride) was developed by the General Electric Company. This material is capable of grinding high-speed steel with ease and accuracy, and is superior to diamond in many applications.

Cubic boron nitride is about twice as hard as aluminum oxide and is capable of withstanding high grinding temperatures up to 2500°F (1370°C) before breaking down. CBN is cool cutting and chemically resistant to all inorganic salts and organic compounds. Because of the extreme hardness of this material, grinding wheels made of CBN are capable of maintaining very close tolerances. These wheels require very little dressing and are capable of removing a constant amount of material across the face of a large work surface without having to compensate for wheel wear. Because of the cool cutting action of CBN wheels, there is little or no surface damage to the work surface. (See N-5.)

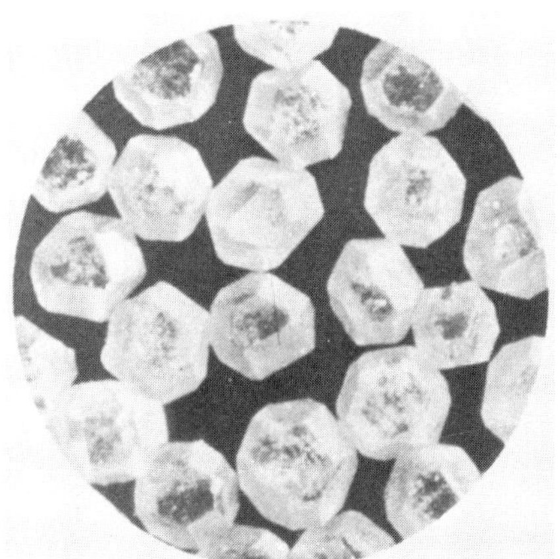

**Fig. 61-4C** Type MBS is a very tough larger crystal used in metal-bonded saws. (*Courtesy The General Electric Company.*)

## Manufactured Diamonds

Diamond, the hardest substance known to human beings, was primarily used in machine shop work for truing and dressing grinding wheels. Because of the high cost of natural diamonds, industry began to look to cheaper, more reliable sources. In 1954, the General Electric Company, after 4 years of research, produced Man-Made Ⓣ diamonds in their laboratory. In 1957, the General Electric Company, after more researching and testing, began the commercial production of these diamonds.

Because the temperature, pressure, and catalyst-solvent can be varied, it is possible to produce diamonds of various sizes, shapes, and crystal structure best suited to a particular need.

### Type RVG Diamond

This manufactured diamond is an elongated, friable crystal with rough edges (Fig. 61-4A). The letters RVG indicate that this type may be used with a resinoid or vitrified bond and is used for grinding ultrahard materials such as tungsten carbide, silicon carbide, and space-age alloys. The RVG diamond may be used for wet and dry grinding.

### Type MBG-II Diamond

This tough, blocky-shaped crystal is not as friable as the RVG type and is used in metal-bonded grinding wheels (Fig. 61-4B). It is used for grinding cemented carbides, sapphires, and ceramics as well as in electrolytic grinding.

### Type MBS Diamond

This is a blocky, extremely tough crystal with a smooth, regular surface which is not very friable (Fig. 61-4C). It is used in metal bonded saws (*MBS*) to cut concrete, marble, tile, granite, stone, and masonry materials.

# ABRASIVE PRODUCTS

After the abrasive has been produced, it is formed into products such as grinding wheels, coated abrasives, polishing and lapping powders, and abrasive sticks, all of which are used extensively in machine shops. (See N-6.)

## Coated Abrasives

Coated abrasives (Fig. 61-5) consist of a flexible backing (cloth or paper) to which abrasive grains have been bonded. Garnet, flint, and emery (natural abrasives) are being replaced by

**Fig. 61-5** A variety of coated abrasives.

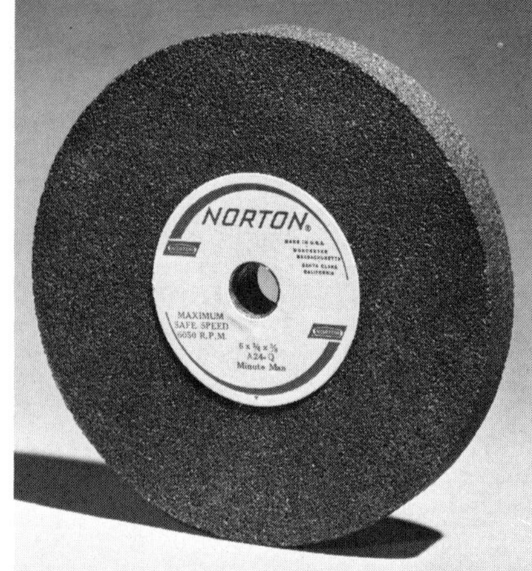


**Fig. 61-6** A type 1 straight grinding wheel. (*Courtesy The Norton Company.*)

N-7 A grinding wheel consists of abrasive grains held together by a bond.

aluminum oxide and silicon carbide in the manufacture of coated abrasives. This is due to the greater toughness and the more uniform grain size and shape of manufactured abrasives.

Coated abrasives serve two purposes in the machine shop: metal grinding and polishing.

Metal grinding, which may be done on a belt or disc grinder, is a rapid, nonprecision method of removing metal. Coarse-grit coated abrasives are used for rapid removal of metal, whereas fine grits are used for polishing.

Emery, a natural abrasive which is black in appearance, is used to manufacture coated abrasives, such as emery cloth and emery paper. Since the grains are not as sharp as artificial abrasives, emery is generally used for polishing metal by hand.

## Selection of Coated Abrasives

### Aluminum Oxide

Aluminum oxide, brown in appearance, is used for high-tensile-strength materials, such as steels, alloy steels, high-carbon steels, and tough bronzes. Aluminum oxide is characterized by the long life of its cutting edges.

For *hand operations,* 60 to 80 grit is used for fast cutting (roughing), while 120 to 180 grit is recommended for finishing operations.

For *machine operations,* such as on belt and disc grinders, 36 to 60 grit is used for roughing, while 80 to 120 grit is recommended for finishing operations.

### Silicon Carbide

Silicon carbide, bluish-black in appearance, is used for low-tensile-strength materials such as cast iron, aluminum, brass, copper, glass, and plastics. The selection of the grit size for hand and machine operations is the same as for aluminum oxide coated abrasives.

# GRINDING WHEELS

Grinding wheels, the most important product made from abrasives, are composed of abrasive material held together with a suitable bond (Fig. 61-6). The basic functions of grinding wheels in a machine shop are:

**1.** Generation of cylindrical, flat, and curved surfaces
**2.** Removal of stock
**3.** Production of highly finished surfaces
**4.** Cutting-off operations
**5.** Production of sharp edges and points

For grinding wheels to function properly, they must be hard and tough, and the wheel surface must be capable of gradually breaking down to expose new sharp cutting edges to the material being ground. (See N-7.)

The material components of a grinding wheel are the *abrasive grain* and the *bond*; however, there are other physical charcteristics, such as *grade* and *structure,* that must be considered in grinding wheel manufacture and selection.

## Grain Size

After crushing, the abrasive grains are sized by passing them over screens having a certain number of meshes per linear inch. The grain size is determined by the number of openings per linear inch in the screen through which the abrasive particles just pass (Fig. 61-7A to C). Grain sizes ranging from 4 to 240 grit are the most common sizes used in the manufacture of grinding wheels. Sizes finer than 240 grit

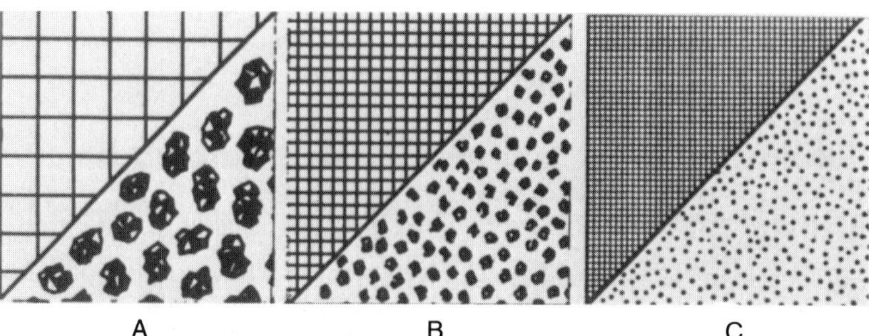


**Fig. 61-7** Relative grain size. (A) 8-grain size; (B) 24-grain size; (C) 60-grain size.

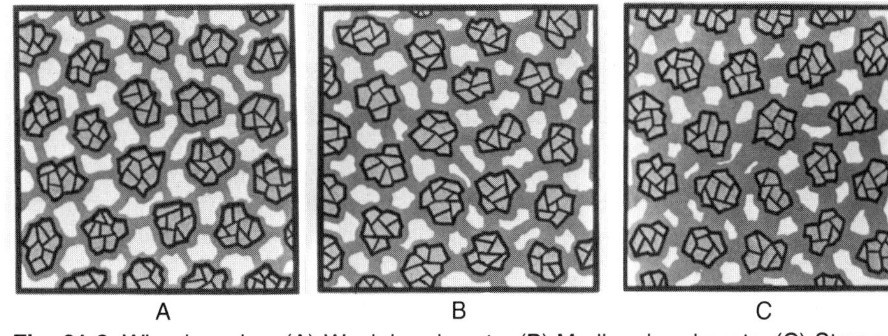


**Fig. 61-8** Wheel grades. (A) Weak bond posts. (B) Medium bond posts. (C) Strong bond posts. (*Courtesy The Carborundum Company.*)

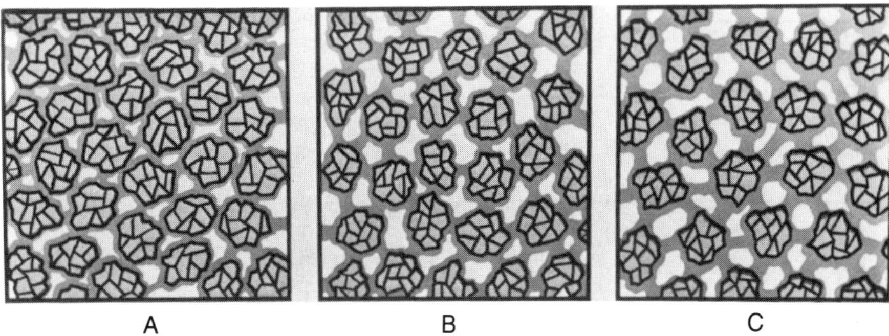


**Fig. 61-9** (A) Dense grinding wheel structure. (B) Medium grinding wheel structure. (C) Open grinding wheel structure.

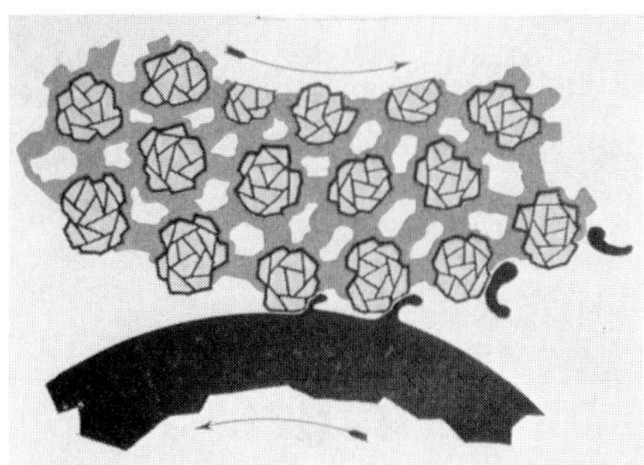

**Fig. 61-10** An open-structure wheel provides greater chip clearance. (*Courtesy The Carborundum Company.*)

N-8 Bond is the agent that holds the abrasive grains together in the form of a wheel or other shape.

N-9 Grade refers to the strength with which the bond holds the grains together.

N-10 Structure refers to the amount of space between abrasive grains.

are separated by hydraulic flotation and are commonly referred to as flour.

Abrasive grains are classified into several ranges: very coarse (6 to 12 grit), coarse (14 to 24 grit), medium (30 to 60 grit), fine (70 to 120 grit), and very fine (150 to 240 grit). Flour sizes used for polishing range from 280 to 600 grit.

## Bonds

Grinding wheels are composed of abrasive grains held together by a bond. The amount of bond used determines the hardness (grade) of the wheel. Standard grinding wheels may use one of the following bonds: vitrified, resinoid, rubber, shellac, silicate, and oxychloride. Vitrified, resinoid, and rubber bonds are the most common and are used in over 90 percent of the grinding wheels manufactured. (See N-8.)

## Wheel Grade and Structure

Although the abrasive grains and the bond are the only tangible components used in the grinding wheel, there are other characteristics which greatly affect the grinding wheel. These are the *grade* and the *structure* of the wheel.

### Grade

The grade indicates the strength with which the bond holds the abrasive particles together in the wheel (Fig. 61-8A to C). In other words, the grade refers to the hardness of the wheel. (See N-9.)

The grade is indicated by a letter ranging from A for the softest wheel (weakest bond posts) to Z for the hardest wheel (strongest bond posts).

### Structure

The structure of a grinding wheel indicates the density (grams per cubic inch) of the wheel, or the relationship of the abrasive grain and bonding material to the voids that separate them (Fig. 61-9A to C). The primary function of the void or space is to provide chip clearance for the metal removed from the work by a grain. (See N-10.)

The variations in wheel structure or density range from 1 (dense) to 15 (open). Although this may seem like a wide range, in reality it is relatively small. A wheel with an open structure has more chip clearance than one with a dense structure (Fig. 61-10).

The structure or porosity of the wheel can be controlled by the mold pressures during the initial wheel forming. This induced porosity, which

**TABLE 61-1 COMMON GRINDING WHEEL SHAPES AND APPLICATIONS**

| Shape | Name | Applications |
|---|---|---|
| | Straight (Type 1) | Cylindrical, centerless, internal, cutter, surface, and offhand grinding operations |
| | Cylinder (Type 2) | Surface grinding on horizontal and vertical spindle grinders |
| | Tapered (both sides) (Type 4) | Snagging operations; tapered sides lessen the chance of the wheel's breaking |
| | Recessed (one side) (Type 5) | Cylindrical, centerless, internal, and surface grinders; recess provides clearance for the mounting flange |
| | Straight cup (Type 6) | Cutter and tool grinder and surface grinding on vertical and horizontal spindle machines |
| | Recessed (both sides) (Type 7) | Cylindrical, centerless, and surface grinders; recesses provide clearance for mounting flanges |
| | Flaring cup (Type 11) | Cutter and tool grinder; used mainly for sharpening milling cutters and reamers |
| | Dish (Type 12) | Cutter and tool grinder; its thin edge permits it to be used in narrow slots |
| | Saucer (Type 13) | Saw gumming, gashing milling cutter teeth |

does not break grain contact, is generally controlled by the addition of ground moth crystals (paradichlorobenzene). These crystals burn out during the firing process to leave voids in the wheel. In the past, sawdust, coke, or crushed nut shells were used for this purpose; however, their use had declined greatly in recent years.

## STANDARD GRINDING WHEEL SHAPES

Nine standard grinding wheel shapes have been established by the United States Department of Commerce, the Grinding Wheel Manufacturers, and the Grinding Machine Manufacturers. Dimensional sizes for each of the shapes have also been standardized. Each of these nine shapes is identified by a number as shown in Table 61-1.

## GRINDING WHEEL MARKINGS

The standard marking system chart (Fig. 61-11) is used by the manufacturers to identify grinding wheels. This information is found on the blotter of all small- and medium-sized grinding wheels. It is stenciled on the side of larger wheels.

The six positions shown in the standard sequence are followed by all manufacturers of grinding wheels. The prefix shown is a manufacturer's symbol and is not always used by all grinding wheel producers. *Note:* This marking system is used only for aluminum oxide and silicon carbide wheels; it is not used for diamond wheels.

## SELECTING A GRINDING WHEEL FOR A SPECIFIC JOB

From the foregoing information, the machinist should be able to select the proper wheel for the job required.

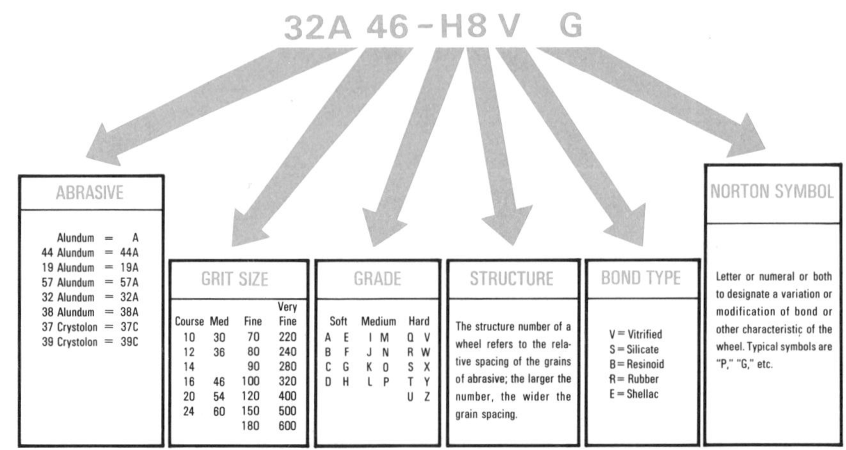


**Fig. 61-11** Straight-grinding-wheel marking system chart. (*Courtesy The Norton Company of Canada.*)

N-11 Grinding wheels are fragile and should be handled with care.

N-12 Never use a grinding wheel which has been dropped unless it has been thoroughly tested for soundness.

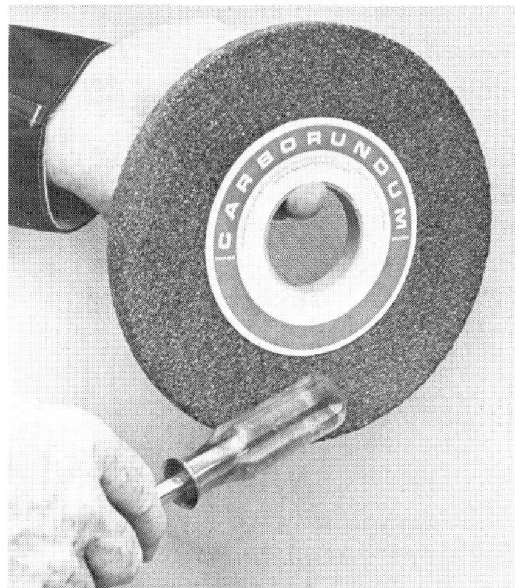

Fig. 61-12 Testing a grinding wheel for cracks. (*Courtesy Kostel Enterprises Ltd.*)

**Example:**

It is required to rough surface grind a piece of SAE 1045 steel using a straight wheel. Coolant is to be used.

*Type of Abrasive:* Because steel is to be ground, *aluminum oxide* should be used.

*Size of grain:* Since the surface is not precision-finished, a medium grain can be used—about *46* grit.

*Grade:* A medium-grade wheel which will break down reasonably well should be selected. Use grade *J*.

*Structure:* Since this steel is of medium hardness, the wheel should be of medium density—about *7*.

*Bond type:* Since the operation is standard surface grinding and since coolant is to be used, a *vitrified* bond should be selected.

After the various factors have been considered, an A46-J7-V grinding wheel should be selected to rough grind SAE 1045 steel. *Note:* These specifications do not include the manufacturer's prefix or the manufacturer's records.

## DIAMOND WHEELS

Diamond wheels are used for grinding cemented carbides and hard vitreous materials, such as glass and ceramics.

Diamond wheels are manufactured in a variety of shapes, such as straight, cup, dish, and thin cut-off wheels.

Wheels of ½ in. (12.7 mm) diameter or less have diamond particles throughout the wheel. Wheels larger than ½ in. (12.7 mm) are made with a diamond surface on the grinding face only. The diamonds for this purpose are made in grain sizes ranging from 100 to 400. The proportions of the diamond and bond mixture vary with the application. This diamond concentration is identified by the letters A, B, or C. "C" concentration will contain four times the number of diamonds of a grinding wheel with an A concentration.

## HANDLING AND STORAGE OF GRINDING WHEELS

Since all grinding wheels are breakable, proper handling and storage is important. Damaged wheels can be dangerous if used; therefore, the following rules should be observed. (See N-11.)

**1.** Do not drop or bump grinding wheels. (See N-12.)

**2.** Always store wheels properly on shelves or in bins provided. Flat and tapered wheels can be stored on edge, while large cup wheels and cylindrical wheels should be stored on the flat sides with a suitable layer of packing between each wheel.

**3.** Thin organic (resinoid and rubber) bonded wheels should be laid flat on a horizontal surface, away from excessive heat, to prevent warping.

**4.** Small cup and small internal grinding wheels may be stored separately in boxes, bins, or drawers.

## INSPECTION OF WHEELS

After wheels have been received, they should be inspected to see that they have not been damaged in transit.

For further assurance that wheels have not been damaged, they should be suspended and tapped lightly with a screwdriver handle for small wheels (Fig. 61-12) or with a wooden mallet for larger wheels. If vitrified or silicate wheels are sound, they give a clear, metallic ring. Organic-bonded wheels give a duller ring, and cracked wheels do not produce a ring. Wheels must be dry and free of sawdust before testing; otherwise the sound will be deadened.

## REVIEW QUESTIONS

**1.** Define an abrasive grain.

**2.** List four abrasive products.

**3.** Name four natural abrasives.

**4.** Name and state the purpose of four manufactured abrasives.

**5.** For what purpose are coated abrasives used?

**6.** Name two material components and two physical characteristics of a grinding wheel.

**7.** Name the three most common bonds used to manufacture grinding wheels.

**8.** How is the grade of a wheel indicated?

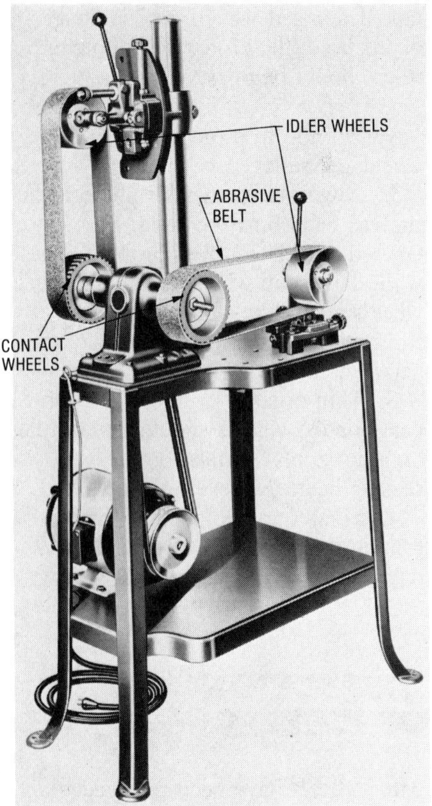


**Fig. 62-1** An abrasive belt grinder with a horizontal and a vertical belt. (*Courtesy Walker-Turner Division Rockwell Manufacturing Co.*)

N-1 Use a coarse grit abrasive belt for rough grinding, a fine grit for finishing.

N-2 Always check the abrasive belt for tension and alignment over the pulleys before starting the machine.

**9.** List five shapes of grinding wheels which can be used on surface grinders.
**10.** Identify the following grinding wheel markings: 57A 60 H 7V.
**11.** Why is it advisable to inspect each grinding wheel before use?

# UNIT 62

# GRINDING MACHINES

Grinding is one of the fastest-growing areas in the machine trade. Improved grinding-machine construction has permitted the production of parts to extremely fine tolerances with improved surface finishes and accuracy. Because of the dimensional accuracy obtained by grinding, interchangeable manufacture has become commonplace in most industries.

Grinding has also, in many cases, eliminated the need for conventional machining. Often the rough part is finished in one grinding operation, thus eliminating the need for other machining processes. The role of grinding machines has changed over the years; initially they were used on hardened work and for truing hardened parts which had been distorted by heat treating. Today, grinding is applied extensively to the production of unhardened parts where high accuracy and surface finish are required. In many cases, modern grinding machines permit the manufacture of intricate parts faster and more accurately than other machining methods.

## OBJECTIVES

After completing this unit, you should be able to:

**1.** Know the purpose and applications of abrasive belt grinders
**2.** State six general grinding wheel rules which apply to all grinding processes
**3.** Name and state the purpose of four types of surface grinders
**4.** Recognize the type of work which can be performed on cylindrical, centerless, and cutter and tool grinders.

## ABRASIVE BELT GRINDER

Abrasive belt grinders have provided industry with a fast, easy, and economical method of finishing flat or contour work. In most cases, this machine has replaced the old hand method of using a file and abrasive cloth. Work that would take hours to finish by hand can be finished in a few minutes on an abrasive belt grinder.

The abrasive belt grinder, whether horizontal or vertical, consists of a motor, a contact wheel, an idler wheel, and an endless abrasive belt (Fig. 62-1).

### Abrasive Belts

*Aluminum oxide* and *silicon carbide* abrasive belts are used on belt grinders. Aluminum oxide abrasive belts should be used for grinding high-tensile-strength materials (all steels and tough bronze). Silicon carbide abrasive belts should be used for grinding materials with low-tensile strength (cast iron, aluminum, brass, copper, glass, plastic). A 60 to 80 grit belt may be used for general-purpose work. For fine finishes a 120 to 220 grit belt is recommended. (See N-1.)

### Safety Precautions

**1.** Always wear eye protectors when grinding.
**2.** Run the abrasive belts in the direction indicated by the arrows stamped on their backs. (See N-2.)
**3.** Never grind on the up side of an abrasive belt (i.e., with the belt rotating toward rather than away from you).
**4.** If much grinding or polishing is required on a workpiece, cool it frequently in a suitable medium.
**5.** Sharp corners or edges should be brought in contact with the belt lightly, otherwise these rough edges will tear the belt.

### Finishing Flat Surfaces

Whenever flat surfaces are to be finished, a hard, flat platen (Fig. 62-2)

Fig. 62-2 A platen under the abrasive belt allows flat surfaces to be finished on an abrasive belt grinder. (*Courtesy The Norton Company.*)

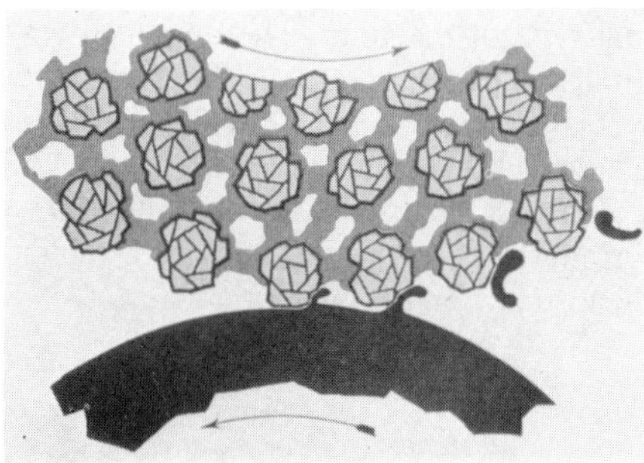

Fig. 62-3 The cutting action of an abrasive grain. (*Courtesy The Carborundum Co.*)

N-3 Grind only on the face of a straight wheel; use disk-shaped wheels for side grinding.

N-4 Most surface grinding in toolrooms is performed on a horizontal spindle reciprocating grinder.

should be mounted on the underside of the belt. This platen will prevent the belt from giving, thereby ensuring truly flat work.

### Finishing Contour Surfaces

Concave and convex surfaces are easily finished by mounting a soft or formed contact wheel on the machine. The work is then held against the contact wheel which conforms to the shape of the part. This method makes it possible to finish intricate forms or sharp radii.

## THE GRINDING PROCESS

In the grinding process the workpiece is brought into contact with a revolving grinding wheel. Each small abrasive grain on the periphery of the wheel acts as an individual cutting tool and removes a chip of metal (Fig. 62-3). As the abrasive grains become dull, the pressure and heat created between the wheel and the workpiece cause the dull face to break away, leaving new sharp cutting edges.

Regardless of the grinding method used, whether it be cylindrical, centerless, or surface grinding, the grinding process is the same and certain general rules will apply in all cases.

**1.** Use a silicon carbide wheel for low-tensile-strength materials and an aluminum oxide wheel for high-tensile-strength materials. (See N-3.)

**2.** Use a hard wheel on soft materials and a soft wheel on hard materials.

**3.** If the wheel is too hard, increase the speed of the work or decrease the speed of the wheel to make it act as a softer wheel.

**4.** If the wheel appears too soft or wears rapidly, decrease the speed of the work or increase the speed of the wheel, but not above its recommended speed.

**5.** A glazed wheel will affect the finish, accuracy, and metal removal rate. The main causes of wheel glazing are:

(a) The wheel speed is too fast.
(b) The work speed is too slow.
(c) The wheel is too hard.
(d) The grain is too small.
(e) The structure is too dense, which causes the wheel to load.

**6.** If a wheel wears too quickly, the cause may be any of the following.

(a) The wheel is too soft.
(b) The wheel speed is too slow.
(c) The work speed is too fast.
(d) The feed rate is too great.
(e) The face of the wheel is too narrow.
(f) The surface of the work is interrupted by holes or grooves.

## SURFACE GRINDER

The surface grinder, used primarily for the grinding of flat surfaces, has become an important machine tool. It can be used to grind either hardened or unhardened workpieces to the close tolerances and high surface finishes required for many jobs.

### Types of Surface Grinders

There are four distinct types of surface grinding machines, all of which provide a means of holding the metal and bringing it into contact with the grinding wheel.

The *horizontal spindle grinder with a reciprocating table* (Fig. 62-4) is probably the most common surface grinder for toolroom work. The work is reciprocated under the grinding wheel, which is fed down to provide the desired depth of cut. Feed is obtained by a transverse movement of the table at the end of each stroke. (See N-4.)

The *horizontal spindle grinder with a rotary table* (Fig. 62-5) is often found in toolrooms for the grinding of flat circular parts. The surface pattern

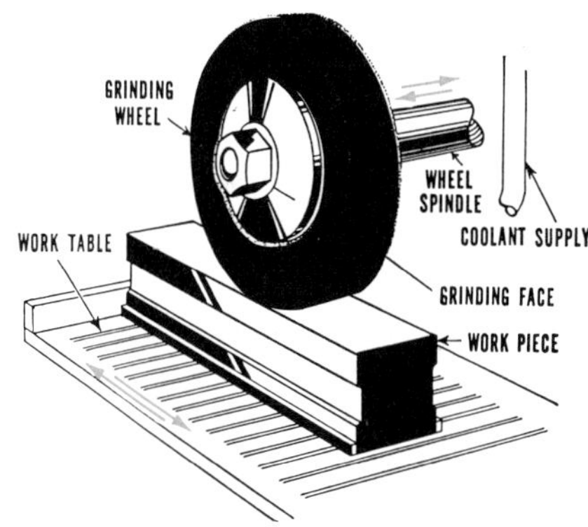


Fig. 62-4 A horizontal spindle reciprocating table surface grinder passes a workpiece under a revolving grinder wheel. (*Courtesy AVCO Bay State Abrasives.*)

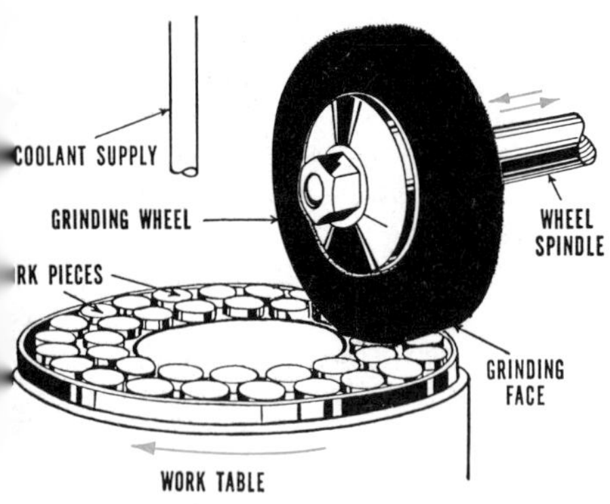

**Fig. 62-5** Horizontal spindle with a rotary table. (*Courtesy AVCO Bay State Abrasives.*)

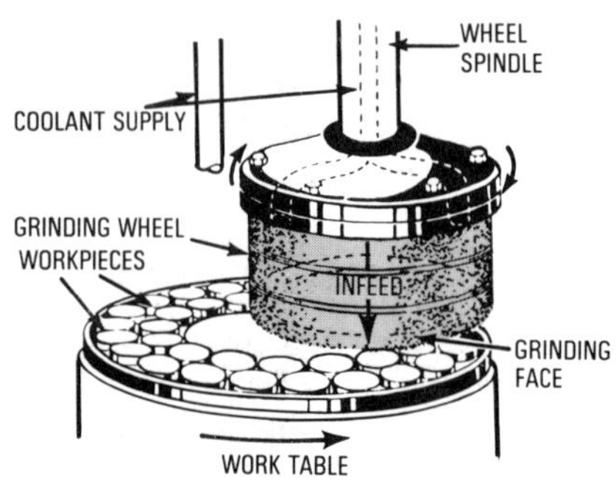

**Fig. 62-6** Vertical spindle with a rotary table. (*Courtesy The Carborundum Co.*)

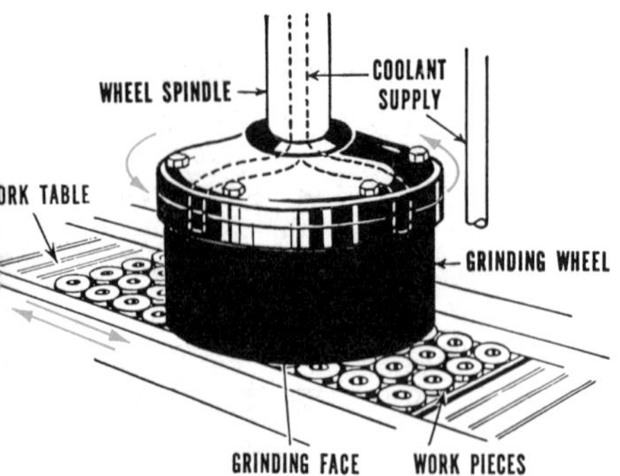

**Fig. 62-7** Vertical spindle with a reciprocating table. (*Courtesy AVCO Bay State Abrasives.*)

N-5 The surface pattern created on a horizontal spindle, rotary table grinder is useful for mating parts.

N-6 The vertical spindle, rotary table grinder is noted for high metal-removal rates.

N-7 The vertical spindle, reciprocating table grinder is capable of heavy cuts.

N-8 Center-type cylindrical grinders are commonly found in toolrooms.

it produces makes it particularly suitable for grinding parts which must rotate in contact with each other. The work is held on the magnetic chuck of a rotating table and passed under a grinding wheel. Feed is obtained by the transverse movement of the wheelhead. This type of machine permits faster grinding of circular parts since the wheel is always in contact with the workpiece. (See N-5.)

The *vertical spindle grinder with a rotary table* (Fig. 62-6) produces a finished surface by grinding with the face of the wheel rather than the periphery, as in horizontal spindle machines. (See N-6.) The surface pattern appears as a series of intersecting arcs. Vertical spindle grinders have a higher metal removal rate than the horizontal-type spindle machines. It is probably the most efficient and accurate form of grinder for the production of flat surfaces.

The *vertical spindle grinder with a reciprocating table* (Fig. 62-7) grinds on the face of the wheel while the work is moved back and forth under the wheel. Because of its vertical spindle and greater area of contact between the wheel and the work, this machine is capable of heavy cuts. (See N-7.)

## CYLINDRICAL GRINDERS

When the diameter of a workpiece must be ground accurately to size and to a high surface finish, it may be done on a *cylindrical* grinder. There are two types of machines suitable for cylindrical grinding—the *center type* and the *centerless type*—each with its special applications. (See N-8.)

### Center-Type Cylindrical Grinders

Work which is to be finished on a cylindrical grinder is generally held between centers, but it may also be held in a chuck for certain types of work.

There are two types of cylindrical grinders (center type): the *plain* and the *universal*. The plain-type grinder (Fig. 62-8) is generally a manufacturing type of machine. The universal cylindrical grinder (Fig. 62-9) is more versatile, since both the wheelhead and headstock may be swiveled.

The principal parts of a cylindrical grinder are the base, wheelhead, table, headstock, and footstock.

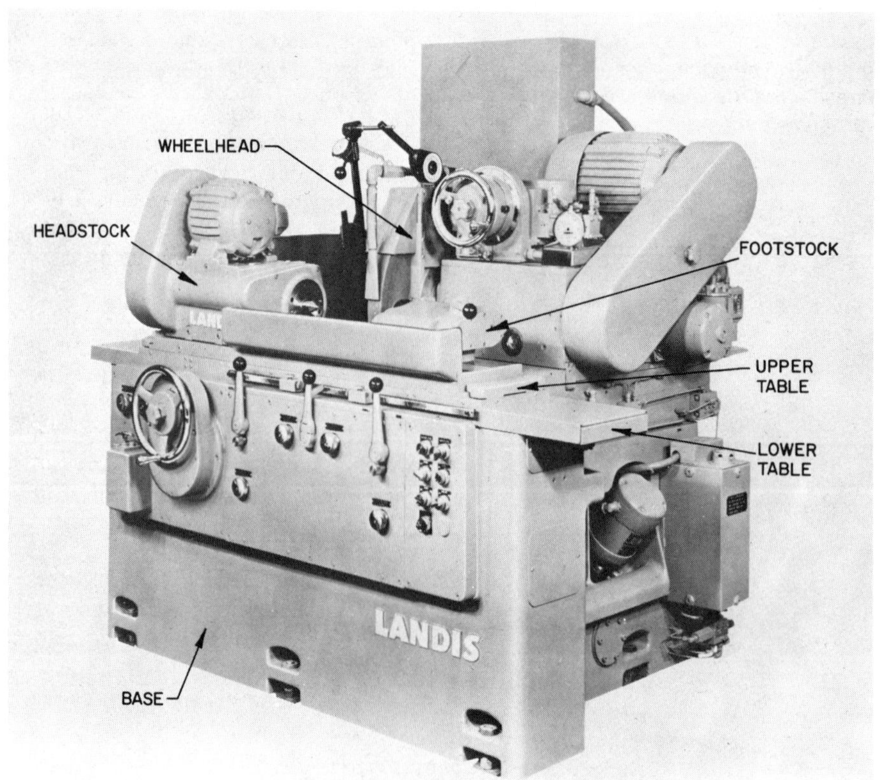

**Fig. 62-8** A plain hydraulic cylindrical grinder. (*Courtesy Landis Tool Co.*)

Fig. 62-9 A universal cylindrical grinder. (*Courtesy Cincinnati Milacron Co.*)

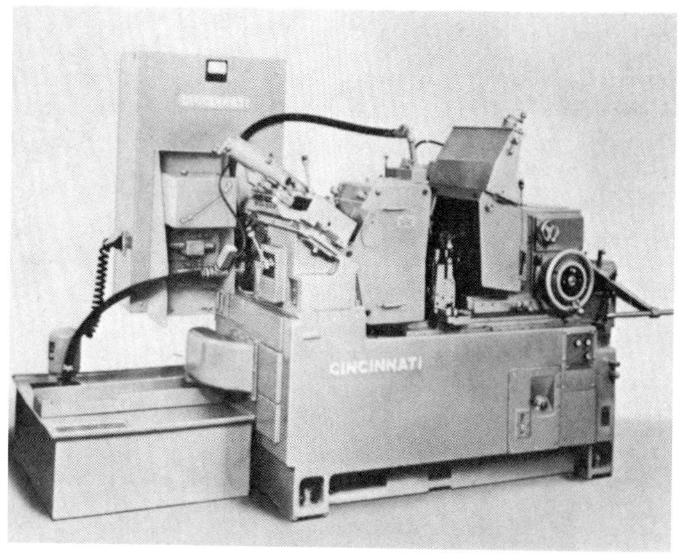

Fig. 62-10 A centerless grinder. (*Courtesy Cincinnati Milacron Co.*)

N-9 Centerless grinding is used to mass-produce cylindrical, tapered, and multidiameter workpieces.

## CENTERLESS GRINDERS

The production of cylindrical, tapered, and multi-diameter workpieces may be achieved on a centerless grinder (Fig. 62-10). As the name suggests, the work is not supported on centers but rather by a work rest blade, a regulating wheel, and a grinding wheel (Fig. 62-11).

On a centerless grinder, the work is supported on the work rest blade, which is equipped with suitable guides for the type of workpiece. The rotation of the grinding wheel forces the workpiece onto the work rest blade and against the regulating wheel, while the regulating wheel controls the speed of the work and the longitudinal feed movement. To provide longitudinal feed to the work, the regulating wheel is set at a slight angle. The rate of feed may be varied by changing the angle and the speed of the regulating wheel. The regulating and grinding wheels rotate in the same direction, and the center heights of these wheels are fixed. Because the centers are fixed, the diameter of the workpiece is controlled by the distance between the grinding and regulating wheels. The cylindrical surface is ground as the work is fed by the regulating wheel past the grinding wheel. The speed at which the work is fed across the grinding wheel is controlled by the speed and angle of the regulating wheel. (See N-9.)

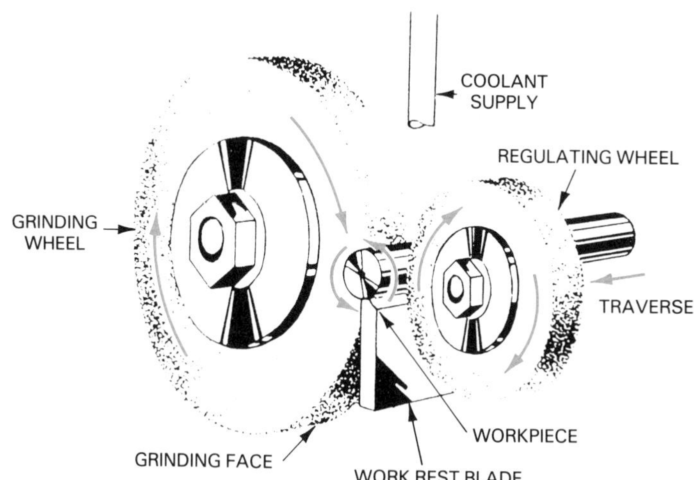


Fig. 62-11 Principle of a centerless grinder. (*Courtesy The Carborundum Co.*)

## THE UNIVERSAL CUTTER AND TOOL GRINDER

The universal cutter and tool grinder (Fig. 62-12) is designed primarily for the grinding of cutting tools such as milling cutters, reamers, and taps. Its universal feature and various attachments permit a variety of other grinding operations to be performed. Other operations which may be performed are internal (Fig. 62-13A), cylindrical (Fig. 62-13B), taper and surface grinding, single-point tool grinding (Fig. 62-13C), and cutting-off operations. Most of the latter operations require additional attachments or accessories.

Fig. 62-12 A universal cutter and tool grinder. (*Courtesy Cincinnati Milacron Co.*)

Fig. 62-13A Setup for internal grinding on a cutter and tool grinder. (*Courtesy Cincinnati Milacron Co.*)

Fig. 62-13B Setup for cylindrical grinding on a cutter and tool grinder. (*Courtesy Cincinnati Milacron Co.*)

Fig. 62-13C A surface grinding attachment being used to sharpen a carbide tool. (*Courtesy Cincinnati Milacron Co.*)

## REVIEW QUESTIONS

**1.** List three reasons why grinding is so important in the machine trade.

**2.** What two abrasives are used for belt grinders?

**3.** What grit size should be used for fine finishes?

**4.** List three safety precautions which should be observed when operating a belt grinder.

**5.** What type of grinding wheel should be used for:

(a) Low-tensile materials
(b) High-tensile materials

**6.** What should be done when the wheel acts

(a) Too hard
(b) Too soft

**7.** Name three types of surface grinders.

**8.** What is the purpose of a

(a) Plain cylindrical grinder
(b) Universal cylindrical grinder

**9.** What part of the centerless grinder supports the workpiece?

**10.** List four operations that can be done on a cutter and tool grinder.

# UNIT 63

# SURFACE GRINDING

Surface grinding is a technical term referring to the production of flat, contoured, and irregular surfaces on a piece of work which is passed under a revolving grinding wheel. The most common grinder used for surface grinding is the horizontal-spindle reciprocating-table machine. The work, usually held on a rectangular magnetic table, is moved back and forth under the abrasive wheel. The table on most surface grinders can be operated manually or automatically.

## OBJECTIVES

After completing this unit, you should be able to:

**1.** Mount and balance a grinding wheel

**2.** True and dress a grinding wheel

**3.** Grind a flat surface to within ± .0005 in. (0.01 mm) of thickness.

**4.** Grind the edges of a workpiece square and parallel

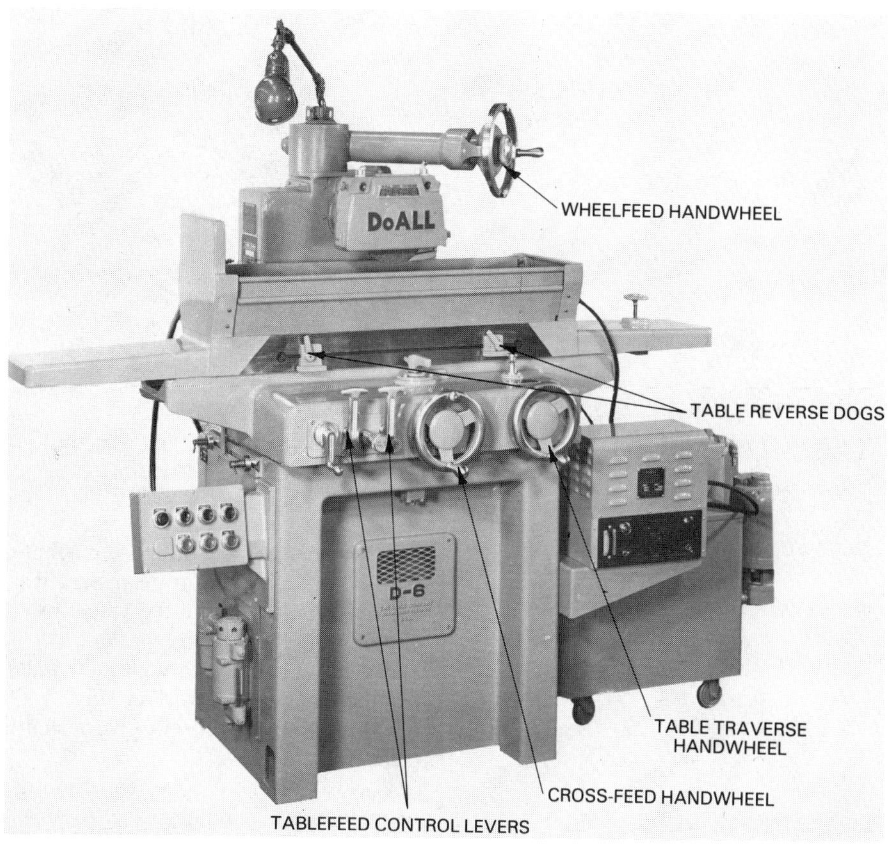


**Fig. 63-1** A hydraulic surface grinder with the main parts labeled. (*Courtesy The DoAll Company.*)

N-1 Always test a grinding wheel for cracks before mounting. It could avoid a serious accident.

# THE HORIZONTAL-SPINDLE RECIPROCATING-TABLE SURFACE GRINDER

The horizontal-spindle reciprocating-table surface grinder is the most commonly used, and will be discussed in detail. Machines of this type may be operated hydraulically (Fig. 63-1) or by hand.

## Parts of a Hydraulic Surface Grinder

The *wheel feed handwheel* moves the grinding wheel up or down to set the depth of cut.

The *table traverse handwheel* moves the table back and forth (longitudinally) under the grinding wheel. It can be operated by hand or automatically.

The *crossfeed handwheel* is used to move the table in or out (transversely). Either hand or automatic crossfeed can be used.

The *table reverse dogs* are used to regulate the length of table travel.

The *table traverse reverse lever* is used to reverse the direction of the table travel.

## Grinding Wheel Care

In order to ensure the best results in any surface grinding operation, proper care of the grinding wheel must be taken. (See N-1.)

**1.** When not in use, all grinding wheels should be properly stored.
**2.** Wheels should be tested for cracks prior to use.
**3.** Select the proper type of wheel for the job.
**4.** Grinding wheels should be properly mounted and operated at the recommended speed.

# MOUNTING A GRINDING WHEEL

A Type 1 aluminum oxide straight wheel (Fig. 63-2) is generally used for most surface grinding operations. Be-

Fig. 63-2 A type 1 straight grinding wheel. (*Courtesy The Norton Company.*)

Fig. 63-3 Testing a grinding wheel to make sure it is not cracked. (*Courtesy Kostel Enterprises Ltd.*)

fore a wheel is mounted on a machine, it should be checked to make sure it is not defective. Suspend the wheel by slipping one finger through the hole and gently tap the side with the handle of a hammer or screwdriver (Fig. 63-3). A good wheel will give a sharp, clear ring.

Care should be used when handling or mounting a grinding wheel to prevent it from being damaged. A wheel that has been misused or damaged may shatter and cause damage to the grinder; this could cause a serious accident.

N-2 Use a rag when holding a grinding wheel to avoid hand injury.

## To Mount a Wheel on a Straight Spindle

**1.** Check that the wheel is not cracked, by tapping it at four points about 90° apart with a plastic- or wooden-handled screwdriver (Fig. 63-3). A good wheel will give a sharp, clear ring.

**2.** Clean the inner flange on the machine and the hole in the grinding wheel (Fig. 63-4).

**3.** Check the wheel to make sure there is a wheel blotter on each side. If not, secure two blotters the same size as the flanges and place one on each side of the wheel.

**4.** Slide the wheel on to the grinder spindle. The wheel should go on freely without binding.

**5.** Clean and place the outer flange against the wheel.

**6.** Hold the wheel and tighten the spindle nut against the flange enough to hold the wheel firmly (Fig. 63-5). (See N-2.)

**7.** Replace the grinding wheel guard on the machine.

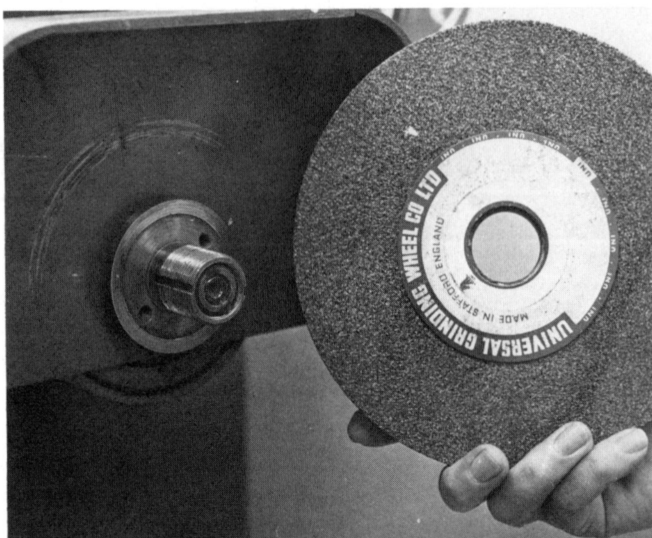


Fig. 63-4 Inspecting the bore of the wheel and the spindle for dirt or burrs. (*Courtesy Kostel Enterprises Ltd.*)

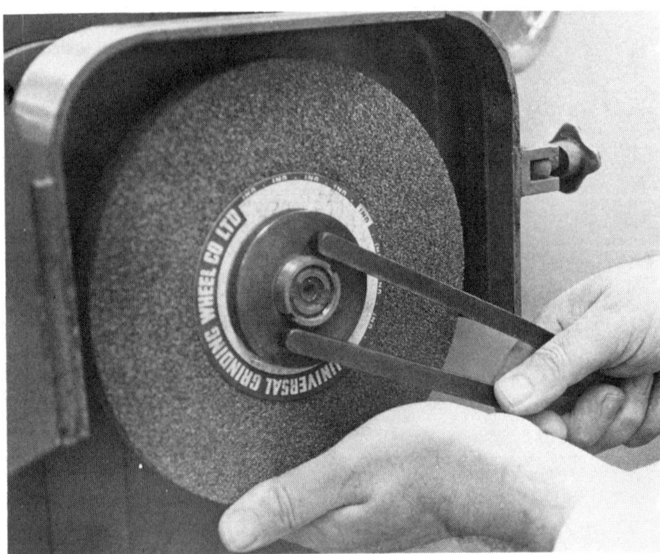


Fig. 63-5 Tightening a grinding wheel on a straight spindle. (*Courtesy Kostel Enterprises Ltd.*)

Fig. 63-6A A grinding wheel balancing stand. (*Courtesy The DoAll Company.*)

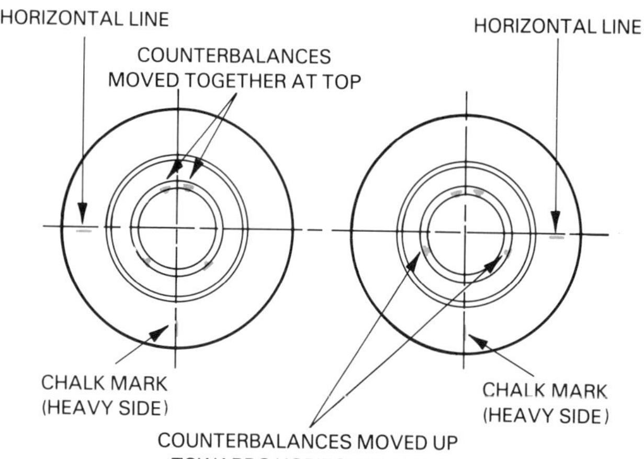


Fig. 63-6B Methods of adjusting the counterbalances to balance a grinding wheel. (*Courtesy Cincinnati Milacron Co.*)

## BALANCING A GRINDING WHEEL

Proper balance of a mounted grinding wheel is very important, since improper balance will greatly affect the surface finish and accuracy of the work. Excessive inbalance creates vibrations which will damage the spindle bearings. (See N-3.)

N-3 Balanced grinding wheels will run with a minimum of vibration and cut better.

After the wheel has been mounted on the adaptor it should be balanced, if provision is made in the adaptor for balancing.

### To Balance a Grinding Wheel

**1.** Mount the wheel and adaptor on the surface grinder and true the wheel with a diamond dresser (Fig. 63-7).

**2.** Remove the wheel assembly and mount a special tapered balancing arbor in the hole of the adaptor.

**3.** Place the wheel and arbor on a balancing stand (Fig. 63-6A) which has been leveled.

**4.** Allow the wheel to rotate until it stops. This will indicate that the heavy side is at the bottom. Mark this point with chalk.

**5.** Rotate the wheel to three positions, one-quarter, one-half, and three-quarters of a turn, to check the balance. If the wheel moves from any of these positions, it is not balanced.

**6.** Loosen the set screws in the wheel counterbalances, in the grooved recess of the flange, and move the counterbalances opposite the chalk mark (Fig. 63-6B).

**7.** Check the wheel in the four positions mentioned in steps 4 and 5.

**8.** Move the counterbalances around the groove an equal amount on each side of the chalk mark and check for balance again.

**9.** Continue to move the counterbalances away from the heavy side until the wheel remains stationary at any position.

**10.** Tighten the counterbalances in place.

### Safety Precautions

**1.** Never run a grinding wheel faster than the speed recommended on the wheel blotter. (See N-4.)

N-4 Wheel r/min $= \frac{CS \times 4}{D}$

**2.** Always have the wheel guard covering at least one-half of the grinding wheel.

**3.** Before starting a grinder, *always* make sure that the magnetic chuck has been turned on, by trying to remove the work from the chuck.

**4.** See that the wheel clears the work before starting the grinder.

**5.** Stand to one side of the wheel before starting the machine.

**6.** Never attempt to clean the magnetic chuck, or mount and remove work, until the wheel has completely stopped. (See N-5.)

N-5 Make sure that the grinding wheel is completely stopped before mounting or removing work.

**7.** *Always* wear safety goggles while grinding.

## TRUING AND DRESSING A GRINDING WHEEL

After mounting a grinding wheel, it is necessary to *true* the wheel to ensure that it will be concentric with the spindle. *Truing* a wheel is the operation of

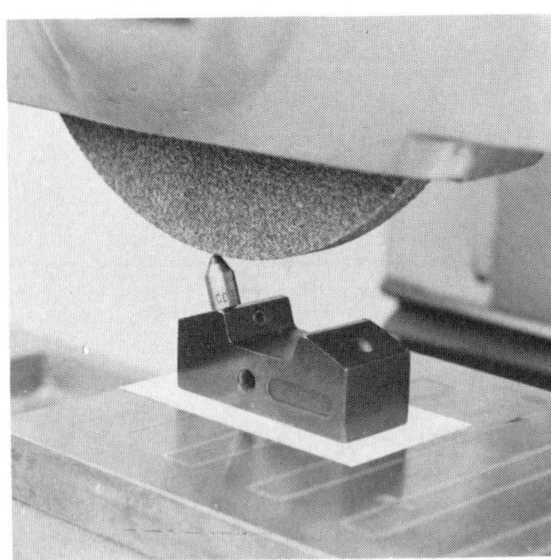

**Fig. 63-7** A diamond dresser positioned under a grinding wheel. (*Courtesy Kostel Enterprises Ltd.*)

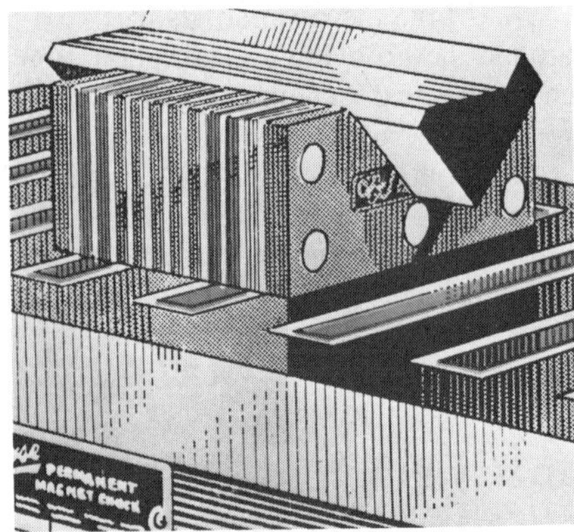

**Fig. 63-8** Work held in a magnetic V-block. (*Courtesy James Neill & Co.*)

N-6 For better wheel dressing, rotate the diamond in the holder occasionally.

N-7 Always wear safety glasses when dressing a grinding wheel.

removing any high spots on the wheel, causing it to run concentric with the spindle. Proper grinding practice requires that a wheel be trued before use.

*Dressing* is the operation of removing the dull grains and metal particles from the cutting surface of a grinding wheel. This operation exposes sharp cutting edges of the abrasive grains to make the wheel cut better. A dull or glazed wheel should be dressed for the following reasons.

**1.** To reduce the heat generated between the work surfaces and the grinding wheel.
**2.** To reduce the strain on the grinding wheel and the machine.
**3.** To improve the surface finish and accuracy of the work.
**4.** To increase the rate of metal removal.

An industrial diamond, mounted in a suitable holder on the magnetic chuck, is generally used to true and dress a grinding wheel (Fig. 63-7).

### Procedure

**1.** Check the diamond for wear and, when necessary, turn it in the holder to expose a new point. (See N-6.)
**2.** The diamond is canted in the holder at a 10° angle. This helps prevent chattering and a tendency to dig in during the dressing operation.
**3.** Clean the magnetic chuck thoroughly with a cloth and then wipe over it with the palm of the hand.
**4.** Place the diamond on the last two magnets, on the left-hand end of the magnetic chuck. Paper should be placed between the diamond holder and the chuck to prevent scratching or marring the chuck surface when removing the diamond holder.
**5.** The point of the diamond should be offset about ½ in. (12.7 mm) to the left of the grinding wheel centerline.
**6.** Make sure that the diamond clears the wheel; then start the grinder. (See N-7.)
**7.** Lower the wheel until it touches the diamond.
**8.** Move the diamond *slowly* across the face of the wheel.
**9.** Take light cuts .001 in. (0.02 mm) until the wheel is clean and sharp and is running true.
**10.** Take a finish pass .0005 in. (0.01 mm) across the face of the grinding wheel.

## WORK-HOLDING DEVICES

### The Magnetic Chuck

In some surface grinding operations the work may be held in a vise, on V-blocks (Fig. 63-8), or bolted directly to the table. However, most of the ferrous work gound on a surface grinder is held on a *magnetic chuck* which is clamped to the table of the grinder.

Magnetic chucks may be of two types: the *electromagnetic chuck* and the *permanent magnetic chuck.*

The *electromagnetic chuck* (Fig. 63-9) uses electromagnets to provide the holding power. It has the following advantages.

**1.** The holding power of the chuck may be varied to suit the area of contact and the thickness of the work.
**2.** A special switch neutralizes the residual magnetism in the chuck, permitting the work to be removed easily from the chuck.

The *permanent magnetic chuck* (Fig. 63-10) provides a convenient means of holding most workpieces to be ground. The holding power is provided by means of permanent magnets. The principle of operation is the same for both electromagnetic and permanent magnetic chucks.

### Magnetic-Chuck Accessories

Often it is not possible to hold all work on the chuck. The size, shape, and type of work will dictate how the work should be held for surface grinding. The holding power of a magnetic chuck is dependent on the size of the workpiece, the area of contact, and the thickness of the workpiece. A highly finished piece will be held better than a poorly machined workpiece.

Very thin workpieces will not be held too securely on the face of a magnetic chuck because there are too few magnetic lines of force entering the workpiece.

An *adapter plate* (Fig. 63-11) is used to securely hold thin work [less than ¼ in. (6.35 mm)]. The plate's alternate layers of steel and brass convert the wider pole spacing of the chuck to finer spacing with more but *weaker* flux paths. This method is particularly suited to small, thin pieces and reduces

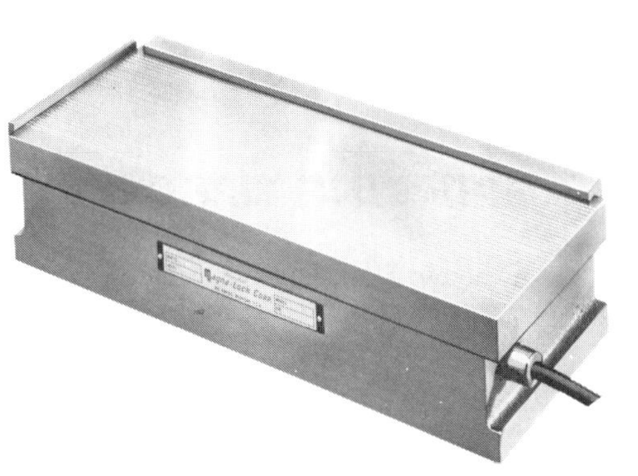

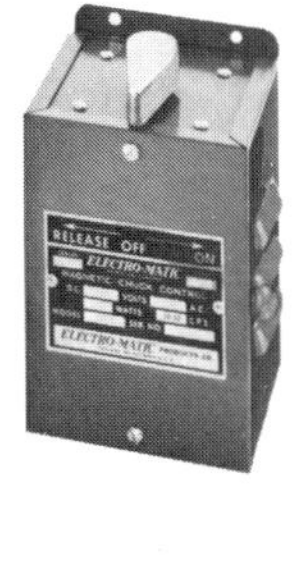

Fig. 63-9 The holding power of an electromagnetic chuck may be varied to suit the work. (*Courtesy Magna-Lock Corp.*)

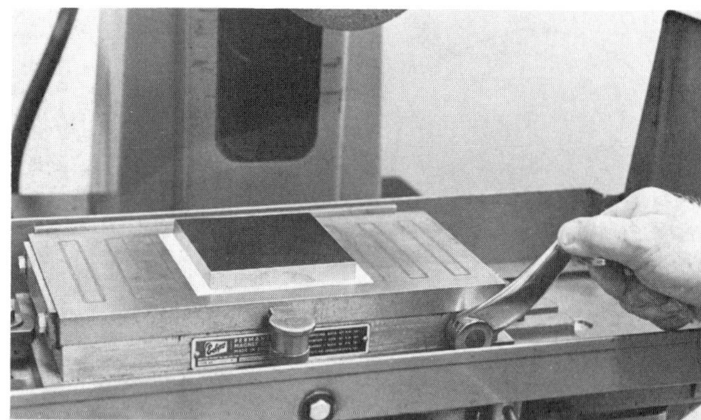

Fig. 63-10 With the handle in the "ON" position, the lines of flux pass through the work to hold it on the chuck face. (*Courtesy Kostel Enterprises Ltd.*)

Fig. 63-11 Thin work held on an adapter plate. (*Courtesy Kostel Enterprises Ltd.*)

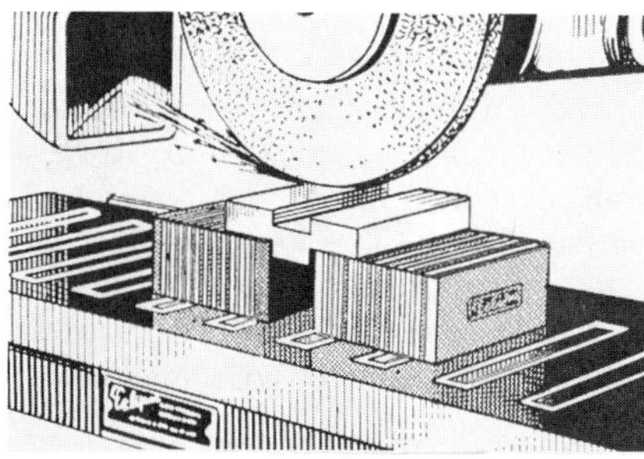

Fig. 63-12 Magnetic chuck blocks can be used to hold small thin pieces for grinding. (*Courtesy James Neill & Co. (Sheffield) Ltd.*)

N-8 Paper between the work and magnetic chuck prevents marring the chuck face when work is removed.

the possibility of distortion when grinding thin work.

*Magnetic chuck blocks* (Figs. 63-12) provide a means of extending the flux paths to hold workpieces that cannot be held securely on the chuck face. V-blocks may be used to hold round or square stock for *light* grinding.

## GRINDING A FLAT SURFACE

The most common operation performed on a surface grinder is that of grinding a flat or horizontal surface. To obtain the best results, the correct type of wheel properly dressed should be used.

### To Grind a Flat (Horizontal) Surface

**1.** Mount the work on a clean chuck, placing a piece of paper between the chuck and the workpiece. (See N-8.)

**2.** Check to see that the work is held firmly (Fig. 63-13).

**3.** Set the table reverse dogs so that the center of the grinding wheel clears each end of the work by approximately 1 in. (25.4 mm) (Fig. 63-14).

**4.** Set the crossfeed for the type of grinding operation—roughing cuts .030 to .050 in. (0.75 to 1.25 mm); finishing cuts .005 to .020 in. (0.15 to 0.50 mm).

**5.** Bring the work under the grinding wheel by hand, *having about 1/8 in. (3.17 mmm) of the wheel over the work* (Fig. 63-15).

**6.** Start the grinder and lower the wheelhead until the wheel just sparks the work.

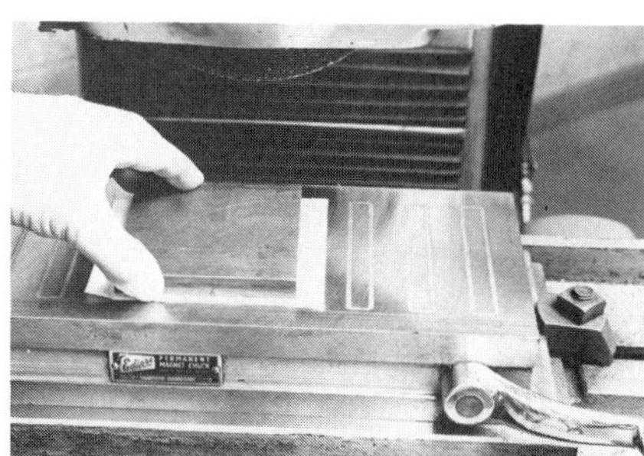

Fig. 63-13 As a safety precaution, after the chuck has been energized, try to remove the workpiece.(*Courtesy Kostel Enterprises Ltd.*)

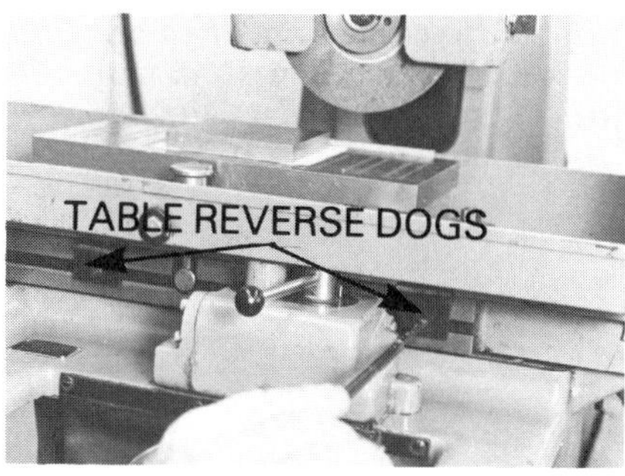


Fig. 63-14 Setting the reverse dogs to obtain the length of table travel. (*Courtesy Kostel Enterprises Ltd.*)

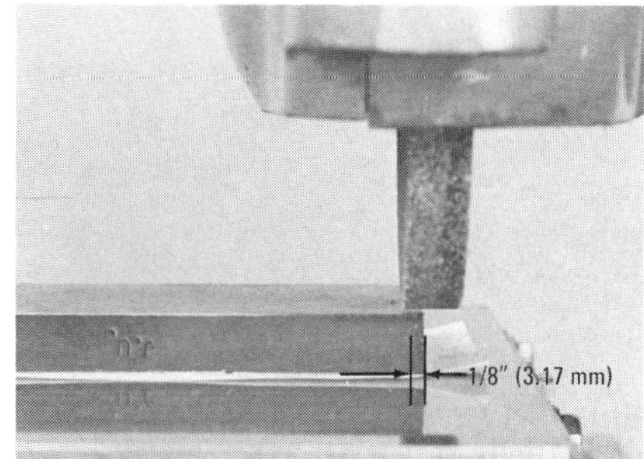


Fig. 63-15 Setting the grinding wheel over 1/8 inch (3.17 mm) of the workpiece. (*Courtesy Kostel Enterprises Ltd.*)

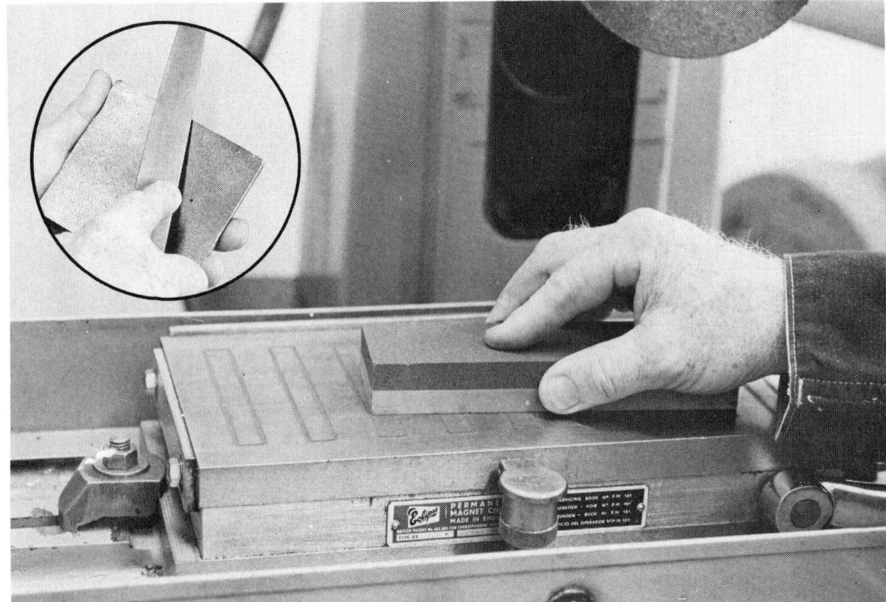

**Fig. 63-16** Removing burrs from the surface of the magnetic chuck with a honing stone before the workpiece is mounted. (*Courtesy Kostel Enterprises Ltd.*)

N-9 Turn the wheelfeed handle one-half turn backward and then bring it back to within .005 in. (0.15 mm) of the original setting.

7. The wheel may have been set on a low spot of the work. (See N-9.) It is good practice, therefore, to always raise the wheel about .005 in. (0.15 mm).

8. Start the table traveling automatically and feed the entire width of the work under the wheel to check for high spots.

9. Lower the wheel for every cut until the surface is completed. Roughing cuts .005 to .015 in. (0.15 to 0.40 mm), finishing cuts .001 to .002 in. (0.02 to 0.05 mm).

10. Release the magnet and remove the workpiece, by raising one edge, to break the magnetic attraction. This will prevent scratching the chuck surface.

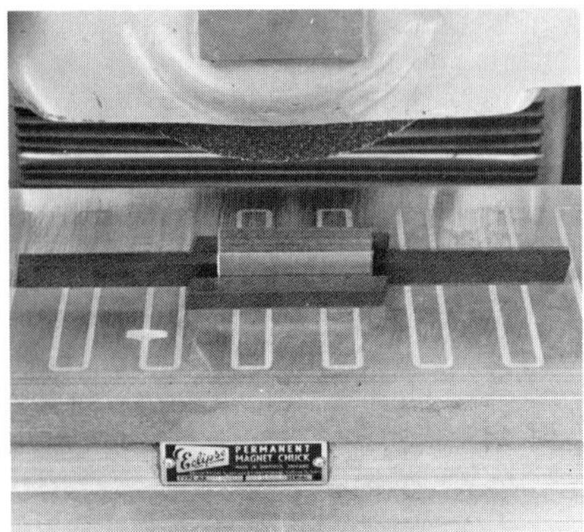

**Fig. 63-17** Steel blocks or parallels are placed around short work to prevent it from moving during grinding. (*Courtesy Kostel Enterprises Ltd.*)

### Securing Short Workpieces on a Magnetic Chuck

1. Remove all burrs from the workpiece and chuck face (Fig. 63-16).

2. Thoroughly clean the face of the magnetic chuck.

3. Place a clean piece of paper on the magnetic chuck.

4. Mount the workpiece(s) and straddle as many inserts as possible.

5. Place steel parallels or blocks that are thinner than the workpiece against the edges of the work to prevent it from moving under the grinding force (Fig. 63-17).

6. Move the magnetic chuck handle into the "ON" position (Fig. 63-10).

7. Check to see that the work is held securely (Fig. 63-13).

## GRINDING FLUIDS

Although work is ground dry in many cases, most machines have provision for applying grinding fluids or coolants. Grinding fluids serve four purposes.

1. *Reduction of grinding heat,* which affects work accuracy, surface finish, and wheel wear.

2. *Lubrication* of the surface between the workpiece and the grinding wheel, which results in a better surface finish.

3. *Removal of swarf* (small metal chips and abrasive grains) from the cutting area.

4. *Control of grinding dust,* which may present a health hazard.

### Grinding the Edges of a Workpiece

Much work machined on a surface grinder must have the edges ground square and parallel so that these edges may be used for further layout or machining operations.

Work to be ground all over should be machined to about .010 in. (0.25 mm) over the finished size for each surface that must be ground. The large, flat surfaces are usually ground first, which then permits them to be used as reference surfaces for additional setups.

When the four edges of a workpiece must be ground, clamp the work to an angle plate so that two adjacent sides may be ground square without moving the workpiece.

### Setting Up the Workpiece

1. Clean and remove all burrs from the workpiece, the angle plate, and the magnetic chuck (Fig. 63-16).

2. Place a piece of paper which is slightly larger than the angle plate on the magnetic chuck.

3. Place one end of the angle plate on the paper (Fig. 63-18).

4. Place a flat-ground surface of the workpiece against the angle plate so that the top and one edge of the workpiece project about ½ in. (12.7 mm)

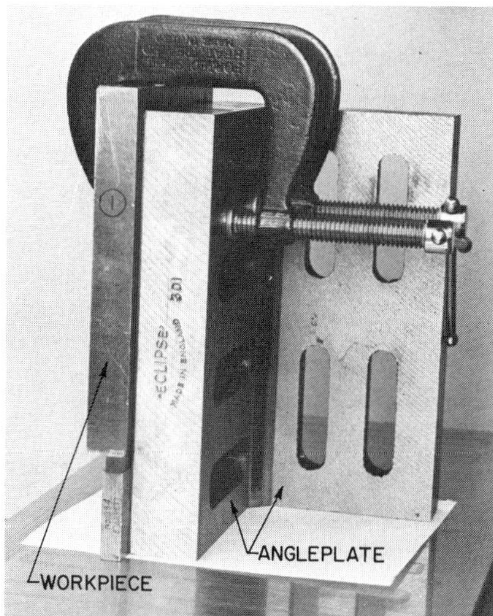

**Fig. 63-18** The workpiece may be clamped to an angle plate for grinding the edges. (*Courtesy Kostel Enterprises Ltd.*)

**Fig. 63-19** The workpiece is set up to grind the first edge. (*Courtesy Kostel Enterprises Ltd.*)

N-10 Be sure that one edge of the work does not extend beyond the base of the angle plate.

N-11 Turn the wheelfeed handle one-half turn backward and then bring it back to within .005 in. (0.15 mm) of the original setting.

beyond the edges of the angle plate (Fig. 63-18). (See N-10.)

**5.** Hold the work firmly against the angle plate and turn on the magnetic chuck.

**6.** Clamp the work securely to the angle plate and set the clamps so that they will not interfere with the grinding operation.

**7.** Turn off the magnetic chuck and carefully place the base of the angle plate on the magnetic chuck (Fig. 63-19).

**8.** Carefully place two more clamps on the end of the workpiece to hold the work securely.

## Grinding the Edges of a Workpiece Square and Parallel

After the work has been properly set up on the magnetic chuck, the following procedure should be followed for grinding the four edges of the workpiece.

**1.** Raise the wheelhead so that it is about ½ in. (12.7 mm) above the top of the work.

**2.** Set the table reverse dogs so that each end of the work clears the grinding wheel by about 1 in. (25.4 mm).

**3.** With the work under the center of the wheel, turn the crossfeed handle until *about ⅛ in. (3.17 mm) of the wheel overlaps the edge of the work* (Fig. 63-15).

**4.** Start the grinding wheel and lower the wheelhead until the wheel just sparks the work.

**5.** Move the work clear of the wheel with the crossfeed handle.

**6.** Raise the wheel about .005 to .010 in. (0.10 to 0.25 mm) in case the wheel had been set to a low spot on the work. (See N-11.)

**7.** Check for further high spots by feeding the table by hand so that the entire length of the work passes under the wheel. Raise the wheel if necessary.

**8.** Engage the table reverse lever and grind the surface until all marks are removed. The depth cut should be .003 to .007 in. (0.07 to 0.17 mm) for roughing cuts and .0005 to .001 in. (0.01 to 0.02 mm) for finishing cuts.

**9.** Stop the machine and remove the clamp from the right-hand end of the work.

**Fig. 63-20** Angle plate and work set for grinding the second edge of the workpiece at 90° to the first edge. (*Courtesy Kostel Enterprises Ltd.*)

**10.** Turn off the magnetic chuck and remove the angle plate and workpiece as one unit. Be careful not to jar the work setup.

**11.** Clean the chuck and the angle plate.

**12.** Place the angle plate (with the attached workpiece) on its end so that the surface to be ground is at the top (Fig. 63-20).

**13.** Fasten two clamps to the right-hand side of the workpiece and the angle plate.

**14.** Remove the original clamps from the top of the setup.

**15.** Repeat steps 1 to 8 and grind the second edge.

**16.** Remove the assembly from the chuck and remove the workpiece from the angle plate.

## Grinding the Third and Fourth Edges

When two adjacent sides have been ground, they are then used as reference surfaces to grind the other two sides square and parallel.

**1.** Clean the workpiece, the angle plate, and the magnetic chuck thoroughly and remove any burrs.

**2.** Place a clean piece of paper on the magnetic chuck.

**3.** Place a ground edge of the workpiece on the paper (Fig. 63-21).

(a) If the workpiece is at least 1 in. (25.4 mm) thick and long

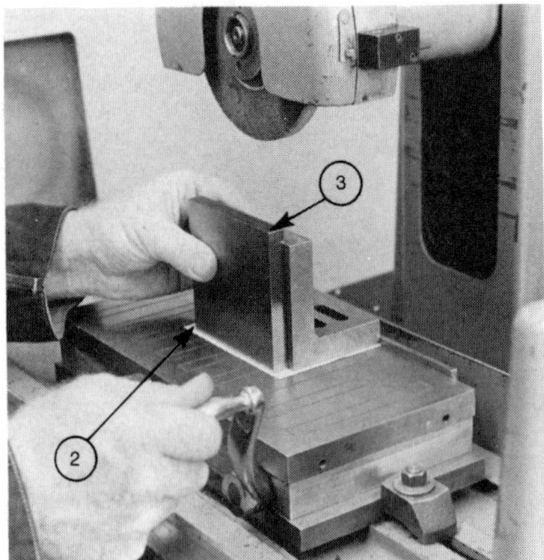


**Fig. 63-21** Place the work against the angle plate with side 2 down on the paper. (*Courtesy Kostel Enterprises Ltd.*)

**Fig. 63-22** Work which has sufficient bearing surface may be set on the chuck for finishing the remaining edges. (*Courtesy Kostel Enterprises Ltd.*)

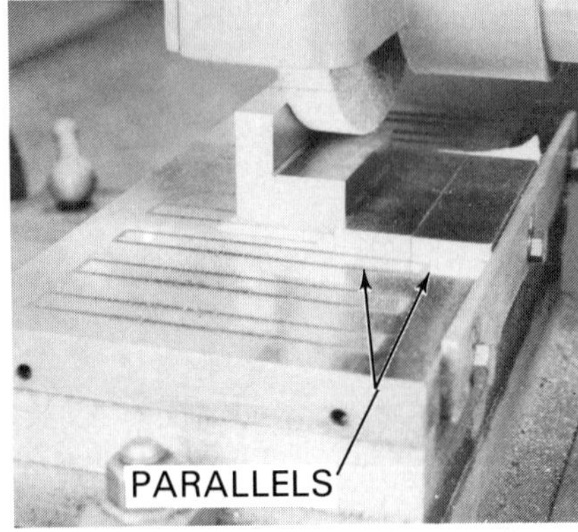


**Fig. 63-23** Setup to grind a vertical surface.

N-12 Always try to remove the piece from the magnetic chuck before grinding.

**Fig. 63-24A** Work may be set to a sine bar when an angle must be ground accurately. (*Courtesy Kostel Enterprises Ltd.*)

enough to span three magnetic poles on the chuck, no angle plate is required (Fig. 63-22).

(b) If the work is less than 1 in. (25.4 mm) thick and does not span three magnetic poles, it should be fastened to an angle plate.

i. Place a ground edge on the paper and place an angle plate no higher than the workpiece against the workpiece.

ii. Turn on the chuck and carefully clamp the work to the angle plate. (See N-12.)

**4.** Grind the third edge to the required size.

**5.** Repeat operations 1 to 3 and grind the fourth edge.

## To Grind a Vertical Surface

Although most grinding performed on a surface grinder is the grinding of flat horizontal surfaces, it is often necessary to grind a vertical surface (Fig. 63-23). Extreme care must be taken in the setup of the workpiece when grinding the vertical surface. Before grinding a vertical surface, it is necessary to relieve the corner of the work to ensure clearance for the edge of the wheel.

## To Grind an Angular Surface

When it is necessary to grind an angular surface, the work may be held at an angle by a sine bar and angle plate (Fig. 63-24A), a sine chuck (Fig. 63-24B), or an adjustable angle vise (Fig. 63-24C). When work is held by any of these methods, grinding is done with a flat dressed wheel.

**Fig. 63-24B** Grinding an angle using a sine chuck.

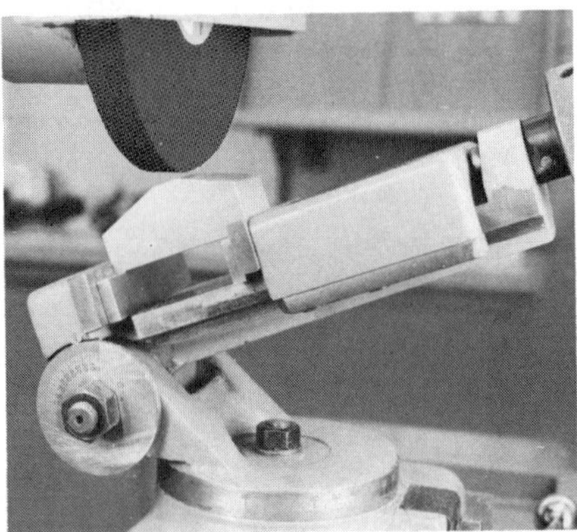

**Fig. 63-24C** Work held in an adjustable angle vise.

# REVIEW QUESTIONS

**1.** What is the purpose of the following surface grinder parts?
   (a) Wheelfeed handwheel
   (b) Table traverse handwheel
   (c) Table reverse dogs

**2.** Why should a grinding wheel be tested for cracks prior to use?

**3.** What effects would an improperly balanced grinding wheel have?

**4.** What precaution should be taken to ensure that the magnetic chuck has been turned on?

**5.** Define truing and dressing.

**6.** How should the diamond be positioned for dressing a surface grinder wheel?

**7.** How deep should cuts be taken when dressing a grinding wheel?

**8.** On what principle do permanent magnetic and electromagnetic chucks operate?

**9.** What can be used to hold thin workpieces for grinding?

**10.** Why should paper be used between the work and magnetic chuck?

**11.** How should a grinding wheel be set to the work surface?

**12.** Name four purposes of grinding fluids.

**13.** How much should be left on a surface for grinding?

**14.** How can the work be set up for grinding the first two edges square?

# UNIT 64

# CYLINDRICAL GRINDING

Cylindrical grinding is used to grind the external or internal diameter of a rigidly supported and rotating workpiece. Although the term cylindrical grinding may also be applied to centerless grinding, it generally refers to work which is ground in a chuck or between supporting centers. Cylindrical grinders can be used to grind all types of hard or soft workpieces to a high degree of accuracy and very-fine surface finishes. In many cases, work which was previously machined on a lathe can now be roughed out and finish ground to eliminate a machining operation.

## OBJECTIVES

After completing this unit, you should be able to:

**1.** Mount, balance, and dress a cylindrical grinder wheel

**2.** Set up and grind external diameters to within ±.001 in. (0.02 mm) of size

**3.** Set up and grind internal diameters to within ±.001 in. (0.02 mm) of size

**4.** Grind in a lathe with a toolpost grinder

## THE UNIVERSAL CYLINDRICAL GRINDER

The universal cylindrical grinder (Fig. 64-1) is the most widely used cylindrical grinder in school shops and toolrooms. These machines can perform a wide variety of external grinding operations such as on straight (cylindrical) and tapered surfaces, shoulders, fillets, and contours. They can also handle internal grinding of straight and tapered holes, face grinding, and some cutter grinding operations.

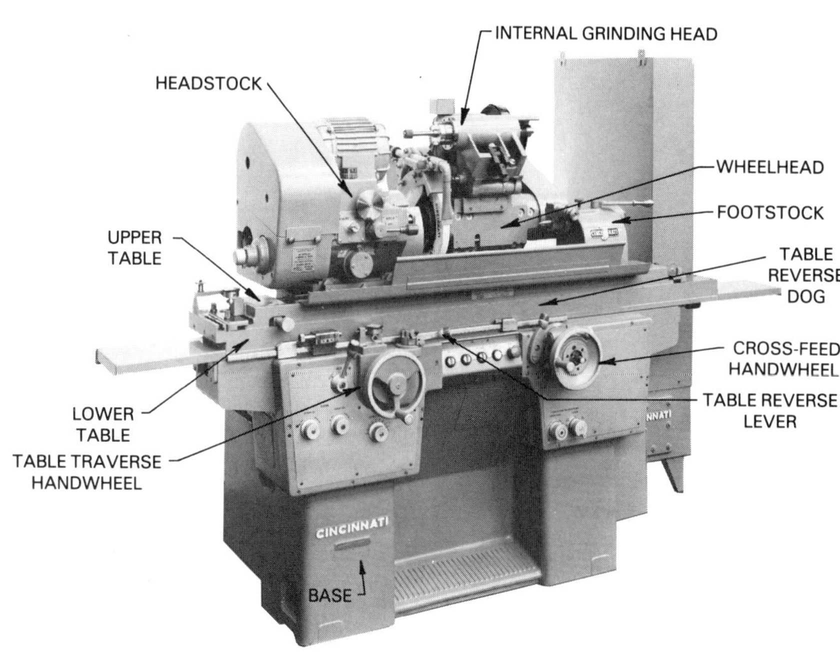


**Fig. 64-1** The main parts of a universal cylindrical grinder. (*Courtesy Cincinnati Milacron Co.*)

**Fig. 64-2** The wheelhead swiveled for grinding a steep taper on work between centers. (*Courtesy Cincinnati Milacron Co.*)

**Fig. 64-3** The grinder set up for internal grinding. (*Courtesy Cincinnati Milacron Co.*)

## PARTS OF THE UNIVERSAL CYLINDRICAL GRINDER

The *base* is of heavy cast-iron construction to provide rigidity. The top of the base is machined to form the *ways* for the table.

The *wheelhead* is mounted on a cross slide at the back of the machine. The ways on which it is mounted are at right angles to the table ways, permitting the wheelhead to be fed toward the table and the work to be fed either automatically or by hand. On universal machines, the wheelhead may be swiveled to permit the grinding of steep tapers by plunge grinding (Fig. 64-2).

The *table,* mounted on the ways, is driven back and forth by hydraulic or mechanical means. The reversal of the table is controlled by *trip dogs.* The table is composed of the *lower table,* which rests on the ways; and the *upper table,* which may be swiveled for grinding tapers and alignment purposes.

The *headstock* is mounted on the left end of the table and contains a motor for rotating the work. A dead center is mounted in the headstock spindle. When work is mounted between centers, it is rotated on *two dead centers* by means of a dog and a drive plate which is attached to, and which revolves with, the headstock spindle. (See N-1.) The purpose of a dead center in the headstock is to overcome any spindle inaccuracies (looseness, burrs, etc.) which may be transferred to the workpiece. Grinding work on two dead centers results in truer diameters which are concentric with the centerline of the work. Work may also be held in a chuck, which is mounted on the spindle nose of the headstock.

N-1 The headstock can be swiveled to grind steep tapers.

The *footstock* supports the right end of the work and is adjustable along the length of the table. The dead center, on which the work is mounted, is spring-loaded to provide the proper center tension on the workpiece.

An *internal grinding attachment* (Fig. 64-3) may be mounted on the wheelhead on most machines for internal grinding. It is usually driven by a separate motor.

## PREPARING THE GRINDER

Since the cylindrical grinder is a precision machine, it is important to prepare it properly. Factors such as selecting the grinding wheel, bal-

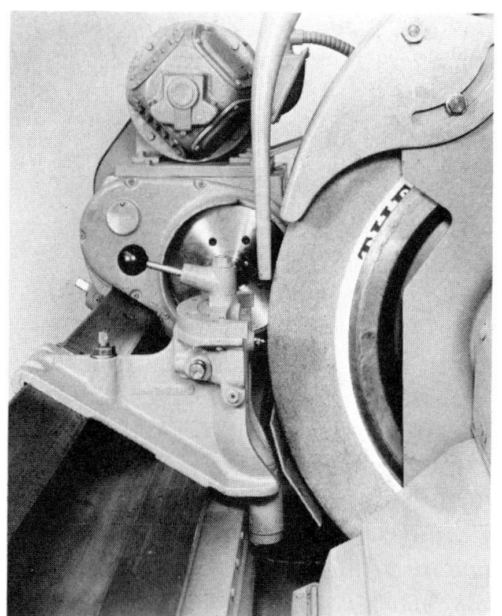

Fig. 64-4 Dressing a grinding wheel with the diamond mounted on the table. (*Courtesy Cincinnati Milacron Co.*)

N-2 As the wheel gets smaller, increase the spindle speed to maintain 6500 surface feet per minute.

N-3 Use a good flow of coolant when grinding.

N-4 True and dress a grinding wheel in order to make it run true and cut better.

N-5 Too slow a dressing feed will cause the wheel to burn the work; too fast a feed will cause a poor surface finish.

ancing and truing, wheel speeds and feeds, and using coolants are very important to the production of accurate work.

*Wheel selection* The metal-removal rate and surface finish are determined by the grinding wheel and therefore it is important to select the correct wheel for the job.

**1.** High-tensile-strength materials use aluminum oxide wheels.
**2.** Low-tensile-strength materials use silicon carbide wheels.
**3.** Rough grinding uses coarse-grain, hard-grade, open-structure wheels.
**4.** Finish grinding uses fine-grain, medium- or soft-grade, close-structure wheels.

*Wheel speed* The speed has a direct relationship to the cutting action of a grinding wheel. Most cylindrical grinding wheels are designed to operate best at 6500 surface feet per minute.

**1.** Fast wheel speed causes the wheel to act harder.
**2.** Slower wheel speed causes the wheel to act softer.

*Wheel feed* The wheel feed on cylindrical grinders should be constant for best results. Irregular or rapid feed may cause the grinding wheel to break, wear unevenly, and produce inaccurate work and poor surface finish. (See N-2.)

*Grinding fluids* Whenever possible, use grinding fluids or coolants to keep the wheel clean and reduce the amount of heat generated. Insufficient coolant will cause the grinding wheel to load, heat the work, and cause distortion and out-of-roundness. (See N-3.)

*Balancing* A wheel which is out of balance can cause excessive vibration, damage to spindle bearings, poor surface finish, and inaccurate work. See Unit 63 for the procedure to balance a grinding wheel.

*Truing and dressing* Truing is a shaping operation which forms or shapes the wheel and makes its grinding surface run concentric with the grinder spindle axis. Dressing is the process of reconditioning a wheel to make it cut better. (See N-4.)

## To True and Dress the Wheel

**1.** Start the grinding wheel to permit the spindle bearings to warm up.
**2.** Mount the proper diamond in the holder and clamp it to the table (Fig. 64-4). The diamond should be mounted at an angle of 10 to 15° to the wheelface and should be held on or slightly below the centerline of the wheel (Fig. 64-5).
**3.** Adjust the wheel until the diamond almost touches the high point of the wheel, which is usually in the center of the wheelface.
**4.** Turn on the coolant if this is to be used for the grinding operation.
**5.** Feed the wheel into the diamond about .001 in. (0.02 mm) per pass, and move it back and forth across the wheelface at a medium rate until the wheelface has been completely dressed.
**6.** Finish dress the wheel by using an infeed of .0005 in. (0.01 mm) and a slow traverse feed. When a very fine finish is required, the diamond should be traversed slowly across the wheelface for a few passes without any additional infeed. (See N-5.)

# GRINDING A PARALLEL DIAMETER

Care must be exercised in preparing and setting up the workpiece to ensure satisfactory results. Factors such as the condition of the center holes, methods of supporting the work, work speed, and table traverse rates must be considered.

## To Parallel Grind an Outside Diameter

**1.** Lubricate the machine as required.
**2.** Start the grinding wheel to warm up the spindle bearings. This

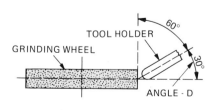

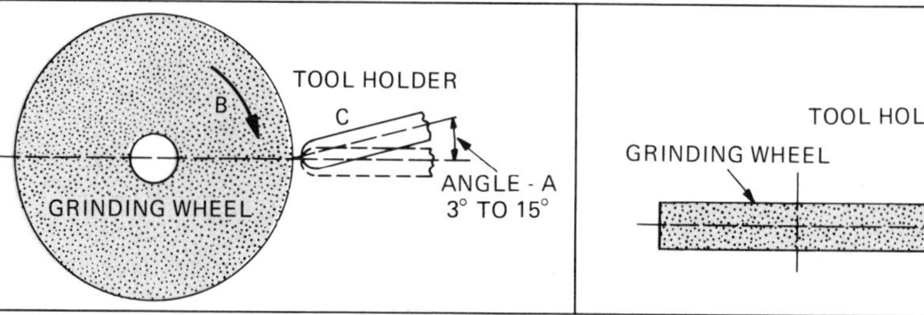


Fig. 64-5 The proper setting of a diamond dresser for cylindrical grinding wheels.

**Fig. 64-6** Dressing a grinding wheel with a diamond mounted on the footstock casting. (*Courtesy Landis Tool Co.*)

**Fig. 64-7** Work mounted between the centers on a cylindrical grinder. (*Courtesy AVCO Bay State Abrasives.*)

N-6 If the centers are not aligned, tapered work will result.

N-7 Slow work speed causes heating and distortion; high work speed makes the wheel act softer and break down quickly.

N-8 Never attempt to measure or make any adjustments to the work until the wheel and the work is stopped.

will ensure the utmost accuracy when grinding.

**3.** True and dress the grinding wheel if required. (Fig. 64-6.)

**4.** Clean the machine centers and the center holes of the work. If the grinder centers are damaged, they must be reground. On hardened steel workpieces, the center holes should be honed or lapped to ensure the utmost accuracy.

**5.** Align the headstock and footstock centers with a test bar and indicator. (See N-6.)

**6.** Lubricate the center holes with a suitable lubricant.

**7.** Set the headstock and footstock for the proper length of work so that the center of the work will be over the center of the table.

**8.** Mount the work between centers with the dog mounted loosely on the left end of the work.

**9.** Tighten the dog on the end of the work and engage the drive plate pin in the fork of the dog (Fig. 64-7).

**10.** Adjust the table dogs so that the wheel will overrun each end of the work by about one-third the width of the wheelface. If grinding must be done up to a shoulder, the table traverse must be carefully set so that it reverses just before the wheel touches the shoulder.

**11.** Set the grinder to the proper speed for the wheel being used. Some machines are provided with a means of increasing the wheel speed as the wheel becomes smaller. If this is not done, the wheel will act softer and wear quickly.

**12.** Set the work speed for the diameter and type of material being ground. Proper work speed is very important. (See N-7.)

**13.** Set the headstock spindle to rotate the work in an opposite direction to that of the grinding wheel. When grinding, the sparks should be directed down toward the table.

**14.** If the machine is so equipped, set the automatic infeed for each table reversal. Also set the dwell, or "tarry," time which permits the wheel to clear itself at each end of the stroke.

**15.** Select and set the desired table traverse or speed. This should be such that the table will move one-half to two-thirds of the wheel width per revolution of the work. Finish grinding is done at a slower rate of table traverse.

**16.** Start the table and move the wheel up to the revolving workpiece until it just sparks.

**17.** Engage the wheel-feed clutch lever and grind until the work is cleaned up.

**18.** Check the work for taper and adjust the table if necessary.

**19.** Determine the amount of material to be removed and set the feed index (if the machine is so equipped) for this amount. The infeed of the wheel will stop automatically when the work is at the proper diameter.

**20.** Stop the machine with the wheel clear of the work and measure the size of the workpiece. If necessary, make a correction on the index setting and grind the piece to size. (See N-8.)

## INTERNAL GRINDING ON A UNIVERSAL CYLINDRICAL GRINDER

Although the universal cylindrical grinder is not designed primarily as an internal grinder, it is used extensively in toolrooms for this purpose. On most universal cylindrical grinders, the internal grinding attachment is mounted on the wheelhead column and is easily swung into place when required. One advantage of this machine is that the outside and inside diameters of a workpiece may often be finished in one setup (Fig. 64-8). Although the grade of the wheel used for internal grinding will depend on the type of work and the rigidity of the machine, the wheels used for internal grinding are generally softer than those for external grinding, for the following reasons.

**1.** There is a larger area of contact between the wheel and the workpiece during the internal grinding operation.

**2.** A soft wheel requires less pressure to cut than a hard wheel; thus the spindle pressure and spring is reduced.

### To Grind a Parallel Internal Diameter on a Universal Cylindrical Grinder

**1.** Mount the workpiece in a universal chuck or a collet chuck or on a face plate. Care must be taken not to distort thin workpieces.

**Fig. 64-8** External and internal diameters can be ground in one setup on a universal cylindrical grinder. (*Courtesy Landis Tool Co.*)

**Fig. 64-9** A toolpost grinder mounted on a lathe for cylindrical grinding.

N-9 For the best grinding results, use the largest spindle and the shortest overhang possible.

N-10 Prevent bell-mouthing the hole by making sure that no more than half the wheel width overlaps the end of the work.

N-11 Grinding dust can get into the machine slides and cause rapid wear.

N-12 Always wear safety glasses when performing any grinding operation.

N-13 Heavy cuts cause work to heat quickly and distort.

**2.** Swing the internal grinding attachment into place and mount the proper spindle in the quill. (See N-9.)

**3.** Mount the proper grinding wheel (as large as possible) for the job.

**4.** Adjust the spindle height until its center is in line with the center axis of the hole in the workpiece.

**5.** True and dress the grinding wheel.

**6.** Set the wheelspeed to 5000 to 6500 surface feet per minute (1525 to 1980 m/min).

**7.** Set the workspeed to 150 to 200 surface feet per minute (45 to 60 m/min).

**8.** Adjust the table dogs so that *only* one-third of the wheel width overlaps the ends of the work at each end of the stroke. On blind holes, the dog should be set to reverse the table just as the wheel clears the undercut at the bottom of the hole. (See N-10.)

**9.** Start the work and the grinding wheel.

**10.** Touch the grinding wheel to the diameter of the hole.

**11.** Turn on the coolant.

**12.** Grind until the hole just cleans up, feeding the wheel in no more than .002 in. (0.05 mm) per table reversal.

**13.** Check the hole size and set the automatic infeed (if the machine is so equipped) to disengage when the work is roughed to within .001 in (0.02 mm) of size.

**14.** Reset the automatic infeed to .0002 in. (0.005 mm) per table reversal.

**15.** Finish grind the work and let the wheel spark out.

**16.** Move the table longitudinally and withdraw the wheel and spindle from the workpiece.

**17.** Check the hole diameter and finish grind if necessary.

## GRINDING ON A LATHE

Cylindrical and internal grinding may be done on a lathe if a proper grinding machine is not available. A toolpost grinder, (Fig. 64-9) mounted on a lathe, may be used for cylindrical and taper grinding as well as angular grinding of lathe centers. An internal grinding attachment for the toolpost grinder permits the grinding of straight and tapered holes. Grinding should only be done on a lathe when no other machine is available, or when the cost of performing a small grinding operation on a part would not warrant setting up a regular grinding machine. Since the work should rotate in an opposite direction to the grinding wheel, the lathe must be equipped with a reversing switch.

### To Cylindrical Grind in a Lathe

**1.** Thoroughly clean the ways of the lathe.

**2.** Cover the ways with cloth or canvas to protect them from the grinding dust. (See N-11.)

**3.** Adjust the tailstock center as required for parallel or taper grinding. When grinding tapers, the taper attachment may be set as required.

**4.** Mount a toolpost grinder on the compound rest and adjust the spindle to center height.

**5.** Mount the proper grinding wheel on the grinder.

**6.** Place a small pan of water directly under the grinding wheel and workpiece to catch as much grinding dust as possible.

**7.** True and dress the grinding wheel. (See N-12.)

**8.** Set a fairly slow spindle speed (depending on the diameter of the workpiece).

**9.** Set the carriage feed to about .060 to .080 in. (1.52 to 2.03 mm), depending on the width of the wheel.

**10.** Mount the work in the lathe.

**11.** Start the lathe and be sure that the work is revolving in reverse.

**12.** Start the grinder and carefully bring it up to the revolving workpiece until it sparks lightly.

**13.** Slowly feed the carriage along the work, by hand, to remove any high spots.

**14.** Move the carriage until the wheel is opposite the right-hand end of the work.

**15.** Feed the grinder in .001 to .003 in. (0.02 to 0.07 mm) using the crossfeed. (See N-13.)

**16.** Engage the automatic feed and grind the workpiece for the required length.

**17.** At the end of the cut, reverse the feed and take a cut in the opposite direction.

**18.** Check the diameter with a micrometer. Do not measure work when it is hot.

**Fig. 64-10** The compound rest set at 30° to grind a lathe center.

N-14 Make sure the work is cool before taking the finishing cuts.

N-15 If a tapered hole is to be ground with the compound rest, set the compound rest to the proper angle before mounting the grinder.

**19.** Grind the workpiece to within .002 in. (0.05 mm) of size.

**20.** Redress the grinding wheel.

**21.** Finish grind the work. After the last cut, let the grinding wheel move back and forth across the length of the ground surface without changing the feed. This will let the wheel spark out. (See N-14.)

### To Grind a Lathe Center

**1.** Remove the chuck or drive plate from the lathe spindle.

**2.** Mount the lathe center to be ground in the headstock spindle.

**3.** Set a slow spindle speed.

**4.** Swing the compound rest to 30° (Fig. 64-10) with the centerline of the lathe.

**5.** Protect the ways of the lathe with cloth or canvas and place a pan of water below the lathe center.

**6.** Mount the toolpost grinder and adjust the center of the grinding spindle to lathe center height.

**7.** Mount the proper grinding wheel, true, and dress.

**8.** Start the lathe, with the spindle revolving in reverse.

**9.** Start the grinder and adjust the grinding wheel until it sparks lightly against the revolving center.

**10.** Lock the carriage in this position.

**11.** Feed the grinding wheel in .001 in. (0.02 mm) using the crossfeed handle.

**12.** Move the grinder along the face of the center using the compound rest feed.

**13.** Check the angle of the center using a center gage and adjust the compound rest if necessary.

**14.** Finish grind the center.

## INTERNAL GRINDING ON A LATHE

Internal grinding may be performed on the lathe using a toolpost grinder with an internal grinding attachment. Internal grinding of tapered holes may be done with the taper attachment or with the compound rest feed handle.

**1.** Mount the work in a chuck or on a faceplate.

**2.** Mount the toolpost grinder and internal attachment at center height on the compound rest. (See N-15.)

**3.** Cover the ways of the lathe with a cloth.

**4.** Start the grinding wheel, true, and dress.

**5.** Start the lathe with the spindle rotating in the same direction as for turning.

**6.** Move the grinding wheel into the hole and carefully adjust it out until it just touches the surface of the hole.

**7.** Move the crossfeed out .001 in. (0.02 mm) and take a pass across the work. Do not overlap the ends of the hole with more than half the width of the wheel to avoid bell-mouthing.

**8.** Take light cuts and continue to grind until the work is within .002 in. (0.05 mm) of size.

**9.** Take a few passes through the hole to allow the wheel to spark out.

**10.** Finish grind to size.

## REVIEW QUESTIONS

**1.** What type of cylindrical grinder is most widely used in school shops and toolrooms?

**2.** What is the purpose of the following grinder parts?

(a) Wheelhead
(b) Table
(c) Headstock
(d) Footstock

**3.** What type of wheel should be used to grind hardened tool steel?

**4.** What purpose do grinding fluids serve?

**5.** How should the diamond be set for dressing a cylindrical grinding wheel?

**6.** What would result if the headstock and footstock centers were not aligned?

**7.** How would a slow work speed affect the grinding action?

**8.** In which direction should the sparks go when grinding external diameters?

**9.** Why are internal grinding wheels generally softer than those used for external grinding?

**10.** How can bell-mouthing be prevented when internal grinding?

**11.** In which direction should the work rotate when internal grinding in a lathe?

**12.** What is the purpose of having a small pan of water under the grinding wheel?

**13.** At what angle should the compound rest be set for grinding a center in a lathe?

**14.** How should the lathe be set up for grinding a tapered hole?

# SECTION

# METALS

The study of metals, starting with the ore, progressing through manufacture, and altering structures to achieve certain qualities or properties all fall into the category of metallurgy. The field of metallurgy may be divided into three categories:

1. The mining and processing of ore into metals
2. The improvement of metals and the development of alloys to suit the needs of industry
3. The heat treatment of metals and alloys to improve their qualities

These topics will be dealt with in this book in the manner outlined above, concentrating on iron, steel, and their alloys.

# UNIT 65

# MANUFACTURE OF IRON AND STEEL

Fig. 65-1 Open-pit mining of iron ore. (*Courtesy Dominion Foundries & Steel Ltd.*)

In ancient times iron was a rare and precious metal. Today steel, a purified form of iron, has become one of our most useful servants. Nature supplied the basic raw products of iron ore, coal, and limestone, and our ingenuity has converted them into a countless number of products. Steel can be made hard enough to cut glass, pliable as the steel in a paper clip, springy as the steel in springs, or strong enough to withstand a pull of 500,000 pounds per square inch. It can be drawn into wire one thousandth of an inch thick or fabricated into giant girders for buildings and bridges. Steel can also be made resistant to heat, rust, and chemical action. Steel is truly our most versatile metal.

## OBJECTIVES

On completion of this unit you should know:

**1.** The procedure used to produce pig iron
**2.** How cast iron is produced in a cupola furnace
**3.** The construction and operation of three common furnaces used to produce steel
**4.** How steel is processed after it is produced in the furnaces

## RAW MATERIALS

The raw materials of steel making, *iron ore, coal,* and *limestone,* must be brought together, often from great distances, and smelted in a blast furnace to produce *pig iron,* which is used to make steel.

### Iron Ore

Iron ore is the chief raw material used in the manufacture of iron and steel. The main sources of iron ore in the United States are the Great Lakes regions of Michigan, Minnesota, and Wisconsin. In Canada, large ore deposits are found in the Steep Rock and Michipicoten districts on the north shore of Lake Superior and in the Ungava district near the Quebec-Labrador border.

### Mining Iron Ore

When layers of iron ore are near the earth's surface, the surface material consisting of sand, gravel, and boulders is first removed, then the iron ore is scooped up by power shovels and loaded into trucks or railway cars (Fig. 65-1). This is called *open-pit mining.* About 75 percent of the ore mined is removed by this method.

When layers of iron ore are too deep in the earth to make open-pit mining economical, *underground* or *shaft mining* is used. Shafts are sunk into the earth, and passageways are cut into the ore body. The ore is blasted loose and brought to the surface by shuttle cars or conveyor belts.

### Types of Iron Ore

Some of the most important types of iron ore are:

- *Hematite*, a rich ore containing about 70 percent iron. It ranges in color from a gray to a bright red.
- *Limonite*, a high-grade brown ore containing water which must be removed before the ore can be shipped to steel mills.
- *Magnetite*, a rich gray to black magnetic ore containing over 70 percent iron.
- *Taconite*, a low-grade ore containing only about 20 to 30 percent iron which is uneconomical to use without further treatment.

### Pelletizing Process

Low-grade iron ores are uneconomical to use in the blast furnace and as a result go through a pelletizing process where most of the rock is removed and the ore is brought to a higher iron concentration. Some steelmaking firms are now pelletizing most of their ores to reduce transportation costs and the problems of pollution and slag disposal at the steel mills. (See N-1.)

N-1 Iron ore pellets reduce transportation and waste-disposal costs.

The crude ore is crushed and ground into a powder and passed through magnetic separators where the iron

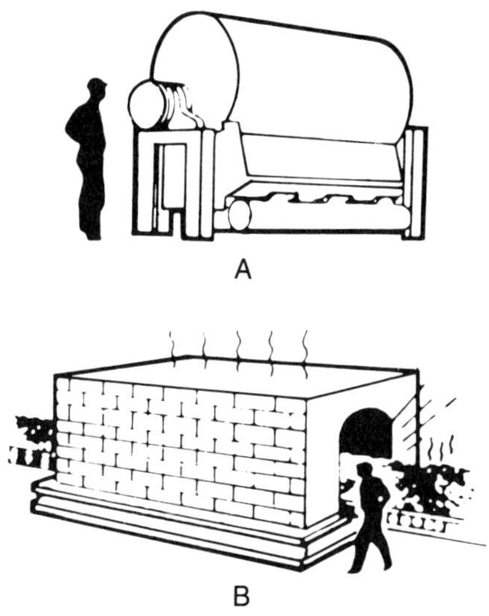


Fig. 65-2 Two steps in the manufacture of iron ore pellets (A) Magnetic separator, (B) pellet-hardening furnace.

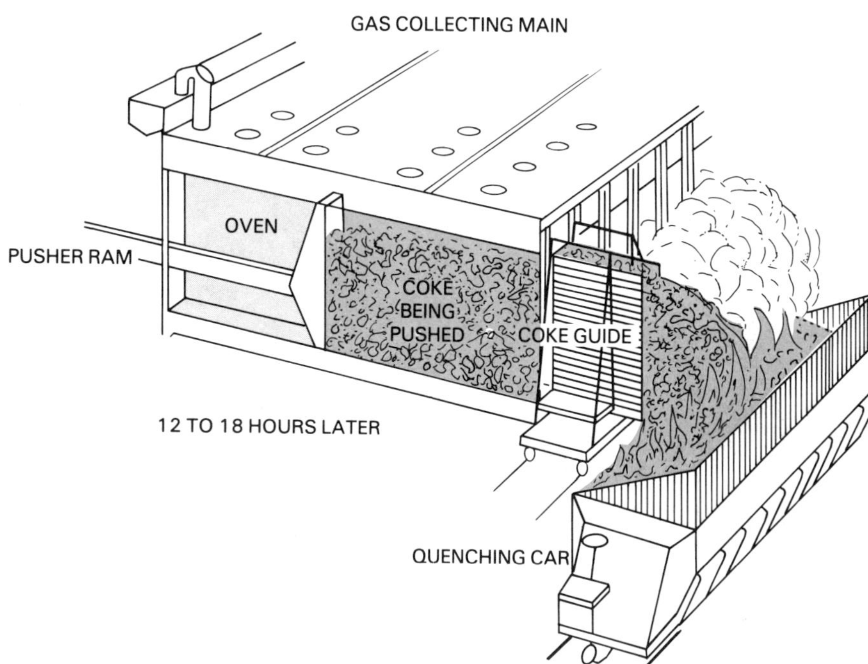


**Fig. 65-3** Coal is converted into coke in long, narrow coking ovens. (*Courtesy The American Iron and Steel Institute.*)

content is increased to about 65 percent (Fig. 65-2A). This high-grade material is mixed with clay and formed into pellets about ½ in. to ¾ in. (12.7 to 19.05 mm) in diameter in a pelletizer. The pellets are then covered with coal dust and sintered (baked) at 2350°F (1290°C) (Fig. 65-2B). The resultant hard, highly concentrated pellets will remain intact during transportation and loading into the blast furnace.

### Coal

Coke, which is used in the blast furnace, is made from a special grade of soft coal that contains small amounts of phosphorus and sulfur. The main sources of this coal are mines in West Virginia, Pennsylvania, Kentucky, and Alabama. Coal may be mined by either strip or underground mining. Before being converted into coke, the coal is crushed and washed. It is then loaded into the top of long, high, narrow ovens which are tightly sealed to exclude air. The coal is baked at 2100°F (1150°C) for about 18 hours, then dumped into railcars and quenched with water (Fig. 65-3).

The gases formed during the coking process are distilled into valuable by-products such as tar, ammonia, and light oils. These by-products are used in the manufacture of fertilizers, nylon, synthetic rubber, dyes, plastic, explosives, aspirin, and sulpha drugs.

### Limestone

Limestone, a gray rock consisting mainly of calcium carbonate, is used in the blast furnace as a flux to fuse and remove the impurities from the iron ore. It is also used as a purifier in the steelmaking furnaces. Limestone is usually found fairly close to steelmaking centers. It is generally mined by open-pit quarrying where the rock is removed by blasting. It is crushed to size before shipment to the steel mill.

## MANUFACTURE OF PIG IRON

N-2 A ferrous metal is one which contains iron.

The first step in the manufacture of any ferrous metal is the production of pig iron in the blast furnace. (See N-2.) The blast furnace (Fig. 65-4), about 130 ft. (40 m) high, is a huge steel shell lined with heat-resistant brick. Once started, the blast furnace runs continuously until the brick lining needs renewal or the demand for the pig iron drops.

Iron ore, coke, and limestone are measured out carefully and transported to the top of the furnace in a *skip car*

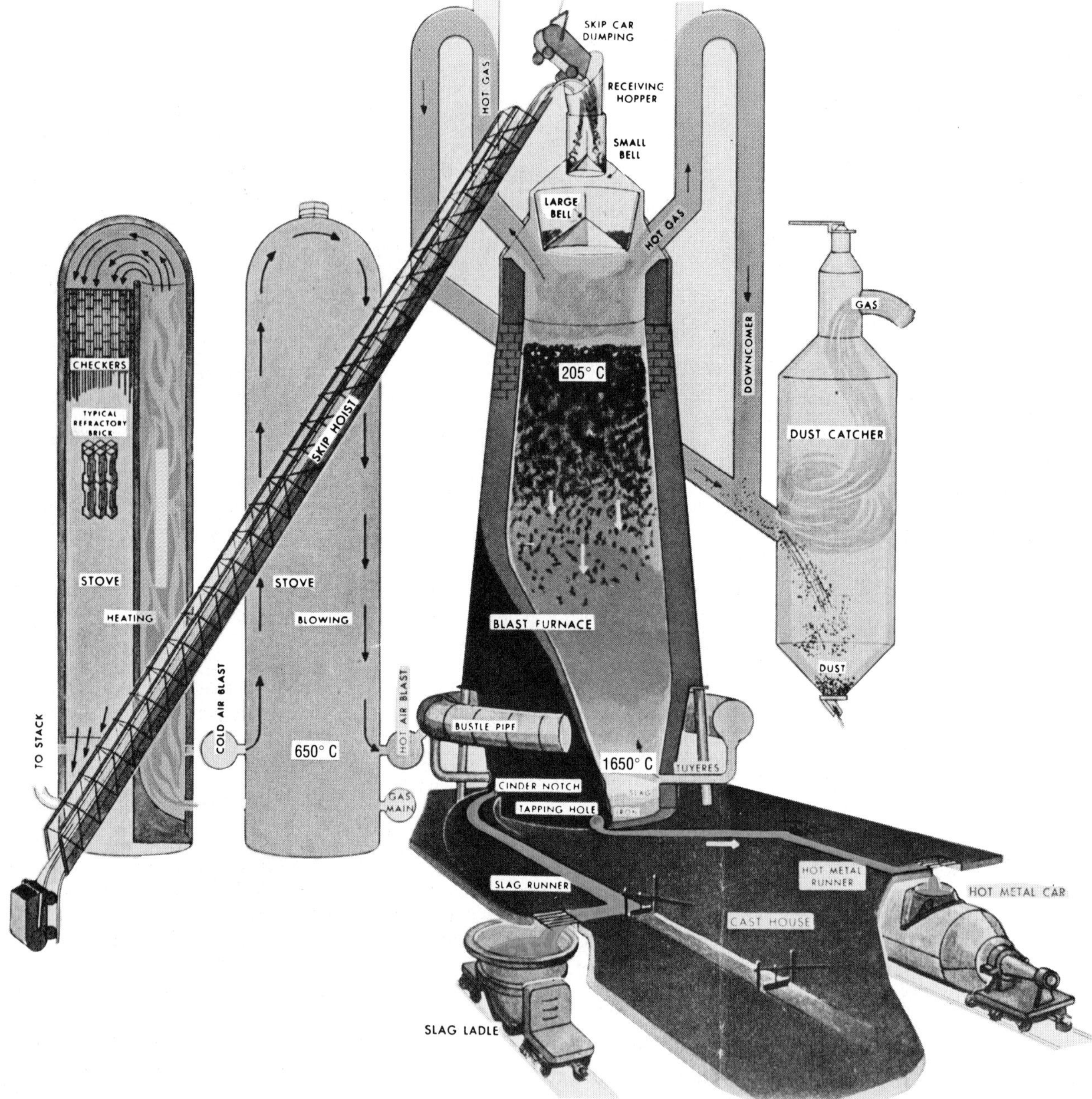


**Fig. 65-4** Schematic view of a blast furnace used to make pig iron. (*Courtesy The American Iron and Steel Institute.*)

(Fig. 65-4). Each ingredient is dumped separately into the furnace through the bell system, forming layers of coke, limestone, and iron ore in the top of the furnace. After the furnace has been charged, the coke is ignited and a continuous blast of hot air from the stoves at 1200°F (650°C) passes through the *bustle pipe* and *tuyeres* causing the coke to burn vigorously. The temperature at the bottom of the furnace reaches about 3000°F (1650°C) or higher. The carbon of the coke unites with the oxygen of the air to form carbon monoxide, which removes the oxygen from the iron ore and frees the metallic iron. The molten iron trickles through the charge and collects in the bottom of the furnace.

The intense heat also melts the limestone which combines with the impurities from the iron ore and coke to form a scum called slag. The slag also seeps down to the bottom of the charge and floats on top of the molten pig iron.

Every 4 to 5 hours the furnace is tapped and the molten iron [up to 350 tons (315 metric tons)] flows into a *hot*

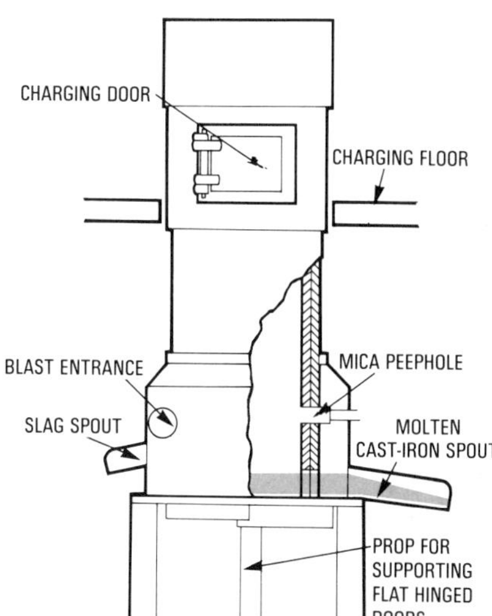


**Fig. 65-5** Cupola furnace used to make cast iron.

N-3 Molten metal is drawn out of a furnace when it is tapped.

N-4 An alloy is a metal which is made up of two or more metals.

N-5 A malleable material is one which is capable of being hammered or rolled without breaking.

N-6 The limestone unites with the impurities and rises to the top as slag.

N-7 Oxygen speeds the burning process, raises the furnace temperature, and decreases the time required to make a heat of steel.

*metal* or *bottle* car and is taken to the steelmaking furnaces. (See N-3.) Sometimes the pig iron is cast directly into *pigs* which are used in foundries for making cast iron. At more frequent intervals, the slag is drawn off into a *slag car* or *ladle* and is used later for making mineral wool insulation, building blocks, and other products.

## MANUFACTURE OF CAST IRON

Most of the pig iron manufactured in a blast furnace is used to make steel. However, a considerable amount is cast into ingots, or "pigs," and later used to manufacture cast-iron products. Cast iron is manufactured in a cupola furnace which resembles a huge stovepipe (Fig. 65-5).

Layers of coke, solid pig iron, scrap iron, steel, and limestone are charged into the top of the furnace. After the furnace is charged, the coke is ignited and air is forced in near the bottom to aid combustion. When the iron is melted, it settles to the bottom of the furnace and is then tapped into ladles.

The molten iron is poured into sand molds of the required shape and the metal assumes the shape of the mold. After the metal has cooled, the castings are removed from the molds.

The principal types of cast iron castings are:

*Gray iron castings,* made from a mixture of pig iron and steel scrap, are the most widely used. They are made into a wide variety of products including parts for automobiles, locomotives, and machinery.

*Chilled iron castings* are made by pouring molten metal into metal molds so that the surface cools very rapidly. The surface of such castings becomes very hard and difficult to machine, and they are used for crusher rolls or other products requiring a hard, wear-resistant surface.

*Alloyed castings* contain certain amounts of alloys such as chromium, molybdenum, nickel, etc. Castings of this type are used extensively by the automobile industry. (See N-4.)

*Malleable castings* are made from a special grade of pig iron and foundry scrap. (See N-5.) After these castings have solidified, they are annealed for several days in special furnaces. This makes the iron malleable and resistant to shock.

## MANUFACTURE OF STEEL

Before molten pig iron from the blast furnace can be converted into steel, some of its impurities must be burned out. This is done in one of three different types of furnace: the *open-hearth furnace,* the *basic oxygen furnace,* or the *electric furnace.*

### Open-Hearth Process

Open-hearth furnaces (Fig. 65-6) are rectangular brick structures resembling huge ovens 40 to 50 ft (12 to 15 m) long and 15 to 17 ft (4.5 to 5.2 m) wide.

The hearth is a large dish-shaped structure about 2 to 3 ft (609 to 914 mm) deep. Above each end of the hearth is a burner and an inlet for hot air from the checkers, which are large heat-storage chambers filled with firebrick stacked in a checkerboard fashion. Some open hearth furnaces have a capacity of over 50 tons (45.37 t).

When a furnace is charged, limestone is put in first as a flux, then steel scrap is added. (See N-6.) After the scrap has been melted down, molten pig iron is poured in. Tongues of flame and hot air from the checkers sweep back and forth over the materials, creating temperatures of about 3000°F (1650°C) melting and burning out impurities. (See N-6.) About every 15 minutes, the direction of the flame and hot air reverses over the metal. This utilizes the heat stored in the checkers at the other end of the furnace.

After a few hours, samples are taken from the bubbling metal for chemical analysis. On the basis of the laboratory reports on these analyses, alloying materials are added to bring the molten metal to the required chemical composition. When the oxygen lance is used, a heat of 230 tons (209 t) of steel can be produced in approximately 6 hours; without the oxygen lance, the time would be about 7½ hours. (See N-7.)

The molten steel is teemed or poured into large cast-iron forms

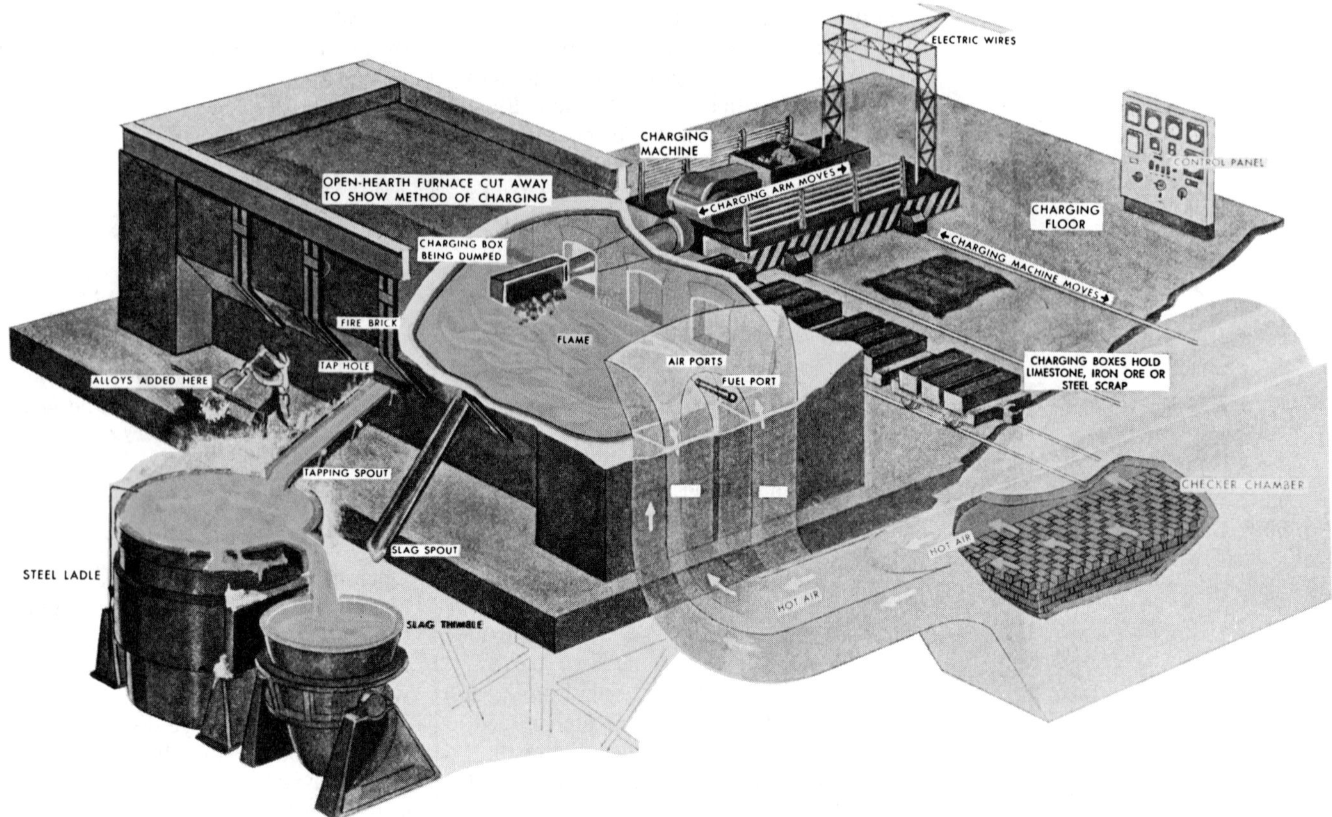


**Fig. 65-6** A schematic view of an open-hearth furnace. (*Courtesy The American Iron and Steel Institute.*)

called *ingot molds*. Here the steel is allowed to solidify and then the molds are removed. The *ingots* are then transferred to soaking pits where they are brought to a uniform temperature throughout, prior to being rolled into the desired shapes and sizes.

## BASIC OXYGEN PROCESS

For many years, about 90 percent of all steel produced in North America was made by *open-hearth furnaces*. The remainder of the steel was produced by *Bessemer converters* and *electric furnaces*.

With the introduction of the *basic oxygen process* in 1955, the emphasis on steelmaking processes shifted. Steelmakers found that the addition of oxygen to any steelmaking process speeded production. The basic oxygen furnace (Fig. 65-7) was developed as a result. Today more than 50 percent of all steel produced is made in basic oxygen furnaces, with much of the remainder being produced in open hearth furnaces modified by the addition of oxygen lances. Special tool steels are still made by electric furnaces, but the production of steel by the Bessemer process has become practically nonexistent.

The basic oxygen furnace (Fig. 65-7) is a cylindrical brick-lined furnace with a dished bottom and cone-shaped top. It may be tilted in both directions for charging and tapping, but it is kept in the vertical position during the steelmaking process.

With the furnace tilted forward, scrap steel (30 to 40 percent of the total charge) is loaded into the furnace (Fig. 65-7A). Molten pig iron (60 to 70 percent of the total charge) is then added (Fig. 65-7B). The furnace is then moved to the vertical position where the fluxes, (mainly burnt lime) are added (Fig. 65-7C). When the furnace is still in the upright position, a water-cooled oxygen lance is lowered into the furnace until the top of the lance is the required height above the molten metal. High-pressure oxygen is blown into the furnace causing a churning, turbulent action during which the undesirable elements are burned out of the steel (Fig. 65-7D).

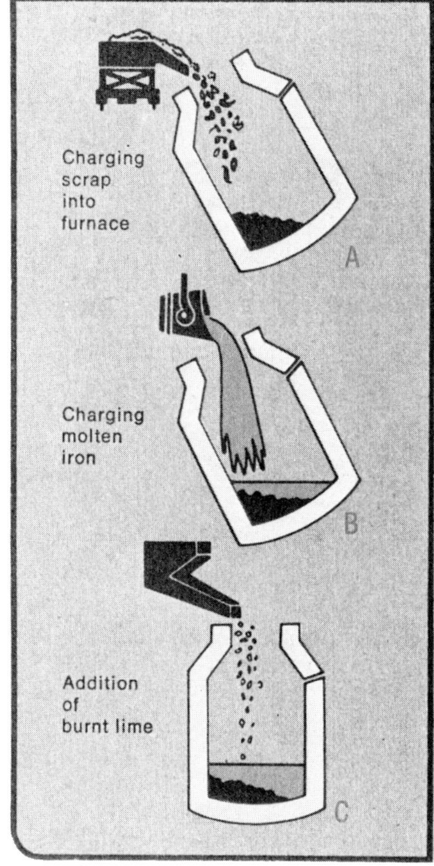


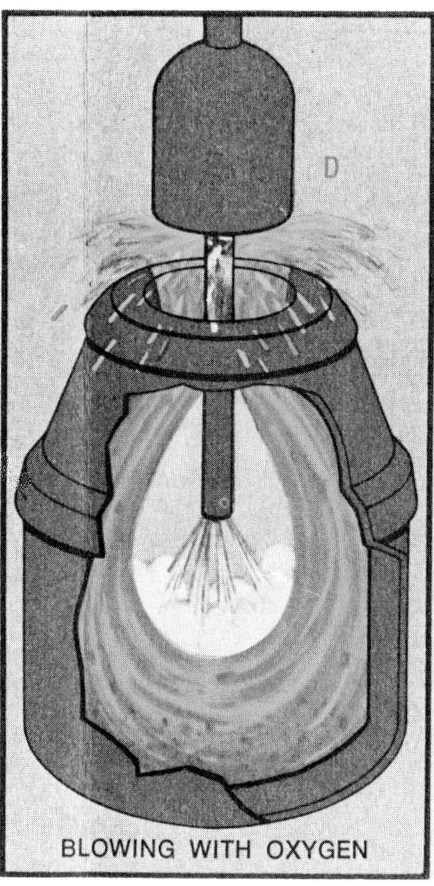

**Fig. 65-7** The operation of a basic oxygen furnace. (*Courtesy Inland Steel Corp.*)

N-8 If the metal sample is not correct, the proper elements are added.

N-9 Limestone is added during the melting process to prevent damage to the furnace lining.

The oxygen blow lasts for about 20 minutes. The lance is then removed and the furnace is tilted to a horizontal position. The temperature of the metal is taken and samples of the metal are tested. (See N-8.) If the temperature and the samples are correct, the furnace is tilted to the tapping position (Fig. 65-7E) and the metal is tapped into a ladle. Alloying elements are added at this time to give the steel its desired properties. An alloy is a combination of two or more metals designed to give desired properties. After tapping, the furnace is tilted in the opposite direction and to an almost vertical position to dump the slag into a slag pot (Fig. 65-7F). About 300 tons (272t) of steel can be produced in 1 hour in a large basic oxygen furnace.

## ELECTRIC FURNACE

The electric furnace (Fig. 65-8) is used primarily to make fine alloy and tool steels. The heat, the amount of oxygen, and the atmospheric condition can be regulated at will in the electric furnace; this furnace is therefore used to make steels that cannot be readily produced in any other way.

Carefully selected steel scrap, containing smaller amounts of the alloying elements than are required in the finished steel along with the required fluxes, is loaded into the saucer-shaped base of the furnace. (See N-9.) The three carbon electrodes are lowered until an arc jumps from them to the scrap. The heat generated by the electric arcs gradually melts all the steel scrap. Alloying materials such as chromium, nickel, tungsten, etc., are then added to make the type of alloy steel required. Depending on the size of the furnace, it takes from 4 to 12 hours to make a heat of steel. When the metal is ready to be tapped, the furnace is tilted forward and the steel flows into a large ladle. From the ladle, the steel is teemed into ingots.

## STEEL PROCESSING

After the steel has been properly refined in any of the furnaces, it is

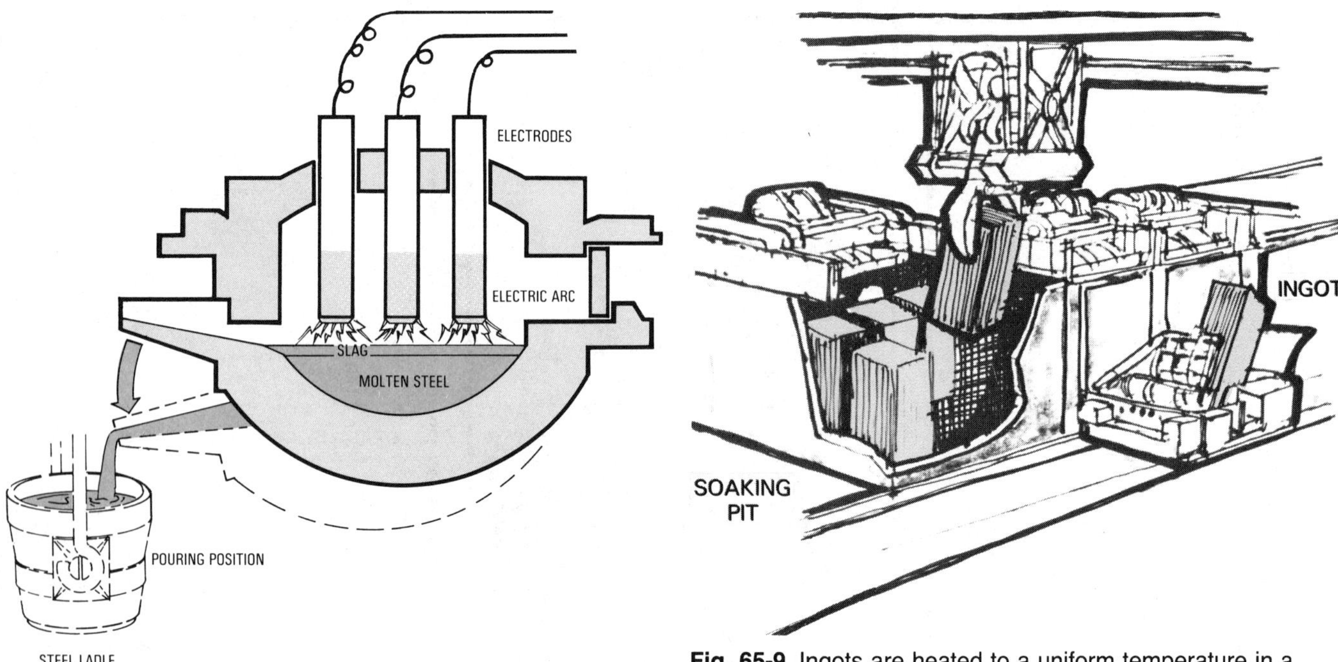


**Fig. 65-8** Electric furnaces are used to make fine alloy and tool steels. (*Courtesy United States Steel Corporation.*)

**Fig. 65-9** Ingots are heated to a uniform temperature in a soaking pit before being rolled into shape.

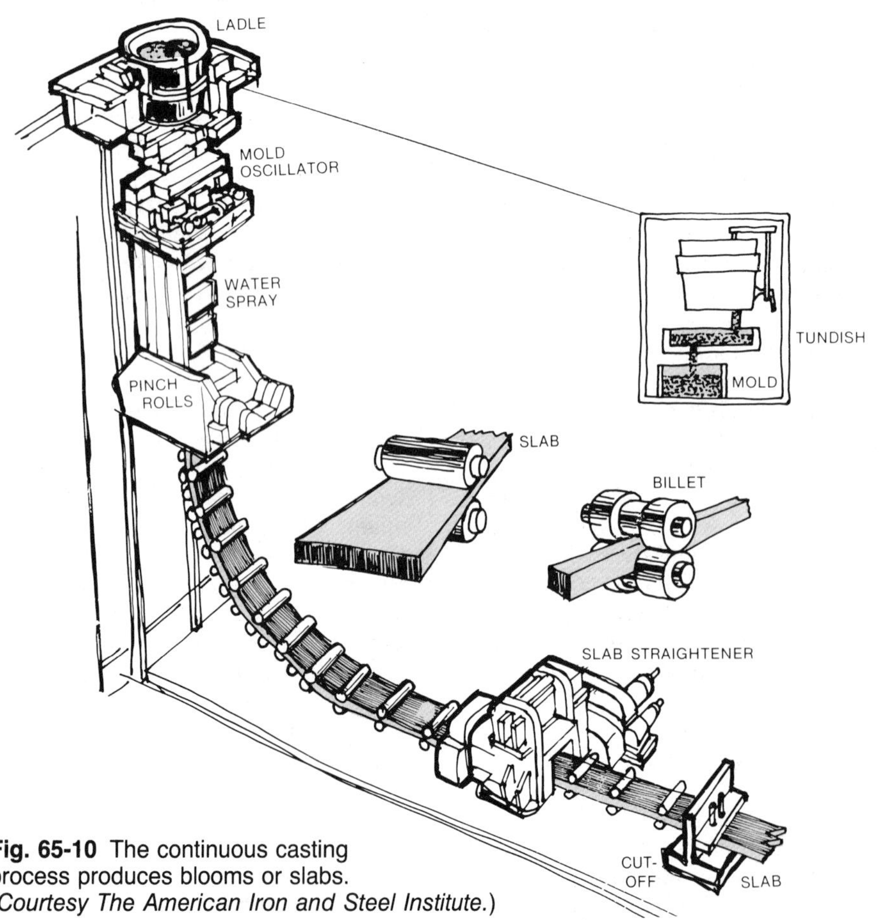


**Fig. 65-10** The continuous casting process produces blooms or slabs. (*Courtesy The American Iron and Steel Institute.*)

tapped into ladles where alloying elements and deoxidizers may be added. the molten steel may then be teemed (poured) into *ingots* weighing as much as 20 tons (18.14 t) and then rolled into billets or slabs, or it may be formed directly into slabs by the *continuous casting process*.

## PRODUCTION OF INGOTS

After the furnace has been tapped, the molten steel is teemed (poured) into ingot molds and allowed to solidify. The ingot molds are then removed or *stripped* and the hot ingots are placed in soaking pits at 2200°F (1204.4°C) to make them a uniform temperature throughout (Fig. 65-9). The ingots are then sent to the rolling mills where they are rolled and reduced in cross section to form *blooms, billets,* or *slabs*.

## CONTINUOUS CASTING PROCESS

The molten steel may also be converted into slabs or blooms by the *con-*

**Fig. 65-11** Various shapes of steel may be produced by rolling. (*Courtesy Kaiser Steel Corporation.*)

*tinuous casting process* (Fig. 65-10). Here, up to 200 tons (181.4t) are poured into a tundish or reservoir at the top of the machine. The molten metal flows from the tundish to an oscillating mold, where a solid skin is formed on the outside of the metal due to the cooling action of the mold. The metal moves down through the cooling system and the skin becomes thicker. As the metal proceeds from the mold, further cooling action solidifies it throughout and the slab is moved to a straightener by means of pinch rolls. The continuous slab is then cut into desired lengths by a traveling acetylene torch.

After the metal has been made into blooms or slabs by either of the forementioned processes, it is further rolled into billets and then, while still hot, into the desired shape, such as round, flat, square, hexagonal, etc. (Fig. 65-11). These rolled products are known as *hot rolled steel* and are easily identified by the bluish black scale on the outside.

Hot rolled steel may be further processed into *cold rolled* or *cold drawn steel* by removing the scale in an acid bath and passing the metal through rolls or dies of the desired shape and size.

## Vacuum Processing of Molten Steel

Steel used in space and nuclear projects is often processed and solidified in a vacuum to remove oxygen, nitrogen, and hydrogen, and thus produce a high-quality steel. High-quality steel, particularly stainless steel, may also be heat-treated in a vacuum and quenched by introducing liquid nitrogen into the furnace. This process produces a cleaner surface, higher hardness, and better corrosion resistance in the metal.

# REVIEW QUESTIONS

**1.** Name the three raw materials used to make steel.

**2.** Name three types of high-grade iron ore.

**3.** List three purposes of pelletizing.

**4.** Briefly describe the coke-making process.

**5.** What type of furnace is used to manufacture pig iron?

**6.** List the raw materials used to manufacture pig iron.

**7.** Briefly describe the operation of the blast furnace.

**8.** In what type of furnace is cast iron manufactured?

**9.** What raw materials are used in the manufacture of cast iron?

**10.** What is the purpose of the checkers in an open hearth furnace?

**11.** How are the impurities removed from the molten metal in an open-hearth furnace?

**12.** What modification has been made to most open hearth furnaces and what has been its result?

**13.** List the main steps in the operation of a basic oxygen furnace.

**14.** What types of steel are made in the electric furnace?

**15.** Explain why the electric furnace is used to produce these steels.

**16.** What is the purpose of soaking the ingot?

**17.** Briefly describe the continuous casting process.

**18.** How may hot rolled steel be identified?

**19.** Briefly describe how hot rolled steel is made into cold rolled steel.

**20.** How is steel used in space and nuclear projects generally processed?

# UNIT 66

# PROPERTIES, COMPOSITION, AND IDENTIFICATION OF METALS

Understanding the properties and heat treatment of metals has become increasingly important to machinists during the past two decades. Study of metal properties and development of new alloys have facilitated reduction in mass and increase in strength of machines, automobiles, aircraft, and many present-day commodities.

The most commonly used metals today are ferrous metals, or those which contain iron. The composition and properties of ferrous materials may be changed by the addition of various alloying elements during manufacture, to impart the desired qualities to the material. Cast iron, machine steel, carbon steel, alloy steel, and high-speed steel are all ferrous metals, each having different properties.

To better understand the use of the various metals, one should be familiar with certain metallurgical terms.

## OBJECTIVES

This unit will help you understand:

**1.** Six important terms related to physical properties of metals
**2.** The difference between machine, carbon, and alloy steels
**3.** The effects of alloying elements in steel
**4.** The properties and uses of nonferrous metals

## PHYSICAL PROPERTIES OF METALS

*Brittleness* (Fig. 66-1A) is that property of a metal which permits no permanent distortion before breaking.

Cast iron is a brittle metal; it will break rather than bend under shock or impact.

**Fig. 66-1A** Brittle metals will not bend but break easily under strain or impact.

**Fig. 66-1B** Ductile metals are easily bent or deformed.

*Ductility* (Fig. 66-1B) is the ability of the metal to be permanently deformed or bent without breaking. Metals such as copper and machine steel, which may be drawn into wire, are ductile materials.

**Fig. 66-1C** Elastic metals return to their original shape after the load is removed.

**Fig. 66-1D** Hard metals resist penetration.

**Fig. 66-1E** Malleable metals may easily be formed or shaped.

**Fig. 66-1F** Tough metals can withstand shock or impact.

**Fig. 66-2** Manganese adds ductility to a metal.

**Fig. 66-3** Chromium adds hardness qualities and wear resistance to a metal.

**Fig. 66-4** Vanadium increases the tensile strength of a metal.

N-1 Chemical elements provide steel with various qualities.

*Elasticity* (Fig. 66-1C) is the ability of a metal to return to its original shape after any force acting upon it has been removed. Properly heat-treated springs are good examples of elastic materials.

*Hardness* (Fig. 66-1D) may be defined as the resistance of a metal to forceable penetration or bending.

*Malleability* (Fig. 66-1E) is that property of a metal which permits it to be hammered or rolled into other sizes and shapes.

*Toughness* (Fig. 66-1F) is the property of a metal to withstand shock or impact. Toughness is the opposite condition to brittleness.

*Fatigue* is the prolonged effect of repeated straining action which causes the metal to strain or break.

## CHEMICAL COMPOSITION AND CLASSIFICATION OF STEEL

Although the steel manufacturers try to produce as pure a steel as possible, it is almost impossible to efficiently make a pure machine steel. When most machine steels are produced, certain chemical elements such as carbon, phosphorus, and sulfur remain in the steel, usually in such small quantities that it does not affect the properties of the steel.

While steel is still in the molten state, elements such as carbon, manganese, or silicon may be added to produce certain desirable qualities in the steel.

Alloy steels are formed by the addition of such elements as chromium, molybdenum, nickel, tungsten, or vanadium in addition to carbon.

### Chemical Elements in Steel

*Carbon* in steel may vary from 0.01 to 1.7%. The amount of carbon will determine the brittleness, hardness, and strength of the steel.

*Manganese* in low-carbon steel makes the metal ductile and gives it good bending quality (Fig. 66-2). In high-speed steel manganese toughens the metal. Manganese content usually varies from 0.30 to 0.80%, but may run higher in special steels.

*Phosphorus* is an undesirable element which makes steel brittle and reduces its ductility. In good-quality steels the phosphorus content should not exceed 0.05%. (See N-1.)

*Silicon* is added to steel in order to remove gases and oxides, thus preventing the steel from becoming porous and oxidizing. It makes the steel harder and tougher. Low carbon steel contains about 0.20% silicon.

*Sulfur,* an undesirable element, causes crystallization of steel when the metal is heated to a red color. Good-quality steel should not contain more than 0.04% sulfur. However, sulfur or lead is added to certain steels to give them a free cutting and good machining quality.

### Alloying Elements in Steel

Alloying elements may be added during the steelmaking process to produce certain qualities in the steel. Some of the more common alloying elements are the following.

*Chromium* in steel imparts hardness and wear resistance (Fig. 66-3). It gives steel a deeper hardness penetration than other alloying metals. It also increases resistance to corrosion.

*Molybdenum* allows cutting tools to retain their hardness when hot. Because molybdenum improves steel's physical structure, it gives steel a greater ability to harden.

*Nickel* in steel improves its toughness, resistance to fatigue failure, impact properties, and resistance to corrosion.

*Tungsten* increases the strength and toughness of steel and its ability to harden. It also gives cutting tools the ability to maintain a cutting edge even at a red heat.

*Vanadium* in amounts up to 0.20% will increase steel's tensile strength (Fig. 66-4) and produce a finer grain structure in steel. Vanadium steel is usually alloyed with chromium to make such parts as springs, gears, wrenches, car axles, and many drop-forged parts.

### Types of Steel

*Low carbon steel,* commonly called machine or mild steel, contains from 0.10 to 0.30% carbon. This steel which is easily forged, welded, and machined, is used for making chains, rivets, bolts, shafting, etc.

*Medium carbon steel* contains from 0.30 to 0.60% carbon and is used for heavy forgings, car axles, rails, etc.

*High carbon steel* commonly called tool steel, contains from 0.60 to 1.7% carbon and can be hardened and tempered. Hammers, crowbars, etc., are made from steel having 0.75% carbon. Cutting tools, such as drills, taps, reamers, etc., are made from steel having 0.90 to 1.00% carbon.

*Alloy steels* are steels which have certain metals, such as chromium, nickel, tungsten, vanadium, etc., added to them to give the steel certain new characteristics. By the addition of various alloys, steel can be made resistant to rust, corrosion, heat, abrasion, shock, and fatigue. (See N-2.)

N-2 An alloy steel is one which contains two or more metals.

*High-speed steels* contain various amounts and combinations of tungsten, chromium, vanadium, cobalt, and molybdenum. Cutting tools made of such steels are used for machining hard materials at high speeds and for taking heavy cuts. High-speed steel cutting tools are noted for maintaining a cutting edge at temperatures where most steels would break down. (See N-3.)

N-3 Many cutting tools are made of high-speed steel.

## High-Strength Low-Alloy Steels

A recent development in the steelmaking industry is that of the high-strength low alloy (H.S.L.A.) steels. These steels, containing a maximum carbon content of 0.28% and small amounts of vanadium, columbium, copper, and other alloying elements, offer many advantages over the regular low carbon construction steels. (See N-4.) These advantages are:

N-4 High-strength low alloy steels can be machined, formed, or welded as easily as low-carbon steels.

**1.** Higher strength than medium carbon steels
**2.** Less expensive than other alloy steels
**3.** Strength properties "built into" the steel and no further heat treating needed
**4.** Bars of smaller cross sections can do the work of larger regular carbon steel bars
**5.** Higher hardness, toughness (impact strength), and higher fatigue failure limits than carbon steel bars
**6.** May be used unpainted because it develops a protective oxide coating on exposure to the atmosphere

# CLASSIFICATION OF STEEL

In order to ensure that the composition of various types of steel remains constant and that a certain type of steel will meet the required specifications, the Society of Automotive Engineers and the American Iron and Steel Institute have devised methods of identifying different types of steel. These numerical systems are similar and both are widely used in industry. (See N-5.)

N-5 The SAE and AISI numbers help to identify the type and composition of a steel.

## The SAE-AISI Classification Systems

The systems designed by the Society of Automotive Engineers and the American Iron and Steel Institute are similar in most respects (Fig. 66-5). They both use a series of four or five numbers to designate the type of steel. The AISI system also indicates the steelmaking process by a letter prefix preceding the number.

The *first digit* in these series indicates the predominant alloying element. For example: 1 indicates carbon steel, 2 nickel steel, etc.

The *second digit* indicates the approximate percentage by weight of the alloying element. Example: 2540 would indicate a nickel steel with about 5% nickel.

The *last two digits* indicate the average carbon content in points (hundredths of one percent). Again 2340 would indicate a nickel steel containing about 3% nickel and 40 points of carbon (40/100 of 1%).

## Examples of Steel Identification

Determine the types of steel indicated by the following numbers: 1015, A2360, 4170.

*1015*—**1** indicates plain carbon steel
—**0** indicates there are no major alloying elements
—**15** indicates that there is between 0.10 and 0.20% carbon content
*Note:* Steel would naturally also contain small quantities of manganese, phosphorus, and sulfur

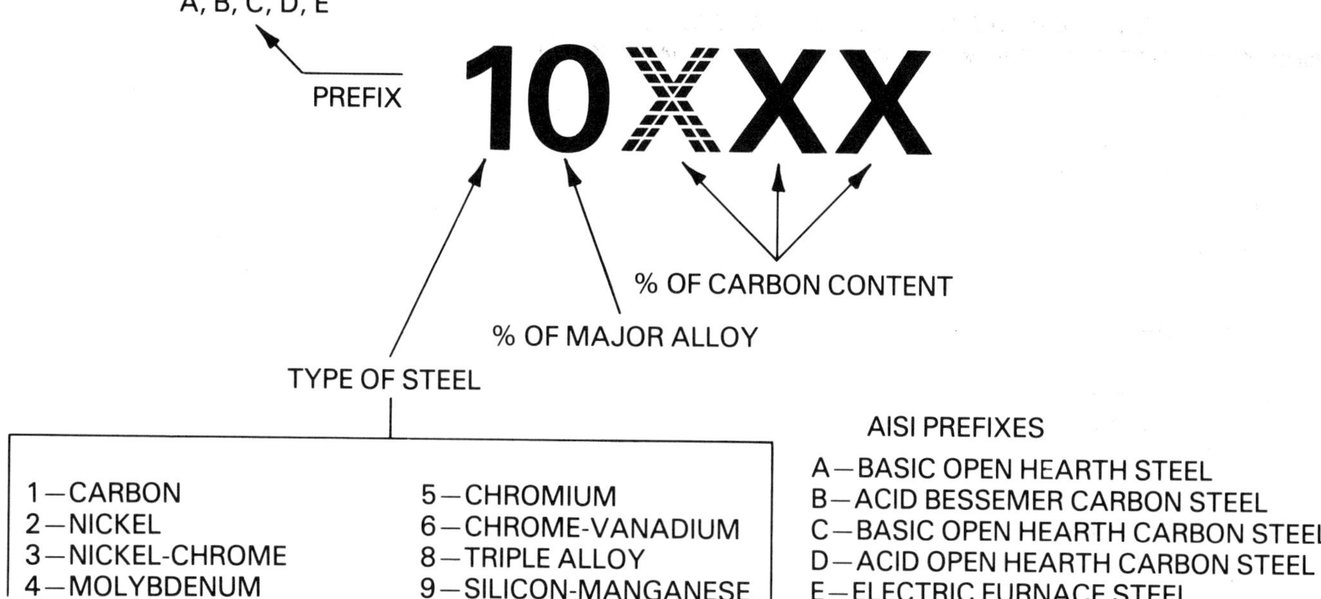


**Fig. 66-5** The SAE and AISI system of classifying steel.

*A2360*—**A** indicates an alloy steel made by the basic open-hearth process
—**23** indicates the steel contains 3.5% nickel (see Table 66-1).
—**60** indicates 0.60% carbon content

*4170*—**41** indicates a chromium-molybdenum steel
—**70** indicates 0.70% carbon content

# IDENTIFICATION OF METALS

The machinist's work consists of machining metals, and therefore it is advantageous to learn as much as possible about the various metals used in the trade. (See N-6.) It is often necessary to determine the type of metal being used by its physical appearance. Some of the more common machine shop metals, their appearance, use, etc., are found in Table 66-2. Metals are usually identified by one of four methods:

N-6 Choose the right metal for each job, otherwise time will be wasted and the part will not function properly.

1. Their appearance
2. Spark testing
3. A manufacturer's stamp
4. A code color painted on the bar

The latter two methods are most commonly used and are probably the most reliable. Each manufacturer, however, may use their own system of stamps or code colors.

## Spark Testing

Any ferrous metal, when held in contact with a grinding wheel, will give off characteristic sparks. Small particles of metal, heated to red or yellow heat, are hurled into the air, where they come in contact with oxygen and oxidize or burn. An element such as carbon burns rapidly, resulting in a bursting of the particles. Spark bursts vary in color, intensity, size, shape, and the distance they fly, according to the makeup of the metal that is being ground. (See N-7.)

N-7 Different colors and shapes of sparks are produced by various elements in a metal.

*Low carbon or machine steel* (Fig. 66-6A) shows sparks in long, light yellow streaks with a little tendency to burst.

*Medium carbon steel* (Fig. 66-6B) is similar to low-carbon steel but has more sparks which burst with a sparkler effect because of the greater percentage of carbon in the steel.

*High carbon or tool steel* (Fig. 66-6C) shows numerous little yellow stars bursting very close to the grinding wheel.

*High-speed steel* (Fig. 66-6D) produces several interrupted spark lines with a dark red, ball-shaped spark at the end.

**TABLE 66-1 SAE CLASSIFICATION OF STEELS**

| | |
|---|---|
| **Carbon steels** | 1xxx |
| Plain carbon | 10xx |
| Free-cutting (resulphurized screw stock) | 11xx |
| Free-cutting, manganese | X13xx |
| **High-manganese** | T13xx |
| **Nickel steels** | 2xxx |
| 0.50% nickel | 20xx |
| 1.50% nickel | 21xx |
| 3.50% nickel | 23xx |
| 5.00% nickel | 25xx |
| **Nickel-chromium steels** | 3xxx |
| 1.25% nickel, 0.60% chromium | 31xx |
| 1.75% nickel, 1.00% chromium | 32xx |
| 3.50% nickel, 1.50% chromium | 33xx |
| 3.00% nickel, 0.80% chromium | 34xx |
| Corrosion- and heat-resisting steels | 30xxx |
| **Molybdenum steels** | 4xxx |
| Chromium-molybdenum | 41xx |
| Chromium-nickel-molybdenum | 43xx |
| Nickel-molybdenum | 46xx and 48xx |
| **Chromium steels** | 5xxx |
| Low-chromium | 51xx |
| Medium-chromium | 52xxx |
| **Chromium-vanadium steels** | 6xxx |
| **Triple-alloy steels** (nickel, chromium, molybdenum) | 8xxx |
| **Manganese-silicon steels** | 9xxx |

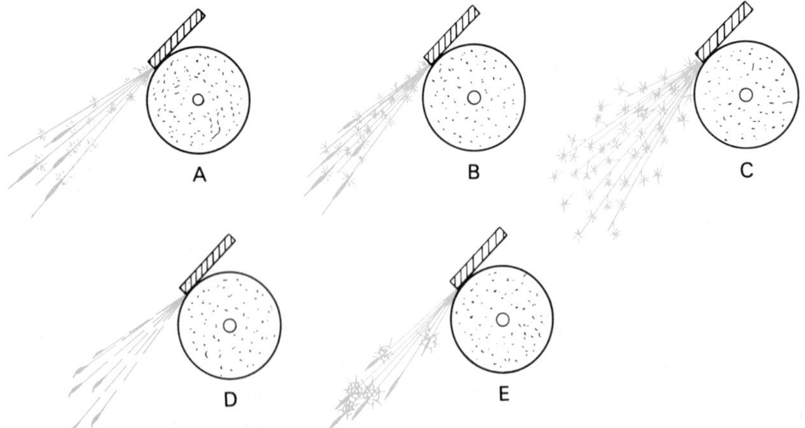

**Fig. 66-6** Spark testing may be used to identify ferrous metals.

*Cast iron* (Fig. 66-6E) shows a definite torpedo-shaped spark with a featherlike effect near the end. It changes from a dark red to a gold color.

# NONFERROUS METALS AND ALLOYS

N-8 Nonferrous metals contain no iron, do not rust, and are nonmagnetic.

Nonferrous metals contain little or no iron. They are resistant to corrosion and are nonmagnetic. In machine shop work, nonferrous metals are used where ferrous metals would be unsuitable. The most commonly used nonferrous metals are aluminum, copper, lead, nickel, tin, and zinc.

*Aluminum* is made from an ore called *bauxite*. It is a white, soft metal used where a light, noncorrosive metal is required. Aluminum is usually alloyed with other metals to increase its strength and stiffness. It is used extensively in aircraft manufacture because it is only one-third as heavy as steel.

*Copper* is a soft, ductile, malleable metal which is very tough and strong. It is reddish in color and second only to silver as an electric conductor. Copper forms the basis of brasses and bronzes. (See N-8.)

*Lead* is a soft, malleable, heavy metal which has a melting point of about 620°F (327°C). It is corrosion-resistant, used for lining vats and tanks and for covering cables. It is also used for making such alloys as babbitt and solder.

*Nickel* is a very hard, corrosion-resistant metal. It is used as a plating agent on steel and brass, and it is added to steel to increase strength and toughness.

*Tin* is a soft white metal having a melting point of about 450°F (232°C). It is very malleable and corrosion resistant. It is used in the manufacture of tin plate and tin foil. It is also used for making such alloys as babbitt, bronze, and solder.

*Zinc* is a bluish-white element which is fairly hard and brittle. It has a melting point of about 788°F (420°C) and is used mainly to galvanize iron and steel.

## Nonferrous Alloys

A nonferrous alloy is a combination of two or more nonferrous metals completely dissolved in each other. Nonferrous alloys are made when certain qualities of both original metals are desired. Some common nonferrous alloys are the following.

*Brass* is an alloy of approximately two-thirds copper and one-third zinc. Sometimes 3% lead is added to make it easy to machine. Its color is normally a bright yellow, but this varies slightly according to the amounts of alloys it contains. Brass is widely used for small bushings, plumbing and radiator parts,

**TABLE 66-2 IDENTIFICATION OF METALS**

| Metal | Carbon content, % | Appearance | Method of processing | Uses |
|---|---|---|---|---|
| Cast iron (C.I.) | 2.5–3.5 | Grey, rough sandy surface | Molten metal poured into sand moulds | Parts of machines, such as lathe beds, etc. |
| Machine steel (M.S.) | 0.10–0.30 | Black, scaly surface | Put through rollers while hot | Bolts, rivets, nuts, machine parts |
| Cold rolled or cold drawn (C.R.S.); (C.D.S.) | 0.10–0.30 | Dull silver, smooth surface | Put through rollers or drawn through dies while cold | Shafting, bolts, screws, nuts |
| Tool steel (T.S.) | 0.60–1.5 | Black, glossy | Same as machine steel | Drills, taps, dies, tools |
| High-speed steel (H.S.S.) | Alloy steel | Black, glossy | Same as machine steel | Dies, tools, taps, drills, toolbits |
| Brass | ............ | Yellow (various shades); rough if cast, smooth if rolled | Same as cast iron, or rolled to shape | Bushings, pump parts, ornamental work |
| Copper | ............ | Red-brown; rough if cast, smooth if rolled | Same as cast iron, or rolled to shape | Soldering irons, electric wire, water pipes |

fittings for water-cooling systems, and miscellaneous castings.

*Bronze* is an alloy composed mainly of copper, tin, and zinc. Some types of bronze contain such additions as lead, phosphorus, manganese, and aluminum to give them special qualities. Bronze is harder than brass and resists surface wear. It is used for machine bearings, gears, propellers, and miscellaneous castings.

*Babbitt* is a soft grayish-white alloy of tin and copper. Antimony may be added to make it harder, and lead is usually added if a softer alloy is required. Babbitt is used in bearings of many reciprocating engines.

## REVIEW QUESTIONS

**1.** How may the properties of a ferrous metal be changed?

**2.** Dcfinc
   (a) Brittleness
   (b) Hardness
   (c) Toughness

**3.** Name three chemical elements that remain in most machine steel.

**4.** Name any four elements that may be added to produce alloy steels.

**5.** What chemical element in steel will determine its hardness and strength?

**6.** What effect does the addition of manganese have on
   (a) Low carbon steel
   (b) High-speed steel

**7.** Why is phosphorus an undesirable element in steel?

**8.** What effect does silicon have when added to steel?

**9.** Name one quality that the following alloying elements imparts to steel.
   (a) Chromium
   (b) Molybdenum
   (c) Nickel
   (d) Tungsten
   (e) Vanadium

**10.** What is the carbon content of
   (a) Low carbon steel
   (b) Medium carbon steel
   (c) High carbon steel

**11.** What is an alloy steel?

**12.** Why is high-speed steel suitable for cutting tools?

**13.** Name four advantages of high-strength low alloy steels over low carbon steels.

**14.** Define the following steels:
   (a) SAE 4395
   (b) AISI 52100

**15.** Define nonferrous metal.

**16.** Briefly describe and give one use for:
   (a) Aluminum
   (b) Copper
   (c) Nickel
   (d) Tin

# UNIT 67

# HEAT TREATMENT OF STEEL

Understanding the properties and heat treatment of metals has become increasingly more important to machinists. The development of new alloys has resulted in an increase in the strength of metals used in the manufacture of machines, automobiles, aircraft, and many of our present-day commodities. The performance of a steel part depends on the following factors:

**1.** The selection of the steel from which the part is made.
**2.** The heat treating operation performed on it
**3.** The testing of the part to ensure that the part possesses the desired qualities

## OBJECTIVES

After completing this unit, you should be able to:

**1.** Understand the changes which occur in carbon steel when it is heat-treated
**2.** Harden, temper, and anneal a piece of tool steel
**3.** Case harden a piece of low carbon steel
**4.** Test the hardness of a metal using the Rockwell hardness tester

## HARDENING CARBON STEEL

Hardening is the process of heating metal uniformly to its proper temperature and then quenching or cooling it in water, oil, air, or a refrigerated area. Hardening produces a fine-grain structure, increases the tensile strength of the metal, and decreases its ductility. (See N-1.)

Carbon tool steel may be hardened by heating it to the critical temperature, which produces a bright cherry-red color, approximately 1450 to 1525°F (790 to 830°C). (See N-2.) When the steel, which is made up of *ferrite* (iron) and *pearlite* (layers of iron and carbon) (Fig. 67-1), is heated to the *critical temperature,* a chemical and physical change takes place. The pearlite combines with the ferrite to form a fine-grain structure called *austenite*. When austenite is quenched in the proper medium (water, oil, or air) it transforms into *martensite,* which is very hard and brittle with a fine-grain structure.

### To Harden Carbon Tool Steel (Fig. 67-3)

**1.** Light the furnace and preheat the steel slowly.
**2.** Heat the steel to its recommended temperature, checking either by color or by pyrometer (Table 67-1).
**3.** Quench in water, brine, or oil, depending on the type of steel or manufacturer's recommendation. (See N-3.)
**4.** Move the work about in the quenching medium in a figure 8 motion to allow the steel to cool quickly and evenly.
**5.** Test for hardness with the edge of a file or a hardness tester.

## TEMPERING

After hardening, the steel is brittle and may break with the slightest tap due to the stresses caused by quenching. (See N-4.) To overcome this brittleness, the steel is tempered; that is, it is reheated until it is brought to the desired temperature or color and then quenched. This structure is called *tempered martensite* (Fig. 67-2). Tempering toughens the steel and makes it less brittle, although a little of the hardness is lost. As steel is heated it changes color, and these colors indicate various tempering temperatures (Table 67-2).

### To Temper Carbon Tool Steel

**1.** Clean off all the surface scale from the work with abrasive cloth.
**2.** Select the heat or color desired (Table 67-2).
**3.** Heat the steel slowly and evenly.
**4.** When the steel reaches the desired color or temperature, quench it quickly in the same cooling medium used for hardening.

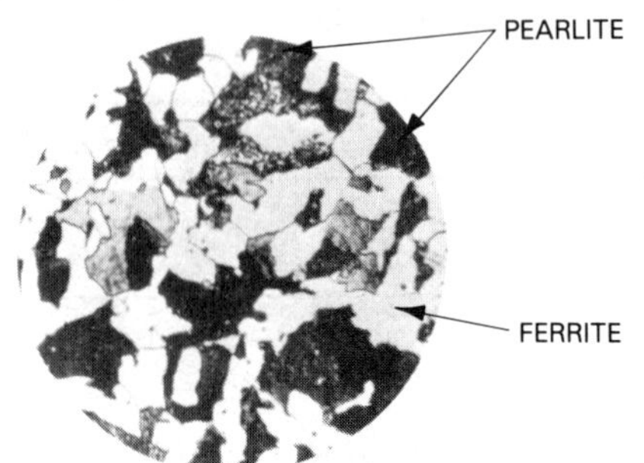


**Fig. 67-1** Steel is made up of ferrite and pearlite.

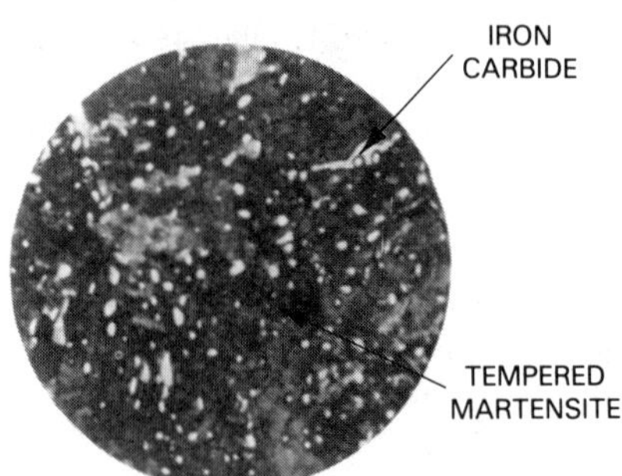


**Fig. 67-2** After heating and rapid cooling, ferrite and pearlite change into martensite.

N-1 Approximate temperatures for hardening and tempering carbon steel may be found in Table 67-1.

N-2 The critical temperature is the temperature to which carbon steel must be heated for the maximum hardness and smallest grain size.

N-3 Long, slender pieces should be quenched vertically to avoid warping.

N-4 Tempering reduces the brittleness in hardened steel.

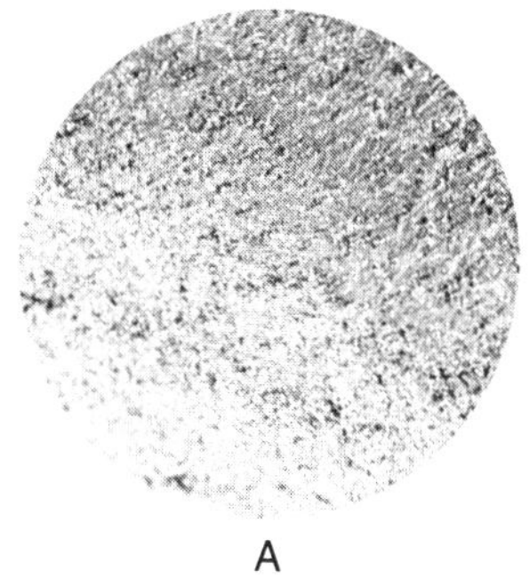

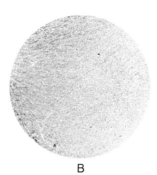

A

B

**Fig. 67-3** The grain structure of carbon tool steel. (A) Before hardening; (B) after hardening. (*Courtesy Kostel Enterprises Ltd.*)

**TABLE 67-1 TEMPERATURE CHART (FAHRENHEIT DEGREES) HEAT TREATMENT**

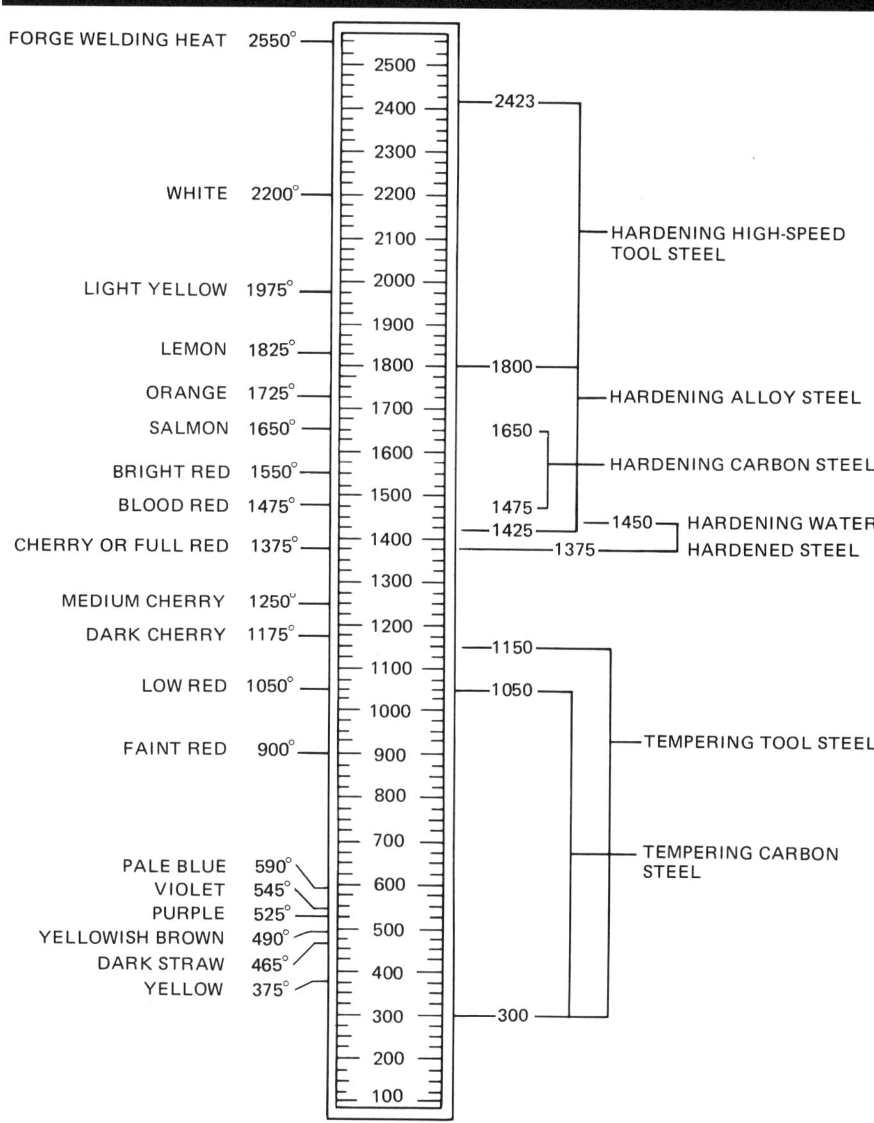


**TABLE 67-2 TEMPERING COLORS**

| Color | Degrees fahrenheit | Degrees celsius | Tools |
|---|---|---|---|
| Faint straw | 430 | 220 | Toolbits, drills, and taps |
| Medium straw | 460 | 240 | Punches and dies, and milling cutters |
| Dark straw | 490 | 255 | Shear blades and hammer faces |
| Purple | 520 | 270 | Axes, wood chisels, and tools |
| Dark blue | 570 | 300 | Knives and steel chisels |
| Light blue | 600 | 320 | Screwdrivers and springs |

## ANNEALING

N-5 Annealing softens hardened steel.

When hardened steel has to be machined, it is generally *annealed*, or softened. Annealing is the process of relieving internal strains and softening the steel by heating it above its critical temperature (see Fig. 67-4) and allowing it to cool slowly in a closed furnace or in ashes, lime, asbestos, or vermiculite. (See N-5.)

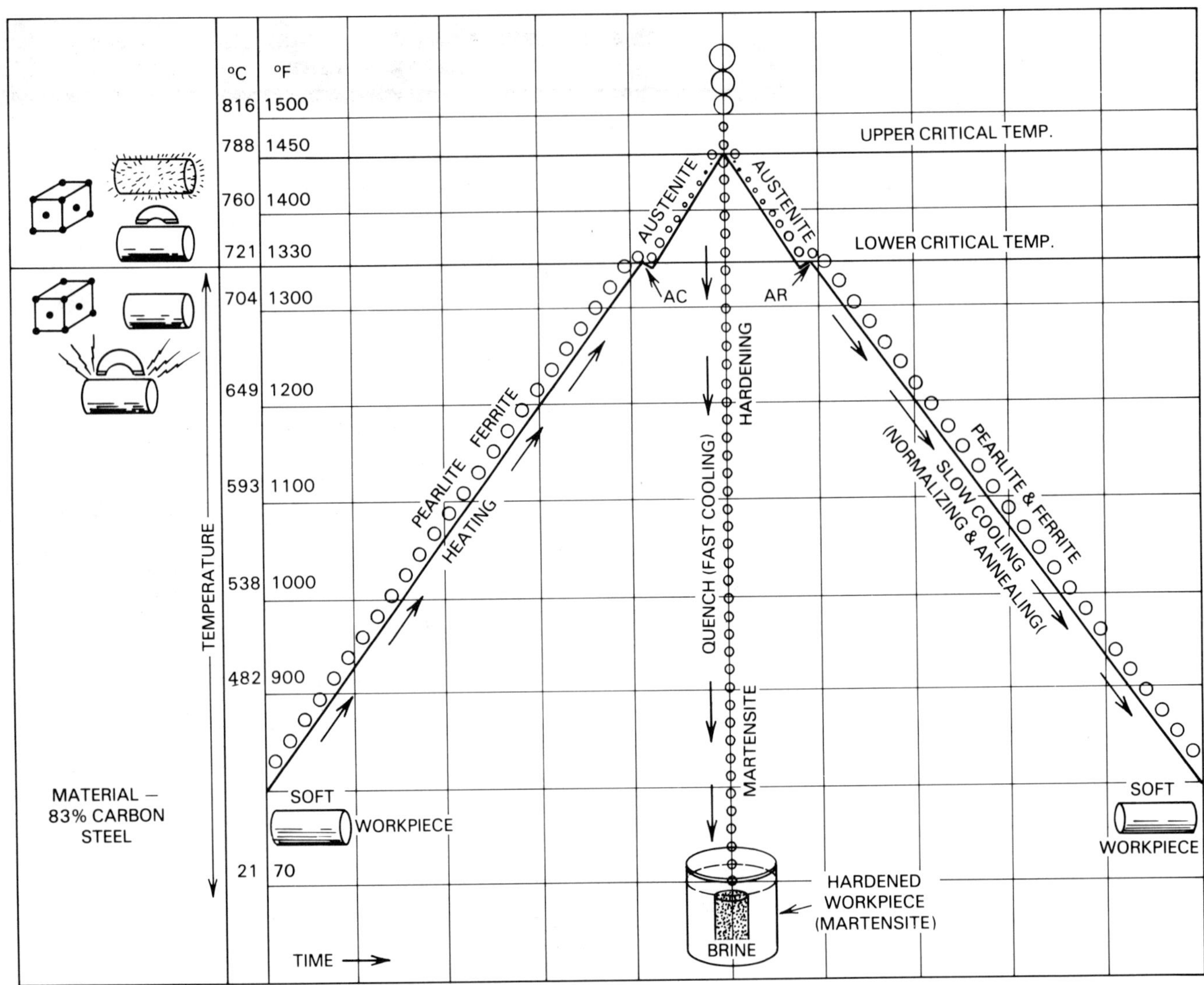


**Fig. 67-4** Steel when quenched at the upper critical temperature retains the smallest grain size.

N-6 Case-hardening produces a hard outer skin on low carbon steel while leaving the center core soft.

## CASE-HARDENING METHODS

Case-hardening is a method used to harden the outer surface of low carbon steel while leaving the center or core soft and ductile (Fig. 67-5A & B). As carbon is the hardening agent, some method must be used to increase the carbon content of low carbon steel before it can be hardened. Case-hardening involves heating the metal to its critical temperature while in contact with some carbonaceous material. Three common methods used for case-hardening are the pack method, the liquid bath method, and the gas method. (See N-6.)

A

B

HARD OUTER SURFACE

SOFT CORE

**Fig. 67-5** The grain structure of low-carbon steel. (A) Before case-hardening; (B) after case-hardening.

### Pack Carburizing

The *pack method,* also called *carburizing,* consists of packing low carbon steel in a closed box with a carbonaceous material and heating it to a temperature of 1650 to 1700°F (900 to 927°C) for 4 to 6 hours.

During this time, carbon from the packing (carbonaceous) material penetrates the surface of the workpiece forming a thin layer of carbon steel around the outside of it (Fig. 67-5B). The longer the packed workpiece is left in the furnace, the deeper will be the carbon penetration.

The steel may then be removed from the box and quickly quenched in water or brine. To avoid excessive warping, it is sometimes better to allow the box to cool, remove the steel, reheat to 1400 to 1500°F (760 to 815°C), and then quench.

In school shops Kasenit, a nonpoisonous coke compound, is often used for case hardening the workpiece.

**1.** Heat the piece of steel to about 1400 to 1500°F (760 to 815°C), which will produce a bright cherry-red color.

**2.** Remove from the furnace and cover the work with powdered Kasenit. (See N-7.) Replace metal in the furnace and leave it there until the Kasenit appears to boil; then quench it in cold water. Note: *This will only give from .010- to .015-in. (0.25- to 0.40-mm) penetration.*

### Liquid Bath Carburizing

In the *liquid bath method,* the steel to be case hardened is immersed in a liquid cyanide bath containing up to 25% sodium cyanide. Potassium cyanide may also be used, but its fumes are dangerous. (See N-8.) The temperature is held at 1550°F (845°C) for 15 minutes to 1 hour depending on the depth of case required. At this temperature the steel will absorb both carbon and nitrogen from the cyanide. The steel is then quickly quenched in water or brine.

### Gas Carburizing

With the *gas method,* carburizing gases are used to case harden low carbon steel. The steel is placed in an inner drum. Carburizing gas is introduced into the drum and the furnace is held between 1650 and 1750°F (900 to 927°C). After a predetermined time, the carburizing gas is shut off and the furnace allowed to cool. The steel is then removed and reheated to 1400 to 1500°F (760 to 815°C) and quickly quenched in water and brine.

## HARDNESS TESTING

The hardness of a metal may be defined as its resistance to forceful penetration or deformation. A steel workpiece may be tested for hardness by the file test, whereby the edge of the metal is nicked with the corner of a file. This method will indicate only whether the metal is hard or soft; it will not indicate the degree of hardness and the ultimate strength of the metal. To measure the degree of hardness of a metal, a hardness tester is used. (See N-9.)

### Hardness Testers

The three most commonly used hardness testers are the following.

*Rockwell.* The Rockwell hardness tester (Fig. 67-6) measures the amount of penetration caused by forcing a diamond point penetrator into the metal using a standard force or load. The greater the penetration, the softer the metal and the lower the Rockwell hardness number. The degree of hardness, or the Rockwell hardness number, is indicated on the dial of the tester. This type of tester is therefore called a direct-reading testing machine. Several different anvils are available for testing different-shaped objects.

*Brinell.* The Brinell hardness tester presses a 10-mm hardened steel ball under a load of 6600 lb (3000 kg) into the surface of the specimen. (See N-10.) The diameter of the impression is measured with a special microscope. This measurement is converted into a Brinell hardness number (Bhn) which defines the hardness of the metal. Harder metals have higher Bhn numbers.

*Shore Scleroscope.* This tester uses a diamond-tipped hammer that is dropped on the metal from a given height. The amount of rebound indicates the degree of hardness. The higher the rebound, the harder the material.

### To Perform a Rockwell C Hardness Test

Although the various makes of Rockwell-type testers may differ slightly in construction, they all operate on the same principle.

*Note:* For accurate results, two or three readings should be taken and av-

N-7 Always wear safety glasses when quenching hot metals during any hardening or case-hardening process.

N-8 Never inhale the poisonous cyanide fumes or allow water to come into contact with molten cyanide because explosive splattering may occur.

N-9 Generally the harder the metal is, the stronger it will be.

N-10 $\text{Brinell hardness number} = \frac{\text{area of impression}}{\text{applied load}}$

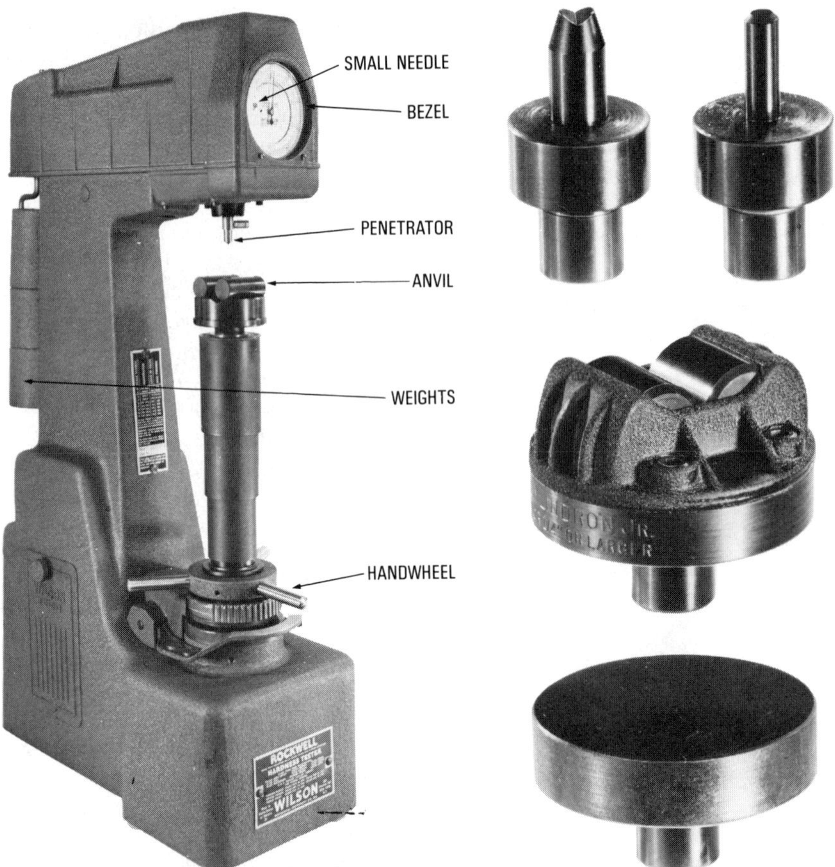


**Fig. 67-6** The Rockwell hardness tester showing various anvils used for supporting different-shaped workpieces. (*Courtesy Wilson Instrument Division, American Chain and Cable Company.*)

eraged. If either the brale or the anvil has been removed and replaced, a "dummy run" should be made to seat these parts properly before performing a test. A test piece of known hardness should be tested occasionally to check the accuracy of the instrument.

**1.** Select the proper penetrator for the material to be tested. Use a 120° diamond for hardened materials. Use a 1/16-in. ball for soft steel, cast iron, and nonferrous metals.

**2.** Mount the proper anvil (Fig. 67-6) for the shape of the part being tested.

**3.** Remove the scale or oxidation from the surface on which the test is to be made. Usually an area of about ½ in. diameter is sufficient. (See N-11.)

N-11 Scale or dirt on a surface will produce a false or incorrect hardness reading.

**4.** Place the workpiece on the anvil and apply the minor load (10 kg) by turning the handwheel, until the small needle is in line with the red dot on the dial.

**5.** Adjust the bezel (outer dial) to zero.

**6.** Apply the major load (150 kg).

**7.** After the large hand stops, remove the major load.

**8.** When the hand ceases to move backward, note the hardness reading on the C scale (black). This reading indicates the difference in penetration of the brale between the minor and major loads (Fig. 67-7).

**9.** Release the minor load and remove the specimen.

## REVIEW QUESTIONS

**1.** Define hardening.

**2.** (a) At what temperature should carbon tool steel be quenched for hardening?

(b) What color would the steel be at this temperature?

(c) What is the name of the steel structure at this temperature?

**3.** (a) After the metal is quenched in a suitable medium (water or oil) what changes take place in the steel?

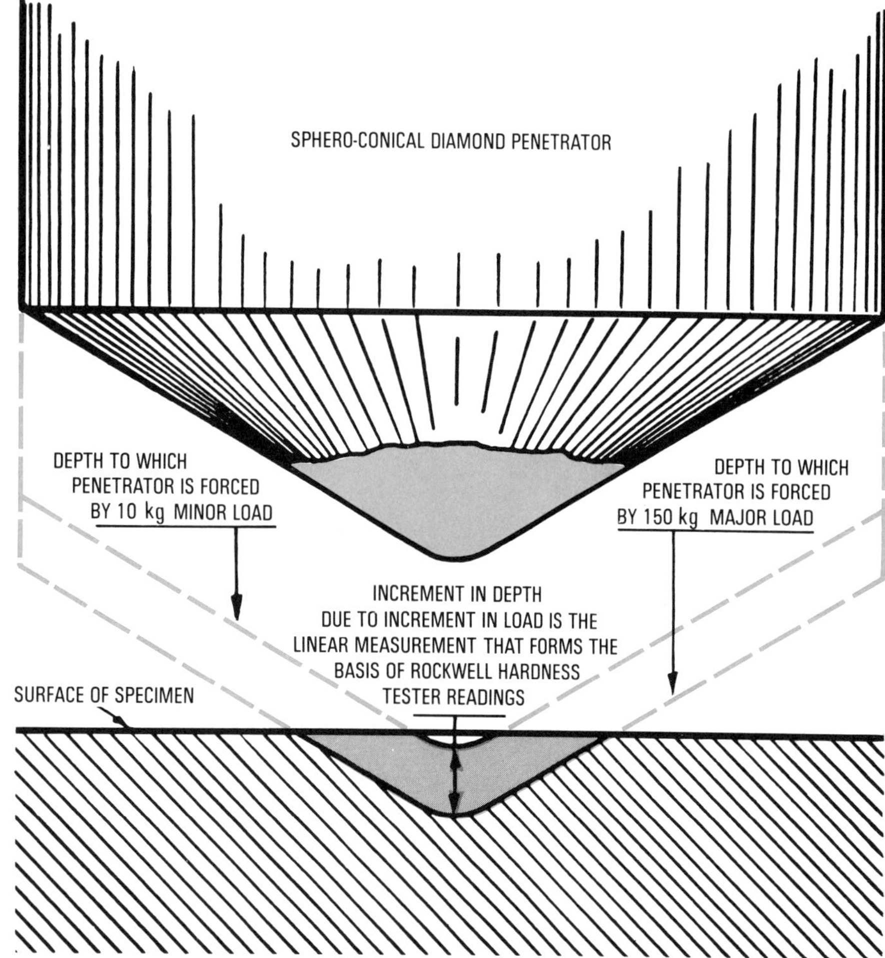


**Fig. 67-7** The operating principle of a Rockwell hardness tester—diamond cone type. (*Courtesy Wilson Instrument Division, American Chain and Cable Company.*)

(b) What is the name of this new structure?

**4.** How should long, slender pieces of steel be quenched when hardening? Explain why.

**5.** How is the metal cooled quickly and evenly in the quenching media?

**6.** What is the purpose of tempering a piece of hardened tool steel?

**7.** Briefly list the steps required to temper a piece of hardened tool steel.

**8.** Define critical temperature.

**9.** What is the purpose of annealing?

**10.** How is metal annealed?

**11.** Define case-hardening.

**12.** Name three methods of case-hardening.

**13.** Briefly describe how a workpiece may be case hardened by the Kasenit method.

**14.** List three precautions to observe when case hardening steel by the liquid bath method.

**15.** Define hardness.

**16.** (a) Name three types of commonly used hardness testers.

(b) Which one is not a direct-reading tester?

**17.** Describe the principle of the Rockwell hardness tester.

**18.** Describe the principle of the Brinell hardness tester.

**19.** On what principle does the Scleroscope operate?

# SECTION

# SPECIAL PROCESSES

The space race has spurred the development of many new metals and alloys to meet the requirements of flight in space. These space-age metals had to be lighter in weight, tougher, stronger, and more heat-resistant than previous metals. The new metals created to meet these requirements presented special problems for those who had to cut and form them into the required parts. Since many of these metals were tough, conventional machining methods such as turning, milling, and drilling could not machine the metals properly or efficiently.

*Electric machining processes* such as electrochemical machining, electrical discharge machining, and electrolytic grinding made it possible to machine intricate forms in exotic and difficult-to-machine metals with accuracy and ease. The development of *numercial control* provided industry with a means of accurately and automatically controlling the movement of machine tools by programmed numerical data in order to produce parts accurately and automatically. *Powder metallurgy* has made it possible to form intricate parts to accurate size and shape by a forming and sintering process, thereby eliminating any need for conventional machining operations to finish the part. The *laser* is expected to play an ever-increasing role in the machining and inspection of parts. *High-energy forming* such as explosive and electromagnetic forming is now used to form metals and shapes which were very difficult or impossible to form by other methods.

There is continual research and development in order to improve metal-removal techniques and to develop better and more accurate machine tools. New metal-removal processes and the ever-increasing use of computers to control and operate machine tools will make the future in this field more exciting than ever.

# UNIT 68

# ELECTRIC-MACHINING PROCESSES

With the introduction of unique mechanisms and exotic materials in recent years, it has been found necessary to develop new methods of efficiently machining metals. Parts made out of cemented carbide or difficult-to-machine metals were previously shaped by the costly process of diamond wheel grinding. *Electrochemical machining, electrical discharge machining,* and *electrolytic grinding* are three methods which have been developed recently. All remove metal by some form of electric discharge.

## OBJECTIVES

After completing this unit you will:

1. Recognize the importance of the electric machining process
2. Understand the principles of:
   (a) Electrochemical machining
   (b) Electrical discharge machining
   (c) Electrolytic grinding

## ELECTROCHEMICAL MACHINING

Electrochemical machining, more commonly called ECM, differs from conventional metal-cutting techniques in that electric and chemical energy are used as the cutting tool. (See N-1.) This process removes metal easily, regardless of the work hardness, and is characterized by its ''chipless'' operation. A nonrotating tool the shape of the cavity required is the *cutting tool*; therefore square or difficult-to-machine shapes can easily be cut in a workpiece. The wear on the cutting tool is hardly noticeable since the tool *never* touches the work. Electrochemical machining is particularly suitable for producing round through holes, square through holes, round or square blind holes, simple cavities which have straight, parallel sides, and planing operations. ECM is especially valuable when metals that exceed a hardness of 42 Rockwell C (400 BHN) are machined. Sharp corners, flat bottom sections, or true radii are difficult to maintain because of the slight overcut which occurs during this process. A significant advantage of ECM is that the surfaces and edges of workpieces are not deformed and are left burr-free. (See N-2.)

N-1 Electrochemical machining combines electric and chemical energy to remove metal.

N-2 The cutting tool (electrode) never comes into contact with the workpiece.

### The Process

For years, metal in a solution has been transferred from one metal to another by means of electroplating baths. Since ECM evolved from this process, it may be wise to examine the electroplating principle (Fig. 68-1).

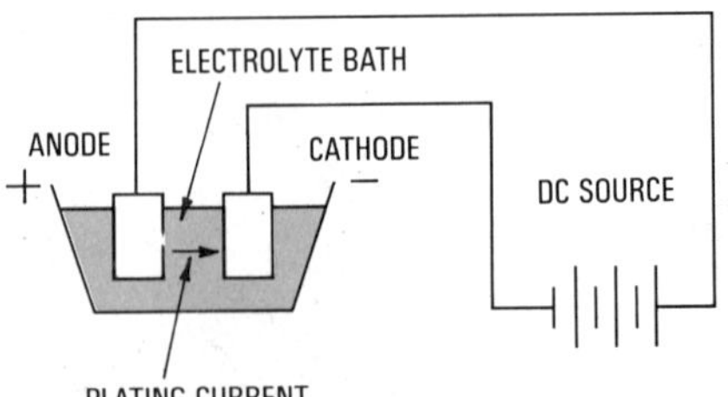

**Fig. 68-1** The principle of the electroplating process.

1. Two bars of unlike metal are immersed in an electrolyte solution.
2. One bar is fastened to a negative lead on a battery while the other is fastened to the positive lead.
3. When the circuit is closed, direct current passes through the electrolyte between the two bars of metal.
4. Chemical reaction transfers metal from one bar to the other.

Electrochemical machining differs from the plating process in that an electrochemical reaction dissolves metal *from a workpiece* into an electrolyte solution. A direct current is passed through an electrolyte solution between the electrode tool (the shape of the cavity desired), which is negative (−), and the workpiece, which is positive (+). This causes metal to be removed ahead of the electrode tool as the tool is fed toward the work. Chemical reaction caused by the direct current in the electrolyte dissolves the metal from the workpiece (Fig. 68-2).

The *electrode* for ECM is not a simple bar of metal, but a precision-insulated hollow tool which has been made to a specific shape and exact size, and through which the electrolyte is forced (Fig. 68-3). The electrode (tool) and workpiece, although located within .002 to .003 in. (0.05 to 0.07 mm), never contact each other. The *electrolyte solution* is a controlled, swiftly flowing stream which carries the current. (See N-3.) The *direct current* used may at times be as high as 10,000 A/in.$^2$ (1550 A/cm$^2$) of work material, and the rate of metal removal is directly proportional to the current

N-3 The electrolyte is generally a very corrosive salt solution.

passing between the tool and the workpiece. A high current will result in a high rate of metal removal; a low current produces a low rate of removal.

## Advantages of ECM

Electrochemical machining is one of the metal-cutting processes which has contributed to the machining of space-age metals. Some of its characteristics and advantages are the following.

**1.** Metal of any hardness can be machined.

**2.** No heat is created during the machining process; therefore, there is no work distortion.

**3.** ECM machines metal without tool rotation.

**4.** Tool wear is insignificant because the tool never touches the work.

**5.** Because the tool never touches the work, thin fragile sections can be machined without distortion.

**6.** The workpiece is left burr-free.

**7.** Intricate forms, difficult to machine by other processes, can be produced easily (Fig. 68-4).

**8.** ECM is suitable for production-type work where multiple holes or cavities may be machined at the same time.

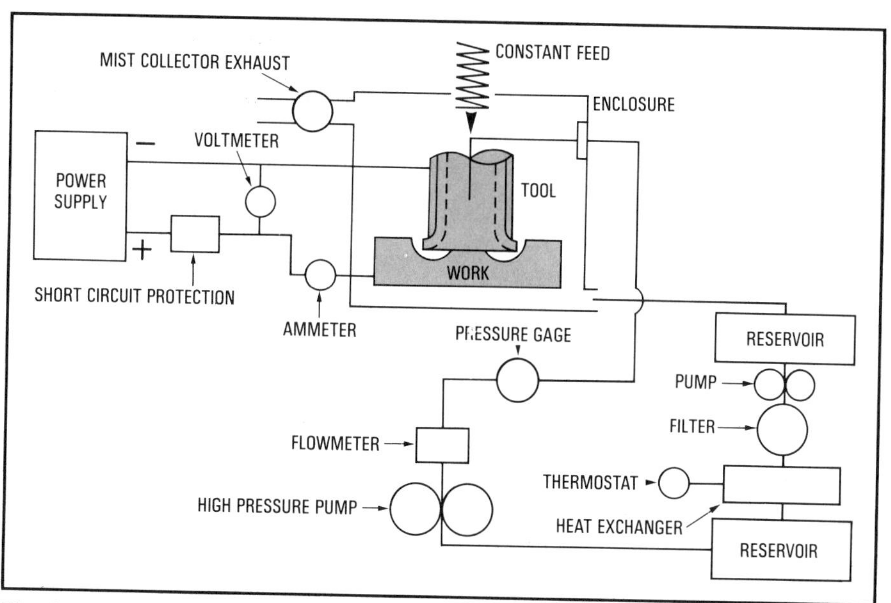


**Fig. 68-2** A schematic diagram of a typical electrochemical machining system.

# ELECTRICAL DISCHARGE MACHINING

*Electrical discharge machining,* commonly known as EDM, is a process that is used to remove metal through the action of an electric discharge of short duration and high current density between the tool and the workpiece (Fig. 68-5). The EDM process can be compared with a miniature version of a lightning bolt striking a surface, creating a localized intense heat, and melting away the work surface.

Electrical discharge machining has proved especially valuable in the machining of super-tough, electrically conductive materials such as the new space-age alloys. These metals would have been difficult to machine by conventional methods, but EDM has made it relatively simple to machine intricate

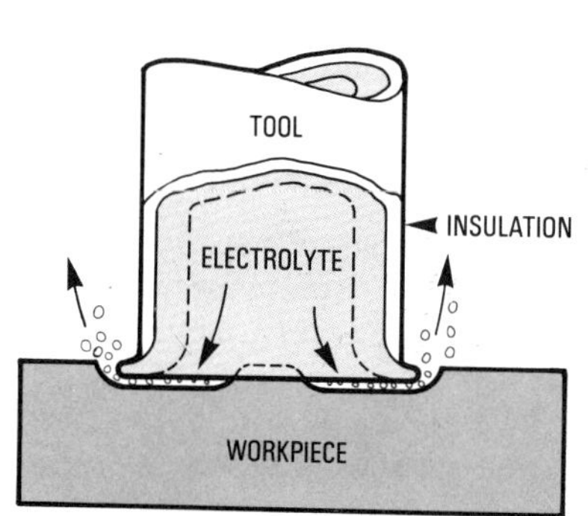


**Fig. 68-3** Electrolyte flows through the tool (electrode) during the electrochemical process.

**Fig. 68-4** An example of electrochemical machining applications. (*Courtesy The Cincinnati Milacron Inc.*)

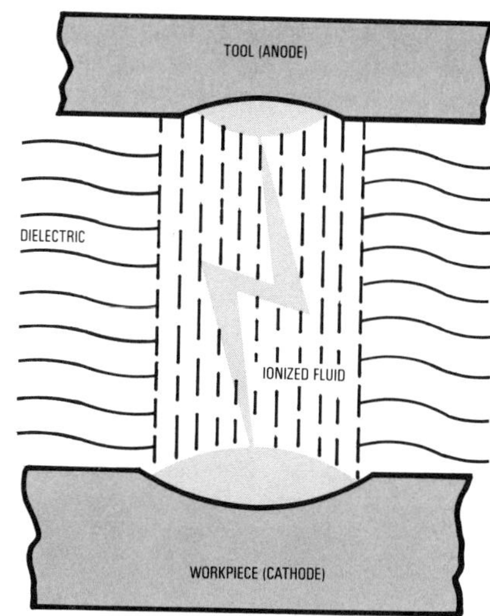


**Fig. 68-5** A controlled spark removes metal during electric discharge machining.

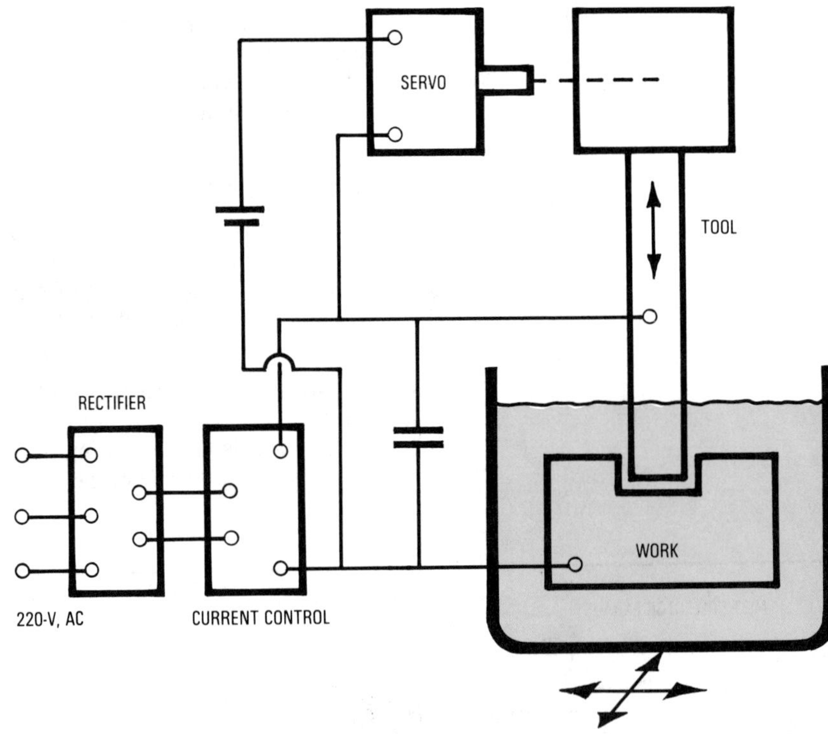


**Fig. 68-6** Basic elements of an electric-discharge system.

shapes that would be impossible to produce with conventional cutting tools.

## Principle of EDM

Electrical discharge machining is a controlled metal-removal technique whereby an electric spark is used to cut (erode) the workpiece by means of the cutting tool or electrode (Fig. 68-6). The *cutting tool (electrode)* is made from electrically conductive material, usually carbon. The electrode, made to the shape of the cavity required, and the workpiece are both submerged in a *dielectric fluid,* which is generally a light lubricating oil. This dielectric fluid should be a nonconductor (or poor conductor) of electricity. (See N-4.) A *servomechanism* maintains a gap of about .001 in. (0.02 mm) between the electrode and the work, preventing them from coming into contact with each other. A *direct current* of low voltage and high amperage is delivered to the electrode at the rate of approximately 20,000 electric impulses per second through a capacitor. These electric-energy impulses become sparks which jump the gap (Fig. 68-5) between the electrode and the workpiece through the dielectric fluid. Intense heat is created in the localized area of the spark impact; the metal melts and a small particle of molten metal is expelled from the surface of the workpiece. (See N-5.) The dielectric fluid, which is constantly being circulated, carries away the eroded particles of metal and also assists in dissipating the heat caused by the spark.

N-4 An electrically conductive cutting tool (electrode) and a nonconductive dielectric fluid are required for electrical discharge machining.

N-5 Metal is removed by electric-spark erosion.

## Advantages of the EDM Process

Electrical discharge machining has many advantages over conventional machining processes.

**1.** Any electrically conductive material can be cut, regardless of its hardness. It is especially valuable for cemented carbides and the new supertough space-age alloys that are extremely difficult to cut by conventional means.

**2.** Work can be machined in a hardened state, thereby overcoming the deformation caused by the hardening process.

**3.** Broken taps or drills can readily be removed from workpieces.

**4.** It does not create stresses in the work material, since the tool (electrode) never comes in contact with the work.

Fig. 68-7 Examples of work produced by electric-discharge machining.

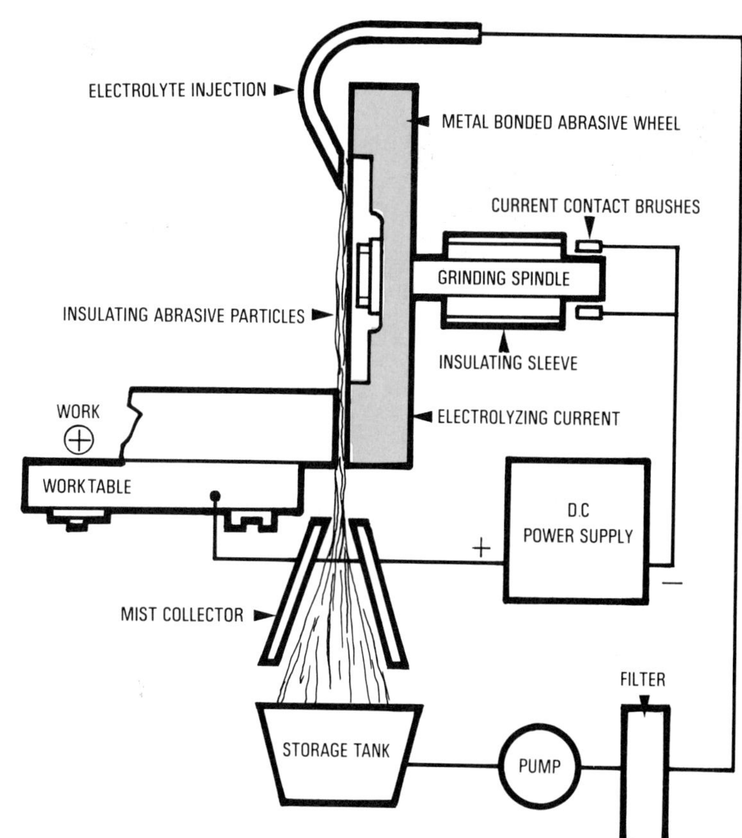

Fig. 68-8 The electrolytic grinding principle.

**5.** The process is burr-free.

**6.** Thin, fragile sections can be easily machined without deforming.

**7.** Secondary finishing operations are generally eliminated for many types of work.

**8.** Intricate shapes, impossible to produce by conventional means, are cut out of a solid with relative ease (Fig. 68-7).

**9.** Better dies and molds can be produced at lower costs.

**10.** A die punch can be used as the electrode to reproduce its shape in the matching die plate, complete with the necessary clearance.

## ELECTROLYTIC GRINDING

*Electrolytic grinding* has been a boon to the machining of thin, fragile metal products and tough, difficult-to-machine space-age alloys. In electrolytic grinding, the metal is removed from the work surface by a combination of electrochemical action and the action of a metal-bonded abrasive grinding wheel. (See N-6.) Approximately 90 percent of the metal removed from the work surface is the result of this electrochemical deplating action, while the remaining 10 percent is "wiped away" by the grinding wheel. The electrolytic grinding process is very similar to the process used for electrochemical machining.

N-6 Metal is removed by a combination of electrical and chemical energy in the electrolytic grinding process.

### The Electrolytic Grinding Process

The metal-bonded electrically conductive grinding wheel and the electrically conductive workpiece are both connected to a *direct current* power supply and are separated by the protruding abrasive particles of the wheel (Fig. 68-8). An electrolytic grinding solution (electrolyte) is injected into the gap between the wheel and the work, completing the electric circuit and producing the necessary deplating action which decomposes the work material. (See N-7.) This decomposed material is removed by the action of the revolving grinding wheel and is washed away in the solution. In electrolytic grinding, the wheel never comes into actual contact with the work material.

N-7 The electrolytic solution is generally very corrosive and machine parts must be protected against this corrosion.

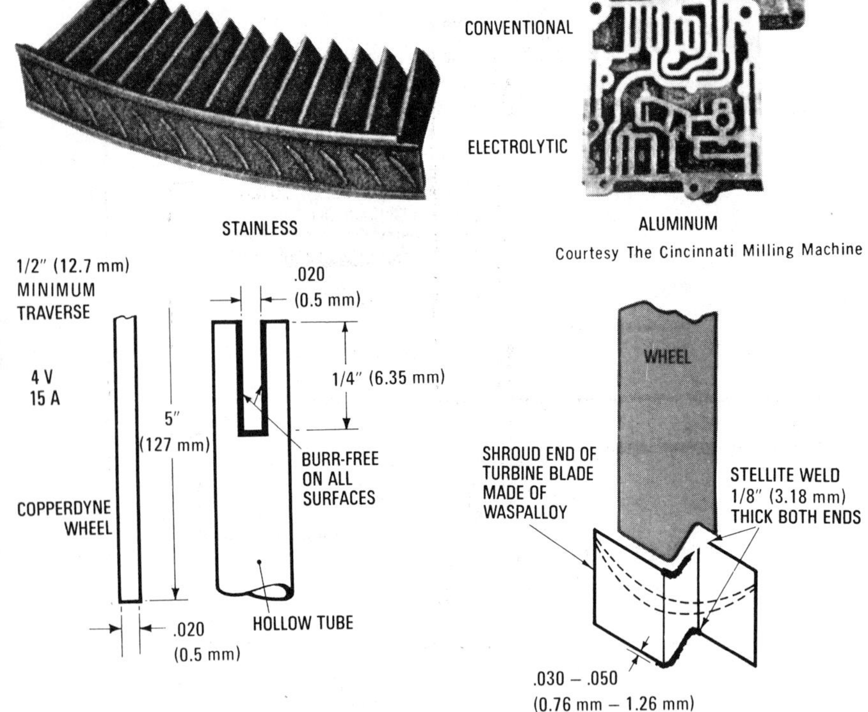


**Fig. 68-9** Examples of electrolytic grinding applications. (*Courtesy Cincinnati Milacron Inc.*)

N-8 Electrolytic grinding metal-removal process is burr-free and does not distort workpieces because the grinding wheel never contacts the work.

### Advantages of Electrolytic Grinding

Electrolytic grinding offers many advantages over conventional machining methods for metal removal.

**1.** It saves wheel costs, especially in metal-bonded diamond wheels, since only 10 percent of the metal is removed by abrasive action.
**2.** There is a high ratio of stock removal in relation to the wheel wear.
**3.** No heat is generated during the grinding operation: therefore, there is no burning or heat distortion of the work.
**4.** The process is burr-free, thereby eliminating deburring operations.
**5.** Thin, fragile workpieces (Fig. 68-9) can be ground without distortion since the wheel never touches the work. (See N-8.)
**6.** Tungsten carbide and supertough alloys can be ground quickly and with ease.
**7.** Exotic materials, such as zirconium, beryllium, etc., can be cut regardless of their hardness, fragility, or thermal sensitivity.
**8.** No stresses are created in the work material.
**9.** No work hardening occurs during this process.

## REVIEW QUESTIONS

**1.** How does electrochemical machining differ from conventional machining?
**2.** List the main steps in the electrochemical machining process.
**3.** State six advantages of electrochemical machining.
**4.** Briefly describe the electrical discharge machining process.
**5.** What purpose does the dielectric serve?
**6.** List six advantages of electrical discharge machining.
**7.** Briefly describe the electrolytic grinding process.
**8.** What type of grinding wheels are used in electrolytic grinding?
**9.** State six advantages of the electrolytic grinding process.

# UNIT 69

# POWDER METALLURGY

Powder metallurgy is a radical departure from the conventional machining process. With conventional machining methods, a piece of steel or bar stock larger than the finished part is selected and the required form is machined. The material cut away in the machining process is in the form of steel chips, which are considered *scrap loss* (Fig. 69-1). In powder metallurgy, the correct types of powders are blended together and formed into the required shape in a die. Since machining is not required, scrap loss is minute as only the required amount of powder is used for each part. Today powder metallurgy is used to produce cams, gears, self-oiling bearings, light filaments, levers, cemented-carbide tools, automotive filters, etc., which were previously produced by conventional machining methods.

## OBJECTIVES

After completing this unit, you should be aware of:

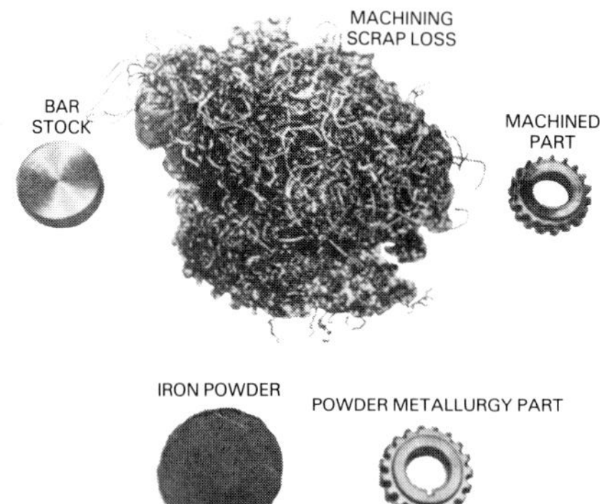


**Fig. 69-1** Comparison of conventional machining and powder-metallurgy processes. (*Courtesy The Powder Metallurgy Parts Mfg. Association.*)

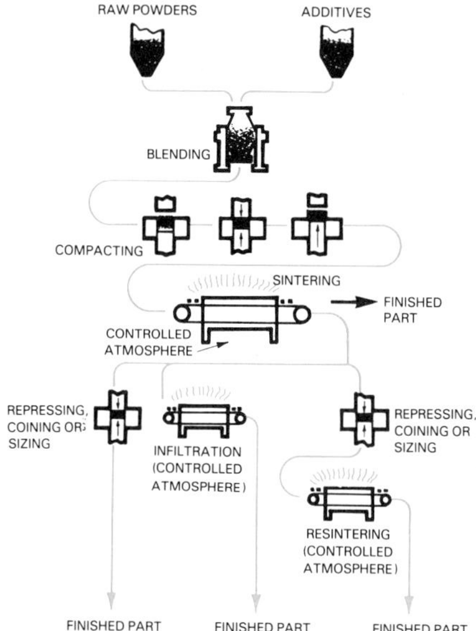


**Fig. 69-2** The steps involved in the powder-metallurgy process. (*Courtesy The Powder Metallurgy Parts Mfg. Association.*)

N-1 Intricate parts may be made to size and shape without machining by the powder metallurgy process.

N-2 Various metal and nonmetal powders can be combined to give the finished part various physical qualities.

N-3 The powders which are blended (mixed) together depend on the requirements of the finished part.

**1.** The principle and operations involved in the powder metallurgy process

**2.** The applications and advantages of the powder metallurgy process

## THE POWDER METALLURGY PROCESS

*Powder metallurgy* is the process of producing metal parts by:

**1.** Blending powdered metals and alloys

**2.** Compressing the powders in a die which is the shape of the part to be produced

**3.** Subjecting the shaped form to elevated temperatures (sintering), which causes the metal particles to "weld" together and form a solid part

A part made by the powder metallurgy process begins with metal powders and goes through three main stages before it becomes a finished product (Fig. 69-2). These include securing raw powders, blending, compacting, and sintering. (See N-1.)

### Raw Powders

Almost any type of metal can be produced in powder form; however, only a few have the desired characteristics and properties necessary for economical production. (See N-2.) Iron and copper base powders are the two main types which lend themselves well to the powder metallurgy process. Metal powders can be produced by atomization, electrolytic deposition, granulation, crushing, reduction, and shotting.

### Blending

The powder for a specific product must be carefully selected to ensure that the finished part will have the required characteristics and that it will be an economical production. (See N-3.)

Once the correct powders have been selected, the mass of each is carefully determined in the proportion required for the finished component, and a die lubricant, such as powdered graphite, zinc stearate, or stearic acid, is added. The purpose of this lubricant is to assist the flow of powder in the die, prevent scoring of the die walls, and permit easy ejection of the compressed part. This composition is then carefully mixed or blended (Fig. 69-2) to ensure homogeneous grain distribution in the finished product.

### Compacting (Briquetting)

The powder blend is then fed into a precision die which is the shape and size of the finished product. The die generally consists of a die shell, an upper punch, and a lower punch (Fig. 69-3). These dies are usually mounted on hydraulic or mechanical presses where pressures from about 3000 lb/in.$^2$ (20.5 MPa) to as much as 200,000 lb/in.$^2$ (1379 MPa) are used to compress the powder. Soft powder particles can be pressed or keyed together readily and therefore do not require as high a pressure as the harder particles. The density and hardness of the finished product increases with the amount of pressure used to compact or briquette the powder.

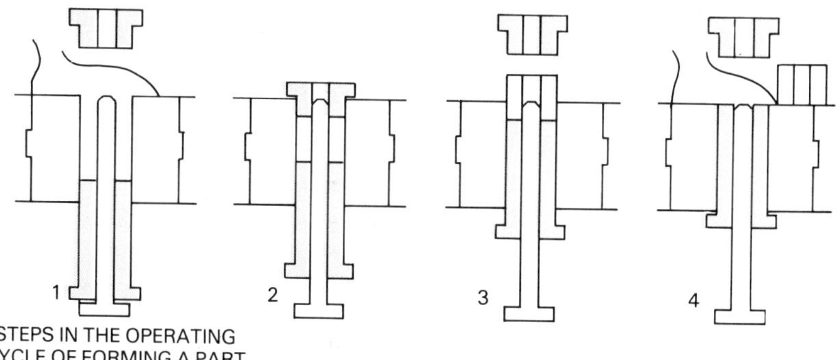


1. THE DIE CAVITY IS FILLED WITH BLENDED POWDER.
2. THE TOP AND BOTTOM PUNCHES COMPRESS THE POWDER TO SHAPE.
3. THE COMPACTED PART IS EJECTED FROM THE DIE BY THE BOTTOM PUNCH.
4. THE COMPACTED PART IS MOVED CLEAR OF THE DIE AREA AND MORE BLENDED POWDER IS FED INTO THE DIE.

**Fig. 69-3** The four steps in compacting metal powders to shape. (*Courtesy The Powder Metallurgy Parts Mfg. Association.*)

N-4 Sintering fuses the part into a solid mass.

N-5 Most powder metallurgy parts are ready for use, but a few require finishing operations to bring them to size or to give them certain desirable qualities.

N-6 Powder metallurgy makes it possible to produce certain intricate parts easily and at lower costs than would be possible with conventional machining methods.

## Sintering

After compacting, the "green part" must be heated sufficiently in order to bond the metal particles into a solid. This operation of heating is known as *sintering*. (See N-4.) The parts are passed through controlled protective atmosphere furnaces which are maintained at a temperature approximately one-third below the melting point of the principal powder. The carefully controlled atmosphere and temperature during the sintering operation prevents oxidation of the pressed powder and permits the particles to bond together. Most sintering is done in a hydrogen and/or nitrogen atmosphere, producing parts with no scaling or discoloration.

The temperature within the furnaces varies from 1600 to 2700°F (871 to 1482°C), depending on the type of metal powder being sintered. The time required to sinter a part varies, depending on its shape and size; normally, however, it is from 15 to 45 minutes.

The purpose of the sintering operation is to bond the powder particles together to form a strong homogeneous part having the desired physical characteristics.

## Finishing Operations

After sintering, most parts are ready for service. However, some parts requiring very close tolerances or other qualities may require some additional operations, such as sizing and repressing, impregnation, infiltration, plating, heat treating, and machining.

### Sizing and Repressing

Parts requiring close tolerances or increased density must be *sized* or *repressed*. This involves putting the part into a die which is similar to the one used for compacting and repressing it. This operation improves the surface finish and the dimensional accuracy, increases the density, and gives added strength to the part.

### Impregnation

The process of filling the pores of a sintered part with a lubricant or nonmetallic material is called *impregnation*. Oilite bearings are a good example of a part impregnated with oil to overcome the necessity for constant lubrication and maintenance. Parts may be impregnated by a vacuum process or by a process that involves soaking the parts in oil for several hours. (See N-5.)

### Infiltration

Filling the pores of a part with a metal or alloy having a lower melting point than the sintered piece is called *infiltration*. The purpose of this operation is to increase the density, strength, hardness, impact resistance, and ductility of the manufactured part.

Slugs of the material to be infiltrated are placed on the compacted parts and passed through the sintering furnace. Because of its lower melting point, the infiltrated material melts and penetrates the pores of the compacted part by capillary action.

### Plating, Heat Treating, and Machining

Depending on the type of material, its application, and requirements, some powder metallurgy parts may be plated, heat-treated, machined, brazed, or welded.

# ADVANTAGES OF POWDER METALLURGY

The use of powder metallurgy is increasing rapidly, with many parts being made better and more economically than with previous manufacturing methods. Some of the advantages of the powder-metallurgy process are listed below. (See N-6.)

**1.** Close dimensional tolerances of plus or minus .001 in. (0.02 mm) can be achieved and smooth finishes can be obtained without costly secondary operations.

**2.** Complex or unusually shaped parts, which would be impractical to obtain by any other method, can be produced.

**3.** Production of porous bearings and cemented carbide tools is possible.

**4.** Pores of a part can be infiltrated with other metals.

**5.** Surfaces of great wear resistance can be produced.

**6.** The porosity of the product can be accurately controlled.

**7.** Products made of extremely pure metals can be produced.

**8.** There is little waste in this process.

**9.** The operation can be automated and employ unskilled labor, keeping costs low.

**10.** The physical properties of the product can be minutely controlled.

**11.** Duplicate parts that are accurate and unvarying can be made.

**12.** Powders made from difficult-to-machine alloys can be formed into parts which would be difficult to produce by machining processes.

## REVIEW QUESTIONS

**1.** Name six products which can be produced by powder metallurgy.

**2.** List the three main steps in the powder metallurgy process.

**3.** Name two main metal powders used in powder metallurgy.

**4.** (a) During the blending process, what material may be added to the base powder?
(b) State three reasons for adding these materials.

**5.** Describe and state the purpose of sintering.

**6.** List four advantages of powder metallurgy.

# UNIT 70

# NUMERICAL CONTROL

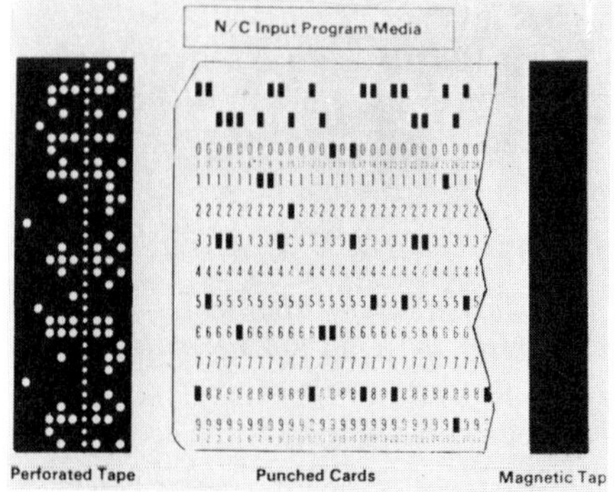


**Fig. 70-1** Types of numerical control input program media. *(Courtesy The Superior Electric Co.)*

N-1 Numerical control guides a machine tool to produce parts accurately by coded information.

Numerical control may be defined as a method of accurately controlling the movement of machine tools by a series of *programmed numerical data* which activates the motors of the machine tool.

A basic form of numerical control has existed since the first print or sketch of a part was dimensioned by a draftsman. The numbers on a print convey the information to the machine operator who proceeds to transform these numbers into manual movements of the machine tool. Numerically controlled machines are supplied with detailed information regarding the part by means of punched cards or tape (Fig. 70-1). The machine decodes this punched information and electronic devices activate the various motors on the machine tool, causing them to follow specific instructions. The measuring and recording devices incorporated into numerically controlled machine tools ensure that the part being manufactured will be accurate. Numerically controlled machines eliminate the possibility of human error which existed before their development.

## OBJECTIVES

After completing this unit you should know:

**1.** The types of systems and controls used in numerical control

**2.** The steps involved in producing a part by numerical control

**3.** The advantages and disadvantages of numerical control

## NUMERICAL CONTROL THEORY

Numerical control is really an efficient method of reading prints and conveying this information to the motors which control the speeds, feeds, and various motions of the machine tool. (See N-1.) The designer's information is punched on a tape which is put into the machine-tool reader. Here holes are scanned by means of small mechanical wires or fingers (Fig. 70-2). Beams of light are sometimes used for this purpose. The fingers or beams of light are connected to an electric circuit. Each time a hole in the tape appears beneath the finger or beam of light, a specific circuit is activated, sending signals to start or stop motors

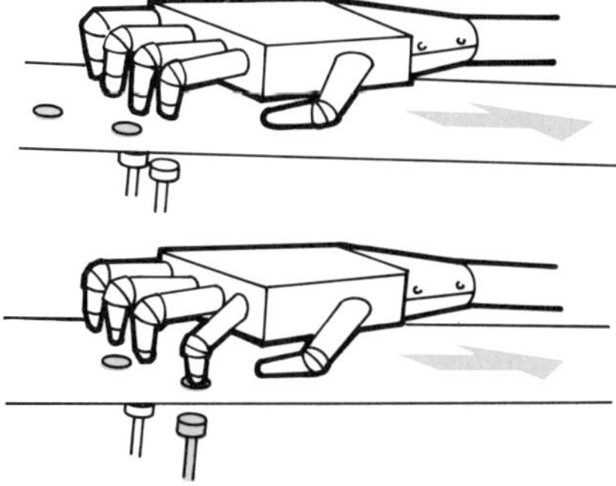

**Fig. 70-2** Mechanical fingers or beams of light read tape information.

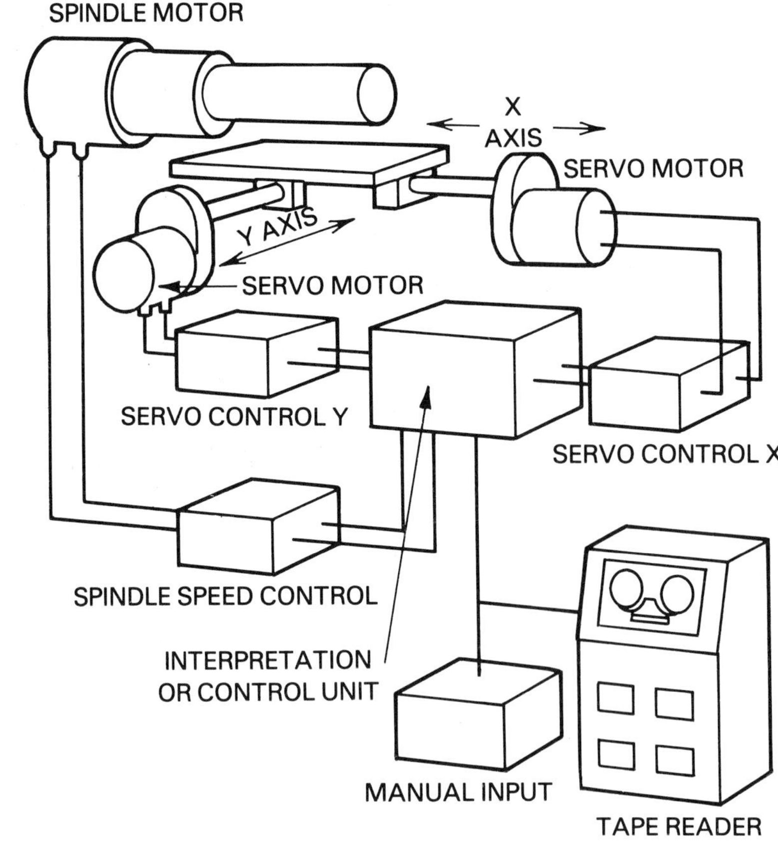


**Fig. 70-3** An open loop numerical control system.

and control various functions of the machine tool.

## TYPES OF NUMERICAL CONTROL SYSTEMS

*Open loop* and *closed loop* are the two main types of control systems used for numerical control machine tools.

### Open Loop System

In the *open loop system* (Fig. 70-3), the tape is fed into a *tape reader* which decodes the information punched on the tape and stores it until the machine is ready to use it and then converts it into electric pulses or signals. (See N-2.) These signals are sent to the *control unit* which energizes the *servo control units*. The servo control units direct the *servomotors* to perform certain functions according to the information supplied by the tape. The amount each servomotor will move the leadscrew of the machine depends on the number of electric pulses it receives from the servo control unit.

N-2 Open loop systems are accurate to .001 in. (0.02 mm) and measurements on prints are shown to three decimal places.

This type of system is fairly simple; however, since there is no means of checking if the servomotor has performed its function correctly, it is not generally used where an accuracy greater than .001 in. (0.02 mm) is required. The open loop system may be compared with a gun crew that has made all the calculations necessary to hit a distant target but does not have an observer to confirm the accuracy of the shot.

### Closed Loop System

The *closed loop system* (Fig. 70-4) can be compared with the same gun crew that now has an observer to confirm the accuracy of the shot. The observer relays the information regarding the accuracy of the shot to the gun crew, which then makes the necessary adjustments to hit the target. (See N-3.)

N-3 Closed loop system transducers permit accuracy of .0001 in. (0.002 mm) and measurements shown on prints are to four decimal places.

The closed loop system is similar to the open loop system with the exception that a *feedback unit* (Fig. 70-4) is introduced in the electric circuit. This feedback unit, generally called a *transducer,* compares the amount the ma-

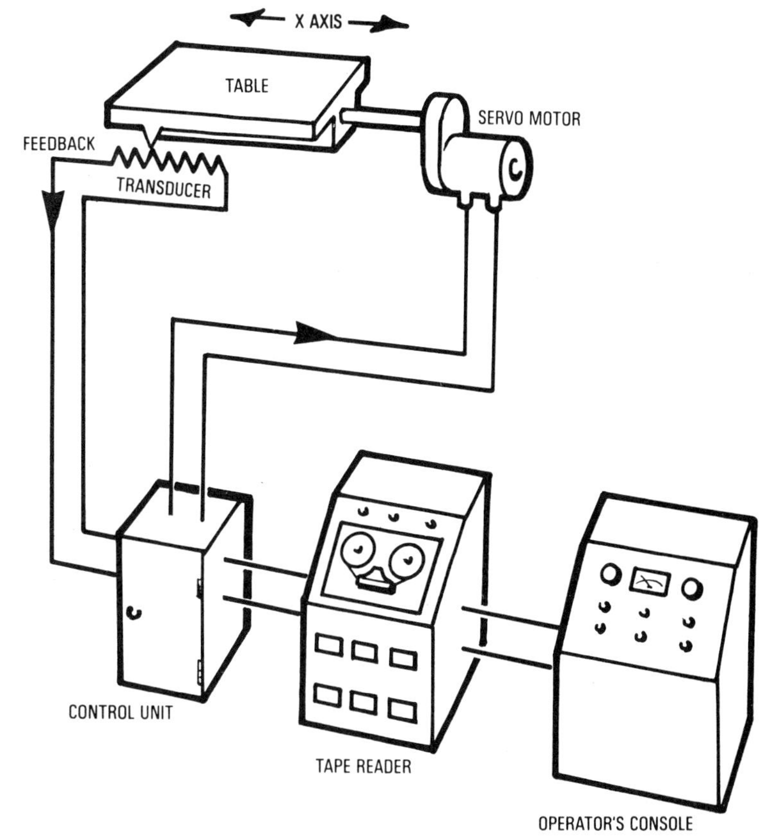


**Fig. 70-4** A closed loop numerical control system showing *x* axis only.

chine table has been moved by the servomotor with the signal sent by the control unit. The control unit instructs the servomotor to make whatever adjustments are necessary until both the signal from the control unit and the one from the servounit are equal. In the closed loop system, 10,000 electric pulses are required to move the machine slide 1 in. (25.4 mm). Therefore, on this type system, one pulse will cause .0001 in. (0.002 mm) movement of the machine slide. Closed loop numerical control systems are very accurate because the accuracy of the command signal is recorded and there is an automatic compensation for error.

## TYPES OF CONTROL

There are two types of controls available which position and control the cutting tool and the workpiece during the machining operation. These are *point-to-point* control and the *continuous-path, or contouring, control*. The type of work to be performed on a machine will determine the type of control that should be purchased.

### Point-to-Point Positioning

With this type of control, the machine slide is instructed to go from point *A* to *B*. The path taken by the machine slide between these two points does not matter, since the operations required must be performed only at points *A* and *B*. The tool slide moves from point *A* to *B* in a series of small steps (Fig. 70-5). The point-to-point commands are ideal for drilling, boring, and tapping operations. This type of positioning is usually found on open loop systems.

### Continuous-Path, or Contouring, Control

When irregular contours are being machined, the control unit must direct the path of the cutting tool at all times. If a machining operation (for example, a straight line cut) must be performed between points *A* and *B*, additional de-

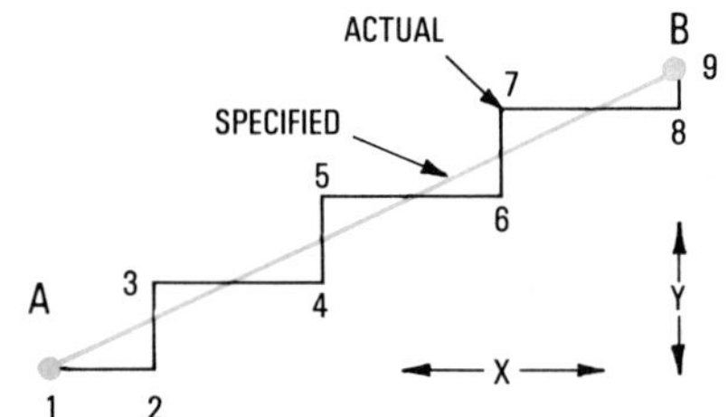


**Fig. 70-5** A comparison of point-to-point control (black) and the straight-cut control (red).

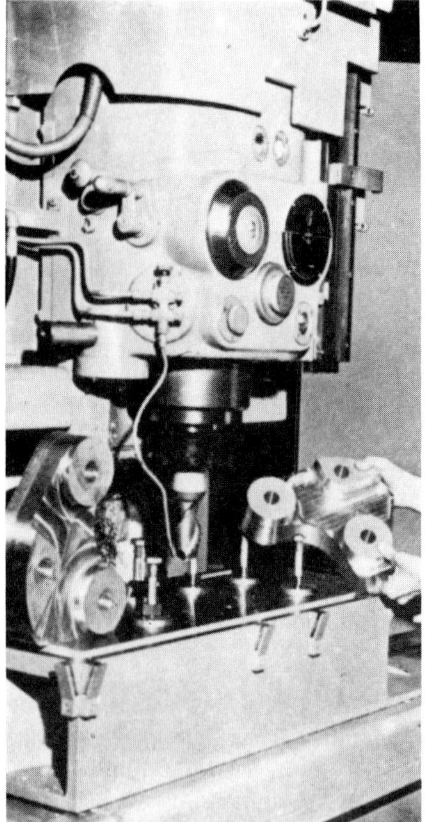

**Fig. 70-6** A complex part produced by a computer-controlled contouring system. (*Courtesy Cincinnati Milacron Inc.*)

N-4 Continuous path or contouring control is used when machining irregular forms or contours.

N-5 The EIA (Electronics Industries Association) and the ASCII (American Standard Code for Information Interchange) tape coding systems are in use. The two systems are very similar but the EIA is the most common.

vices must be added to the system. (See N-4.) Much more detailed signals are sent to the machine slide so that each step it takes to get from point *A* to *B* is minute. The *continuous-path*, or *straight-cut, system* would follow the specified path *AB* illustrated in Fig. 70-5 at a controlled rate, producing a straight-line cut. A *computer* is essential for the continuous-path system because of the large number of coordinates required to guide the machine tool on a straight-line path or on complex profiles, such as is shown in Fig. 70-6.

## Input Media

With the development of numerical control, a variety of input media has been used to convey the information from the print to the machine. The most common types of input media used are magnetic tape, punched cards, and punched tape. Most machines now use a standard 1-in.-wide punched tape which may be made of paper, Mylar, or foil.

## Tape Coding

The Electronics Industries Association adopted the binary-coded decimal system for standardizing information punched on tapes used for numerical control. The standard coding system used for the 1-in.- (25.4-mm-) wide, 8 channel tape is illustrated in Fig. 70-7. (See N-5.)

The numbers 1 to 8 on the top of the tape represent only the number of each channel and have no relationship to the holes punched in the tape.

**1.** *The sprocket holes* between channels 3 and 4 are used to drive the tape through the machine. The sprocket holes are set off center so that it is impossible to put the reverse side of the tape into the machine.

**2.** *Channels 1, 2, 3, and 4* are used for *numerical data,* such as dimensions, speeds, feeds, etc. These channels have numerical values of 1, 2, 4, and 8, respectively.

**3.** *Channel 5,* marked CH, is called the *parity check.* An odd number of holes must appear in each row; otherwise the tape reader will stop the machine. If an even number of holes must be punched in one row to get the correct information on the tape, an additional hole must be punched in channel 5. The odd parity check tests for any errors or mechanical failures in the tape preparation.

**4.** *Channel 6* always represents a "zero."

**5.** *Channel 7,* marked X, is used to select a letter to identify various machine operations and is not used for dimensions. Each letter (a to z) in the left-hand, vertical column represents a certain machine function or operation, such as drilling, boring, reaming, etc. When channels 6 and 7 are punched together, in conjunction with channels 1, 2, 3, or 4, operations "a to i" will

| EIA STANDARD RS-244 KEYBOARD SYMBOLS | ALTERNATE KEYBOARD SYMBOLS | CODE ON TAPE CHANNEL NUMBERS 1 | 2 | 3 | • | 4 | 5 | 6 | 7 | 8 | SYSTEM FUNCTION |
|---|---|---|---|---|---|---|---|---|---|---|---|
| TAPE FEED | SPACE, BUZZ, feed | | | | • | | | | | | LEADER |
| | RWST, %, $ | ● | ● | | • | ● | ● | | ● | | RWS (REWIND STOP) |
| TAB | | | ● | ● | • | ● | ● | ● | | | TAB |
| + | | | | | • | | ● | ● | ● | | + (OPTIONAL) |
| − | | | | | • | | | | ● | | − |
| 1 | | ● | | | • | | | | | | 1 |
| 2 | | | ● | | • | | | | | | 2 |
| 3 | | ● | ● | | • | | ● | | | | 3 |
| 4 | | | | ● | • | | | | | | 4 |
| 5 | | ● | | ● | • | | ● | | | | 5 |
| 6 | | | ● | ● | • | | ● | | | | 6 |
| 7 | | ● | ● | ● | • | | | | | | 7 |
| 8 | | | | | • | ● | | | | | 8 |
| 9 | | ● | | | • | ● | ● | | | | 9 |
| 0 | | | | | • | | | ● | | | 0 |
| CAR. RET. OR EOB | | | | | • | | | | | ● | EOB (END OF BLOCK( |
| DELETE | TAPE FEED | ● | ● | ● | • | ● | ● | ● | ● | | DELETE |
| | | 1 | 2 | 3 | • | 4 | 5 | 6 | 7 | 8 | |

**Fig. 70-7** The EIA (Electronics Industries Association) standard coding for a 1-inch-wide eight-channel tape. (*Courtesy The Superior Electric Co.*)

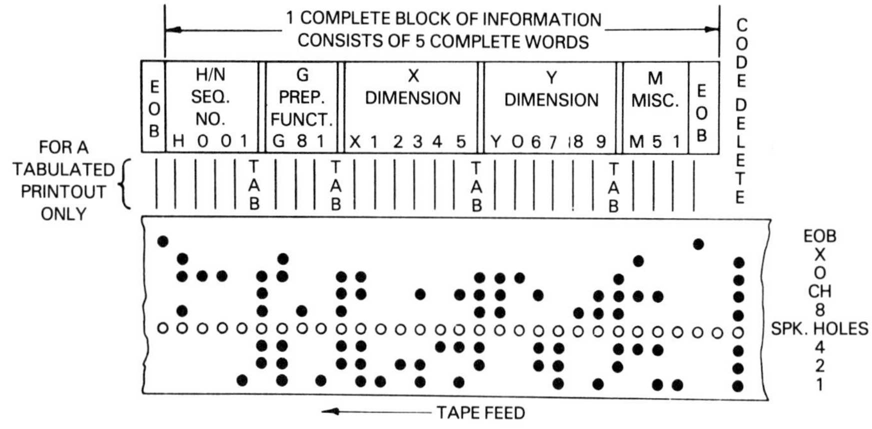


**Fig. 70-8** A complete block of information punched on a tape. (*Courtesy Cincinnati Milacron Inc.*)

be selected. When only channel 7 is punched, in conjunction with channels 1, 2, 3, or 4, operations "j to r" will be selected. If only channel 6 is punched, in conjunction with channels 1, 2, 3, or 4, operations "s to z" will be selected.

**6.** *Channel 8,* marked EL (or EOB), represents the *end of a line or block* of information. It must always be at the beginning of a tape and at the end of each block of information.

The *TAB* code, punched in channels 2, 3, 4, 5, and 6, is used to separate each operation or dimension. The remaining codes in Fig. 70-7 beginning with a period (.) and ending with lower case are all acceptable codes used in numerical control tapes.

## Examples of Punched Information

**1.** Assume that the numerical value of 1 must be recorded. A hole must be punched in channel 1 because it has a numerical value of 1.

**2.** Assume that the numerical value of 5 must be recorded. A hole must be punched in channels 3 and 1 because channel 3 has a numerical value of 4 and channel 1 has a numerical value of 1. These two totaled equal a numerical value of 5.

*Note:* Since only two holes are punched across this row, an extra hole must be punched in the parity check channel (5) because the tape reader will not recognize an even number of holes (Fig. 70-7). If a hole is not punched in the parity check channel, the tape reader will stop the machine.

**3.** Operation "g" would be recorded by punching holes in channels 1, 2, 3, 6, and 7.

## Tape Format

Each block of information must contain five complete words or pieces of information. If five complete words are not included in each block, the tape reader will not recognize the information on the tape, and therefore will not activate the control unit. (See N-6.) Figure 70-8 shows a punched tape containing one complete block of information consisting of five complete words. From left to right, the information punched on the tape is as follows:

N-6 Each block of information must contain five words or information, otherwise the tape reader will stop the machine at that point.

**1.** A hole is punched in channel 8 (EOB), which represents the end or the beginning of a line or block of information.

**2.** The *first word* of the block represents the number of the operation. H001 represents the first operation on the tape. The *letter H* is recorded by punching holes in channels 4, 6, and 7 (Fig. 70-8). The *two zeros* are recorded by punching channel 6 in two successive rows. The 1 is recorded by punching channel 1, which has a numerical value of 1.

**3.** The *TAB* code is used to separate each word into a block of information.

**4.** The *second word* in the block represents the type of operation to be performed. G81 is a drill cycle on Cincinnati Milacron Numerical Control systems. *The letter G* is recorded by punching channels 1, 2, 3, 6, and 7. The *8* is recorded by punching channel 4, which has a numerical value of 8.

The 1 is recorded by punching channel 1, which has a numerical value of 1.

**5.** The *third word* represents the distance the table slide must move from the *x* axis. The information contained in the third word of Fig. 70-8 is X12345. As standard tapes on closed loop systems program all dimensions to ten-thousandths of an inch, the machine will move 1.2345 from the *x* axis.

**6.** The *fourth word* represents the distance the table slide must move from the *y* axis. In Fig. 70-8, Y06789 represents a table movement of 0.6789 from the *y* axis.

**7.** The *fifth word* represents a miscellaneous machining function. M51 would select the proper cam so that the hole in the workpiece is drilled to the required depth.

## Program, Tape, and Machine Preparation

Although numerical control systems differ greatly in detail and complexity according to the manufacturer, all have

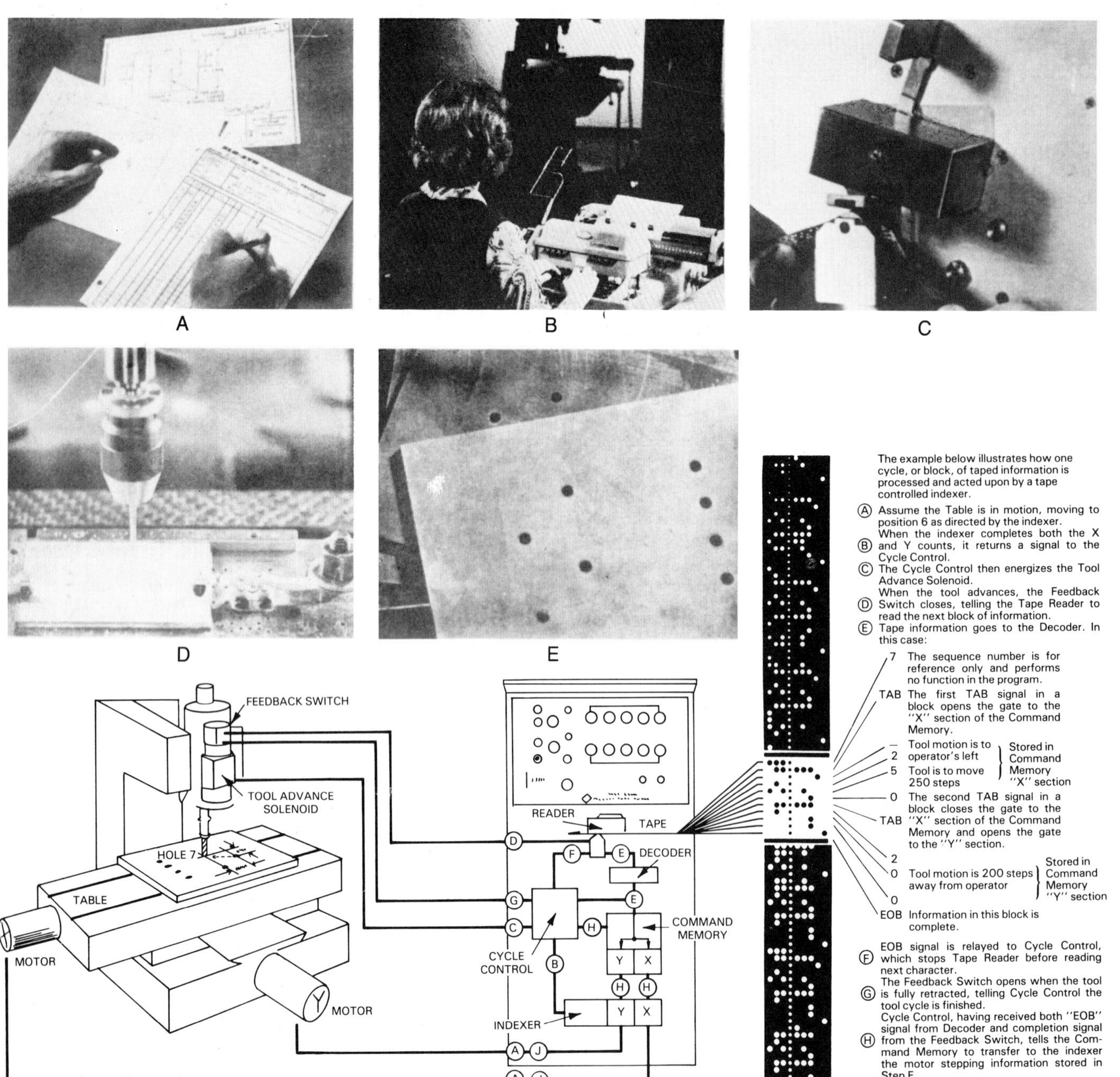


**Fig. 70-9** Numerical control sequence of operations.

N-7

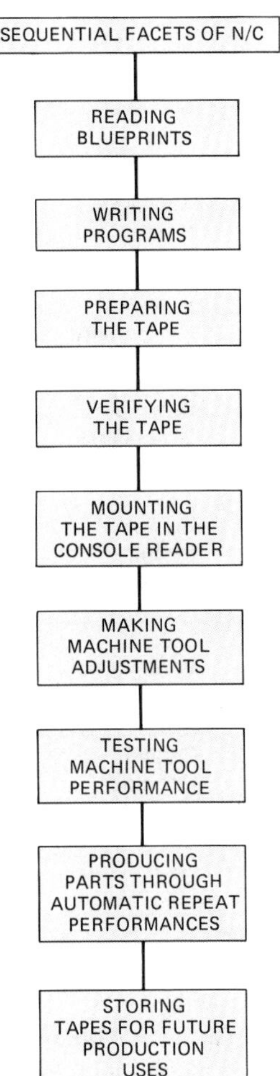


basically the same elements. Regardless of the type of input media used, all operate a machine tool the same way. A complete sequence of operations beginning with the programmer transferring the information from a print to a program sheet until the finished part is removed from the machine is illustrated in Fig. 70-9. (See also N-7.)

**1.** A programmer reviews the part drawing, determines the sequence of operations required, and records this information on a program sheet (Fig. 70-9A).

**2.** A typist transfers the information contained on the program sheet to the tape using a special-tape punching machine (Fig. 70-9B). To ensure that the typist has not made an error in punching the information on the tape, a second typist may also type the same program. These two tapes are then compared, and if they are identical it is assumed the program is correct. On some types of tape-punching machines, the finished tape may be run back through the machine, which makes a duplicate tape and at the same time produces a typed copy of the program. The typed copy of the program can be compared with the original program to determine its accuracy.

**3.** The punched tape and a copy of the program sheet are then handed to the machine operator. After positioning the part to be machined on the machine table according to the instructions contained on the program sheet, he threads the tape into the tape reader (Figs. 70-9C and 70-10).

**4.** The tape reader which automatically advances the tape is then started.

**5.** As each block of information is decoded by the tape reader, it sends the necessary information to the *cycle control*.

**6.** After the end of each machining operation (Fig. 70-9D), the *feedback switch* informs the cycle control that the previous operation has been completed.

**7.** The cycle control instructs the *command memory unit* to transfer the tape instructions for the next operation to the *indexer*.

**8.** The indexer starts the servomotors which in turn move the machine table slides the required amount; and the next machining operation is performed until all machining operations have been completed (Fig. 70-9E).

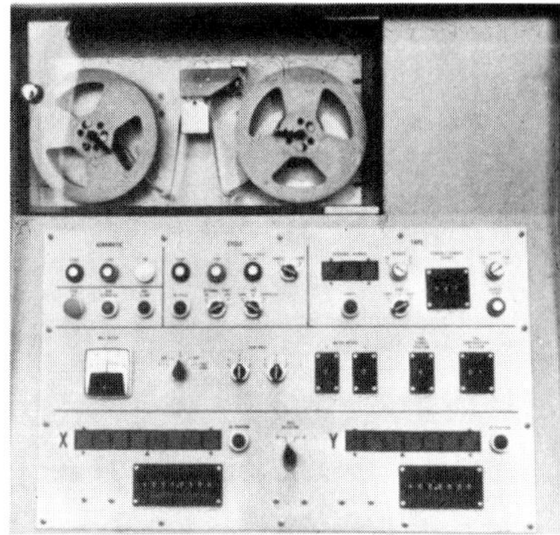

**Fig. 70-10** A Cincinnati Milacron Acromatic tape control reader. (*Courtesy Cincinnati Milacron Inc.*)

## Advantages

Numerical control has been applied to a variety of machines and has gained wide acceptance by industry. Machine tools such as lathes, turret drills, and milling and boring machines are some of the more common types of equipment which employ numerical control. The following are some of the advantages of numerical control equipment.

**1.** The machine has greater flexibility, since one machine can act as a drill, mill, and turret lathe.

**2.** Once the program has been set up, there is usually a 20 to 30 percent increase in production.

**3.** The reliability of the system eliminates the human error associated with manual operation, thereby reducing scrap loss.

**4.** Special jigs and fixtures usually required for positioning are eliminated, since the machine can locate positions quickly and accurately.

**5.** The time required for setting up and locating the workpiece is reduced.

**6.** Complex operations can be performed with ease.

**7.** Single parts or production runs can be made with minimum effort and cost.

**8.** The program can be quickly changed by inserting a new tape in the machine.

**9.** Inspection costs are reduced because of the reliability of the system.

**10.** Once the program and tooling have been tested, the equipment does not require a highly skilled operator.

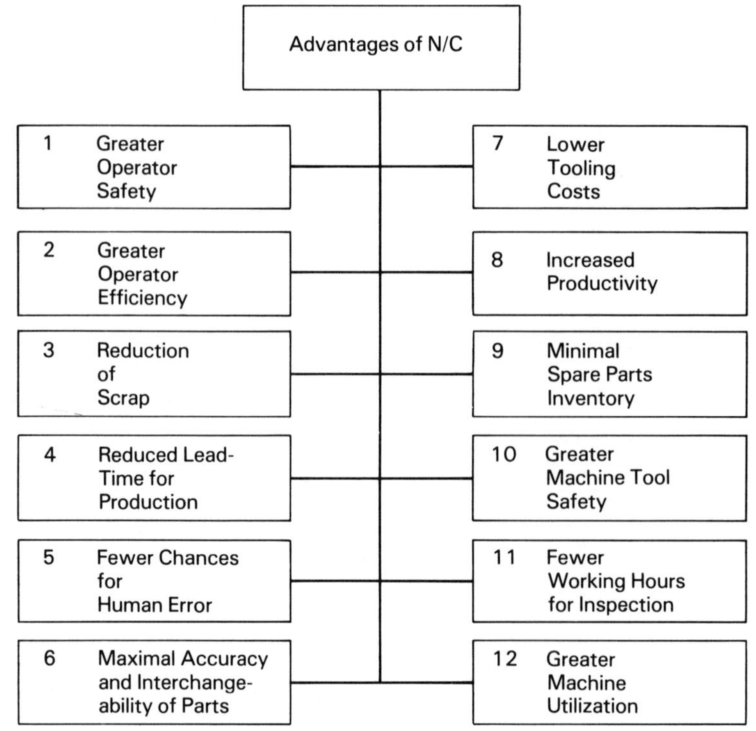


**Fig. 70-11** Advantages of numerical control.

### Disadvantages

**1.** The initial cost of numerical control machines is higher than that of conventional machines.

**2.** Personnel trained in electronics are required to service the equipment.

**3.** Additional floor space is required for this equipment.

**4.** Personnel must be trained in the programming and operation of this equipment.

## REVIEW QUESTIONS

**1.** Define numerical control.

**2.** Briefly describe an open loop system.

**3.** What is the basic difference between an open loop system and a closed loop system?

**4.** What is the accuracy of a machine fitted with

(a) An open loop system

(b) A closed loop system

**5.** Name two types of controls used to position and control the cutting tool.

**6.** Why is a computer essential for the operation of a continuous-path control?

**7.** What feature on a standard tape prevents it from being mounted incorrectly?

**8.** What is *parity check* and what is its purpose?

**9.** Explain the purpose of TAB and EL code.

**10.** List briefly the steps required to produce a part by numerical control.

**11.** List six important advantages of numerical control.

# APPENDIX OF TABLES

## TABLE 1 DECIMAL INCH, FRACTIONAL INCH, AND MILLIMETER EQUIVALENTS

| Decimal inch | Fractional inch | | Millimeter | Decimal inch | Fractional inch | | Millimeter |
|---|---|---|---|---|---|---|---|
| .015625 | | 1/64 | 0.397 | .515625 | | 33/64 | 13.097 |
| .03125 | 1/32 | | 0.794 | .53125 | 17/32 | | 13.494 |
| .046875 | | 3/64 | 1.191 | .546875 | | 35/64 | 13.891 |
| .0625 | 1/16 | | 1.588 | .5625 | 9/16 | | 14.288 |
| .078125 | | 5/64 | 1.984 | .578125 | | 37/64 | 14.684 |
| .09375 | 3/32 | | 2.381 | .59375 | 19/32 | | 15.081 |
| .109375 | | 7/64 | 2.778 | .609375 | | 39/64 | 15.478 |
| .125 | 1/8 | | 3.175 | .625 | 5/8 | | 15.875 |
| .140625 | | 9/64 | 3.572 | .640625 | | 41/64 | 16.272 |
| .15625 | 5/32 | | 3.969 | .65625 | 21/32 | | 16.669 |
| .171875 | | 11/64 | 4.366 | .671875 | | 43/64 | 17.066 |
| .1875 | 3/16 | | 4.762 | .6875 | 11/16 | | 17.462 |
| .203125 | | 13/64 | 5.159 | .703125 | | 45/64 | 17.859 |
| .21875 | 7/32 | | 5.556 | .71875 | 23/32 | | 18.256 |
| .234375 | | 15/64 | 5.953 | .734375 | | 47/64 | 18.653 |
| .25 | 1/4 | | 6.350 | .75 | 3/4 | | 19.05 |
| .265625 | | 17/64 | 6.747 | .765625 | | 49/64 | 19.447 |
| .28125 | 9/32 | | 7.144 | .78125 | 25/32 | | 19.844 |
| .296875 | | 19/64 | 7.541 | .796875 | | 51/64 | 20.241 |
| .3125 | 5/16 | | 7.938 | .8125 | 13/16 | | 20.638 |
| .328125 | | 21/64 | 8.334 | .828125 | | 53/64 | 21.034 |
| .34375 | 11/32 | | 8.731 | .84375 | 27/32 | | 21.431 |
| .359375 | | 23/64 | 9.128 | .859375 | | 55/64 | 21.828 |
| .375 | 3/8 | | 9.525 | .875 | 7/8 | | 22.225 |
| .390625 | | 25/64 | 9.922 | .890625 | | 57/64 | 22.622 |
| .40625 | 13/32 | | 10.319 | .90625 | 29/32 | | 23.019 |
| .421875 | | 27/64 | 10.716 | .921875 | | 59/64 | 23.416 |
| .4375 | 7/16 | | 11.112 | .9375 | 15/16 | | 23.812 |
| .453125 | | 29/64 | 11.509 | .953125 | | 61/64 | 24.209 |
| .46875 | 15/32 | | 11.906 | .96875 | 31/32 | | 24.606 |
| .484375 | | 31/64 | 12.303 | .984375 | | 63/64 | 25.003 |
| .5 | 1/2 | | 12.700 | 1. | 1 | | 25.400 |

## TABLE 2*

| Conversion of inches to millimeters | | | | | | Conversion of millimeters to inches | | | | | |
|---|---|---|---|---|---|---|---|---|---|---|---|
| Inches | Millimeters | Inches | Millimeters | Inches | Millimeters | Millimeters | Inches | Millimeters | Inches | Millimeters | Inches |
| .001 | 0.025 | .290 | 7.37 | .660 | 16.76 | 0.01 | .0004 | 0.35 | .0138 | 0.68 | .0268 |
| .002 | 0.051 | .300 | 7.62 | .670 | 17.02 | 0.02 | .0008 | 0.36 | .0142 | 0.69 | .0272 |
| .003 | 0.076 | .310 | 7.87 | .680 | 17.27 | 0.03 | .0012 | 0.37 | .0146 | 0.70 | .0276 |
| .004 | 0.102 | .320 | 8.13 | .690 | 17.53 | 0.04 | .0016 | 0.38 | .0150 | 0.71 | .0280 |
| .005 | 0.127 | .330 | 8.38 | .700 | 17.78 | 0.05 | .0020 | 0.39 | .0154 | 0.72 | .0283 |
| .006 | 0.152 | .340 | 8.64 | .710 | 18.03 | 0.06 | .0024 | 0.40 | .0157 | 0.73 | .0287 |
| .007 | 0.178 | .350 | 8.89 | .720 | 18.29 | 0.07 | .0028 | 0.41 | .0161 | 0.74 | .0291 |
| .008 | 0.203 | .360 | 9.14 | .730 | 18.54 | 0.08 | .0031 | 0.42 | .0165 | 0.75 | .0295 |
| .009 | 0.229 | .370 | 9.40 | .740 | 18.80 | 0.09 | .0035 | 0.43 | .0169 | 0.76 | .0299 |
| .010 | 0.254 | .380 | 9.65 | .750 | 19.05 | 0.10 | .0039 | 0.44 | .0173 | .077 | .0303 |
| .020 | 0.508 | .390 | 9.91 | .760 | 19.30 | 0.11 | .0043 | 0.45 | .0177 | 0.78 | .0307 |
| .030 | 0.762 | .400 | 10.16 | .770 | 19.56 | 0.12 | .0047 | 0.46 | .0181 | 0.79 | .0311 |
| .040 | 1.016 | .410 | 10.41 | .780 | 19.81 | 0.13 | .0051 | 0.47 | .0185 | 0.80 | .0315 |
| .050 | 1.270 | .420 | 10.67 | .790 | 20.07 | 0.14 | .0055 | .048 | .0189 | 0.81 | .0319 |
| .060 | 1.524 | .430 | 10.92 | .800 | 20.32 | 0.15 | .0059 | 0.49 | .0193 | 0.82 | .0323 |
| .070 | 1.778 | .440 | 11.18 | .810 | 20.57 | 0.16 | .0063 | 0.50 | .0197 | 0.83 | .0327 |
| .080 | 2.032 | .450 | 11.43 | .820 | 20.83 | 0.17 | .0067 | 0.51 | .0201 | 0.84 | .0331 |
| .090 | 2.286 | .460 | 11.68 | .830 | 21.08 | 0.18 | .0071 | 0.52 | .0205 | 0.85 | .0335 |
| .100 | 2.540 | .470 | 11.94 | .840 | 21.34 | 0.19 | .0075 | 0.53 | .0209 | 0.86 | .0339 |
| .110 | 2.794 | .480 | 12.19 | .850 | 21.59 | 0.20 | .0079 | 0.54 | .0213 | 0.87 | .0343 |
| .120 | 3.048 | .490 | 12.45 | .860 | 21.84 | 0.21 | .0083 | 0.55 | .0217 | 0.88 | .0346 |
| .130 | 3.302 | .500 | 12.70 | .870 | 22.10 | 0.22 | .0087 | 0.56 | .0220 | 0.89 | .0350 |
| .140 | 3.56 | .510 | 12.95 | .880 | 22.35 | 0.23 | .0091 | 0.57 | .0224 | 0.90 | .0354 |
| .150 | 3.81 | .520 | 13.21 | .890 | 22.61 | 0.24 | .0094 | 0.58 | .0228 | 0.91 | .0358 |
| .160 | 4.06 | .530 | 13.46 | .900 | 22.86 | 0.25 | .0098 | 0.59 | .0232 | 0.92 | .0362 |
| .170 | 4.32 | .540 | 13.72 | .910 | 23.11 | 0.26 | .0102 | 0.60 | .0236 | 0.93 | .0366 |
| .180 | 4.57 | .550 | 13.97 | .920 | 23.37 | 0.27 | .0106 | 0.61 | .0240 | 0.94 | .0370 |
| .190 | 4.83 | .560 | 14.22 | .930 | 23.62 | 0.28 | .0110 | 0.62 | .0244 | 0.95 | .0374 |
| .200 | 5.08 | .570 | 14.48 | .940 | 23.88 | 0.29 | .0114 | 0.63 | .0248 | 0.96 | .0378 |
| .210 | 5.33 | .580 | 14.73 | .950 | 24.13 | 0.30 | .0118 | 0.64 | .0252 | 0.97 | .0382 |
| .220 | 5.59 | .590 | 14.99 | .960 | 24.38 | 0.31 | .0122 | 0.65 | .0256 | 0.98 | .0386 |
| .230 | 5.84 | .600 | 15.24 | .970 | 24.64 | 0.32 | .0126 | 0.66 | .0260 | 0.99 | .0390 |
| .240 | 6.10 | .610 | 15.49 | .980 | 24.89 | 0.33 | .0130 | 0.67 | .0264 | 1.00 | .0394 |
| .250 | 6.35 | .620 | 15.75 | .990 | 25.15 | 0.34 | .0134 | .... | .... | .... | .... |
| .260 | 6.60 | .630 | 16.00 | 1.000 | 25.40 | | | | | | |
| .270 | 6.86 | .640 | 16.26 | .... | .... | | | | | | |
| .280 | 7.11 | .650 | 16.51 | .... | .... | | | | | | |

*Courtesy Automatic Electric Company.

## TABLE 3 LETTER DRILL SIZES

| Letters | Inches | Millimeters | Letters | Inches | Millimeters | Letters | Inches | Millimeters | Letters | Inches | Millimeters |
|---|---|---|---|---|---|---|---|---|---|---|---|
| A | .234 | 5.94 | H | .266 | 6.76 | N | .302 | 7.67 | T | .358 | 9.09 |
| B | .238 | 6.05 | I | .272 | 6.91 | O | .316 | 8.03 | U | .368 | 9.35 |
| C | .242 | 6.15 | J | .277 | 7.04 | P | .323 | 8.20 | V | .377 | 9.58 |
| D | .246 | 6.25 | K | .281 | 7.14 | Q | .332 | 8.43 | W | .386 | 9.80 |
| E | .250 | 6.35 | L | .290 | 7.37 | R | .339 | 8.61 | X | .397 | 10.08 |
| F | .257 | 6.53 | M | .295 | 7.49 | S | .348 | 8.84 | Y | .404 | 10.26 |
| G | .261 | 6.63 | | | | | | | Z | .413 | 10.49 |

## TABLE 4 NUMBER DRILL SIZES

| Numbers | Inches | Millimeters | Numbers | Inches | Millimeters | Numbers | Inches | Millimeters |
|---|---|---|---|---|---|---|---|---|
| 1 | .2280 | 5.79 | 34 | .1110 | 2.82 | 66 | .0330 | 0.84 |
| 2 | .2210 | 5.61 | 35 | .1100 | 2.79 | 67 | .0320 | 0.81 |
| 3 | .2130 | 5.41 | 36 | .1065 | 2.71 | 68 | .0310 | 0.79 |
| 4 | .2090 | 5.31 | 37 | .1040 | 2.64 | 69 | .0292 | 0.74 |
| 5 | .2055 | 5.22 | 38 | .1015 | 2.58 | 70 | .0280 | 0.71 |
| 6 | .2040 | 5.18 | 39 | .0995 | 2.53 | 71 | .0260 | 0.66 |
| 7 | .2010 | 5.10 | 40 | .0980 | 2.49 | 72 | .0250 | 0.64 |
| 8 | .1990 | 5.05 | 41 | .0960 | 2.44 | 73 | .0240 | 0.61 |
| 9 | .1960 | 4.98 | 42 | .0935 | 2.37 | 74 | .0225 | 0.57 |
| 10 | .1935 | 4.91 | 43 | .0890 | 2.26 | 75 | .0210 | 0.53 |
| 11 | .1910 | 4.85 | 44 | .0860 | 2.18 | 76 | .0200 | 0.51 |
| 12 | .1890 | 4.80 | 45 | .0820 | 2.08 | 77 | .0180 | 0.46 |
| 13 | .1850 | 4.70 | 46 | .0810 | 2.06 | 78 | .0160 | 0.41 |
| 14 | .1820 | 4.62 | 47 | .0785 | 1.99 | 79 | .0145 | 0.37 |
| 15 | .1800 | 4.57 | 48 | .0760 | 1.93 | 80 | .0135 | 0.34 |
| 16 | .1770 | 4.50 | 49 | .0730 | 1.85 | 81 | .0130 | 0.33 |
| 17 | .1730 | 4.39 | 50 | .0700 | 1.78 | 82 | .0125 | 0.32 |
| 18 | .1695 | 4.31 | 51 | .0670 | 1.70 | 83 | .0120 | 0.30 |
| 19 | .1660 | 4.22 | 52 | .0635 | 1.61 | 84 | .0115 | 0.29 |
| 20 | .1610 | 4.09 | 53 | .0595 | 1.51 | 85 | .0110 | 0.28 |
| 21 | .1590 | 4.04 | 54 | .0550 | 1.40 | 86 | .0105 | 0.27 |
| 22 | .1570 | 3.99 | 55 | .0520 | 1.32 | 87 | .0100 | 0.25 |
| 23 | .1540 | 3.91 | 56 | .0465 | 1.18 | 88 | .0095 | 0.24 |
| 24 | .1520 | 3.86 | 57 | .0430 | 1.09 | 89 | .0091 | 0.23 |
| 25 | .1495 | 3.80 | 58 | .0420 | 1.07 | 90 | .0087 | 0.22 |
| 26 | .1470 | 3.73 | 59 | .0410 | 1.04 | 91 | .0083 | 0.21 |
| 27 | .1440 | 3.66 | 60 | .0400 | 1.02 | 92 | .0079 | 0.20 |
| 28 | .1405 | 3.57 | 61 | .0390 | 0.99 | 93 | .0075 | 0.19 |
| 29 | .1360 | 3.45 | 62 | .0380 | 0.97 | 94 | .0071 | 0.18 |
| 30 | .1285 | 3.26 | 63 | .0370 | 0.94 | 95 | .0067 | 0.17 |
| 31 | .1200 | 3.05 | 64 | .0360 | 0.91 | 96 | .0063 | 0.16 |
| 32 | .1160 | 2.95 | 65 | .0350 | 0.89 | 97 | .0059 | 0.15 |
| 33 | .1130 | 2.87 | | | | | | |

**TABLE 5 COMMERCIAL TAP DRILL SIZES (75% of thread depth)**
**AMERICAN NATIONAL AND UNIFIED FORM THREAD**

| NC National Coarse* | | | NF National Fine* | | |
|---|---|---|---|---|---|
| Tap size | Threads per inch | Tap drill size | Tap size | Threads per inch | Tap drill size |
| # 5 | 40 | #38 | # 5 | 44 | #37 |
| # 6 | 32 | #36 | # 6 | 40 | #33 |
| # 8 | 32 | #29 | # 8 | 36 | #29 |
| #10 | 24 | #25 | #10 | 32 | #21 |
| #12 | 24 | #16 | #12 | 28 | #14 |
| 1/4 | 20 | # 7 | 1/4 | 28 | # 3 |
| 5/16 | 18 | F | 5/16 | 24 | I |
| 3/8 | 16 | 5/16 | 3/8 | 24 | Q |
| 7/16 | 14 | U | 7/16 | 20 | 25/64 |
| 1/2 | 13 | 27/64 | 1/2 | 20 | 29/64 |
| 9/16 | 12 | 31/64 | 9/16 | 18 | 33/64 |
| 5/8 | 11 | 17/32 | 5/8 | 18 | 37/64 |
| 3/4 | 10 | 21/32 | 3/4 | 16 | 11/16 |
| 7/8 | 9 | 49/64 | 7/8 | 14 | 13/16 |
| 1 | 8 | 7/8 | 1 | 14 | 15/16 |
| 1 1/8 | 7 | 63/64 | 1 1/8 | 12 | 1 3/64 |
| 1 1/4 | 7 | 1 7/64 | 1 1/4 | 12 | 1 11/64 |
| 1 3/8 | 6 | 1 7/32 | 1 3/8 | 12 | 1 19/64 |
| 1 1/2 | 6 | 1 11/32 | 1 1/2 | 12 | 1 27/64 |
| 1 3/4 | 5 | 1 9/16 | | | |
| 2 | 4 1/2 | 1 25/32 | | | |
| **NPT National Pipe Thread** | | | | | |
| 1/8 | 27 | 11/32 | 1 | 11 1/2 | 1 5/32 |
| 1/4 | 18 | 7/16 | 1 1/4 | 11 1/4 | 1 1/2 |
| 3/8 | 18 | 19/32 | 1 1/2 | 11 1/2 | 1 23/32 |
| 1/2 | 14 | 23/32 | 2 | 11 1/2 | 2 3/16 |
| 3/4 | 14 | 15/16 | 2 1/2 | 8 | 2 5/8 |

The major diameter of an NC or NF number size tap or screw = (N × .013) + .060.
Example: The major diameter of a #5 tap equals (5 × .013) + .060 = .125 diameter.

**TABLE 6 METRIC TAP DRILL SIZES**

| Nominal diameter, mm | Thread pitch, mm | Tap drill size, mm | Nominal diameter, mm | Thread pitch, mm | Tap drill size, mm |
|---|---|---|---|---|---|
| 1.6 | 0.35 | 1.2 | 20.0 | 2.5 | 17.5 |
| 2.0 | 0.40 | 1.6 | 24.0 | 3.0 | 21.0 |
| 2.5 | 0.45 | 2.05 | 30.0 | 3.5 | 26.5 |
| 3.0 | 0.50 | 2.5 | 36.0 | 4.0 | 32.0 |
| 3.5 | 0.60 | 2.9 | 42.0 | 4.5 | 37.5 |
| 4.0 | 0.70 | 3.3 | 48.0 | 5.0 | 43.0 |
| 5.0 | 0.80 | 4.2 | 56.0 | 5.5 | 50.5 |
| 6.0 | 1.00 | 5.3 | 64.0 | 6.0 | 58.0 |
| 8.0 | 1.25 | 6.8 | 72.0 | 6.0 | 66.0 |
| 10.0 | 1.50 | 8.5 | 80.0 | 6.0 | 74.0 |
| 12.0 | 1.75 | 10.2 | 90.0 | 6.0 | 84.0 |
| 14.0 | 2.00 | 12.0 | 100.0 | 6.0 | 94.0 |
| 16.0 | 2.00 | 14.0 | | | |

## TABLE 7 ISO METRIC PITCH AND DIAMETER COMBINATIONS

| Nominal diameter, mm | Thread pitch, mm | Nominal diameter, mm | Thread pitch, mm |
|---|---|---|---|
| 1.6 | 0.35 | 20 | 2.5 |
| 2 | 0.40 | 24 | 3.0 |
| 2.5 | 0.45 | 30 | 3.5 |
| 3 | 0.50 | 36 | 4.0 |
| 3.5 | 0.60 | 42 | 4.5 |
| 4 | 0.70 | 48 | 5.0 |
| 5 | 0.80 | 56 | 5.5 |
| 6.0 | 1.00 | 64 | 6.0 |
| 8 | 1.25 | 72 | 6.0 |
| 10 | 1.50 | 80 | 6.0 |
| 12 | 1.75 | 90 | 6.0 |
| 14 | 2.00 | 100 | 6.0 |
| 16 | 2.00 | | |

## TABLE 8 REVOLUTIONS PER MINUTE FOR WORK OR CUTTING TOOL DIAMETERS

| Diameter | | Cutting speed, ft/min or m/min | | | | | | | |
|---|---|---|---|---|---|---|---|---|---|
| Inches | mm | 40 ft or 12 m | 60 ft or 18 m | 80 ft or 24 m | 90 ft or 27 m | 100 ft or 30 m | 120 ft or 36 m | 150 ft or 45 m | 200 ft or 60 m |
| 1/8 | 3 | 1280 | 1920 | 2560 | 2880 | 3200 | 3840 | 4800 | 6400 |
| 1/4 | 6 | 640 | 960 | 1280 | 1440 | 1600 | 1920 | 2400 | 3200 |
| 3/8 | 9 | 427 | 640 | 853 | 960 | 1067 | 1280 | 1600 | 2133 |
| 1/2 | 13 | 320 | 480 | 640 | 720 | 800 | 960 | 1200 | 1600 |
| 5/8 | 16 | 256 | 384 | 512 | 576 | 640 | 768 | 960 | 1280 |
| 3/4 | 19 | 213 | 320 | 427 | 480 | 533 | 640 | 800 | 1067 |
| 7/8 | 22 | 183 | 274 | 366 | 411 | 457 | 548 | 686 | 914 |
| 1 | 25 | 160 | 240 | 320 | 360 | 400 | 480 | 600 | 800 |
| 1 1/4 | 32 | 128 | 192 | 256 | 288 | 320 | 384 | 480 | 640 |
| 1 1/2 | 38 | 107 | 160 | 213 | 240 | 267 | 320 | 400 | 533 |
| 1 3/4 | 44 | 91 | 137 | 183 | 206 | 229 | 274 | 343 | 457 |
| 2 | 50 | 80 | 120 | 160 | 180 | 200 | 240 | 300 | 400 |
| 2 1/2 | 63 | 64 | 96 | 128 | 144 | 160 | 192 | 240 | 320 |
| 3 | 76 | 53 | 80 | 107 | 120 | 133 | 160 | 200 | 267 |
| 3 1/2 | 89 | 46 | 68 | 91 | 103 | 114 | 136 | 171 | 229 |
| 4 | 100 | 40 | 60 | 80 | 90 | 100 | 120 | 150 | 200 |

## TABLE 9 FORMULA SHORTCUTS

..., block out (cover) the unknown; the remainder is the formula. In each diagram the horizontal line is the division line; ... the multiplication line.

| | | |
|---|---|---|
| $A$ = Area | $L$ = Length | $b$ = Base |
| $C$ = Circumference | $R$ = Radius | $h$ = Height |
| CS = Cutting speed | r/min = Revolutions/minute | m = Meters |
| $D$ = Diameter | S = Strokes/minute | mm = Millimeters |

### Circle

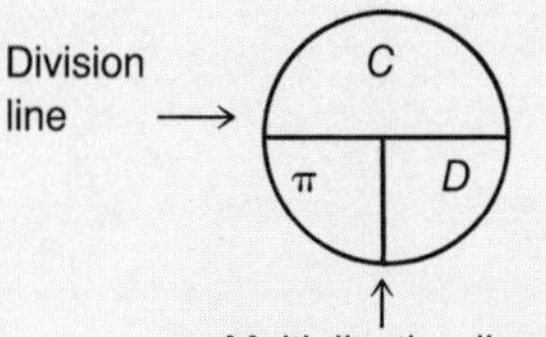


$$C = \pi \times D$$

$$D = \frac{C}{\pi}$$

### Four-element formulas

1. Block out unknown.
2. Cross-multiply diagonally opposite elements.
3. Divide by remaining element.

#### Triangles

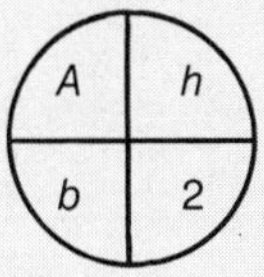


$$A = \frac{b \times h}{2}$$

$$b = \frac{A \times 2}{h}$$

$$h = \frac{A \times 2}{b}$$

#### Shaper speed

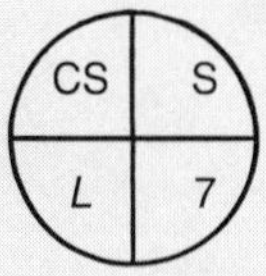


$$CS = \frac{S \times L}{7}$$

$$S = \frac{CS \times 7}{L}$$

$$L = \frac{CS \times 7}{S}$$

### Area

#### Squares and rectangles

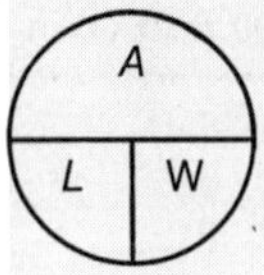


$$A = L \times W$$

$$L = \frac{A}{W}$$

$$W = \frac{A}{L}$$

#### Circles

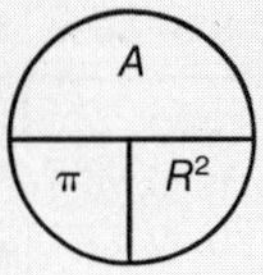


$$A = \pi \times R^2$$

$$R^2 = \frac{A}{\pi}$$

### Revolutions per minute (r/min)

(Lathe, drill, mill, grinder)

#### Inch

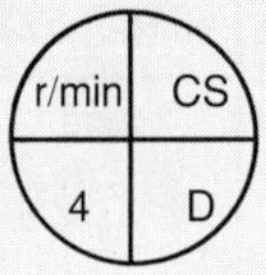


$$\text{r/min} = \frac{\text{CS (ft)} \times 4}{D \text{ (in.)}}$$

$$CS = \frac{\text{r/min} \times D}{4}$$

$$D = \frac{CS \times 4}{\text{r/min}}$$

#### Metric

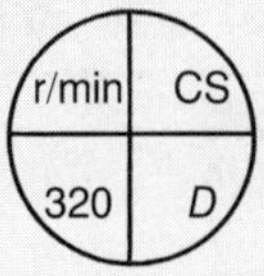


$$\text{r/min} = \frac{\text{CS (m)} \times 320}{D \text{ (mm)}}$$

$$CS = \frac{\text{r/min} \times D}{320}$$

$$D = \frac{CS \times 320}{\text{r/min}}$$

## TABLE 10 COMMONLY USED FORMULAS

Code

c.p.t. = Chip per tooth
CS = Cutting speed
$D$ = Large diameter
$d$ = Small diameter
G.L. = Guide bar length
$N$ = Number of threads per inch
= Number of strokes per minute
= Number of teeth in cutter
O.L. = Overall length of work
$P$ = Pitch
T.D.S. = Tap drill size
T.L. = Taper length
t/ft. = Taper per foot
t/mm = Taper per millimeter
T.O. = Tailstock offset

| Inch | Metric |
|---|---|
| $\text{T.D.S.} = D - \left(\frac{1}{N}\right)$ | $\text{T.D.S.} = D - P$ |
| $\text{r/min} = \frac{\text{CS (ft)} \times 4}{D \text{ (in.)}}$ | $\text{r/min} = \frac{\text{CS (m)} \times 320}{D \text{ (mm)}}$ |
| $\text{t/ft} = \frac{(D - d) \times 12}{\text{T.L.}}$ | $\text{t/mm} = \frac{(D - d)}{\text{T.L.}}$ |
| $\text{T.O.} = \frac{\text{t/ft} \times \text{O.L.}}{24}$ | $\text{T.O.} = \frac{\text{t/mm} \times \text{O.L.}}{2}$ |
| $\text{Guide bar setover} = \frac{(D - d) \times 12}{\text{T.L.}}$ | $\text{Guide bar setover} = \frac{(D - d)}{2} \times \frac{\text{G.L.}}{\text{T.L.}}$ |
| Milling feed (in./min) = N × c.p.t. × r/min | Milling feed (mm/min) = $N$ × c.p.t. × r/min |
| $N \text{ (shaper)} = \frac{\text{CS (ft)} \times 7}{L \text{ (in.)}}$ | $N \text{ (shaper)} = \frac{\text{CS (m)} \times 0.6}{\text{S.L. (m)}}$ |

## TABLE 11 HARDNESS CONVERSION CHART*

| 10-mm ball 3000 kg | 120° cone 150 kg | 1/16-in. ball 100 kg | Model C | 1000 lb.-in.² | 10-mm ball 3000 kg | 120° cone 150 kg | 1/16-in. ball 100 kg | Model C | 1000 lb.-in.² |
|---|---|---|---|---|---|---|---|---|---|
| Brinell | Rockwell C | Rockwell B | Shore Scleroscope | Tensile strength | Brinell | Rockwell C | Rockwell B | Shore Scleroscope | Tensile strength |
| 800 | 72 | ...... | 100 | ...... | 276 | 30 | 105 | 42 | 136 |
| 780 | 71 | ...... | 99 | ...... | 269 | 29 | 104 | 41 | 132 |
| 760 | 70 | ...... | 98 | ...... | 261 | 28 | 103 | 40 | 129 |
| 745 | 68 | ...... | 97 | 367 | 258 | 27 | 102 | 39 | 127 |
| 725 | 67 | ...... | 96 | 357 | 255 | 26 | 102 | 39 | 125 |
| 712 | 66 | ...... | 95 | 350 | 249 | 25 | 101 | 38 | 123 |
| 682 | 65 | ...... | 93 | 337 | 245 | 24 | 100 | 37 | 119 |
| 668 | 64 | ...... | 91 | 326 | 240 | 23 | 99 | 36 | 117 |
| 652 | 63 | ...... | 89 | 318 | 237 | 23 | 99 | 35 | 115 |
| 626 | 62 | ...... | 87 | 306 | 229 | 22 | 98 | 34 | 113 |
| 614 | 61 | ...... | 85 | 299 | 224 | 21 | 97 | 33 | 110 |
| 601 | 60 | ...... | 83 | 292 | 217 | 20 | 96 | 33 | 107 |
| 590 | 59 | ...... | 81 | 290 | 211 | 19 | 95 | 32 | 104 |
| 576 | 57 | ...... | 79 | 281 | 206 | 18 | 94 | 32 | 102 |
| 552 | 56 | ...... | 76 | 270 | 203 | 17 | 94 | 31 | 100 |
| 545 | 55 | ...... | 75 | 268 | 200 | 16 | 93 | 31 | 98 |
| 529 | 54 | ...... | 74 | 259 | 196 | 15 | 92 | 30 | 96 |
| 514 | 53 | 120 | 72 | 254 | 191 | 14 | 92 | 30 | 94 |
| 502 | 52 | 119 | 70 | 247 | 187 | 13 | 91 | 29 | 92 |
| 495 | 51 | 119 | 69 | 244 | 185 | 12 | 91 | 29 | 91 |
| 477 | 49 | 118 | 67 | 233 | 183 | 11 | 90 | 28 | 90 |
| 461 | 48 | 117 | 66 | 227 | 180 | 10 | 89 | 28 | 89 |
| 451 | 47 | 117 | 65 | 223 | 175 | 9 | 88 | 27 | 86 |
| 444 | 46 | 116 | 64 | 219 | 170 | 7 | 87 | 27 | 84 |
| 427 | 45 | 115 | 62 | 209 | 167 | 6 | 87 | 27 | 82 |
| 415 | 44 | 115 | 60 | 204 | 165 | 5 | 86 | 26 | 81 |
| 401 | 43 | 114 | 58 | 196 | 163 | 4 | 85 | 26 | 80 |
| 388 | 42 | 114 | 57 | 191 | 160 | 3 | 84 | 25 | 78 |
| 375 | 41 | 113 | 55 | 184 | 156 | 2 | 83 | 25 | 76 |
| 370 | 40 | 112 | 54 | 182 | 154 | 1 | 82 | 25 | 75 |
| 362 | 39 | 111 | 53 | 179 | 152 | .... | 82 | 24 | 74 |
| 351 | 38 | 111 | 51 | 173 | 150 | .... | 81 | 24 | 74 |
| 346 | 37 | 110 | 50 | 170 | 147 | .... | 80 | 24 | 72 |
| 341 | 37 | 110 | 49 | 168 | 145 | .... | 79 | 23 | 71 |
| 331 | 36 | 109 | 47 | 163 | 143 | .... | 79 | 23 | 70 |
| 323 | 35 | 109 | 46 | 158 | 141 | .... | 78 | 23 | 69 |
| 311 | 34 | 108 | 46 | 153 | 140 | .... | 77 | 22 | 69 |
| 301 | 33 | 107 | 45 | 148 | 135 | .... | 75 | 22 | 67 |
| 293 | 32 | 106 | 44 | 144 | 130 | .... | 72 | 22 | 65 |
| 285 | 31 | 105 | 43 | 140 | ...... | .... | ...... | .... | ...... |

*Courtesy The Greenfield Tap & Die Co.*

## TABLE 12 VALUES OF FRACTIONAL SIZES EXPRESSED IN MILLIMETERS*†

25.4 mm equals 1 inch

| Fractional sizes | | 1 in. | 2 in. | 3 in. | 4 in. | 5 in. | 6 in. |
|---|---|---|---|---|---|---|---|
| | | 25.4 | 50.8 | 76.2 | 101.6 | 127. | 152.4 |
| 1/64 | 0.40 | 25.80 | 51.20 | 76.60 | 102. | 127.39 | 152.79 |
| 1/32 | 0.79 | 26.19 | 51.59 | 76.99 | 102.39 | 127.79 | 153.19 |
| 3/64 | 1.19 | 26.59 | 51.99 | 77.39 | 102.79 | 128.19 | 153.59 |
| 1/16 | 1.59 | 26.99 | 52.39 | 77.79 | 103.19 | 128.59 | 153.98 |
| 5/64 | 1.98 | 27.38 | 52.78 | 78.18 | 103.58 | 128.98 | 154.38 |
| 3/32 | 2.38 | 27.78 | 53.18 | 78.58 | 103.98 | 129.38 | 154.78 |
| 7/64 | 2.77 | 28.17 | 53.58 | 78.98 | 104.37 | 129.78 | 155.18 |
| 1/8 | 3.17 | 28.57 | 53.97 | 79.37 | 104.77 | 130.17 | 155.57 |
| 9/64 | 3.57 | 28.97 | 54.37 | 79.77 | 105.17 | 130.57 | 155.97 |
| 5/32 | 3.97 | 29.37 | 54.77 | 80.17 | 105.57 | 130.97 | 156.37 |
| 11/64 | 4.37 | 29.76 | 55.16 | 80.56 | 105.96 | 131.36 | 156.76 |
| 3/16 | 4.76 | 30.16 | 55.56 | 80.96 | 106.36 | 131.76 | 157.16 |
| 13/64 | 5.16 | 30.56 | 55.96 | 81.36 | 106.76 | 132.16 | 157.56 |
| 7/32 | 5.56 | 30.96 | 56.36 | 81.75 | 107.16 | 132.55 | 157.95 |
| 15/64 | 5.95 | 31.35 | 56.75 | 82.15 | 107.55 | 132.95 | 158.35 |
| 1/4 | 6.35 | 31.75 | 57.15 | 82.55 | 107.95 | 133.35 | 158.75 |
| 17/64 | 6.75 | 32.15 | 57.55 | 82.95 | 108.34 | 133.74 | 159.14 |
| 9/32 | 7.14 | 32.54 | 57.94 | 83.34 | 108.74 | 134.14 | 159.54 |
| 9/64 | 7.54 | 32.94 | 58.34 | 83.74 | 109.14 | 134.54 | 159.94 |
| 5/16 | 7.94 | 33.34 | 58.74 | 84.14 | 109.54 | 134.94 | 160.33 |
| 21/64 | 8.33 | 33.73 | 59.13 | 84.53 | 109.93 | 135.33 | 160.73 |
| 11/32 | 8.73 | 34.13 | 59.53 | 84.93 | 110.33 | 135.73 | 161.13 |
| 23/64 | 9.13 | 34.53 | 59.93 | 85.33 | 110.73 | 136.13 | 161.53 |
| 3/8 | 9.52 | 34.92 | 60.32 | 85.72 | 111.12 | 136.52 | 161.92 |
| 25/64 | 9.92 | 35.32 | 60.72 | 86.12 | 111.52 | 136.92 | 162.32 |
| 13/32 | 10.32 | 35.72 | 61.12 | 86.52 | 111.92 | 137.32 | 162.72 |
| 27/64 | 10.72 | 36.11 | 61.51 | 86.91 | 112.31 | 137.71 | 163.11 |
| 7/16 | 11.11 | 36.51 | 61.91 | 87.31 | 112.71 | 138.11 | 163.51 |
| 29/64 | 11.51 | 36.91 | 62.31 | 87.71 | 113.11 | 138.51 | 163.91 |
| 15/32 | 11.91 | 37.31 | 62.71 | 88.1 | 113.5 | 138.9 | 164.3 |
| 31/64 | 12.3 | 37.7 | 63.1 | 88.5 | 113.9 | 139.3 | 164.7 |

| Fractional sizes | | 1 in. | 2 in. | 3 in. | 4 in. | 5 in. | 6 in. |
|---|---|---|---|---|---|---|---|
| 1/2 | 12.7 | 38.1 | 63.5 | 88.9 | 114.3 | 139.7 | 165.1 |
| 33/64 | 13.10 | 38.49 | 63.90 | 89.3 | 114.69 | 140.09 | 165.49 |
| 17/32 | 13.49 | 38.89 | 64.29 | 89.69 | 115.09 | 140.49 | 165.89 |
| 35/64 | 13.89 | 39.29 | 64.69 | 90.09 | 115.49 | 140.89 | 166.29 |
| 9/16 | 14.29 | 39.69 | 65.09 | 90.49 | 115.89 | 141.29 | 166.68 |
| 37/64 | 14.68 | 40.08 | 65.48 | 90.88 | 116.28 | 141.68 | 167.08 |
| 19/32 | 15.08 | 40.48 | 65.88 | 91.28 | 116.68 | 142.08 | 167.48 |
| 39/64 | 15.48 | 40.88 | 66.28 | 91.68 | 117.08 | 142.48 | 167.88 |
| 5/8 | 15.87 | 41.27 | 66.67 | 92.07 | 117.47 | 142.87 | 168.27 |
| 41/64 | 16.27 | 41.67 | 67.07 | 92.47 | 117.87 | 143.27 | 168.67 |
| 21/32 | 16.67 | 42.07 | 67.47 | 92.87 | 118.27 | 143.67 | 169.07 |
| 43/64 | 17.07 | 42.46 | 67.86 | 93.26 | 118.66 | 144.06 | 169.46 |
| 11/16 | 17.46 | 42.86 | 68.26 | 93.66 | 119.06 | 144.46 | 169.86 |
| 45/64 | 17.86 | 43.26 | 68.66 | 94.06 | 119.46 | 144.86 | 170.26 |
| 23/32 | 18.26 | 43.66 | 69.05 | 94.45 | 119.85 | 145.25 | 170.65 |
| 47/64 | 18.65 | 44.05 | 69.45 | 94.85 | 120.25 | 145.65 | 171.05 |
| 3/4 | 19.05 | 44.45 | 69.85 | 95.25 | 120.65 | 146.05 | 171.45 |
| 49/64 | 19.45 | 44.85 | 70.25 | 95.65 | 121.04 | 146.44 | 171.84 |
| 25/32 | 19.84 | 45.24 | 70.64 | 96.04 | 121.44 | 146.84 | 172.24 |
| 51/64 | 20.24 | 45.64 | 71.04 | 96.44 | 121.84 | 147.24 | 172.64 |
| 13/16 | 20.64 | 46.04 | 71.44 | 96.84 | 122.24 | 147.63 | 173.03 |
| 53/64 | 21.03 | 46.43 | 71.83 | 97.23 | 122.63 | 148.03 | 173.43 |
| 27/32 | 21.43 | 46.83 | 72.23 | 97.63 | 123.03 | 148.43 | 173.83 |
| 55/64 | 21.83 | 47.23 | 72.63 | 98.03 | 123.43 | 148.83 | 174.22 |
| 7/8 | 22.22 | 47.62 | 73.02 | 98.42 | 123.82 | 149.22 | 174.62 |
| 57/64 | 22.62 | 48.02 | 73.42 | 98.82 | 124.22 | 149.62 | 175.02 |
| 29/32 | 23.02 | 48.42 | 73.82 | 99.22 | 124.62 | 150.02 | 175.42 |
| 59/64 | 23.42 | 48.81 | 74.21 | 99.61 | 125.01 | 150.41 | 175.81 |
| 15/16 | 23.81 | 49.21 | 74.61 | 100.01 | 125.41 | 150.81 | 176.21 |
| 61/64 | 24.21 | 49.61 | 75.01 | 100.41 | 125.81 | 151.21 | 176.61 |
| 31/32 | 24.61 | 50.01 | 75.4 | 100.8 | 126.2 | 151.6 | 177. |
| 63/64 | 25. | 50.4 | 75.8 | 101.2 | 126.6 | 152. | 177.4 |

*To use, read down appropriate inch column to the desired fraction line. The number indicated is the size in millimeters.

†*Courtesy The Cleveland Twist Drill Co.*

## TABLE 13 METRIC-ENGLISH CONVERSION TABLE

| Multiply | By | To get equivalent number of |
|---|---|---|
| | Length | |
| Inch | 25.4 | Millimeters (mm) |
| Foot | 0.304 8 | Meters (m) |
| Yard | 0.914 4 | Meters |
| Mile | 1.609 | Kilometers (km) |
| | Area | |
| $Inch^2$ | 645.2 | $Millimeters^2$ ($mm^2$) |
| | 6.45 | $Centimeters^2$ ($cm^2$) |
| $Foot^2$ | 0.092 9 | $Meters^2$ ($m^2$) |
| $Yard^2$ | 0.836 1 | $Meters^2$ |
| | Volume | |
| $Inch^3$ | 16 387. | $mm^3$ |
| | 16.387 | $cm^3$ |
| | 0.016 4 | Liters (l) |
| Quart (U.S.) | 0.946 4 | Liters |
| Quart (imperial) | 1.136 | Liters |
| Gallon (U.S.) | 3.785 4 | Liters |
| Gallon (imperial) | 4.459 | Liters |
| $Yard^3$ | 0.764 6 | $Meters^3$ ($m^3$) |
| | Mass | |
| Pound | 0.453 6 | Kilograms (kg) |
| Ton | 907.18 | Kilograms (kg) |
| Ton | 0.907 | Tonne (t) |
| | Force | |
| Kilogram | 9.807 | Newtons (N) |
| Ounce | 0.278 0 | Newtons |
| Pound | 4.448 | Newtons |
| | Temperature | |
| Degree Fahrenheit | (°F − 32) ÷ 1.8 | Degree Celsius (C) |
| Degree Celsius | (°C × 1.8) + 32 | Degree Fahrenheit (F) |

| Multiply | By | To get equivalent number of |
|---|---|---|
| | Acceleration | |
| $Foot/second^2$ | 0.304 8 | Meter per $second^2$ ($m/s^2$) |
| $Inch/second^2$ | 0.025 4 | Meter per $second^2$ |
| | Torque | |
| Pound-inch | 0.112 98 | Newton-meters (N-m) |
| Pound-foot | 1.355 8 | Newton-meters |
| | Power | |
| Horsepower | 0.746 | Kilowatts (kW) |
| | Pressure or stress | |
| Inches of water | 0.249 1 | Kilopascals (kPa) |
| Pounds/square inch | 6.895 | Kilopascals |
| | Energy or work | |
| BTU | 1 055. | Joules (J) |
| Foot-pound | 1.355 8 | Joules |
| Kilowatthour | 3 600 000. or $3.6 \times 10^6$ | Joules (J = one W's) |
| | Light | |
| Footcandle | 1.076 4 | Lumens per $meter^2$ ($lm/m^2$) |
| | Fuel performance | |
| Miles per gallon | 0.425 1 | Kilometers per liter (km/l) |
| Gallons per mile | 2.352 7 | Liters per kilometers (l/km) |
| | Velocity | |
| Miles per hour | 1.609 3 | Kilometers per hr (km/h) |

# INDEX